Non-linear dynamics and statistical theories for basic geophysical flows

Non-linear dynamics and statistical theories for basic geophysical flows

ANDREW J. MAJDA
New York University

XIAOMING WANG
Florida State University

CAMBRIDGE
UNIVERSITY PRESS

CAMBRIDGE UNIVERSITY PRESS
Cambridge, New York, Melbourne, Madrid, Cape Town, Singapore, São Paulo

CAMBRIDGE UNIVERSITY PRESS
The Edinburgh Building, Cambridge CB2 2RU, United Kingdom

www.cambridge.org
Information on this title: www.cambridge.org/9780521834414

First published 2006

Printed in the UK

A catalogue record for this publication is available from the British Library

ISBN-13 978-0-521-83441-4 hardback
ISBN-10 0-521-83441-4 hardback

Contents

Preface *page* xi

1 Barotropic geophysical flows and two-dimensional fluid flows: elementary introduction **1**
1.1 Introduction 1
1.2 Some special exact solutions 8
1.3 Conserved quantities 33
1.4 Barotropic geophysical flows in a channel domain – an important physical model 44
1.5 Variational derivatives and an optimization principle for elementary geophysical solutions 50
1.6 More equations for geophysical flows 52
References 58

2 The response to large-scale forcing **59**
2.1 Introduction 59
2.2 Non-linear stability with Kolmogorov forcing 62
2.3 Stability of flows with generalized Kolmogorov forcing 76
References 79

3 The selective decay principle for basic geophysical flows **80**
3.1 Introduction 80
3.2 Selective decay states and their invariance 82
3.3 Mathematical formulation of the selective decay principle 84
3.4 Energy–enstrophy decay 86
3.5 Bounds on the Dirichlet quotient, $\Lambda(t)$ 88
3.6 Rigorous theory for selective decay 90
3.7 Numerical experiments demonstrating facets of selective decay 95
References 102

A.1 Stronger controls on $\Lambda(t)$ 103
A.2 The proof of the mathematical form of the selective decay principle in the presence of the beta-plane effect 107

4 Non-linear stability of steady geophysical flows 115
4.1 Introduction 115
4.2 Stability of simple steady states 116
4.3 Stability for more general steady states 124
4.4 Non-linear stability of zonal flows on the beta-plane 129
4.5 Variational characterization of the steady states 133
References 137

5 Topographic mean flow interaction, non-linear instability, and chaotic dynamics 138
5.1 Introduction 138
5.2 Systems with layered topography 141
5.3 Integrable behavior 145
5.4 A limit regime with chaotic solutions 154
5.5 Numerical experiments 167
References 178
Appendix 1 180
Appendix 2 181

6 Introduction to information theory and empirical statistical theory 183
6.1 Introduction 183
6.2 Information theory and Shannon's entropy 184
6.3 Most probable states with prior distribution 190
6.4 Entropy for continuous measures on the line 194
6.5 Maximum entropy principle for continuous fields 201
6.6 An application of the maximum entropy principle to geophysical flows with topography 204
6.7 Application of the maximum entropy principle to geophysical flows with topography and mean flow 211
References 218

7 Equilibrium statistical mechanics for systems of ordinary differential equations 219
7.1 Introduction 219
7.2 Introduction to statistical mechanics for ODEs 221
7.3 Statistical mechanics for the truncated Burgers–Hopf equations 229
7.4 The Lorenz 96 model 239
References 255

8 Statistical mechanics for the truncated quasi-geostrophic equations 256
8.1 Introduction 256
8.2 The finite-dimensional truncated quasi-geostrophic equations 258
8.3 The statistical predictions for the truncated systems 262
8.4 Numerical evidence supporting the statistical prediction 264
8.5 The pseudo-energy and equilibrium statistical mechanics for fluctuations about the mean 267
8.6 The continuum limit 270
8.7 The role of statistically relevant and irrelevant conserved quantities 285
References 285
Appendix 1 286

9 Empirical statistical theories for most probable states 289
9.1 Introduction 289
9.2 Empirical statistical theories with a few constraints 291
9.3 The mean field statistical theory for point vortices 299
9.4 Empirical statistical theories with infinitely many constraints 309
9.5 Non-linear stability for the most probable mean fields 313
References 316

10 Assessing the potential applicability of equilibrium statistical theories for geophysical flows: an overview 317
10.1 Introduction 317
10.2 Basic issues regarding equilibrium statistical theories for geophysical flows 318
10.3 The central role of equilibrium statistical theories with a judicious prior distribution and a few external constraints 320
10.4 The role of forcing and dissipation 322
10.5 Is there a complete statistical mechanics theory for ESTMC and ESTP? 324
References 326

11 Predictions and comparison of equilibrium statistical theories 328
11.1 Introduction 328
11.2 Predictions of the statistical theory with a judicious prior and a few external constraints for beta-plane channel flow 330
11.3 Statistical sharpness of statistical theories with few constraints 346
11.4 The limit of many-constraint theory (ESTMC) with small amplitude potential vorticity 355
References 360

12 Equilibrium statistical theories and dynamical modeling of flows with forcing and dissipation **361**

12.1 Introduction 361

12.2 Meta-stability of equilibrium statistical structures with dissipation and small-scale forcing 362

12.3 Crude closure for two-dimensional flows 385

12.4 Remarks on the mathematical justifications of crude closure 405

References 410

13 Predicting the jets and spots on Jupiter by equilibrium statistical mechanics **411**

13.1 Introduction 411

13.2 The quasi-geostrophic model for interpreting observations and predictions for the weather layer of Jupiter 417

13.3 The ESTP with physically motivated prior distribution 419

13.4 Equilibrium statistical predictions for the jets and spots on Jupiter 423

References 426

14 The statistical relevance of additional conserved quantities for truncated geophysical flows **427**

14.1 Introduction 427

14.2 A numerical laboratory for the role of higher-order invariants 430

14.3 Comparison with equilibrium statistical predictions with a judicious prior 438

14.4 Statistically relevant conserved quantities for the truncated Burgers–Hopf equation 440

References 442

A.1 Spectral truncations of quasi-geostrophic flow with additional conserved quantities 442

15 A mathematical framework for quantifying predictability utilizing relative entropy **452**

15.1 Ensemble prediction and relative entropy as a measure of predictability 452

15.2 Quantifying predictability for a Gaussian prior distribution 459

15.3 Non-Gaussian ensemble predictions in the Lorenz 96 model 466

15.4 Information content beyond the climatology in ensemble predictions for the truncated Burgers–Hopf model 472

15.5 Further developments in ensemble predictions and information theory 478
References 480

16 Barotropic quasi-geostrophic equations on the sphere 482
16.1 Introduction 482
16.2 Exact solutions, conserved quantities, and non-linear stability 490
16.3 The response to large-scale forcing 510
16.4 Selective decay on the sphere 516
16.5 Energy enstrophy statistical theory on the unit sphere 524
16.6 Statistical theories with a few constraints and statistical theories with many constraints on the unit sphere 536
References 542
Appendix 1 542
Appendix 2 546

Index 550

Preface

This book is an introduction to the fascinating and important interplay between non-linear dynamics and statistical theories for geophysical flows. The book is designed for a multi-disciplinary audience ranging from beginning graduate students to senior researchers in applied mathematics as well as theoretically inclined graduate students and researchers in atmosphere/ocean science. The approach in this book emphasizes the serendipity between physical phenomena and modern applied mathematics, including rigorous mathematical analysis, qualitative models, and numerical simulations. The book includes more conventional topics for non-linear dynamics applied to geophysical flows, such as long time selective decay, the effect of large-scale forcing, non-linear stability and fluid flow on the sphere, as well as emerging contemporary research topics involving applications of chaotic dynamics, equilibrium statistical mechanics, and information theory. The various competing approaches for equilibrium statistical theories for geophysical flows are compared and contrasted systematically from the viewpoint of modern applied mathematics, including an application for predicting the Great Red Spot of Jupiter in a fashion consistent with the observational record. Novel applications of information theory are utilized to simplify, unify, and compare the equilibrium statistical theories and also to quantify aspects of predictability in non-linear dynamical systems with many degrees of freedom. No previous background in geophysical flows, probability theory, information theory, or equilibrium statistical mechanics is needed to read the text. These topics and related background concepts are all introduced and developed through elementary examples and discussion throughout the text as they arise. The book is also of wider interest to applied mathematicians and other scientists to illustrate how ideas from statistical physics can be applied in novel ways to inhomogeneous large-scale complex non-linear systems.

The material in the book is based on lectures of the first author given at the Courant Institute in 1995, 1997, 2001, and 2004. The first author thanks Professor

Pedro Embid as well as his former Ph.D. students Professor Pete Kramer and Seuyung Shim for their help with early versions of Chapters 1, 2, 3, 4, and 6 of the present book. Joint research work with Professors Richard Kleeman and Bruce Turkington as well as Majdas former Courant post docs, Professors Marcus Grote, Ilya Timofeyev, Rafail Abramov, and Mark DeBattista have been incorporated into the book; their explicit and implicit contributions are acknowledged warmly. The authors acknowledge generous support of the National Science Foundation and the Office of Naval Research during the development of this book, including partial salary support for Xiaoming Wangs visit to Courant in the spring semester of 2001.

1

Barotropic geophysical flows and two-dimensional fluid flows: elementary introduction

1.1 Introduction

The atmosphere and the ocean are the two most important fluid systems of our planet. The bulk of the atmosphere is a thin layer of air 10 km thick that engulfs the earth, and the oceans cover about 70% of the surface of our planet. Both the atmosphere and the ocean are in states of constant motion where the main source of energy is supplied by the radiation of the sun. The large-scale motions of the atmosphere and the ocean constitute geophysical flows and the science that studies them is geophysical fluid dynamics. The motions of the atmosphere and the ocean become powerful mechanisms for the transport and redistribution of energy and matter. For example, the motion of cold and warm atmospheric fronts determine the local weather conditions; the warm waters of the Gulf Stream are responsible for the temperate climate in northern Europe; the winds and the currents transport the pollutants produced by industries. It is clear that the motions of the atmosphere and the ocean play a fundamental role in the dynamics of our planet and greatly affect the activities of mankind.

It is apparent that the dynamical processes involved in the description of geophysical flows in the atmosphere and the ocean are extremely complex. This is due to the large number of physical variables needed to describe the state of the system and the wide range of space and time scales involved in these processes. The physical variables may include the velocity, the pressure, the density, and, in addition, the humidity in the case of atmospheric motions or the salinity in the case of oceanic motions. The physical processes that determine the evolution of the geophysical flows are also numerous. They may include the Coriolis force due to the earth's rotation; the sun's radiation; the presence of topographical barriers, as represented by mountain ranges in the case of atmospheric flows and the ocean floor and the continental masses in the case of oceanic flows. There may be also dissipative energy mechanisms, for example due to eddy diffusivity or Ekman drag. The ranges of spatial and temporal scales involved in the description of geophysical flows is also very large. The space scales may vary from a few hundred meters to thousands of kilometers. Similarly, the time scales maybe as short as minutes and as long as days, months, or even years.

The above remarks make evident the need for simplifying assumptions regarding the relevant physical mechanisms involved in a given geophysical flow process, as well as the relevant range of space and time scales needed to describe the process. The treatises of Pedlosky (1987) and Gill (1982) are two excellent references to consult regarding the physical foundations of geophysical flows and different simplifying approximations utilized in the study of the various aspects of geophysical fluids. Here we concentrate on large-scale flows for the atmosphere or *mesoscale* flows in the oceans. The simplest set of equations that meaningfully describes the motion of geophysical flows under these circumstances is given by the:

Barotropic quasi-geostrophic equations

$$\begin{aligned} &\frac{Dq}{Dt} = \mathcal{D}(\Delta)\psi + \mathcal{F}(\vec{x}, t) \\ &q = \omega + \beta y + h(x, y), \quad \text{where } \omega = \Delta\psi \\ &\vec{v} = \nabla^{\perp}\psi = \begin{pmatrix} -\dfrac{\partial\psi}{\partial y} \\ \dfrac{\partial\psi}{\partial x} \end{pmatrix}, \end{aligned} \tag{1.1}$$

where $\frac{D}{Dt}$ stands for the advective (or material) derivative

$$\frac{D}{Dt} \equiv \frac{\partial}{\partial t} + v_1\frac{\partial}{\partial x} + v_2\frac{\partial}{\partial y}$$

and Δ denotes the Laplacian operator

$$\Delta = \operatorname{div}\nabla = \frac{\partial^2}{\partial x^2} + \frac{\partial^2}{\partial y^2}.$$

In equation (1.1), q is the potential vorticity, $\vec{v}$ is the horizontal velocity field, ω, is the relative vorticity, and ψ is the stream function. The horizontal space variables are given by $\vec{x} = (x, y)$ and t denotes time. The term βy is called the beta-plane effect from the Coriolis force and its significance will be explained later. The term $h = h(x, y)$ represents the bottom floor topography. The term $\mathcal{D}(\Delta)\psi$ represents various possible dissipation mechanisms. Finally, the term $\mathcal{F}(\vec{x}, t)$ accounts for additional external forcing. The fluid density is set to 1.

Before continuing, we would like to explain briefly, in physical terms and without going into any technical details, the origin of the barotropic quasi-geostrophic equations. The barotropic rotational equations, also called rotating shallow water equations (Pedlosky, 1987), admit two different modes of propagation, slow and fast. The slow mode of propagation corresponds to the motion of the bulk of the fluid by advection. This is the slow motion we see in the weather patterns in the atmosphere, evolving on a time scale of days. The fast mode corresponds to gravity waves, which evolve on a short time scale of the order of several minutes, but do not contribute to the bulk motion of the fluid. The barotropic quasi-geostrophic equations are the result of "filtering out" the fast gravity waves from the rotating barotropic equations. There is also a formal analogy between barotropic quasi-geostrophic equations and incompressible flows; in the theory of compressible fluid flows the incompressible limit is obtained by "filtering out" the "fast" acoustic waves and retaining only the "slow" vortical modes associated to convection by the fluid (Majda, 1984). Indeed, it was this analogy that originally inspired Charney (1949) when he first formulated the quasi-geostrophic equations and thus opened the modern era of numerical weather prediction (Charney, 1949; Charney, Fjörtoft, and von Neumann, 1950).

The full derivation of the rotating barotropic equations and the corresponding barotropic quasi-geostrophic equations is lengthy and will take us too far from our main objective, which is the study of the quasi-geostrophic equations. For a thorough treatment of the barotropic rotational equations the reader is referred to Pedlosky (1987). Formal as well as rigorous derivations of the barotropic quasi-geostrophic equations from the rotating shallow water equations can be found in Majda (2003), Embid and Majda (1996).

Rather than deriving the quasi-geostrophic equations, we would like to explain the physical meaning and significance of the different terms appearing in equation (1.1). For barotropic quasi-geostrophic flows, the potential vorticity q is made of three different contributions. The first term $\omega = \Delta\psi = \operatorname{curl} \vec{v}$ is the fluid vorticity and represents the local rate of rotation of the fluid. The second term βy is the beta-plane effect from the Coriolis force and its appearance will be explained later. The third term $h = h(x, y)$ represents the bottom topography, as given by the ocean floor or a mountain range.

The horizontal velocity field, $\vec{v}$, is determined by the orthogonal gradient of the stream function ψ, $\vec{v} = \nabla^{\perp}\psi$, where the orthogonal gradient of ψ is defined as

$$\nabla^{\perp}\psi = \begin{pmatrix} -\dfrac{\partial\psi}{\partial y} \\ \\ \dfrac{\partial\psi}{\partial x} \end{pmatrix}.$$

In particular, the velocity field $\vec{v}$ is incompressible because

$$\operatorname{div} \vec{v} = \nabla \cdot \vec{v} = \nabla \cdot \nabla^{\perp}\psi = 0.$$

The reason ψ is called the stream function is because at any fixed instant in time the velocity field $\vec{v}$ is perpendicular to the gradient of ψ, i.e. $\vec{v}$ is tangent to the level curves of ψ. Therefore the level curves of ψ represent the streamlines of the fluid. In addition, there is another important interpretation of ψ. Physically ψ represents the (hydrostatic) pressure of the fluid. In this context, the equation $\vec{v} = \nabla^{\perp}\psi$ corresponds to the fact that the flow field is in *geostrophic balance*, and therefore the streamlines also happen to be the *isobars* of the flow. In particular, we conclude that for a steady solution of the quasi-geostrophic equations the fluid flows along the isobars. This is in marked contrast with the situation in non-rotating fluids, where typically the flow is from regions of high pressure to those of low pressure.

The importance of the potential vorticity q is in the fact that it completely determines the state of the flow. Indeed in the barotropic quasi-geostrophic equations, once we know the potential vorticity q, the second equation in equation (1.1) immediately yields the vorticity ω. Since $\omega = \Delta\psi$, we can determine the stream function ψ, and then introduce it into the third equation in equation (1.1), namely $\vec{v} = \nabla^{\perp}\psi$, to determine the advective velocity field.

Next we return to a brief discussion of the beta-plane effect (cf. Pedlosky, 1987). This effect is essentially the result of linearizing the Coriolis force when we consider the motion of the fluid in the tangent plane approximation. More specifically, although the earth is spherical, we assume that the spatial scale of motion is moderate enough so that the region occupied by the fluid can be approximated by a tangent plane (this is certainly the case for mesoscale flows, even for horizontal ranges of the order of 10^3 km). This is what is called the tangent plane approximation. The equations of motion in equation (1.1) are written in terms of horizontal Cartesian coordinates in the tangent plane. In this context, the spatial variable x corresponds to longitude (with positive direction towards the east) and the variable y to latitude (with positive direction towards the north).[1] In fact, throughout this book we often refer to flows pointing in the positive (negative) x-direction as eastward (westward). Since the tangent plane rotates with the earth it becomes a non-inertial frame, and the Coriolis force due to the earth's rotation becomes an important effect in geophysical flows. Moreover, because of the curvature of the earth, the contribution of the Coriolis force depends on the latitude at which the tangent plane is being considered; the Coriolis force will increase from zero at the equator to its maximum value at the poles. Since the tangent plane approximation assumes a moderate range in latitude and longitude, a Taylor expansion approximation of the Coriolis force is permissible; the linear term of this Taylor expansion yields the beta-plane effect βy considered in equation (1.1). For the actual details of the tangent plane approximation and the beta-plane effect, the reader is encouraged to consult Pedlosky (1987) or Gill (1982).

[1] For simplicity we will always assume that the tangent plane approximation is considered in the northern hemisphere

There are many choices of dissipation operator $\mathcal{D}(\Delta)$, ranging from Ekman drag to Newtonian viscosity or hyper-viscosity. We list some commonly used dissipation operators below for later convenience:

(i) Newtonian (eddy) viscosity

$$\mathcal{D}(\Delta)\psi = \nu\Delta^2\psi$$

This form of the diffusion is identical to the ordinary molecular friction in a Newtonian fluid. For geophysical flows, the value of the coefficient is often assumed to be many orders of magnitude larger than that for molecular viscosity, and represents, crudely, smaller-scale turbulence effects. This led to the name, eddy viscosity.

(ii) Ekman drag dissipation

$$\mathcal{D}(\Delta)\psi = -d\Delta\psi,$$

which is common to the large-scale pieces of the geophysical flow. This arises from boundary layer effects in rapidly rotating flows.

(iii) Hyper-viscosity dissipation

$$\mathcal{D}(\Delta)\psi = (-1)^j d_j \Delta^j \psi, \quad j = 3, 4, 5, \cdots$$

This form of the dissipation term is frequently utilized in the study and numerical simulation of geophysical flows, where its role is to introduce very little dissipation in the large scales of the flow but to strongly damp out the small scales. The validity of the use of such hyper-viscous mechanisms is still an open issue among geophysical fluid dynamicists.

(iv) Ekman drag dissipation + Hyper-viscosity

$$\mathcal{D}(\Delta)\psi = -d\Delta\psi + (-1)^j d_j \Delta^j \psi, \qquad d_j \geq 0, \quad d > 0, \quad j > 2.$$

This is a combination of the previous two dissipation mechanisms.

(v) Radiative damping

$$\mathcal{D}(\Delta)\psi = d\psi$$

This represents a crude model for radiative damping when models with stratification are involved. Radiative damping is an unusual dissipation operator since it damps the large scales more strongly than the small scales in contrast to the standard diffusion operators in (i) and (iii) above.

(vi) General dissipation operator

$$\mathcal{D}(\Delta)\psi = \sum_{j=0}^{l} (-1)^j d_j \Delta^j \psi,$$

which encompasses all other forms of dissipation mechanisms previously discussed.

For simplicity we will consider periodic boundary conditions for the flow in both the x and y variables, say with period 2π in both variables

$$\begin{aligned} \vec{v}(x+2\pi, y, t) &= \vec{v}(x, y, t), \\ \vec{v}(x, y+2\pi, t) &= \vec{v}(x, y, t), \end{aligned} \tag{1.2}$$

or in terms of the stream function ψ

$$\psi(x+2\pi, y, t) = \psi(x, y+2\pi, t) = \psi(x, y, t). \tag{1.3}$$

We may also impose the zero average condition

$$\int \psi(x, y, t)\, dxdy = 0, \tag{1.4}$$

since the stream function is always determined up to a constant, and we can choose the constant here so that the average is zero. The assumption of periodicity in both variables is not unreasonable (except near drastic topographical barriers, such as continents). It allows us to use Fourier series and separation of variables as a main mathematical tool (see page 10 for a Fourier series tool kit). Physically, periodicity allows us to avoid other issues such as the appearance of boundary layers or the generation of vorticity at the boundary. However, occasionally we will consider other boundary conditions besides the periodic one. In particular, we will study flows in channel domains or in a rectangular basin which can be treated through minor modification of periodic flows with special geometry.

It is worthwhile to point out that, in the special case where there are no beta-plane effects or bottom floor topography, i.e. $\beta = 0$, $h = 0$, then the potential vorticity q reduces to the vorticity ω, $q = \omega$, and if we assume Newtonian dissipation, then the barotropic quasi-geostrophic equations reduce to the classical Navier–Stokes equations for a two-dimensional flow, written in the vorticity-stream form (Majda and Bertozzi, 2001; Chorin and Marsden, 1993)

Two-dimensional classical fluid flow equations

$$\frac{D\omega}{Dt} = \nu\Delta\omega + \mathcal{F}(\vec{x}, t), \qquad \omega = \Delta\psi, \qquad \vec{v} = \nabla^{\perp}\psi \tag{1.5}$$

and in the case without dissipation we have the classical Euler equations with forcing

$$\frac{D\omega}{Dt} = \mathcal{F}(\vec{x}, t), \qquad \omega = \Delta\psi, \qquad \vec{v} = \nabla^{\perp}\psi. \tag{1.6}$$

One of our objectives of this book is to compare and contrast the barotropic quasi-geostrophic equations and the Navier–Stokes equations to better understand the role of the beta-plane effect and the topography on the behavior of geophysical flows.

Even though we have restricted ourselves to the study of the barotropic quasi-geostrophic equations, it is still not possible for us (and not our intention) to cover all the possible problems associated with these equations. Instead we will focus on various topics that we consider physically interesting, yet mathematically tractable. We are especially interested in geophysical fluid flow phenomena, influenced by the presence of the Coriolis force and topography, combined with the presence of various dissipative and external forcing mechanisms, and on their role in the emergence and persistence of large coherent structures, as observed in mesoscale flows. However, many of the ideas and techniques apply to more complex models for geophysical flows, such as the F-plane equations, two layer models, continuously stratified quasi-geostrophic flow. The final section of this chapter discusses all of these models briefly as well as the inter-relations among them and the basic barotropic model. Generalizations of some of the material in the course to these equations are straightforward, while other material involves subtle current research.

Here we include a list of some of the topics that we will study in subsequent chapters:

(i) Exact solutions showing interesting physics.
There are many interesting patterns in geophysical flows ranging from Rossby waves to jets. One of our tasks here is to present some special exact solutions. They will illustrate simple Rossby wave motion, Taylor vortices, shear flows, simple topographic effects, etc.

(ii) Conserved quantities.
Conserved quantities play an essential role in both the physical understanding and mathematical study of geophysical flows. In this book we will carefully study various conserved quantities. A set of important conserved quantities are summarized, with the geophysical effects and domain geometry effects distinguished. These conserved quantities will play a central role in the subsequent study of non-linear stability of geophysical flows and the statistical theories of large-scale coherent structures.

(iii) Response to large-scale forcing.
We will establish the stability of motion on the ground shell, provided that the forcing is of the largest scale and dissipation is present. This will provide us with an explicit example of stable large coherent structure in damped and driven environment.

(iv) Selective decay.
We will demonstrate various facets of selective decay, both numerically and mathematically. This is an interesting example of how the inverse cascade is observed in two-dimensional flows.

(v) Non-linear stability of certain steady geophysical flows.
Stability is directly related to the issue of whether a specific flow is observable. Here the non-linear stability for certain geophysical flows is discussed, utilizing the Arnold–Kruskal method.

(vi) Large- and small-scale interaction via topographic stress.
Interesting phenomena arise when we introduce topography. An effective new stress, called topographic stress, effectively mediates the energy exchange between the large- and small-scale flows. We will see below that this is an interesting source of explicit examples with chaotic dynamics.

(vii) Equilibrium statistics mechanics and statistical theories for large coherent structures.
Here we develop a self-contained treatment of equilibrium statistical mechanics for geophysical flows in an elementary fashion. We develop elementary models with statistical features of the atmosphere and ocean, and equilibrium statistical theories for large coherent structures. The main perspective in achieving this is through information theory, which is developed in the text in a self-contained fashion.
If we are interested in large coherent structures instead of the small-scale fine structures, equilibrium statistical theory provides a way to predict the large coherent structure without calculating the details of the solutions. Various approaches will be presented. These will include the classical statistical theory with two conserved quantities, theories that attempt to incorporate infinitely many conserved quantities, and the current statistical theory with a few judicious constraints. In addition, a special numerical laboratory is developed and utilized to compare these approaches quantitatively.

(viii) Crude dynamic modeling for geophysical flows.
We will develop novel ways in which the equilibrium statistical theories can represent complex flows with damping and driving. This will include crude closure algorithms for both the classical fluid flows and flows with topography. This study will rely heavily on numerical evidence, but will also be supported with mathematical analysis.

(ix) We will apply the ideas developed in (vii) and (viii) to successfully predict the Great Red Spot of Jupiter in a fashion which is completely self-consistent with the observational record from the Voyager and Galileo space missions.

(x) Barotropic quasi-geostrophic equations on the sphere.
Actual fluid flow in the atmosphere occurs on the sphere and there are important new effects. Here all the previous problems will be reconsidered on the sphere. Some peculiar phenomena arise due to the special spherical symmetry.

(xi) We will show how information theory can be used to quantify predictability for ensemble predictions in geophysical flows. The following issues will be addressed. How important is the mean compared with the variance? When is a prediction reliably bi-modal with two different scenarios?

1.2 Some special exact solutions

Next we introduce and describe several families of special exact solutions of the barotropic quasi-geostrophic equations, equation (1.1). These solutions will be given with increasing levels of complexity, as we add more physical effects into the quasi-geostrophic equations. We will start by considering special steady flows free from beta-plane, topographical, diffusive, and external forcing effects. Even in this simplified situation, we will find a rich family of simple flows with interesting flow topology, which include shear flows, array of eddies, and Taylor vortices. Then we continue by systematically adding beta-plane effects, dissipation, and special external forcing, known as generalized Kolmogorov forcing. In particular, in this situation we will find flows with a large-scale mean flow and Rossby waves. We will also study the effects from the bottom floor topography and how it modifies the vorticity of the flow. This will be followed by examples incorporating the combined effects of the beta-plane and the topography. Finally, we conclude this section with an example of interaction of a large-scale shear flow with beta-plane dynamics. These special exact solutions are invaluable. They help us to build intuition and insight by revealing explicitly the behavior of the flow under the different physical mechanisms. They also provide us with ideal examples to test numerical methods designed to solve the quasi-geostrophic equations, as well as further theories about these geophysical flows.

In general it is far from easy to find exact solutions of the barotropic quasi-geostrophic equations. The difficulty lies in the non-linear character of the equations, through the non-linear advection term

$$\frac{Dq}{Dt} = \frac{\partial q}{\partial t} + \vec{v} \cdot \nabla q.$$

Since the velocity $\vec{v}$ is given by the perpendicular gradient of the stream function ψ, $\vec{v} = \nabla^{\perp}\psi$, we can rewrite this non-linear term as

$$\nabla^{\perp}\psi \cdot \nabla q = \det \begin{pmatrix} \nabla\psi \\ \nabla q \end{pmatrix} = \begin{vmatrix} \frac{\partial\psi}{\partial x} & \frac{\partial\psi}{\partial y} \\ \frac{\partial q}{\partial x} & \frac{\partial q}{\partial y} \end{vmatrix} = J(\psi, q),$$

where $J(\psi, q)$ is the Jacobian determinant of q and ψ. Therefore the potential vorticity equation in equation (1.1) takes the form

$$\frac{\partial q}{\partial t} + J(\psi, q) = \mathcal{D}(\Delta)\psi + \mathcal{F}(\vec{x}, t), \tag{1.7}$$

or equivalently

$$\frac{\partial q}{\partial t} + \nabla^{\perp}\psi \cdot \nabla q = \mathcal{D}(\Delta)\psi + \mathcal{F}(\vec{x}, t),$$

where $q = \Delta\psi + \beta y + h$.

To eliminate the non-linearity we must require the vanishing of the Jacobian determinant $J(\psi, q)$, and this certainly happens if the potential vorticity q and the stream function ψ are functionally dependent, i.e. if $q = F(\psi)$ for some function F.

Although such an assumption makes the potential vorticity equation linear, it also makes the elliptic equation for the stream function ψ non-linear

$$F(\psi) = q = \omega + \beta y + h = \Delta\psi + \beta y + h.$$

Clearly, any q and ψ, functionally related by $q = F(\psi)$ automatically define a steady (time-independent) exact solution of the barotropic quasi-geostrophic equation without damping or external forcing. In later chapters, the reader will find many examples of solutions of this type.

Here we concentrate on finding special exact solutions with both forcing and dissipation with the stronger ansatz: we *assume* that q and ψ are linearly dependent

$$q = \mu\psi, \tag{1.8}$$

then the elliptic equation for ψ now also becomes linear.

Summarizing, *under the linear dependence assumption* $q = \mu\psi$, the solution of the barotropic quasi-geostrophic equations, equation (1.1), is given by the:

Reduced linear system for the stream function ψ

$$\mu\frac{\partial\psi}{\partial t} = \mathcal{D}(\Delta)\psi + \mathcal{F}(\vec{x}, t), \quad \mu\psi = \Delta\psi + \beta y + h(\vec{x}), \tag{1.9}$$

where the velocity field $\vec{v}$, the vorticity ω, and the potential vorticity q are then given in terms of the stream function ψ by

$$\vec{v} = \nabla^{\perp}\psi, \quad \omega = \Delta\psi\ , \quad q = \mu\psi. \tag{1.10}$$

Throughout this book, we study geophysical flows in idealized periodic geometry or on special domains, such as channels or the square, which are related to the periodic geometry though symmetry considerations. Unless noted otherwise, all domains are 2π-periodic in each direction. Since the domain is periodic and the equations for the stream function ψ in equation (1.7) are linear, it is a natural desire to use the Fourier series as the main tool to study them. The only potential impediment may come from the beta-plane term βy, which is not a periodic function. However, we will solve this problem later with the introduction of a suitable large-scale mean flow for the velocity field.

Next we summarize some important basic properties of the Fourier series, which we will use throughout the book, and then discuss some particular solutions of the linear equations in equation (1.7).

Fourier series tool kit

Here we recall a few basic properties of the Fourier series that are used in this book.

For any complex-valued functions $f(x, y)$ and $g(x, y)$, 2π-periodic in each variable:

Inner product: The inner product of f and g is defined as

$$(f, g)_0 = (4\pi^2)^{-1} \int_0^{2\pi} \int_0^{2\pi} f\bar{g}\, dx\, dy. \tag{1.11}$$

Complete orthonormal basis: There exists a complete orthonormal basis for the space of square integrable complex-valued functions on $[0, 2\pi] \times [0, 2\pi]$ given by

$$\vec{e}_{\vec{k}} = e^{i\vec{x}\cdot\vec{k}}, \vec{k} = (k_1, k_2) \in \mathcal{Z}^2, \tag{1.12}$$

where $i = \sqrt{-1}$ and $\mathcal{Z}^2 = \{(k_1, k_2)|k_1, k_2 \text{ are arbitrary integers}\}$. It is orthonormal in the sense that

$$(\vec{e}_{\vec{k}}, \vec{e}_{\vec{l}})_0 = \begin{cases} 0, & \vec{k} \neq \vec{l} \\ 1, & \vec{k} = \vec{l} \end{cases} \tag{1.13}$$

Fourier coefficients: The $\vec{k}$th Fourier coefficient of an integrable complex-valued function f is defined as

$$\hat{f}_{\vec{k}} = (f, e^{i\vec{x}\cdot\vec{k}})_0 = (4\pi^2)^{-1} \int_0^{2\pi} \int_0^{2\pi} f\, e^{-i\vec{x}\cdot\vec{k}}\, dx\, dy. \tag{1.14}$$

Expansion property: For each square integrable complex-valued function f, the following expansion formula holds

$$f = \sum_{\vec{k}\in\mathcal{Z}^2} \hat{f}_{\vec{k}} e^{i\vec{x}\cdot\vec{k}}. \tag{1.15}$$

Moreover, f is real valued if and only if

$$\bar{\hat{f}}_{\vec{k}} = \hat{f}_{-\vec{k}} \quad \text{for all } \vec{k}. \tag{1.16}$$

Parseval's identity: For any square integrable functions f and g, the following Parseval's identity holds

$$(f, g)_0 = \sum_{\vec{k}\in\mathcal{Z}^2} \hat{f}_{\vec{k}} \bar{\hat{g}}_{\vec{k}}. \tag{1.17}$$

Differentiation property: Let $D^\alpha = \frac{\partial^{\alpha_1}}{\partial x^{\alpha_1}} \frac{\partial^{\alpha_2}}{\partial y^{\alpha_2}}$ be a general derivative operator and assume f is sufficiently differentiable, then

$$\widehat{D^\alpha f}_{\vec{k}} = (ik_1)^{\alpha_1} (ik_2)^{\alpha_2} \hat{f}_{\vec{k}}. \tag{1.18}$$

A simple application of the Fourier series toolkit is the Poincaré inequality used throughout the text.

If f is a 2π-periodic function with zero mean, i.e. $\hat{f}_0 = 0$, then

$$\|f\|_0^2 \leq \|\nabla f\|_0^2. \tag{1.19}$$

The proof is an immediate application of (1.17) and (1.18) since

$$\|f\|_0^2 = \sum_{|\vec{k}|\geq 1} |\hat{f}_{\vec{k}}|^2 \leq \sum_{|\vec{k}|\geq 1} |\vec{k}|^2 |\hat{f}_{\vec{k}}|^2 \leq \|\nabla\|_0^2.$$

1.2.1 Exact solutions for the Euler equations

First we consider special solutions of the quasi-geostrophic equations with q and ψ linearly dependent and in the absence of physical effects from beta-plane, topography, dissipation, and external forcing (we remarked earlier that in this case the barotropic quasi-geostrophic equations (1.1) reduce to the Euler equations (1.6)). In this case equation (1.7) for the stream function reduces to

$$\mu \frac{\partial \psi}{\partial t} = 0 \quad \mu\psi = \Delta\psi, \tag{1.20}$$

so that the stream function ψ is steady and corresponds to an eigenfunction of the Laplacian with eigenvalue μ. The periodic eigenfunctions of the Laplacian are readily given by the Fourier modes. Indeed, $\psi = e^{i\vec{k}\cdot\vec{x}}$, where $\vec{k} \in \mathcal{Z}^2 - \{0\}$ is an eigenfunction of the Laplacian with eigenvalue $\mu = -|\vec{k}|^2$ where $\mathcal{Z}^2 = \{(k_1, k_2)|\ k_1, k_2 \text{ integers}\}$ is the two-dimensional integer lattice. Moreover, if we denote $\Lambda_j = -\mu_j$, where μ_j is a given eigenvalue of the Laplacian, then the most general form of the associated eigenfunction ψ_j is

$$\psi_j = \sum_{|\vec{k}|^2=\Lambda_j} \hat{A}_{\vec{k}} e^{i\vec{k}\cdot\vec{x}} + c.c., \tag{1.21}$$

where, as usual, *c.c.* stands for complex conjugate. This solution involves the linear combination of those Fourier modes, where wave numbers $\vec{k}$ are in the same shell $|\vec{k}|^2 = \Lambda_j$. The eigenvalues Λ_j are ordered in increasing magnitude, $0 < \Lambda_1 < \Lambda_2 < \cdots$, with $\Lambda_1 = 1$, $\Lambda_2 = 2$, $\Lambda_3 = 5$, etc.

Next we present several examples that illustrate the diverse variety of flows that can be generated with the stream function ψ_j in equation (1.19), and which display interesting flow topologies and symmetries.

Example 1: Ground state eigenmode flows.

For the first example we choose the ground state eigenvalue $\Lambda_1 = 1$. The associated wave numbers $\vec{k}$ in the shell $|\vec{k}| = 1$ are $\vec{k} = (\pm 1, 0), (0, \pm 1)$, and the stream function ψ in equation (1.19) reduces to

$$\psi = a\sin(x) + b\sin(y) + c\cos(x) + d\cos(y),$$

with the corresponding velocity field given by

$$\vec{v} = \nabla^{\perp}\psi = \begin{pmatrix} -b\cos(y) + d\sin(y) \\ a\cos(x) - c\sin(x) \end{pmatrix}.$$

Next we show how changing the parameters a, b, c, and d produce a variety of interesting simple flow patterns.

Case 1a: Simple shear flow.

If we set $a = A$, $b = c = d = 0$, then the stream function ψ reduces to

$$\psi = A\sin(x),$$

and the corresponding velocity field $\vec{v}$ represents a simple shear flow

$$\vec{v} = \begin{pmatrix} 0 \\ A\cos(x) \end{pmatrix}.$$

The stream function ψ and the its velocity field are depicted in Figure 1.1. In this case, the streamlines are given by the straight lines $x = const.$

Case 1b: 2-D array of swirling eddies.

On the other hand, if we set $a = b = A$, and $c = d = 0$, the stream function ψ now yields

$$\psi = A\left[\sin(x) + \sin(y)\right],$$

and the velocity field $\vec{v}$ becomes

$$\vec{v} = \begin{pmatrix} -A\cos(y) \\ A\cos(x) \end{pmatrix}.$$

The streamlines for this flow are depicted in Figure 1.2. The resulting flow pattern consists of a doubly periodic array of alternating eddies rotating in clockwise and counter-clockwise fashion. These eddy cells are aligned at 45° from the x and y axes, and the flow is confined within each one of the cells. In the period square $[0, 2\pi] \times [0, 2\pi]$ there are four stagnation points, $p_1 = (1/2, 1/2)\pi$, $p_2 = (3/2, 1/2)\pi$, $p_3 = (3/2, 3/2)\pi$, and $p_4 = (1/2, 3/2)\pi$. The stagnation points p_1 and p_3 correspond to the eddy centers, with the fluid locally rotating around these stagnation points. Figure 1.2(b) shows that p_3 is a local maximum of the vorticity and the rotation is counter-clockwise around p_3, whereas p_1 is a local minimum of the vorticity and the rotation is clockwise in this case. On the other hand, the stagnation points p_2 and p_4 correspond to saddle points for the vorticity; around these points the fluid behaves locally as a strain flow. The vorticity ω is depicted in Figure 1.2(b).

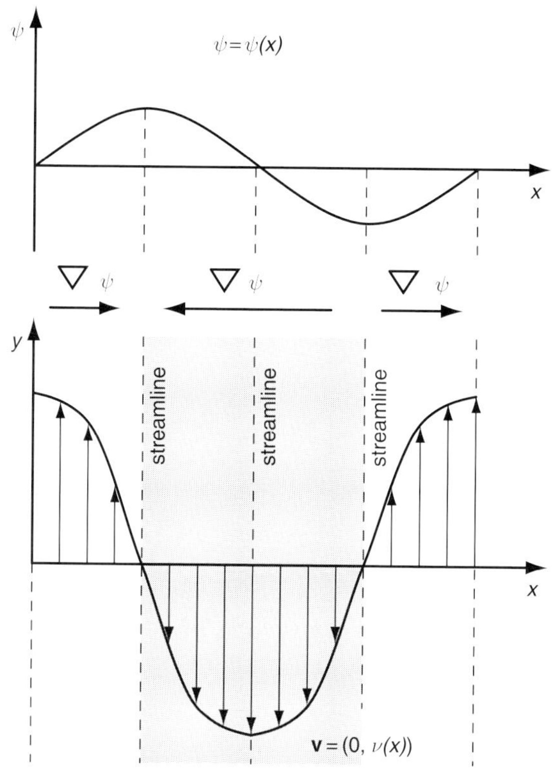

Figure 1.1 Example of a simple shear flow with stream function $\psi = A\sin(x)$ and velocity field $\vec{v} = (0, A\cos(x))$. Notice that the fluid velocity is parallel to the streamlines.

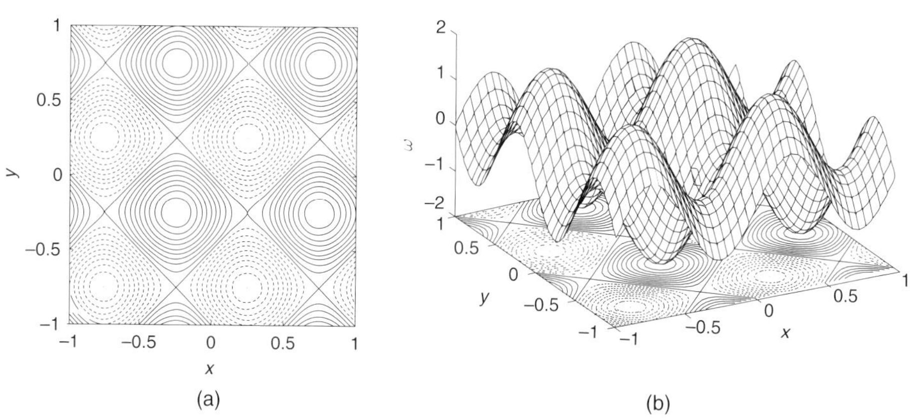

Figure 1.2 2-D array of swirling eddies with stream function $\psi(x, y) = \sin(x) + \sin(y)$ (x and y both normalized by 2π). (a) Plot of the streamlines: counter-clockwise rotation (—); clockwise rotation (- -). (b) Plot of the vorticity ω.

Case 1c: Swirling eddies embedded in a shear-flow stream.

In general, if we let $c = d = 0$ and vary a and b, the resulting flow will display free flowing streamlines as in the shear-flow example in Case 1a, together with embedded pockets of swirling eddies as in the periodic array of eddies in Case 1b. A typical streamline configuration is displayed in Figure 1.3(a), for the selected values of $a_1 = 0.4$ and $a_2 = 0.8$. The associated vorticity is displayed in Figure 1.3(b).

Example 2: Taylor vortices and flows with symmetries.

For this example we choose the second eigenvalue $-\mu = \Lambda_2 = 2$. The wave numbers $\vec{k}$ associated with the shell $|\vec{k}|^2 = \Lambda_2$ are $\vec{k} = (\pm 1, \pm 1), (\pm 1, \mp 1)$, and in this case the stream function ψ in equation (1.10) becomes

$$\psi = a\sin(x)\sin(y) + b\sin(x)\cos(y)$$
$$+c\cos(x)\sin(y) + d\cos(x)\cos(y).$$

In particular, if $a = A$, and $b = c = d = 0$, then the resulting flow pattern is known as the Taylor vortices. This flow pattern consists of a doubly periodic array of eddies, with alternating clockwise and counter-clockwise rotating vortices, all aligned with the x and y axes. The streamlines of the flow pattern are displayed in Figure 1.4(a) and the vorticity is displayed in Figure 1.4(b) for the case where $A = 1$. It is clear from Figure 1.4(a) that there are four vortices in the period box $[0, 2\pi] \times [0, 2\pi]$, two of them rotate clockwise and the other two counter-clockwise. Moreover, the stream function ψ is anti-symmetric under reflection

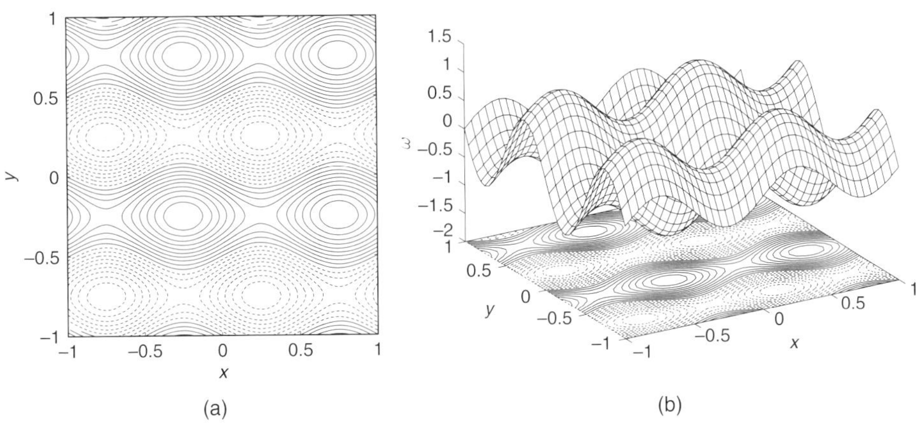

Figure 1.3 2-D array of swirling eddies with stream function $\psi(x, y) = 0.4\sin(x) + 0.8\sin(y)$ (a) Plot of the streamlines: counter-clockwise rotation (—); clockwise rotation (- -). (b) Plot of the vorticity ω.

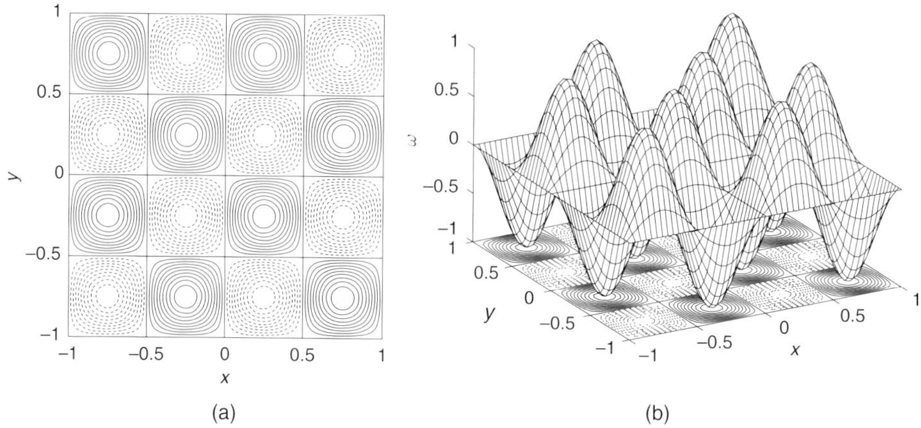

Figure 1.4 Taylor vortices with stream function $\psi(x, y) = \sin(x)\sin(y)$ (x and y both normalized by 2π). (a) Plot of the streamlines: counter-clockwise rotation (—); clockwise rotation (- -). (b) Plot of the vorticity ω.

across the lines given by $C = \{\vec{x}|\psi(\vec{x}) = 0\}$, which include the lines $x = \pi$ and $y = \pi$. In particular, for the lines $x = \pi$ and $y = \pi$ it follows that

$$\psi(\pi - x, y) = -\psi(\pi + x, y), \quad \psi(x, \pi - y) = -\psi(x, \pi + y).$$

Clearly, the Taylor vortices satisfy the no-penetration boundary condition on the level set $C = \{\vec{x}|\psi(\vec{x}) = 0\}$

$$\vec{v} \cdot \vec{n}\Big|_C = \nabla^{\perp}\psi \cdot \frac{\nabla\psi}{|\nabla\psi|}\Bigg|_C = 0.$$

In addition, the vorticity $\omega = -\Lambda_2\psi$ also vanishes on the set C. This implies that the Taylor vortices also satisfy the free-slip boundary condition

$$\frac{\partial}{\partial n}(\vec{v} \cdot \vec{\tau})\Bigg|_C = 0,$$

where $\vec{\tau}$ is the tangent vector to C.

Finally, we also remark that additional flows satisfying the symmetry conditions are also obtained when the stream function ψ is of the form

$$\psi = \sum_{k_1, k_2 = -\infty}^{\infty} a_{k_1 k_2} \sin(k_1 x) \sin(k_2 y).$$

These flows with symmetries include as a particular case the Taylor vortices and for that reason they are called *generalized Taylor vortices*.

1.2.2 Exact solutions with beta-effect and generalized Kolmogorov forcing

In this next example we include the beta-plane effect, dissipation, and external forcing, but assume flat bottom topography so that $h = 0$. Under the assumption of linear dependence of q and ψ, the reduced linear equations for the stream function ψ, equation (1.7), are

$$\begin{aligned} \mu \frac{\partial \psi}{\partial t} &= \mathcal{D}(\Delta)\psi + \mathcal{F}(\vec{x}, t), \\ \mu\psi &= \Delta\psi + \beta y. \end{aligned} \tag{1.22}$$

The inclusion of the non-periodic beta-plane term prevents the direct application of the Fourier analysis as was done before. In fact, we can no longer expect the stream function ψ to be periodic. To circumvent this problem we postulate a non-periodic stream function ψ consisting of a non-periodic large-scale mean flow and a small-scale periodic component

$$\psi = -V_0 y + \psi'(x, y, t), \tag{1.23}$$

where V_0 is the large-scale mean velocity and the small-scale part ψ' is assumed to be 2π-periodic. Note that the velocity field itself will be 2π-periodic. Introducing this form of the stream function ψ into equation (1.21), we see that the mean flow contribution $-V_0 y$ will exactly match the beta-plane term βy, provided that we choose for V_0 the specific value

$$V_0 = -\frac{\beta}{\mu}, \tag{1.24}$$

in which case the reduced linear equations for ψ in equation (1.21) yield the simpler system of equations for the small-scale stream function ψ'

$$\begin{aligned} \mu \frac{\partial \psi'}{\partial t} &= \mathcal{D}(\Delta)\psi' + \mathcal{F}(\vec{x}, t), \\ \mu\psi' &= \Delta\psi'. \end{aligned} \tag{1.25}$$

This last equation can be treated through Fourier analysis as before, provided that we restrict the external forcing $\mathcal{F}$ in such a way that it only excites the modes associated to the eigenvalue μ. This special form of the external forcing is known as *generalized Kolmogorov forcing* and it is of the form

$$\mathcal{F}(\vec{x}, t) = \sum_{|\vec{k}|^2 = \Lambda} \hat{F}_{\vec{k}} e^{i\vec{x}\cdot\vec{k}}. \tag{1.26}$$

With this form of forcing and letting $\Lambda = -\mu$, we assume that the small-scale stream function has the Fourier expansion

$$\psi' = \sum_{|\vec{k}|^2=\Lambda} \hat{A}_{\vec{k}}(t) e^{i\vec{x}\cdot\vec{k}} + c.c. \tag{1.27}$$

Introducing these formulas for the solution ψ' and the external forcing into the equation for ψ' in equation (1.24) and solving for each wave number $\vec{k}$ finally yields the following ordinary differential equation for the Fourier amplitude $\hat{A}_{\vec{k}(t)}$

$$\frac{d\hat{A}_{\vec{k}}}{dt} = \frac{\mathcal{D}(-\Lambda)}{-\Lambda}\hat{A}_{\vec{k}} + \frac{\hat{F}_{\vec{k}}}{-\Lambda}. \tag{1.28}$$

Summarizing, the stream function ψ is given by

$$\psi = -V_0 y + \sum_{|\vec{k}|^2=\Lambda} \hat{A}_{\vec{k}}(t) e^{i\vec{x}\cdot\vec{k}} + c.c.,$$

where the Fourier amplitudes $\hat{A}_{\vec{k}}(t)$ are the solutions of the ODE in equation (1.27) and V_0 is given by $V_0 = -\beta/\mu$. The non-linear stability of these solutions is studied in Chapter 2.

The large-scale component $\psi_0 = -V_0 y$ of the stream function ψ yields the mean flow velocity given by

$$\nabla^{\perp}\psi_0 = \begin{pmatrix} V_0 \\ 0 \end{pmatrix}.$$

Since μ is always negative, then the direction of the wind is determined by the sign of β; for example, if $\beta > 0$ (as in the northern hemisphere), then in this case the beta-plane effect is responsible for the presence of a permanent eastward mean flow since $V_0 = -\beta/\mu > 0$.

Finally, we remark that the ordinary differential equations that govern the Fourier amplitudes $\hat{A}_{\vec{k}}(t)$ of the small-scale flow are of damped and driven type. Clearly they are driven by the external forcing through the forcing amplitudes $\hat{F}_{\vec{k}}$. They are also damped by the dissipative mechanisms, because the coefficient $-\mathcal{D}(-\Lambda)/\Lambda$ satisfies

$$\frac{\mathcal{D}(-\Lambda)}{-\Lambda} = -\frac{\sum_{j=0}^{L} d_j(-1)^j(-\Lambda)^j}{\Lambda} = -\frac{\sum_{j=0}^{L} d_j(\Lambda)^j}{\Lambda} < 0.$$

This is the first mathematical confirmation of the fact that the general operator $\mathcal{D}(\Delta)$ actually represents dissipation. We may also observe that higher frequencies (or wave numbers, thus higher Λ) are damped more efficiently with higher-order dissipation operators, with the order being hyper-viscosity > eddy viscosity > Ekman damping.

1.2.3 Rossby waves

Let us reconsider the previous example. As we saw earlier, the presence of the beta-plane effect forced us to consider a stream function ψ consisting of a large-scale mean flow $-V_0 y$ and a small-scale component ψ'. Here we want to take a second look at the small-scale component ψ'. If we assume at the outset that the stream function has the form

$$\psi = -V_0 y + \psi',$$

and evaluate the Jacobian determinant $J(\psi, q)$

$$J(\psi, q) = V_0 \frac{\partial q}{\partial x} + J(\psi', q),$$

then the potential vorticity equation for q in equation (1.1) becomes

$$\frac{\partial q}{\partial t} + V_0 \frac{\partial q}{\partial x} + J(\psi', q) = \mathcal{D}\psi' + \mathcal{F}. \tag{1.29}$$

The advective term contribution due to the constant mean flow velocity V_0 is easily removed with a Galilean coordinate transformation

$$\tilde{x} = x - V_0 t, \quad \tilde{y} = y, \quad \tilde{t} = t,$$

that is, we are in a coordinate system that moves with the mean flow velocity V_0. Since differentiation in the new variables follows the rules given by

$$\frac{\partial}{\partial x} = \frac{\partial}{\partial \tilde{x}}, \quad \frac{\partial}{\partial y} = \frac{\partial}{\partial \tilde{y}}, \quad \frac{\partial}{\partial t} = \frac{\partial}{\partial \tilde{t}} - V_0 \frac{\partial}{\partial \tilde{x}},$$

it follows that in the moving coordinate system the advective term contribution from V_0 is eliminated, and the potential vorticity q and the small-scale stream function ψ' solve the

Barotropic beta-plane equations

$$\begin{aligned} \frac{\partial q}{\partial t} + J(\psi', q) &= \mathcal{D}\psi' + \mathcal{F}, \\ q &= \Delta\psi' + \beta y, \end{aligned} \tag{1.30}$$

which are the quasi-geostrophic equations in the moving coordinate system.

Since we know that the small-scale solution ψ' in the previous example was given by equation (1.26), we can write this solution in terms of the new coordinates. Keeping in mind that the mean velocity $V_0 = -\beta/\mu = \beta/\Lambda$, we obtain

$$\psi'(\tilde{\vec{x}}, \tilde{t}) = \sum_{|\vec{k}|^2=\Lambda} \hat{A}_{\vec{k}} e^{i(\vec{k}\cdot\tilde{\vec{x}} + k_1 V_0 \tilde{t})} + c.c$$

$$= \sum_{|\vec{k}|^2=\Lambda} \hat{A}_{\vec{k}} e^{i(\vec{k}\cdot\tilde{\vec{x}} + k_1 \frac{\beta}{|\vec{k}|^2} \tilde{t})} + c.c., \tag{1.31}$$

and we recognize that ψ' is a traveling wave of the form $e^{i\vec{k}\cdot\vec{x} - i\omega(\vec{k})t}$, where ω and $\vec{k}$ satisfy the dispersion relation

$$\omega(\vec{k}) = -\frac{k_1\beta}{|\vec{k}|^2}. \tag{1.32}$$

This solution represents a dispersive wave known as a *Rossby wave* (Pedlosky, 1987). Indeed these waves travel horizontally at the speed (the so-called horizontal phase speed) of $\frac{\omega(\vec{k})}{k_1} = -\frac{\beta}{|\vec{k}|^2}$. Thus waves with different wave numbers travel at different speeds and hence are dispersive. See the lecture notes by Majda (2003) for an extensive discussion of dispersive waves in geophysical flows.

In the absence of dissipation and forcing it is interesting to see how the elementary exact solutions of the Euler equations in Subsection 1.2.1 are modified by the beta-effect. They become time-dependent solutions with different phase velocities. Indeed it is easy to see that we have the following:

1. The simple shear in x, $\psi = A\sin(x)$, as presented in Example 1 of Subsection 1.2.1, becomes a Rossby wave $\psi = A\sin(\tilde{x} + \beta t)$ with horizontal phase velocity $= -\beta$.
2. General ground eigenmode flows, $\psi = a\sin(x) + b\cos(x) + c\sin(y) + d\cos(y)$, as presented in Example 1, become the super-position of Rossby waves and zonal shear flows $\psi = a\sin(\tilde{x} + \beta t) + b\cos(\tilde{x} + \beta t) + c\sin(y) + d\cos(y)$ with horizontal phase velocity $= -\beta$.
3. Taylor vortex, $\psi = A\sin(x)\sin(y)$, as presented in Example 2 of Subsection 1.2.1 becomes a Rossby wave $\psi = A\sin(\tilde{x} + \frac{\beta}{2}t)\sin(y)$ with horizontal phase velocity $= -\frac{\beta}{2}$.

We observe that waves with different wave numbers move at different speeds. In fact, waves with higher wave numbers move slower than waves with lower wave numbers in general. This is a feature for geophysical waves, which is different from most of the classical dispersive waves where waves with higher wave number move faster (Majda, 2003).

Let us verify that the Rossby waves are indeed dispersive waves from the mathematical point of view. This is equivalent to checking that the determinant of the Hessian of the dispersive relation does not vanish except for a set of measure zero. The interested reader is encouraged to consult the lecture notes by Majda (2003) for this and much more on dispersive waves. The Hessian $\mathrm{H}(\omega(\vec{k}))$ can be computed easily as

$$\mathrm{H}(\omega(\vec{k})) = -\frac{\beta}{(k_1^2+k_2^2)^3}\begin{pmatrix} 2k_1^3 - 6k_1k_2^2 & 6k_1^2k_2 - 2k_2^3 \\ 6k_1^2k_2 - 2k_2^3 & -2k_1^3 + 6k_1k_2^2 \end{pmatrix},$$

hence we deduce that

$$\begin{aligned}\det \mathrm{H}(\omega(\vec{k})) &= -\frac{4\beta^2}{(k_1^2+k_2^2)^6}((k_1^3 - 3k_1k_2^2)^2 + (3k_1^2k_2 - k_2^3)^2) \\ &= -\frac{4\beta^2}{(k_1^2+k_2^2)^6}(k_1^2(k_1^2 - 3k_2^2)^2 + k_2^2(3k_1^2 - k_2^2)^2).\end{aligned}$$

Since the determinant of the Hessian vanishes only if $k_1^2 = 3k_2^2, 3k_1^2 = k_2^2$, or at the origin, this verifies our claim that Rossby waves are dispersive.

1.2.4 Topographic effect in steady states

In our next example we consider flows that are subject to topography and beta-plane effects, but ignore the effects of dissipation and external forcing. We still assume that the potential vorticity q and the stream function ψ are linearly dependent, as expressed in equation (1.10). Under these circumstances, the reduced linear equations for the stream function ψ in equation (1.7) yield

$$\mu\frac{\partial\psi}{\partial t} = 0, \quad \mu\psi = \Delta\psi + \beta y + h(\vec{x}), \tag{1.33}$$

so that the flow is stationary. The beta-plane effect is again eliminated from equation (1.32), with the assumption that the stream function ψ contains a large-scale mean flow $-V_0y$ and a small-scale component ψ'

$$\psi = -V_0y + \psi', \quad V_0 = -\frac{\beta}{\mu}, \tag{1.34}$$

so that equation (1.32) gets reduced to the following equation for the small-scale stream function ψ'

$$\mu\psi' = \Delta\psi' + h(\vec{x}). \tag{1.35}$$

Immediately we observe how radically different are the solvability conditions for ψ', depending on whether the topography effects are present or absent. More precisely, we recall that for the two examples previously considered there was no topography effects and that μ *had to be* an eigenvalue of the Laplacian with associated eigenfunction given by the small-scale stream function ψ'. On the contrary, if there is topography, then equation (1.34) is in general solvable only if μ *is not* an eigenvalue of the Laplacian. This is a simple consequence of the Fredholm alternative, which we will verify explicitly in the present case through Fourier analysis. For simplicity, we assume that the topography h and the stream function ψ' are periodic functions with zero average and with Fourier expansions given by

$$h = \sum_{|\vec{k}|\neq 0} \hat{h}_{\vec{k}} e^{i\vec{x}\cdot\vec{k}} + c.c., \qquad \psi' = \sum_{|\vec{k}|\neq 0} \hat{\psi}'_{\vec{k}} e^{i\vec{x}\cdot\vec{k}} + c.c.. \tag{1.36}$$

Introducing these expansions into equation (1.34) and solving for each mode by separation of variables yields

$$-(|\vec{k}|^2 + \mu)\hat{\psi}'_{\vec{k}} = -\hat{h}_{\vec{k}}, \tag{1.37}$$

which can be solved in general only if μ is not one of the eigenvalues of the Laplacian ($\mu + |\vec{k}|^2 \neq 0$). In this case we can solve for the Fourier coefficients $\hat{\psi}'_{\vec{k}}$ and obtain the stream function

$$\psi = -V_0 y + \psi' = \frac{\beta}{\mu} y + \sum_{|\vec{k}|\neq 0} \frac{1}{|\vec{k}|^2 + \mu} \hat{h}_{\vec{k}} e^{i\vec{x}\cdot\vec{k}} \tag{1.38}$$

and the corresponding velocity field $\vec{v}$ and (relative) vorticity ω

$$\vec{v} = \nabla^{\perp}\psi = -\begin{pmatrix} \beta/\mu \\ 0 \end{pmatrix} + \sum_{|\vec{k}|\neq 0} \frac{1}{|\vec{k}|^2 + \mu} \hat{h}_{\vec{k}} i e^{i\vec{x}\cdot\vec{k}} \begin{pmatrix} -k_2 \\ k_1 \end{pmatrix}, \tag{1.39}$$

$$\omega = \Delta\psi = -\sum_{|\vec{k}|\neq 0} \frac{|\vec{k}|^2}{|\vec{k}|^2 + \mu} \hat{h}_{\vec{k}} e^{i\vec{x}\cdot\vec{k}}. \tag{1.40}$$

Next we explore some interesting properties of the solutions we just derived. Let us start by considering the situation where there is no beta-plane effect, so that $\beta = 0$, and therefore also the mean flow $V_0 = 0$. From the formula for the velocity field, $\vec{v}$, it readily follows that in this case the kinetic energy E is given by

$$E = \frac{1}{2}\int |\vec{v}|^2 = \frac{1}{2} \sum_{|\vec{k}|\neq 0} \frac{(2\pi)^2 |\vec{k}|^2 |\hat{h}_{\vec{k}}|^2}{(|\vec{k}|^2 + \mu)^2}, \tag{1.41}$$

since the fluid density is 1.

A plot of the energy E versus the parameter μ is given in Figure 1.5. Since in general we may assume that $\hat{h}_{\vec{k}} \neq 0$, then the energy E will become singular when μ is an eigenvalue of the Laplacian, $\mu = -|\vec{k}|^2$, $\vec{k} \in \mathcal{Z}^2 - \{0\}$. In particular, if we make the natural assumption that $\hat{h}_{\vec{k}} \neq 0$ for $|\vec{k}| = 1$, then the energy E varies from some positive value to infinity as μ varies in the range $-1 < \mu < +\infty$.

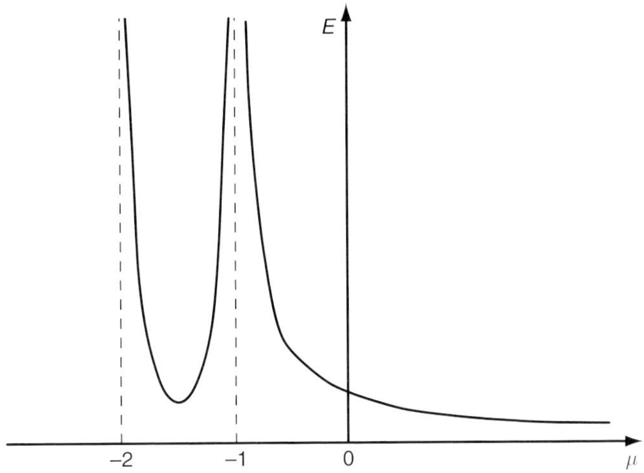

Figure 1.5 Plot of the total energy $E = \frac{1}{2}\sum_{|\vec{k}|\neq 0} \frac{(2\pi)^2|\vec{k}|^2|\hat{h}_{\vec{k}}|^2}{(|\vec{k}|^2+\mu)^2}$ versus μ in the case of no beta-plane effect (and no mean flow V_0). The singularities are given by the eigenvalues of the Laplacian, $\mu = -1, -2, -5, \ldots$

These are somehow robust states. In a later chapter we will see that these states are non-linearly stable. In other words, they are typically observed states without mean flow.

Next we consider the case $\beta \neq 0$. In this case there is a mean flow contribution given by $V_0 = -\beta/\mu$, so that the total kinetic energy has contributions from the large-scale mean flow and the small-scale flow. From the Fourier representation of the velocity field $\vec{v}$ in equation (1.28), it follows that the total kinetic energy E is given by

$$
\begin{aligned}
E_{\text{tot}} &= \frac{1}{2}V_0^2(2\pi)^2 + E_{\text{small scale}} \qquad (1.42)\\
&= \left(\frac{\beta^2}{2\mu^2} + \frac{1}{2}\sum_{|\vec{k}|\neq 0} \frac{|\vec{k}|^2|\hat{h}_{\vec{k}}|^2}{(|\vec{k}|^2+\mu)^2}\right)(2\pi)^2.
\end{aligned}
$$

A plot of the kinetic energy E versus μ for this case is given in Figure 1.6. In this case the presence of the beta-plane effect introduces through the presence of the mean flow an additional singularity at $\mu = 0$, so that in this case the energy E goes to infinity as μ approaches zero, while in the range $0 < \mu < \infty$. Again we expect that the steady state solutions with $\mu > 0$ will be stable. Interestingly, when $\beta > 0$ these steady states have negative mean flow velocity, $V_0 = -\beta/\mu < 0$, and are associated with westward mean flow. This is in marked contrast with the earlier example, where there was beta-plane effect but no topography present, and

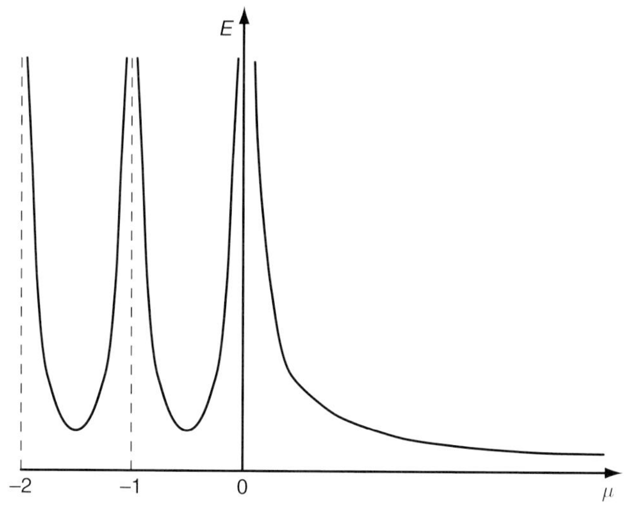

Figure 1.6 Plot of the total energy $E = \left(\frac{\beta^2}{2\mu^2} + \frac{1}{2} \sum_{|\vec{k}|\neq 0} \frac{|\vec{k}|^2 |\hat{h}_{\vec{k}}|^2}{(|\vec{k}|^2+\mu)^2} \right) (2\pi)^2$ versus μ in the case of beta-plane effect (and non-zero mean flow V_0). The singularities are $\mu = 0, -1, -2, -5, \ldots$

the resulting flow had an eastward mean flow. Nevertheless, this conclusion is consistent with the observations of flows in the ocean. The flow in the oceans is much slower than in the atmosphere and the effects due to the topography are more important. Indeed, in many cases the flow will be westward because of the topographic effect.

Next we would like to discuss the effect that the topography has on the vorticity. For example, we can wonder whether a rise in the terrain will increase or decrease the rotation of the flow? It turns out that the vorticity of the flow decreases as it climbs up the terrain and increases as it climbs down. If we ignore the dissipation and external forcing effects, then the equations for the potential vorticity q in (1.1) reduce to

$$\frac{\partial q}{\partial t} + \vec{v} \cdot \nabla q = 0, \qquad q = \omega + \beta y + h.$$

Thus q should be conserved along the particle trajectory $\vec{X}(\alpha, t)$, which satisfies

$$\frac{d\vec{X}(\alpha, t)}{dt} = \vec{v}(\vec{X}(\alpha, t), t).$$

That is

$$q(\vec{X}(\alpha, t), t) = q_0(\vec{X}(\alpha, 0)).$$

Now suppose the flow transports a fluid particle from a location with a lower value of $h+\beta y$ to a location with a higher value of $h+\beta y$, i.e. the particle climbs a mountain. Then the conservation of the potential vorticity tells us that (negative anti-cyclonic) vorticity ω must be produced. The same reasoning applies for the sea valley case. Thus, if we think of steady states as limits at long times of the dynamics, we expect the relative vorticity ω and the topography h to mostly have opposite signs. We could certainly see that this is the case if the vorticity ω and the topography h are anti-linear, $\omega = -\alpha h$ with $\alpha > 0$; under these circumstances an increase in h certainly corresponds to a decrease in ω, and vice versa. However, in general, the vorticity ω and the topography h are not related in such a manner; this is certainly evident in the explicit formulas in equation (1.37) and equation (1.39) for the topography and the vorticity. One way to provide a rigorous quantitative argument supporting the above assertion that an increase in the topography entails a corresponding decrease in the vorticity utilizes the correlation of two functions.

The correlation of two functions f and g, $\mathrm{Corr}(f, g)$, is defined by

$$\mathrm{Corr}(f, g) = \frac{\int fg}{(\int f^2)^{\frac{1}{2}}(\int g^2)^{\frac{1}{2}}}.$$

Intuitively, the correlation function gives a measure of the "angle" between the functions f and g. The correlation function satisfies the following properties:

(i) $\mathrm{Corr}(f, g) = 1$ if and only if $g = \alpha f$ with $\alpha > 0$.
(ii) $\mathrm{Corr}(f, g) = -1$ if and only if $g = \alpha f$ with $\alpha < 0$.
(iii) In general, $-1 \le \mathrm{Corr}(f, g) \le 1$ by the Cauchy–Schwarz inequality.

The strong condition that ω and h are anti-linear, $\omega = -\alpha h$, is now seen to be equivalent to showing that $\mathrm{Corr}(h, \omega) = -1$. Instead, we will show that h and ω are anti-correlated, $\mathrm{Corr}(h, \omega) < 0$ for an appropriate range of μ. Indeed, a straightforward calculation utilizing the Fourier representations of ω and h shows that

$$\int h\omega = \sum_{|\vec{k}|\neq 0} \overline{\hat{h}_{\vec{k}}} \omega_{\vec{k}} = -(2\pi)^2 \sum_{|\vec{k}|\neq 0} \frac{|\vec{k}|^2 |\hat{h}_{\vec{k}}|^2}{|\vec{k}|^2 + \mu} < 0,$$

and therefore we conclude that for $-1 < \mu < 0$

$$\mathrm{Corr}(h, \omega) = \frac{\int h\omega}{(\int h^2)^{\frac{1}{2}}(\int \omega^2)^{\frac{1}{2}}} < 0, \tag{1.43}$$

and this implies that *topography and vorticity are anti-correlated* provided that $\mu > -1$. Thus, the states with $\mu > -1$ satisfy the physical intuition described above.

Although it is clear that the topography h and the vorticity ω are not anti-linear over all scales, we will show next that they are approximately anti-linear at sufficiently small scales. The argument that follows nicely quantifies the effect of topography on various scales of the vorticity.

Since $\Delta\psi = \mu\psi - h$, the unit of μ is $\frac{1}{L^2}$. Let us look at the piece of solution at length scale smaller than the length scale $|\mu|^{-\frac{1}{2}}$ i.e. $|\vec{k}| \gg |\mu|^{\frac{1}{2}}$. Let us define the vorticity on small scale as

$$\begin{aligned}\omega_\Theta &= -\sum_{|\vec{k}|^2>\Theta} \frac{|\vec{k}|^2}{|\vec{k}|^2+\mu} \hat{h}_{\vec{k}} e^{i\vec{x}\cdot\vec{k}}, \qquad \Theta \gg |\mu|^{\frac{1}{2}} \\ &\cong -\sum_{|\vec{k}|^2>\Theta} 1 \times \hat{h}_{\vec{k}} e^{i\vec{x}\cdot\vec{k}} = -h_{\text{small scale}},\end{aligned}$$

thus showing that at small scales the vorticity and the topography are roughly opposite to each other. Therefore, when the flow goes over a sea mountain ($h > 0$), the small-scale vorticity decreases, i.e. small-scale flow becomes anti-cyclonic. On the other hand, when the flow goes over a sea valley, the small-scale vorticity increases, i.e. small-scale flow becomes more cyclonic. We also remark that our result is exact for $\mu = 0$.

1.2.5 A dynamical solution with beta-plane and layered topography

Here we present one more example which has the beta-plane effect (hence Rossby waves) and topography. These solutions, a set of linear ODE, provide convenient elementary test problems for the effect of various dissipations on the Rossby waves.

We first rewrite the equation for the potential vorticity equation (1.7), in terms of the vorticity $\omega = \Delta\psi$ rather than the potential vorticity q. Since

$$\begin{aligned}J(\psi, q) &= J(\psi, \Delta\psi) + J(\psi, \beta y) + J(\psi, h) \\ &= J(\psi, \omega) + \beta\psi_x + \nabla^\perp\psi \cdot \nabla h, \\ \frac{\partial q}{\partial t} &= \frac{\partial \omega}{\partial t},\end{aligned}$$

the equation (1.7) for the potential vorticity becomes

$$\frac{\partial \omega}{\partial t} + J(\psi, \omega) + \beta\psi_x + \nabla^\perp\psi \cdot \nabla h = \mathcal{D}(\Delta)\psi + \mathcal{F}. \tag{1.44}$$

We assume a special topography called layered topography. Let $\vec{k}_0 = (k_{0,x}, k_{0,y})$ be fixed and assume

$$h(\vec{x}) = \sum_{j\neq 0} \hat{h}_j e^{ij\vec{k}_0\cdot\vec{x}}. \tag{1.45}$$

We look for solutions with the same layered structure, i.e. we try

$$\psi = \sum_{j \neq 0} \hat{\psi}_j(t) e^{ij\vec{k}_0 \cdot \vec{x}}, \tag{1.46}$$

$$\omega = \sum_{j \neq 0} (-j^2 |\vec{k}_0|^2) \hat{\psi}_j(t) e^{ij\vec{k}_0 \cdot \vec{x}}. \tag{1.47}$$

The advantage of choosing layered topography and stream function is that the non-linear term vanishes. Indeed

$$\begin{aligned} J(\psi, \omega) &= \psi_x \omega_y - \psi_y \omega_x \\ &= \sum_{j_1 \neq 0, j_2 \neq 0} (j_1 k_{0,x} j_2^3 k_{0,y} |\vec{k}_0|^2 \hat{\psi}_{j_1}(t) \hat{\psi}_{j_2}(t) \\ &\qquad - j_1 k_{0,y} j_2^3 k_{0,x} |\vec{k}_0|^2 \hat{\psi}_{j_1}(t) \hat{\psi}_{j_2}(t)) e^{i(j_1 + j_2)\vec{k}_0 \cdot \vec{x}} \\ &= 0, \\ \nabla^{\perp} \psi \cdot \nabla h &= \psi_x h_y - \psi_y h_x \\ &= \sum_{j_1 \neq 0, j_2 \neq 0} (j_1 k_{0,x} j_2 k_{0,y} \hat{\psi}_{j_1}(t) \hat{h}_{j_2}(t) \\ &\qquad - j_1 k_{0,y} j_2 k_{0,x} \hat{\psi}_{j_1}(t) \hat{h}_{j_2}(t)) e^{i(j_1 + j_2)\vec{k}_0 \cdot \vec{x}} \\ &= 0. \end{aligned}$$

Hence we are left to solve the reduced linear equation

$$\frac{\partial \omega}{\partial t} + \beta \psi_x = \mathcal{D}(\Delta)\psi + \mathcal{F}. \tag{1.48}$$

By assuming a layered forcing of the form

$$\mathcal{F} = \sum_{j \neq 0} \hat{\mathcal{F}}_j(t) e^{ij\vec{k}_0 \cdot \vec{x}},$$

and using the identities

$$\begin{aligned} \frac{\partial \omega}{\partial t} &= \sum_{j \neq 0} (-j^2 |\vec{k}_0|^2) \frac{d\hat{\psi}_j(t)}{dt} e^{ij\vec{k}_0 \cdot \vec{x}}, \\ \beta \psi_x &= \sum_{j \neq 0} ij\beta k_{0,x} \hat{\psi}_j(t) e^{ij\vec{k}_0 \cdot \vec{x}}, \\ \mathcal{D}(\Delta)\psi &= \sum_{j \neq 0} \mathcal{D}(-j^2 |\vec{k}_0|^2) \hat{\psi}_j(t) e^{ij\vec{k}_0 \cdot \vec{x}}, \end{aligned}$$

we can change the linear PDE (1.48) to the system of ODE for the Fourier coefficients $\hat{\psi}_j(t)$

$$\frac{d\hat{\psi}_j(t)}{dt} - \frac{i\beta k_{0,x}}{j|\vec{k}_0|^2}\hat{\psi}_j = \frac{\mathcal{D}(-j^2|\vec{k}_0|^2)}{-j^2|\vec{k}_0|^2}\hat{\psi}_j(t) - \frac{\hat{\mathcal{F}}_j(t)}{j^2|\vec{k}_0|^2}. \tag{1.49}$$

This linear ODE system provides us with an excellent example to study the effect of various dissipative mechanisms on Rossby wave propagation, both numerically and analytically. In Chapter 5 we will discuss the chaotic dynamics of solutions of this type interacting with a large-scale mean flow.

1.2.6 Beta-plane dynamics with large-scale shear flow

For realistic geophysical flows, if we average wind over a month at mid-latitude, we will see the average wind velocity varies according to its latitude. So large-scale mean flow is typically sheared in this situation. In this section we study the barotropic quasi-geostrophic motion riding on a large-scale shear flow. Our model for this flow is the following:

Beta-plane equations with mean shear flow

$$\begin{aligned} q &= \Delta\psi + \beta y, \qquad \Delta\psi \equiv \omega, \\ \vec{v} &= \begin{pmatrix} \gamma y \\ 0 \end{pmatrix} + \nabla^{\perp}\psi, \quad \psi|_{t=0} \text{ is periodic}, \\ \frac{\partial\omega}{\partial t} &+ \gamma y\frac{\partial\omega}{\partial x} + J(\psi,\omega) + \beta\psi_x = \mathcal{D}(\Delta)\psi. \end{aligned} \tag{1.50}$$

We remark that the velocity field is *not* spatially periodic in this case except for special flow configurations. The coefficient γ measures the strength of the mean shear flow. We do not assume ψ is periodic for all time in a fixed reference frame, but periodic in the Lagrangian coordinates frame following the large-scale shear flow.

In fact, the equation (1.50) has interesting exact solutions, which generalize those just presented in previous subsections to the situation with an ambient uniform shear flow provided that topography and external forcing are ignored. For the purpose of deriving the exact solution, we introduce Lagrangian coordinates following the mean flow

$$\frac{dx}{dt'} = \gamma y,\ x(0) = X; \quad \frac{dy}{dt'} = 0,\ y(0) = Y; \quad \frac{dt}{dt'} = 1,\ t(0) = 0.$$

Hence we have

$$\begin{pmatrix} x \\ y \\ t \end{pmatrix} = \begin{pmatrix} X+\gamma Y t' \\ Y \\ t' \end{pmatrix}, \qquad \begin{pmatrix} \frac{\partial}{\partial x} \\ \frac{\partial}{\partial y} \\ \frac{\partial}{\partial t} \end{pmatrix} = \begin{pmatrix} \frac{\partial}{\partial X} \\ \frac{\partial}{\partial Y} - \gamma t' \frac{\partial}{\partial X} \\ \frac{\partial}{\partial t'} - \gamma Y \frac{\partial}{\partial X} \end{pmatrix}$$

and set

$$\vec{x} = (x, y), \quad \vec{X} = (X, Y), \quad \omega(X, Y, t) = \omega(x(X, Y), Y, t).$$

We then have

$$\begin{aligned} J_{\vec{x}}(\psi, \omega) &= \det \begin{pmatrix} \nabla_{\vec{x}}\psi \\ \nabla_{\vec{x}}\omega \end{pmatrix} = \det \begin{pmatrix} 1 & 0 \\ -\gamma t' & 1 \end{pmatrix} \times \det \begin{pmatrix} \nabla_{\vec{X}}\psi \\ \nabla_{\vec{X}}\omega \end{pmatrix} \\ &= \det \begin{pmatrix} \nabla_{\vec{X}}\psi \\ \nabla_{\vec{X}}\omega \end{pmatrix} = J_{\vec{X}}(\psi, \omega), \end{aligned}$$

and hence the transformed equations in the Lagrangian coordinates take the form

$$\mathcal{L}_{\vec{X}}(t)\psi = \omega, \qquad \mathcal{L}_{\vec{X}}(t) = \frac{\partial^2}{\partial X^2} + \left(\frac{\partial}{\partial Y} - \gamma t \frac{\partial}{\partial X}\right)^2, \quad \Delta_{\vec{x}} = \mathcal{L}_{\vec{X}}(t),$$

$$\frac{\partial \omega}{\partial t} + J_{\vec{X}}(\psi, \omega) + \beta \psi_X = \mathcal{D}(\mathcal{L}_{\vec{X}}(t))\psi. \tag{1.51}$$

Now we *assume* that the stream function is periodic in the Lagrangian coordinates following the background large shear flow. Applying the Fourier representation in Lagrangian coordinates, we can write

$$\begin{aligned} \psi(X, Y, t) &= \sum_{|\vec{k}|\neq 0} \hat{\psi}_{\vec{k}} e^{i\vec{k}\cdot\vec{X}}, \quad \overline{\hat{\psi}_{\vec{k}}} = \hat{\psi}_{-\vec{k}}, \\ \omega(X, Y, t) &= \sum_{|\vec{k}|\neq 0} (-\Lambda_{\vec{k}}(t))\hat{\psi}_{\vec{k}} e^{i\vec{k}\cdot\vec{X}}, \\ \Lambda_{\vec{k}}(t) &= k_1^2 + (k_2 - \gamma t k_1)^2. \end{aligned} \tag{1.52}$$

In order to find exact solutions to the system, we further *assume* that we have layered structure, i.e. there exists $\vec{k}_0 = (k_{01}, k_{02}) \neq 0$ such that

$$\omega(\vec{X}, t) = \sum_{j\neq 0} A_j e^{ij\vec{k}_0\cdot\vec{X}} + c.c. \tag{1.53}$$

We then have, for the stream function

$$\psi(\vec{X}, t) = \sum_{j\neq 0} \frac{-A_j}{\Lambda_j(t)} e^{ij\vec{k}_0\cdot\vec{X}} + c.c., \tag{1.54}$$

where

$$\Lambda_j(t) = \Lambda_{j\vec{k}_0} = j^2 \Lambda_{\vec{k}_0}. \tag{1.55}$$

Following the same reasoning already developed in the previous example of layered structure, we conclude that the non-linear terms drop out and we have the reduced linear dynamics of

$$\frac{d}{dt} A_j(t) - ij\beta \frac{k_{0,1}}{\Lambda_j(t)} A_j(t) = -\frac{\mathcal{D}(-\Lambda_j(t))}{\Lambda_j(t)} A_j(t) \tag{1.56}$$

and the solutions are

$$A_j(t) = A_j(0) \exp\left(\int_0^t \left(\frac{ij\beta k_{01}}{\Lambda_{j\vec{k}_0}(s)} - \frac{\mathcal{D}(-\Lambda_{j\vec{k}_0}(s))}{\Lambda_{j\vec{k}_0}(s)} \right) ds \right) \tag{1.57}$$

and hence we have:

Plane wave solutions in a uniform shear flow

$$\omega(t) = \sum_{j\neq 0} A_j(0) \exp\left(-\int_0^t \frac{\mathcal{D}(-\Lambda_{j\vec{k}_0}(s))}{\Lambda_{j\vec{k}_0}(s)} ds \right) \times \exp\left(ij\vec{k}_0 \cdot \vec{X} + i\int_0^t \frac{j\beta k_{01}}{\Lambda_{j\vec{k}_0}(s)} ds \right) \tag{1.58}$$

$$= \sum_{j\neq 0} A_j(0) \exp\left(-\int_0^t \frac{\mathcal{D}(-\Lambda_{j\vec{k}_0}(s))}{\Lambda_{j\vec{k}_0}(s)} ds \right)$$

$$\times \exp\left(ij\vec{k}_0 \cdot \vec{x} - ijk_{01}\gamma y t + i\int_0^t \frac{j\beta k_{01}}{\Lambda_{j\vec{k}_0}(s)} ds \right). \tag{1.59}$$

Notice that Λ_j changes as time changes. We may interpret these waves as Rossby waves riding on a shear flow. The above solutions provide a very interesting quantitative test for the behavior of various dissipative mechanisms in more realistic flows, which often consist of Rossby waves embedded in shear flows.

A test problem for dissipative mechanisms

Recall that without background shear flows, hyper-viscosity has more damping power than the eddy viscosity, which in turn has more damping power than the Ekman drag at least on layered exact solutions for the enstrophy. We will see that this picture can be completely changed in the presence of the background large shear flow. It is easy to see that the enstrophy takes the form

$$\mathcal{E}(t) = \sum_{j\neq 0} |A_j(0)|^2 \exp\left(-2\sum_{m=1}^{l} d_m \int_0^t \Lambda_{j\vec{k}_0}(s)^{m-1} ds \right). \tag{1.60}$$

Thus the enstrophy is *conserved* if there is no dissipation. For the purpose of illustration, we will focus on a plane wave solution in a uniform shear flow with one single mode solution, i.e.

$$\omega = A_{\vec{k}} e^{i\vec{k}\cdot\vec{X}} + c.c.$$

This is allowed in the exact solution with $j = 1$, $\vec{k} = \vec{k}_0$. We will consider three types of dissipations: Ekman damping only, eddy viscosity only, and hyper-viscosity only. In this case of one mode solutions

$$\mathcal{E}(t) = \mathcal{E}(0)\mathcal{D}, \tag{1.61}$$

$$\mathcal{D} = \exp\left(-2d_m \int_0^t \Lambda_{\vec{k}}(s)^{m-1} ds\right). \tag{1.62}$$

We will use the log of the decay factor ($\ln \mathcal{D}$) of the enstrophy to illustrate the point. After a simple integration we have:

- Ekman damping, $m = 1$

$$\ln \mathcal{D}_{\text{Ek}} = -2d_1 t. \tag{1.63}$$

- Eddy diffusivity, $m = 2$

$$\ln \mathcal{D}_{\text{Ed}} = -2d_2\left(\frac{1}{3}\gamma^2 k_1^2 t^3 - \gamma k_1 k_2 t^2 + \left(k_1^2 + k_2^2\right) t\right). \tag{1.64}$$

- Hyper-viscosity, $m = 3$

$$\begin{aligned}\ln \mathcal{D}_{\text{Hp}} = -2d_3 \Big(&\frac{1}{5}\gamma^4 k_1^4 t^5 + a_4\gamma^3 t^4 + a_3\gamma^2 t^3 \\ &-2\gamma k_1 k_2 \left(k_1^1 + k_2^2\right) t^2 + \left(k_1^2 + k_2^2\right)^2 t\Big),\end{aligned} \tag{1.65}$$

where $a_3 = 2k_1^2\left(k_2^2 + \frac{1}{3}k_1^2\right)$, $a_4 = -k_1^3 k_2$.

We first discuss how each type of the dissipation operator is affected if γ is changed. For this purpose we take the partial derivatives of the log of the decay factor with respect to γ and we deduce

$$\frac{\partial}{\partial\gamma}\ln \mathcal{D}_{\text{Ek}} = 0, \tag{1.66}$$

$$\frac{\partial}{\partial\gamma}\ln \mathcal{D}_{\text{Ed}} = -2d_2\left(\frac{2}{3}\gamma k_1^2 t^3 - k_1 k_2 t^2\right), \tag{1.67}$$

$$\begin{aligned}\frac{\partial}{\partial\gamma}\ln \mathcal{D}_{\text{Hp}} = -2d_3 \Big(&\frac{4}{5}\gamma^3 k_1^4 t^5 + 3a_4\gamma^2 t^4 \\ &+2a_3\gamma t^3 - 2k_1 k_2 \left(k_1^1 + k_2^2\right) t^2\Big).\end{aligned} \tag{1.68}$$

Hence we may conclude:

Large uniform shear effects on dissipation: sensitivity on γ, one mode solution

(i) Long time
- Case Ekman damping: no effect.
- Case Eddy viscosity: the increase of γ will enhance dissipation if $\gamma k_1^2 > 0$ and decrease dissipation if $\gamma k_1^2 < 0$.
- Case hyper-viscosity: the increase of γ will enhance dissipation if $\gamma k_1^2 > 0$ and decrease dissipation if $\gamma k_1^2 < 0$.

(ii) Short time
- Case Ekman damping: no effect.
- Case Eddy viscosity: the increase of γ will enhance dissipation if $k_1 k_2 < 0$ and decrease dissipation if $k_1 k_2 > 0$.
- Case hyper-viscosity: the increase of γ will enhance dissipation if $k_1 k_2 < 0$ and decrease dissipation if $k_1 k_2 > 0$.

We now turn to compare the different effects of the background large shear flows on different dissipations when γ is fixed.

It is easy to see from the explicit formula for the decay factors that the order of dissipations is preserved at the long time. In fact the higher-order dissipation (eddy diffusivity and hyper-viscosity) are further enhanced if $k_1 \neq 0$. Of course when $k_1 = 0$ (i.e. the flow is a zonal flow), the background shear flow has no effect on the dynamics and the order of dissipation is preserved as well.

Next we turn to the regime where time is short. In order to have a fair comparison, we will assume that initially (at time zero) they have the same decay rate. This implies that there exists a $d > 0$ such that

$$d_3 = d, \quad d_2 = (k_1^2 + k_2^2)d, \quad d_1 = \left(k_1^2 + k_2^2\right)^2 d.$$

Thus for short time we have

$$\ln \mathcal{D}_{\text{Ed}} \doteq \ln \mathcal{D}_{\text{Ek}} + 2d\gamma k_1 k_2 \left(k_1^2 + k_2^2\right) t^2 \tag{1.69}$$

$$\ln \mathcal{D}_{\text{Hp}} \doteq \ln \mathcal{D}_{\text{Ed}} + 2d\gamma k_1 k_2 \left(k_1^2 + k_2^2\right) t^2. \tag{1.70}$$

Hence we may conclude:

Large uniform shear effects on dissipation: sensitivity on operator and wave number, one mode solution

(i) Long time. The order of dissipation is maintained. In fact:
- Case Ekman damping: no effect.
- Case eddy viscosity: γ enhances the dissipation with a higher-order dissipation proportional to $\gamma^2 k_1^2 t^3$.
- Case hyper-viscosity: γ enhances the dissipation with a higher-order dissipation proportional to $\gamma^4 k_1^4 t^5$.

(ii) Short time. The order of dissipation could be completely reversed depending on the wave number and γ. More specifically:

- The order of dissipation is maintained, i.e. hyper-viscosity > eddy viscosity > Ekman damping, if $\gamma k_1 k_2 < 0$.
- The order of dissipation is completely reversed, i.e. hyper-viscosity < eddy viscosity < Ekman damping, if $\gamma k_1 k_2 > 0$.

We want to point out here that we assumed periodicity in the Lagrangian coordinates following the large-scale background shear. This is not the same as periodicity in the Eulerian coordinates in general. However, for the special case of channel geometry, these two notions are the same, as will be clear later in the chapter. Unfortunately, the exact solutions that we built above do not fit the channel geometry in general unless we assume zonal flows only. Of course, there is no influence by the background large-scale zonal shear on zonal flows; hence it is not interesting for our purpose.

1.3 Conserved quantities

Conserved quantities play a decisive role both in physics and in mathematics. They are especially important in studying general properties of solutions. They are of great importance in the non-linear stability theory and statistical theory. Thus we can hardly over-estimate the importance of conserved quantities. They will be used throughout the remainder of this book.

1.3.1 Conservation of energy

The conservation of energy is especially important in physics. The traditional way to show kinetic energy conservation is to use the momentum equation. But for the geophysical fluid, it is often more convenient to use the potential vorticity equation.

Conservation of energy for periodic flows

For the sake of simplicity, we consider first the simple case without mean flow.

Keeping

$$\vec{v} = \nabla^{\perp}\psi, \quad q = \Delta\psi + \beta y + h \tag{1.71}$$

in mind, we see the kinetic energy has a natural candidate

$$E = \frac{1}{2}\int |\vec{v}|^2 = \frac{1}{2}\int |\nabla\psi|^2, \tag{1.72}$$

since the fluid density is 1.

The traditional way of measuring the kinetic energy change is to use the evolution equation for the velocity (the momentum equation). However, for geophysical flows, the equation for potential vorticity actually defines the dynamics and is often more convenient. Using the dynamic equation

$$\frac{\partial q}{\partial t} + \nabla^{\perp}\psi \cdot \nabla q = \mathcal{D}(\Delta)\psi + \mathcal{F}(\vec{x}, t),$$

and differentiating the kinetic energy in time, we get the energy change rate

$$\begin{aligned}\frac{dE}{dt} &= \int \nabla\psi \cdot \nabla\psi_t = -\int \psi \frac{\partial}{\partial t}\Delta\psi = -\int \psi \frac{\partial q}{\partial t} \\ &= \left\{\int \psi\nabla^{\perp}\psi \cdot \nabla q\right\} + \left\{-\int \psi\mathcal{D}(\Delta)\psi\right\} + \left\{-\int \psi\mathcal{F}(\vec{x}, t)\right\} \\ &= \{1\} + \{2\} + \{3\}.\end{aligned} \tag{1.73}$$

We will see that $\{1\}$ represents contribution from advection, $\{2\}$ represents contribution from dissipation, and $\{3\}$ represents contribution from external forcing.

Indeed, we have:

Claim The net contribution from advection is zero, i.e. $\{1\} = 0$.
Because the potential vorticity q, $q = \Delta\psi + \beta y + h$, is not a periodic function in the presence of βy, the actual calculation becomes subtle under periodic geometry.

Since

$$\begin{aligned}\psi\nabla^{\perp}\psi \cdot \nabla q &= \psi\nabla^{\perp}\psi \cdot \nabla(\omega + h) + \beta\psi\psi_x \\ &= \frac{1}{2}\nabla^{\perp}(\psi^2) \cdot \nabla(\omega + h) + \frac{\beta}{2}(\psi^2)_x,\end{aligned}$$

we get

$$\begin{aligned}\{1\} &\equiv \int_{[0,2\pi]\times[0,2\pi]} \psi\nabla^{\perp}\psi \cdot \nabla q \\ &= \frac{1}{2}\int_{[0,2\pi]\times[0,2\pi]} \nabla^{\perp}(\psi^2) \cdot \nabla(\omega + h) + \frac{\beta}{2}\int_{[0,2\pi]\times[0,2\pi]} (\psi^2)_x \\ &= \frac{1}{2}\int_{\partial[0,2\pi]\times[0,2\pi]} (\omega + h)\nabla^{\perp}(\psi^2) \cdot \vec{n} + \frac{\beta}{2}\left(\int_{x=2\pi} (\psi^2) - \int_{x=0} (\psi^2)\right) \\ &= 0,\end{aligned}$$

where $\vec{n}$ denotes the unit outer-normal to the periodic domain $[0, 2\pi] \times [0, 2\pi]$. The boundary terms drop out since ψ, ω, and h are periodic.

Claim $\{2\}$ represents dissipation, i.e. $\{2\} \leq 0$.
We first recall that we have already seen that the dissipation operator really represents dissipation in some exact solutions. Now we prove the same property for general solutions.

Indeed

$$\{2\} = -\int \psi \mathcal{D}(\Delta)\psi = -\int \psi \sum_{j=0}^{l} d_j(-1)^j \Delta^j \psi.$$

Notice for $j = 2k$ even, we have

$$\int \psi(-1)^j \Delta^j \psi = \int \psi \Delta^{2k} \psi = \int \Delta^k \psi \Delta^k \psi = \int \Delta^{j/2} \psi \Delta^{j/2} \psi.$$

Likewise, if $j = 2k + 1$ is odd, we have

$$\int \psi(-1)^j \Delta^j \psi = -\int \psi \Delta^{2k+1} \psi = -\int \Delta^k \psi \Delta^{k+1} \psi = \int \nabla \Delta^k \psi \cdot \nabla \Delta^k \psi$$

$$= \int \nabla \Delta^{(j-1)/2} \psi \cdot \nabla \Delta^{(j-1)/2} \psi,$$

where we utilized the classical integration by parts formula

$$\int (\Delta F) G = -\int \nabla F \cdot \nabla G = \int F \Delta G$$

for periodic functions F and G.

Hence we have

$$\{2\} = -\sum_{j=0,\, j:\text{even}}^{l} d_j \int |\Delta^{j/2} \psi|^2 - \sum_{j=1,\, j:\text{odd}}^{l} d_j \int |\nabla \Delta^{(j-1)/2} \psi|^2.$$

Claim $\{3\}$ We will assume that $\int \mathcal{F} = 0$. This means that we can rewrite $\mathcal{F}$ in the alternative form of $\mathcal{F} = \text{curl} \vec{F}$, $\vec{F} = \nabla^\perp \Phi$ or $\mathcal{F} = \Delta \Phi$. More specifically, if $\mathcal{F}$ takes the Fourier expansion

$$\mathcal{F} = \sum_{\vec{k} \neq 0} \hat{\mathcal{F}}_{\vec{k}} e^{i\vec{k}\cdot\vec{x}},$$

then Φ is given by

$$\Phi = \sum_{\vec{k} \neq 0} \frac{-\hat{\mathcal{F}}_{\vec{k}}}{|\vec{k}|^2} e^{i\vec{k}\cdot\vec{x}}$$

and we have the desired alternative forms. It then follows that

$$\{3\} = -\int \psi \mathcal{F} = -\int \psi \,\text{curl}\, \vec{F} = \int \nabla^\perp \psi \cdot \vec{F} = \int \vec{v} \cdot \vec{F}$$

or

$$\{3\} = -\int \psi \mathcal{F} = -\int \psi \Delta \Phi = -\int \Delta \psi \Phi = -\int \omega \Phi.$$

Combining our estimates on the three terms and putting them back into the equation (1.74), we obtain the results on conservation of energy in the absence of the mean velocity field.

Theorem 1.1 (Kinetic energy identity) *The solutions of the barotropic quasi-geostrophic equations (1.1) with periodic boundary conditions satisfy the kinetic energy identity*

$$\frac{d}{dt}\frac{1}{2}\int |\vec{v}|^2 = -\mathcal{D}_E + \int \vec{v}\cdot\vec{F}, \tag{1.74}$$

where

$$\mathcal{D}_E = \sum_{j=0,\, j:\text{even}}^{l} d_j \int |\Delta^{j/2}\psi|^2 + \sum_{j=1,\, j:\text{odd}}^{l} d_j \int |\nabla\Delta^{(j-1)/2}\psi|^2$$

$$\operatorname{curl} \vec{F} = \mathcal{F}.$$

Corollary 1.1 *In the absence of dissipation and forcing, the kinetic energy of periodic flow is conserved.*

Remarks on the dissipation

Case 1: Newtonian (eddy) viscosity ($j = 2$ only, $d_2 = \nu$).

In this case the dissipation takes the form

$$\begin{aligned} -\nu\int |\Delta\psi|^2\,dx &= -\nu\int\left\{\left(\frac{\partial^2\psi}{\partial x^2}\right)^2 + \left(\frac{\partial^2\psi}{\partial y^2}\right)^2 + 2\frac{\partial^2\psi}{\partial x^2}\frac{\partial^2\psi}{\partial y^2}\right\} \\ &= -\nu\int\left\{\left(\frac{\partial^2\psi}{\partial x^2}\right)^2 + \left(\frac{\partial^2\psi}{\partial y^2}\right)^2 + 2\left(\frac{\partial^2\psi}{\partial x\partial y}\right)^2\right\} \\ &= -\nu\int |\nabla\vec{v}|^2\,dx, \end{aligned} \tag{1.75}$$

where we have performed integration by parts and utilized the fact that $\vec{v} = \nabla^{\perp}\psi$.

Case 2: Ekman drag dissipation ($j = 1$ only, $d_1 = d$).

In this case the dissipation takes the form

$$-d\int |\nabla\psi|^2 = -d\int |\nabla^{\perp}\psi|^2 = -d\int |\vec{v}|^2. \tag{1.76}$$

The dissipation depends on the velocity itself, not on its gradient. This property justifies the name *drag dissipation*.

1.3.2 Large-scale and small-scale flow interaction via topography

In the previous subsection we have discussed the energy conservation of periodic flow. We now turn to the case where there is a non-trivial time-dependent mean flow. The question we would like to address is whether the energy is conserved and if there is any way of the energy going from small scale (or periodic) to large scale (or mean) or vice versa. Here the stream function is given by

$$\psi = -V(t)y + \psi',$$

where ψ' is periodic and $V(t)$ is an evolving mean flow. We will see that there is natural large-scale and small-scale flow interaction via topography.

For simplicity, we will assume no dissipation or external forcing, i.e.

$$\mathcal{D} = 0, \mathcal{F} = 0.$$

With

$$q = \Delta\psi + h + \beta y, \quad \frac{Dq}{Dt} = 0$$

the small-scale part of the kinetic energy is given by

$$E_{\text{small scale}} = \frac{1}{2}\int |\nabla\psi'|^2.$$

Hence

$$\begin{aligned} \frac{d}{dt}E_{\text{small scale}} &= -\int \frac{\partial q}{\partial t}\psi' = \int \psi'\nabla^{\perp}\psi' \cdot \nabla q + \int V(t)\frac{\partial q}{\partial x}\psi' \\ &= V(t)\int \frac{\partial \Delta\psi'}{\partial x}\psi' + V(t)\int \frac{\partial h}{\partial x}\psi'. \end{aligned} \tag{1.77}$$

Notice that

$$\int \Delta\frac{\partial\psi'}{\partial x}\psi' = -\int \frac{\partial}{\partial x}(\nabla\psi')\cdot\nabla\psi' = -\int \frac{\partial}{\partial x}(\frac{1}{2}|\nabla\psi'|^2) = 0;$$

we deduce

$$\frac{d}{dt}E_{\text{small scale}} = V(t)\int \frac{\partial h}{\partial x}\psi'. \tag{1.78}$$

The latter integral is called *topographic stress* for reasons which will be explained later.

An important thing we should observe here is that the large-scale mean flow can change the energy of the small-scale flow through topographic stress. A natural thing is to ask where the small-scale energy is going. Since there is no dissipation and forcing, it is natural to postulate the assumption that the total energy is conserved. In this way they get the evolution equation for the mean flow from the total energy conservation. The topography mediates the energy transfer between large scales and small scales. We may quantify this remarkable fact in the following way

$$\begin{aligned} E_{\text{total}} &\equiv E_{\text{mean}}(t) + E_{\text{small scale}}(t) \\ &\equiv A_R \frac{1}{2} V^2(t) + \frac{1}{2} \int |\nabla \psi'|^2 \\ &\equiv \text{constant}, \end{aligned} \tag{1.79}$$

where A_R is the area of the domain of interest, i.e. $A_R = 4\pi^2$ for the standard periodic domain. The factor A_R occurs to keep correct dimensional physical units. Hence

$$\begin{aligned} \frac{d}{dt} E_{\text{total}} &= A_R V(t) \frac{d}{dt} V(t) + \frac{d}{dt} E_{\text{small scale}} \\ &= A_R V(t) \frac{d}{dt} V(t) + V(t) \int \frac{\partial h}{\partial x} \psi' \\ &= 0. \end{aligned} \tag{1.80}$$

This implies, in the case $V \neq 0$

$$\frac{d}{dt} V(t) = - ⨍ \frac{\partial h}{\partial x} \psi', \tag{1.81}$$

where here and throughout the rest of this book

$$⨍_\Omega f = \frac{1}{A_R} \int_\Omega f,$$

with A_R given by the area of Ω. So we can derive the mean flow evolution equation from the conservation principle. It is a good piece of wisdom that we can often derive dynamic equations from conservation principles in physics. This is a serendipity of mathematics and physics.

We summarize the results in this subsection as:

Equations for large-scale and small-scale flow interaction via topographic stress

$$\text{(A)} \quad q = \Delta \psi' + h + \beta y, \tag{1.82}$$

$$\text{(B)} \quad \psi = -V(t) y + \psi', \tag{1.83}$$

$$\text{(C)} \quad \frac{\partial q}{\partial t} + J(\psi, q) = 0, \tag{1.84}$$

$$\text{(D)} \quad \frac{d}{dt} V(t) = - ⨍ \frac{\partial h}{\partial x} \psi'. \tag{1.85}$$

Now we explain why the integral

$$\int \frac{\partial h}{\partial x}\psi' = -\int h\frac{\partial \psi'}{\partial x}$$

is called *topographic stress*. This is because ψ' is the pressure (due to geostrophic balance) and hence $\frac{\partial \psi'}{\partial x}$ is the pressure gradient, and thus $h\frac{\partial \psi'}{\partial x}$ represents topographic stress.

Next we check that the steady states with mean flow and topography described in (1.33) and (1.34) remain steady states of the interacting system in (1.81)–(1.84) above. Recall that these steady states are given by the formulas

$$\psi = -V_0 y + \psi', \quad V_0 = -\frac{\beta}{\mu}, \quad \mu\psi' = \Delta\psi' + h.$$

In order to establish our claim, we need to show that the topographic stress vanishes for these solutions. Indeed, we have

$$\begin{aligned}\int \frac{\partial h}{\partial x}\psi' &= -\int h\frac{\partial \psi'}{\partial x} = \int \Delta\psi'\frac{\partial \psi'}{\partial x} - \mu\int \psi'\frac{\partial \psi'}{\partial x}\\ &= \frac{1}{2}\int \frac{\partial}{\partial x}(|\nabla\psi'|^2 - \mu(\psi')^2) = 0,\end{aligned}$$

since ψ' is a periodic function and the integral of a derivative of a periodic function vanishes. We study the non-linear stability of these steady states in Chapter 4.

In Chapter 5, we will consider a special type of solutions to (1.82)–(1.85). This special type of solutions consists of a large-scale Mean Flow component and a Rossby Wave component. The two components interact through topographic stress. We will find very rich examples of integrable and chaotic dynamics.

The equations of large-scale and small-scale interaction via topographic stress enjoy another conserved quantity, the large-scale enstrophy $Q = \beta V(t) + \frac{1}{2}⨍(\omega + h)^2$, for general periodic flows. Indeed

$$\begin{aligned}\frac{d}{dt}\frac{1}{2}\int(\omega+h)^2 &= \int(\omega+h)\frac{\partial\omega}{\partial t}\\ &= -\int(\omega+h)\nabla^{\perp}\psi'\cdot\nabla(\omega+h+\beta y) - V(t)\int(\omega+h)\frac{\partial}{\partial x}(\omega+h+\beta y)\\ &= -\frac{1}{2}\int\nabla^{\perp}\psi'\cdot\nabla(\omega+h)^2 - \beta\int(\omega+h)\frac{\partial\psi'}{\partial x} - \frac{V(t)}{2}\int\frac{\partial}{\partial x}(\omega+h)^2\\ &= -\beta\int\Delta\psi'\frac{\partial\psi'}{\partial x} - \beta\int h\psi'_x = \frac{\beta}{2}\int\frac{\partial}{\partial x}|\nabla\psi'|^2 + \beta\int\frac{\partial h}{\partial x}\psi'\\ &= -\beta A_R\frac{dV(t)}{dt},\end{aligned}$$

since ψ', ω, h are periodic. Thus Q is also conserved.

When we study barotropic flows on the sphere in a later chapter, we will see that similar equations as in (1.81)–(1.84) arise automatically as changes in angular momentum through "mountain torque," the spherical analogue of topographic stress.

1.3.3 Infinite number of conserved quantities – generalized enstrophy

Here we explore the additional physically conserved quantities besides energy, which might be conserved with various geophysical effects in different geometry.

Ignoring the dissipation, external forcing, and the large mean flow, we may rewrite the evolution equations as

$$q = \Delta\psi + \beta y + h, \tag{1.86}$$

$$\vec{v} = \nabla^{\perp}\psi, \tag{1.87}$$

$$\frac{\partial q}{\partial t} + \vec{v}\cdot\nabla q = 0. \tag{1.88}$$

Since q is conserved along the particle trajectory, a good guess is that the integral of any function of q, i.e $\int_{[0,2\pi]\times[0,2\pi]} G(q)$, is conserved. In particular, $\int_{[0,2\pi]\times[0,2\pi]} \frac{1}{2}q^2$ is called *potential enstrophy*, whereas $\int_{[0,2\pi]\times[0,2\pi]} \frac{1}{2}\omega^2$ is called *the enstrophy*. The quantities $\int_{[0,2\pi]\times[0,2\pi]} G(q)$ are called *the generalized enstrophies*. We will see the guess is true if $\beta = 0$. But not necessarily true if $\beta \neq 0$. Indeed

$$\begin{aligned}
\frac{d}{dt}\int_{[0,2\pi]\times[0,2\pi]} G(q) &= \int_{[0,2\pi]\times[0,2\pi]} G'(q)\frac{\partial q}{\partial t} \\
&= -\int_{[0,2\pi]\times[0,2\pi]} (\vec{v}\cdot\nabla q)G'(q) \\
&= -\int_{[0,2\pi]\times[0,2\pi]} (\vec{v}\cdot\nabla)G(q) \\
&= -\int_{[0,2\pi]\times[0,2\pi]} \operatorname{div}(\vec{v}G(q)) \\
&= -\int_{\partial[0,2\pi]\times[0,2\pi]} (\vec{v}\cdot\vec{n})G(q) \\
&= -\int_{y=2\pi} \psi_x G(\Delta\psi + h + \beta y) + \int_{y=0} \psi_x G(\Delta\psi + h) \\
&= -\int_0^{2\pi} \psi_x(x,0,t)(G(\Delta\psi(x,0,t) + h(x,0) + 2\pi\beta) \\
&\quad -G(\Delta\psi(x,0,t) + h(x,0))dx).
\end{aligned} \tag{1.89}$$

From this we can read off that the generalized enstrophy $\int_{[0,2\pi]\times[0,2\pi]} G(q)$ is not conserved in general, and it is conserved if one of the following is true:

- $\beta = 0$, i.e. there is no beta-effect.
- G is periodic with period $2\pi\beta$.
- $\psi_x(x, 0, t) \equiv 0$, i.e. there exists special symmetry in the flow.

It is easily checked that the situation remains unchanged even if we take into consideration the large-scale mean flow $V(t)$.

Thus we conclude:

Case 1: $\beta = 0$, i.e. $q = \Delta\psi' + h$.

All generalized enstrophies $\int_{[0,2\pi]\times[0,2\pi]} G(q)$ are conserved quantities.

Case 2: $\beta \neq 0$.

The generalized enstrophies $\int_{[0,2\pi]\times[0,2\pi]} G(q)$ are not conserved in general. They are conserved in the special case when G is periodic with period $2\pi\beta$, or when the flow has extra symmetry so that $\psi_x(= v_2) = 0$ at both $y = 0$, $y = 2\pi$.

A natural question to ask is then if there exists any subspace of the stream function which has the symmetry and is invariant under the non-linear dynamics. The answer is yes and there is a rich family of these kind of subspaces. The first example is related to channel flow, as we shall see later in the chapter

$$\psi' = \sum_{k\geq 1} (a_{jk} \cos jx + b_{jk} \sin jx) \sin ky. \tag{1.90}$$

It is easy to verify that this is an invariant subspace of the stream function space, provided that the topography also takes the above form. Indeed it is straightforward to see that the linear terms of the evolution equation do not break the symmetry. The only problem is on the non-linear term $\nabla^{\perp}\psi \cdot \nabla q'$, where $q' = \Delta\psi + h$. But a routine calculation establishes that the non-linear term also preserves the symmetry of the flow.

Of course, we can think of flow which has more symmetries. For instance the swimming pool geometry

$$\psi' = \sum_{k\geq 1, j\geq 1} a_{jk} \sin jx \sin ky. \tag{1.91}$$

This is invariant under the non-linear dynamics, provided that $\beta = 0$ and the topography h has the same kind of symmetry. Actually, the ground state of this flow corresponds to the so-called *Taylor vortex* described earlier in Subsection 1.2.1.

Another interesting case arises if we consider the beta-plane channel flow, i.e. we will assume periodicity in the zonal (longitude) direction, but we will replace the azimuthal (latitude) periodicity of the velocity by the solid wall boundary condition. More precisely we impose

$$\vec{v}(x+2\pi, y, t) = \vec{v}(x, y, t)$$
$$v_2(x, 2\pi, t) = v_2(x, 0, t) = 0.$$

This also leads to infinite number of conserved quantities. We will discuss this case later in this chapter.

1.3.4 Several conserved quantities

We proved above that all of the generalized enstrophies are not necessarily conserved quantities if $\beta \neq 0$ in general periodic geometry. Still, there is the possibility of there existing several significant conserved quantities. For this purpose we take a more careful study of these conserved quantities under various conditions.

Case 1: $\beta \neq 0, h \equiv 0, V(t) \equiv 0$.

We recall the evolution equations

$$q = \omega + \beta y = \Delta\psi + \beta y, \tag{1.92}$$
$$\frac{\partial \omega}{\partial t} + J(\psi, \omega) + \beta\psi_x = 0. \tag{1.93}$$

In this case, we have the following conserved quantities:

- Energy: $\frac{1}{2}\int |\nabla\psi|^2$.
- Enstrophy (not potential enstrophy): $\frac{1}{2}\int \omega^2$.

The conservation of energy is derived in subsection 1.3.1. As for the conservation of enstrophy, we have

$$\begin{aligned}\frac{d}{dt}\left(\frac{1}{2}\int \omega^2\right) = \int \omega\frac{\partial \omega}{\partial t} &= -\int (\nabla^{\perp}\psi\cdot\nabla\omega)\omega - \beta\int \omega\psi_x \\ &= -\frac{1}{2}\int \nabla^{\perp}\psi\cdot\nabla(\omega^2) - \beta\int \Delta\psi\,\psi_x \\ &= \beta\int \nabla\psi\cdot\nabla\psi_x = \frac{\beta}{2}\int \frac{\partial}{\partial x}|\nabla\psi|^2 \\ &= 0,\end{aligned}$$

since ψ, ω are periodic.

Case 2: $\beta \neq 0,\ h \neq 0,\ V(t) = 0$.

In this case, there is only one conserved quantity unless there is special symmetry in the flow. The conserved quantity is:

- Energy: $\frac{1}{2}\int |\nabla\psi|^2$.

We should notice that $\int|\omega|^2$, $\int|q|^2$ are not conserved in general. $\int|\omega|^2$ is not conserved in general, for there is an additional $J(\psi, h)$ term in the evolution equation. In fact

$$\frac{d}{dt}\left(\frac{1}{2}\int\omega^2\right) = \int J(\psi, h)\omega = -\int J(\psi, \omega)h \neq 0$$

in general. The potential enstrophy is not conserved in general due to (1.88):

Case 3: $\beta \neq 0$, $h \neq 0$, $V(t) \neq 0$.

We assume that the mean flow moves according to the large-scale and small-scale flow interaction equations in (1.81)–(1.84). Then we have the following two conserved quantities as discovered before:

- Total energy: $E = \frac{1}{2}A_R V^2(t) + \frac{1}{2}\int|\nabla\psi|^2$.
- Large-scale enstrophy: $Q = \beta V(t) + \frac{1}{2}\fint(\omega + h)^2$.

Of course, for flows with the symmetries in (1.89) the same calculation as in (1.88) above yields the conservation of all the generalized enstrophies $\int G(q)$ for $\beta \neq 0$ as well as $\beta = 0$.

1.3.5 Summary of conserved quantities: periodic geometry

Here we summarize the conserved quantities for the:

Equations for large-scale and small-scale flow interaction via topographic stress

$$q = \Delta\psi' + h + \beta y, \qquad \omega = \Delta\psi', \tag{1.94}$$

$$\psi = -V(t)y + \psi', \qquad \vec{v} = \nabla^{\perp}\psi = \begin{pmatrix} -\frac{\partial\psi}{\partial y} \\ \frac{\partial\psi}{\partial x} \end{pmatrix}, \tag{1.95}$$

$$\frac{\partial q}{\partial t} + J(\psi, q) = 0, \tag{1.96}$$

$$\frac{d}{dt}V(t) = -\fint \frac{\partial h}{\partial x}\psi'. \tag{1.97}$$

Case I: Full periodic geometry $[0, 2\pi] \times [0, 2\pi]$ **and** $\beta = 0$.

Assume $\beta = 0$ and h arbitrary:

(i) Total mean energy: $E \equiv \frac{1}{2}V^2(t) + \frac{1}{2}\fint|\nabla\psi|^2$.
(ii) Generalized enstrophy: $\mathcal{G}(q) = \int_0^{2\pi}\int_0^{2\pi} G(q)\,dx\,dy$, for any function $G(q)$.

Case II: Subtle issues in periodic geometry if $\beta \neq 0$.
In this case, far fewer conserved quantities exist.

Sub-case A: $\beta \neq 0,\ h \equiv 0,\ V(t) = 0$.

(i) Energy: $E \equiv \frac{1}{2} \int |\nabla\psi|^2$.
(ii) Enstrophy: $\frac{1}{2} \int \omega^2$.
(iii) Non-robust conserved quantities: $\int_0^{2\pi} \int_0^{2\pi} G(q)\, dx\, dy$, if $G(q)$ is periodic with period $\beta 2\pi$.

Sub-case B: $\beta \neq 0,\ h \neq 0,\ V(t) \equiv 0$.
We neglect the large mean equation in this case, since we have an independent dynamics for the small scale.

(i) **Single** robust conserved quantity energy: $E \equiv \frac{1}{2} \int |\nabla\psi|^2$.
(ii) Neither enstrophy $\frac{1}{2} \int \omega^2$ nor potential enstrophy $\frac{1}{2} \int q^2$ are conserved.
(iii) Non-robust conserved quantities: $\int_0^{2\pi} \int_0^{2\pi} G(q)\, dx\, dy$, provided $G(q)$ is periodic with period $\beta 2\pi$.

Sub-case C: $\beta \neq 0,\ h \neq 0,\ V(t) \neq 0$.

(i) Total mean energy: $E \equiv \frac{1}{2} V^2 + \frac{1}{2} ⨍ |\nabla\psi'|^2$.
(ii) Large-scale enstrophy: $\beta V(t) + \frac{1}{2} ⨍ (\omega + h)^2$.
(iii) Non-robust conserved quantities: $\int_0^{2\pi} \int_0^{2\pi} G(q)\, dx\, dy$, provided $G(q)$ is periodic with period $\beta 2\pi$.

It is tempting to conclude that the extrema (maximum and minimum) of the potential vorticity are conserved, since the potential vorticity is advected along the stream lines. However this is not true, since the potential vorticity is not periodic if $\beta \neq 0$, and the stream lines may escape the periodic domain. It is easy to cook up counter-examples where the extrema are not conserved. Another way to understand this is to view the problem as flow on the whole plane with periodic structure. Then the potential vorticity is an unbounded quantity on the whole plane. We will see later in this chapter that it will be very different in the case of channel geometry.

1.4 Barotropic geophysical flows in a channel domain – an important physical model

Here we give a more complete discussion of the formulation for geophysical flows on a channel domain mentioned earlier in the chapter. We assume for simplicity here that the units for the problem have been chosen so that the channel domain is given by the region

$$0 < x < 2\pi, \quad 0 < y < \pi. \tag{1.98}$$

It is an easy exercise for the reader to adapt the discussion presented here to channel domain with a different aspect ratio of length to height.

In such a channel domain, it is natural to consider velocity fields which are periodic in x and satisfy solid wall boundary conditions, $\vec{v}\cdot\vec{n}|_{y=0,\pi}=0$, i.e.

$$\begin{aligned} &\vec{v}(x+2\pi, y, t)=\vec{v}(x, y, t), \\ &v_2(x, \pi, t)=v_2(x, 0, t)=0 \end{aligned} \tag{1.99}$$

on the fundamental domain in (1.98), where $y=0(y=\pi)$ is the bottom (top) of the channel.

In atmospheric modeling, channel domains are utilized to model a span of mid-latitudes, for example, $25°N-75°N$, around the entire globe within the beta-plane approximation, so that the periodicity in x is a very natural boundary condition. In oceanography, the most powerful large-scale current in the world is the Antarctic circumpolar current (ACC), which completes an entire circular route around Antarctica with the narrowest point occurring at the Drake Passage at the tip of South America. Once again, the simplest model with qualitative geometric features of the ACC involves barotropic flow on a channel domain, with natural periodicity in x and finite extent in y. In both these physical models, the role of the beta-effect and topography are very important. Note also that, in these physical models, standard viscous boundary layers do not play a significant role, since we are considering inviscid models here and the boundaries are somewhat artificial, especially for the atmosphere. For general flow in a channel domain with solid wall boundary conditions, the second requirement in (1.99) implies that the stream function $\psi(\vec{v}=\nabla^{\perp}\psi)$ satisfies

$$\psi(x, \pi, t)=A(t) \quad \psi(x, 0, t)=B(t), \tag{1.100}$$

where $A(t)$, $B(t)$ are time-varying constants. Since a stream function is always determined within an arbitrary constant, without loss of generality we set

$$B(t)\equiv 0. \tag{1.101}$$

The most general stream function satisfying these boundary conditions has the form

$$\psi(x, y)=-Vy+\psi'(x, y), \quad \psi'|_{y=0,\pi}=0, \tag{1.102}$$

where $V=-A/\pi$.

In order to make a connection with the periodic geometry, we extend ψ' oddly in the y direction to the box $[0, 2\pi]\times[-\pi, \pi]$, i.e.

$$\tilde{\psi}'(x, y)=\begin{cases} \psi'(x, y) & \text{for } y\geq 0, \\ -\psi'(x, -y) & \text{for } y<0. \end{cases} \tag{1.103}$$

Thanks to the odd extension in y and the zero boundary condition of ψ' at the wall ($y=0$, $y=\pi$), both $\tilde{\psi}'$ and the small-scale velocity $\nabla^{\perp}\tilde{\psi}'$ are periodic functions in x and y. (The small-scale velocity will not be periodic for an even extension in y in general.) Since it is an odd function in y, the Fourier representation of $\tilde{\psi}'$ cannot contain even terms (the cosines) in y. Hence we have, after restricted back to the original domain, the following Fourier expansion for ψ'

$$\psi'(x,y)=\sum_{k\geq 1}\sum_{j}(a_{jk}\cos(jx)+b_{jk}\sin(jx))\sin(ky). \tag{1.104}$$

For simplicity, we ignore dissipation and forcing in the discussion below. With the general stream function given in the above form under the above discussion, the dynamic equations are basically identical to those derived earlier for the periodic geometry, namely:

Barotropic quasi-geostrophic equations in a channel with topographic stress

$$\begin{aligned} &q=\Delta\psi'+h+\beta y, \quad \psi=-V(t)y+\psi',\\ &\frac{\partial q}{\partial t}+J(\psi,q)=0, \quad \frac{d}{dt}V(t)=-⨍\frac{\partial h}{\partial x}\psi'. \end{aligned} \tag{1.105}$$

We note here that the area for the domain is $2\pi^2$ and hence $2\pi^2$ is now the one utilized in the normalized integral in the dynamic equation for the large mean V.

From our previous discussions in Subsections 1.3.3–1.3.5 on conserved quantities, it follows that the system (1.105) has the following:

Conserved quantities

(i) Total mean energy

$$E=\frac{1}{2}V^2(t)+\frac{1}{2}⨍|\nabla\psi'|^2. \tag{1.106}$$

(ii) Large-scale enstrophy

$$Q=\beta V(t)+\frac{1}{2}⨍(\omega+h)^2. \tag{1.107}$$

(iii) Generalized enstrophies

$$\int G(q) \tag{1.108}$$

for an arbitrary function $G(q)$, where $q=\Delta\psi'+h+\beta y$.

The proofs are pretty much the same as in the periodic case and the conservation of generalized enstrophies follows from the special channel symmetry (1.104).

The conservation of the infinitely many generalized enstrophies has some interesting implications. In particular, it implies the conservation of extrema of the potential vorticity. This is physically reasonable: since the potential vorticity is advected along the stream lines, it is periodic in the zonal (x) direction and the vertical boundaries are impermeable.

Mathematically we let m be the maximum of the initial potential vorticity in the channel

$$m = \max q(\cdot, 0) \tag{1.109}$$

and we define

$$G(q) = \begin{cases} (q-m)^2, & \text{if } q \geq m \\ 0, & \text{if } q < m. \end{cases} \tag{1.110}$$

Then G is differentiable and hence $\mathcal{G} = \int G(q)$ is conserved. Notice that $\mathcal{G}(0) = 0$ since m is the maximum of the initial potential vorticity. Thus $\mathcal{G}(t) = 0$ for all t, which further implies

$$q(\cdot, t) \leq m \tag{1.111}$$

for all t. On the other hand, for any $\varepsilon > 0$ we may define

$$G_\varepsilon(q) = \begin{cases} (q-m+\varepsilon)^2, & \text{if } q \geq m-\varepsilon \\ 0, & \text{if } q < m-\varepsilon. \end{cases} \tag{1.112}$$

Then G_ε is differentiable and hence $\mathcal{G}_\varepsilon = \int G_\varepsilon(q)$ is conserved. Notice that $\mathcal{G}_\varepsilon(0) > 0$ since m is the maximum of the initial potential vorticity and $\varepsilon > 0$. Thus $\mathcal{G}_\varepsilon(t) > 0$ for all t, which further implies

$$\max q(\cdot, t) \geq m - \varepsilon \tag{1.113}$$

for all t and all $\varepsilon > 0$. Hence

$$\max q(\cdot, t) \geq m, \tag{1.114}$$

which leads to

$$\max q(\cdot, t) = m = \max q(\cdot, 0). \tag{1.115}$$

The conservation of the minimum can be verified the same way.

Note that this argument also shows that we have conservation of the extrema for general periodic flows with $\beta = 0$.

1.4.1 The impulse and conserved quantities

For a channel domain, a very natural quantity to consider is the impulse, I, defined by

$$I = \fint y\omega, \quad \text{where } \omega = \Delta\psi'. \tag{1.116}$$

Note that I has the units of velocity. Physically, the impulse represents the mean (zonal) x-momentum induced by the relative vorticity in the fluid. The physical interpretation will be clear after our discussion below.

For $\beta \neq 0$, the standard choice of the conserved potential enstrophy, $G(q) = \frac{1}{2}q^2$, yields the conserved quantity

$$\frac{1}{2}\fint(\omega+h)^2 + \beta I, \tag{1.117}$$

since $q = \omega + h + \beta y$. The difference of this conserved quantity and the conserved large-scale enstrophy Q yields the fact that for $\beta \neq 0$

$$V(t) - I(t) \quad \text{is conserved in time.} \tag{1.118}$$

Thus, the production of mean x-momentum (horizontal flux rate) in time is exactly associated with the changes of impulse in time. In particular, for a zonal topography, the impulse

$$I, \quad \text{is conserved in time, if } h = h(y). \tag{1.119}$$

The elementary discussion we have just presented relied on requiring $\beta \neq 0$. However, direct calculations utilizing integration by parts also yield the two general conservation principles for impulses discussed above. This is left as an exercise for the interested reader.

The conservation of impulse for channel flow is in marked contrast to periodic flow for which the impulse is not conserved in general, even in the absence of topography.

1.4.2 Conservation of circulation

Another interesting quantity for the channel is the so-called circulation, which is defined as

$$\Gamma = \int \omega, \quad \text{where } \omega = \Delta\psi'. \tag{1.120}$$

This measures the total vorticity (or the global rotation of the fluid) in the domain. It is called circulation since

$$\begin{aligned}\Gamma &= \int \left(-\frac{\partial v_1}{\partial y} + \frac{\partial v_2}{\partial x}\right) = \int_{\partial\Omega} (-n_2 v_1 + v_2 n_1) ds \; = \; \int_{\partial\Omega} \vec{v}\cdot\vec{\tau}\, ds \\ &= -\Gamma_{\text{top}} + \Gamma_{\text{bottom}},\end{aligned}$$

where $\vec{\tau}$ is the unit counter-clockwise tangent vector at the boundary of the channel and $\Gamma_{\text{top}} = \int_{y=\pi} v_1\, dx$, $\Gamma_{\text{bottom}} = \int_{y=0} v_1\, dx$.

The conservation of circulation is a special case of the conservation of generalized enstrophy with $G(q) = q$. In general the circulation in a channel is not zero. This is different from the periodic geometry where the circulation is identically zero.

In the case of zonal topography (thus the impulse is conserved), we may derive the separated conservation of the circulation on the top (Γ_{top}) and on the bottom (Γ_{bottom}). Indeed

$$A_R I = \int y\omega = \int y\Delta\psi = \int_{y=\pi} \pi \frac{\partial\psi}{\partial y} - \int \frac{\partial\psi}{\partial y} = -\pi\Gamma_{\text{top}}. \tag{1.121}$$

Hence Γ_{top} is conserved, which further implies the conservation of Γ_{bottom} by the conservation of total circulation.

1.4.3 *Summary of conserved quantities: channel geometry*

Here we summarize the conserved quantities discussed above for the barotropic quasi-geostrophic equations in a channel (1.105).

Conserved quantities in the channel geometry

(i) Total mean energy

$$E = \frac{1}{2}V^2(t) + \frac{1}{2}⨍|\nabla\psi'|^2. \tag{1.122}$$

(ii) Large-scale enstrophy

$$Q = \beta V(t) + \frac{1}{2}⨍(\omega + h)^2. \tag{1.123}$$

(iii) Generalized enstrophies

$$\int G(q) \tag{1.124}$$

for an arbitrary function $G(q)$, where $q = \Delta\psi' + h + \beta y$.

(iv) Mean velocity–impulse difference

$$V - I, \tag{1.125}$$

where the impulse I is defined as

$$I = ⨍ y\omega, \quad \text{where } \omega = \Delta\psi'. \tag{1.126}$$

Moreover, in the case of zonal topography, i.e. $h = h(y)$, the impulse I is conserved in time.

(v) Circulation (special case of conservation of generalized enstrophy with $G(q) = q$)

$$\Gamma = \int \omega, \quad \text{where } \omega = \Delta\psi'. \tag{1.127}$$

Moreover, in the case of zonal topography, i.e. $h = h(y)$, we have separated conservation of circulation

$$\Gamma_{\text{top}} = \int_{y=\pi} v_1\, dx, \quad \Gamma_{\text{bottom}} = \int_{y=0} v_1\, dx. \tag{1.128}$$

(vi) Potential vorticity extrema (special case of conservation of generalized enstrophies)

$$\max q(\cdot, t), \quad \min q(\cdot, t), \tag{1.129}$$

where $q = \Delta\psi + h + \beta y$.

1.5 Variational derivatives and an optimization principle for elementary geophysical solutions

For an ordinary differentiable function F from R^N to R, the gradient, ∇F, is the unique vector field, so that the directional derivatives satisfy

$$\lim_{\varepsilon\to 0} \frac{F(\vec{x}+\varepsilon\vec{y}) - F(\vec{x})}{\varepsilon} \stackrel{def}{=} \nabla F(\vec{x})\cdot\vec{y}, \tag{1.130}$$

where $\vec{a}\cdot\vec{b}$ is the usual Euclidean inner product.

Throughout this book, we will often calculate directional derivatives of functionals, $\mathcal{F}$, from an infinite-dimensional Hilbert space equipped with an inner product into the real numbers. Let H denote this Hilbert space with the inner product $\langle f, g\rangle$. In most of our applications in this book, the inner product space will be ordinary square integrable functions on a two-dimensional domain, i.e.

$$\langle f, g\rangle = (f, g)_0 = \int fg. \tag{1.131}$$

Given a functional, $\mathcal{F}(u)$, from H to R, the variational derivative of $\mathcal{F}$, denoted $\frac{\delta\mathcal{F}}{\delta u}$, is the function satisfying

$$\lim_{\varepsilon\to 0} \frac{\mathcal{F}(u+\varepsilon\delta u) - \mathcal{F}(u)}{\varepsilon} \stackrel{\text{def}}{=} \left\langle \frac{\delta\mathcal{F}}{\delta u}, \delta u\right\rangle, \tag{1.132}$$

for any δu. Thus, the variational derivative, $\frac{\delta\mathcal{F}}{\delta u}$, plays the analogous notion to that of the gradient in finite dimensions from (1.130). We will use this notion in (1.132) in a purely formal fashion throughout the book as is usually done by physicists; however, in every context in this book, such manipulations can be justified rigorously, although we will not do that here.

1.5.1 Some important variational derivatives

There are some elementary variational derivatives which play an important role throughout the book. We develop these general formulas briefly here and give one simple application in the next subsection.

First, we calculate the variational derivative of the kinetic energy with respect to the vorticity. Recall that for periodic functions, ω, the (small-scale) mean kinetic energy is the quadratic functional

$$E(\omega) = \frac{1}{2} \fint |\vec{v}|^2 = \frac{1}{2} \fint |\nabla\psi|^2 = -\frac{1}{2} \fint \psi\omega, \tag{1.133}$$

where $\omega = \Delta\psi$.

Given a directional variation $\delta\omega$, define $\delta\psi$ by

$$\Delta(\delta\psi) = \delta\omega. \tag{1.134}$$

We calculate explicitly that

$$\frac{E(\omega+\varepsilon\delta\omega) - E(\omega)}{\varepsilon} = -\frac{1}{2} \fint (\psi\delta\omega + \delta\psi\omega) - \varepsilon\frac{1}{2} \fint \delta\psi\delta\omega \tag{1.135}$$

so that

$$\begin{aligned} \lim_{\varepsilon\to 0} \frac{E(\omega+\varepsilon\delta\omega) - E(\omega)}{\varepsilon} &= -\frac{1}{2} \fint (\psi\delta\omega + \delta\psi\omega) = -\frac{1}{2} \fint (\psi\delta\omega + \delta\psi\Delta\psi) \\ &= -\fint \psi\delta\omega = (-\psi, \delta\omega)_0 \end{aligned}$$

with the inner product in (1.131). Thus, with the definition in (1.132), we have the formula

$$\frac{\delta E}{\delta\omega} = -\psi, \quad \Delta\psi = \omega. \tag{1.136}$$

Other important functionals, where it is easy to calculate the variational derivatives, are given by

$$\mathcal{G}(\omega) = \fint G(\omega) \tag{1.137}$$

for a differentiable function G. Here it is easy to see that

$$\lim_{\varepsilon\to 0} \frac{\mathcal{G}(\omega+\varepsilon\delta\omega) - \mathcal{G}(\omega)}{\varepsilon} = \fint G'(\omega)\delta\omega \tag{1.138}$$

so that

$$\frac{\delta\mathcal{G}}{\delta\omega} = G'(\omega). \tag{1.139}$$

The two elementary formulas of functional derivatives calculated above in (1.136) and (1.139) will occur throughout the book.

1.5.2 An optimization principle for elementary geophysical solutions

As a simple application of the ideas to a calculus problem in infinite dimensions, we consider the averaged enstrophy

$$\mathcal{E}(\omega) = \frac{1}{2} \fint (\omega + h)^2 \tag{1.140}$$

and ask the following question:

Which functions, ω, minimize the enstrophy for a fixed value of the energy, $E(\omega) = E_0$?

Intuitively, according to the Lagrange multiplier principle, the function ω should satisfy the Euler–Lagrange equation

$$\frac{\delta \mathcal{E}}{\delta \omega} = -\mu \frac{\delta E}{\delta \omega}, \tag{1.141}$$

where μ is a Lagrangian multiplier. The formula in equation (1.141) is the generalization of the standard Lagrange multiplier principle with variational derivatives replacing gradients. Since $\frac{\delta E}{\delta \omega} = -\psi$, and $\frac{\delta \mathcal{E}}{\delta \omega} = \omega + h$, the Euler–Lagrange equation becomes

$$\Delta \psi + h = \mu \psi. \tag{1.142}$$

Thus, we are led to the same exact solutions with and without topography that we introduced in Sections 1.3 and 1.4 above. Of course, as in ordinary calculus, not every solution of the equation (1.142) is necessarily a minimizer of the enstrophy, and some of the solutions in (1.142) are saddle points. In fact, for non-zero topography h, only the solutions with μ satisfying $-1 < \mu < +\infty$ are minimizers. See Section 4.5 below.

1.6 More equations for geophysical flows

So far we have focused our attention on the simplest dynamic equations for geophysical flows, namely the barotropic quasi-geostrophic equations (1.1). There are other interesting models that govern the dynamics of basic geophysical flows. In this section we introduce a few more models for basic geophysical flows. They include the F-plane equation, two layer models, and continuously stratified quasi-geostrophic models. We will also discuss briefly the inter-relations among these models and the basic barotropic model. As indicated in the introduction, some of the material in the book can be generalized to these equations (models) in a straightforward fashion, while other material involves subtle current research.

1.6.1 The models

In this subsection we recall a few more models for basic geophysical flows. The reader is referred to the book of Pedlosky (1987) for the physical background as well as derivations for these models.

We first introduce the F-plane model, which is the simplest model next to the barotropic model:

F-plane equations

$$\frac{\partial q}{\partial t} + J(\psi, q) = \mathcal{D}(\Delta)\psi + \mathcal{F}, \tag{1.143}$$

$$q = \Delta\psi - F^2\psi + h + \beta y, \tag{1.144}$$

where

$$F = \frac{L}{L_R}, \tag{1.145}$$

with L the characteristic horizontal length of the flow and $L_R = \sqrt{gH_0}/f_0$ the Rossby deformation radius, H_0 the typical depth of the fluid layer and f_0 the rotation rate of the fluid. In another words, F measures the relative strength of stratification over the rotation. The reader is referred to Pedlosky (1987) for a formal derivation and Majda (2003) for a formal and rigorous derivation. Notice that the F-plane model has no vertical structure and hence it is a one-layer model. This is one of the simplest models that takes into consideration the effect of stratification.

The next model we introduce is a quasi-geostrophic model that has full vertical structure.

Continuously stratified quasi-geostrophic equations

$$\frac{\partial}{\partial t}\left(\Delta\psi + F^2\frac{\partial^2\psi}{\partial z^2}\right) + J\left(\psi, \Delta\psi + F^2\frac{\partial^2\psi}{\partial z^2}\right) + \beta\frac{\partial\psi}{\partial x} = \mathcal{D}_3\psi + \mathcal{F}. \tag{1.146}$$

Here $\psi(x, y, z, t)$ is the stream function with *fully periodic geometry* in the domain

$$[0, 2\pi] \times [0, 2\pi] \times [0, 2\pi\Theta],$$

with period 2π in x and y and period $\frac{2\pi}{\Theta}$ in z for the sake of simplicity, Δ, $\nabla^\perp$, ∇ are the horizontal Laplacian, perpendicular gradient and gradient operators, and $\mathcal{D}_3$ is some three-dimensional dissipation operator, F, is a non-dimensional constant which measures the relative strength of stratification over rotation as defined above. The motivated reader may consult the book by Majda (2003) for a formal and rigorous derivation of these equations from the Boussinesq equations, or Pedlosky (1987) for a formal derivation and more.

Next we introduce a model with complexity between the simple F-plane model and the more complex continuously stratified model. We consider two layers of fluids with one on top of the other. This is one of the simplest models that has a vertical structure. It has enjoyed surprising success in explaining various phenomena in the atmosphere and ocean. The dynamic equations take the form:

Two-layer model

$$\frac{\partial}{\partial t}(\Delta\psi_1 - F_1(\psi_1-\psi_2)) + J(\psi_1, \Delta\psi_1 - F_1(\psi_1-\psi_2)) + \beta\frac{\partial}{\partial x}\psi_1 = \mathcal{F}_1,$$

$$\frac{\partial}{\partial t}(\Delta\psi_2 + F_2(\psi_1-\psi_2)) + J(\psi_2, \Delta\psi_2 + F_2(\psi_1-\psi_2)) + \beta\frac{\partial}{\partial x}\psi_2 = \mathcal{F}_2, \quad (1.147)$$

where $\psi_1(x,y,t)(\psi_2(x,y,t))$ is the stream function of the top (bottom) layer of fluid respectively, and

$$F_j = \frac{f_0^2 L^2}{g(\delta\rho/\rho)D_j}, j = 1, 2 \quad (1.148)$$

with f_0 being the rotation rate, L being the horizontal length scale, g being the gravity constant, $\delta\rho$ being the difference of fluid densities in the two layers, ρ being the reference density, and D_j being the depth of the jth fluid layer.

1.6.2 Relationships between various models

The models that we have introduced are all related, with the continuously stratified model (1.146) being the most complex, two-layer model (1.147) the second, F-plane model (1.143) the third, and the barotropic model (1.1) the simplest.

In this subsection we briefly discuss the inter-relationships among these models. In fact we will demonstrate (formally) that the two-layer model (1.147) (with equal depths) can be derived as a two vertical modes truncation of the continuously stratified model (1.146); the F-plane model (1.143) can be derived from the two-layer model (1.147) when the relative depth of the bottom layer approaches infinity; the barotropic model (1.1) is the limit of the F-plane model as the Rossby deformation radius approaches infinity (or equivalently, $F \to 0$). We may also derive the one-layer model (1.143) directly from the continuously stratified model (1.146) via a one vertical model truncation.

Derivation of the barotropic one-layer model from the continuously stratified model

In order to derive the one-layer model we propose the following one mode vertical truncation of the continuously stratified model

$$\psi(x, y, z, t) = \psi_b(x, y, t). \quad (1.149)$$

In another word we look for solutions with no vertical structure (hence barotropic). Substitute this into the continuously stratified model (1.146), and, assuming the forcing has no vertical structure, we deduce

$$\frac{\partial \Delta \psi_b}{\partial t} + \beta \frac{\partial \psi_b}{\partial x} + J(\psi_b, \Delta \psi_b) = \mathcal{D}\psi_b + \mathcal{F}, \tag{1.150}$$

which is exactly the barotropic quasi-geostrophic model (1.1) without topography.

Derivation of the two-layer model from the continuously stratified model

Next we derive the two-layer quasi-geostrophic equations with equal depth (thus $F_1 = F_2$) by taking a two mode vertical Galerkin truncation of the continuously stratified model (1.146). Such models are prominent in studies of the dynamics of the atmosphere and ocean (see Pedlosky, 1987). The intuition behind them is that the z independent vertical mode gives the barotropic mode and thus we need a vertical shear to incorporate the baroclinic mode associated with the transport of heat. For simplicity in exposition, we neglect dissipation and external forcing in the derivation below.

Without loss of generality, we assume $\Theta = 1$ and we approximate the stream function ψ with the first two modes in the vertical Fourier expansion

$$\psi(x, y, z, t) = \psi_b(x, y, t) + \psi_t(x, y, t)\sqrt{2}\sin z. \tag{1.151}$$

Substitute this into the continuously stratified model (1.146) and notice for the Jacobian we have

$$\begin{aligned} J\left(\psi, \Delta\psi + F^2 \frac{\partial^2 \psi}{\partial z^2}\right) &= J(\psi_b + \psi_t \sqrt{2}\sin z, \Delta\psi_b + (\Delta\psi_t - F^2\psi_t)\sqrt{2}\sin z) \\ &= J(\psi_b, \Delta\psi_b) + \sqrt{2}\sin z(J(\psi_t, \Delta\psi_b) \\ &\quad + J(\psi_b, \Delta\psi_t - F^2\psi_t)) + 2\sin^2 z J(\psi_t, \Delta\psi_t - F^2\psi_t). \end{aligned}$$

The projection of the right-hand side on to the first two vertical Fourier bases $\{1, \sqrt{2}\sin z\}$ yields, following components for each modes

$$\frac{\partial}{\partial t}\Delta\psi_b + J(\psi_b, \Delta\psi_b) + J(\psi_t, \Delta\psi_t - F^2\psi_t) + \beta\frac{\partial}{\partial x}\psi_b = 0, \tag{1.152}$$

$$\frac{\partial}{\partial t}(\Delta\psi_t - F^2\psi_t) + J(\psi_t, \Delta\psi_b) + J(\psi_b, \Delta\psi_t - F^2\psi_t) + \beta\frac{\partial}{\partial x}\psi_t = 0. \tag{1.153}$$

These two equations can be re-arranged (addition and subtraction) into the following form

$$\frac{\partial}{\partial t}(\Delta(\psi_b+\psi_t)-F^2\psi_t)+J(\psi_b+\psi_t,\Delta(\psi_b+\psi_t)-F^2\psi_t)$$
$$+\beta\frac{\partial}{\partial x}(\psi_b+\psi_t)=0, \tag{1.154}$$
$$\frac{\partial}{\partial t}(\Delta(\psi_b-\psi_t)+F^2\psi_t)+J(\psi_b-\psi_t,\Delta(\psi_b-\psi_t)+F^2\psi_t)$$
$$+\beta\frac{\partial}{\partial x}(\psi_b-\psi_t)=0. \tag{1.155}$$

The well-known (inviscid) two-layer model then follows, once we define the stream function ψ_i for each layer as follows

$$\psi_1=\psi_b+\psi_t, \quad \psi_2=\psi_b-\psi_t, \tag{1.156}$$

and the *two-layer model* takes the form

$$\frac{\partial}{\partial t}(\Delta\psi_1-F'(\psi_1-\psi_2)+J(\psi_1,\Delta\psi_1-F'(\psi_1-\psi_2))$$
$$+\beta\frac{\partial}{\partial x}\psi_1=0, \tag{1.157}$$
$$\frac{\partial}{\partial t}(\Delta\psi_2+F'(\psi_1-\psi_2))+J(\psi_2,\Delta\psi_2+F'(\psi_1-\psi_2))$$
$$+\beta\frac{\partial}{\partial x}\psi_2=0, \tag{1.158}$$

where

$$F'=\frac{F^2}{2}.$$

This is the two-layer model with equal depth, no dissipation and no external forcing.

Derivation of the one- and one-half-layer model from the two-layer model

We consider the two-layer model (1.147) with the bottom layer much thicker than the top layer. Physically this could happen in many situations; for instance, in the ocean with a relatively thin (mixing) layer sitting on a much thicker (inertial) layer. Since $F_1(F_2)$ are inversely proportional to the depth of the layer with the other parameters fixed (see (1.148)), we may quantify this assumption in the following relationship for the two-layer model (1.147)

$$F_2=\varepsilon F_1, \tag{1.159}$$

where ε is a small quantity. We may then rewrite the second equation in (1.147) as

$$\frac{\partial}{\partial t}(\Delta\psi_2+\varepsilon F_1(\psi_1-\psi_2))+J(\psi_2,\Delta\psi_2+\varepsilon F_1(\psi_1-\psi_2))+\beta\frac{\partial}{\partial x}\psi_2=0.$$

Formally setting $\varepsilon=0$ in the above equation, we deduce that the bottom layer satisfies the barotropic quasi-geostrophic equations

$$\frac{\partial}{\partial t}(\Delta\psi_2)+J(\psi_2,\Delta\psi_2)+\beta\frac{\partial}{\partial x}\psi_2=\mathcal{F}_2(\vec{x}). \tag{1.160}$$

Thus the limiting dynamics of the bottom layer is independent of the top layer. Now for any fixed (time-independent) flow configuration $\psi_2(\vec{x})$, it is a solution to the above limiting dynamic equation for the bottom layer with a forcing term $\mathcal{F}_2(\vec{x})$ dictated by this given stream function. Let

$$h(\vec{x})=F_1\psi_2(\vec{x}), \tag{1.161}$$

then the limiting dynamics of the top layer is governed by

$$\frac{\partial}{\partial t}(\Delta\psi_1-F_1\psi_1+h)+J(\psi_1,\Delta\psi_1-F_1\psi_1+h)+\beta\frac{\partial}{\partial x}\psi_1=\mathcal{F}_1(\vec{x}), \tag{1.162}$$

or

$$\frac{\partial}{\partial t}(q)+J(\psi_1,q)=\mathcal{F}_1(\vec{x}), \tag{1.163}$$

where $q=\Delta\psi_1-F_1\psi_1+h+\beta y$. In the case of $\mathcal{F}=0$, we recover the F-plane equation without dissipation or external forcing. We observe that $h(\vec{x})=F_1\psi_2(\vec{x})$ serves as an *effective topography*. The interested reader may consult Majda and Wang (2005) for a rigorous justification of this limit.

Derivation of the barotropic quasi-geostrophic model from the F-plane model

This is achieved by setting the Rossby deformation radii L_R equal to ∞; or equivalently setting $F=0$ in the F-plane equation (1.143), we arrive at the barotropic quasi-geostrophic model (1.1) with effective topography.

The interested reader is encouraged to consult the books of Pedlosky (1987), Gill (1982), and Majda (2003) for more information on these models.

It is not difficult to make lists of conserved quantities for the models that we introduced in Subsection 1.6.1, either in the full periodic geometry or the channel geometry. The derivation of most of the conserved quantities here is straightforward generalization of our derivation for the barotropic quasi-geostrophic model and it is left as an exercise for the interested reader.

References

Charney, J. (1949), On the scale of atmospheric motions. *Geofys. Pub. Oslo* **17**(2), 1–17.

Charney, J., Fjörtoft, R., and Von Neumann, J. (1950), Numerical integration of the barotropic vorticity equation. *Tellus* **2**, 237–254.

Chorin, A. and Marsden, J. (1993), *A Mathematical Introduction to Fluid Mechanics*, 3rd edn. New York: Springer-Verlag.

Embid, P. and Majda, A. (1996), Averaging over fast gravity waves for geophysical flows with arbitrary potential vorticity. *Comm. Partial Diff. Eqs.* **21**, 619–658.

Folland, G. (1976), *Introduction to Partial Differential Equations*. Princeton, NJ: Princeton University Press.

Gill, A. (1982), *Atmosphere-Ocean Dynamics*. San Diego: Academic Press.

Majda, A. (1984), *Compressible Fluid Flow and Systems of Conservation Laws in Several Space Variables*. New York: Springer-Verlag.

Majda, A. (2003), *Introduction to P.D.E.'s and Waves for the Atmosphere and Ocean*, *Courant Institute Lecture Notes Series*, Vol. 9. Amer. Math. Soc..

Majda, A. and Bertozzi, A. (2001), *Vorticity and Incompressible Flow*. Cambridge: Cambridge University Press.

Majda, A. and Wang, X. (2005), Validity of the one and one-half layer model and effective topography. Comm. P.D.E., in press.

Pedlosky, J. (1987), *Geophysical Fluid Dynamics*, 2nd edn. New York: Springer-Verlag.

2

The response to large-scale forcing

2.1 Introduction

It is an apparent fact in many instances, for example in the mesoscale and large-scale motions of the atmosphere and the ocean, that the fluid develops large-scale, coherent, and essentially two-dimensional flow patterns. These flows are the result of many competing effects, including dissipation and external forcing. Therefore, a fundamental problem is to understand how these large-scale flow structures develop through energy transfer to large scales for a fluid with dissipation and external forcing. In this chapter we study the existence and stability of large-scale flow structures for two-dimensional flows. In the simplest situation we assume that the flow is homogeneous (there is no density stratification) and it is described by the barotropic quasi-geostrophic equations with external forcing in the absence of topography and mean flow

$$\frac{\partial \omega}{\partial t} + \beta \frac{\partial \psi}{\partial x} + J(\psi, \omega) = \mathcal{D}(\Delta)\psi + \mathcal{F}(\vec{x}, t). \tag{2.1}$$

We will assume that the stream function is doubly periodic, i.e.

$$\psi(x+2\pi, y) = \psi(x, y+2\pi) = \psi(x, y) \tag{2.2}$$

and hence the velocity $\vec{v}$ is doubly periodic as well. The case of channel geometry with periodicity in the longitude direction

$$\psi(x+2\pi, y) = \psi(x, y) \tag{2.3}$$

and free-slip boundary condition (for the Newtonian viscosity case) at the latitude direction

$$\frac{\partial \psi}{\partial x}(x, H) = \frac{\partial \psi}{\partial x}(x, 0) = \omega(x, H) = \omega(x, 0) = 0 \tag{2.4}$$

can be treated in exactly the same fashion, where H denotes the height of the channel.

We will assume a fairly general dissipation operator, which includes Ekman drag, Newtonian viscosity, hyper-viscosity, or their combination

$$\mathcal{D}(\Delta) = \sum_{j=1}^{l} d_j(-\Delta)^j,\ d_j \geq 0. \tag{2.5}$$

In the special case of having generalized Kolmogorov forcing

$$\mathcal{F}(\vec{x}, t) = \sum_{|\vec{k}|^2=\Lambda} \hat{\mathcal{F}}_{\vec{k}}(t) e^{i\vec{x}\cdot\vec{k}} + c.c, \tag{2.6}$$

where Λ is one of the eigenvalues of the Laplacian operator, the set $\{|\vec{k}|^2, \vec{k} \in \mathcal{Z}^2\}$ in this case, and $c.c$ represents the complex conjugate so that the forcing term could take the real value, we have exact solutions of the form (see for instance Section 1.3)

$$\psi = \sum_{|\vec{k}|^2=\Lambda} A_{\vec{k}}(t) e^{i(\vec{k}\cdot\vec{x}-\omega(\vec{k})t)} + c.c; \quad \omega(\vec{k}) = -\frac{k_1\beta}{|\vec{k}|^2}, \tag{2.7}$$

and the coefficients are solved via the following ODE

$$\frac{dA_{\vec{k}}}{dt} = \frac{\mathcal{D}(-\Lambda)}{-\Lambda} A_{\vec{k}} + \frac{\hat{\mathcal{F}}_{\vec{k}}(t) e^{i\omega(\vec{k})t}}{-\Lambda}, \tag{2.8}$$

and the solutions takes the form

$$A_{\vec{k}}(t) = A_{\vec{k}}(0) e^{-\frac{\mathcal{D}(-\Lambda)}{\Lambda}t} - \frac{1}{\Lambda}\int_0^t e^{-\frac{\mathcal{D}(-\Lambda)}{\Lambda}(t-s)} \hat{\mathcal{F}}_{\vec{k}}(s) e^{i\omega(\vec{k})s} ds. \tag{2.9}$$

The purpose of this chapter is to show that these solutions are, in a very precise sense, non-linearly stable under large-scale forcing (corresponding to ground state Fourier modes) and genuine dissipation. Roughly speaking, if we wait for a long time, any initial perturbation on the system will die out and the long time behavior of the system will be our exact solutions as long as there is genuine dissipation, i.e. $\sum_{j=1}^{l} d_j > 0$. We also present a non-linear stability result of the solution (in a restricted sense) in the inviscid (no dissipation, $d_j = 0$) case.

The result here gives us an explicit example of a stable large coherent structure in the presence of forcing and dissipation. In the case of a steady forcing, we also have an example of absence of turbulence with an arbitrarily large Reynolds number. This tells us that in order to generate turbulence or complex chaotic behavior we not only need the amplitude of the external forcing to be large but also need some structure on the external forcing, even in this toy model of geophysical flows.

The approach here is a unified approach for classical fluid dynamics and the special geophysical flows considered here. The introduction of the Coriolis term (the beta-plane effect) changes the long time behavior, i.e. the final states, in a significant manner, as can be inferred from the exact solutions in (2.7)–(2.9) above. For instance, for the case without damping ($C(\vec{d}) = 0$) and no external forcing ($\mathcal{F} = 0$), the long time behavior of the barotropic quasi-geostrophic equations is those Rossby wave type solutions (see Section 1.3 for more details), while the long time behavior is those steady states when there is no geophysical effects.

A remarkable identity

At the basis of the analysis presented here lies an important observation concerning flows with generalized Kolmogorov forcing. We already know that the kinetic energy $E(t)$ and the enstrophy $\mathcal{E}(t)$ are quadratic functionals whose evolution has already been described in the previous chapter. The remarkable fact about flows with general Kolmogorov forcing is that:

There is a quadratic functional $\mathcal{W}(t)$ for the flow whose evolution does not involve the external forcing which is:

(i) *exactly conserved without dissipation,*
(ii) *decreasing in time for suitable dissipation in general whenever $\mathcal{W}(t)$ is non-negative.*

Indeed, let us assume a generalized Kolmogorov forcing, i.e.

$$\Delta\mathcal{F} = -\Lambda_p \mathcal{F} \tag{2.10}$$

or equivalently

$$\Delta\Phi = -\Lambda_p \Phi, \quad \text{where } \mathcal{F} = \Delta\Phi. \tag{2.11}$$

We recall the energy identity

$$\frac{d}{dt}\frac{1}{2}\int |\vec{v}|^2 = -\int \psi D(\Delta)\psi - \int \omega\Phi, \tag{2.12}$$

and the enstrophy identity

$$\frac{d}{dt}\frac{1}{2}\int \omega^2 = \int \omega\frac{\partial\omega}{\partial t} = \int \omega\mathcal{D}(\Delta)\psi + \int \omega\Delta\Phi. \tag{2.13}$$

Thanks to the assumption on generalized Kolmogorov forcing, we notice that the terms involving external forcing in the energy and enstrophy identity have the following simple relationship

$$\int \omega\Delta\Phi = -\Lambda_p \int \omega\Phi. \tag{2.14}$$

This inspires us to consider the following quadratic functional

$$\mathcal{W}(\omega(t)) \equiv \mathcal{E}(\omega(t)) - \Lambda_p E(\omega(t)), \tag{2.15}$$

and consider the dynamics of this quantity using the energy and enstrophy identity. Then we deduce the following **remarkable identity**

$$\begin{aligned} \frac{d}{dt}\mathcal{W}(\omega(t)) &= \frac{d}{dt}\mathcal{E}(\omega(t)) - \Lambda_p \frac{d}{dt}E(\omega(t)) \\ &= \Lambda_p \int \psi \mathcal{D}(\Delta)\psi + \int \omega \mathcal{D}(\Delta)\psi = \mathbf{D}, \end{aligned} \tag{2.16}$$

where we have used the identity $\int \omega\Delta\Phi = -\Lambda_p \int \omega\Phi$.

The significance of this identity lies in the fact that the external forcing term is not involved explicitly in the dynamics of the quadratic functional $\mathcal{W}$.

2.2 Non-linear stability with Kolmogorov forcing

Now we restrict ourselves to the special case of large-scale forcing, i.e. the forcing is on the ground shell, or $\Lambda = 1$, the first eigenvalue of the Laplacian operator on the 2π torus, and $\vec{k} = \{(\pm 1, 0), (0, \pm 1)\}$.

Since the forcing is on one energy shell, a naive guess would be that any perturbation off that energy shell will eventually die out, since those modes are not forced directly. This is not true in general, since the non-linear term tends to transfer energy between different modes. However this is true when the external forcing is on the ground energy shell, with genuine dissipation under the doubly periodic boundary condition or the free-slip channel boundary condition. The quadratic functional $\mathcal{W}$ introduced above will play a central role in the study below.

The analysis here borrows and generalizes ideas of Marchioro (1986), Constantin, Foias, and Temam (1988).

2.2.1 Non-linear stability in restricted sense

Since we suspect that the dynamics on the ground energy shell have a distinguished position, we would like to introduce the following convenient decomposition of the potential vorticity into the grand energy shell and higher modes using Fourier expansion.

Notation

$$P_{\Lambda_1}\omega = \sum_{|\vec{k}|^2=1} \hat{\omega}_{\vec{k}} e^{i\vec{x}\cdot\vec{k}}, \tag{2.17}$$

$$\omega' = \sum_{|\vec{k}|^2 \geq 2} \hat{\omega}_{\vec{k}} e^{i\vec{x}\cdot\vec{k}} = (1 - P_{\Lambda_1})\omega, \tag{2.18}$$

$$\nabla^\perp \psi' = \vec{v}' = \sum_{|\vec{k}|^2 \geq 2} \frac{i}{|\vec{k}|^2} \begin{pmatrix} -k_2 \\ k_1 \end{pmatrix} \hat{\omega}_{\vec{k}} e^{i\vec{x}\cdot\vec{k}}. \tag{2.19}$$

Here P_{Λ_1} projects the solution onto the finite-dimensional ground state modes (or large-scale modes). Notice that $\omega = P_{\Lambda_1}\omega + \omega'$, $P_{\Lambda_1}\omega$ and ω' are orthogonal.

The main result of this subsection is the following non-linear stability result in the restricted sense.

Theorem 2.1 *Under the above hypothesis, i.e. forcing on the ground energy shell, we have*

$$\frac{1}{2}\int (\omega')^2(t) \leq Ce^{-C(\vec{d})t}\frac{1}{2}\int (\omega')^2(0), \tag{2.20}$$

where

$$C(\vec{d}) = \sum_{j=1}^{l} 2^{j-1} d_j.$$

It is noted that, although the growth of non-ground state modes are controlled by the theorem, this theorem tells us nothing about the growth of ground modes. In fact the ground modes could grow in the absence of dissipation ($d_j = 0$) due to resonance. Thus this theorem proves only restricted stability.

An immediate corollary of the theorem is the following fact:

Corollary 2.1 *If there is no dissipation mechanism, i.e.* $d_j \equiv 0$, *we have*

$$\frac{1}{4}\int (\omega')^2(0) \leq \frac{1}{2}\int (\omega')^2(t) \leq \int (\omega')^2(0). \tag{2.21}$$

Proof of the theorem: We invoke the Fourier representation of the vorticity and stream function. Recall

$$\frac{1}{2}\fint \omega^2 = \frac{1}{2}\sum_{|\vec{k}|^2 \geq 1} |\hat{\omega}_{\vec{k}}|^2, \tag{2.22}$$

$$\frac{1}{2}\fint |\nabla^\perp \psi|^2 = \frac{1}{2}\sum_{|\vec{k}|^2 \geq 1} |\vec{k}|^{-2}|\hat{\omega}_{\vec{k}}|^2, \tag{2.23}$$

$$\begin{aligned} \frac{1}{4\pi^2}\mathcal{W}(\omega) &= \frac{1}{2}\fint \omega^2 - \Lambda_p \frac{1}{2}\fint |\nabla^\perp \psi|^2 \\ &= \frac{1}{2}\sum_{|\vec{k}|^2 \geq 1} |\hat{\omega}_{\vec{k}}|^2 (1 - \Lambda_p |\vec{k}|^{-2}). \end{aligned} \tag{2.24}$$

For the purpose of stability argument, it would be convenient if we have a non-negative $\mathcal{W}(\omega)$. This is the case with Kolmogorov forcing on the ground energy shell. Indeed setting $\Lambda_p = 1$, we can see

$$\frac{1}{4\pi^2}\mathcal{W}(\omega) = \frac{1}{2}\sum_{|\vec{k}|^2\geq 2} |\hat{\omega}_{\vec{k}}|^2(1-|\vec{k}|^{-2}) \geq 0. \tag{2.25}$$

In fact we have, for all time

$$\frac{1}{2}\fint|\omega'|^2 \geq \frac{1}{4\pi^2}\mathcal{W}(\omega) \geq \frac{1}{4}\fint|\omega'|^2, \tag{2.26}$$

since $|\vec{k}|^{-2} \leq \frac{1}{2}$. This immediately implies the corollary for the inviscid case. Indeed $\mathcal{W}$ is a conserved quantity in this case of no dissipation, thanks to the identity (2.16) for $\mathcal{W}$. Thus

$$\begin{aligned}\frac{1}{4}\int \omega'^2(0) \leq \mathcal{W}(\omega(0)) = \mathcal{W}(\omega(t)) \leq \frac{1}{2}\int \omega'^2(t) \leq 2\mathcal{W}(\omega(t))\\ = 2\mathcal{W}(\omega(0)) \leq \int \omega'^2(0).\end{aligned} \tag{2.27}$$

This is exact the corollary that we want to prove.

We now continue with the proof of the theorem in the general case with dissipation. We are thus forced to deal with the dissipation, or the right-hand side of the identity for $\mathcal{W}(\omega(t))$.

Recall that

$$\begin{aligned}\mathcal{D}(\Delta)\psi &= \sum_{|\vec{k}|^2\geq 1}\sum_{j=1}^{l} d_j(-1)^j\Delta^j(e^{i\vec{x}\cdot\vec{k}}\hat{\psi}_{\vec{k}}) = \sum_{|\vec{k}|^2\geq 1}\sum_{j=1}^{l} d_j|\vec{k}|^{2j}e^{i\vec{x}\cdot\vec{k}}\hat{\psi}_{\vec{k}}\\ &= -\sum_{|\vec{k}|^2\geq 1}\sum_{j=1}^{l} d_j|\vec{k}|^{2j-2}e^{i\vec{x}\cdot\vec{k}}\hat{\omega}_{\vec{k}},\end{aligned} \tag{2.28}$$

since $\hat{\psi}_{\vec{k}} = -|\vec{k}|^{-2}\hat{\omega}_{\vec{k}}$; thus we have, thanks to Parseval's identity and the assumption that $\Lambda_p = 1$

$$\begin{aligned}\frac{1}{4\pi^2}\mathbf{D} &= \Lambda_p\fint\psi\mathcal{D}(\Delta)\psi + \fint\omega\mathcal{D}(\Delta)\psi\\ &= -\sum_{|\vec{k}|^2\geq 1}|\hat{\omega}_{\vec{k}}|^2\left(\sum_{j=1}^{l} d_j|\vec{k}|^{2j-2} - \Lambda_p\sum_{j=1}^{l} d_j|\vec{k}|^{2j-4}\right)\\ &= -\sum_{|\vec{k}|^2\geq 1}|\hat{\omega}_{\vec{k}}|^2\left(\sum_{j=1}^{l} d_j|\vec{k}|^{2j-2} - \sum_{j=1}^{l} d_j|\vec{k}|^{2j-4}\right)\end{aligned}$$

$$= -\sum_{|\vec{k}|^2\geq 2} |\hat{\omega}_{\vec{k}}|^2 \left(\sum_{j=1}^{l} d_j |\vec{k}|^{2j-2} - \sum_{j=1}^{l} d_j |\vec{k}|^{2j-4} \right)$$

$$= -\sum_{|\vec{k}|^2\geq 2} \sum_{j=1}^{l} d_j |\vec{k}|^{2j-2} (1 - |\vec{k}|^{-2}) |\hat{\omega}_{\vec{k}}|^2. \tag{2.29}$$

Now we define

$$C(\vec{d}) \equiv \sum_{j=1}^{l} d_j 2^{j-1}; \tag{2.30}$$

we have

$$C(\vec{d}) \leq \sum_{j=1}^{l} d_j |\vec{k}|^{2j-2}, \text{ provided } |\vec{k}|^2 \geq 2. \tag{2.31}$$

Combining the above inequalities, we deduce

$$\begin{aligned} \frac{1}{4\pi^2} C(\vec{d}) \mathcal{W}(\omega) &= \sum_{j=1}^{l} d_j 2^{j-1} \frac{1}{2} \sum_{|\vec{k}|^2\geq 2} (1 - |\vec{k}|^{-2}) |\hat{\omega}_{\vec{k}}|^2 \\ &\leq \frac{1}{2} \sum_{|\vec{k}|^2\geq 2} \sum_{j=1}^{l} d_j |\vec{k}|^{2j-2} (1 - |\vec{k}|^{-2}) |\hat{\omega}_{\vec{k}}|^2 \\ &= -\frac{1}{4\pi^2} \frac{1}{2} \mathbf{D}. \end{aligned} \tag{2.32}$$

This implies, together with the dynamic equation (2.15) for $\mathcal{W}(t)$

$$\frac{d}{dt} \mathcal{W}(t) \leq -2C(\vec{d}) \mathcal{W}(t). \tag{2.33}$$

Here and elsewhere in the book we will use the *Gronwall Inequality* (see for instance Hartman, 1964). This inequality states that, if functions $f(t)$ and $g(t)$ satisfy the inequality

$$\frac{d}{dt} g(t) \leq c g(t) + f(t), \tag{2.34}$$

where c is a constant, then

$$g(t) \leq e^{c(t-s)} g(s) + \int_s^t e^{c(t-\tau)} f(\tau) d\tau. \tag{2.35}$$

Thus, thanks to Gronwall's inequality, (2.26) and (2.33), we have

$$4\pi^2 \times \frac{1}{4} \|\omega'(t)\|_{L^2}^2 \leq \mathcal{W}(t) \leq e^{-2C(\vec{d})t} \mathcal{W}(0) \leq e^{-2C(\vec{d})t} 4\pi^2 \times \frac{1}{2} \|\omega'(0)\|_{L^2}^2. \tag{2.36}$$

With this we finish the proof of this theorem.

2.2.2 Finite-dimensional dynamics on the ground modes and non-linear stability

In the previous subsection, we proved that perturbation off the ground modes will decay exponentially in the presence of genuine dissipation ($C(\vec{d}) > 0$). The question of stability is then reduced to the study of the finite-dimensional dynamics on the ground modes. It is then natural to compare the dynamics on the ground modes with the exact solution of the equation with a new initial value which equals the projection of the original initial value to the ground modes. More precisely, we introduce the exact solution as

$$\psi_{\text{exact}} = \sum_{|\vec{k}|^2=1} A_{\vec{k}}(t) e^{i(\vec{k}\cdot\vec{x}-\omega(\vec{k})t)} + c.c; \quad \omega(\vec{k}) = -\frac{k_1\beta}{|\vec{k}|^2}, \tag{2.37}$$

$$\omega_{\text{exact}} = -\psi_{\text{exact}}, \tag{2.38}$$

where the coefficients are solved via the following ODE

$$\frac{dA_{\vec{k}}}{dt} = -\mathcal{D}(-1)A_{\vec{k}} - \hat{\mathcal{F}}_{\vec{k}}(t)e^{i\omega(\vec{k})t}, \tag{2.39}$$

and the solutions takes the form

$$A_{\vec{k}}(t) = A_{\vec{k}}(0)e^{-\mathcal{D}(-1)t} - \int_0^t e^{-\mathcal{D}(-1)(t-s)}\hat{\mathcal{F}}_{\vec{k}}(s)e^{i\omega(\vec{k})s}\,ds, \tag{2.40}$$

where $A_{\vec{k}}(0)$ are the Fourier coefficients of $\psi(0)$, the initial data of the problem with $|\vec{k}|^2 = 1$. Alternatively, we may characterize ω_{exact} as the solution to the following linear equation

$$\frac{\partial\omega}{\partial t} + \beta\frac{\partial\psi}{\partial x} = \mathcal{D}(\Delta)\psi + \mathcal{F}(\vec{x}, t). \tag{2.41}$$

The initial data here are not that important, since we have a linear dissipated system and we are interested in long time behavior only. It is an easy exercise to verify that the influence of the initial data dies out exponentially, as we can observe from (2.40).

Our goal is to estimate the difference between ω_{exact} and ω. According to the result in the previous subsection, it suffices to consider the difference between ω_{exact} and $P_{\Lambda_1}\omega$ if we neglect exponentially decaying terms. We are interested in the stability property of the solution ω. Now, if we could prove that ω_{exact} and $P_{\Lambda_1}\omega$ stay close together, the stability would follow, since the dynamics on the ground modes, or ω_{exact} are non-linearly stable, as we can see from (2.37)–(2.40).

Here is our main result of this subsection:

Theorem 2.2 *Assume that there is genuine dissipation, i.e.* $C(\vec{d}) > 0$, *and assume that* $\mathcal{F}_{\vec{k}}(t)$ *grow sub-exponentially, i.e. for any* $\alpha > 0$, *there exists* c_α *such that*

$$\sum_{|\vec{k}|^2=1} |\mathcal{F}_{\vec{k}}|^2 \le c_\alpha e^{\alpha t}, \tag{2.42}$$

and therefore the exact solutions on the ground energy shell also grow sub-exponentially, i.e. $A_{\vec{k}}$ *also grow sub-exponentially*

$$\sum_{|\vec{k}|^2=1} |A_{\vec{k}}|^2 \le \bar{c}_\alpha e^{\alpha t}. \tag{2.43}$$

Let $\omega(\vec{x}, t)$ *be any solution of the barotropic quasi-geostrophic equations with* $\mathcal{F}$ *given by (2.6) with* $\Lambda = 1$ *and initial condition* $\omega(\vec{x}, t)|_{t=0} = \omega_0(\vec{x})$. *Then there exist constants* $c_1, c_2 > 0$, *such that*

$$\|\omega_{\text{exact}}(\vec{x}, t) - \omega(\vec{x}, t)\|_0^2 \le c_1 e^{-c_2 t}.$$

This theorem implies that, for arbitrary initial conditions, the flow relaxes exponentially on a dissipative time scale to the exact solution ω_{exact}.

Proof of the theorem: We first recall that

$$\|\omega'\|_0^2 = \|(Id - P_{\Lambda_1})\omega\|_0^2 \le c e^{-c' t},$$

for some positive constants c, c' by our earlier result. Thus the theorem is equivalent to proving

$$\|\omega_{\text{exact}}(\vec{x}, t) - P_{\Lambda_1}\omega(\vec{x}, t)\|_0^2 \le c_1 e^{-c_2 t},$$

by the usual triangle inequality.

Since we already have the dynamic equation for the exact solution ω_{exact} (2.41), it is natural to find the equation satisfied by $P_{\Lambda_1}\omega$ and then consider the difference of ω_{exact} and $P_{\Lambda_1}\omega$ by studying the equation satisfied by $\omega_{\text{exact}} - \omega$. In view of the equation of ω_{exact}, we would like to write the non-linear interaction term as the external forcing term for the equation for $P_{\Lambda_1}\omega$.

Fourier representation for the dynamic equations

We recall the Fourier representation of the solutions for the damped and forced barotropic quasi-geostrophic equations

$$\frac{\partial \omega}{\partial t} + \nabla^\perp \psi \cdot \nabla \omega + \beta \frac{\partial \psi}{\partial x} = \mathcal{D}(\Delta)\psi + \mathcal{F}(\vec{x}, t).$$

with

$$\omega = \sum_{|\vec{k}|^2 \geq 1} e^{i\vec{x}\cdot\vec{k}} \hat{\omega}_{\vec{k}}(t), \quad \psi = \sum_{|\vec{k}|^2 \geq 1} \frac{-\hat{\omega}_{\vec{k}}(t)}{|\vec{k}|^2} e^{i\vec{x}\cdot\vec{k}}.$$

The terms in the equation can be represented using the Fourier representation

$$\nabla^\perp \psi = -\sum_{|\vec{m}|^2 \geq 1} \frac{i}{|\vec{m}|^2} \begin{pmatrix} -m_2 \\ m_1 \end{pmatrix} \hat{\omega}_{\vec{m}}(t) e^{i\vec{x}\cdot\vec{m}}, \tag{2.44}$$

$$\nabla\omega = \sum_{|\vec{l}|^2 \geq 1} i \begin{pmatrix} l_1 \\ l_2 \end{pmatrix} \hat{\omega}_{\vec{l}}(t) e^{i\vec{x}\cdot\vec{l}}, \tag{2.45}$$

$$\nabla^\perp \psi \cdot \nabla\omega = \sum_{|\vec{k}|^2 \geq 1} \hat{P}_{\vec{k}} e^{i\vec{x}\cdot\vec{k}}, \tag{2.46}$$

where

$$\hat{P}_{\vec{k}} = \sum_{\vec{m}+\vec{l}=\vec{k}, |\vec{m}|^2 \neq |\vec{l}|^2} \frac{\vec{m}^\perp \cdot \vec{l}}{|\vec{m}|^2} \hat{\omega}_{\vec{m}}(t) \hat{\omega}_{\vec{l}}(t), \tag{2.47}$$

and we have used the fact that

$$\sum_{\vec{m}+\vec{l}=\vec{k}, |\vec{m}|^2 = |\vec{l}|^2} \frac{\vec{m}^\perp \cdot \vec{l}}{|\vec{m}|^2} \hat{\omega}_{\vec{m}}(t) \hat{\omega}_{\vec{l}}(t) = 0,$$

which can be verified by changing the role of $\vec{l}$ and $\vec{m}$ in the summation, and using the identity $\vec{m}^\perp \cdot \vec{l} = -m_2 l_1 + m_1 l_2 = -(-l_2 m_1 + l_1 m_2) = -\vec{l}^\perp \cdot \vec{m}$.

Thus the equivalent β-plane equation in the Fourier representation takes the form

Fourier series representation of barotropic quasi-geostrophic equations

$$\frac{d\hat{\omega}_{\vec{k}}(t)}{dt} + \sum_{\vec{m}+\vec{l}=\vec{k}, |\vec{m}|^2 \neq |\vec{l}|^2} \frac{\vec{m}^\perp \cdot \vec{l}}{|\vec{m}|^2} \hat{\omega}_{\vec{m}}(t) \hat{\omega}_{\vec{l}}(t) - \frac{i\beta k_1}{|\vec{k}|^2} \hat{\omega}_{\vec{k}}(t)$$
$$= -\frac{\mathcal{D}(-|\vec{k}|^2)}{|\vec{k}|^2} \hat{\omega}_{\vec{k}}(t) + \hat{\mathcal{F}}_{\vec{k}}(t). \tag{2.48}$$

We now restrict $\vec{k}$ to the ground state modes $|\vec{k}|^2 = 1$ (since these are the object that we are interested in) and assume external forcing on the ground states only.

We are interested in finding out the contributions from the quadratic term to the ground energy shell ($|\vec{k}| = 1$). It is apparent that the contribution comes from either the interaction between the ground modes and higher modes, i.e.

$$\vec{l} + \vec{m} = \vec{k},\ |\vec{l}| = 1,\ |\vec{k}| = 1$$

or from interaction among higher modes, i.e.

$$|\vec{l}|^2 \geq 2,\ |\vec{m}|^2 \geq 2,\ \vec{l} + \vec{m} = \vec{k}.$$

In the case of ground modes/higher modes interaction, it is clear that $|\vec{m}| \leq 2$. By elementary number theory, or simply exhausting all possibility, we see that $|\vec{m}|^2 \neq 3$. For the case $|\vec{m}|^2 = 4$, we must have that $\vec{l}$ and $\vec{m}$ co-linear and hence $\vec{m}^\perp \cdot \vec{l} = 0$. We see that the only non-trivial contribution in the case $|\vec{l}| = 1$ is when $|\vec{m}|^2 = 2$. Hence we deduce, for $|\vec{k}| = 1$

$$\begin{aligned}\frac{d\hat{\omega}_{\vec{k}}(t)}{dt} - \frac{i\beta k_1}{|\vec{k}|^2}\hat{\omega}_{\vec{k}}(t) &= -\frac{\mathcal{D}(-|\vec{k}|^2)}{|\vec{k}|^2}\hat{\omega}_{\vec{k}}(t) + \hat{\mathcal{F}}_{\vec{k}}(t) \\ &\quad - \sum_{|\vec{l}|^2\geq 2, |\vec{m}|^2\geq 2, \vec{l}+\vec{m}=\vec{k}} \frac{\vec{m}^\perp \cdot \vec{l}}{|\vec{m}|^2}\hat{\omega}_{\vec{m}}(t)\hat{\omega}_{\vec{l}}(t) \\ &\quad - \sum_{|\vec{l}|^2=1, |\vec{m}|^2=2, \vec{l}+\vec{m}=\vec{k}} \frac{\vec{m}^\perp \cdot \vec{l}}{2}\hat{\omega}_{\vec{m}}(t)\hat{\omega}_{\vec{l}}(t).\end{aligned} \tag{2.49}$$

The last line in the last equation represents interaction between ground modes and higher modes through quadratic non-linear terms in the equation. And a ground mode can interact with other ground modes in the presence of higher modes. For instance for $\vec{k} = (1, 0)$, $\vec{m} = (1, -1)$, $\vec{l} = (0, 1)$, they satisfy $\vec{k} = \vec{l} + \vec{m}$, $|\vec{k}| = 1$, $|\vec{l}| = 1$, $|\vec{m}|^2 = 2$. These terms can be viewed as scattering terms.

Denoting

$$C_{\vec{k},\vec{l}} = \sum_{\vec{m}+\vec{l}=\vec{k}, |\vec{m}|^2=2} \frac{\vec{m}^\perp \cdot \vec{l}}{2}\hat{\omega}_{\vec{m}},$$

and

$$H_{\vec{k}} = \sum_{|\vec{l}|^2\geq 2, |\vec{m}|^2\geq 2, \vec{l}+\vec{m}=\vec{k}} \frac{\vec{m}^\perp \cdot \vec{l}}{|\vec{m}|^2}\hat{\omega}_{\vec{m}}\hat{\omega}_{\vec{l}},$$

we get the full non-linear ODEs for the ground state modes $|\vec{k}| = 1$

$$\frac{d\hat{\omega}_{\vec{k}}(t)}{dt} - i\beta k_1\hat{\omega}_{\vec{k}}(t) = -\mathcal{D}(-1)\hat{\omega}_{\vec{k}}(t) + \hat{\mathcal{F}}_{\vec{k}}(t) - \sum_{|\vec{l}|=1} C_{\vec{k},\vec{l}}\hat{\omega}_{\vec{l}}(t) - H_{\vec{k}}. \tag{2.50}$$

The above equation is almost the same as the one for ω_{exact}, except that it has the new forcing terms due to the interaction between the ground state modes and higher modes or interaction between higher modes. Now the name of the game is to prove the smallness, more precisely exponential decay, of these extra forces.

It is quite clear that

$$|C_{\vec{k},\vec{l}}(t)| \leq Ce^{-\frac{1}{2}C(\vec{d})t}, \tag{2.51}$$

since each term in the sum has the decay rate $Ce^{-\frac{1}{2}C(\vec{d})t}$ and it is a finite sum. We will prove that the contribution from higher modes is exponentially small. We state this as a claim and will prove this fact at the end of this subsection.

Claim

$$|H_{\vec{k}}(t)| \leq Ce^{-\frac{C(\vec{d})}{2}t}. \tag{2.52}$$

This implies the result of the theorem once we take the difference of ω_{exact} and $P_{\Lambda_1}\omega$ and apply the usual energy method, i.e. take the inner product of the difference equation with $\omega_{\text{exact}} - P_{\Lambda_1}\omega$ and apply the usual Gronwall inequality. Indeed, in terms of Fourier representation we have, for

$$\delta\omega = \omega_{\text{exact}} - P_{\Lambda_1}\omega,$$

$$\frac{d\delta\hat{\omega}_{\vec{k}}(t)}{dt} - i\beta k_1 \delta\hat{\omega}_{\vec{k}}(t) + \sum_{|\vec{l}|=1} C_{\vec{k},\vec{l}}\delta\hat{\omega}_{\vec{l}} = -\mathcal{D}(-1)\delta\hat{\omega}_{\vec{k}}(t) - \sum_{|\vec{l}|=1} C_{\vec{k},\vec{l}}\hat{\omega}_{\text{exact},\vec{l}} - H_{\vec{k}}.$$

Taking the inner product with $\delta\omega$ and taking the real part, noticing that

$$\text{Re}(i\beta k_1 \delta\hat{\omega}_{\vec{k}}, \delta\hat{\omega}_{\vec{k}}) = 0,$$

and

$$\left|\left(\sum_{|\vec{l}|=1} C_{\vec{k},\vec{l}}\delta\hat{\omega}_{\vec{l}}, \delta\hat{\omega}_{\vec{k}}\right)\right| \leq Ce^{-\frac{C(\vec{d})}{2}t}\|\delta\omega(t)\|_0^2,$$

also by the assumption on the exact solution and the claim

$$\left|\left(\sum_{|\vec{l}|=1} C_{\vec{k},\vec{l}}\hat{\omega}_{\text{exact},\vec{l}}, \delta\hat{\omega}_{\vec{k}}\right)\right| \leq C_\alpha e^{-\frac{C(\vec{d})}{2}t+\alpha t}\|\delta\omega(t)\|_0,$$

$$|(H_{\vec{k}}, \delta\hat{\omega}_{\vec{k}})| \leq Ce^{-\frac{C(\vec{d})}{2}t}\|\delta\omega(t)\|_0,$$

we deduce, with $\alpha = C(\vec{d})/4$

$$
\begin{aligned}
\frac{d\|\delta\omega\|_0^2}{dt} &\le \left(-2\mathcal{D}(-1) + Ce^{-\frac{C(\vec{d})}{2}t}\right)\|\delta\omega\|_0^2 + C'_\alpha e^{-\frac{C(\vec{d})}{2}t+\alpha t}\|\delta\omega(t)\|_0 \\
&= \left(-2\mathcal{D}(-1) + Ce^{-\frac{C(\vec{d})}{2}t}\right)\|\delta\omega\|_0^2 + C'_\alpha e^{-\frac{C(\vec{d})}{4}t}\|\delta\omega(t)\|_0 \\
&\le \left(-2\mathcal{D}(-1) + Ce^{-\frac{C(\vec{d})}{2}t} + \frac{C'_\alpha}{4}e^{-\frac{C(\vec{d})}{4}t}\right)\|\delta\omega\|_0^2 + C'_\alpha e^{-\frac{C(\vec{d})}{4}t}.
\end{aligned}
$$

Now choosing T large so that

$$
-2\mathcal{D}(-1) + Ce^{-\frac{C(\vec{d})}{2}t} + \frac{C'_\alpha}{4}e^{-\frac{C(\vec{d})}{4}t} \le -\mathcal{D}(-1), \quad \text{for all } t \ge T
$$

and denoting

$$
\kappa_1 = \mathcal{D}(-1), \quad \kappa_2 = C'_\alpha, \quad \kappa_3 = \frac{C(\vec{d})}{4}
$$

we have

$$
\frac{d\|\delta\omega\|_0^2}{dt} \le -\kappa_1\|\delta\omega\|_0^2 + \kappa_2 e^{-\kappa_3 t}, \text{ for all } t \ge T.
$$

Without loss of generality we may assume that $\kappa_1 > \kappa_3$ (otherwise just replace κ_3 with a smaller positive number), and we deduce, thanks to the Gronwall inequality

$$
\|\delta\omega(t)\|_0^2 \le e^{-\kappa_1(t-s)}\|\delta\omega(s)\|_0^2 + \frac{\kappa_2}{\kappa_1-\kappa_3}(1-e^{-(\kappa_1-\kappa_3)(t-s)})e^{-\kappa_3 t},
$$

for $t \ge s \ge T$. This proves the theorem.

Now it remains to prove the claim. Since

$$
\vec{l} + \vec{m} = \vec{k} \quad \text{and} \quad |\vec{k}| = 1,
$$

we have

$$
|\vec{m}| - 1 \le |\vec{l}| \le |\vec{m}| + 1.
$$

This implies

$$
\left|\frac{\vec{m}^\perp \cdot \vec{l}}{|\vec{m}|^2}\right| \le C, \qquad \frac{|\vec{l}|}{|\vec{m}|} \le C,
$$

provided that $|\vec{m}|^2 \ge 2$, $|\vec{l}|^2 \ge 2$.

Utilizing the above fact, we can see that

$$
\begin{aligned}
|H_{\vec{k}}(t)| &= \left| \sum_{|\vec{l}|^2\geq 2, |\vec{m}|^2\geq 2, \vec{l}+\vec{m}=\vec{k}} \frac{\vec{m}^{\perp}\cdot\vec{l}}{|\vec{m}|^2} \hat{\omega}_{\vec{m}}(t)\hat{\omega}_{\vec{l}}(t) \right| \\
&\leq C \sum_{|\vec{m}|^2\geq 2, |\vec{k}|=1, |-\vec{m}+\vec{k}|^2\geq 2} |\hat{\omega}_{\vec{m}}(t)|\,|\hat{\omega}_{-\vec{m}+\vec{k}}(t)|
\end{aligned}
$$

(by the Cauchy–Schwarz inequality)

$$
\begin{aligned}
&\leq C \left(\sum_{|\vec{m}|^2\geq 2, |\vec{k}|=1} |\hat{\omega}_{\vec{m}}(t)|^2 \right)^{\frac{1}{2}} \left(\sum_{|\vec{m}|^2\geq 2, |\vec{k}|=1, |-\vec{m}+\vec{k}|^2\geq 2} |\hat{\omega}_{-\vec{m}+\vec{k}}(t)|^2 \right)^{\frac{1}{2}} \\
&\leq C \sum_{|\vec{m}|^2\geq 2} |\hat{\omega}_{\vec{m}}(t)|^2 \leq Ce^{-C(\vec{d})t}\|\omega(0)\|^2.
\end{aligned}
$$

This completes the proof of the claim and hence the theorem.

2.2.3 *Counter-example of unstable ground state modes dynamics for truncated inviscid flows*

The non-linear stability result here depends on the existence of genuine dissipation in an essential way. In the absence of dissipation, a situation similar to Euler's equations, we have non-linear stability in the restricted sense, as was stated in Corollary 2.1. The natural question to ask is if non-linear instability could occur for inviscid dynamics of ground state modes. This is possible indeed. In the last chapter of this book we will present an exact solution to the barotropic quasi-geostrophic equations on the sphere which is stable in the restricted sense but non-linearly unstable due to resonance.

For the periodic geometry, we will consider Galerkin truncation of the inviscid dynamics and demonstrate that the ground state modes' dynamics are unstable in general for the truncated dynamics.

Consider a truncation of the stream function of the form

$$
\psi(\vec{x}, t) = \sum_{|\vec{k}|^2\leq\Lambda} \hat{\psi}_{\vec{k}}(t)e^{i\vec{k}\cdot\vec{x}} \tag{2.53}
$$

and satisfying the Galerkin truncated barotropic quasi-geostrophic equations

$$
\frac{\partial\Delta\psi}{\partial t} + \beta\frac{\partial\psi}{\partial x} + P(J(\psi, \Delta\psi)) = \mathcal{D}(\Delta)\psi + P(\mathcal{F}(\vec{x}, t)), \tag{2.54}
$$

where P is the projection onto the subspace expanded by Fourier modes with the square of the wave number less than or equal to Λ, i.e.

$$P(\phi) = \sum_{|\vec{k}|^2 \leq \Lambda} \hat{\phi}_{\vec{k}}(t) e^{i\vec{k}\cdot\vec{x}}, \tag{2.55}$$

$$\phi = \sum_{|\vec{k}|^2 \neq 0} \hat{\phi}_{\vec{k}}(t) e^{i\vec{k}\cdot\vec{x}}. \tag{2.56}$$

The Galerkin truncated equations in (2.54) are the physical space equations, which can be derived equivalently by truncating the Fourier representation (2.48) to all modes satisfying $|\vec{k}|^2 \leq \Lambda$ in (2.48).

Suppose that we still have Kolmogorov forcing, i.e.

$$\Delta\mathcal{F} = -\Lambda_1\mathcal{F},$$

it is then easy to see that the quadratic functional

$$\mathcal{W}_\Lambda(t) = \mathcal{E}(\omega(t)) - E(\omega(t))$$

is a conserved quantity in time for the truncated dynamics (2.54) where $\omega = \Delta\psi$. Thus we still have stability for the ground state modes' dynamics in the restricted sense, as we proved in Corollary 2.1. We will show however that these ground state modes' dynamics are in general non-linearly unstable due to resonance. To fix the idea, we consider the case with $\Lambda = \Lambda_2 = 2$, i.e. we consider Galerkin truncation on to the first two energy shells. For simplicity let us consider the special case of $\beta = 0$ (thus Euler dynamics) and consider the invariant subspace of the stream function of the form

$$\psi(\vec{x}, t) = x_1(t)\cos(x) + y_1(t)\cos(y) + x_2(t)\cos(x-y) + y_2(t)\cos(x+y). \tag{2.57}$$

In terms of the complex-valued Fourier representation, this amounts to saying

$$\hat{\psi}_{(1,0)}(t) = \hat{\psi}_{(-1,0)}(t) = \frac{1}{2}x_1(t), \quad \hat{\psi}_{(0,1)}(t) = \hat{\psi}_{(0,-1)}(t) = \frac{1}{2}y_1(t),$$
$$\hat{\psi}_{(1,-1)}(t) = \hat{\psi}_{(-1,1)}(t) = \frac{1}{2}x_2(t), \quad \hat{\psi}_{(1,1)}(t) = \hat{\psi}_{(-1,-1)}(t) = \frac{1}{2}y_2(t).$$

A straightforward calculation yields

$$P_{\Lambda_2}(J(\psi, \Delta\psi)) = 0 \tag{2.58}$$

$$P_{\Lambda_1}(J(\psi, \Delta\psi)) = (x_2 - y_2)y_1\cos(x) - (x_2 - y_2)x_1\cos(y). \tag{2.59}$$

This implies that this form of stream function is invariant under the truncated Euler dynamics, provided that the Kolmogorov forcing takes the form

$$\begin{aligned}\mathcal{F}(\vec{x},t) &= F_1(t)\cos(x)+F_2(t)\cos(y)\\ &= \frac{1}{2}\{F_1(t)e^{ix}+F_1(t)e^{-ix}\}+\frac{1}{2}\{F_2(t)e^{iy}+F_2(t)e^{-iy}\}. \qquad (2.60)\end{aligned}$$

The exact dynamics for these four modes is then given by the following ODE, which is the real coefficient of the projection on the modes with $|\vec{k}|^2=1$, $|\vec{k}|^2=2$ in (2.48).

$$\frac{dx_1}{dt} = -(x_2-y_2)y_1-F_1, \quad x_1(0)=x_{10}, \qquad (2.61)$$

$$\frac{dy_1}{dt} = (x_2-y_2)x_1-F_2, \quad y_1(0)=y_{10}, \qquad (2.62)$$

$$\frac{dx_2}{dt} = 0, \quad x_2(0)=x_{20}, \qquad (2.63)$$

$$\frac{dy_2}{dt} = 0, \quad y_2(0)=y_{20}. \qquad (2.64)$$

Denoting

$$z(t)=x_1(t)+iy_1(t), \quad F(t)=F_1(t)+iF_2(t), \qquad (2.65)$$

then z satisfies the (complex-valued) ODE

$$\frac{dz}{dt}=i(x_2-y_2)z-F, \qquad (2.66)$$

thus we deduce that the solution is given by

$$x_2(t) = x_{20}, \qquad (2.67)$$

$$y_2(t) = y_{20}, \qquad (2.68)$$

$$z(t) = e^{i(x_{20}-y_{20})t}z(0)-e^{i(x_{20}-y_{20})t}\int_0^t e^{-i(x_{20}-y_{20})\tau}F(\tau)d\tau. \qquad (2.69)$$

The exact solution of the ground state modes $(\cos(x),\cos(y))$ with Fourier coefficients $x_1(t)$ and $y_1(t)$, and initial data x_0, y_0 is given by

$$z_{\text{exact}}(t)=x_1(t)+iy_1(t)=z_{\text{exact}}(0)-\int_0^t F(\tau)d\tau = x_0+iy_0-\int_0^t F(\tau)d\tau. \quad (2.70)$$

This implies non-linear instability of the exact solution for general perturbation with non-zero second energy shell projection satisfying

$$x_{20}-y_{20}\neq 0.$$

We consider three special types of forcing – zero, constant (non-zero), and periodic in time – to illustrate the instability.

(i) Case $F(t) \equiv 0$, i.e. we have inviscid Euler without forcing, we then have

$$z(t) = e^{i(x_{20}-y_{20})t} z(0).$$

Thus

$$\begin{aligned} &|z(t) - z_{\text{exact}}(t)| \\ &\geq |(e^{i(x_{20}-y_{20})t} - 1) z_{\text{exact}}(0)| - |z(0) - z_{\text{exact}}(0)| \\ &\geq |z_{\text{exact}}(0)| - |z(0) - z_{\text{exact}}(0)| \quad (\text{when } |e^{i(x_{20}-y_{20})t} - 1| \geq 1). \end{aligned}$$

Thus small perturbation $(|z(0) - z_{\text{exact}}(0)|, x_{20}, y_{20})$ of the initial data lead to order one error provided $z_{\text{exact}}(0) \neq 0$ which is a generic situation. This is instability. A remark is that this is the strongest instability that we can expect in such situation, since the energy and enstrophy are conserved and thus no error could grow without bound. In the case, with $z_{\text{exact}}(0) = 0$, we are then studying the non-linear stability of the zero (rest) state. It is of course non-linearly stable by the conservation of energy and enstrophy.

(ii) Case $F(t) \equiv C \neq 0$. In this case

$$\begin{aligned} z_{\text{exact}}(t) &= z_{\text{exact}}(0) - Ct, \\ z(t) &= e^{i(x_{20}-y_{20})t} z(0) + \frac{C}{i(x_{20} - y_{20})}(1 - e^{i(x_{20}-y_{20})t}). \end{aligned}$$

Hence the exact dynamics on the ground state modes is unbounded, while the perturbed dynamics is bounded. This is clear instability.

(iii) Case $F(t) = Ce^{i\alpha t}$, i.e. we have periodic in time Kolmogorov forcing. There are two distinct sub-cases:

- Sub-case $\alpha = (x_{20} - y_{20})$, i.e. we have resonance in time frequency of internal rotation and external forcing. Then

$$\begin{aligned} z(t) &= e^{i(x_{20}-y_{20})t} z(0) - Cte^{i(x_{20}-y_{20})t}, \\ z_{\text{exact}}(t) &= z_{\text{exact}}(0) - \frac{C}{i\alpha}(e^{i\alpha t} - 1). \end{aligned}$$

  Hence the exact dynamics on the ground modes is bounded, while the perturbed dynamics is unbounded. This qualitative difference clearly demonstrates instability.

- Sub-case $\alpha \neq (x_{20} - y_{20})$. Then

$$\begin{aligned} z(t) &= e^{i(x_{20}-y_{20})t} z(0) - \frac{C}{i(\alpha - x_{20} + y_{20})}(e^{i\alpha t} - e^{i(x_{20}-y_{20})t}), \\ z_{\text{exact}}(t) &= z_{\text{exact}}(0) - \frac{C}{i\alpha}(e^{i\alpha t} - 1). \end{aligned}$$

Thus

$$
\begin{aligned}
&|z(t) - z_{\text{exact}}(t)| \\
&\geq \left|(e^{i(x_{20}-y_{20})t} - 1)\left(z_{\text{exact}}(0) + \frac{C}{i\alpha}\right)\right| - \left|\frac{C}{i(\alpha - x_{20} + y_{20})} - \frac{C}{i\alpha}\right| \\
&\quad - \left|e^{i(x_{20}-y_{20})t}\left(z(0) + \frac{C}{i(\alpha - x_{20} + y_{20})} - z_{\text{exact}}(0) - \frac{C}{i\alpha}\right)\right| \\
&\geq \left|(e^{i(x_{20}-y_{20})t} - 1)\left(z_{\text{exact}}(0) + \frac{C}{i\alpha}\right)\right| \\
&\quad -|z(0) - z_{\text{exact}}(0)| - 2\left|\frac{C}{i(\alpha t - x_{20} + y_{20})} - \frac{C}{i\alpha}\right|
\end{aligned}
$$

It is easy to see that the last two terms are small since we have small perturbation. On the other hand, the first term could be of order one for those times t such that $|e^{i(x_{20}-y_{20})t} - 1| \geq 1$ and provided that $z_{\text{exact}}(0) - \frac{C}{i\alpha} \neq 0$, which is a generic situation. Again this implies that small perturbation leads to order one error, which is instability.

The case with the beta-effect is more involved, since β generates Rossby waves on the second energy shell, which in turn drives the ground energy shell. We then have a linear ODE system with time periodic coefficients for the ground state modes in this time. In general, exact solutions are hard to find. Nevertheless, the stability issues can be easily verified by combining Floquet theory with elementary numerical computation. An elementary computation for a spectrally truncated system of size say $|\vec{k}|^2 \leq 5$ indicates there is non-linear instability due to the resonance effect. In Chapter 16, we will see a fully non-linear example of an exact solution which is stable in the restricted sense but non-linearly unstable due to the resonance effect with the beta-effect without any truncation.

2.3 Stability of flows with generalized Kolmogorov forcing

We have discussed the nonlinear stability of flows for a system with Kolmogorov forcing. A natural question to ask is then, what happens for generalized Kolmogorov forcing. We notice that the key observation posted in the introduction to the chapter is still valid. In fact, we have already gone through some of the computations in the previous section.

We assume a generalized Kolmogorov forcing

$$\mathcal{F} = \Delta\Phi = -\Lambda_p \Phi. \tag{2.71}$$

Even though the result is true with hyper-viscosity as well, for simplicity we assume there is no hyper-viscosity, i.e.

$$d_j = 0, \quad j \geq 3. \tag{2.72}$$

We consider the quadratic functional $\mathcal{W} = \mathcal{E} - \Lambda_p E$ as before. The dynamics of this quantity is given by, thanks to the remarkable identity (2.16)

$$\frac{d\mathcal{W}}{dt} + 2d_1\mathcal{W} + d_2(\|\nabla\Delta\psi\|_0^2 - \Lambda_p\|\Delta\psi\|_0^2) = 0. \tag{2.73}$$

Since we would like to have decay on $\mathcal{W}$, we try to combine the last two terms into a positive definite quantity. Indeed we have the following *key identity* (this trick will be used in the next chapter as well)

$$\begin{aligned}\|\nabla\Delta\psi\|_0^2 - \Lambda_p\|\Delta\psi\|_0^2 &= \|\nabla(\Delta+\Lambda_p)\psi\|_0^2 - 2(\nabla\Delta\psi, \Lambda_p\nabla\psi)_0 \\ &\quad -\Lambda_p^2\|\nabla\psi\|^2 - \Lambda_p\|\Delta\psi\|_0^2 \\ &= \|\nabla(\Delta+\Lambda_p)\psi\|_0^2 + \Lambda_p\|\Delta\psi\|_0^2 - \Lambda_p^2\|\nabla\psi\|^2 \\ &= \|\nabla(\Delta+\Lambda_p)\psi\|_0^2 + 2\Lambda_p\mathcal{W},\end{aligned} \tag{2.74}$$

thus we deduce

$$\frac{d\mathcal{W}}{dt} = -2(d_1 + d_2\Lambda_p)\mathcal{W} - \|\nabla(\Delta+\Lambda_p)\psi\|_0^2. \tag{2.75}$$

This implies the decay of the quadratic functional $\mathcal{W}$

$$\mathcal{W}(t) \leq e^{-2(d_1+d_2\Lambda_p)t}\mathcal{W}(0), \tag{2.76}$$

by simply dropping the term $\|\nabla(\Delta+\Lambda_p)\psi\|_0^2$ on the right-hand side of the equation and applying Gronwall's inequality.

An important difference between the kinetic energy $E(t)$ (or the enstrophy $\mathcal{E}(t)$) and the quadratic functional $\mathcal{W}(t)$ is that the former is positive definite, whereas $\mathcal{W}(t)$ is in principle an indefinite quadratic functional. Because of this fact, the previous estimate is not as powerful as its counterparts for kinetic energy and enstrophy. However, in some instances we can ascertain that $\mathcal{W}(t)$ is always non-negative and then the above estimate guarantees exponential decay for $\mathcal{W}(t)$. For example, one case where $\mathcal{W}(t) \geq 0$ occurs if $\Lambda_p = \Lambda_1$, where Λ_1 is the first eigenvalue of the Laplacian operator sustained by the specific geometry; for instance, it is 1 for doubly periodic geometry on 2π torus. Indeed, in this case the Poincaré inequality guarantees that

$$\frac{\mathcal{E}(t)}{E(t)} = \frac{\|\nabla\vec{v}\|_0^2}{\|\vec{v}\|_0^2} \geq \Lambda_p, \tag{2.77}$$

which is equivalent to $\mathcal{W}(t) \geq 0$.

Otherwise, for general flows the quadratic functional $\mathcal{W}(t)$ can change sign and all we can conclude is:

Corollary 2.2

1. *If* $\mathcal{W}(0) > 0$, *then* $\limsup_{t\to\infty} \mathcal{W}(t) \leq 0$.
2. *If* $\mathcal{W}(0) \leq 0$, *then* $\mathcal{W}(t) \leq 0$ *for all* t.

Proof: This is a direct consequence of the decay estimate on $\mathcal{W}$.

The quadratic functional $\mathcal{W}(t)$ can be viewed as a functional $\mathcal{W}(\omega)$ of the vorticity ω. In fact, we can provide an explicit formula for $\mathcal{W}(\omega)$ in terms of the Fourier modes $\hat{\omega}_{\vec{k}}$ of the vorticity and the relation that $\hat{\psi}_{\vec{k}} = -\frac{1}{|\vec{k}|^2}\hat{\omega}_{\vec{k}}$. In terms of the Fourier expansion, we have

$$\mathcal{W}(\omega(t)) = \mathcal{E}(t) - \Lambda_p E(t) = 4\pi^2 \times \frac{1}{2}\sum_{\vec{k}\neq 0}\left(1 - \frac{\Lambda_p}{|\vec{k}|^2}\right)|\widehat{\omega}_{\vec{k}}|^2(t). \tag{2.78}$$

An interesting situation arises when $\mathcal{W}(0) \leq 0$. In this case we know from the previous corollary that $\mathcal{W}(t) \leq 0$ for all time. We notice that the quadratic functional $\mathcal{W}(\omega)$ has in general negative and positive terms; the low modes with $|\vec{k}|^2 < \Lambda_p$ give negative contributions, and the remaining high modes with $|\vec{k}|^2 > \Lambda_p$ give positive contributions to $\mathcal{W}(\omega)$. This observation together with the fact that $\mathcal{W}(t) \leq 0$ for all times permit us to derive an upper bound for all the high modes in terms of the few low modes

$$\begin{aligned}\left(1 - \frac{\Lambda_p}{\Lambda_{p+1}}\right)\sum_{|\vec{k}|^2>\Lambda_p}|\widehat{\omega}_{\vec{k}}|^2(t) &\leq \sum_{|\vec{k}|^2>\Lambda_p}\left(1 - \frac{\Lambda_p}{|\vec{k}|^2}\right)|\widehat{\omega}_{\vec{k}}|^2(t) \leq \\ \sum_{|\vec{k}|^2<\Lambda_p}\left(\frac{\Lambda_p}{|\vec{k}|^2} - 1\right)|\widehat{\omega}_{\vec{k}}|^2(t) &\leq \left(\frac{\Lambda_p}{\Lambda_{p-1}} - 1\right)\sum_{|\vec{k}|^2<\Lambda_p}|\widehat{\omega}_{\vec{k}}|^2(t).\end{aligned} \tag{2.79}$$

In conclusion, we proved:

Proposition 2.1 (Slaving principle) *If* $\mathcal{W}(0) \leq 0$, *then* $\mathcal{W}(t) \leq 0$ *for all* t, *and the high modes are dominated by the low modes by the upper bound*

$$\sum_{|\vec{k}|^2>\Lambda_p}|\widehat{\omega}_{\vec{k}}|^2(t) \leq C\sum_{|\vec{k}|^2<\Lambda_p}|\widehat{\omega}_{\vec{k}}|^2(t), \tag{2.80}$$

where C *is the constant* $C = \Lambda_{p+1}(\Lambda_p - \Lambda_{p-1})/\Lambda_{p-1}(\Lambda_{p+1} - \Lambda_p)$.

The above proposition says that when $\mathcal{W}(t) \leq 0$ the infinitely many high modes are "slaved" by a finite number of low modes. In effect, this slaving principle implies control on the enstrophy $\mathcal{E}(t)$ and control on $\|\nabla\vec{v}\|_0^2 = \|\omega\|_0^2$, provided that only the finite-dimensional amplitudes on the right-hand side of (2.80) are bounded. Therefore by Rellich's theorem the velocity field $\vec{v}$ is relatively compact, i.e. $\vec{v}$ is almost finite dimensional.

For general generalized Kolmogorov forcing we cannot expect a non-linear stability result, such as the case for large-scale (ground state modes) forcing, treated in the previous section. Indeed for p large enough, an exact solution of the form $\sin(\vec{k} \cdot \vec{x})$, $|\vec{k}|^2 = \Lambda_p$ becomes unstable for the Navier–Stokes equation with corresponding generalized Kolmogorov forcing, see for instance Yudovich (1965) or Babin and Vishik (1992). Their method can be carried over with minor changes to this case with beta-plane effect. So the above results are in a sense the best that we could expect. Numerical experiments with the case, $\beta = 0$, show that in these circumstances there can be remarkable homoclinic chaos, with bursts and quiescent periods for the vorticity (Nicolaenko and She, 1991).

References

Babin, A. V. and Vishik, M. I. (1992), *Attractors of Evolution Equations*. Amsterdam: North-Holland.

Constantin, P., Foias, C., and Temam, R. (1988), On the dimension of the attractors in two-dimensional turbulence. *Physica D* **30**, 284–296.

Hartman, P. (1964), *Ordinary Differential Equations*. New York: Wiley.

Marchioro, C. (1986), An example of absence of turbulence for any Reynolds number. *Com. Math. Phy.* **105**, 99–106.

Nicolaenko, B., and Z. She, 1991, Symmetry-breaking homoclinic chaos and vorticity bursts in periodic Navier-Stokes flows. *Eur. J. Mech.*, B/Fluids **10** (2-suppl.), 67–74.

Yudovitch, V. I. (1965), Example of the generation of a secondary stationary or periodic flow when there is loss of stability of the laminar flow of a viscous incompressible fluid. *J. Appl. Math. Mech.* **29**(3), 453–467.

3

The selective decay principle for basic geophysical flows

3.1 Introduction

In the last chapter we have considered the case when the system is damped and driven at the large scales. We discovered that a large coherent structure emerges. Now we consider the case without external forcing, i.e. the freely decaying quasi-geostrophic equation

$$\frac{\partial \omega}{\partial t} + \beta \frac{\partial \psi}{\partial x} + J(\psi, \omega) = \mathcal{D}(\Delta)\psi, \tag{3.1}$$

where

$$\mathcal{D}(\Delta) = \sum_{j=1}^{k} d_j (-\Delta)^j,$$

with

$$\sum_{j=2}^{k} d_j > 0 \quad \text{and } d_j \geq 0 \quad \text{for all } j,$$

i.e. we assume the presence of Newtonian viscosity, or hyper-viscosity, or both.

It is apparent that all solutions converge to zero as time approaches infinity. To illustrate this point, we consider the special case of (3.1) with only Newtonian viscosity, $\mathcal{D}(\Delta) = d_2 \Delta \omega$. Then the energy $E(t) = \frac{1}{2}\|\vec{v}\|_0^2$ and the enstrophy $\mathcal{E}(t) = \frac{1}{2}\|\omega\|_0^2$ satisfy

$$\frac{dE(t)}{dt} \leq -d_2\|\omega\|_0^2 = -d_2\|\nabla\vec{v}\|_0^2 \leq -2d_2 E(t),$$

$$\frac{d\mathcal{E}(t)}{dt} \leq -d_2\|\nabla\omega\|_0^2 \leq -2d_2\mathcal{E}(t),$$

where we have applied Poincaré's inequality, $2\|f\|_0^2 \leq \|\nabla f\|_0^2$, for periodic functions with zero mean in the above. Now, Gronwall's inequality yields

$$E(t) \leq e^{-2d_2 t} E(0), \qquad \mathcal{E}(t) \leq e^{-2d_2 t}\mathcal{E}(0),$$

which demonstrates the exponential decay of solutions. Here, we would like to study in detail how solutions decay. We are particularly interested in the emergence of large-scale coherent structures.

Early numerical investigation of the evolution of coherent structures for freely decaying 2-D Navier–Stokes flows indicated that the enstrophy decays much more rapidly than the energy. This suggests that we might find a suitable intermediate time scale over which the energy changes slightly, so as to be regarded as nearly conserved, while the enstrophy sweeps down much more sharply. This led physicists to hypothesize the following selection principle to characterize the large time asymptotic states of the flow:

Physicist's selective decay principle: *After a long time, solutions of the quasi-geostrophic equations and/or the two-dimensional incompressible Navier–Stokes equations approach those states which minimize the enstrophy for a given energy.*

The appeal of such a principle is that it reduces the calculation of the asymptotic states of the quasi-geostrophic equations and/or the Navier–Stokes system to a simpler problem in the calculus of variations as discussed briefly earlier in Section 1.5. However, from the mathematical point of view, the selective decay principle stated above is somewhat imprecise; the solution of the quasi-geostrophic equations and/or the Navier–Stokes equations may not approach only a minimizing state, but rather some critical point of the enstrophy. On the other hand, the minimizing state is the only stable state. In any event, we include in the discussion all rather than just the enstrophy minimizers at constant energy.

For conciseness, we refer to such a critical point of the enstrophy at constant energy as a *selective decay state*. We do not know *a priori* that selective decay states, so defined, are *bona fide* solutions to (3.1) or are invariant for the underlying equations. At the moment, a selective decay state is just a velocity or vorticity profile satisfying the variational principle. The selective decay hypothesis is that the arbitrary initial stream function somehow approaches the flow configuration of a selective decay state in the long time limit. It is hard to imagine the selective decay principle being valid unless the selective decay state turned out to be invariant to the Navier–Stokes equations and/or the quasi- geostrophic equations. More precisely, we are interested in asking if, under time evolution via the quasi-geostrophic equations or the Navier–Stokes equations, a selective decay state continues to minimize the enstrophy at its energy level at later times. Such a property will at least render the selective decay principle meaningful. If such selective states are invariant under the barotropic quasi-geostrophic dynamics, we then need to check if an arbitrary solution converges to some selective decay state. This is the key part of the selective decay principle. We also need to explain the numerical fact that all flows converge to the largest structure allowed by the geometry. This is most likely to be explained using a stability argument. We also need to identify the geophysical effects (in our case it is the beta-plane and the artificial hyper-viscosity), since we are interested in geophysical applications. To summarize, we need to address the following issues:

- Invariance. This is the one that make the selective decay state meaningful.
- Convergence. This is needed to justify the selective decay principle.
- Stability. This is useful in interpreting the numerical results which indicate all flows converge to some ground states.
- Beta-plane effect, hyper-viscosity effect. This is useful since we would like to identify the geophysical effects.

In the remaining sections, we will study the questions of existence and stability of the selective decay states for the quasi-geostrophic equations and the Navier–Stokes equations with periodic boundary conditions and follow similar ideas as those developed by Foias and Saut (1984) for the classical fluid flow case and Majda–Shim–Wang (2001) for the geophysical flow case. The chapter ends with a series of numerical experiments illustrating important facets of selective decay with and without geophysical effects.

Even though the selective decay principle holds for both geophysical and classical two-dimensional flows, the presence of the geophysical effect, i.e. the beta-plane, alters the final dynamics in a fundamental way. We will see that the selective decay states for the classical fluid flows are steady state generalized Taylor vortices, while the selective decay states for geophysical flows with the beta-effect are the superposition of generalized Taylor vortices and those Rossby waves. We will also explain the role of hyper-viscosity. Roughly speaking, it enhances the process of selective decay. Heuristically, this is because the hyper-viscosity damps higher frequency modes much more efficiently than lower frequency modes.

3.2 Selective decay states and their invariance

Recall that the selective decay states $\psi_*(E_0)$ are critical points of the enstrophy functional

$$\mathcal{E}(\psi) = \frac{1}{2}\|\omega\|^2 = \frac{1}{2}\|\Delta\psi\|_0^2 \tag{3.2}$$

under the following energy constraint

$$E(\psi) = \frac{1}{2}\fint|\nabla^\perp\psi|^2 = \frac{1}{2}\|\nabla^\perp\psi\|_0^2 = \frac{1}{2}\|\nabla\psi\|_0^2 = E_0. \tag{3.3}$$

According to the Lagrange multiplier method we deduce the following simultaneous functional relations

$$E(\psi_\star) = E_0, \qquad \frac{\delta\mathcal{E}}{\delta\psi}|_{\psi_\star} = \Lambda\frac{\delta E}{\delta\psi}|_{\psi_\star}.$$

Recall that the variational derivatives of the quadratic energy and enstrophy functionals are calculated as in Section 1.5 as

$$\frac{\delta E}{\delta \psi} = -\Delta \psi, \qquad \frac{\delta \mathcal{E}}{\delta \psi} = \Delta^2 \psi.$$

Hence, we end up with a simultaneous system

$$E(\psi_\star) = E_0, \qquad \Delta^2 \psi_\star = -\Lambda \Delta \psi_\star.$$

This implies, since we assumed zero mean for all stream functions and hence the Laplacian is invertible, that the stream function $\psi_\star$ for the selective decay state must satisfy

$$E(\psi_\star) = E_0, \tag{3.4}$$

$$-\Delta \psi_\star = \Lambda \psi_\star. \tag{3.5}$$

Hence, it must be one of the eigenfunctions of the Laplace operator.

It is interesting to note that as discussed later in the book such eigenvalue–eigenfunction problems also emerge from classical energy–enstrophy statistical mechanics in predicting the most probable states (see Chapters 6 and 8 among others).

The eigenfunctions of the Laplacian in the periodic setting can be easily calculated as the generalized Taylor vortices

$$\psi_j = \sum_{|\vec{k}|^2 = \Lambda_j} A_{\vec{k}} e^{i\vec{k}\cdot\vec{x}}, \tag{3.6}$$

where $\Lambda_j = |\vec{k}|^2$, $\vec{k} \in Z^2$ are the eigenvalues.

It is easy to see, using the Fourier representation (3.6)

$$\mathcal{E}(\psi_j) = \Lambda_j E(\psi_j). \tag{3.7}$$

Hence we may conclude:

(i) The *ground state* is the actual *minimizer* of the enstrophy with given energy. Other Taylor vortices are saddle points.
(ii) All flows will approach a Taylor vortex of the *lowest* eigenvalue $\Lambda = \Lambda_\star$ permitted by the symmetries if the physicists' selective decay principle is true.

We now check the first issue for selective decay states, i.e. the invariance of these states under the barotropic quasi-geostrophic dynamics.

The first thing we notice is that the non-linear term drops for selective decay states, i.e.

$$J(\psi, \Delta\psi) = 0,$$

since

$$\Delta\psi = -\Lambda_j\psi.$$

Hence as discussed earlier in Section 1.3, we end up with a linear equation

$$-\Lambda_j\frac{\partial\psi}{\partial t} + \beta\frac{\partial\psi}{\partial x} = \mathcal{D}(-\Lambda_j)\psi.$$

The solutions can be computed explicitly as

$$\psi(t) = \sum_{|\vec{k}|^2=\Lambda_j} A_{\vec{k}}(0)e^{i\vec{k}\cdot\vec{x}-ik_1t\frac{\beta}{\Lambda_j}}e^{-\frac{\mathcal{D}(-\Lambda_j)}{\Lambda_j}t} \tag{3.8}$$

for the stream function ψ. The vorticity can be calculated correspondingly.

We notice that after renormalization, the selective states are nothing but superposition of generalized Taylor vortices and Rossby waves in the presence of the beta-effect. Without the beta-effect we have steady states (generalized Taylor vortices) only. We may further *speculate*, based on this representation

- The beta-plane effect generates Rossby waves at large times.
- The Rossby waves degenerate into generalized Taylor vortices at vanishing beta-effect.
- Ekman drag ($\mathcal{D}(-\Lambda_j) = d_1\Lambda_j$) has no selective decay effect since all wave numbers decay at the same rate.

Of course, these statements still need to be verified.

We also see that our selective decay states preserve their form under the quasi-geostrophic dynamics, and therefore remain selective decay states at all later times. As we have said in the previous section, this shows that the selective decay principle at least makes sense. Physically, it predicts that an arbitrary initial flow will appear as one of the selective decay states, a decaying Taylor vortex in a moving frame, in the long time limit.

3.3 Mathematical formulation of the selective decay principle

Since we are studying freely decaying flows, it is natural to study the *normalized (in H^1) stream function*

$$\tilde{\psi}(t) = \frac{\psi(t)}{\|\nabla\psi(t)\|_0}, \tag{3.9}$$

so that the energy is normalized to one.

As the numerical results suggest consistently more rapid decay of the enstrophy over the energy, we introduce the quotient of the enstrophy over energy, i.e. the *Dirichlet quotient* $\Lambda(t)$

$$\Lambda(t) = \frac{\mathcal{E}(t)}{E(t)}. \tag{3.10}$$

We now can state the mathematical formulation of the selective decay principle:

Theorem 3.1 (Mathematical form of selective decay) *For a fluid with positive viscosity or hyper-viscosity, in the absence of external forces, the following selective decay principle holds for the freely decaying barotropic quasi- geostrophic equations (3.1): for arbitrary initial data, the generalized Dirichlet quotient $\Lambda(t)$ monotonically decreases to Λ_j for an eigenvalue Λ_j of the Laplace operator, i.e.*

$$\lim_{t\to\infty}\Lambda(t) = \Lambda_j. \tag{3.11}$$

There exists an exact solution $\xi_j(t)$ of the barotropic quasi-geostrophic equation with motion restricted on the jth shell given explicitly by (3.8) with $\mathcal{D}=0$, i.e. a superposition of a zonal flow and Rossby waves, such that

$$\left\| \nabla\tilde{\psi}(t) - \nabla\xi_j(t) \right\|_0 \to 0 \text{ as } t\to\infty. \tag{3.12}$$

The Rossby waves degenerate into generalized Taylor vortices in the absence of the geophysical beta-plane effect.

This partially justifies the physicist's selective decay principle; we do see that all flows eventually tend toward a selective decay state, a critical point of the enstrophy on the momentary energy manifold. But this theorem does not say that the flows must tend to the *minimum* enstrophy point; in fact, this statement is in a strict mathematical sense false. The exact solutions in (3.8) with $\Lambda_k > 2$ provide the counter-examples with $\Lambda(t) \equiv \Lambda_k > 2$. Later, however, when we consider stability of the various selective decay states, we shall see that only the minimizing enstrophy state (subject to imposed symmetries) is stable. So for practical purposes, the physicist's selective decay principle may be said to be valid.

In the remainder of the main text of this chapter we prove a weaker form of the preceding theorem. The proof of the strong form stated in Theorem 3.1, can be found in the second appendix to the chapter.

Theorem 3.2 (Weak form of selective decay principle) *For any initial data with finite energy, there exists an eigenvalue Λ_j of the Laplacian such that (3.11) holds. Moreover, for any increasing sequence of times $\{t_j\}_{j=1}^{\infty}$ such that $t_j\to\infty$, there exists a subsequence $\{t_{j_k}\}_{k=1}^{\infty}$ and a selective decay state ψ_* corresponding to the eigenvalue Λ_j, so that*

$$\left\| \nabla\tilde{\psi}(t_{j_k}) - \nabla\psi_* \right\|_0 \to 0 \text{ as } k\to\infty. \tag{3.13}$$

This weak version does not tell us how limits of different subsequences are related. In particular, for the case of Navier–Stokes equations it does not rule out the possibility that a given state might fluctuate between different selective decay states corresponding to the same Laplacian eigenvalue $-\Lambda_*$ even though this is not allowed according to the strong form. The weak form does not tell us about the geophysical effect neither.

3.4 Energy–enstrophy decay

We begin the proof of this theorem by first showing that the enstrophy decays more rapidly than the energy. This was the numerically observed phenomenon which spawned the selective decay principle. Note that we are *not* saying that the actual value of the enstrophy as a function of time $\mathcal{E}(t)$ decays *much* more rapidly than the energy $E(t)$. In fact, the upper bounds on the decay of these two functionals are identical and sharp, as we shall see below. What we shall demonstrate below is the decay of the Dirichlet quotient, which serves as an effective Lyapunov functional.

Proposition 3.1 *The Dirichlet quotient $\Lambda(t)$ defined in equation (3.10) satisfies*

$$\frac{d\Lambda(t)}{dt} = -\frac{1}{E(t)} \sum_{j=2}^{k} \sum_{l=0}^{j-2} d_j \Lambda^l(t) \|(-\Delta)^{\frac{j+1-l}{2}} \psi - \Lambda(t)(-\Delta)^{\frac{j-1-l}{2}} \psi\|_0^2. \qquad (3.14)$$

In particular, $\frac{d\Lambda(t)}{dt} \leq 0$ with equality obtained only for the selective decay states $(\Delta\psi_j = -\Lambda_j \psi_j)$.

Remark: We notice that the proposition suggests that hyper-viscosity enhances selective decay since it enhances the decay rate of $\Lambda(t)$. It also indicates that Ekman damping (d_1) has no selective decay effect.

Proof of Proposition 3.1: Now we establish the validity of equation (3.14) for the decay of $\Lambda(t)$.

For convenience we recall the definition of fractional power of the Laplacian operator in the current periodic setting (see for instance Folland, 1976, or Foias and Temam, 1989). For $\psi = \sum_{\vec{k}} A_{\vec{k}} e^{i\vec{k}\cdot\vec{x}}$

$$(-\Delta)^s \psi = \sum_l \sum_{|\vec{k}|^2=\Lambda_l} \Lambda_l^s A_{\vec{k}} e^{i\vec{k}\cdot\vec{x}}. \qquad (3.15)$$

This immediately implies

$$\|\nabla\psi\|_0 = \|\nabla^\perp\psi\|_0 = \|(-\Delta)^{\frac{1}{2}}\psi\|_0. \qquad (3.16)$$

Our first step is to write down the dynamic equations satisfied by the enstrophy and the energy. Indeed, multiplying equation (3.1) by $\psi(\Delta\psi)$ and integrating over the domain, we deduce

$$\frac{dE(t)}{dt} = -\sum_{j=1}^{k} d_j \|(-\Delta)^{\frac{j}{2}}\psi\|_0^2,$$
$$\frac{d\mathcal{E}(t)}{dt} = -\sum_{j=1}^{k} d_j \|(-\Delta)^{\frac{j+1}{2}}\psi\|_0^2. \tag{3.17}$$

The next step is a straightforward differentiation, substituting (3.17) for the enstrophy and energy time derivatives into the time derivative of the Dirichlet quotient

$$\begin{aligned}
\frac{d\Lambda(t)}{dt} &= \frac{d}{dt}\left(\frac{\mathcal{E}(t)}{E(t)}\right) \\
&= \frac{\dot{\mathcal{E}}(t)}{E(t)} - \frac{\mathcal{E}(t)\dot{E}(t)}{E^2(t)} \\
&= \frac{1}{E(t)}(\dot{\mathcal{E}}(t) - \Lambda(t)\dot{E}(t)) \\
&= -\frac{1}{E(t)}\left(\sum_{j=1}^{k} d_j \|(-\Delta)^{\frac{j+1}{2}}\psi\|_0^2 - \Lambda(t)\sum_{j=1}^{k} d_j \|(-\Delta)^{\frac{j}{2}}\psi\|_0^2\right) \\
&= -\frac{1}{E(t)}\sum_{j=1}^{k} d_j (\|(-\Delta)^{\frac{j+1}{2}}\psi\|_0^2 - \Lambda(t)\|(-\Delta)^{\frac{j}{2}}\psi\|_0^2).
\end{aligned} \tag{3.18}$$

Next we turn our attention to the numerator of equation (3.18). The idea here is to write the right hand as sum of negative definite terms. The key observations are the following identities, which are similar to those used earlier in (2.74)

$$\begin{aligned}
&\|(-\Delta)^{\frac{j+1}{2}}\psi\|_0^2 - \Lambda(t)\|(-\Delta)^{\frac{j}{2}}\psi\|_0^2 \\
&\quad = \|(-\Delta)^{\frac{j+1}{2}}\psi - \Lambda(t)(-\Delta)^{\frac{j-1}{2}}\psi\|_0^2 + 2\Lambda(t)((-\Delta)^{\frac{j+1}{2}}\psi, (-\Delta)^{\frac{j-1}{2}}\psi) \\
&\qquad - \Lambda^2(t)\|(-\Delta)^{\frac{j-1}{2}}\psi\|_0^2 - \Lambda(t)\|(-\Delta)^{\frac{j}{2}}\psi\|_0^2 \\
&\quad = \|(-\Delta)^{\frac{j+1}{2}}\psi - \Lambda(t)(-\Delta)^{\frac{j-1}{2}}\psi\|_0^2 + 2\Lambda(t)((-\Delta)^{\frac{j}{2}}\psi, (-\Delta)^{\frac{j}{2}}\psi) \\
&\qquad - \Lambda^2(t)\|(-\Delta)^{\frac{j-1}{2}}\psi\|_0^2 - \Lambda(t)\|(-\Delta)^{\frac{j}{2}}\psi\|_0^2 \\
&\quad = \|(-\Delta)^{\frac{j+1}{2}}\psi - \Lambda(t)(-\Delta)^{\frac{j-1}{2}}\psi\|_0^2 + \Lambda(t)(\|(-\Delta)^{\frac{j}{2}}\psi\|_0^2 \\
&\qquad - \Lambda(t)\|(-\Delta)^{\frac{j-1}{2}}\psi\|_0^2),
\end{aligned} \tag{3.19}$$

and

$$\| -\Delta\psi\|_0^2 - \Lambda(t)\|(-\Delta)^{\frac{1}{2}}\psi\|_0^2 = 0. \tag{3.20}$$

Thus, we have, by simple iteration

$$\begin{aligned} &\|(-\Delta)^{\frac{j+1}{2}}\psi\|_0^2 - \Lambda(t)\|(-\Delta)^{\frac{j}{2}}\psi\|_0^2 \\ &\quad = \sum_{l=0}^{j-2} \Lambda^l(t)\|(-\Delta)^{\frac{j+1-l}{2}}\psi \;-\Lambda(t)(-\Delta)^{\frac{j-1-l}{2}}\psi\|_0^2, \end{aligned} \tag{3.21}$$

and the proposition follows immediately.

3.5 Bounds on the Dirichlet quotient, $\Lambda(t)$

With the preceding proposition, we collect three facts leading to the mathematical justification for the selective decay principle:

(i) The proposition states that $\Lambda(t)$ is monotone non-increasing. In fact, $\Lambda(t)$ is strictly decreasing unless the stream function is one of the selective decay states $\psi = \psi_j$, in which case $\Lambda(t) = \Lambda_j$ for all succeeding times.

(ii) By Poincare's inequality, $\Lambda(t)$ is bounded below by the first eigenvalue of the Laplacian $\Lambda_1 = 1$.

(iii) Since $\Lambda(t)$ is a decreasing function bounded from below, the limit

$$\lim_{t\to\infty} \Lambda(t) = \Lambda_* \geq \Lambda_1 \tag{3.22}$$

must exist.

Next, we want to demonstrate that Λ_* must be an eigenvalue of the Laplacian, i.e.

$$\Lambda_* = \Lambda_j \text{ for some } j. \tag{3.23}$$

Without loss of generality we assume that $d_2 > 0$, i.e. there is Newtonian dissipation. We begin by noting that

$$\begin{aligned} &2d_2 \int_0^\infty \frac{\|(-\Delta)^{\frac{3}{2}}\psi - \Lambda(t)(-\Delta)^{\frac{1}{2}}\psi\|_0^2}{\|\nabla^\perp\psi\|_0^2}\, dt \\ &\quad \leq -\int_0^\infty \frac{d\Lambda}{dt} = \Lambda(0) - \lim_{t\to\infty}\Lambda(t) = \Lambda(0) - \Lambda_* < \infty \end{aligned}$$

so the integral

$$\int_0^\infty \frac{\|(-\Delta)^{\frac{3}{2}}\psi - \Lambda(t)(-\Delta)^{\frac{1}{2}}\psi\|_0^2}{\|\nabla^\perp\psi\|_0^2}\, dt < \infty \tag{3.24}$$

must converge.

Now let

$$\delta = \min_j(|\Lambda_* - \Lambda_j|). \tag{3.25}$$

Then we have, for large t

$$\min_j(|\Lambda(t) - \Lambda_j|) \geq \frac{\delta}{2} \tag{3.26}$$

since $\Lambda(t) \to \Lambda_*$.

On the other hand, we observe, after invoking Fourier representation for the stream function, that

$$\frac{\|(-\Delta)^{\frac{3}{2}}\psi - \Lambda(t)(-\Delta)^{\frac{1}{2}}\psi\|_0^2}{\|\nabla^\perp\psi\|_0^2} = \frac{\sum_{\vec{k}\neq 0}(|\vec{k}|^2 - \Lambda(t))^2|\vec{k}|^2|\hat{\psi}_{\vec{k}}|^2}{\sum_{\vec{k}\neq 0}|\vec{k}|^2|\hat{\psi}_{\vec{k}}|^2}$$

$$\geq \min_{\vec{k}\in Z^2}\{||\vec{k}|^2 - \Lambda(t)|\} \geq \frac{\delta}{2}.$$

This, together with the convergence of the indefinite integral (3.24), implies that $\delta = 0$. Thus, there exists a j such that

$$\lim_{t\to\infty}\Lambda(t) = \Lambda_* = \Lambda_j \tag{3.27}$$

as desired.

Because $\Lambda(t)$ is monotone and bounded between Λ_* and $\Lambda(0)$ for all $t \in (0, \infty)$, we are tempted to deduce that

$$0 \geq \frac{d\Lambda(t)}{dt} \geq -2d_2\frac{\|(-\Delta)^{\frac{3}{2}}\psi - \Lambda(t)(-\Delta)^{\frac{1}{2}}\psi\|_0^2}{\|\nabla^\perp\psi\|_0^2}$$

$$= -2d_2\frac{\|(-\Delta - \Lambda(t))\nabla\psi\|_0^2}{\|\nabla\psi\|_0^2} \longrightarrow 0 \tag{3.28}$$

as $t \to \infty$. But we can easily construct smooth (C^∞) monotone functions of bounded variation with the property that their derivatives do not have a limit as $t \to \infty$ (for example, consider an infinite sequence of smooth bumps of non-decreasing height over the positive real axis, such that the total area under the graph is finite. Then the indefinite integral of such a function gives a desired counter-example). At this point of the mathematical argument, we only know that $\frac{d\Lambda(t)}{dt}$ converges to zero in some measure-theoretic sense (so that its integral over $[0, \infty)$ converges). But we really want (3.28) to hold in the usual sense; otherwise we would not be able to handle those sequences of times $\{t_j\}$ for which

$$\lim_{j\to\infty}\frac{\left\|(-\Delta - \Lambda(t_j))\nabla\psi(t_j)\right\|_0^2}{\left\|\nabla\psi(t_j)\right\|_0^2}$$

does not exist.

In fact, equation (3.28) can be shown to be valid after a little more computation. The details are given in the first appendix at the end of the chapter. Let us proceed under the presumption that (3.28) has been verified.

3.6 Rigorous theory for selective decay

3.6.1 Convergence to an asymptotic state

Now choose an increasing sequence of times $\{t_j\}_{j=1,\infty}$ such that $\lim_{j\to\infty} t_j = \infty$. We will show that, under a suitable rescaling of amplitudes, the stream function for a certain subsequence of these times will look like a selective decay state in the long time limit; that is, the stream function will take on the structure of a generalized Taylor vortex.

Recall, the normalized stream function is defined as

$$\tilde{\psi}(t) = \frac{\psi(t)}{\|\nabla\psi(t)\|_0}. \tag{3.29}$$

Mathematically, we wish to establish the existence of a subsequence $\{t_{j_k}\}_{k=1,\infty}$ such that $\nabla\tilde{\psi}(t_{j_k}) \stackrel{L^2}{\rightarrow} \nabla\psi_*$, with ψ_* a selective decay state. Such a convergence of subsequence is usually accomplished via some kind of compactness argument. The desired compactness is usually achieved by showing the boundedness of the subsequence in a higher Sobolev space.

In order to obtain the desired compactness, we notice that the normalization (3.29) implies the relations

$$\left\|\nabla\tilde{\psi}(t)\right\|_0 = 1, \qquad \left\|\Delta\tilde{\psi}(t)\right\|_0^2 = \Lambda(t).$$

We can thus deduce that

$$\left\|\nabla\tilde{\psi}(t)\right\|_0^2 + \left\|\Delta\tilde{\psi}(t)\right\|_0^2 = 1 + \Lambda(t) \le 1 + \Lambda(0). \tag{3.30}$$

Therefore, the normalized stream function $\tilde{\psi}$ is bounded in the Sobolev space H^2 independent of time, since we assumed all stream functions have zero mean. Since the sequence $\tilde{\psi}(t_j)$ is bounded in H^2, and bounded sets in a Hilbert space are weakly precompact (Reed and Simon, 1980), there exists a subsequence of times $\{t_{j_k}\}_{k=1,\infty} \subseteq \{t_j\}_{j=1,\infty}$ such that $\tilde{\psi}(t_{j_k})$ converges *weakly* in H^2 to some $\psi_* \in H^2$. In addition, by Rellich's lemma (Folland, 1976) H^2 is compactly embedded in H^1, so that the sequence $\tilde{\psi}(t_{j_k})$ converges strongly to ψ_* in H^1. Thus we have

$$\tilde{\psi}(t_{j_k}) \to \psi_* \text{ weakly in } H^2; \quad \tilde{\psi}(t_{j_k}) \to \psi_* \text{ strongly in } H^1. \tag{3.31}$$

This limiting state ψ_* is not a zero flow, since

$$\|\nabla\psi_*\|_0 = \lim_{k\to\infty}\left\|\nabla\tilde{\psi}(t_j)\right\|_0 = 1 \tag{3.32}$$

thanks to the strong convergence.

3.6.2 Convergence to the selective decay state

Our next goal is to show that ψ_* is in fact an eigenfunction corresponding to the eigenvalue $\Lambda_* = \Lambda_j$.

For this purpose, let ϕ be an arbitrary test function. We notice that

$$\begin{aligned}((-\Delta - \Lambda_*)\psi_*, \phi)_0 &\equiv (\psi_*, (-\Delta - \Lambda_*)\phi)_0 \\ &= \lim_{k\to\infty}(\tilde{\psi}(t_{j_k}), (-\Delta - \Lambda(t_{j_k}))\phi)_0 \\ &= \lim_{k\to\infty}((-\Delta - \Lambda(t_{j_k}))\tilde{\psi}(t_{j_k}), \phi)_0 \\ &= 0,\end{aligned}$$

since, by assumption (3.28)

$$\lim_{k\to\infty}\left\|(-\Delta - \Lambda(t_{j_k}))\nabla\tilde{\psi}(t_{j_k})\right\|_0^2 = 0.$$

Therefore, since the test function is arbitrary

$$(-\Delta - \Lambda_*)\psi_* = 0, \tag{3.33}$$

i.e. ψ_* is a selective decay state corresponding to the eigenvalue $\Lambda_* = \Lambda_j$ as expected.

3.6.3 Stability of the selective decay states

One physically interesting question is whether all the generalized Taylor vortices, corresponding to various eigenvalues Λ_j, are equally likely asymptotic states. Observationally, scientists have found the ground states (the actual *minimizer* of the enstrophy rather than just a critical point) to be the actual "selected" decay states, but our theorem does not yet indicate why. Let us now address this point by first appending a direct corollary:

Corollary 3.1 *Suppose that the initial value of the Dirichlet quotient $\Lambda(0)$ is less than the second lowest eigenvalue allowed by any imposed symmetries. Then*

$$\lim_{t\to\infty}\Lambda(t) = \Lambda_* \tag{3.34}$$

will be the lowest eigenvalue permitted by the symmetries. Consequently, the flow will cluster near ground states in the long time limit.

This observation follows immediately from the monotonic decay of $\Lambda(t)$ and the fact that the limiting value must be an eigenvalue. For the case of flows without symmetries, the ground state is four dimensional (Chapter 1). Our theorem cannot say that a given initial flow ultimately settles down to any one state in the lowest eigenspace. Furthermore, shear flows are associated with Λ_1, and, if our given flows approached shear flows, we would not observe a nice, large, coherent vortex. Hence we consider now flows with swimming pool symmetry (see (1.90))

$$\psi = \sum_{\vec{k}} A_{\vec{k}} \sin(k_1 x)\sin(k_2 y). \tag{3.35}$$

Corollary 3.2 *Consider a flow with swimming pool symmetry. If our initial data satisfy* $\Lambda(0) < 5 = \Lambda_2$, *then*

$$\lim_{t\to\infty} \Lambda(t) = \Lambda_* = 2 = \Lambda_1. \tag{3.36}$$

Also, for every increasing sequence of times $\{t_j\}_{j=1}^{\infty}$ *such that* $t_j \to \infty$, *there exists a subsequence* t_{j_k} *such that*

$$\left\| \nabla\tilde{\psi}(t_{j_k}) - \nabla\psi_* \right\|_0 \to 0, \tag{3.37}$$

where ψ_* *is one of the two anti-parallel selective decay states*

$$\psi_* = \pm\frac{1}{\pi}\sin(x)\sin(y). \tag{3.38}$$

Proof: The assertion that $\Lambda_* = 2$ follows from the monotonic decay of $\Lambda(t)$ and the fact that the lowest eigenvalue permitted by the symmetry is 2, while the next lowest is 5. The rest of the corollary follows from the observation that the two possible ψ_* listed are the only ground state Taylor vortices satisfying the symmetry and the normalization $\|\nabla\psi_*\|_0 = 1$.

Remark: Notice however such symmetry cannot be preserved under the presence of the beta-plane effect. Hence we are considering flows without geophysical effects here in this corollary.

This corollary gives a justification for the heuristic selective decay principle; if the enstrophy–energy ratio begins small enough, then the asymptotic form of the flow resembles a unique Taylor vortex, in fact the one which minimizes the enstrophy (for given energy). The structure of this selective decay state (3.38) is a coherent vortex, of the form typically observed. But our corollary is not yet really satisfactory, for it does not explain why typical initial data, with $\Lambda(0)$ possibly above 5, ought to approach this ground state. We need to add a stability analysis for all the selective decay states given in the theorem for a good mathematical understanding of why only the ground state (respecting imposed symmetries) is actually observed. The next proposition does this, showing that the flows approaching higher states are unstable under small perturbations from the lower eigenmodes.

Proposition 3.2 *All selective decay states in E_k, the eigenspace associated with eigenvalue Λ_k, with $k > 1$ are unstable in the sense that there exist arbitrarily small perturbations such that the resulting stream function asymptotically approaches selective decay states in a strictly lower eigenspace.*

Proof: Start with a selective decay state $\psi_k \in E_k$. Consider any $\psi_j \in E_j$ with $j < k$, such that $\left\|\nabla\psi_j\right\|_0 = 1$, and consider the flow $\psi_k + \varepsilon\psi_j$, where $\varepsilon > 0$ is an arbitrary perturbation strength parameter

$$\left\|\nabla(\psi_k + \varepsilon\psi_j) - \nabla\psi_k\right\|_0 = \varepsilon\left\|\nabla\psi_j\right\|_0 = \varepsilon.$$

Next we compute the Dirichlet quotient of the perturbed stream function. Utilizing the assumption that $\left\|\nabla\psi_j\right\|_0 = 1$ and the orthogonality of the eigenspaces, we have

$$\Lambda(\psi_k + \varepsilon\psi_j) = \frac{\Lambda_k \left\|\nabla\psi_k\right\|_0^2 + \varepsilon^2\Lambda_j}{\left\|\nabla\psi_k\right\|_0^2 + \varepsilon^2} = \Lambda_k + \frac{\varepsilon^2}{\left\|\nabla\psi_k\right\|_0^2 + \varepsilon^2}(\Lambda_j - \Lambda_k) < \Lambda_k. \quad (3.39)$$

Hence, since the Dirichlet quotient is monotonically decreasing in time

$$\Lambda_*(\psi_k + \varepsilon\psi_j) < \Lambda_k \quad (3.40)$$

and thus $\Lambda_*(\psi_k + \varepsilon\psi_j)$ must be a lower eigenvalue. Thus, the asymptotic structure of the perturbed flow, for any arbitrarily small ε, will resemble selective decay states from a lower eigenspace.

Another way to see the instability of the selective decay states against perturbations from lower eigenstates is to recall the formula for the decay of generalized Taylor vortices

$$\left\|\nabla\psi_j(t)\right\|_0 = e^{-2\frac{\mathcal{D}(-\Lambda_j)}{\Lambda_j}t}\left\|\nabla\psi_j(0)\right\|_0. \quad (3.41)$$

We observe that generalized Taylor vortices from lower eigenspaces decay slower, and we might imagine that perturbations from lower eigenspaces would thus persist and in fact grow in relative magnitude to the original Taylor vortex. This is not a proof, since it does not take into account nonlinear interactions, but it gives a partial understanding.

Hence our mathematical theory gives plausibility to the selective decay hypothesis, that all physical flows eventually evolve to the enstrophy-minimizing configuration. An interesting question we have not answered is how the flow actually approaches the selective decay state. Intriguing dynamics can arise; for example, a flow might approach a certain "higher" selective decay state, hover there for a while, then cascade down to the lower decay states. Such a phenomenon is known as a *coarsening transition*.

3.6.4 Underlying simplifying mechanisms

Now that we have established the mathematical theory, we can attempt to cut through the technical details of the arguments and find the fundamental mechanism responsible for the striking simplicity of the long-term flows. First, we criticize the original physical notion that selective decay is due to the enstrophy decaying *much more* quickly than the energy. We have shown this assertion to be false; the fact that the quotient Λ of enstrophy to energy approaches a constant positive value means that the energy and the enstrophy are really decaying on approximately the same time scale. But there is still a crucial element of truth to the physical hypothesis; the monotonic decrease of $\Lambda(t)$ means that the enstrophy is *consistently* decaying more rapidly than the energy. Even though the rates of decay of energy and enstrophy become ever closer, the enstrophy is at all times decaying more quickly (except for the special case of an exact selective decay state). This premise itself might be enough to go through the heuristic motivation for the selective decay principle, arguing that the asymptotic states ought to be the minimizers of enstrophy on constant energy surfaces. Our mathematical proof has added rigor to *this* line of reasoning.

The monotonic decrease of $\Lambda(t) = \frac{\mathcal{E}(t)}{E(t)}$ leads to another, possibly clearer, physical understanding of the selective decay process. Writing $\Lambda(t)$ in terms of the Fourier coefficients of the stream function

$$\Lambda(t) = \frac{\|\Delta\psi\|_0^2}{\|\nabla\psi\|_0^2} = \frac{\sum_{\vec{k}\in Z^2,\vec{k}\neq 0} |\hat{\psi}_{\vec{k}}|^2|\vec{k}|^4}{\sum_{\vec{k}\in Z^2,\vec{k}\neq 0} |\hat{\psi}_{\vec{k}}|^2|\vec{k}|^2} \tag{3.42}$$

we discover that $\Lambda(t)$ is nothing more than the mean- squared wave number, averaged with respect to the spectral distribution of the energy $E = \sum_{\vec{k}\in Z^2,\vec{k}\neq 0} |\hat{\psi}_{\vec{k}}|^2|\vec{k}|^2$. So we have in fact proven that the energy spectrum is flowing to lower wave numbers, and thus larger scales, contributing to the appearance of coherent structures. This flow is opposite to the normal turbulent cascade toward higher wave numbers, and is thus called an *inverse cascade*. On the other hand, the enstrophy spectrum seems to shift higher wave numbers and smaller scales, where it is dissipated. Since dissipation is quadratically more effective as wave number magnitude increases, we can understand why the enstrophy is decaying more rapidly than the energy.

An alternative way which partially explains the selective decay procedure is the following. Since everything is decaying exponentially fast, the quadratic non-linear term plays a less and less significant role in the evolution. Thus, the long time behavior of the solutions is like a linearized equation. For these almost linear equation, the Newtonian viscosity or hyper-viscosity dissipates high wave numbers much more efficiently than the ground modes. This leads to the inverse cascade and selective decay. A word of caution is that for exponentially decaying flows, the non-linear term is not necessarily negligible for the long time dynamics of the normalized flow. For instance, the Euler flow with Ekman damping decays exponentially, while the long time dynamics of the normalized flow is exactly the Euler flow after appropriate scaling in time (see Majda and Holen, 1998).

Finally, we note the key role viscosity has played in our proofs. Intuitively, it acts to kill off the enstrophy flowing to high wave numbers, thereby leaving behind only the energy in low wave number, large-scale structures. Nonetheless, we observe that the coherent structures appear more quickly than on a viscous time scale, and so self-organization does not seem to be an entirely viscous phenomenon. In later lectures we will present other statistical theories applicable to inviscid flows.

3.7 Numerical experiments demonstrating facets of selective decay

Here we present various numerical simulations which verify our selective decay principle and illustrate interesting facets of the selective decay process with geophysical effects. In particular, we will see that, even though the selective decay states are the same with or without the beta-plane, β often strongly influences the geometric nature of selective decay states.

For the numerical results reported here, a standard pseudo-spectral method in space and an explicit fourth-order Runge–Kutta method with adaptive time stepping is utilized. For the non-linear term (Jacobian), the code computes $\vec{v} = \nabla^{\perp}\psi$ and $\nabla\Delta\psi$ in Fourier space with an exponential filter for de-aliasing high frequencies, and then computes the non-linear product $\vec{v}\cdot\nabla\Delta\psi = J(\psi, \Delta\psi)$ in physical space. The reader is referred to the work of Majda and Holen (1997) or Grote and Majda (1997) for more details on the numerical method.

There are two classes of initial data that we shall consider here: perturbations of exact solutions to the sinh-Poisson equations and some initial data used in earlier simulation by Rhines (1975).

3.7.1 Measure of anisotropy

We are particularly interested in the beta-effect. It seems that β often changes the geometric feature of the long time asymptotic dynamics. In order to give a quantitative measurement of the anisotropy in the flow, we recall the Rhines (1975) measure of anisotropy which is defined as

$$\mathcal{R} = \frac{\|v_x\|_0^2}{\|v_x\|_0^2 + \|v_y\|_0^2} = \frac{\|\frac{\partial\psi}{\partial y}\|_0^2}{\|\nabla^\perp\psi\|_0^2}. \tag{3.43}$$

For $\mathcal{R} = 1$, we have a purely zonal flow along the x- axis, while for $\mathcal{R} = 0$ we have a shear flow in the latitude (y) direction.

3.7.2 Explicit solutions of the sinh–Poisson equation

We begin with a one-parameter family of analytic solutions of the sinh–Poisson equations

$$\bar{\omega}_{c,1} = \Delta\bar{\psi}_{c,1} = -c\sinh(\bar{\psi}_{c,1}). \tag{3.44}$$

These equations arise naturally from empirical statistical theories for fluid flows as we shall encounter in later chapters. The parameter $c = c(k)$ depends on a parameter k taking values in

$$0 \le k \le \left(\frac{1}{2}\right)^{\frac{1}{2}}, \quad c(k) = \frac{2}{\pi^2}(1-2k^2)(\mathcal{K}(k))^2, \tag{3.45}$$

where $\mathcal{K}(k)$ is the complete elliptic integral of the first kind (Abramowitz and Stegun, 1970). The parameter $c(k)$ ranges from 0 to 2.07, i.e. $0 \le c(k) \le 2.07$ as k varies over the interval in (3.45) and $c(0) = c_0 = 2$.

A class of special exact solutions to the sinh–Poisson equation is derived in Ting, Chen, and Lee (1987). The exact formula for $\bar{\psi}_{c,1}$ is

$$\bar{\psi}_{c,1} = 4\operatorname{arctanh}(\mathcal{C}_0(k)P_k(x)P_k(y)), \tag{3.46}$$

where

$$\mathcal{C}_0(k) = k(1-k^2)^{-\frac{1}{2}}, \quad P_k(x) = \operatorname{cn}\left(4\left(\frac{x-\pi}{2\pi}\right)\mathcal{K}(k), k\right), \tag{3.47}$$

with cn being the Jacobi cnoidal function. We note that the cnoidal function $\operatorname{cn}(z,k)$ is periodic in z with period $4\mathcal{K}(k)$, so that P_k always has period 2π. By construction, these flows define exact steady solutions for flow on the square, $\Omega = \{(x,y)|0 < x < \pi, 0 < y < \pi\}$ and satisfy the swimming pool symmetry

$$\psi = \sum_{k_1,k_2=-\infty}^{\infty} a_{k_1k_2}\sin(k_1x).\sin(k_2y) \tag{3.48}$$

We get a two parameter family of exact solutions by rescaling those in (3.46), (3.47), i.e. we define $\bar{\psi}_{\frac{c}{b},b} = \frac{\bar{\psi}_{c,1}}{b}$, $\bar{\omega}_{\frac{c}{b},b} = \frac{\bar{\omega}_{c,1}}{b}$ and with (3.44), these functions satisfy

$$\bar{\omega}_{\frac{c}{b},b} = \Delta\bar{\psi}_{\frac{c}{b},b} = -\left(\frac{c}{b}\right)\sinh\left(b\bar{\psi}_{\frac{c}{b},b}\right), \tag{3.49}$$

which explains our notation.

All of the above functions, $\bar{\psi}_{\frac{c}{b},b}$, have the same symmetries as the Taylor vortices

$$\psi = A\sin x \sin y. \tag{3.50}$$

Furthermore, as $c(k)$ tends to zero, $\bar{\omega}_{c,1}$ tends to a point vortex with positive circulation at the center of the square, Ω, while as $k \to 0$, $c(k) \to 2$, and $\bar{\psi}_{c,1}$ looks like the stream function for a Taylor vortex

$$\bar{\psi}_{c,1} \approx 2k \sin x \sin y. \tag{3.51}$$

The quantity, $|\bar{\omega}_{c,1}|_\infty$ where $|\omega|_\infty = \max_{(x,y)\in T^2}|\omega(x,y)|$ is a monotone decreasing function of c for $0 < c < 2.07$ and ranges from $+\infty$ to zero (see Majda and Holen, 1998).

3.7.3 Numerical examples

In this subsection we describe five types of numerical simulations. For all numerical simulations, there will be no hyper-viscosity or Ekman drag. We assume Newtonian viscosity only. The Reynolds number of the problem is then defined as

$$Re = \frac{L\max|\vec{v}|}{d_2} = \frac{L\max|\nabla^\perp\psi|}{d_2}, \tag{3.52}$$

where L is the typical length which is 2π in the periodic case and π in the swimming pool symmetry case.

The initial data for the first simulation are a perturbation by two vortices of the exact solution to the sinh–Poisson equations described in equation (3.46). The two vortices are chosen to break the swimming pool symmetry. More specifically the initial data take the form

$$\Delta\psi(\vec{x},0) = \omega_0 = \omega_{c,1} + \sum_{j=1}^{2} A_j b_r(|\vec{x}-\vec{x}_j|), \tag{3.53}$$

where

$$b_r(s) = \begin{cases} \left(1-\left(\frac{s}{r}\right)^2\right)^2 & \text{for } s \le r \\ 0 & \text{for } s > r \end{cases} \tag{3.54}$$

and

$$\vec{x}_1 = (\pi, \pi), \quad \vec{x}_2 = \left(\frac{\pi}{2}, \pi\right); \quad A_1 = 5, \quad A_2 = -5 \tag{3.55}$$

$$r = \pi/10, \tag{3.56}$$

where $\omega_{c,1}$ is the vorticity of the sinh–Poisson solution. The initial Reynolds number is about 260. Figure 3.1 illustrates the numerical results on energy, enstrophy, and Dirichlet quotient decay, Rhines measure for anisotropy and selective decay states with and without β. We emphasize the difference in the measure of anisotropy and the difference in the geometric feature of the decay states. For the case with $\beta = 5$, we observe complete anisotropy with zonal flow in the limit in contrast to $\beta = 0$.

The initial data for the second simulation are a perturbation by nine random vortices of the same solution to the sinh–Poisson equation. The random vortices are of the same form and same magnitudes with random location and random radius. The initial Reynolds number is about 179. Figure 3.2 illustrates the numerical results on energy, enstrophy, and Dirichlet quotient decay, Rhines measures for anisotropy and selective decay states with and without β. This time the selective decay states with β look like the horizontal translation of the selective decay states without β, and, for these initial data, the selective decay states show no effect of β except translation in phase.

The initial data for the third simulation are the following stream function used by Rhines (1975)

$$\begin{aligned}\psi(\vec{x}, 0) = {} & \cos(x+0.3) + 0.9\sin(3(y+1.8)+2x) \\ & +0.87\sin(4(x-0.7)+(y+0.4)+0.815) \\ & +0.8\sin(5(x-4.3)+0.333)+0.7\sin(7y+0.111).\end{aligned} \tag{3.57}$$

Notice that the various modes superimposed here result in a broad spectrum of the stream function. The initial Reynolds number is about 8046. Figure 3.3 illustrates the numerical results on the decay of energy, Dirichlet quotient (the difference in enstrophy decay is almost indistinguishable), Rhines measure of anisotropy for different β, and the difference in the geometric nature of the selective decay states for different values of β. For $\beta = 3$, the final state is roughly a 75 % zonal flow and has a large zonal jet. The cases with $\beta = 0$ and $\beta = 5$ are more isotropic. These results point to very subtle development of zonal jets in the final state.

In the fourth test example for selective decay, the initial data are a perturbation of (3.46), which keeps the swimming pool symmetry (3.48). The perturbation in each symmetry quadrant consisted of six random vortices with amplitudes equal to 50% of the size of the maximum vorticity of the underlying sinh–Poisson flow, $\bar{\omega}_{c,1}$. The viscosity is set to $\nu = 0.001$ and the initial Reynolds

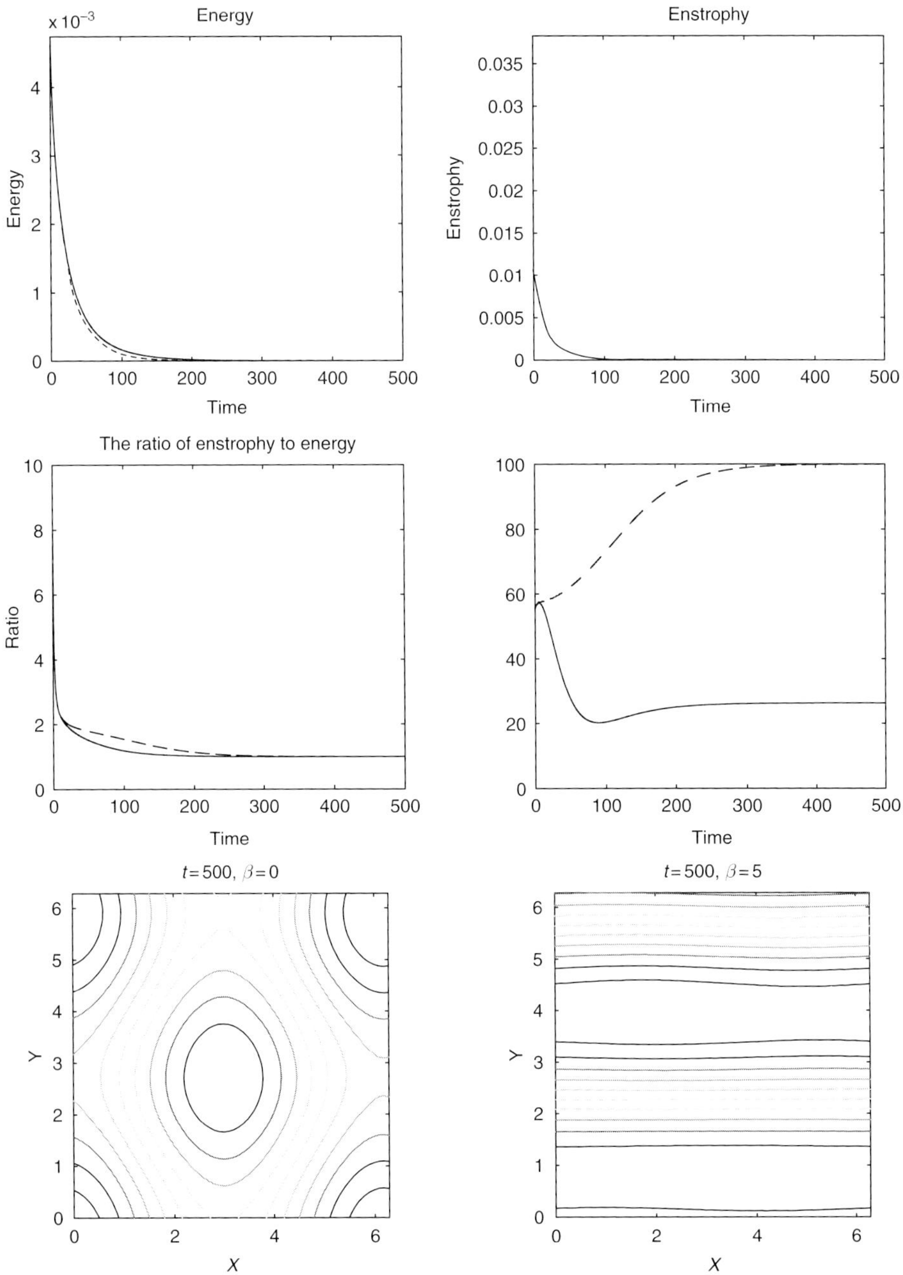

Figure 3.1 Decay of an exact solution of the sinh–Poisson equation perturbed by two vortices. Solid lines: $\beta = 0$. Dashed lines: $\beta = 5$. Upper left: energy v. time ($t = 0 \sim 500$). Upper right: enstrophy vs. time ($t = 0 \sim 500$). Center left: $\Lambda(t)$ v. time ($t = 0 \sim 500$). Center right: anisotropy measure v. time ($t = 0 \sim 500$). Lower left: snapshot of the stream function at $t = 500$, with $\beta = 0$. Lower right: snapshot of the stream function at $t = 500$, with $\beta = 5$.

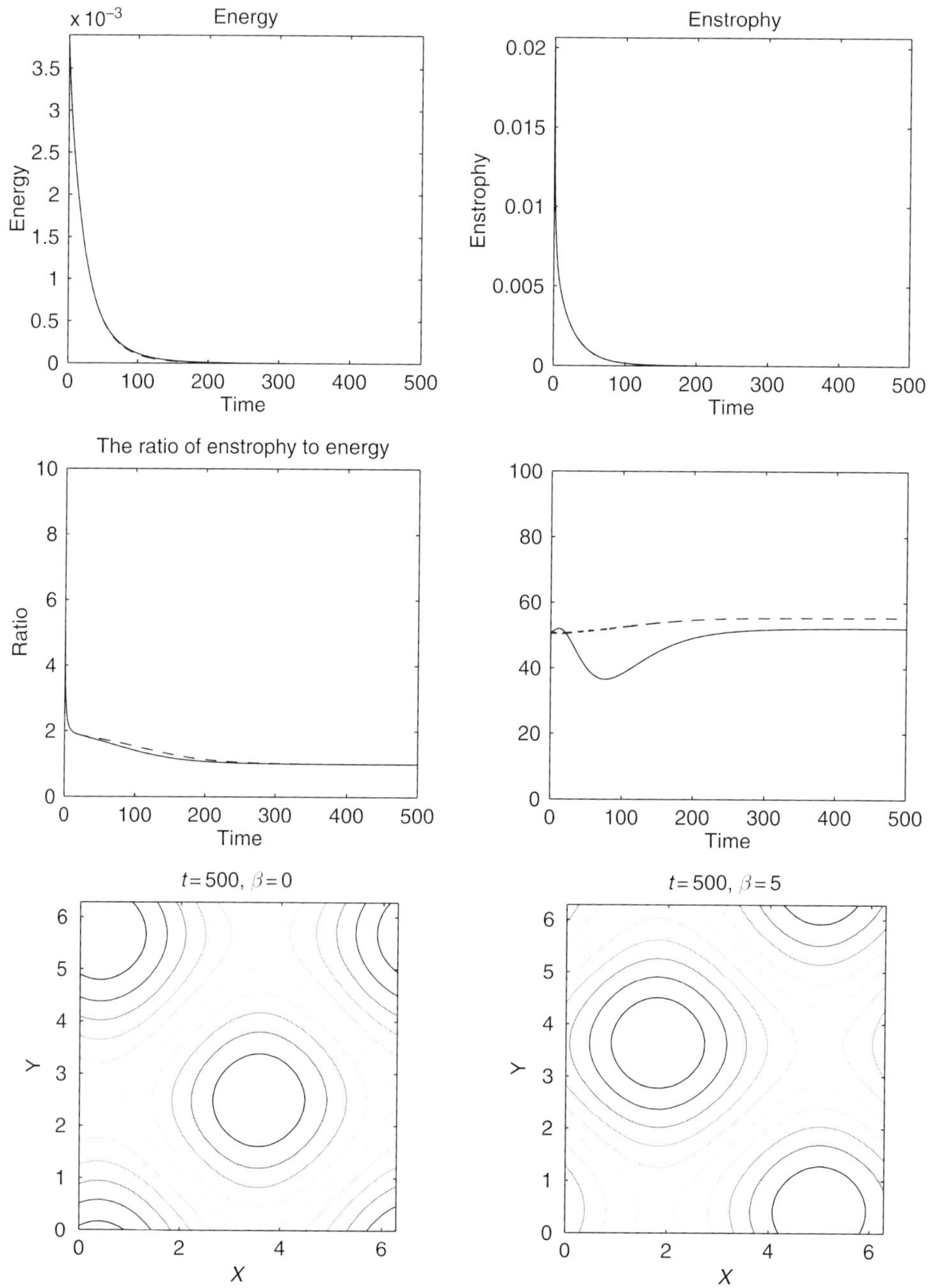

Figure 3.2 Decay of an exact solution of the sinh–Poisson equation perturbed by nine vortices. Solid lines: $\beta = 0$. Dashed lines: $\beta = 5$. Upper left: energy v. time ($t = 0 \sim 500$). Upper right: enstrophy v. time ($t = 0 \sim 500$). Center left: $\Lambda(t)$ v. time ($t = 0 \sim 500$). Center right: anisotropy measure v. time ($t = 0 \sim 500$). Lower left: snapshot of the stream function at $t = 500$, with $\beta = 0$. Lower right: snapshot of the stream function at $t = 500$, with $\beta = 5$.

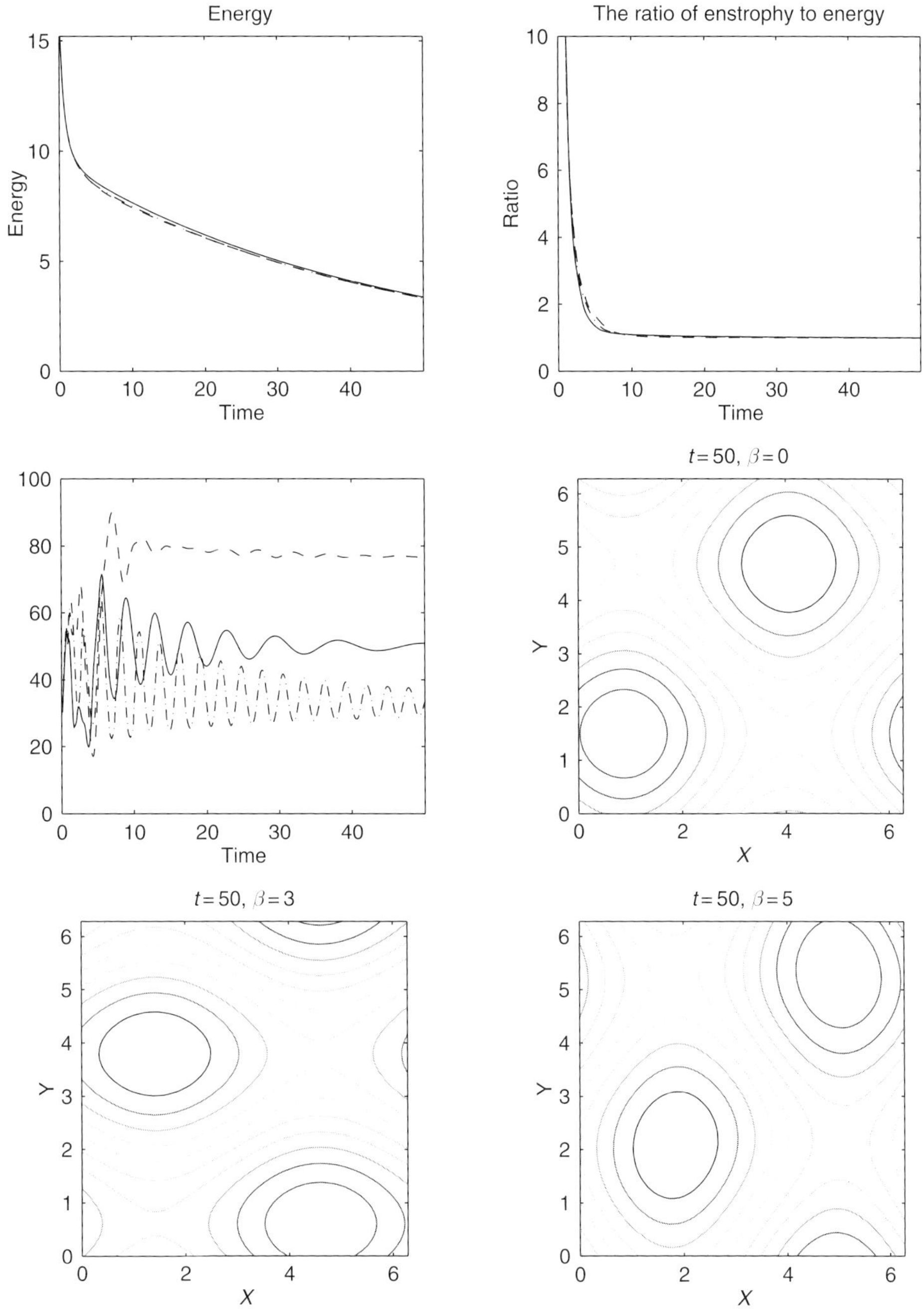

Figure 3.3 Decay of the QG flow specified in (3.57). Solid lines: $\beta = 0$. Dashed lines: $\beta = 3$. Dashdot lines: $\beta = 5$. Upper left: energy v. time ($t = 0 \sim 50$). Upper right: $\Lambda(t)$ v. time ($t = 0 \sim 50$). Center left: anisotropy measure v. time ($t = 0 \sim 50$). Center right: snapshot of the stream function at $t = 50$, with $\beta = 0$ Lower left: snapshot of the stream function at $t = 50$, with $\beta = 3$. Lower right: snapshot of the stream function at $t = 50$, with $\beta = 5$.

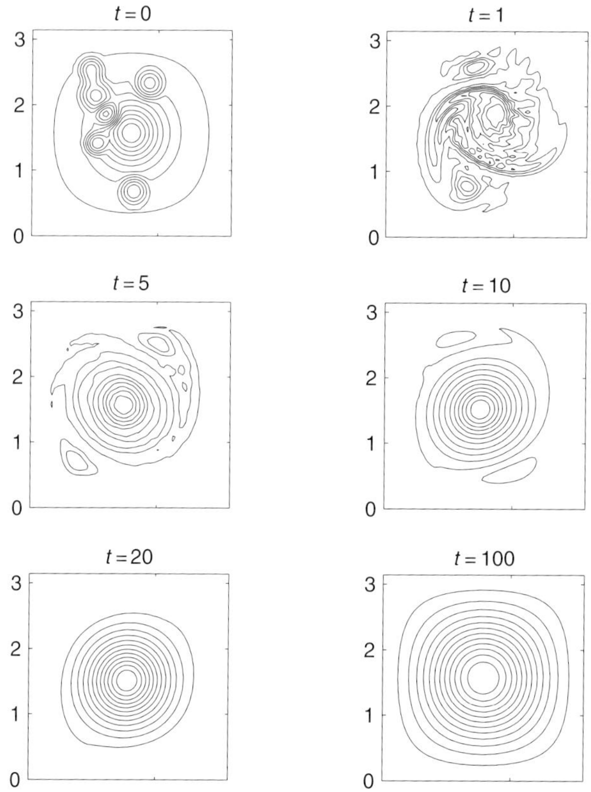

Figure 3.4 Snapshots of the vorticity at $t = 0, 1, 5, 10, 20, 100$ during pure decay, with Newtonian dissipation $\nu = 0.001$.

number is $Re = 25\,000$. Figure 3.4 contains snapshots of the vorticity at times $t = 0, 1, 5, 10, 20, 100$. The small-scale vortices, entrained by the large background vortex, interact and merge rapidly to form a large vortex, which spreads to the boundary of the domain. Hence Figure 3.4 confirms qualitatively the theory of selective decay, which predicts the emergence of a Taylor vortex as the long time solution of the 2D Navier–Stokes equations in a periodic geometry with symmetries.

References

Abramowitz, M. and Stegun, I. (1970), *Handbook of Mathematical Functions*. New York: Dover.

Brezis, H. and Gallouet, T. (1980), Nonlinear Schrödinger evolution equations, *Nonlinear Anal., TMA* **4**, 677–681.

Foias, C. and Saut, J.-C. (1984), Asymptotic behaviour, as $t \to \infty$ of solutions of Navier–Stokes equations and non-linear spectral manifolds. *Indiana Univ. Math. J.* **33**, 459–477.

Foias, C. and Temam, R. (1989), Gevrey class regularity for the solutions of the Navier–Stokes equations. *J. Funct. Anal.* **87**, 359–369.
Folland, G. (1976), *Introduction to Partial Differential Equations*. Princeton, NJ: Princeton University Press.
Grote, M. and Majda, A. (1997), Crude closure dynamics through large scale statistical theory. *Physics of Fluids* **9**, 3431–3442.
Majda, A. and Bertozzi, A. (2001), *Vorticity and Incompressible Flow*. Cambridge: Cambridge University Press.
Majda, A. and Holen, M. (1998), Dissipation, topography, and statistical theories for large scale coherent structure. *CPAM* **50**, 1183–1234.
Majda, A., Shim, S.-Y., and Wang X. (2001), Selective decay for geophysical flows. *Methods and Applications of Analysis* **7**(3), 511–554.
Reed, M. and Simon, B. (1980), *Methods of Modern Mathematical Physics*. New York: Academic Press.
Rhines, P. (1975), Waves and turbulence on a β-plane. *J. Fluid Mech.* **69**, 417–443.
Ting, A. C., Chen, H. H., and Lee, Y. C. (1987), Exact solutions of a nonlinear boundary value problem: the vortices of the two-dimensional sinh–Poisson equation. *Physica D* **26**, 37–66.

Appendix 1 Stronger controls on $\Lambda(t)$

In Section 3.5, we wanted to establish the fact in (3.28)

$$\frac{d\Lambda(t)}{dt} \longrightarrow 0, \tag{3.58}$$

given that:

(i) $\Lambda(t)$ is monotone non-increasing, and
(ii) $\Lambda(0) \geq \Lambda(t) \geq \Lambda_*$ for all times t.

It is sufficient to prove that

$$f(t) = \frac{\|(\Delta - \Lambda(t))\nabla\psi\|_0^2}{\|\nabla\psi\|_0^2} \longrightarrow 0. \tag{3.59}$$

Notice, after a straightforward calculation

$$f = f_1 + f_2, \tag{3.60}$$

$$f_1(t) = \frac{D(t)}{E(t)}, \tag{3.61}$$

$$f_2(t) = 3\Lambda^2(t), \tag{3.62}$$

where

$$D(t) = \frac{1}{2}\|(-\Delta)^{\frac{3}{2}}\psi(t)\|_0^2 = \frac{1}{2}\|\nabla\omega\|_0^2. \tag{3.63}$$

Thanks to our existing estimates on the Dirichlet quotient, we have

$$f \in L^1(\mathbf{R}_+), \tag{3.64}$$

$$f_2' \in L^1(\mathbf{R}_+). \tag{3.65}$$

It is obvious that the contribution from f_2 is manageable, since it has a limit and the derivative of it is integrable. However, we lack control on f_1 for the moment.

The missing link is a *uniform* Lipschitz condition. In fact, we only need a uniform "lower Lipschitz" condition

$$\frac{g(t)-g(s)}{t-s} \geq M \tag{3.66}$$

for some prespecified constant M and for all $t, s \geq 0$. For smooth functions, this is equivalent to a uniform lower bound on the derivative. An upper Lipschitz condition is defined analogously.

Lemma 3.1 *If $f \in L^1(\mathbf{R}_+)$, $f = f_1 + f_2$, $f_2' \in L^1(\mathbf{R}_+)$, and f_1 satisfies a uniform lower (or upper) Lipschitz condition, then*

$$\lim_{t\to\infty} f(t) = 0 \tag{3.67}$$

in the usual sense.

Proof: The idea is simple: the Lipschitz condition forces the width of any bump to be at least proportional to the height, and this does away with our counter-example of ever-thinner bumps tending to infinity. The modification with a function whose derivative is integrable is minimal.

We proceed with a by-way-of-contradiction proof. Without loss of generality, we assume that f_1 satisfies a uniform lower Lipschitz condition with constant M, and that there exists some $\delta > 0$ and a sequence $\{t_j\}_{j=1}^{\infty}$ such that $t_j \to \infty$ and $f(t_j) \geq \delta$. Let us omit enough terms in the sequence so that

$$|t_j - t_{j+1}| > \frac{2\delta}{|M|} \tag{3.68}$$

for all j. The lower Lipschitz condition forces

$$f_1(t) - f_1(t_j) \geq -|M|(t-t_j) \text{ for } t \geq t_j, \tag{3.69}$$

also

$$f_2(t) - f_2(t_j) \geq -\int_{t_j}^{t} |f_2'(s)|ds \text{ for } t \geq t_j. \tag{3.70}$$

Adding the two inequalities and utilizing the assumption $f(t_j) \geq \delta$ we deduce

$$f(t) \geq \delta - |M|(t-t_j) - \int_{t_j}^{t} |f_2'(s)|ds, \tag{3.71}$$

and therefore

$$\int_{t_j}^{t_j+\frac{\delta}{|M|}} f(t) \geq \frac{\delta^2}{2|M|} - \int_{t_j}^{t_j+\frac{\delta}{|M|}} |f_2'(s)|ds \tag{3.72}$$

for each j. And thus by the non-overlapping condition (3.68), we have

$$\int_0^\infty |f(t)| \geq \sum_{j=1}^\infty \int_{t_j}^{t_j+\frac{\delta}{|M|}} f(t)dt \geq \sum_{j=1}^\infty \frac{\delta^2}{2|M|} - \int_0^\infty |f_2'(t)|dt = \infty, \tag{3.73}$$

and hence f is not integrable over the positive real axis. This is a contradiction. Our lemma is proven for this case.

A simple modification gives this result for f with a uniform upper Lipschitz condition. Considering now $f = -g$ proves the lemma for the case where there exists some $\delta > 0$ and a sequence $\{t_j\}_{j=1}^\infty$ such that $t_j \to \infty$ and $f(t_j) \leq -\delta$. This completes the proof of the lemma.

Now we know what to aim for: either a lower or an upper uniform Lipschitz condition on f_1. It turns out that a uniform upper Lipschitz condition is more readily available.

Lemma 3.2 $\dfrac{D(t)}{E(t)}$ *and its first time derivative are bounded above uniformly on* $\mathbf{R}_+$.

Proof: Clearly $D(t)$ is playing a central role in this analysis, so we proceed by studying its time evolution. Taking the inner product of the quasi-geostrophic equations with $-\Delta^2\psi$ we have

$$\frac{dD(t)}{dt} \leq c\|\nabla^2\psi(t)\|_{L^\infty} D(t) - \sum_{j=1}^k d_j\|(-\Delta)^{j/2+1}\psi(t)\|_0^2, \tag{3.74}$$

where we utilized the following estimate on the non-linear term

$$|(J(\psi,\Delta\psi),\Delta^2\psi)| \leq |(\nabla J(\psi,\Delta\psi),\nabla\Delta\psi)| \leq c\|\nabla^2\psi\|_{L^\infty} D(t).$$

Now we consider the evolution of the time derivative of f_1

$$\begin{aligned}
\frac{df_1}{dt} &= \frac{1}{E}\frac{dD}{dt} - \frac{D}{E^2}\frac{dE}{dt} \\
&\leq \frac{c\|\nabla^2\psi\|_{L^\infty} D}{E} - \frac{\sum_{j=1}^k d_j\|(-\Delta)^{j/2+1}\psi\|_0^2}{E} + \frac{D\sum_{j=1}^k d_j\|(-\Delta)^{j/2}\psi\|_0^2}{E^2} \\
&= \frac{c\|\nabla^2\psi\|_{L^\infty} D}{E} - \\
&\quad \sum_{j=1}^k d_j \frac{\|(-\Delta)^{j/2+1}\psi\|_0^2\|(-\Delta)^{1/2}\psi\|_0^2 - \|(-\Delta)^{j/2}\psi\|_0^2\|(-\Delta)^{3/2}\psi\|_0^2}{E^2}.
\end{aligned} \tag{3.75}$$

To estimate the summation in the above inequality we investigate each term in the sum. We first make the:

Claim

$$\|(-\Delta)^{j/2+1}\psi\|_0^2\|(-\Delta)^{1/2}\psi\|_0^2 \geq \|(-\Delta)^{j/2}\psi\|_0^2\|(-\Delta)^{3/2}\psi\|_0^2, \quad \text{for } j \geq 2.$$

Assuming the claim is true we then deduce

$$\frac{df_1}{dt} \leq \frac{c\|\nabla^2\psi\|_{L^\infty} D}{E} = c\|\nabla^2\psi\|_{L^\infty} f_1. \tag{3.76}$$

Invoking the Gronwall inequality we have

$$f_1(t) \leq f_1(0)\exp\left(c\int_0^\infty \|\nabla^2\psi(\tau)\|_{L^\infty} d\tau\right) = \text{constant} < \infty. \tag{3.77}$$

This verifies the lemma, since $\|\nabla^2\psi(\tau)\|_{L^\infty}$ decay exponentially fast. Indeed, by Agmon type inequality

$$\|\nabla^2\psi\|_{L^\infty} \leq c\|\nabla^2\psi\|_0^{\frac{1}{2}}\|\nabla^4\psi\|_0^{\frac{1}{2}}.$$

It is well-known that $\|\nabla^4\psi\|_0 = \|\Delta\omega\|_0$ remains uniformly bounded in time, even in the presence of smooth steady state forcing (see for instance Foias and Temam, 1989). This combined with the known fact that $\|\nabla^2\psi\|_0 = \|\omega\|_0$ decays exponentially, we deduce the exponential decay of the uniform norm of the vorticity.

It remains to prove the claim. For this purpose we invoke the Fourier expansion for the stream function again

$$\psi(\vec{x}) = \sum_l \sum_{|\vec{k}|=\Lambda_l} A_{\vec{k}} e^{i\vec{k}\cdot\vec{x}},$$

and hence

$$\begin{aligned}
\|(-\Delta)^{j/2+1}\psi\|_0^2\|(-\Delta)^{1/2}\psi\|_0^2 &= \left(\sum_l \sum_{|\vec{k}_l|=\Lambda_l} \Lambda_l^{j+2}|A_{\vec{k}_l}|^2\right)\left(\sum_m \sum_{|\vec{k}_m|=\Lambda_m} \Lambda_m|A_{\vec{k}_m}|^2\right)\\
&= \frac{1}{2}\sum_l \sum_{|\vec{k}_l|=\Lambda_l} \sum_m \sum_{|\vec{k}_m|=\Lambda_m} \left(\Lambda_l^{j+2}\Lambda_m + \Lambda_m^{j+2}\Lambda_l\right)|A_{\vec{k}_l}|^2|A_{\vec{k}_m}|^2;
\end{aligned}$$

$$\begin{aligned}
\|(-\Delta)^{j/2}\psi\|_0^2\|(-\Delta)^{3/2}\psi\|_0^2 &= \left(\sum_l \sum_{|\vec{k}_l|=\Lambda_l} \Lambda_l^{j}|A_{\vec{k}_l}|^2\right)\left(\sum_m \sum_{|\vec{k}_m|=\Lambda_m} \Lambda_m^3|A_{\vec{k}_m}|^2\right)\\
&= \frac{1}{2}\sum_l \sum_{|\vec{k}_l|=\Lambda_l} \sum_m \sum_{|\vec{k}_m|=\Lambda_m} \left(\Lambda_l^{j}\Lambda_m^3 + \Lambda_m^{j}\Lambda_l^3\right)|A_{\vec{k}_l}|^2|A_{\vec{k}_m}|^2.
\end{aligned}$$

Thus the claim hinges on the following elementary inequality

$$\Lambda_l^{j+2}\Lambda_m + \Lambda_m^{j+2}\Lambda_l \geq \Lambda_l^j\Lambda_m^3 + \Lambda_m^j\Lambda_l^3,$$

which is equivalent to, after dividing both sides of the inequality by $\Lambda_l\Lambda_m^{j+2}$ and denoting $\lambda = \frac{\Lambda_l}{\Lambda_m}$

$$\lambda^{j+1} - \lambda^{j-1} - \lambda^2 + 1 \geq 0 \quad \text{for } \lambda \geq 0.$$

The above inequality can be verified easily.

This completes the proof of the claim.

Appendix 2 The proof of the mathematical form of the selective decay principle in the presence of the beta-plane effect

The purpose of this appendix is to give a rigorous proof of the strong form of the selective decay principle (Theorem 3.1) in the presence of the beta-plane effect.

Some of the arguments here are a little bit technical and assume familiarity with basic knowledge of functional analysis and Sobolev spaces. The reader may safely skip this part.

Since

$$\lim_{t\to\infty} \Lambda(t) = \Lambda_j$$

(the limit exists, see previous argument in Section 3.5), we suspect that the dynamics is concentrated on the jth energy shell. To verify our intuition, we introduce the following notations. Let

$$E_j = \text{span}\{e^{i\vec{k}\cdot\vec{x}} | |\vec{k}|^2 = \Lambda_j\} \tag{3.78}$$

be the eigenspace of the Laplacian operator corresponding to the eigenvalue Λ_j. Let

$$P_{\Lambda_j} = \text{the orthogonal projection from } L^2 \text{ on to } E_j, \tag{3.79}$$

$$Q_{\Lambda_j^-} = \text{the orthogonal projection from } L^2 \text{onto span}\{E_l, l < j\}, \tag{3.80}$$

$$Q_{\Lambda_j^+} = I - P_{\Lambda_j} - Q_{\Lambda_j^-}, \tag{3.81}$$

where I denotes the identity operator.

In terms of the normalized stream function

$$\tilde{\psi} = \frac{\psi}{\|v\|_0}$$

our main theorem is equivalent to the combination of the following three properties

$$Q_{\Lambda_j^-}\tilde{\psi}(t) \to 0 \quad \text{in } H^1 \quad \text{as } t \to \infty, \tag{3.82}$$

$$Q_{\Lambda_j^+}\tilde{\psi}(t) \to 0 \quad \text{in } H^1 \quad \text{as } t \to \infty, \tag{3.83}$$

and there exists a solution $\tilde{\xi}_j$ of the linearized quasi-geostrophic equation

$$\frac{\partial \Delta \tilde{\xi}_j}{\partial t} + \beta \frac{\partial \tilde{\xi}_j}{\partial x} = 0$$

with

$$(I - P_{\Lambda_j})\tilde{\xi}_j = 0$$

and

$$P_{\Lambda_j}\tilde{\psi}(t) - \tilde{\xi}_j(t) \to 0 \quad \text{in } H^1 \qquad \text{as } t \to \infty. \tag{3.84}$$

For the sake of simplicity we will assume Newtonian viscosity only, i.e.

$$\mathcal{D}(\Delta) = \nu \Delta^2.$$

The case with more general damping mechanism can be treated similarly (see the previous appendix) at the expense of losing brevity.

For the purpose of deriving the desired estimates, it is necessary to derive better estimates on the decay rates of the velocity and vorticity fields for each solution. For this purpose, we take the inner product of the freely decaying quasi-geostrophic equation (3.1) with ψ and we deduce

$$\frac{1}{2}\frac{d}{dt}\|\vec{v}(t)\|_0^2 + \nu \|\omega(t)\|_0^2 = 0.$$

This implies

$$\frac{d}{dt}\|\vec{v}(t)\|_0^2 = -2\nu\|\omega(t)\|_0^2 = -2\nu\Lambda(t)\|\vec{v}(t)\|_0^2 \leq -2\nu\Lambda_j\|\vec{v}(t)\|_0^2.$$

Hence we deduce

$$\|\vec{v}(t)\|_0 \leq \|\vec{v}(t_0)\|_0 e^{-\nu\Lambda_j(t-t_0)}. \tag{3.85}$$

This further implies that

$$\|\omega(t)\|_0 \leq \|\omega(t_0)\|_0 e^{-\nu\Lambda_j(t-t_0)}, \tag{3.86}$$

since

$$\|\omega(t)\|_0^2 = \Lambda(t)\|\vec{v}(t)\|_0^2 \leq \Lambda(t_0)\|\vec{v}(t_0)\|_0^2 e^{-2\nu\Lambda_j(t-t_0)} = \|\omega(t_0)\|_0^2 e^{-2\nu\Lambda_j(t-t_0)}.$$

A lower bound on the energy decay rate can be derived as well. Indeed, since the Dirichlet quotient is non-increasing, we have, thanks to the energy equation

$$\frac{1}{2}\frac{d}{dt}\|\vec{v}(t)\|_0^2 + \nu\Lambda(t_0)\|\vec{v}(t)\|_0^2 \geq 0, \text{ for all } t \geq t_0. \tag{3.87}$$

Hence we have

$$\|\vec{v}(t)\|_0^2 \geq \|\vec{v}(t_0)\|_0^2 e^{-2\nu\Lambda(t_0)(t-t_0)}, \ \ \forall\, t \geq t_0. \tag{3.88}$$

Next we derive a decay rate for $\nabla\Delta\psi$, which is also needed in proving the rigorous selective decay principle.

For this purpose, we take the inner product of the freely decaying quasi-geostrophic equation (3.1) with $\Delta^2\psi$. Notice

$$|\int J(\psi, \Delta\psi)\Delta^2\psi| \leq \|\nabla\psi\|_{L^\infty}\|\nabla\Delta\psi\|_0\|\Delta^2\psi\|_0$$

(thanks to Agmon's inequality)

$$\leq c\|\nabla\psi\|_0^{\frac{1}{2}}\|\nabla\Delta\psi\|_0^{\frac{3}{2}}\|\Delta^2\psi\|_0$$

(thanks to interpolation inequality)

$$\leq c\|\nabla\psi\|_0^{\frac{1}{2}}\|\Delta\psi\|_0^{\frac{3}{4}}\|\Delta^2\psi\|_0^{\frac{7}{4}}$$

(thanks to Hölder's inequality)

$$\leq \frac{\nu}{2}\|\Delta^2\psi\|_0^2 + c\|\nabla\psi\|_0^4\|\Delta\psi\|_0^6$$

(thanks to (3.85) and (3.86)

$$\leq \frac{\nu}{2}\|\Delta^2\psi\|_0^2 + c\|\nabla\psi(t_0)\|_0^4\|\Delta\psi(t_0)\|_0^6 e^{-10\nu\Lambda_j(t-t_0)}$$

we deduce

$$\frac{d}{dt}\|\nabla\Delta\psi\|_0^2 + \nu\|\Delta^2\psi\|_0^2 \leq c\|\nabla\psi(t_0)\|_0^4\|\Delta\psi(t_0)\|_0^6 e^{-10\nu\Lambda_j(t-t_0)},$$

which implies, thanks to Poincaré's inequality

$$\frac{d}{dt}\|\nabla\Delta\psi\|_0^2 + \nu\Lambda_1\|\nabla\Delta\psi\|_0^2 \leq c\|\nabla\psi(t_0)\|_0^4\|\Delta\psi(t_0)\|_0^6 e^{-10\nu\Lambda_j(t-t_0)},$$

which further implies, after applying the Gronwall inequality

$$\|\nabla\Delta\psi(t)\|_0^2 \leq e^{-\nu\Lambda_1(t-t_0)}\|\nabla\Delta\psi(t_0)\|_0^2 + c\|\nabla\psi(t_0)\|_0^4\|\Delta\psi(t_0)\|_0^6 e^{-\nu\Lambda_1(t-t_0)}. \tag{3.89}$$

We now derive the equation satisfied by the normalized stream function. For this purpose we notice

$$\frac{d}{dt}\Delta\tilde{\psi}(t) = \frac{1}{\|\vec{v}(t)\|_0}\frac{d}{dt}\Delta\psi(t) - \frac{\Delta\psi(t)}{\|\vec{v}(t)\|_0^3}\frac{1}{2}\frac{d}{dt}\|\vec{v}(t)\|_0^2$$

(thanks to the dynamic equations for the stream function and energy)

$$= \frac{1}{\|\vec{v}(t)\|_0}\left(-\beta\frac{\partial\psi}{\partial x} - J(\psi,\Delta\psi) + \nu\Delta^2\psi + \frac{\Delta\psi(t)}{\|\vec{v}(t)\|_0^2}\nu\|\omega\|_0^2\right)$$

$$= -\beta\frac{\partial\tilde{\psi}}{\partial x} - \|\vec{v}(t)\|_0 J(\tilde{\psi},\Delta\tilde{\psi}) + \nu\Delta^2\tilde{\psi} + \nu\Lambda(t)\Delta\tilde{\psi}(t).$$

Hence we arrive at the equation for the normalized stream functions

$$\frac{d}{dt}\Delta\tilde{\psi}(t) + \beta\frac{\partial\tilde{\psi}}{\partial x} + \|\vec{v}(t)\|_0 J(\tilde{\psi},\Delta\tilde{\psi}) - \nu\Delta^2\tilde{\psi} - \nu\Lambda(t)\Delta\tilde{\psi}(t) = 0. \tag{3.90}$$

We now proceed to prove the decay, in H^1, of $Q_{\Lambda_j^+}\tilde{\psi}$. For this purpose we take the inner product of the normalized equation (3.90) with $Q_{\Lambda_j^+}\tilde{\psi}$. Notice for the non-linear term we have

$$\begin{aligned} \left|\int J(\tilde{\psi},\Delta\tilde{\psi})Q_{\Lambda_j^+}\tilde{\psi}\right| &= \left|\int J(\tilde{\psi},Q_{\Lambda_j^+}\tilde{\psi})\Delta\tilde{\psi}\right| \\ &\le \|\nabla\tilde{\psi}\|_{L^4}^2\|\Delta\tilde{\psi}\|_0 \end{aligned}$$

(thanks to Sobolev imbedding and interpolation inequality)

$$\begin{aligned} &\le c\|\nabla\tilde{\psi}\|_0\|\Delta\tilde{\psi}\|_0^2 \quad (\text{since}\|\nabla\tilde{\psi}\|_0 = 1) \\ &= c\|\Delta\psi\|_0^2\|\vec{v}(t)\|_0^{-2} \\ &= c\Lambda(t). \end{aligned}$$

Hence we deduce, since $\Lambda(t)$ is uniformly bounded

$$\begin{aligned} \frac{1}{2}\frac{d}{dt}\|\nabla Q_{\Lambda_j^+}\tilde{\psi}\|_0^2 + \nu\|\Delta Q_{\Lambda_j^+}\tilde{\psi}\|_0^2 - \nu\Lambda(t)\|\nabla Q_{\Lambda_j^+}\tilde{\psi}\|_0^2 &\le c\Lambda(t)\|\vec{v}(t)\|_0 \\ &\le ce^{-\nu\Lambda_j(t-t_0)}. \end{aligned}$$

Since the smallest wave number in the range of $Q_{\Lambda_j^+}$ is Λ_{j+1}, we have

$$\|\Delta Q_{\Lambda_j^+}\tilde{\psi}\|_0^2 \ge \Lambda_{j+1}\|\nabla Q_{\Lambda_j^+}\tilde{\psi}\|_0^2.$$

Therefore

$$\frac{1}{2}\frac{d}{dt}\|\nabla Q_{\Lambda_j^+}\tilde{\psi}\|_0^2 + \nu(\Lambda_{j+1} - \Lambda(t))\|\nabla Q_{\Lambda_j^+}\tilde{\psi}\|_0^2 \le ce^{-\nu\Lambda_j(t-t_0)}.$$

Recall that $\lim_{t\to\infty}\Lambda(t) = \Lambda_j$, thus we may take t_0 large enough so that

$$\Lambda(t) \le \frac{1}{2}(\Lambda_{j+1} + \Lambda_j), \qquad \text{for all} \quad t \ge t_0.$$

This implies

$$\frac{d}{dt}\|\nabla Q_{\Lambda_j^+}\tilde{\psi}\|_0^2+\nu(\Lambda_{j+1}-\Lambda_j)\|\nabla Q_{\Lambda_j^+}\tilde{\psi}\|_0^2\le ce^{-\nu\Lambda_j(t-t_0)},\quad \text{for all } t\ge t_0,$$

which further implies, after applying Gronwall-type inequality

$$\begin{aligned}\|\nabla Q_{\Lambda_j^+}\tilde{\psi}(t)\|_0^2 &\le e^{-\nu(\Lambda_{j+1}-\Lambda_j)(t-t_0)}\|\nabla Q_{\Lambda_j^+}\tilde{\psi}(t)\|_0^2\\ &\quad+c(e^{-\nu\Lambda_j(t-t_0)}+e^{-\nu(\Lambda_{j+1}-\Lambda_j)(t-t_0)}).\end{aligned}\tag{3.91}$$

Letting t approach infinity we have the estimate (3.82).

Next we prove the decay, in H^1, of $Q_{\Lambda_j^-}\tilde{\psi}$. For this purpose, we take the inner product of the normalized equation (3.90) with $Q_{\Lambda_j^-}\tilde{\psi}$. Notice for the non-linear term we have

$$\begin{aligned}|\int J(\tilde{\psi},\Delta\tilde{\psi})Q_{\Lambda_j^-}\tilde{\psi}| &= |\int J(\tilde{\psi},Q_{\Lambda_j^-}\tilde{\psi})\Delta\tilde{\psi}|\le\|\nabla\tilde{\psi}\|_{L^4}^2\|\Delta\tilde{\psi}\|_0\\ &\le c\|\nabla\tilde{\psi}\|_0\|\Delta\tilde{\psi}\|_0^2=c\|\Delta\psi\|_0^2\|\vec{v}(t)\|_0^{-2}\\ &= c\Lambda(t).\end{aligned}$$

Hence we deduce

$$\begin{aligned}\frac{1}{2}\frac{d}{dt}\|\nabla Q_{\Lambda_j^-}\tilde{\psi}\|_0^2+\nu\|\Delta Q_{\Lambda_j^-}\tilde{\psi}\|_0^2-\nu\Lambda(t)\|\nabla Q_{\Lambda_j^-}\tilde{\psi}\|_0^2 &\ge -c\Lambda(t)\|\vec{v}(t)\|_0\\ &\ge -ce^{-\nu\Lambda_j(t-t_0)},\end{aligned}$$

where we have used the uniform boundedness of $\Lambda(t)$.

It is obvious, thanks to the definition of $Q_{\Lambda_j^-}$, that

$$\|\Delta Q_{\Lambda_j^-}\tilde{\psi}\|_0^2\le\Lambda_{j-1}\|\nabla Q_{\Lambda_j^-}\tilde{\psi}\|_0^2.$$

We then deduce

$$\begin{aligned}\frac{1}{2}\frac{d}{dt}\|\nabla Q_{\Lambda_j^-}\tilde{\psi}\|_0^2 &\ge \nu(\Lambda(t)-\Lambda_{j-1})\|\nabla Q_{\Lambda_j^-}\tilde{\psi}\|_0^2-ce^{-\nu\Lambda_j(t-t_0)}\\ &\ge \nu(\Lambda_j-\Lambda_{j-1})\|\nabla Q_{\Lambda_j^-}\tilde{\psi}\|_0^2-ce^{-\nu\Lambda_j(t-t_0)}\end{aligned}$$

or

$$\frac{d}{dt}(e^{-2\nu(\Lambda_j-\Lambda_{j-1})t}\|\nabla Q_{\Lambda_j^-}\tilde{\psi}(t)\|_0^2)\ge -ce^{-\nu(3\Lambda_j-2\Lambda_{j-1})t}.$$

Integrating this inequality from t to T (with $T>t>t_0$) we have

$$\begin{aligned}&e^{-2\nu(\Lambda_j-\Lambda_{j-1})T}\|\nabla Q_{\Lambda_j^-}\tilde{\psi}(T)\|_0^2\\ &\quad\ge e^{-2\nu(\Lambda_j-\Lambda_{j-1})t}\|\nabla Q_{\Lambda_j^-}\tilde{\psi}(t)\|_0^2-ce^{-\nu(3\Lambda_j-2\Lambda_{j-1})t}\end{aligned}$$

or equivalently

$$e^{-2\nu(\Lambda_j-\Lambda_{j-1})(T-t)}\|\nabla Q_{\Lambda_j^-}\tilde{\psi}(T)\|_0^2 \geq \|\nabla Q_{\Lambda_j^-}\tilde{\psi}(t)\|_0^2 - ce^{-\nu\Lambda_j t}. \tag{3.92}$$

Letting T approach infinity in (3.92) we deduce

$$\|\nabla Q_{\Lambda_j^-}\tilde{\psi}(t)\|_0^2 \leq ce^{-\nu\Lambda_j t}, \tag{3.93}$$

which further implies the estimate (3.83).

We are now left with the analysis of the dynamics on the jth shell. Our goal here is to construct $\tilde{\xi}_j$ and prove the estimate (3.84). For this purpose we apply P_{Λ_j} to the freely decaying quasi-geostrophic equation (3.1), denoting $P_{\Lambda_j}\psi = \psi_j$, we have

$$-\Lambda_j \frac{d}{dt}\psi_j + \beta\frac{\partial\psi_j}{\partial x} + P_{\Lambda_j}J(\psi, \Delta\psi) = \nu\Lambda_j^2\psi_j$$

or equivalently

$$\frac{d}{dt}\psi_j - \frac{\beta}{\Lambda_j}\frac{\partial\psi_j}{\partial x} - \frac{1}{\Lambda_j}P_{\Lambda_j}J(\psi, \Delta\psi) + \nu\Lambda_j\psi_j = 0, \tag{3.94}$$

which is equivalent to

$$\frac{d}{dt}\psi_j - \frac{\beta}{\Lambda_j}\frac{\partial\psi_j}{\partial x} + \nu\Lambda_j\psi_j = e^{-\nu\Lambda_j t}f, \tag{3.95}$$

where

$$f = f = \frac{e^{\nu\Lambda_j t}}{\Lambda_j}P_{\Lambda_j}J(\psi, \Delta\psi). \tag{3.96}$$

The equation can be written in the following form

$$\frac{d}{dt}(e^{\nu\Lambda_j t}\psi_j) - \frac{\beta}{\Lambda_j}\frac{\partial(e^{\nu\Lambda_j t}\psi_j)}{\partial x} = f. \tag{3.97}$$

It is easy to see that f satisfies the following estimate

$$\|f(t)\|_0 \leq ce^{\nu\Lambda_j t}\|\nabla\psi\|_{L^\infty}\|\nabla\Delta\psi\|_0$$

(thanks to a Brezis–Gallouet (1980) inequality)

$$\leq ce^{\nu\Lambda_j t}\|\Delta\psi\|_0\left(1+\log\frac{\|\nabla\Delta\psi\|_0^2}{\Lambda_1\|\Delta\psi\|_0^2}\right)^{\frac{1}{2}}\|\nabla\Delta\psi\|_0$$

(thanks to (3.89) and lower bound on the decay rate of ψ)

$$\leq ce^{\nu\Lambda_j t}\|\Delta\psi\|_0\left(1+\log\frac{e^{-\nu\Lambda_1 t}}{e^{-2\nu\Lambda(0)t}}\right)^{\frac{1}{2}}e^{-\nu\Lambda_1 t/2}$$

(thanks to (3.86))

$$\leq c(2\Lambda(0)-\Lambda_1)^{\frac{1}{2}} t^{\frac{1}{2}} e^{-\nu\Lambda_1 t/2}.$$

This implies that $f \in L^1$.

Now we invoke the Fourier series representation

$$\psi_j = \sum_{|\vec{k}|^2=\Lambda_j} \hat{\psi}_{\vec{k}} e^{i\vec{k}\cdot\vec{x}}, \quad f = \sum_{|\vec{k}|^2=\Lambda_j} \hat{f}_{\vec{k}} e^{i\vec{k}\cdot\vec{x}},$$

and we consider

$$\phi_j = \sum_{|\vec{k}|^2=\Lambda_j} \hat{\psi}_{\vec{k}} e^{(\nu\Lambda_j - \frac{i\beta k_1}{\Lambda_j})t} e^{i\vec{k}\cdot\vec{x}}.$$

We have

$$\frac{d}{dt}\phi_j = g = \sum_{|\vec{k}|^2=\Lambda_j} e^{-\frac{i\beta k_1}{\Lambda_j}t} \hat{f}_{\vec{k}} e^{i\vec{k}\cdot\vec{x}}.$$

It is easy to see that $g \in L^1$ since $f \in L^1$ and hence we may define

$$\phi_\infty = \phi_j(t_0) + \int_{t_0}^{\infty} g(s)\, ds \tag{3.98}$$

and we have

$$\|\phi_j(t) - \phi_\infty\|_0 \leq ct^{\frac{1}{2}} e^{-\nu\Lambda_1 t/2}. \tag{3.99}$$

Suppose that ϕ_∞ takes the form

$$\phi_\infty = \sum_{|\vec{k}|^2=\Lambda_j} \hat{\phi}_{\infty,\vec{k}} e^{i\vec{k}\cdot\vec{x}}. \tag{3.100}$$

We then define

$$\xi_j(t) = \sum_{|\vec{k}|^2=\Lambda_j} e^{\frac{i\beta k_1}{\Lambda_j}t} \hat{\phi}_{\infty,\vec{k}} e^{i\vec{k}\cdot\vec{x}}. \tag{3.101}$$

It is easy to see that ξ_j solves the equation

$$\frac{d}{dt}\Delta\xi_j + \beta\frac{\partial\xi_j}{\partial x} = 0$$

and satisfies

$$(I - P_{\Lambda_j})\xi_j = 0,$$

and

$$\|e^{\nu\Lambda_j t}\psi_j(t) - \xi_j(t)\|_0 \to 0. \tag{3.102}$$

It remains to prove (3.84), or equivalently, since all norms are equivalent on the finite-dimensional space E_j

$$\left\| \frac{\psi_j(t)}{\|\nabla\psi(t)\|_0} - \xi_j(t) \right\|_0 \to 0. \tag{3.103}$$

Thanks to our estimates (3.91) and (3.93) we have

$$\|\nabla\psi\|_0 = \|\nabla\psi_j\|_0 + \text{l.o.t.},$$

where l.o.t. denotes lower-order terms. On the other hand, thanks to (3.102) and the equivalence of norms on E_j

$$e^{\nu\Lambda_j t}\|\nabla\psi_j(t)\|_0 \to \lim_{t\to\infty} \|\nabla\xi_j(t)\|_0 = \|\nabla\phi_\infty\|_0,$$

hence

$$e^{\nu\Lambda_j t}\|\nabla\psi\|_0 = \|\nabla\phi_\infty\|_0 + \text{l.o.t.},$$

which implies

$$\frac{\nabla\psi_j(t)}{\|\nabla\psi(t)\|_0} = \frac{e^{\nu\Lambda_j t}\nabla\psi_j(t)}{e^{\nu\Lambda_j t}\|\nabla\psi\|_0} \to \frac{\nabla\xi_j(t)}{\|\nabla\phi_\infty\|_0}.$$

This completes the proof of (3.84) with

$$\tilde{\xi}_j = \frac{\xi_j(t)}{\|\nabla\phi_\infty\|_0}. \tag{3.104}$$

4

Non-linear stability of steady geophysical flows

4.1 Introduction

The observation of the geophysical flows in the atmosphere and the ocean reveal the existence of large-scale coherent flow structures. Examples of these structures are the atmospheric cyclonic and anti-cyclonic flow patterns, mesoscale ocean eddies, currents, and jets. These structures develop under fairly broad conditions and are characterized by their essentially steady nature, as well as their robustness and persistence in time. Possibly the most dramatic example of such coherent flow structures is exemplified by the Great Red Spot of Jupiter, discovered by Robert Hooke in 1664, which has persisted for at least 300 years.

From a dynamical point of view, such robust and persistent steady states must be non-linearly stable; small but finite initial perturbations of the steady states must remain small in time for the coherent flow structures to be observable. It is therefore clear that a fundamental problem is the study of the non-linear dynamical stability of the steady geophysical flows under small initial perturbations of the flow. This chapter and the next are devoted to the study of non-linear stability or instability of several classes of steady flows introduced earlier in Chapter 1. This study considers geophysical flows with topography and beta-plane effects, but without external forcing and dissipation mechanisms. In particular, we are interested in gaining a better understanding of what is the role played in the stability of the steady states by the beta-plane effect, and by the non-linear interaction of the large-scale mean flow and the small-scale flow through topographic stress.

In this chapter we will discuss mainly steady flows that are non-linearly stable. Section 4.2 is devoted to the non-linear stability of simple steady state solutions with linear potential vorticity–stream function relation. In Subsection 4.2.1 we first introduce the rigorous definition of non-linear stability and the energy method to prove non-linear stability. In Subsection 4.2.2 we establish the non-linear stability of simple steady states with no mean flow and linear q–ψ relation, $q=\mu\psi$, provided that the parameter μ is above the threshold value given by the first eigenvalue of the Laplacian, $\mu>-\Lambda_1$. The non-linear stability of steady flows with mean flow and beta effects is considered in Subsection 4.2.3. Again it is assumed that the steady flows satisfy a linear q–ψ relation; here we show that these flows are non-linearly stable for large-scale westward mean flows, i.e. $\mu > 0$. The non-linear stability analysis presented here is based on the extension of the energy method for conservative systems with one degree of freedom to fluid systems with infinitely many degrees of freedom; this approach was originally utilized by Bernstein *et al.* (1958) in the context of plasma physics, followed by Arnold (1967) for the Euler equations, and by Blumen (1968) for the quasi-geostrophic equations.

On the other hand, in Chapter 5 we discuss other unstable steady states with a linear q–ψ relation which exhibit a rich dynamic behavior through the competing effects of topographic stress and the beta-plane. Even with a simple layered topography in a fixed direction, the resulting flows can be surprisingly complex, including chaotic dynamic behavior and topographic blocking flow patterns; a key mechanism being the non-linear coupling of the large-scale mean flow and the small-scale flow through the topographic stress.

In Section 4.3, we establish the non-linear stability of more general steady state flows given by a non-linear q–ψ relation, $\psi = g(q)$. These steady states play an important role in some equilibrium statistical theories for the prediction of the most probable state of the flow, which will be discussed in later chapters. The purpose of Section 4.4 is to illustrate the stabilizing effect of β on zonal flows. This offers a partial explanation of the zonal tendency of quasi-geostrophic flows on a beta-plane.

In addition, studies of 2D turbulence above topography (Bretherton and Haidvogel, 1976) show the emergence of steady states; during this process the kinetic energy remains at large scales and is essentially constant, whereas the enstrophy cascades to the small scales, where it is dissipated. In this context, the resulting steady states become enstrophy minimizers at constant kinetic energy. In Section 4.5, we conclude by showing that the stable steady states discussed earlier in Sections 4.2 and 4.3 are generalized enstrophy minimizers at fixed kinetic energy, thus providing an alternative variational characterization of these states. This alternative characterization as generalized enstrophy minimizer will be utilized later in the study of statistical sharpness etc.

4.2 Stability of simple steady states

4.2.1 Non-linear stability and the energy method

We start by recalling the quasi-geostrophic barotropic equations with mean flow and small-scale flow interaction through topographic stress

$$\begin{aligned}
&\frac{\partial q}{\partial t} + J(\psi, q) = 0, \qquad \frac{dV}{dt}(t) = -\fint \frac{\partial h}{\partial x}\psi', \\
&q = \Delta\psi + h + \beta y \quad \psi = -V(t)y + \psi',
\end{aligned} \tag{4.1}$$

where $q(x, y, t)$ is the potential vorticity, $\psi(x, y, t)$ is the stream function, $V(t)$ is the large-scale mean flow, and $h(\vec{x})$ represents the topography. Also notice that here we assumed that there is no external forcing or damping. The mechanisms involved in the energy transfer between large and small scales are the topographic stress, the beta-plane effect, and the non-linear advection term.

In the above equations the flow is decomposed in terms of the large-scale mean flow $V(t)$ and the small-scale stream function ψ'

$$\psi(x, y, t) = -V(t)y + \psi'(x, y, t), \tag{4.2}$$

so that the velocity field $\vec{v}$ is

$$\vec{v} = \nabla^{\perp}\psi = \begin{pmatrix} V(t) \\ 0 \end{pmatrix} + \nabla^{\perp}\psi'.$$

Similarly, we split the potential vorticity q in terms of the large-scale βy and the small-scale component q'

$$q = q' + \beta y, \quad q' = \Delta\psi' + h, \tag{4.3}$$

and finally the relative vorticity ω is given in terms of the small-scale stream function ψ' by

$$\omega = \Delta\psi'. \tag{4.4}$$

Throughout this chapter we assume that $\vec{v}$ and h are 2π-periodic functions unless specified otherwise, and the bar across the integral sign in (4.1) indicates that the integral has been normalized by the area of the domain of integration, in this case $(2\pi)^2$.

Assume now that $q = \overline{q}(x, y)$ and $V(t) = \overline{V}$ is a steady state solution of the barotropic quasi-geostrophic equations (4.1). Usually a study of the stability of the flow is based on a linearized stability analysis, which only applies in principle to infinitesimal perturbations of the flow. Here we want to bound the perturbations without resorting to the linearization of the equations, i.e. we want to consider small but finite perturbations of the flow. Intuitively, the steady state solution $q = \overline{q}$, $V = \overline{V}$ is non-linearly stable if for sufficiently small but finite initial perturbations $\delta q_0(x, y)$ and δV_0, the resulting perturbation $\delta q(x, y, t)$ and $\delta V(t)$ remains small as time evolves.

A convenient measure of the size of the perturbations is given by the L^2-norm

$$\|(q, V)\|_0^2 = \|q\|_0^2 + \|V\|_0^2 = \fint |q|^2 + |V|^2.$$

Next we introduce the precise concept of non-linear stability in the L^2 sense.

Definition 4.1 (Non-linear stability) The steady state $(\overline{q}, \overline{V})$ of the barotropic quasi-geostrophic equations (4.1) is *non-linearly stable* in the L^2 sense if there are constants $C>0$ and $R>0$ so that given any initial perturbation δq_0 of $\overline{q}$ and any initial perturbation δV_0 of $\overline{V}$

$$q|_{t=0}=\overline{q}+\delta q_0, \quad V|_{t=0}=\overline{V}+\delta V_0,$$

with $\|(\delta q_0, \delta V_0)\|_0 \leq R$, then the resulting perturbed solution $(q(x, y, t), V(t))$ of (4.1)

$$q(x, y, t)=\overline{q}+\delta q(x, y, t), \quad V(t)=\overline{V}+\delta V(t),$$

where the perturbations δq and δV satisfy

$$\|(\delta q(t), \delta V(t))\|_0 \leq C\|(\delta q_0, \delta V_0)\|_0$$

for any time $t \geq 0$.

The approach to establish that a steady solution of the barotropic quasi-geostrophic equations (4.1) is non-linearly stable is an extension of the well-known energy method (or Lyapunov method) for conservative systems with finite degrees of freedom, to systems with infinitely many degrees of freedom. For example, we recall that for a conservative Hamiltonian system in the plane

$$\frac{dx}{dt}=-\frac{\partial H}{\partial y}, \quad \frac{dy}{dt}=\frac{\partial H}{\partial x},$$

the Hamiltonian $H=H(x, y)$ is a conserved quantity. If the equilibrium point $(x, y)=(\overline{x}, \overline{y})$ corresponds to a local minimum of the Hamiltonian, then it is well-known that such a equilibrium point is non-linearly stable. Next we want to generalize this result to systems with infinite degrees of freedom. We start with the following definition:

Definition 4.2 We say that a non-linear functional $\mathcal{W}(\delta q, \delta V)$ is *locally positive* definite (with respect to the L^2 norm) provided that there is $R_0>0$ and $C \geq 1$ so that if $\|(\delta q, \delta V)\|_0 \leq R_0$ then

$$C^{-1}\|(\delta q, \delta V)\|_0^2 \leq \mathcal{W}(\delta q, \delta V) \leq C\|(\delta q, \delta V)\|_0^2. \tag{4.5}$$

Proposition 4.1 *Assume that* $\mathcal{W}(\delta q, \delta V)$ *be a conserved functional for the quasi-geostrophic equations* (4.1), *with the property that it is locally positive definite. Then the steady state* $(\overline{q}(x, y), \overline{V})$ *is non-linearly stable.*

Proof: Assume that the initial perturbation satisfies $\|(\delta q_0, \delta V_0)\|_0 \leq R=\frac{R_0}{2C}$. We may then deduce that future perturbations cannot leave the ball of radius $R_0/2$ and hence the inequalities (4.5) are always satisfied. Indeed, suppose that future perturbation $(\delta q(t), \delta V(t))$ satisfies $\|(\delta q(t), \delta V(t))\|_0 \leq R_0$ for all $t \leq T$, we then have, since $\mathcal{W}(\delta q, \delta V)$ is a conserved quantity for the flow and it satisfies the inequalities in (4.5)

$$C^{-1}\,\|(\delta q(t), \delta V(t))\|_0^2 \le \mathcal{W}(\delta q(t), \delta V(t)) = \mathcal{W}(\delta q_0, \delta V_0) \le C\,\|(\delta q_0, \delta V_0)\|_0^2 .$$

From these inequalities it follows that

$$\|(\delta q(t), \delta V(t))\|_0 \le C\|(\delta q_0, \delta V_0)\|_0 \le R_0/2.$$

Thus we may conclude that (4.5) applies for all time and hence the steady state $(\overline{q}(x, y), \overline{V})$ is non-linearly stable.

Therefore to prove the non-linear stability of a steady state it is sufficient to produce a conserved functional $\mathcal{W}(\delta q, \delta V)$ satisfying the inequalities (4.5). From Section 1.4 we have at our disposal two conserved functionals, namely the total energy $E(q, V)$

$$E(q(t), V(t)) = \frac{1}{2}V^2(t) + \frac{1}{2}⨍|\nabla^\perp\psi'|^2 = \frac{1}{2}V^2(t) - \frac{1}{2}⨍\psi'\omega, \tag{4.6}$$

and the large-scale enstrophy $\mathcal{E}_L(q, V)$

$$\mathcal{E}_L(q(t), V(t)) = \beta V(t) + \frac{1}{2}⨍|q'|^2. \tag{4.7}$$

The idea is to combine the total energy $E(q, V)$ and the (large-scale) enstrophy $\mathcal{E}_L(q, V)$ to construct a conserved functional $\mathcal{W}(\delta q, \delta V)$ for the perturbation $(\delta q, \delta V)$, with the property that $\mathcal{W}(\delta q, \delta V)$ satisfies the inequalities in (4.5). This program will be carried out in the examples below. We point out that in general the functional $\mathcal{W}$ depends on the underlying background steady state flow $(\overline{q}(x, y), \overline{V})$; therefore whether or not this functional satisfies the inequalities in (4.5) will also depend on the specific steady state utilized.

4.2.2 Simple states with topography, but no mean flow or beta-effect

In this first example we consider steady state solutions of the barotropic quasi-geostrophic equations (4.1) with no mean flow, $V(t) \equiv 0$, and no beta-plane effect, $\beta = 0$. In this situation the quasi-geostrophic equations (4.1) reduce to

$$\frac{\partial q}{\partial t} + J(\psi, q) = 0, \quad q = \Delta\psi + h, \tag{4.8}$$

and in this case there is only the small-scale flow

$$q = q', \quad \psi = \psi', \quad \omega = \Delta\psi', \tag{4.9}$$

and for convenience we will drop the primes in the sequel. In Chapter 1 we constructed a family of steady state solutions of (4.8) with linear q–ψ relation. In terms of the Fourier expansion for the topography $h(x, y)$ the solution $q = \overline{q}(x, y)$ is given by

$$\begin{aligned} \overline{q} &= \mu\overline{\psi} \\ h &= \sum_{\vec{k}\neq 0} \widehat{h}_{\vec{k}}\, e^{i\vec{x}\cdot\vec{k}} \\ \overline{\psi} &= \sum_{\vec{k}\neq 0} \frac{1}{|\vec{k}|^2+\mu}\widehat{h}_{\vec{k}}\, e^{i\vec{x}\cdot\vec{k}}, \end{aligned} \tag{4.10}$$

provided that $\mu \neq -|\vec{k}|^2$, $\vec{k} \in \mathcal{Z}^2 \setminus \{0\}$. Next we show that these steady states are non-linearly stable as long as the parameter μ is in the range $-1 < \mu < \infty$.

Theorem 4.1 *The steady state $q = \overline{q}(x, y)$ in* (4.10) *is a non-linearly stable steady state solution of quasi-geostrophic equations* (4.8) *provided that $-1 < \mu < \infty$.*

Proof: Following Proposition 4.1, we want to construct a conserved functional $\mathcal{W}(\delta q)$ for the perturbation δq which is positive definite with respect to the L^2 norm. We already know that the energy $E(q)$ and the enstrophy $\mathcal{E}(q)$ are conserved functionals. In the case, where there is no large-scale mean flow, the expression for the energy in (4.6) reduces to

$$E(q) = -\frac{1}{2}\fint \psi'\omega = -\frac{1}{2}\fint (q-h)\Delta^{-1}(q-h), \tag{4.11}$$

and, similarly, the expression for the enstrophy $\mathcal{E}(q)$ becomes

$$\mathcal{E}(q) = \frac{1}{2}\fint |q|^2. \tag{4.12}$$

Next we expand these quadratic functionals around the steady state $q = \overline{q}$. If we let $q = \overline{q} + \delta q$, then as explained in Section 1.5 the energy $E(q)$ in (4.11) has the expansion

$$E(\overline{q}+\delta q) = E(\overline{q}) - \fint \overline{\psi}\delta q - \frac{1}{2}\fint \delta q \Delta^{-1}\delta q, \tag{4.13}$$

and the enstrophy $\mathcal{E}(q)$ has the expansion

$$\mathcal{E}(\overline{q}+\delta q) = \mathcal{E}(\overline{q}) + \fint \overline{q}\delta q + \frac{1}{2}\fint |\delta q|^2. \tag{4.14}$$

We observe that both the expansion for the energy $E(q)$ and the expansion for the enstrophy $\mathcal{E}(q)$ contain linear terms. In order to eliminate these linear contributions, we consider the following linear combination of $E(q)$ and $\mathcal{E}(q)$, denoted by $\mathcal{W}_\mu(\delta q)$

$$\begin{aligned}\mathcal{W}_\mu(\delta q) &= [\mathcal{E}(\overline{q}+\delta q)-\mathcal{E}(\overline{q})]+\mu[E(\overline{q}+\delta q)-E(\overline{q})] \\ &= ⨍(\overline{q}-\mu\overline{\psi})\delta q+\frac{1}{2}⨍|\delta q|^2-\frac{\mu}{2}⨍\delta q\Delta^{-1}\delta q,\end{aligned} \tag{4.15}$$

and, since $\overline{q}=\mu\overline{\psi}$, then the linear terms in (4.15) cancel and $\mathcal{W}_\mu(\delta q)$ reduces to

$$\mathcal{W}_\mu(\delta q)=\frac{1}{2}⨍|\delta q|^2-\frac{\mu}{2}⨍\delta q\Delta^{-1}\delta q. \tag{4.16}$$

Next we want to verify that the functional $\mathcal{W}_\mu(\delta q)$ is positive definite. Here it is convenient to utilize Fourier expansions in order to diagonalize the quadratic functional $\mathcal{W}_\mu(\delta q)$. If the Fourier expansion of δq is given by

$$\delta q=\sum_{\vec{k}\neq 0}\widehat{\delta q}_{\vec{k}}e^{i\vec{k}\cdot\vec{x}}, \tag{4.17}$$

then Parseval's identity yields the following representation of the functional $\mathcal{W}_\mu(\delta q)$ in terms of the Fourier coefficients of δq

$$\mathcal{W}_\mu(\delta q)=\frac{1}{2}\sum_{\vec{k}\neq 0}\left(1+\frac{\mu}{|\vec{k}|^2}\right)\left|\widehat{\delta q}_{\vec{k}}\right|^2. \tag{4.18}$$

Now everything boils down to studying the signature of the quadratic form in (4.18). Clearly each term in the quadratic form is positive definite for $\mu>-1$. It is straightforward to check that for $1\leq s<\infty$ and $-1<\mu<\infty$ the following inequality holds

$$C_\mu^{-1}\leq\frac{1}{2}\left(1+\frac{\mu}{s}\right)\leq C_\mu, \tag{4.19}$$

where the constant $C_\mu\geq 1$ is given by

$$C_\mu=\begin{cases}2/(1+\mu) & \text{if } -1<\mu\leq 0\\ \max\{(1+\mu)/2,2\} & \text{if } 0\leq\mu<\infty\end{cases} \tag{4.20}$$

(notice that the constant C_μ blows up as μ approaches the value $\mu=-1$). Introducing the estimates in (4.19) back into the expansion of the quadratic functional $\mathcal{W}_\mu(\delta q)$ in (4.18) yields

$$C_\mu^{-1}\|\delta q\|_0^2\leq\mathcal{W}_\mu(\delta q)\leq C_\mu\|\delta q\|_0^2, \tag{4.21}$$

and this proves that the quadratic functional $\mathcal{W}_\mu(\delta q)$ is positive definite. Finally invoking Proposition 4.1 we conclude that the steady state $q=\overline{q}(x,y)$ in (4.10) is a non-linearly stable solution of the quasi-geostrophic equations (4.8).

4.2.3 Simple states with topography, mean flow, and beta-effect

Next we consider steady state solutions of the barotropic quasi-geostrophic equations (4.1) with linear q–ψ relation, with mean flow, and beta-plane effects, i.e. $V(t) \neq 0$, and $\beta > 0$. In Chapter 1 we derived a family of steady state solutions of the barotropic quasi-geostrophic (4.1). These solutions $(q, V) = (\overline{q}(x, y), \overline{V})$ involve a large-scale mean flow in addition to the small-scale flow, and are given by

$$\begin{aligned} \overline{q} = \mu\overline{\psi}, \quad \overline{V} = -\frac{\beta}{\mu}, \quad h = \sum_{\vec{k}\neq 0} \widehat{h}_{\vec{k}}\, e^{i\vec{x}\cdot\vec{k}} \\ \overline{\psi} = \frac{\beta}{\mu}y + \sum_{\vec{k}\neq 0} \frac{1}{|\vec{k}|^2+\mu}\widehat{h}_{\vec{k}}\, e^{i\vec{x}\cdot\vec{k}} \end{aligned} \tag{4.22}$$

provided that $\mu \neq -|\vec{k}|^2$, $\vec{k} \in \mathcal{Z}^2$. The large-scale mean flow velocity $\vec{V} = (-\beta/\mu, 0)$ corresponds to a uniform zonal flow, which points westward if $\mu > 0$ and eastward if $\mu < 0$. Next we show that these steady state flows are non-linearly stable for westward mean flows, i.e. when $\mu > 0$.

Theorem 4.2 *The steady state $(q, V) = (\overline{q}(x, y), \overline{V})$ given by (4.22) is a non-linearly stable steady state solution of quasi-geostrophic equations (4.1) provided that $0 < \mu < \infty$.*

Proof: The proof follows along the same lines as Proposition 4.2. Again, following Proposition 4.1, we want to construct a conserved functional $\mathcal{W}(\delta q, \delta V)$ for the perturbation $(\delta q, \delta V)$, which is positive definite with respect to the L^2 norm. We already know that the energy $E(q, V)$ in (4.6) and the enstrophy $\mathcal{E}(q, V)$ in (4.7) are conserved functionals. Next we expand these quadratic functionals around the steady state $(q, V) = (\overline{q}(x, y), \overline{V})$. Let $q = \overline{q} + \delta q$ and $V = \overline{V} + \delta V$. The energy $E(q, V)$ in (4.6) has the expansion

$$E(\overline{q}+\delta q, \overline{V}+\delta V) = E(\overline{q}, \overline{V}) + \overline{V}\delta V - \fint \overline{\psi}\delta q + \frac{1}{2}(\delta V)^2 - \frac{1}{2}\fint \delta q \Delta^{-1}\delta q, \tag{4.23}$$

and the enstrophy $\mathcal{E}(q, V)$ in (4.7) has the expansion

$$\mathcal{E}(\overline{q}+\delta q, \overline{V}+\delta V) = \mathcal{E}(\overline{q}, \overline{V}) + \beta\delta V + \fint \overline{q}\delta q + \frac{1}{2}\fint |\delta q|^2. \tag{4.24}$$

Once more we observe that both the expansion for the energy $E(q, V)$ and the expansion for the enstrophy $\mathcal{E}(q, V)$ contain linear terms. In order to eliminate these linear contributions we consider the following linear combination of $E(q, V)$ and $\mathcal{E}(q, V)$, denoted by $\mathcal{W}_\mu(\delta q, \delta V)$

$$\begin{aligned}\mathcal{W}_\mu(\delta q, \delta V) =& [\mathcal{E}(\overline{q}+\delta q, \overline{V}+\delta V) - \mathcal{E}(\overline{q}, \overline{V})] \\ &+ \mu[E(\overline{q}+\delta q, \overline{V}+\delta V) - E(\overline{q}, \overline{V})] \\ =& (\beta+\mu\overline{V})\delta V + \fint(\overline{q} - \mu\overline{\psi})\delta q + \\ &+ \frac{\mu}{2}(\delta V)^2 + \frac{1}{2}\fint|\delta q|^2 - \frac{\mu}{2}\fint \delta q \Delta^{-1}\delta q.\end{aligned} \tag{4.25}$$

Since the steady state satisfies $\overline{q} = \mu\overline{\psi}$ and $\overline{V} = -\beta/\mu$, then the linear terms in (4.25) cancel and $\mathcal{W}_\mu(\delta q, \delta V)$ reduces to

$$\mathcal{W}_\mu(\delta q, \delta V) = \frac{\mu}{2}(\delta V)^2 + \frac{1}{2}\fint|\delta q|^2 - \frac{\mu}{2}\fint \delta q \Delta^{-1}\delta q. \tag{4.26}$$

Next we want to verify that the functional $\mathcal{W}_\mu(\delta q, \delta V)$ is positive definite. Again we utilize Fourier expansions in order to diagonalize the quadratic functional $\mathcal{W}_\mu(\delta q, \delta V)$. With the the Fourier expansion of δq given by (4.17), and utilizing Parseval's identity, it follows that the quadratic functional $\mathcal{W}_\mu(\delta q, \delta V)$ has the representation

$$\mathcal{W}_\mu(\delta q, \delta V) = \frac{\mu}{2}(\delta V)^2 + \frac{1}{2}\sum_{\vec{k}\neq 0}\left(1 + \frac{\mu}{|\vec{k}|^2}\right)|\widehat{\delta q}_{\vec{k}}|^2. \tag{4.27}$$

Next, we want to study the signature of the quadratic form for $\mathcal{W}_\mu(\delta q, \delta V)$ in (4.27). Again, it is straightforward to verify that for $1 \leq s < \infty$ and $0 < \mu < \infty$ the following inequalities hold

$$C_\mu^{-1} \leq \frac{1}{2}\left(1 + \frac{\mu}{s}\right) \leq C_\mu, \quad \text{and} \quad C_\mu^{-1} \leq \frac{\mu}{2} \leq C_\mu, \tag{4.28}$$

where the constant $C_\mu \geq 1$ is given by

$$C_\mu = \begin{cases} 2/\mu & \text{if } 0 < \mu < 1 \\ \max\{(1+\mu)/2, 2\} & \text{if } 1 \leq \mu < \infty \end{cases} \tag{4.29}$$

(notice that the constant C_μ blows up as μ approaches the value $\mu = 0$; this is consistent with the fact that the coefficient $\mu/2$ for the perturbation of the mean flow goes to zero). Introducing the estimates in (4.28) back into the expansion of the quadratic functional $\mathcal{W}_\mu(\delta q)$ in (4.27) yields

$$C_\mu^{-1}\|(\delta q, \delta V)\|_0^2 \leq \mathcal{W}_\mu(\delta q, \delta V) \leq C_\mu\|(\delta q, \delta V)\|_0^2, \tag{4.30}$$

and this proves that the quadratic functional $\mathcal{W}_\mu(\delta q, \delta V)$ is positive definite. Finally, from Proposition 4.1 we conclude that the steady state $(q, V) = (\overline{q}(x, y), \overline{V})$ in (4.22) is a non-linearly stable solution of the quasi-geostrophic equations (4.1).

Are the steady states with $-1 < \mu < 0$, which are stable with topography alone, dynamically unstable with the addition of the mean flow interaction? In Subsection 5.3.3, we provide explicit examples establishing that this is typically the case. Thus, the technique developed above is not merely a mathematical one, but sharply delineates non-linear stability from instability in these specific examples.

4.3 Stability for more general steady states

In the last section we studied the non-linear stability of steady states satisfying a linear q–ψ relation. Next we want to consider the non-linear stability problem for steady states of the barotropic quasi-geostrophic equations which satisfy such a nonlinear q–ψ relation of the form

$$\overline{\psi} = g(\overline{q}), \tag{4.31}$$

where $g = g(q)$ is a non-linear function. Such steady states were introduced in Chapter 1. To simplify the problem we assume that there is no large-scale/small-scale interaction through topographic stress, i.e. either no mean flow $V \equiv 0$, or zonal topography $h = h(y)$, so that the barotropic quasi-geostrophic equations (4.1) reduce to

$$\frac{\partial q}{\partial t} + J(\psi, q) = 0, \quad q = \Delta\psi + h + \beta y. \tag{4.32}$$

In this case a steady state solution $q = \overline{q}(x, y)$ with non-linear q–ψ relation is given by the solution of the non-linear elliptic equation

$$g^{-1}(\overline{\psi}) = \Delta\overline{\psi} + h + \beta y, \tag{4.33}$$

where g^{-1} is the inverse function of g provided that g is invertible. Such non-linear elliptic equations arise naturally as the mean field equations for the most probable states predicted by various statistical theories in later chapters in this book. For a given $g(g^{-1})$, the existence (or non-existence) of solutions to such non-linear elliptic equations is non-trivial. Nevertheless, if such a solution exists, it then solves the barotropic quasi-geostrophic equations with large-scale mean/small-scale interaction via topography (4.1). Indeed, it is easy to see that the topographic stress vanishes since

$$\fint \frac{\partial\overline{\psi}}{\partial x} h = \fint \frac{\partial\overline{\psi}}{\partial x}(g^{-1}(\overline{\psi}) - \Delta\overline{\psi} - \beta y) = 0$$

by the periodicity in the longitude (x) direction.

According to Proposition 4.1, to establish the non-linear stability of the steady state $q = \overline{q}(x, y)$, we need to construct a conserved functional $\mathcal{W}(\delta q, \delta V)$ that is positive definite. Notice that there is no mean flow/small-scale interacting through topography; the perturbation to the large-scale mean remains a constant in time. Thus we need only to find a conserved functional for the small-scale perturbation δq that is positive definite. In the examples discussed in the previous section the conserved functional $\mathcal{W}$ was always quadratic, and it was given by a linear combination of the kinetic energy E and the enstrophy $\mathcal{E}$. However, in the present case the functional relationship between $\overline{q}$ and $\overline{\psi}$ is non-linear and we cannot expect to find a quadratic functional $\mathcal{W}$. Nevertheless, below we will show that, under certain circumstances, it is possible to construct non-linear conserved functional $\mathcal{W}$ that is locally positive definite, thus establishing the non-linear stability of the steady state. To construct this functional $\mathcal{W}$ we must utilize the generalized enstrophy $\mathcal{G}(q)$, defined by

$$\mathcal{G}(q) = \fint G(q), \tag{4.34}$$

where $G = G(q)$ is an arbitrary function of q. From Chapter 1 we know that the generalized enstrophy $\mathcal{G}(q)$ is conserved under the following assumptions:

(A) $\beta = 0$, $V \equiv 0$, i.e. there is no beta-plane effect and no large-scale mean flow, or
(B) $\beta \neq 0$, $V \equiv 0$, but the flow has channel symmetry without large-scale mean

$$\psi = \sum_{n\geq 1, j\geq 1} (a_{nj} \cos nx + b_{nj} \sin nx) \sin jy,$$

or
(C) $\beta \neq 0$, zonal topography ($h = h(y)$) with channel geometry

$$\psi = -Vy + \sum_{n\geq 1, j\geq 1} (a_{nj} \cos nx + b_{nj} \sin nx) \sin jy. \tag{4.35}$$

In what follows we will tacitly assume that either one of the assumptions in (4.35) is satisfied.

Notice that both energy and total energy are conserved since there is no large mean flow/small-scale flow interaction through topographic stress. Next we expand the energy $E(q)$ and the generalized enstrophy $\mathcal{G}(q)$ around the steady state $q = \overline{q}$. let $q = \overline{q} + \delta q$. The expansion for the energy $E(q)$ is

$$E(\overline{q} + \delta q) = E(\overline{q}) - \fint \overline{\psi} \delta q - \frac{1}{2} \fint \delta q \Delta^{-1} \delta q. \tag{4.36}$$

On the other hand, the expansion for the generalized enstrophy is given by Taylor's formula with remainder

$$\mathcal{G}(\overline{q} + \delta q) = \mathcal{G}(\overline{q}) + \fint G'(\overline{q}) \delta q + \frac{1}{2} \fint |\delta q|^2 a(\overline{q}, \delta q), \tag{4.37}$$

where the function $a(\overline{q}, \delta q)$ is given by

$$a(\overline{q}, \delta q) = 2\int_0^1 (1-s)G''(\overline{q}+s\delta q)\, ds. \tag{4.38}$$

We observe that both the expansion of $E(\overline{q}+\delta q)$ in (4.36) and the expansion of $\mathcal{G}(\overline{q}+\delta q)$ in (4.37) contain linear terms. In order to eliminate the linear terms we consider the non-linear functional $\mathcal{W}_{\mathcal{G}}(\delta q)$ defined by

$$\begin{aligned}\mathcal{W}_{\mathcal{G}}(\delta q) &= [\mathcal{G}(\overline{q}+\delta q) - \mathcal{G}(\overline{q})] + [E(\overline{q}+\delta q) - E(\overline{q})] \\ &= \fint (G'(\overline{q}) - \overline{\psi})\delta q - \frac{1}{2}\fint \delta q \Delta^{-1}\delta q + \frac{1}{2}\fint |\delta q|^2 a(\overline{q}, \delta q).\end{aligned} \tag{4.39}$$

Now it is clear that the linear contribution in (4.39) is eliminated if we choose the non-linear function $G(q)$ to be an anti-derivative of $g(q)$ from (4.13)

$$G'(q) = g(q). \tag{4.40}$$

With this choice of $G(q)$ the functional $\mathcal{W}_{\mathcal{G}}(\delta q)$ reduces to

$$\mathcal{W}_{\mathcal{G}}(\delta q) = -\frac{1}{2}\fint \delta q \Delta^{-1}\delta q + \frac{1}{2}\fint |\delta q|^2 a(\overline{q}, \delta q), \tag{4.41}$$

where a is determined in (4.38).

Next we want to investigate when the conserved functional $\mathcal{W}_{\mathcal{G}}(\delta q)$ is locally positive definite. From (4.41) it is apparent that we need to control the non-linear coefficient $a(\overline{q}, \delta q)$. For this purpose we recall from Section 1.4 that the extrema of the potential vorticity are conserved quantities in the special case of $\beta = 0$ or channel geometry. Thus we have the following lemma, thanks to (4.35) and the conservation of extrema of potential vorticity

Lemma 4.1 *Let $q = q(x, y, t)$ be a solution of the barotropic quasi-geostrophic equation (4.1) with initial data $q = q_0(x, y)$. Assume that the initial data satisfy the bounds*

$$a \le q_0(x, y) \le b, \tag{4.42}$$

then the solution $q = q(x, y, t)$ also satisfies

$$a \le q(x, y, t) \le b \tag{4.43}$$

for all time t.

Now we are ready to establish sufficient conditions for the conserved functional $\mathcal{W}_{\mathcal{G}}(\delta q)$ to be locally positive definite and, consequently, for the steady state $q = \overline{q}(x, y)$ to be non-linearly stable.

Theorem 4.3 *Let $q = \overline{q}(x, y)$ be a steady state solution of the barotropic quasi-geostrophic equation (4.1) and satisfying*

$$\overline{\psi} = g(\overline{q}), \quad \overline{q} = \Delta\overline{\psi} + h + \beta y, \quad \overline{a} \leq \overline{q} \leq \overline{b},$$

let the initial perturbation $\delta q_0(x, y)$ have the bounds $\tilde{\alpha} \leq \delta q_0(x, y) \leq \tilde{\beta}$, and denote $a = \overline{a} + \tilde{\alpha}$ and $b = \overline{b} + \tilde{\beta}$. Now assume that there exist positive constants c_1, c_2, c_3, c_4 such that the non-linear function $g(q)$ satisfies either

$$\begin{aligned} &(1) \quad 0 < c_1 \leq g'(q) \leq c_2, \quad a \leq q \leq b, \text{ or} \\ &(2) - c_3 \leq g'(q) \leq -c_4 < -\Lambda_1^{-1}, \quad a \leq q \leq b. \end{aligned} \tag{4.44}$$

Then the steady state $q = \overline{q}(x, y)$ is non-linearly stable with the norm $\|\delta q\|_0^2$ provided one of the conditions in (4.35) holds.

Recall that we studied the stability of steady states with a linear q–ψ relation in Theorem 4.1. Hence we suspect that for a linear function g, corresponding to a linear q–ψ relation, Theorem 4.3 should be related to Theorem 4.1. Indeed, for the special case of linear g, i.e. $g(q) = cq$, with no beta-effect and no large-scale mean, we have

$$\overline{\psi} = c(\Delta\overline{\psi} + h)$$

or equivalently

$$\Delta\overline{\psi} + h = \mu\overline{\psi}, \quad \mu = \frac{1}{c}.$$

It is then clear that this is a special case of Theorem 4.1 with case (1) in (4.44) corresponding to the assumption $\mu > 0$ and case (2) corresponding to the assumption $-\Lambda_1 < \mu < 0$.

Proof: In view of Proposition 4.1, it is enough to exhibit a conserved functional that is positive definite. Since there is no mean flow/small-scale interaction through topography, the perturbation to the mean flow in the case of case (C) in (4.35) will remain a constant in time. Thus it is sufficient to construct a conserved functional that is positive definite for the small-scale perturbation δq only. For this purpose we consider the functional $\mathcal{W}_{\mathcal{G}}(\delta q)$ in (4.41)

$$\mathcal{W}_{\mathcal{G}}(\delta q) = -\frac{1}{2}⨍\delta q \Delta^{-1}\delta q + \frac{1}{2}⨍|\delta q|^2 a(\overline{q}, \delta q). \tag{4.45}$$

We will prove the result in two separate cases, depending on whether the non-linear function $g(q)$ satisfies assumptions (1) or (2) in (4.44). For (1) we will show that $\mathcal{W}_{\mathcal{G}}(\delta q)$ is positive definite, and for (2) we will show that $-\mathcal{W}_{\mathcal{G}}(\delta q)$ is positive definite.

Case 1: $g(q)$ satisfies assumption (1).

Here we assume $c_1 \le g'(q) \le c_2$ for $a \le q \le b$. Consider the term $a(\overline{q}, \delta q)$ in (4.45)

$$a(\overline{q}, \delta q) = 2\int_0^1 (1-s)G''(\overline{q}+s\delta q)\, ds. \tag{4.46}$$

The function $G''(q) = g'(q)$ satisfies assumption (1). In addition, $q = \overline{q}+\delta q$ satisfies $a \le q(x, y, t) \le b$ by Lemma 4.1. From here it follows that $a(\overline{q}, \delta q)$ satisfies the bounds

$$c_1 \le a(\overline{q}, \delta q) \le c_2, \tag{4.47}$$

and now we deduce that $\mathcal{W}_{\mathcal{G}}(\delta q)$ is bounded by

$$-\frac{1}{2}⨍\delta q\Delta^{-1}\delta q + \frac{c_1}{2}⨍|\delta q|^2 \le \mathcal{W}_{\mathcal{G}}(\delta q) \le -\frac{1}{2}⨍\delta q\Delta^{-1}\delta q + \frac{c_2}{2}⨍|\delta q|^2. \tag{4.48}$$

Now we can verify that the functional $\mathcal{W}_{\mathcal{G}}(\delta q)$ is positive definite. If we introduce the Fourier expansion of δq in (4.17) into (4.48) and apply Parseval's identity, it follows that

$$\frac{1}{2}\sum_{\vec{k}\neq 0}\left(c_1 + \frac{1}{|\vec{k}|^2}\right)|\widehat{\delta q}_{\vec{k}}|^2 \le \mathcal{W}_{\mathcal{G}}(\delta q) \le \frac{1}{2}\sum_{\vec{k}\neq 0}\left(c_2 + \frac{1}{|\vec{k}|^2}\right)|\widehat{\delta q}_{\vec{k}}|^2. \tag{4.49}$$

It is straightforward to verify the following inequality for $1 \le s < \infty$

$$C^{-1} \le \frac{1}{2}\left(c_1 + \frac{1}{s}\right) < \frac{1}{2}\left(c_2 + \frac{1}{s}\right) \le C, \tag{4.50}$$

where $C = \max\{(1+c_2)/2, 2/c_1\}$. Introducing this inequality in (4.49) we obtain

$$C^{-1}\|\delta q\|_0^2 \le \mathcal{W}_{\mathcal{G}}(\delta q) \le C\|\delta q\|_0^2. \tag{4.51}$$

This proves that $\mathcal{W}_{\mathcal{G}}(\delta q)$ is positive definite in Case 1.

Case 2: $g(q)$ satisfies assumption (2).

In this case we assume that $-c_3 \le g'(q) \le -c_4 \le -\Lambda_1^{-1}$ for $a \le q \le b$, where $\Lambda_1^{-1} = 1$ is the fundamental eigenvalue of the Laplacian with 2π-periodic boundary conditions. In this case we want to show that $-\mathcal{W}_{\mathcal{G}}(\delta q)$ is positive definite. First consider the term $a(\overline{q}, \delta q)$ in (4.45)

$$a(\overline{q}, \delta q) = 2\int_0^1 (1-s)G''(\overline{q}+s\delta q)\, ds. \tag{4.52}$$

Now the function $G''(q) = g'(q)$ satisfies assumption (2) and $q = \overline{q}+\delta q$ satisfies $a \le q(x, y, t) \le b$ by Lemma 4.1. From here it follows that $a(\overline{q}, \delta q)$ satisfies the bounds

$$-c_3 \le a(\overline{q}, \delta q) \le -c_4. \tag{4.53}$$

Introducing this resulting (4.45) we deduce that $-\mathcal{W}_{\mathcal{G}}(\delta q)$ satisfies the bounds

$$\frac{c_4}{2}\fint|\delta q|^2+\frac{1}{2}\fint\delta q\Delta^{-1}\delta q\le-\mathcal{W}_{\mathcal{G}}(\delta q)\le\frac{c_3}{2}\fint|\delta q|^2+\frac{1}{2}\fint\delta q\Delta^{-1}\delta q, \quad (4.54)$$

where $0 < 1 = \Lambda_1^{-1} < c_4 < c_3$. Next we verify that the functional $-\mathcal{W}_{\mathcal{G}}(\delta q)$ is positive definite. Introducing the Fourier expansion of δq in (4.17) into (4.54) and applying Parseval's identity, it follows that

$$\frac{1}{2}\sum_{\vec{k}\neq 0}\left(c_4-\frac{1}{|\vec{k}|^2}\right)|\widehat{\delta q}_{\vec{k}}|^2\le-\mathcal{W}_{\mathcal{G}}(\delta q)\le\frac{1}{2}\sum_{\vec{k}\neq 0}\left(c_3-\frac{1}{|\vec{k}|^2}\right)|\widehat{\delta q}_{\vec{k}}|^2. \quad (4.55)$$

Again, it is straightforward to verify the inequality

$$C^{-1}\le\frac{1}{2}\left(c_4-\frac{1}{s}\right)<\frac{1}{2}\left(c_3-\frac{1}{s}\right)\le C \quad (4.56)$$

for $1\le s<\infty$, where $C=\max\{c_3/2, 2/(c_4-\Lambda_1^{-1})\}$. Introducing this inequality in (4.55) we conclude

$$C^{-1}\|\delta q\|_0^2\le-\mathcal{W}_{\mathcal{G}}(\delta q)\le C\|\delta q\|_0^2. \quad (4.57)$$

This proves that $-\mathcal{W}_{\mathcal{G}}(\delta q)$ is positive definite in case (2).

In both Cases 1 and 2 we obtained a positive definite conserved functional. Therefore from Proposition 4.1 it follows that in both cases the steady state is non-linearly stable. This concludes the proof of the proposition.

4.4 Non-linear stability of zonal flows on the beta-plane

It is observed that the suitable (say around 30 days) average of weather systems are mostly zonal flows, and it is the perturbations and oscillations about these states that determine the weather. Thus a very important and natural question to ask is which zonal steady state is non-linearly stable. In contrast to the previous sections where the beta-plane did not play an active role in the stability of those steady states, we will see that the beta-plane has a stabilizing effect on zonal jets.

We start by recalling the barotropic quasi-geostrophic equations on a beta-plane with no damping, forcing, or topography

$$\frac{\partial q}{\partial t}+\nabla^{\perp}\psi\cdot\nabla q=0, \quad (4.58)$$

where

$$q=\Delta\psi+\beta y.$$

For simplicity we will assume periodic geometry. The channel geometry can be treated in a similar fashion with slight modification.

It is easy to see that the equations possess many steady states that are zonal shear flows. More precisely, for any stream function that is a function of the latitude (y) only, i.e.

$$\psi = \Psi(y) \tag{4.59}$$

the velocity field is given by

$$\vec{v} = \begin{pmatrix} -\Psi_y(y) \\ 0 \end{pmatrix}, \tag{4.60}$$

hence these states are called zonal shear flows (zonal jets) since the velocity is in the zonal (x) direction and it varies (shears) according to the latitudes. For such a zonal shear flow, the corresponding potential vorticity is given by

$$q = Q(y) = \frac{\partial^2}{\partial y}\Psi(y) + \beta y = \Psi_{yy}(y) + \beta y. \tag{4.61}$$

These states are steady states of the quasi-geostrophic equation since the stream function and the potential vorticity are functionally dependent and hence the non-linear term vanishes.

Following an argument of Shepherd (1987), we will see that β has some stabilizing effects on these kind of zonal jets. More precisely we have

Theorem 4.4 *Given any zonal jet $q = Q(y)$ in periodic geometry, if β is large enough in the sense that*

$$Q_y = \Psi_{yyy} + \beta > 0, \tag{4.62}$$

then this zonal flow is non-linear stable. In fact

$$⨍(\delta q(t))^2\,dxdy \le \frac{|Q_y|_{\max}}{|Q_y|_{\min}} ⨍(\delta q(0))^2\,dxdy. \tag{4.63}$$

We may further quantify this largeness assumption of β in the following immediate corollary:

Corollary 4.1 *Let*

$$\Psi(y) = \sum_{k\ge 1}(a_k \sin(ky) + b_k \cos(ky)) \tag{4.64}$$

be a given zonal flow in the periodic geometry. Define

$$A = \sum_{k\ge 1} |k|^3(|a_k| + |b_k|). \tag{4.65}$$

Then the zonal flow is stable if

$$\beta > A. \tag{4.66}$$

How do we prove the stability of these zonal jets? The idea again is to utilize conserved quantities.

It is observed that the zonal steady states are invariant under zonal translation and hence the equation for the perturbation is also invariant under zonal translation. This implies that there are new conserved quantities for the perturbation by Hamiltonian formulation of the equations and the so-called Noether's theorem. Instead of following the Hamiltonian approach, we will construct the conserved quantity in a straightforward manner. Our strategy is:

(A) Find a non-linear conserved quantity for perturbation.
(B) Show that the conserved quantity is positive definite.

For this purpose we let δq be the perturbation to the potential vorticity, i.e.

$$\begin{aligned} q &= Q(y) + \delta q, \\ \psi &= \Psi(y) + \delta\psi, \\ \Delta\delta\psi &= \delta q. \end{aligned} \tag{4.67}$$

We consider Q as a function defined on the real line. It is clear that the range of Q is the whole real line. Since Q_y is positive, Q is invertible. Let $Y_0(q)$ be the inverse function on the real line and hence

$$Y_0(Q(y)) = y, \quad Y_0' = (Q_y)^{-1}. \tag{4.68}$$

We now consider the following "morally quadratic" quantity for the perturbation

$$A(Q, \delta q) = \int_0^{\delta q} \{Y_0(Q + \tilde{q}) - Y_0(Q)\} d\tilde{q}. \tag{4.69}$$

This is the density of the pseudo-momentum or non-linear wave activity. It arises from the translation invariance in x of these zonal flows. See Shepherd (1990) for the derivation via Hamiltonian dynamics.

Claim

$$\frac{D}{Dt} A(Q, q) = -\psi_x(\psi_{xx} + \psi_{yy}) = -\frac{1}{2}((\psi_x)^2 - (\psi_y)^2)_x - (\psi_x\psi_y)_y, \tag{4.70}$$

where

$$\frac{D}{Dt} = \frac{\partial}{\partial t} + \nabla^{\perp}\psi \cdot \nabla.$$

The claim (4.70) indicates that the advection of the density of pseudo-momentum along the streamline is a perfect divergence. This identity is established at the end of this section. Hence, we naturally integrate (4.70) in space and conclude the conservation of pseudo-momentum. This is clear for the channel geometry. However, there is subtlety involved for the periodic geometry. Indeed, from

Section 1.3 we know that the integration of a quantity that is advected by the flow is not necessarily conserved in the periodic geometry unless the quantity is periodic. Thus, we need to verify that the density of pseudo- momentum is periodic in the periodic geometry before we can conclude the conservation of the pseudo-momentum. Notice that A is periodic in x. It is not hard to verify the periodicity of A in y. Indeed, for any $\tilde{q}$, since the range of Q is the whole real line and Q is monotone, there exists a unique y' such that $\tilde{q} = Q(y)$. Thus

$$\begin{aligned} Y_0(\tilde{q}+2\pi\beta) &= Y_0(Q(y')+2\pi\beta) = Y_0(Q(y'+2\pi)) = y'+2\pi \\ &= Y_0(\tilde{q})+2\pi. \end{aligned}$$

This implies

$$\begin{aligned} Y_0(Q(y+2\pi)+\tilde{q}) - Y_0(Q(y+2\pi)) &= Y_0(Q(y)+2\pi\beta+\tilde{q}) - Y_0(Q(y)+2\pi\beta) \\ &= Y_0(Q(y)+\tilde{q}) - Y_0(Q(y)), \end{aligned}$$

which further implies the periodicity of A in y when combined with the periodicity of the perturbation δq. Thus, we deduce:

Corollary 4.2 *The quantity*

$$\mathcal{A}(Q, \delta q) = \fint A(Q(y), \delta q(x, y, t))\, dx\, dy \tag{4.71}$$

is conserved for the quasi-geostrophic equation in the periodic geometry.

The theorem then follows from this corollary. Indeed we have

$$\begin{aligned} A(Q, \delta q(t)) &\le \int_0^{\delta q} \max(|Y_0'(Q)|)\tilde{q}\, d\tilde{q} = \frac{1}{2}(\delta q)^2 |Y_0'(Q)|_{\max} \\ &= \frac{1}{2}(\delta q)^2 (|Q_y|_{\min})^{-1}, \end{aligned}$$

where the max (min) is taken over the real line, which is the same as over a period in the latitude direction since Q_y (and hence Y_0' by (4.68)) is periodic although Q is not. Similarly we have

$$A(Q, \delta q(t)) \ge \frac{1}{2}(\delta q)^2 (|Q_y|_{\max})^{-1}.$$

Thus, after integrating in space, we have

$$\mathcal{E}(\delta q)(|Q_y|_{\max})^{-1} \le \mathcal{A}(Q, \delta q) \le \mathcal{E}(\delta q)(|Q_y|_{\min})^{-1}, \tag{4.72}$$

where

$$\mathcal{E}(\delta q) = \frac{1}{2} \fint (\delta q)^2.$$

Combining this with the conservation of $\mathcal{A}$ we deduce

$$(|Q_y|_{\max})^{-1}\mathcal{E}(\delta q(t)) \leq \mathcal{A}(Q, \delta q(t)) = \mathcal{A}(Q, \delta q(0)) \leq (|Q_y|_{\min})^{-1}\mathcal{E}(\delta q(0)), \tag{4.73}$$

or

$$\mathcal{E}(\delta q(t)) \leq \frac{|Q_y|_{\max}}{|Q_y|_{\min}}\mathcal{E}(\delta q(0)).$$

This ends the proof of Theorem 4.4. except the proof of the claim in (4.70).

Now we move on to the proof of the claim.

Applying the chain rule and utilizing the equations for Q and δq we have

$$\begin{aligned}\frac{D}{Dt}(A(Q, \delta q)) &= \frac{\partial A}{\partial Q}\frac{DQ}{Dt} + \frac{\partial A}{\partial \delta q}\frac{D\delta q}{Dt} = \delta\psi_x Q_y \frac{\partial A}{\partial Q} - \delta\psi_x Q_y \frac{\partial A}{\partial \delta q} \\ &= \delta\psi_x Q_y \left(\frac{\partial A}{\partial Q} - \frac{\partial A}{\partial \delta q}\right).\end{aligned}$$

Utilizing (4.69) we may deduce

$$\frac{\partial A}{\partial Q} = Y_0(Q+\delta q) - Y_0(Q) - Y_0'(Q)\delta q; \qquad \frac{\partial A}{\partial \delta q} = Y_0(Q+\delta q) - Y_0(Q).$$

Hence

$$\frac{D}{Dt}(A(Q, \delta q)) = -\delta\psi_x Q_y Y_0'(Q)\delta q = -\delta\psi_x \delta q = -\delta\psi_x(\delta\psi_{xx} + \delta\psi_{yy}),$$

where we have used the fact that Y_0 is the inverse function of Q and hence $Q_y Y_0'(Q) = 1$. This ends the proof of the claim. Notice that this is the only place we use our choice of Y_0 as the inverse of Q. Hence we may start with a general Y_0, impose the conservation of the "morally quadratic" quantity $\mathcal{A}(Q, \delta q)$, to arrive at the conclusion that Y_0 must be a constant multiple of the inverse of Q. This provides a means to generate the new conserved quantity for the perturbation without invoking the Hamiltonian formulation.

The channel domain can be treated in a similar fashion. The main difference in the channel case is that we need to extend Q as a linear function in y outside the channel.

4.5 Variational characterization of the steady states

So far in this chapter we have discussed the non-linear stability of steady states $(q, V) = (\overline{q}, \overline{V})$ of the barotropic quasi-geostrophic equations (4.1) satisfying a q–ψ relation of the form

$$\overline{\psi} = g(\overline{q}). \tag{4.74}$$

As was mentioned in the introduction, numerical studies of two-dimensional turbulence over topography show the emergence of large-scale coherent steady states (Bretherton and Haidvogel, 1976), where it is advocated that these steady states correspond to enstrophy minimizers at fixed kinetic energy. In addition, in Chapter 3 we saw that according to the selective decay principle, solutions of the quasi-geostrophic equations with dissipation approach states which minimize the enstrophy for a given energy. These selective decay states obey a linear q–ψ relation. Here we will show that this variational characterization applies to the non-linearly stable steady states discussed in Section 4.3.

First we introduce the definition of generalized selective decay states that encompasses large-scale mean flows and generalized enstrophy minimizers. These definitions naturally have to incorporate the appropriate energy or enstrophy depending on the context

Definition 4.3 Generalized selective decay states

(1) In the case of flows with large-scale mean flow, a *selective decay state* (q^*, V^*) is a minimizer of the large-scale enstrophy $\mathcal{E}_L(q, V)$

$$\mathcal{E}_L(q, V) = \beta V + \frac{1}{2} ⨍ q'^2$$

subject to the total energy $E(q, V)$ constraint

$$\mathcal{S} = \{(q, V) | E(q, V) = -\frac{1}{2} ⨍ \psi\omega + \frac{1}{2} V^2 = E_0\}.$$

(2) In the case of flows without large-scale mean flow, a *selective decay state* q^* is a minimizer of the generalized enstrophy functional

$$\mathcal{G}(q) = ⨍ G(q)$$

with the energy constraint

$$\mathcal{S} = \{q \,|\, E(q) = E_0\}.$$

As in Section 1.5, we apply the Lagrange multiplier method to solve these variational problems. Let us start with case (1) corresponding to flows with large-scale mean flow. The variational derivatives of the total energy E and the large-scale enstrophy $\mathcal{E}_L$ are given by

$$\frac{\delta E}{\delta q} = -\psi, \tag{4.75}$$

$$\frac{\delta E}{\delta V} = V, \tag{4.76}$$

$$\frac{\delta \mathcal{E}_L}{\delta q} = \Delta\psi + h, \tag{4.77}$$

$$\frac{\delta \mathcal{E}_L}{\delta V} = \beta. \tag{4.78}$$

According to the Lagrange multiplier principle, a minimizer of the large-scale enstrophy $\mathcal{E}_L$ with the constraint of fixed total energy E must satisfy the equations

$$\Delta\psi + h = \mu\psi, \tag{4.79}$$

$$\beta = -\mu V, \tag{4.80}$$

where μ is the Lagrange multiplier. But now we recognize that these steady states (q^*, V^*) are precisely the steady states studied in Section 4.2. Therefore the Lagrange multiplier principle recovers the steady states with non-zero large-scale mean flow as critical points for the large-scale enstrophy at fixed total energy.

Next we consider the steady states in case (2), which are given by a non-linear q–ψ relation of the form $\overline{\psi} = g(\overline{q})$. In this case the variational derivative of the energy E and generalized enstrophy $\mathcal{G}$ is given by, according to Section 1.5

$$\frac{\delta E}{\delta q} = -\psi, \tag{4.81}$$

$$\frac{\delta \mathcal{G}}{\delta q} = G'(q). \tag{4.82}$$

Now, according to the Lagrange multiplier principle, a minimizer of the generalized enstrophy $\mathcal{E}_G$, subject to the constraint of constant energy E, must satisfy the equation

$$\frac{\delta \mathcal{G}}{\delta q} = -\mu \frac{\delta E}{\delta q}, \tag{4.83}$$

where μ is a Lagrange multiplier. From here we conclude that

$$\overline{\psi} = g(\overline{q}), \tag{4.84}$$

where

$$g(\overline{q}) \equiv \mu^{-1} G'(\overline{q}).$$

Again, the Lagrange multiplier method recovers the non-linear q–ψ steady states as critical points for the generalized enstrophy at fixed energy.

A natural question to ask is whether these selective decay states are real minimizers of the (total) enstrophy with given (total) energy. The affirmative answer is given in the next proposition.

Proposition 4.2 *(Variational characterization of selective decay states)*

(A) Let $\overline{q} = \mu\overline{\psi}$ be a selective decay state of barotropic quasi-geostrophic equations (4.1):

1. *If $h(\vec{x}) \neq 0, \beta \equiv 0, V(t) \equiv 0$, then the steady state $\overline{q} = \mu\overline{\psi}$ given by (4.79) is the minimizer of the enstrophy with fixed energy if and only if $\mu \geq -1$, and it is the unique minimizer if $\mu > -1$.*

2. *If $h(\vec{x}) \neq 0$, $\beta \neq 0$, $V(t)$ general, then the steady state $\overline{q} = \mu\overline{\psi}$ given by (4.79) and (4.80) is the minimizer of the large-scale enstrophy with fixed energy if and only if $\mu \geq 0$, and it is the unique minimizer if $\mu > 0$.*
3. *If $h(\vec{x}) \equiv 0$, then the steady state $\overline{q} = \mu\overline{\psi}$ by (4.79) is the minimizer of the enstrophy with fixed energy if and only if $\mu = -|\vec{k}_0|^2 = -\Lambda_1 = -1$.*

(B) Let $\overline{q}$ be a bounded generalized selective decay state of the barotropic quasi-geostrophic equations (4.1) without mean flow satisfying

$$\overline{\psi} = g(\overline{q}), \quad \overline{q} = \Delta\overline{\psi} + h + \beta y, \quad \overline{a} \leq \overline{q} \leq \overline{b}.$$

Assume there is a bounded perturbation $\delta\omega$ added, i.e. $\tilde{\alpha} \leq \delta\omega \leq \tilde{\beta}$, and denote $a = \overline{a} + \tilde{\alpha}$ and $b = \overline{b} + \tilde{\beta}$. We further assume either
Case (1)

$$0 < c_1 \leq g'(r) \leq c_2, \qquad a \leq r \leq b$$

or
Case (2)

$$-c_3 \leq g'(q) \leq -c_4 < -\Lambda_1^{-1}, \qquad a \leq q \leq b.$$

Then either (1) or (2) implies the generalized selective decay state is a local minimizer of the enstrophy with fixed energy in the sense that

$$\mathcal{G}(q + \delta q) \leq \mathcal{G}(q)$$

with the perturbation $\delta q (= \delta\omega)$ satisfying the above assumptions plus the energy constraint

$$E(q + \delta q) = E(q).$$

Proof: The proof is a simple variation of the proofs of the techniques utilized in Sections 4.2 and 4.3, and we will prove only the result for the case (A.2). Let us show that in this case the large-scale enstrophy $\mathcal{E}_L(q, V)$ is minimized at fixed energy $E(q, V)$. Since the energy is constant then $E(q + \delta q, V + \delta V) - E(q, V) = 0$, therefore we have

$$\begin{aligned}\mathcal{E}_L(\overline{q} + \delta q, \overline{V} + \delta V) - \mathcal{E}_L(\overline{q}, \overline{V}) &= [\mathcal{E}_L(\overline{q} + \delta q, \overline{V} + \delta V) - \mathcal{E}_L(\overline{q}, \overline{V})] + \\ &+ \mu[E(\overline{q} + \delta q, \overline{V} + \delta V) - E(\overline{q}, \overline{V})] = \mathcal{W}_\mu(\delta q, \delta V).\end{aligned}$$

Hence according to (4.27) for $\mathcal{W}_\mu(\delta q, \delta V)$ we have

$$\mathcal{E}_L(\overline{q} + \delta q, \overline{V} + \delta V) - \mathcal{E}_L(\overline{q}, \overline{V}) = \frac{\mu}{2}(\delta V)^2 + \frac{1}{2}\sum_{\vec{k}\neq 0}\left(1 + \frac{\mu}{|\vec{k}|^2}\right)\left|\widehat{\delta q}_{\vec{k}}\right|^2$$

and this last expression is non-negative for $\mu \geq 0$, and strictly positive for $\mu > 0$. This shows that the steady state $(q, V) = (\overline{q}(x, y), \overline{V})$ minimizes the large-scale enstrophy $\mathcal{E}_L(q, V)$ at fixed energy $E(q, V)$. The remaining cases can be verified in similar fashions.

References

Arnold, V. (1965), Conditions for nonlinear stability of stationary plane curvilinear flows of an ideal fluid. *Dokl. Akad. Nauk SSSR* **162**, 975–978.

Bernstein, I. Frieman, E., Kruskal, M., and Kulsrud, R. (1958), An energy principle for hydromagnetic stability problems. *Proc. Roy. Soc.* **A244**, 17–40.

Blumen, W. (1968), On the stability of quasi-geostrophic flow. *J. Atmos. Sci.* **25**, 929–931.

Bretherton, F. and Haidvogel, D. (1976), Two-dimensional turbulence above topography. *J. Fluid Mech.* **78(1)**, 129–154.

Carnevale, G.F. and Frederiksen, J.S. (1987), Nonlinear stability and statistical mechanics of flow over topography. *J. Fluid Mech.* **175**, 157–181.

Majda, A. and Bertozzi, A. (2001), *Vorticity and Incompressible Flow*. Cambridge: Cambridge University Press.

Shepherd, T.G. (1987), Non-ergodicity of inviscid two-dimensional flow on a beta-plane and on the surface of a rotating sphere. *J. Fluid Mech.* **184**, 289–302.

Shepherd, T.G. (1990), Symmetries, conservation laws, and Hamiltonian structure in geophysical fluid dynamics. *Adv. Geophysics* **32**.

5

Topographic mean flow interaction, non-linear instability, and chaotic dynamics

5.1 Introduction

In this chapter we continue to study the dynamic behavior of the barotropic quasi-geostrophic equations in the absence of dissipation and external forcing, paying special attention to the non-linear interaction of the large-scale mean flow and the small-scale flow through topographic stress. Situations of obvious importance in atmosphere and ocean science occur when smaller-scale motions have a significant feedback and interaction with a larger-scale mean flow. One prototype situation of this sort occurs in the interaction of large-scale and small-scale components of barotropic flow over topography via topographic stress. In two influential papers, Charney and DeVore (1979) and Hart (1979) studied the multiple equilibrium states of this system with dissipation and single mode topography, and suggested their possible importance as model states for atmospheric blocking (see also Carnevale and Frederiksen, 1987; Vallis, 1985 for further developments).

In oceanography, in the special case of single mode topography as well as damping and driving, these equations have been used as a model for large-scale mean flow modification through topographic stress for flow along a continental shelf with smaller-scale topographic ridges (Allen *et al.*, 1991; Samelson and Allen, 1987); also recently Holloway (Holloway, 1987; Edy and Holloway, 1994) has emphasized the possible dynamical significance of topographic stress in modifying coastal currents in many oceanographic contexts.

An ideal model to study the complex non-linear interaction of the large-scale and the small-scale flow and the role of the topography is given by the equations for large-scale mean flow and small-scale flow interaction through topographic stress introduced in Chapter 1, Subsection 1.3.1

$$\frac{\partial q}{\partial t} + \nabla^{\perp}\psi \cdot \nabla q + V(t)\frac{\partial q}{\partial x} + \beta\frac{\partial \psi}{\partial x} = 0, \tag{5.1}$$

$$q = \Delta\psi + h, \tag{5.2}$$

$$\frac{dV}{dt} = \fint h\frac{\partial \psi}{\partial x}. \tag{5.3}$$

Here the small-scale flow is given in terms of the stream function ψ, and q is the small-scale potential vorticity (for simplicity we have omitted the primes in the the small-scale variables in equations (5.1)–(5.3)). The large-scale velocity field only has the zonal component $V(t)$, and the topography is given by the function $h = h(x, y)$. The parameter $\beta > 0$ is the contribution from the beta-plane effect. Both the small-scale potential vorticity q and the small-scale stream function ψ, as well as the topography h, are assumed to be 2π-periodic functions in both variables x and y, with zero average. The large-scale velocity $V(t)$ is strongly coupled with the small-scale flow through equation (5.3), where the bar across the integral sign indicates that the integral has been normalized by the area of the domain of integration.

We recall from Chapter 1, Section 1.3, the existence of conserved quantities for flows given by equations (5.1)–(5.3). One important conserved quantity is the total energy E given by

$$E = \frac{1}{2}V^2(t) + \frac{1}{2}\fint |\nabla^{\perp}\psi|^2 = \frac{1}{2}V^2(t) - \frac{1}{2}\fint \psi\omega, \tag{5.4}$$

where $\omega = \Delta\psi$ is the vorticity; the other important conserved quantity is the total enstrophy $\mathcal{E}$ given by

$$\mathcal{E} = \beta V(t) + \frac{1}{2}\fint |q|^2. \tag{5.5}$$

In Chapter 1, Section 1.2.4, we identified special steady states of (5.1)–(5.3) with constant mean flow V, and with the potential vorticity q and the stream function ψ linearly related

$$q = \mu\psi, \quad V = -\frac{\beta}{\mu}, \quad q = \psi + h, \tag{5.6}$$

and from the equation above it is clear that these steady states represent a westward flow when $\mu > 0$ and an eastward flow when $\mu < 0$. These steady states are non-linearly stable for $\mu > 0$ and typically unstable for $\mu < 0$. In fact, in Section 4.2, the non-linear stability of the steady states in (5.6) was established rigorously when $\mu > 0$, i.e. for westward mean flow, $V < 0$. In that case the proof relied on the construction of a positive definite invariant functional $\mathcal{W}$, utilizing the total energy E and the total enstrophy $\mathcal{E}$ in equations (5.4)–(5.5). In Subsection 5.3.3 below, we present rigorous examples demonstrating the linear and non-linear instability of these special states for $-1 < \mu < 0$ through explicit interaction of the mean flow and small-scale flow through topographic stress.

In spite of the rigorous non-linear stability results established in Chapter 4, much remains to be understood about the dynamics of the equations (5.1)–(5.3) and the non-linear interaction of the large and small flow scales, as well as the role of topography. Here, through a combination of theory and numerical experiments, we study the dynamic behavior of solutions of the equations for mean flow and small-scale interaction via topographic stress without any dissipation or forcing, given by equations (5.1)–(5.3). We do this to emphasize the remarkably rich intrinsic chaos which occurs in these systems and substantially alters the mean flow through the energy conserving interaction with the small scales via topographic stress. Here we will show that this intrinsic chaos occurs for fixed layered topography with at least two non-zero modes and with a non-zero variation in the Coriolis parameter β, which is typical for flows in the atmosphere and ocean. In fact, if β vanishes, that is $\beta = 0$, these equations are completely integrable and do not possess intrinsic chaos, although they can have a complex heteroclinic structure (see Subsection 5.3.2). Below, we use a combination of theory and numerical experiments (see Subsection 5.5.2) to produce prototype topographic blocking patterns, with topographic stress as the only transfer mechanism; these patterns nevertheless have time series which resemble those in recent laboratory experiments on topographic blocking (Weeks *et al.*, 1997). Next, we outline the contents of the remainder of this chapter.

In Section 5.2, we present the important special case involving layered anisotropic topography, which is the main focus of this section. In particular, we show that, in this situation, there is a Lagrangian form of the basic equations (see (5.19)–(5.25) below), which has a Hamiltonian structure as well as action-angle variables. This provides the starting point for the theoretical developments in Sections 5.3 and 5.4. In Section 5.3, we discuss the dynamics of the layered (anisotropic) equations with topographic stress in two interesting regimes with completely integrable behavior: the first involves setting the Coriolis parameter β to zero, while the second involves single mode topography with non-zero β. The structure presented in Section 5.2 plays an important role in these developments.

In Section 5.4, we develop a parameter regime for the layered model where we utilize theory to predict intrinsic Hamiltonian chaos including situations with global stochastic behavior. In this regime, the mean flow, the topography, and the beta-effect are all small in a suitable distinguished limit, but the topography is much smaller than both the mean flow and the beta-effect. At the beginning of Section 5.4, we show that the formal limiting dynamics in this regime reduces to equations identical to the motion of a single particle in a one-dimensional spectrum of longitudinal waves, which have been studied extensively (Zaslavsky and Filonenko, 1968; Zaslavsky *et al.*, 1991; Chirikov, 1979; Escande, 1985); furthermore, in a suitable formal sense, this limiting process involves a regular perturbation. In Subsection 5.4.2, we utilize this structure to provide a theoretical framework for strong resonance and global stochasticity in the layered equations via topographic stress with at least two topographic modes of comparable size; these results provide the theoretical underpinning for the prototype blocking patterns described through numerical experiments in Subsection 5.5.2. In Subsection 5.4.3, we establish by Melnikov's method that weak stochasticity also occurs in this regime for suitable topography through homoclinic chaos.

In Section 5.5, we present a series of numerical experiments involving the basic equations with mean flow and small-scale interaction via topographic stress. In Subsection 5.5.1, we describe the behavior of solutions involving random perturbations of single mode topography for a wide range of initial mean flows and values of β. In Subsection 5.5.2, we describe prototype chaotic blocking regimes with two-mode topography motivated by the theoretical developments in Subsection 5.4.2. Numerical experiments with random perturbations of multi-mode topography are described in Subsection 5.5.3. Finally, in Subsection 5.5.4, we present some preliminary results on symmetry breaking perturbations of the prototype topographic blocking patterns.

We end this introduction by remarking that, as established in Sections 5.2 through 5.4, the equations for mean flow and small-scale interaction via topographic stress with layered topography provide a very rich infinite-dimensional Hamiltonian with complex heteroclinic structure (see McLaughlin and Overman, 1995 for whiskered tori in related developments for soliton equations), where rigorous finite-dimensional truncation can be achieved. Even in the integrable case in Section 5.2 with $\beta = 0$, the phase space structure is extremely complex since the integrable two-dimensional Hamiltonian has coefficients which depend on the initial data and, as the initial data vary complex bifurcations in phase space occur, much like the "trouser, pant leg" constructions for whiskered tori (McLaughlin and Overman, 1995).

5.2 Systems with layered topography

We shall now construct special solutions to the full non-linear system, which inherit the non-linear coupling of the small-scale flow with the large-scale mean flow via topographic stress. To do so, we restrict ourselves to topography layered in the fixed direction, $\vec{l} = (\ell_x, \ell_y)$, and assume that both ψ and q only depend on $\xi = \vec{l} \cdot \vec{x}$, with $\vec{x} = (x, y)$. As a consequence, the small-scale non-linear term in (5.1), $\nabla\psi \cdot \nabla^{\perp} q$, is identically zero. However, the non-linear coupling due to topographic stress in (5.3) remains and is responsible for much of the complex behavior presented below. For this special type of solution of the full system in (5.1)–(5.3), the expansions of ψ and h in Fourier series are

$$\psi(x, y, t) = \sum_{k \neq 0} \psi_k(t) e^{ik\vec{l} \cdot \vec{x}}, \tag{5.7}$$

$$h(x, y) = \sum_{j \neq 0} h_k e^{ik\vec{l} \cdot \vec{x}}, \quad \vec{x} = (x, y). \tag{5.8}$$

Next, we introduce equations (5.7) and (5.8) into (5.1)–(5.3), which leads to a system of ordinary differential equations for $\psi_k(t)$, only coupled through the small-scale mean flow interaction by topographic stress.

Layered topographic equations in Fourier form

$$\frac{d\psi_k}{dt} = ik\ell_x \left(\frac{\beta}{|\vec{l}|^2 k^2} - V \right) \psi_k + ik\ell_x \frac{h_k}{|\vec{l}|^2 k^2} V, \tag{5.9}$$

$$\frac{dV}{dt} = -\ell_x \sum_{k \neq 0} ikh_k (\psi_k)^*. \tag{5.10}$$

We note that $h_{-k} = (h_k)^*$ and $\psi_{-k} = (\psi_k)^*$ since both h and ψ are real, where $*$ denotes complex conjugation.

5.2.1 Hamiltonian structure

By rescaling the equations and changing the coordinate system to align the x-component with $\vec{l}$, it is always possible to reduce the equations in (5.9), (5.10) with $\ell_x \neq 0$, to the same convenient form as in the special situation of layered topography aligned with the North–South (or y) direction. Indeed, with the change of variables

$$q' = \frac{q}{|\vec{l}|}, \quad \psi' = |\vec{l}|\psi, \quad h' = \frac{h}{|\vec{l}|}, \quad \beta' = \frac{\beta}{|\vec{l}|^2}, \quad V' = V, \quad t' = l_x t,$$

we recover (5.9), (5.10) with the special case $\vec{l} = (1, 0)$. Hence any choice of $l_x \neq 0$ and l_y can be reduced to the special situation $l_x = 1$ and $l_y = 0$ via the change of variables delineated above. Therefore, without loss of generality, we restrict ourselves to the particular situation of layered topography orthogonal to the zonal mean flow, so that q, ψ, and h are functions of (x, t) only. With all of these special assumptions the equations in (5.1)–(5.3) become the

Layered topographic equations

$$\frac{\partial q}{\partial t} + V\frac{\partial q}{\partial x} + \beta\frac{\partial \psi}{\partial x} = 0, \tag{5.11}$$

$$q = \frac{\partial^2 \psi}{\partial x^2} + h, \tag{5.12}$$

$$\frac{dV}{dt} = \fint h\frac{\partial \psi}{\partial x}. \tag{5.13}$$

To exhibit the Hamiltonian structure inherent in the equations from (5.11)–(5.13), we rewrite them in terms of large-scale "advected coordinates". Thus, we define the new variables

$$x' = x - \Lambda(t), \quad t' = t, \quad \Lambda(t) = \int_0^t V(s)ds, \tag{5.14}$$

and

$$\tilde{q}(x', t') = q(x, t), \quad \tilde{\psi}(x', t') = \psi(x, t). \tag{5.15}$$

We note that q and $\tilde{q}$ coincide at $t = 0$ when $\Lambda(0) = 0$. Setting the initial value of Λ to a value different from zero corresponds to shifting the topography in x by that same amount relative to the inertial frame. In these new coordinates, equations (5.11) and (5.12) become

$$\frac{\partial \tilde{q}}{\partial t'} + \beta\frac{\partial \tilde{\psi}}{\partial x'} = 0, \tag{5.16}$$

$$\tilde{q} = \frac{\partial^2 \psi}{\partial x'^2} + h(x' + \Lambda(t)). \tag{5.17}$$

We differentiate (5.16) with respect to x' and use (5.17) to obtain

$$\frac{\partial^2 \tilde{q}}{\partial t' \partial x'} + \beta\tilde{q} - \beta h(x' + \Lambda(t)) = 0.$$

To further simplify the notation we define

$$H(x) = \int_0^x h(s)ds. \tag{5.18}$$

The function $H(x)$ is periodic because the average of $h(x)$ is zero. After dropping the primes in notation, the above manipulation leads to the following set of equations, which is equivalent to the basic equations in (5.11)–(5.13).

Lagrangian layered topographic equations

$$\frac{\partial^2 \tilde{q}}{\partial t \partial x}(t, x) + \beta \tilde{q}(t, x) - \beta h(x + \Lambda) = 0, \tag{5.19}$$

$$\dot{V} = -\int_0^{2\pi} H(x) \tilde{q}(t, x - \Lambda(t))\, dx, \tag{5.20}$$

$$\dot{\Lambda} = V. \tag{5.21}$$

In terms of the Fourier series representation of $\tilde{q}$ and h

$$\tilde{q}(t, x) = \sum_{k \neq 0} \tilde{q}_k(t) \exp(ikx), \quad h(x) = \sum_{k \neq 0} h_k \exp(ikx), \tag{5.22}$$

the equations in (5.19)–(5.21) become the following infinite set of ordinary differential equations

$$\dot{\tilde{q}}_k = -\frac{\beta}{ik} \left[\tilde{q}_k - h_k \exp(ik\Lambda)\right], \tag{5.23}$$

$$\dot{V} = -\sum_{k \neq 0} \frac{h_k}{ik} (\tilde{q}_k)^* \exp(ik\Lambda), \tag{5.24}$$

$$\dot{\Lambda} = V. \tag{5.25}$$

Now with $\tilde{q}_k = a_k + ib_k$ and $h_k = |h_k| \exp(i\theta_k)$, $k \geq 1$, the equations in (5.23)–(5.25) determine an infinite-dimensional Hamiltonian system over the space $\{V, \Lambda, a_1, b_1, \ldots\}$, whose Hamiltonian is given by the total energy E in equation (5.4)

$$E = \frac{1}{2}V^2 - \frac{1}{2}\int \psi\omega =$$

$$E = \frac{1}{2}V^2 + \sum_{k>0} \frac{1}{k^2} \left[a_k^2 + b_k^2 - 2a_k |h_k| \cos(k\Lambda + \theta_k) \right. \tag{5.26}$$
$$\left. -2b_k |h_k| \sin(k\Lambda + \theta_k) + |h_k|^2\right].$$

Indeed with $Z = [V, \Lambda, a_1, b_1, \ldots]$, we can rewrite (5.23)–(5.25) in the Hamiltonian form

$$\dot{Z} = J\nabla E.$$

Here J is an infinite symplectic block-diagonal matrix in the space $\{V, \Lambda, a_1, b_1, \ldots\}$, whose diagonal consists of 2×2 block matrices $J_0, J_1, J_2, \ldots,$ with

$$J_0 = \begin{pmatrix} 0 & -1 \\ 1 & 0 \end{pmatrix} \quad J_k = \frac{k\beta}{2} \begin{pmatrix} 0 & -1 \\ 1 & 0 \end{pmatrix}, \quad k \geq 1.$$

This Hamiltonian system also conserves the total enstrophy $\mathcal{E}$ in equation (5.5)

$$\mathcal{E} = \beta V + \sum_{k>0} |\tilde{q}_k|^2. \tag{5.27}$$

We can also write the equations (5.23)–(5.25) in terms of "action-angle" coordinates, with

$$\tilde{q}_k = \sqrt{2I_k}\exp i\phi_k.$$

This leads to the equivalent system

$$\dot{I}_k = \frac{\beta}{k}\sqrt{2I_k}|h_k|\sin(k\Lambda - \phi_k + \theta_k), \tag{5.28}$$

$$\dot{\phi}_k = \frac{\beta}{k} - \frac{\beta}{k}\frac{|h_k|}{\sqrt{2I_k}}\cos(k\Lambda - \phi_k + \theta_k), \tag{5.29}$$

$$\dot{V} = -\frac{2\sqrt{2I_k}|h_k|}{k}\sin(k\Lambda - \phi_k + \theta_k), \tag{5.30}$$

$$\dot{\Lambda} = V. \tag{5.31}$$

5.3 Integrable behavior

Here we discuss several interesting regimes where the *Lagrangian layered topographic equations* in (5.19)–(5.21) are completely integrable.

5.3.1 The case $h = 0$

In the absence of topography, $h = 0$, the equations (5.23)–(5.25) trivially reduce to

$$V = const, \quad \tilde{q}(t,x) = \sum_{k\neq 0} \tilde{q}_k(0)\exp\left[i\frac{\beta}{k}t + ikx\right], \tag{5.32}$$

and the solution is a system of independent Rossby waves, whose frequency is proportional to β. Hence, as is well known, β/k determines the frequency of oscillation for each of the Rossby waves.

5.3.2 The case $\beta = 0$

Next, we consider the special case $\beta = 0$. Then the equation (5.19) implies that

$$\tilde{q}(t,x) = \tilde{q}(0,x) = const,$$

and both V and Λ decouple from $\tilde{q}$. With $\tilde{q}_k = |\tilde{q}_k| \exp(i\phi_k)$, it is straightforward to show from (5.24) that

$$\dot{V} = -2 \sum_{k>0} \frac{|h_k||\tilde{q}_k|}{k} \sin(k\Lambda - \phi_k + \theta_k).$$

From (5.20), (5.21) we observe that the dynamics of V and Λ are determined by the two-dimensional Hamiltonian system

$$\dot{V} = -\frac{\partial}{\partial \Lambda} \Phi(\Lambda), \tag{5.33}$$

$$\dot{\Lambda} = V. \tag{5.34}$$

Here Φ, given by

$$\Phi(\Lambda) = -\fint_0^{2\pi} H(x) W(x - \Lambda) dx, \quad W(x) = \int_0^x \tilde{q}(0, s)\, ds, \tag{5.35}$$

plays the role of a "mechanical potential" and depends on the initial data in the small-scale field.

For fixed topography and initial conditions in the small scales, all solutions of (5.33)–(5.34) can be visualized by displaying the contours of the Hamiltonian $H(V, \Lambda) = V^2/2 + \Phi(\Lambda)$. In Figure 5.1, the solutions are shown in the (V, Λ)-plane for single mode topography. We observe the emergence of a large set of solutions, whose mean velocity V oscillates about the average value $V = 0$. In particular for these solutions, the mean flow does not show any preferred direction and periodically reverses its direction from east to west, and vice-versa. The maximal amplitude of these oscillations is determined by solutions heteroclinic to unstable equilibria. These separatrices divide the phase space (V, Λ) into two distinct regions, which exhibit a different qualitative behavior; either the time average of the mean flow V is zero or it lies strictly above a threshold different from zero, with $|V(t)| > 0$ for all time.

What if we perturb the topography by adding more scales? In Figure 5.2 we show solutions after a few smaller scales were added to the same basic topographic mode. The effect on the phase portrait is small and the qualitative behavior barely changes. However, as we increase the amplitudes of the second and third mode in the topography and the initial conditions, new steady states clearly appear as depicted in Figure 5.3. Instead of one we now see the emergence of three separatrices per period, heteroclinic to the unstable equilibria. These heteroclinic orbits vary with the initial data and can be quite complicated for several mode topography.

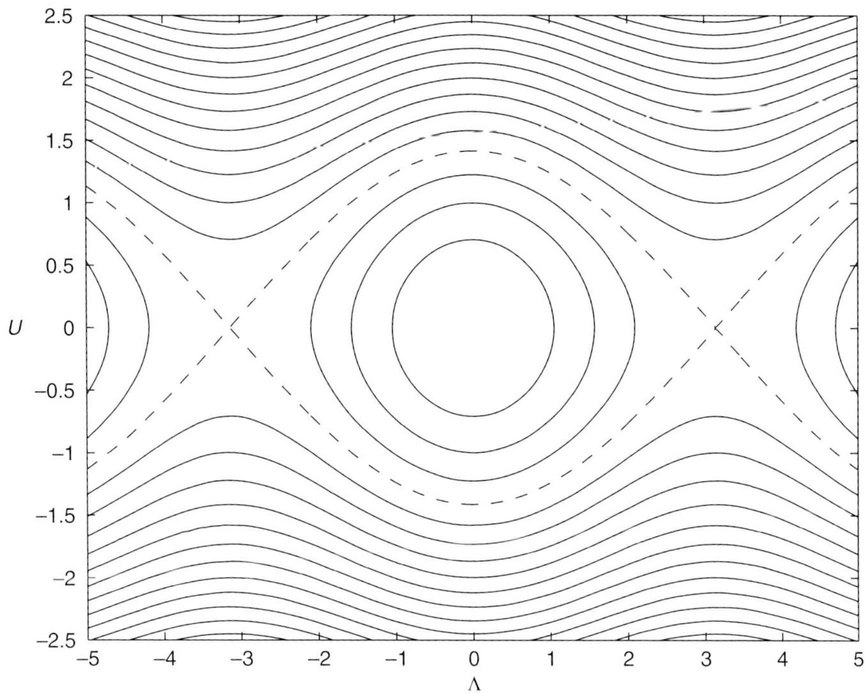

Figure 5.1 Contour lines of the Hamiltonian $H(\Lambda, U)$ for single-mode topography are shown for $\beta = 0$. The heteroclinic orbits, denoted by dashed lines, separate the phase plane (Λ, U) into two distinct regions: either the average mean is zero or it is strictly eastward (or westward) for all time. The stable equilibrium lies at $(0, 0)$ and the unstable equilibrium at $(\pm\pi, 0)$.

In Figure 5.3, the phase portrait is not symmetric about $\Lambda = 0$ because we have chosen random phase shifts ϕ_k and θ_k. Whenever a trajectory lingers about a state with a small mean, i.e. V remains close to zero, the flow configuration tends to appear blocked. As a solution trapped between the two outer separatrices meanders along the heteroclinic orbit, it first reverses the mean flow direction from eastward to westward; then, the mean remains westward, while the trajectory approaches the two inner unstable equilibria, and the flow lingers in a blocked state with a small westward mean component; finally, the mean reverses again and the now eastward flow completes the loop to regain its initial zonal configuration with a rather large mean.

Despite the existence of complicated heteroclinic orbits, all solutions behave in a regular periodic fashion, unless dissipation or forcing are added. Indeed, as we shall see in Sections 5.4 and 5.5, the crucial ingredient for chaotic behavior without dissipation, non-zero β, is still missing in this system.

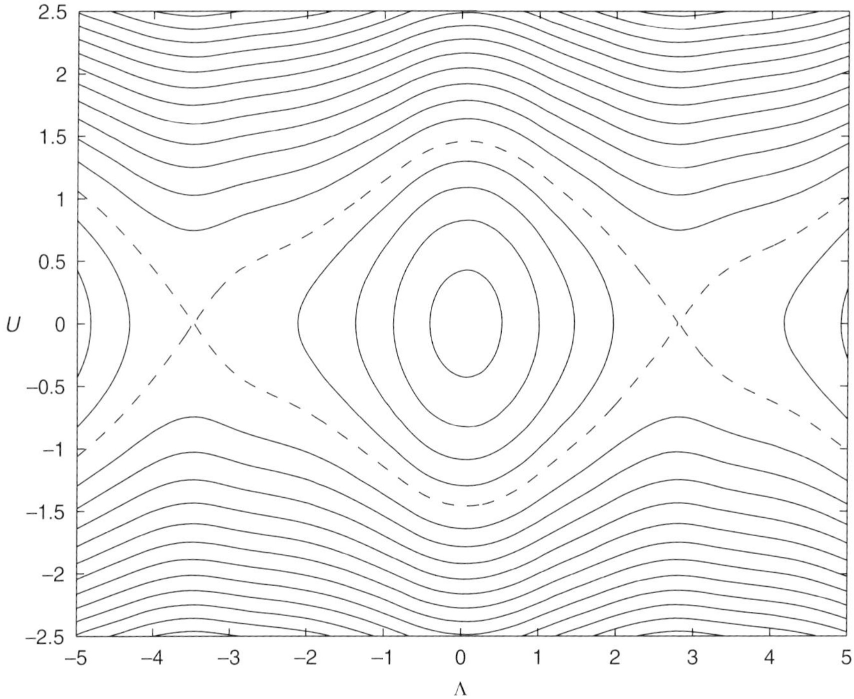

Figure 5.2 Contour lines of the Hamiltonian $H(\Lambda, U)$ for peturbed single-mode topography with added small scales are shown for $\beta = 0$. The heteroclinic orbits (or separatrices) are denoted by dashed lines.

5.3.3 Single mode topography

To study the complex interaction between β, the topography, and the mean flow, we first consider the special case of single mode topography (Hart, 1979; Samelson and Allen, 1987; Vallis, 1985). Let h_k be the unique non-vanishing mode in the Fourier expansion of the topography in (5.22). Then all other modes, $j \neq k$, with $h_j = 0$, decouple into an infinite set of independent linear oscillators

$$w_j(t) = w_j(0) \exp\left[i \frac{\beta}{j} t \right], \quad j \neq k.$$

Thus all $\tilde{q}_j, j \neq k$, are independent Rossby waves which, if initially zero, will remain zero for all time, regardless of the mean flow. Only $\tilde{q}_k$, which corresponds to the non-vanishing topographic mode h_k, interacts with the mean flow and will be excited through topographic stress, even if zero initially.

Specifically, we consider the single mode solution

$$\psi(x, y, t) = a(t) \sin(kx) + b(t) \cos(kx), \tag{5.36}$$

$$h(x, y) = H \sin(kx), \quad k \neq 0, \tag{5.37}$$

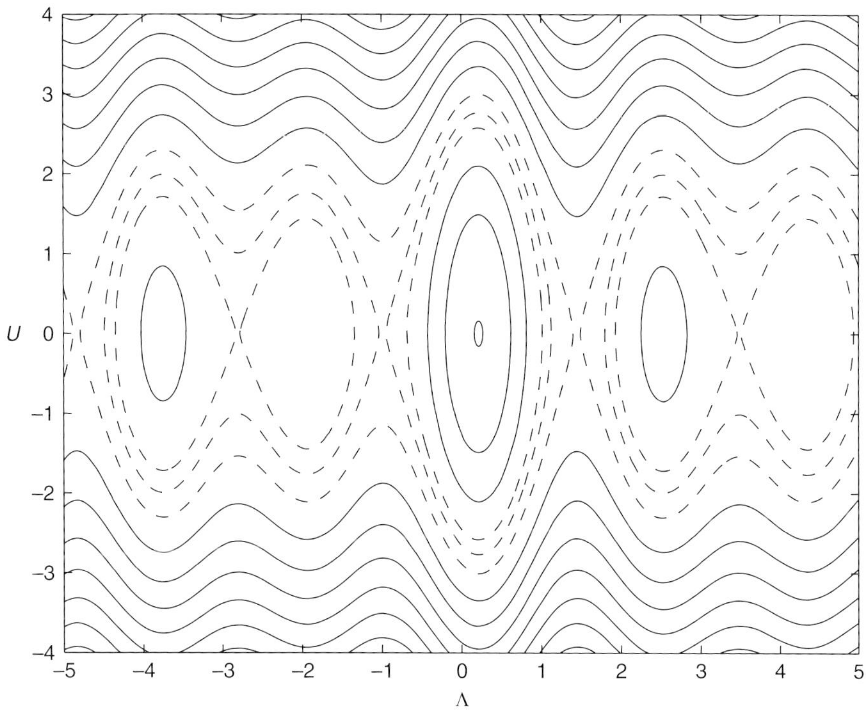

Figure 5.3 Contour lines of the Hamiltonian $H(\Lambda, U)$ for strongly perturbed multi-mode topography are shown for $\beta = 0$. The unstable equilibria lie at the intersections of the heteroclinic orbits, denoted by dashed lines. The stable equilibria are located at the centers of the circular orbits.

where $H \neq 0$ denotes the height of the topography. For these special exact solutions, the full set of non-linear equations in (5.11)–(5.13) simplifies to the three-dimensional dynamical system

$$\dot{a} = \left(V - \frac{\beta}{k^2}\right) kb, \quad a(0) = \bar{a}, \tag{5.38}$$

$$\dot{b} = -\left(V - \frac{\beta}{k^2}\right) ka + \frac{H}{k} V, \tag{5.39}$$

$$\dot{V} = -\frac{kHb}{2}. \tag{5.40}$$

The energy and the enstrophy, E and $\mathcal{E}$, are now given by

$$E = \frac{V^2}{2} + \frac{k^2(a^2 + b^2)}{4}, \tag{5.41}$$

$$\mathcal{E} = \beta V + \frac{k^4(a^2 + b^2) + H^2 - 2k^2 aH}{4}. \tag{5.42}$$

Since E and $\mathcal{E}$ are both conserved in time, the solutions of (5.38)–(5.40) lie on the intersections of the two manifolds $E = const$ and $\mathcal{E} = const$. The surface defined by $E = const$ is a prolate spheroid (ellipsoid of circular cross-section), whereas $\mathcal{E} = const$ is a concave surface and defines a one-to-one mapping from (a, b) to V for a fixed value of $\mathcal{E}$. Next we utlize the elementary examples of exact solutions in (5.38)–(5.40) to rigorously demonstrate what we claimed at the end of Section 4.2 on non-linear stability. Namely, that the interaction with the mean flow destablizes these steady states in general for $-1 \le \mu < 0$.

What are the steady states of this system? Since $H \neq 0$, we see from (5.40) that b must be identically zero, which immediately implies that $a(t) = \bar{a}$ is constant in time. Hence any steady state takes the form

$$x^* = (\bar{a}, 0, V), \qquad V = -\frac{\beta}{H/\bar{a} - k^2}, \tag{5.43}$$

where the value of V was obtained by setting $\dot{b} = 0$ in (5.39). For small topography, $H < k^2\bar{a}$, the steady state corresponds to eastward mean flow, but once $H > \bar{a}k^2$, the steady mean reverses and points westward. These steady-states coincide with those introduced in Chapter 1, Subsection 1.2.4 under the assumption of a linear q–ψ relation, and given the physical parameters H and β they are fully determined by μ

$$\mu = \frac{H}{\bar{a}} - k^2, \qquad V = -\frac{\beta}{\mu}. \tag{5.44}$$

To determine the linear stability of any steady state, $x^* = (H/(\mu + k^2), 0, -\beta/\mu)$, we compute the eigenvalues of the Jacobian of the equations on the right-hand side of (5.38)–(5.40)

$$DF\,|_{x=x^*} = \begin{pmatrix} 0 & -\beta(\mu+k^2)/\mu & 0 \\ \beta(\mu+k^2)/\mu & 0 & H\mu/(k(\mu+k^2)) \end{pmatrix} \tag{5.45}$$

The eigenvalues λ_i are

$$\lambda_1 = 0, \qquad \lambda_{2,3} = \pm\left[-\frac{\mu H^2}{2(\mu+k^2)} - \frac{\beta^2(\mu+k^2)^2}{\mu^2}\right]^{1/2}. \tag{5.46}$$

The steady state x^* is linearly stable if λ_2 and λ_3 are imaginary. Thus, instability occurs when the term in brackets in (5.46) is positive, which can only happen for $\mu < 0$. Since the second term involving β is strictly negative, the beta-effect always tends to stabilize the steady state. In fact for $\beta \neq 0$ and without topography, $H = 0$, the equilibrium is stable for all μ. On the other hand for $\beta = 0$, but in the presence of topography, $H \neq 0$, the equilibrium is stable if μ satisfies either $\mu > 0$ or $\mu < -k^2$ with instability for $-k^2 < \mu < 0$. The combined effect of topography and β is determined by the superposition of these two extreme cases. In the situation with both $H \neq 0$ and $\beta \neq 0$, the steady states are stable for all $\mu > 0$ and for all $\mu < -k^2$. However inside the interval $-k^2 < \mu < 0$, the stabilizing effect of β and the destabilizing effect of the topography compete. Let μ_c denote the unique μ for which the expression in square brackets in (5.46) vanishes. Then to the left of μ_c the equilibrium is unstable, whereas to the right it is stable. In summary for $\beta \neq 0$ and $H \neq 0$, the steady state is stable for $\mu < k^2$ and $\mu > \mu_c$, and it is unstable for $-k^2 < \mu < \mu_c$. The location of μ_c depends on the relative magnitude of β versus H. As β increases, μ_c shifts towards $-k^2$, thereby further reducing the instability region. Physically the unstable equilibria correspond to eastward mean flows, where the topography is large enough to overcome the stabilizing effect of β but not quite large enough to create a situation of topographically "locked" flow. The stable branch $\mu < -k^2$ is likely to be unstable in a fully non-linear regime (Carnevale and Frederiksen, 1987).

To determine the number of possible equilibria in any given situation, we compute the total enstrophy at each steady state x^* as a function of μ

$$\mathcal{E} = -\frac{\beta^2}{\mu} + \frac{H^2\mu^2}{4(\mu+k^2)^2}. \tag{5.47}$$

In Figure 5.4, the total enstrophy is shown in the special case $\beta = H = k = 1$. We note that for $\mathcal{E} < 0$, there is a unique stable equilibrium, which corresponds to westward flow. As $\mathcal{E}$ crosses zero, a unique steady state exists up to a critical threshold, which we denote by $\mathcal{E}_c$. This bifurcation point corresponds to the previous limit μ_c between stable and unstable regions. For $\mathcal{E} > \mathcal{E}_c$, two new equilibria appear, one stable (elliptic) and one unstable (hyperbolic). Figure 5.5 displays contour lines of constant energy in the (a, b)-plane for a fixed value of the enstrophy $\mathcal{E} > \mathcal{E}_c$. The two homoclinic orbits, which are connected to the unstable equilibrium, isolate the two stable equilibria from each other in the (a, b)-plane. All three equilibria correspond to eastward mean flow.

Next, we focus on the dynamic behavior of the mean flow and switch to the 'action-angle' variable form (5.28)–(5.31) for single mode topography. We reduce

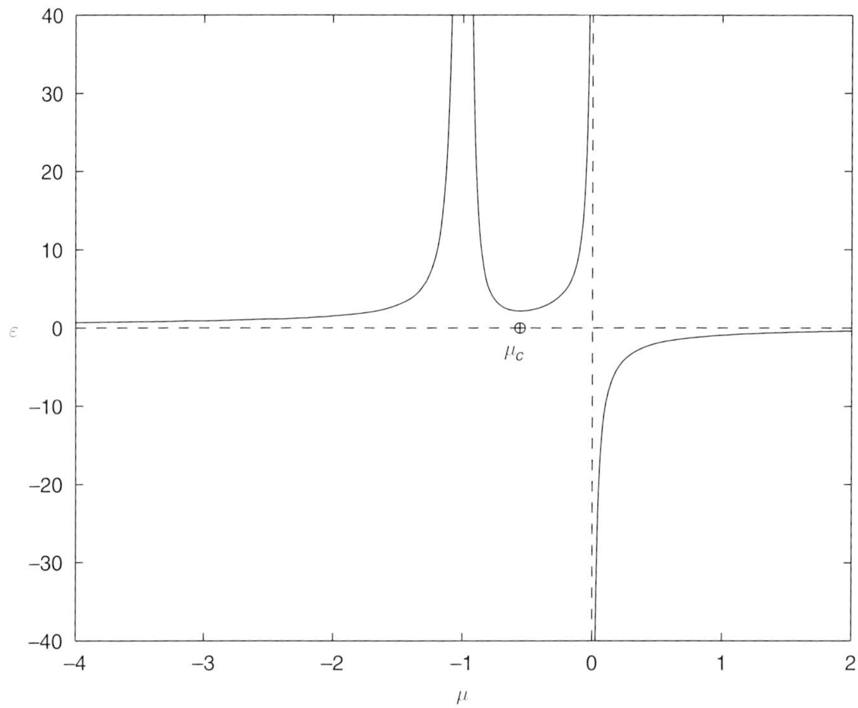

Figure 5.4 The enstrophy $\mathcal{E}(\mu)$ in (5.47) is shown at each steady state for $\beta = H = k = 1$. The unstable equilibria correspond to the interval $-1 < \mu < \mu_c$.

its dimensionality by rewriting this dynamical system in the new set of variables $(\mathcal{E}, \phi_k', V, \Lambda')$

$$I_k = \frac{\mathcal{E} - \beta V}{2}, \quad \Lambda = \Lambda' + \phi_k', \quad \phi_k = k(\Lambda' + \phi_k') - \Lambda',$$

where neither $\mathcal{E}$ nor V were modified. Then the equations for V and Λ' decouple from that of ϕ_k'; moreover, since the potential enstrophy, $\mathcal{E} = \beta V + 2I_k$, is conserved in time, we obtain the two-dimensional dynamical system

$$\dot{V} = -2\frac{\sqrt{\mathcal{E} - \beta V}|h_k|}{k} \sin(\Lambda' + \theta_k), \tag{5.48}$$

$$\dot{\Lambda}' = kV - \frac{\beta}{k} + \frac{\beta}{k}\frac{|h_k|}{\sqrt{\mathcal{E} - \beta V}} \cos(\Lambda' + \theta_k). \tag{5.49}$$

Its phase space is 2π-periodic in Λ' and is restricted to the region

$$V \leq \frac{\mathcal{E}}{\beta} \quad -\pi \leq \Lambda' < \pi.$$

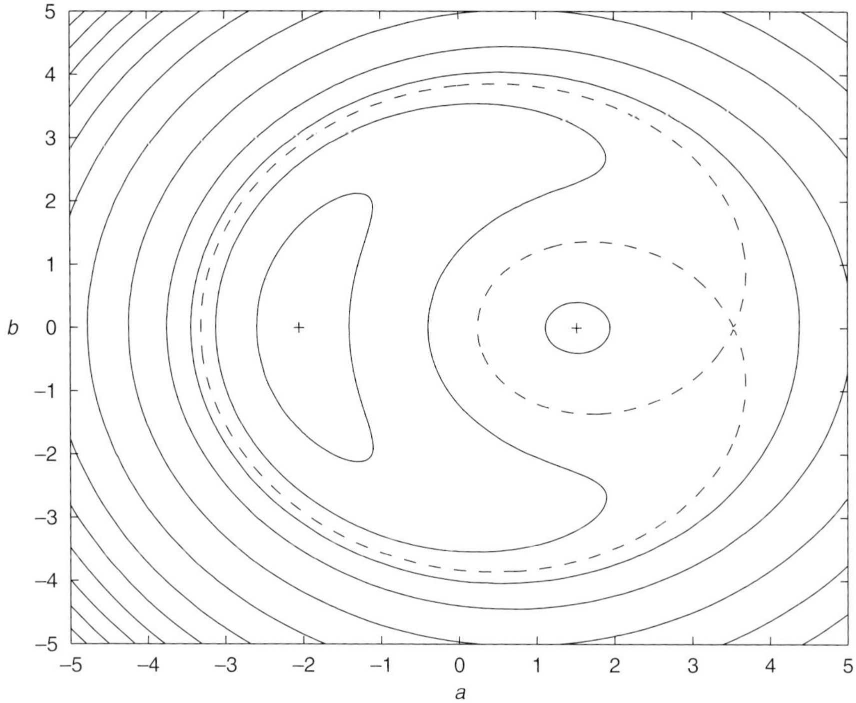

Figure 5.5 Contour lines of constant energy E are shown in the (a, b) plane for fixed enstrophy, $\mathcal{E} = 3$, with $\beta = H = k = 1$. The two stable equilibria are denoted by plus signs and the unstable equilibrium lies at the intersection of the two heteroclinic orbits, denoted by dashed lines.

To draw the phase portrait of (5.48)–(5.49), we use the following first integral of the system, related to the conservation of energy

$$\mathcal{H} = \frac{k}{2}\left(V - \frac{\beta}{k^2}\right)^2 - 2\frac{|h_k|\sqrt{\mathcal{E} - \beta V}}{k}\cos(\Lambda' + \theta_k). \tag{5.50}$$

As we vary the enstrophy, we observe a change in the topology of the phase portrait when $\mathcal{E}$ crosses the critical threshold $\mathcal{E}_c$

$$\mathcal{E}_c = 3\left(\frac{\beta^2|h_k|}{2k^2}\right)^{2/3} + \left(\frac{\beta}{k}\right)^2. \tag{5.51}$$

In Figure 5.6, the phase portrait in the (V, Λ')-plane is shown for $\mathcal{E} < 0$ on the left and for $\mathcal{E} > \mathcal{E}_c$ on the right. In the left frame, we find a single stable equilibrium, which corresponds to a westward mean. In the right frame, two new equilibria have appeared, one stable and the other unstable, located at the intersection of the two homoclinic orbits. The two stable equilibria correspond to strong and weak eastward mean flow, respectively.

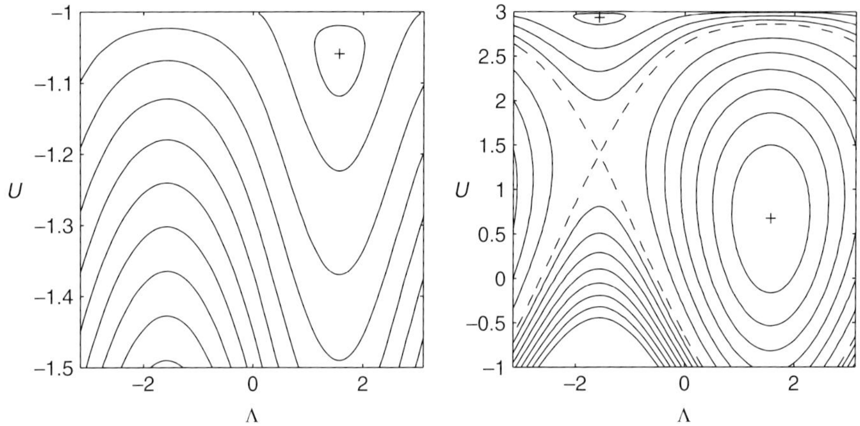

Figure 5.6 Contour lines of $\mathcal{H}(\Lambda, U)$, which is defined by (5.50), are shown in the (Λ, U) plane for fixed enstrophy $\mathcal{E}$, with $\beta = H = k = 1$. Left: $\mathcal{E} = -1$, the stable equilibrium, denoted by a plus sign, corresponds to westward flow; right: $\mathcal{E} = 3$, the two stable equilibria are denoted by plus signs and correspond to strong and weak eastward mean flow. The unstable equilibrium lies at the intersection of the two heteroclinic orbits, denoted by dashed lines, and also corresponds to eastward mean flow.

5.4 A limit regime with chaotic solutions

We wish to gain a mathematical understanding of the combined effect of β and multi-mode topography. In Section 5.5, we will discuss numerical solutions of the equations with layered topography for a wide range of parameter regimes. Nevertheless such a numerical case study is necessarily limited in its nature. On the other hand, with the loss of integrability, much of the underlying mathematical structure required for analysis has vanished. To gain a global mathematical insight, we will study a special limit in this section, which greatly reduces the complexity of the dynamical system, while preserving an infinite-dimensional signature of this system.

To begin we introduce the scaling parameter, γ, into the *Lagrangian layered equations* in (5.19)–(5.21). We set

$$h = \gamma h', \quad \beta = \gamma^{1/2}\beta', \quad t = \gamma^{-1/2}t', \quad V = \gamma^{1/2}V', \tag{5.52}$$

and rewrite the equations in (5.19)–(5.21) in these new variables

$$\frac{\partial^2 \tilde{q}}{\partial t \partial x}(t, x), + \beta' \tilde{q}(t, x) - \gamma \beta' h'(x + \Lambda) = 0, \tag{5.53}$$

$$\dot{V}' = -\int_0^{2\pi} H'(x)\tilde{q}(t, x - \Lambda)\, dx, \tag{5.54}$$

$$\dot{\Lambda} = V'. \tag{5.55}$$

In the remainder of this section we will study the dynamics generated by the above equation in the formal limit $\gamma \to 0$. A small value of γ corresponds to the physical situation where h, β, and V are small, the small-scale velocity $\nabla^{\perp}\psi$ is of order one, and the time oscillations of V have a long period. In this case, most of the energy of the system is stored in the small-scale part of the velocity field. From (5.52) it follows that for $\gamma \ll 1$ the topography is small relative to both β and the mean flow V.

To simplify the notation, we drop the primes from the new variables. Next, we set $\gamma = 0$ in (5.53), which leads to

$$\frac{\partial^2 \tilde{q}}{\partial t \partial x}(t,x) + \beta \tilde{q}(t,x) = 0. \tag{5.56}$$

By using the Fourier representation of $\tilde{q}$ and h in (5.22), we rewrite (5.56) as

$$i\dot{\tilde{q}}_k = -\frac{\beta}{k}\tilde{q}_k, \quad k \geq 1.$$

The solution of equation (5.53) is then given by a superposition of Rossby waves

$$\tilde{q}(t,x) = \sum_{k \neq 0} |\tilde{q}_k(0)| \exp[ikx + i\phi_k(t)], \quad \phi_k(t) = \frac{\beta}{k}t + \phi_k(0), \quad k \geq 1. \tag{5.57}$$

Next, we substitute (5.57) into (5.54) to obtain the reduced dynamics

$$\dot{V} = -2\sum_{k>0} \frac{|h_k||\tilde{q}_k|}{k} \sin(k\Lambda + \theta_k - \phi_k), \tag{5.58}$$

$$\dot{\Lambda} = V. \tag{5.59}$$

Here $|\tilde{q}_k(t)| = |\tilde{q}_k(0)| = const$, ϕ_k is given by (5.57), $h_k = |h_k|\exp(i\theta_k)$. Thus for $\gamma = 0$, the evolution of V is dictated by a single, time-dependent, second-order ordinary differential equation, similar to that for the motion of a single particle in a one-dimensional spectrum of longitudinal waves, which has been studied extensively (Zaslavsky and Filonenko, 1968; Zaslavsky *et al.*, 1991; Chirikov, 1979; Escande, 1985).

Although the equations in (5.53)–(5.55) remain valid for $\gamma = 0$, the original change of variables defined by (5.52) becomes meaningless in the limit. Nevertheless, we shall see that much of the behavior of V for $\gamma = 0$ extends to the case $\gamma \neq 0$, and we will see this confirmed dramatically in the numerical experiments presented in Subsection 5.5.2. To develop our understanding of the limit $\gamma \to 0$, we first consider the special situation of single mode topography.

5.4.1 Single mode topography

From Subsection 5.3.3 we recall that the equations in (5.53)–(5.55) can be completely integrated for single mode topography, that is when $h_j = 0$ for all $j \neq k$ and $h_k \neq 0$. We shall now study the limit $\gamma \to 0$ in this special case to get a better understanding of the limiting process. The analogues of (5.48) and (5.49) in terms of the rescaled variables (5.52) are

$$\dot{V} = -2\frac{\sqrt{\mathcal{E}_r - \gamma\beta V}|h_k|}{k}\sin(\Lambda + \theta_k) \tag{5.60}$$

$$\dot{\Lambda} = kV - \frac{\beta}{k} + \frac{\gamma\beta}{k}\frac{|h_k|}{\sqrt{\mathcal{E}_r - \gamma\beta V}}\cos(\Lambda + \theta_k), \tag{5.61}$$

where the reduced enstrophy, $\mathcal{E}_r$, is defined by

$$\mathcal{E}_r = \gamma\beta V + 2I_k.$$

Next, we rewrite the first integral of (5.60)–(5.61) in the new variables, which yields

$$\mathcal{H} = \frac{k}{2}\left(V - \frac{\beta}{k^2}\right)^2 - 2\frac{|h_k|\sqrt{\mathcal{E} - \gamma\beta V}}{k}\cos(\Lambda' + \theta_k). \tag{5.62}$$

The analogue of $\mathcal{E}_c$ defined in (5.51) is

$$\mathcal{E}_c = 3\left(\frac{\gamma^2\beta^2|h_k|}{2k^2}\right)^{2/3} + \gamma\left(\frac{\beta}{k}\right)^2.$$

For γ sufficiently small, $\mathcal{E}_r > \mathcal{E}_c$, which implies that (5.60) and (5.61) have three equilibria (V_l, Λ_l), $l = 1, 2, 3$, as described in Subsection 5.3.3 and illustrated in Figure 5.6. That figure shows two non-linear resonance regions, one in which U oscillates about V_1 and another in which it oscillates about V_2. When γ is small these non-linear resonances correspond to the cases where the reduced potential enstrophy is dominated either by $\tilde{q}_k$ (region V_1) or by V (region V_2). As $\gamma \to 0$, $V_1 \to \beta/k^2$, $V_3 \to \beta/k^2$, and $V_2 \to +\infty$. With $\gamma = 0$, the system reduces to a simple pendulum shifted in momentum space by β/k^2. Then the equilibria (V_1, Λ_1) and (V_3, Λ_3) are the usual elliptic and hyperbolic equilibria, respectively, of this modified pendulum. Moreover, the time average of V for any solution in the resonance region is exactly β/k^2. The analyticity of $\mathcal{H}$ in the region $V < \mathcal{E}_r/\gamma\beta$ clearly implies that the orbits of (5.60), (5.61) for small γ are close to those for $\gamma = 0$, as long as these orbits are contained in some region where V is not too large. Thus, as to be expected, the limit $\gamma \to 0$ is regular and well-behaved in this integrable case.

5.4.2 Interaction of non-linear resonances

Our goal in this section is to study the behavior of the large-scale velocity V when γ is small and there are at least two large modes in the topography. Here most of our statements will not be mathematically rigorous, but instead are based on previous numerical investigations and heuristic arguments (most of them can be found in Escande, 1985; Chirikov, 1979). However, these ideas lead to quantitative prediction and explanation for the interesting topographic blocking phenomenon presented in Subsection 5.5.2 for $\gamma \ll 1$.

Our starting point is the system of ordinary differential equations (5.58), (5.59), which we obtained from the equations (5.53)–(5.55) by setting $\gamma = 0$. To understand the basic features of the equations in (5.58), (5.59), we will concentrate on the simplest non-trivial case of two modes in the topography (neither of them needs to be small). Hence, we assume that the topography is given by

$$h(x) = |h_k|[\exp(ikx + i\theta_k) + \exp(-ikx - i\theta_k)] + |h_l|[\exp(ilx + i\theta_l) + \exp(-ilx - i\theta_l)],$$

where $h_k \neq 0$.

In this case, the dynamics is determined by

$$\dot{V} = -2\frac{|h_k||\tilde{q}_k|}{k}\sin(k\Lambda + \theta_k - \phi_k) - 2\frac{|h_l||\tilde{q}_l|}{l}\sin(l\Lambda + \theta_l - \phi_l), \quad (5.63)$$

$$\dot{\Lambda} = V. \quad (5.64)$$

Here, $\phi_k = (\beta/k)t + \phi_k(0)$ and $\phi_l = (\beta/l)t + \phi_l(0)$. Our goal is to exhibit the basic mechanism for the onset of large regions of stochastic dynamics. To do so we will study the dynamics of the Poincaré map induced by the time-T flow of the above system, where T is $2\pi/\beta$ times the lowest common multiple of k and l. The following analysis can be found, for instance in Escande (1985) and Chirikov (1979).

We begin by defining resonance zones. If $|h_l| = 0$ and $|h_k| \neq 0$, we recover the situation discussed in Subsection 5.4.1 and depicted in Figure 5.6. We define the k-resonance zone as the region in the (Λ, V) space inside the separatrix. We note that the k-resonance zone is centered at the elliptic fixed point with $V = \beta/k^2$. The width of the k-resonance zone is

$$D_k = \frac{\sqrt{32|h_k||\tilde{q}_k|}}{k}$$

In a similar way, for $|h_k| = 0$ and $|h_l| \neq 0$, we define the l-resonance zone. It is centered at the elliptic fixed point with $V = \beta/l^2$, and its width is

$$D_l = \frac{\sqrt{32|h_l||\tilde{q}_l|}}{l}.$$

When both h_k and h_l are different from zero, the geometric complexity greatly increases. It is then possible to prove that the Poincaré map has hyperbolic fixed points corresponding to the modes k and l. The hyperbolic fixed points associated with k have stable and unstable manifolds that necessarily intersect each other. In this case, the k-resonance zone is defined as the region inside the closed curve formed by pieces of these manifolds (see Meiss, 1992; section VIII-A, for a detailed discussion of this point). The proofs of the claims in this paragraph, at least for generic cases, can be found in Denzler (1987), and Mather (1993).

If both $|h_k|$ and $|h_l|$ are different from zero but are sufficiently small, equations (5.63) and (5.64) can be considered as a perturbation of a free particle: $\dot{V}=0$ and $\dot{\Lambda}=V$. In this case, the k-resonance zone resides inside a strip of width D_k centered at $V=\beta/k^2$; its overall shape essentially unchanged from the case $|h_l|=0$. Similarly, we find the l-resonance zone centered near $V=\beta/l^2$. Between the two resonance zones the dynamics of the Poincaré map is rather simple: most orbits remain close to those of a free particle system and reside inside invariant curves, which lie between the two resonance zones.

As the magnitude of $|h_k|$ and $|h_l|$ increases, the appearance of the resonance zones changes continuously, and the stochastic layers which surround the separatrices increase in thickness – see Figure 5.7 in Subsection 5.4.3. With the increase in width of these stochastic layers, they can occupy almost entirely the resonance regions; it then becomes difficult to define the boundaries of the resonance regions. Although, the size of the stochastic regions surrounding the resonances k and l can be large, they might be sufficiently far from each other to remain

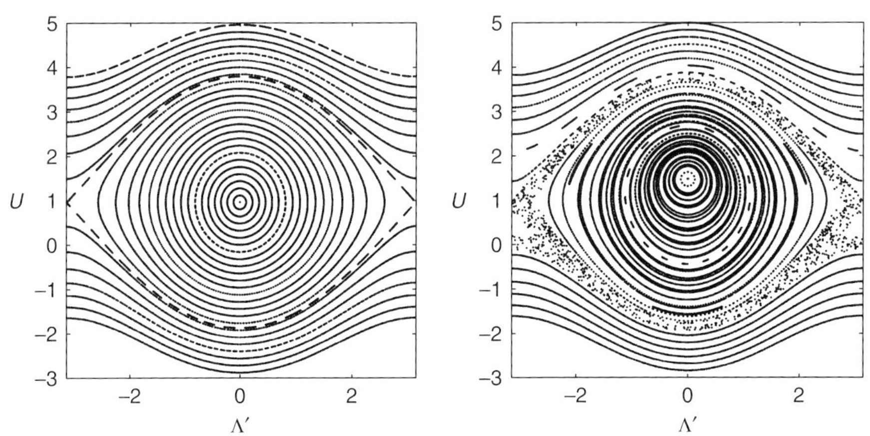

Figure 5.7 Iterations of Poincaré maps of (5.78), (5.79) are shown in the (Λ', U) plane for $\beta=h_k=\tilde{q}_k=1$. Left: $h_l=0$; right: $h_l=0.1$ and $\tilde{q}_l=1$. Note on the right the stochastic layer appearing around the unperturbed separatrix, denoted by dashed lines on the left.

isolated, separated by the invariant curves between them. Therefore the solutions of (5.63) and (5.64) can display highly erratic behavior, yet with a time average value of V close to β/k^2 or β/l^2. Moreover, the V component of such a solution will never increase beyond some threshold V_*, strictly between β/k^2 and β/l^2. In other words, solutions to equations (5.63) and (5.64) can display erratic behavior, while remaining trapped within a single resonance region.

Finally, if the two resonance zones overlap so that no invariant circle separates them, the Poincaré map contains a stochastic region connecting the two resonance regions, which leads to a situation of global stochasticity. Then solutions to (5.63), (5.64) can display irregular behavior with large excursions in V, while visiting different resonance regions.

The transition to global stochasticity is related to the disappearance of the last invariant circle between two resonance zones. After the break-up of the last invariant curve, invariant Cantor sets (or Mather sets'), which in some sense replace the invariant curves appear in the (V, Λ) space. Especially right after the onset of global stochasticity, the gap of these Cantor sets remains small enough to act as partial barriers and constrain the motion in phase space. A solution may then remain for a long time within a certain region of phase space, say near β/k^2, suddenly escape to another region, say near β/l^2, stay there for a long time, and so forth. This behavior probably corresponds to the prototype topographic blocking events, which we document in Section 5.2 for numerical solutions of the *Layered topographic equations in Fourier form* from (5.9), (5.10).

Determining the set of parameters which correspond to the threshold of global stochasticity in equations (5.63) and (5.64) is a complicated problem. An extensive discussion on this issue in the context of (5.63), (5.64) can be found in Escande (1985). A simple, heuristic, well-known approach is the "Chirikov resonance overlap criterion." It involves verifying whether the k-resonance zone obtained by setting $|h_l| = 0$ overlaps the l-resonance zone obtained by setting $|h_k| = 0$. In other words, the Chirikov criterion reduces to verifying whether the distance between the centers of the resonance regions is larger than the sum of their half widths, when either resonance is considered independently. To simplify the notation when verifying the Chirikov criterion in our model, we define the parameter

$$
\begin{aligned}
s &= 2\sqrt{2}\left(\frac{\sqrt{|\tilde{q}_k||h_k|}}{k} + \frac{\sqrt{|\tilde{q}_l||h_l|}}{l}\right)\left|\frac{\beta}{l^2} - \frac{\beta}{k^2}\right|^{-1} \\
&= 2\sqrt{2}\frac{l^2 P_k + k^2 P_l}{|l^2 - k^2|}. \qquad (5.65)
\end{aligned}
$$

Here P_k (and similarly for P_l with k replaced by l) is defined by

$$P_k = \frac{k}{\beta}\sqrt{|h'_k||\tilde{q}_k|}, \tag{5.66}$$

which remains unchanged in terms of the original unprimed variables (here denoted by an asterisk, recall (5.52))

$$P_k = \frac{k}{\beta^*}\sqrt{|h^*_k||\tilde{q}_k|}. \tag{5.67}$$

The physical interpretation of P_k is quite interesting. In the absence of topography, the Rossby wave frequency of oscillation related to the wave number k is given by

$$\frac{\beta^*}{k}. \tag{5.68}$$

Thus, β^*/k determines a time scale for the oscillations of the k-mode Rossby wave in the absence of topography (if β^*/k increases by a factor of two, so does the Rossby wave frequency). However, with $\beta = 0$ and a single topographic mode, $h_k \neq 0$, the time scale of the oscillations of U is determined by

$$\sqrt{|h_k||\tilde{q}_k|}. \tag{5.69}$$

Indeed, if $\beta = 0$ and h_k is the unique topographic mode, the dynamics of U is governed by

$$\dot{V} = -2\frac{|h_k||\tilde{q}_k|}{k}\sin(k\Lambda - \phi_k + \theta_k), \tag{5.70}$$

$$\dot{\Lambda} = V. \tag{5.71}$$

By introducing the new variables

$$t' = \sqrt{|h_k||\tilde{q}_k|}\,t, \quad V' = kV, \quad \Lambda' = k\Lambda - \phi_k + \theta_k,$$

we rewrite (5.70), (5.71) as

$$\dot{V}' = -2\sin(\Lambda')$$
$$\dot{\Lambda}' = V',$$

which does not depend on the kth mode anymore. Therefore, the frequencies of two solutions of (5.70), (5.71), for two different sets of values $|h_{k1}||\tilde{q}_{k1}|$ and $|h_{k2}||\tilde{q}_{k2}|$ which are similar (identical, except for possible space–time dilation), are related through their ratio $\sqrt{|h_{k1}||\tilde{q}_{k1}|}/\sqrt{|h_{k2}||\tilde{q}_{k2}|}$. Hence the quantity in (5.69) sets the time scale of V for $\beta = 0$ and a single topographic mode, whereas P_k

is the ratio between the time scale set by topography (5.69) and that set by the beta-effect on the kth mode (5.68).

In terms of s in (5.65), the Chirikov criterion predicts global stochasticity for $s > s_c$, where $s_c \simeq 1$. For equations (5.63) and (5.64), when $|h_k||\tilde{q}_k|/k^2$ and $|h_l||\tilde{q}_l|/l^2$ have the same order of magnitude, it predicts the value of s within a 30% error margin (Escande, 1985). The Chirikov criterion provides a rough estimate for parameter values that result in irregular dynamics in certain parts of the phase space. This estimate can be extrapolated for small $\gamma > 0$, and thus can still produce reasonable quantitative prediction, as shown in Subsection 5.5.2. Even more important than the quantitative prediction of the Chirikov criterion is its implication that the "amount" of irregular dynamical behavior increases with P_k and P_l.

What happens when $s \to \infty$, that is when $\beta \to 0$ (or $k, l \to \infty$) for fixed values of h_k, h_l, $\tilde{q}_k$, and $\tilde{q}_l$? The previous argument suggests that equations (5.63) and (5.64) contain a "large amount of stochasticity in phase space," although we know from Subsection 5.3.2 that the system is integrable for $\beta = 0$! This puzzling situation is commonly called adiabatic chaos in the physics literature (Menyuk, 1985; Kaper and Kovacic, 1994). The term adiabatic derives from the slow temporal variation of the right-hand sides of (5.63), (5.64) for small β, and describes the situation when the theory of adiabatic invariants of classical mechanics applies (Arnold *et al.*, 1997).

To develop our understanding of this seemingly paradoxical situation, we consider the following special topography. In our model, the length scale for the horizontal variation of the topography is determined by its dominant mode k. Let us suppose that k is large and hence that the topographic variations occur at a much smaller spatial scale than the characteristic length scale of the flow (essentially 2π in our model). Then we consider the case when the topography consists only of the two modes, k and $l = k + 1$; in addition for simplicity we assume that

$$\theta_k = \theta_l = \phi_k(0) = \phi_l(0) = 0, \quad |h_k \tilde{q}_k| = |h_l \tilde{q}_l| = \frac{a}{4}.$$

We introduce the new variables

$$V' = kV, \quad \Lambda' = k\Lambda - \frac{\beta t}{k}$$

and rewrite (5.63), (5.64) as

$$\begin{aligned}
\dot{V}' &= -\frac{a}{2}\sin(\Lambda') - \frac{a}{2}[1 + \mathcal{O}(\varepsilon^{1/2})]\sin[(1 + \varepsilon^{1/2})\Lambda' + (2\varepsilon + \mathcal{O}(\varepsilon^{3/2}))t], \\
\dot{\Lambda}' &= V' - \beta\varepsilon,
\end{aligned}$$

with $\varepsilon = k^{-2}$. For large k we may approximate the above system by (we drop the prime sign)

$$\dot{V} = -a\sin(\Lambda)\cos(b), \tag{5.72}$$

$$\dot{\Lambda} = V, \tag{5.73}$$

$$b = \varepsilon\beta t. \tag{5.74}$$

It is associated with the Hamiltonian function

$$H_a = \frac{V^2}{2} - \cos(b)\cos\Lambda, \qquad b = \varepsilon\beta t. \tag{5.75}$$

Equations (5.72) and (5.73) coincide with those in Menyuk (1985) and provide a good model for the phenomenon of adiabatic chaos.

For small ε, the dynamics of (5.72), (5.73) during a time interval, which is long yet much smaller than $1/\varepsilon$, approximate those obtained by setting b constant. Initially the dynamics of (5.72) and (5.73) resemble those of a simple pendulum obtained by setting $b = 0$; then for $t \simeq \pi/(4\varepsilon)$, the dynamics resemble that of a pendulum obtained by setting $b = \pi/4$, and so forth. In particular both situations contain trajectories whose Λ and V components oscillate about zero. Subsequently, for $t \simeq \pi/(2\varepsilon)$ and hence $\cos(b) \simeq 0$, equations (5.72) and (5.73) behave like a free particle $\ddot{\Lambda} = 0$ for which the sign of V never changes. Due to this mixed oscillatory–nonoscillatory behavior around $V = 0$, solutions of (5.72), (5.73) separate into two classes, depending on their long time behavior: first, solutions where the sign of V changes at most a finite number of times and Λ increases unboundedly, and, second, solutions where the sign of V changes an infinite number of times. In the terminology of Menyuk (1985) the first type is called "untrapped" and the second type "trapped." A numerical investigation of (5.72), (5.73) showed that the untrapped solutions are typically well-behaved and quasi-periodic (Menyuk, 1985), which corroborates previous results on the perpetual conservation of adiabatic invariants obtained using the KAM theorem (Arnold, 1989). It also showed that the dynamic behavior of trapped solutions is usually irregular and "chaotic" (Menyuk, 1985), although it appears periodic over very long time intervals. The V component of these trapped solutions oscillates about zero whenever b is close to $(2n+1)\pi/2$, $n = 0, 1, 2, \ldots$. The sequence of intervals where the behavior of V alternates from oscillatory to positive or negative can appear in any order.

Although we have concentrated on the situation with two modes in the topography, some aspects of the above analysis, such as the interaction of nearby resonances, certainly apply when h contains an arbitrary number of modes and γ is small.

5.4.3 Two modes in the topography: a perturbative Melnikov analysis

We recall from the previous section that, if $\gamma = 0$, $|h_l| \neq 0$, and $|h_k| \neq 0$, the only non-trivial part of the dynamics is that of V given by

$$\dot{V} = -2\frac{|h_k||\tilde{q}_k|}{k}\sin(k\Lambda + \theta_k - \phi_k) - 2\frac{|h_l||\tilde{q}_l|}{l}\sin(l\Lambda + \theta_l - \phi_l), \quad (5.76)$$

$$\dot{\Lambda} = V, \quad (5.77)$$

where $\phi_k = (\beta/k)t + \phi_k(0)$ and $\phi_l = (\beta/l)t + \phi_l(0)$. It is well-known (Zaslavsky and Filonenko, 1968; Guckenheimer and Holmes, 1990; Moser, 1973) that this system has invariant sets of solutions where the dynamics is highly erratic (or "chaotic") for $h_k \neq 0$ and $h_l \neq 0$. Indeed, for most parameters values in (5.76), (5.77), regular ("quasi-periodic") and irregular behavior coexist in different parts of the phase space, as we discussed earlier in Subsection 5.5.2. In this section we shall only consider the case, for which a more or less complete mathematical theory exists, namely when $|h_l|$ is small while $|h_k|$ is $\mathcal{O}(1)$.

To begin, we introduce the new variable

$$\Lambda' = k\Lambda - \frac{\beta}{k}t - \phi_k(0) + \theta_k.$$

In terms of Λ', the equations in (5.76) and (5.77) become

$$\dot{V} = -2\frac{|h_k||\tilde{q}_k|}{k}\sin(\Lambda') - 2\frac{|h_l||\tilde{q}_l|}{l}\sin(c_1\Lambda' + c_2 t + c_3), \quad (5.78)$$

$$\dot{\Lambda}' = kV - \frac{\beta}{k}, \quad (5.79)$$

where

$$c_1 = \frac{l}{k}, \quad c_2 = \left[\frac{l}{k^2} - \frac{1}{l}\right]\beta, \quad c_3 = \theta_l - \phi_l(0) + \frac{(\phi_k(0) - \theta_k)l}{k}. \quad (5.80)$$

If we set $|h_l| = 0$, all solutions of (5.78), (5.79) are contained in the level sets of the function

$$\frac{k}{2}\left(V - \frac{\beta}{k}\right)^2 - 2\frac{|h_k||\tilde{q}_k|}{k}\cos(\Lambda'), \quad (5.81)$$

defined in the extended phase space (V, Λ', t). Since this function is a first integral, the system is integrable. For $|h_l|$ sufficiently small the KAM theorem (Arnold, 1989) applies and implies that most but not necessarily all solutions of (5.78), (5.79) are contained inside invariant surfaces as in the case $|h_l| = 0$. The dynamics remain regular (quasi-periodic with two frequencies) in most of the phase space. However, the KAM theory does not provide any information on certain regions in phase space, such as described below.

For $|h_l| = 0$, the equations (5.78), (5.79) possess a hyperbolic equilibrium at $(V, \Lambda') = (\beta/k^2, \pi)$. There exists a pair of homoclinic orbits or separatrices, which connect this equilibrium to itself (here we identify points whose Λ'-coordinate differs by a multiple of 2π). The presence of this hyperbolic equilibrium for $|h_l| = 0$ implies that for $|h_l| \neq 0$ there must be a hyperbolic periodic solution of period $2\pi/c_2$, which is contained in a small ball centered at $(V, \Lambda') = (\beta/k^2, \pi)$. Usually the two-dimensional stable and unstable manifolds of this periodic orbit are not contained inside the same surface in the extended phase space. In particular, it can be shown that, if these two manifolds intersect transversally, there exists an invariant set (a topological "Smale Horseshoe") where the dynamics is highly irregular (Moser, 1973). Then the equations (5.78), (5.79) do not admit a nonconstant global analytic first integral, that is a function of (V, Λ', t) whose level surfaces are invariant under the flow of the equations (5.78), (5.79) (Moser, 1973). The region near the original separatrix, where the irregular behavior occurs, is usually called the "stochastic layer." In Figure 5.7, we show sets of iterations of Poincaré maps of (5.78) and (5.79) for two different values of $|h_l|$: $|h_l| = 0$ and $|h_l| = 10^{-1}$. From these pictures it is clear that the dynamics is regular in most of the phase space except for the stochastic layer surrounding the unperturbed separatrix.

To prove rigorously that the stable and unstable manifolds of the periodic orbit intersect transversally, we need to show that the Melnikov function $M(t_0)$ of the real variable t_0 has a non-degenerate zero (Guckenheimer and Holmes, 1990). In Appendix A, we show that the Melnikov function associated with (5.78), (5.79) vanishes at a certain t_*. Moreover, for $l = k/2$ the equations (5.93), (5.94) imply that

$$\frac{dM(t_*)}{dt} = \dot{M}(t_*) = 32\pi|\tilde{q}_l|\frac{c_2}{\alpha}\frac{\exp[-c_2 3\pi/(2\alpha)]}{1-\exp(-2\pi c_2/\alpha)}, \tag{5.82}$$

with

$$\frac{c_2}{\alpha} = \frac{\beta}{\sqrt{|h_k||\tilde{q}_k|}}\frac{1}{k}\left(-\frac{3}{2\sqrt{2}}\right).$$

Alternatively, for $l = 2k$, the equations (5.93), (5.5.4) imply that

$$\dot{M}(t_*) = \frac{8}{3}\pi|\tilde{q}_l|\frac{c_2^2}{\alpha^2}\left[-2+\frac{c_2^2}{\alpha^2}\right]\frac{\exp[-c_2 3\pi/(2\alpha)]}{1-\exp(-2\pi c_2/\alpha)}, \tag{5.83}$$

with

$$\frac{c_2}{\alpha} = \frac{\beta}{\sqrt{|h_k||\tilde{q}_k|}} \frac{1}{k} \left(\frac{3}{2\sqrt{2}}\right).$$

It appears difficult to derive an expression for $\dot{M}(t_*)$ valid for arbitrary k and l. The values $l = k/2$ and $l = 2k$ represent the cases when a single mode in the topography is perturbed by a sub-harmonic or a harmonic mode, respectively.

The size of $|\dot{M}(t_*)|$ is related to the growth rate of the width of the stochastic layer with respect to the variation of $|h_l|$ (Zaslavsky and Filonenko, 1968; Zaslavsky, 1994; Escande, 1985), and it increases with the growth rate. We observe that $|\dot{M}(t_*)|$ is exponentially small with respect to β, as $\beta \to \infty$, provided all other parameters remain constant. This fact is consistent with the excellent averaging property of fast oscillatory Rossby waves discussed in Subsection 5.5.1 and Appendix 2.

In (5.82) and (5.83), $|\dot{M}(t_*)|$ is (but for the multiplicative constant $|\tilde{q}_l|$) only a function of

$$P_k = \frac{k\sqrt{|h_k||\tilde{q}_k|}}{\beta},$$

which we introduced previously in Subsection 5.4.2. Whether $l = k/2$ or $l = 2k$, $|\dot{M}(t_*)|$ tends to zero exponentially fast as $1/P_k \to \infty$, yet in the sub-harmonic case, $l = k/2$, the decay is less rapid. In the sub-harmonically perturbed case, $|\dot{M}(t_*)| \to 16|\tilde{q}_l|$ as $P_k \to \infty$, while in the harmonically perturbed case $|\dot{M}(t_*)| \to 0$ as $P_k \to \infty$. Thus, when P_k is large, sub-harmonic perturbations tend to produce much larger stochastic layers than harmonic perturbations. Although the results in this section were obtained for $\gamma = 0$, they remain valid for γ small but different from zero. By using the scaling defined in (5.52), we can rewrite the equations in (5.23)–(5.25) in the current situation as

$$\begin{aligned}
\tilde{q}_j(t) &= \tilde{q}_j(0) \exp(i\beta t/j), \qquad j \neq k, l, \\
i\dot{\tilde{q}}_k &= -\frac{\beta}{k}\tilde{q}_k + \gamma\frac{\beta}{k} h_k \exp(ik\Lambda), \\
i\dot{\tilde{q}}_l &= -\frac{\beta}{l}\tilde{q}_l + \gamma\frac{\beta}{l} h_l \exp(il\Lambda), \\
\dot{V} &= -\sum_{m=\pm k, \pm l} \frac{h_m}{im} \tilde{q}_{-m} \exp(im\Lambda), \\
\dot{\Lambda} &= V.
\end{aligned} \tag{5.84}$$

This system of differential equations for $\tilde{q}_k$, $\tilde{q}_l$, V, and Λ, is Hamiltonian with three degrees of freedom. It admits as first integrals the reduced energy and the enstrophy

$$H_r = \frac{\gamma}{2}V^2 + \frac{1}{2}\sum_{k=\pm k,\pm l}\frac{1}{k^2}\left[|\tilde{q}_k|^2 - 2\gamma\tilde{q}_{-k}h_k\exp(i\Lambda k)\right],$$

$$\mathcal{E}_r = \gamma\beta V + \sum_k |\tilde{q}_k|^2.$$

If $\gamma = 0$, the gradients of H_r and $\mathcal{E}_r$ are linearly independent for $|\tilde{q}_k| \neq 0$ and $|\tilde{q}_l| \neq 0$. Next we *assume that* $|\tilde{q}_k| > a$ *and* $|\tilde{q}_l| > a$, *for some given* $a > 0$. The linear independence of H_r and $\mathcal{E}_r$ then implies that these functions remain linearly independent for γ sufficiently small. Together with the Hamiltonian character of the flow, this implies for γ sufficiently small that the dimension can be reduced from six down to three – the first integral $\mathcal{E}_c$ reduces the system's dimensionality by one, and the Hamiltonian function of this new system reduces it even further. For $\gamma = 0$, $|h_l|$ sufficiently small, and every pair $(|\tilde{q}_k|, |h_l|)$, the reduced system given by (5.76), (5.77) has a hyperbolic periodic orbit, as shown above. Moreover the unstable and stable manifold of this periodic orbit intersect transversally inside the reduced phase space, as demonstrated by the Melnikov calculation in Appendix A. The hyperbolicity and transversality, in addition to the system's and its first integrals' smooth dependence on γ in (5.84), imply the existence of a hyperbolic periodic orbit with a transversal homoclinic orbit, for every pair $(H_r, \mathcal{E}_r)$ and for $\gamma \neq 0$ sufficiently small. Therefore, the dynamical system in (5.84) contains invariant sets (topological horseshoes) with chaotic dynamics; moreover, it does not admit a third analytic first integral independent of H_r and $\mathcal{E}_r$ (Moser, 1973).

To conclude we summarize the two main results in this section. First, the present results are perturbative and show that the dominant dynamic behavior of the perturbed system is essentially identical to that of the unperturbed integrable system. The key parameter that determines the 'amount' of irregular behavior is P_k, which also played an important role in Subsection 5.4.2. Second, the non-integrability result determined in this section suggests that there may be no additional "simple" conserved quantities for the original system in (5.11)–(5.13) besides the energy and the enstrophy. Any additional conserved quantity (for arbitrary h and β) would have to be of a special form to be compatible with the non-integrability result delineated above. Allen *et al.* (1991) have also utilized Melnikov's method to study chaos in the layered equations with topographic stress. These authors study single mode topography with damping and driving as perturbations. Here we utilize the Melnikov method in a completely different context involving two-mode topography in the distinguished regime with $\gamma \ll 1$.

5.5 Numerical experiments

Here we present a numerical study for a wide range of parameter regimes of the *Layered topographic equations in Fourier form* from (5.9)–(5.10). The numerical integration is performed using a fifth-order Runge–Kutta method with adaptive time stepping. Although these solutions are obtained numerically, numerical artifacts are kept to a minimum since the computations involve neither spatial discretization nor artificial smoothing. Both the energy and enstrophy were monitored in all computations and remained constant within a 1% error margin.

5.5.1 Perturbation of single mode topography

By restricting the underlying topography to a single mode, we obtained a global description of the dynamics for $\beta \neq 0$ in Subsection 5.3.3. Here we will study the effect of adding small-scale random perturbations to the system via the numerical solution of the equations in (5.9)–(5.10). Thus, we consider the same basic topographic mode, but superimpose small-scale random perturbations as

$$h(x) = \sin(x) - \frac{i}{2} \sum_{2 \leq k \leq 10} \frac{e^{i(kx+\theta_k)}}{k^p} + \text{c.c.} \tag{5.85}$$

The θ_k are random phase shifts uniformly distributed in $[0, 2\pi)$, whereas p, which controls the smoothness of the small-scale features, is set to $p = 2$. In (5.85) and below, c.c. denotes the complex conjugate.

How well do our previous conclusions for single mode topography with nonzero β extend to a multi-modal situation? A partial answer to this question is provided by the study of the effect of perturbations on the steady states of the unperturbed system (5.43). As in Subsection 5.3.3, we set $\beta = 1$ but perturb the steady state by adding random small-scale features to the initial flow. Thus, the initial conditions are

$$\psi(x, 0) = \bar{a} \sin(x) - \frac{i}{2} \sum_{2 \leq k \leq 10} \frac{e^{i(kx+\phi_k)}}{k^p} + \text{c.c.}, \quad p = 2. \tag{5.86}$$

Akin to the perturbed topography, the ϕ_k are uniformly chosen random phase shifts; the rapid decay of the Fourier components, $\mathcal{O}(k^{-2})$, guarantees a smooth initial flow configuration.

We begin with the two stable steady states shown in Figure 5.5. The first, eastward, corresponds to $\mu \simeq -0.34$, whereas the second, westward, corresponds to $\mu \simeq 0.94$; both $\bar{a}$ and V_0 are determined by μ through (5.44). By adding small scales, we perturb the steady states and observe the evolution of the mean $V(t)$ shown in the top of Figure 5.8. The two mean flows hover about their previously steady average and the effect of small scales remains negligible.

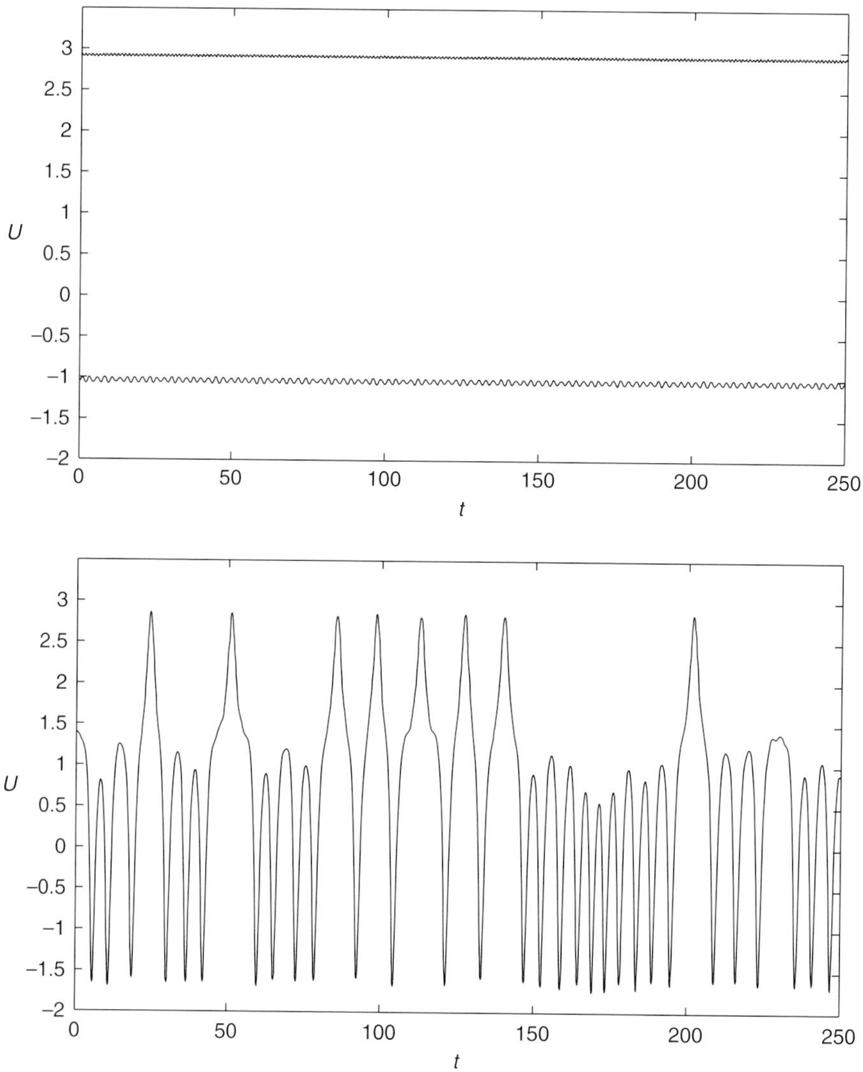

Figure 5.8 The mean $U(t)$ is shown for the randomly perturbed topography in (5.85) and the random initial conditions in (5.86), with $\beta = 1$. Top: perturbed stable equilibria (eastward and westward); bottom: perturbed unstable equilibrium.

Next, we consider the unstable eastward flow configuration from the right frame of Figure 5.6, which corresponds to $\mu \simeq -0.72$. As expected, the added small-scale perturbations throw the mean off its unstable (hyperbolic) equilibrium. The lower frame in Figure 5.8 displays the large deviations of $V(t)$ as it completely reverses the flow direction at random.

To investigate the combined effect of β and multi-mode topography, we consider the same topography (5.85), but with initially eastward flow, $V_0 = 1$. In Figure 5.9 the evolution of the mean is shown for $\beta = 0.1, 1, 5, 25$. Although $\psi(x, y, t)$ was initially set to zero, the interaction of the mean with the topography excites all the smaller scales in the flow, which then feed back into the mean via topographic stress (5.10). For small $\beta = 0.1$, the mean does not display any preferred flow direction and remains almost symmetric about the $V = 0$ axis. As we increase β to $\beta = 1$, the symmetry is broken and $V(t)$ shows a clear eastward bias, while the amplitude of oscillation decreases. For $\beta = 5$, however, smaller oscillations appear in the evolution of the mean, while the oscillation amplitude of the slowest mode, $k = 1$, is greatly reduced so that the flow does not reverse anymore. The depletion of the lowest topographic mode is a typical effect due to larger β. Indeed, for small values of β, most of the energy exchange between the mean and the smaller scales tends to occur via the largest topographic mode (here $k = 1$). Larger values of β typically stimulate the energy exchange among the smallest scales, while the bulk of the energy shifts into the mean (mode $k = 0$).

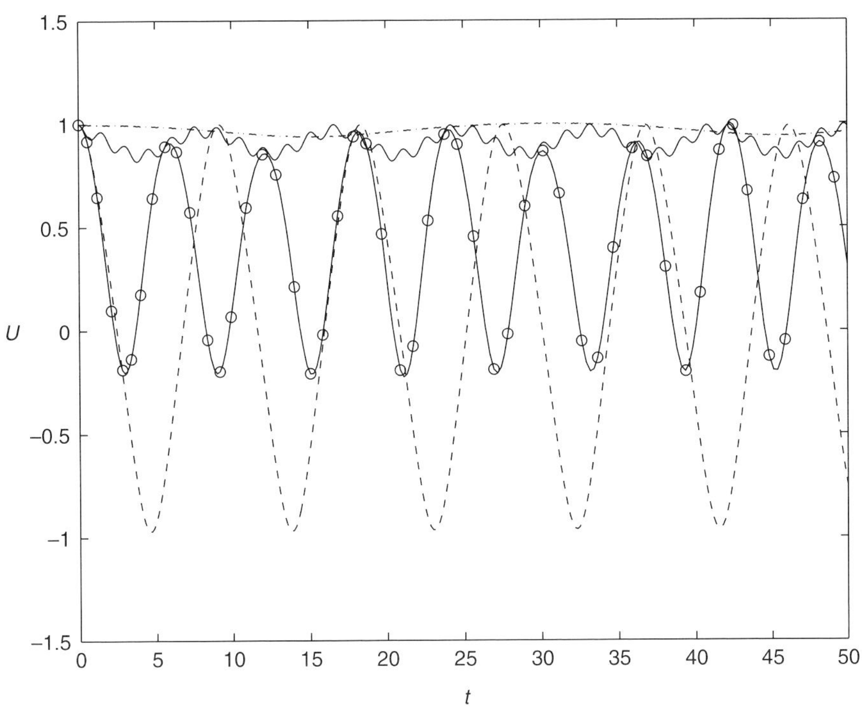

Figure 5.9 The effect of β on the mean $U(t)$ is shown for the randomly perturbed topography in (5.85), with the initial conditions set to $U_0 = 1$ and $\psi = 0$: $\beta = 0.1$ (dash), $\beta = 1$ (circles), $\beta = 5$ (solid), and $\beta = 1$ (dash–dot).

Finally when $\beta = 25$, the solution becomes very stiff and V barely departs off its initial value $V_0 = 1$. This behavior corroborates the theoretical prediction in the limit $\beta \to \infty$, which is derived in Appendix 2. The small-scale mean flow interaction via topographic stress is now suppressed and V remains essentially constant.

5.5.2 Two-mode layered topography and topographic blocking events

Here we show how the reduced dynamics equations in (5.58), (5.59), developed in Section 5.4, can be used to obtain quantitative prediction of chaotic behavior in the full system in (5.9)–(5.10). We recall that the reduced dynamics were obtained after the change of variables (5.52) in the limit $\gamma \to 0$.

We fix $\beta' = 5$ and consider the following two-mode topography and initial conditions

$$h'(x) = \cos x + 2\cos 2x, \quad \tilde{q}(x, 0) = 2\cos x + 2\cos 2x.$$

Iterates of the Poincaré map of the reduced dynamics equations (5.58), (5.59) are shown in Figure 5.10. Two resonance zones clearly appear, centered at $V' = \beta'/k^2$ with $k = 1$ and $k = 2$, respectively. They are surrounded by thin stochastic layers, which break up with increasing distance from the center and merge with the region of global stochasticity. For initial westward flow, $V' < 0$, or strong eastward flow, $V' > 7$, the dynamics is regular and barely affected by the topography; the mean displays quasi-periodic behavior with small departures from its average.

Next, we recast the primed variables into the original variables for a small but finite value of γ

$$h(x) = \gamma h'(x) = \gamma\cos x + 2\gamma\cos 2x, \tag{5.87}$$

$$\psi(x, 0) = (\gamma - 2)\cos x + \frac{\gamma - 1}{2}\cos 2x. \tag{5.88}$$

We choose $\gamma = 0.1$ and calculate the corresponding values of $\beta = \sqrt{\gamma}\beta'$ and $V(0) = \sqrt{\gamma}V'(0)$, given any $V'(0)$. In Figure 5.11, we show the evolution of $V(t)$, obtained by numerical integration of the layered topographic equations (5.9)–(5.10) for three different initial values of the mean $V_0 = V'(0)\sqrt{0.1}$.

In the upper frame, the mean flow starts in the middle of the upper resonance zone at $V_0 = 5\sqrt{0.1}$. As predicted by the reduced dynamics for small γ, in the full system the mean flow displays small oscillatory behavior about the mean and remains trapped in the upper resonance zone for all time. Next, we choose the initial value $V_0' = 3.5$, which lies on the fringe of the upper stochastic region. For quite a long time the mean remains within a stochastic layer centered about the upper resonance zone at $V = \sqrt{\gamma}V'/k^2 \simeq 1.6$, $k = 1$, until it eventually escapes

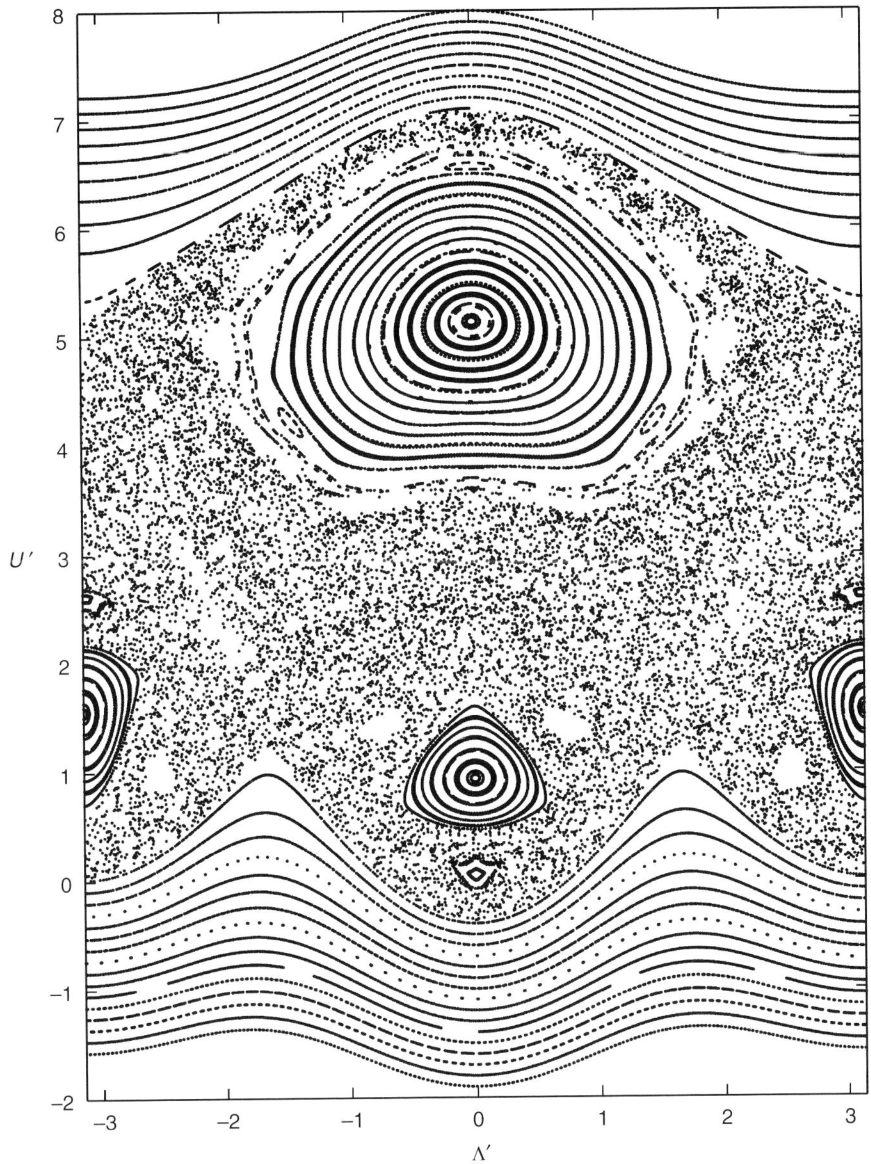

Figure 5.10 Iterates of the Poincaré map of the reduced dynamics equations in (5.58), (5.59) are shown in the (Λ', U') plane, for two-mode topography (5.87) and $\beta' = 5$. The two resonant zones are centered about $U' = \beta'/k^2$, with $k = 1$ for the upper and $k = 2$ for the lower resonance.

into the surrounding stochastic region. It then rapidly reaches the lower resonance zone, centered about $V = \sqrt{0.1}U_0'/k^2 \simeq 0.4$, with $k = 2$.

To reproduce erratic transitions between blocked and unblocked flow configurations with statistics resembling those observed in Weeks *et al.* (1997), we

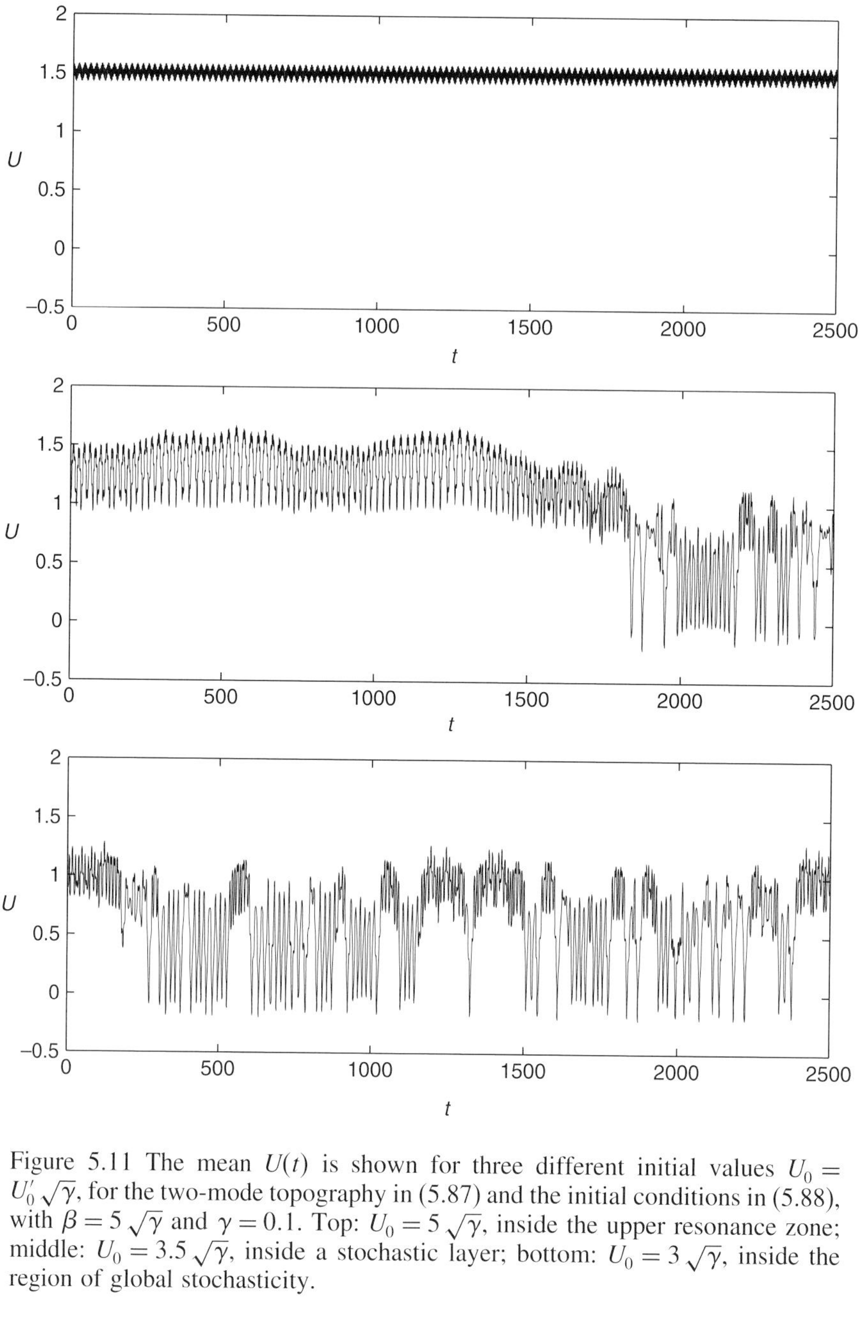

Figure 5.11 The mean $U(t)$ is shown for three different initial values $U_0 = U_0' \sqrt{\gamma}$, for the two-mode topography in (5.87) and the initial conditions in (5.88), with $\beta = 5\sqrt{\gamma}$ and $\gamma = 0.1$. Top: $U_0 = 5\sqrt{\gamma}$, inside the upper resonance zone; middle: $U_0 = 3.5\sqrt{\gamma}$, inside a stochastic layer; bottom: $U_0 = 3\sqrt{\gamma}$, inside the region of global stochasticity.

choose the initial value $V' = 3$ well inside the stochastic region. Now the mean flow clearly exhibits erratic transitions between blocked and unblocked states. Whenever the mean lingers about the lower resonance zone, $k = 2$, the flow is in a blocked state because the mean component is small. On the other hand, whenever

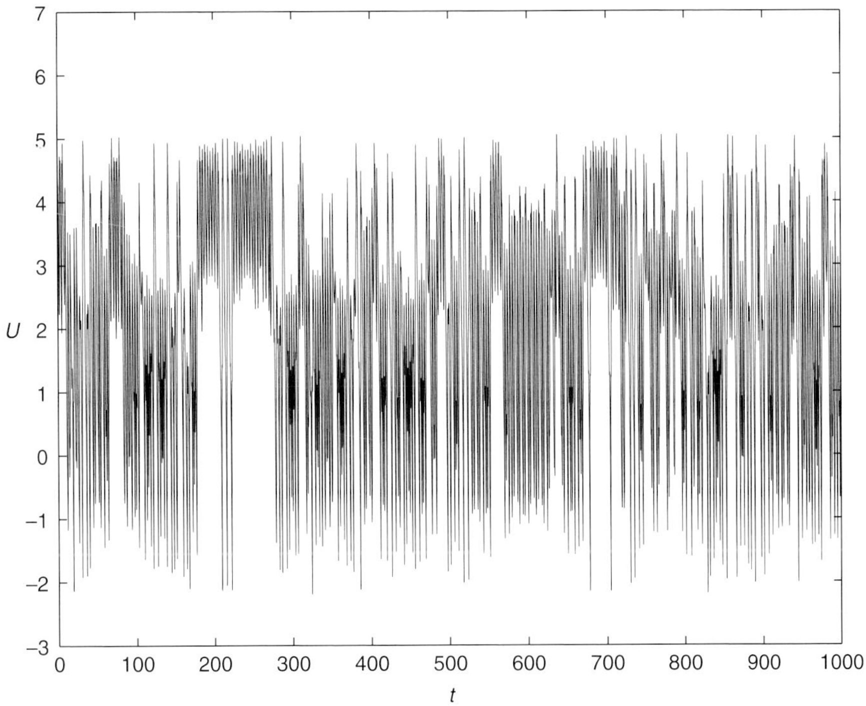

Figure 5.12 The mean $U(t)$ is shown with the initial value $U_0 = 5\sqrt{\gamma}$, $\gamma = 1$, for the two-mode topography in (5.87) and the initial conditions in (5.88), with $\beta = 5$.

the mean hovers about the upper resonance zone, $k = 1$, the flow is in a zonal (unblocked) configuration.

What if we choose a large value of γ? In that case we do not expect the reduced dynamics, obtained in the limit of small γ, to predict the behavior of the full system. Indeed, in Figure 5.12 we show the evolution of $V(t)$ for the same initial conditions as in the upper frame of Figure 5.11, but with $\gamma = 1$, and hence $\beta = 5$ and $V_0 = 5$. Instead of the previously trapped behavior with $\gamma = 0.1$, which confirmed the prediction of the reduced dynamics, the mean now exhibits random oscillatory behavior of very large amplitude. At this stage all stochastic layers that shield the resonance zones have broken apart and merged with the surrounding stochastic region.

5.5.3 Random perturbations with multi-mode topography

From the previous numerical study we might be led to believe that two equally strong modes in the topography are necessary for the onset of erratic transitions between blocked and zonal flow configurations. To show that this is not the case,

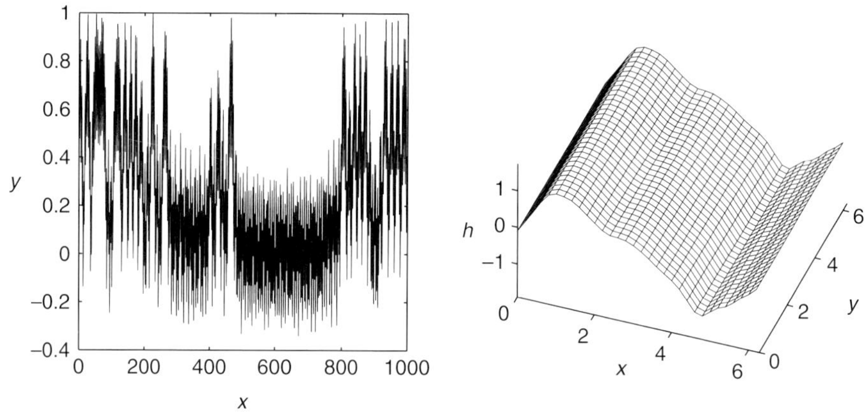

Figure 5.13 Left, the mean $U(t)$ is shown for the initial flow configuration defined by (5.89) and $U_0 = 1$, with $\beta = 2$; right, the underlying random multi-mode topography in (5.85).

we consider a ubiquitous smooth topography with ten random initial modes, as described in (5.85) with $p = 2$. It is shown in the left frame of Figure 5.13. The initial conditions are chosen in a similar fashion with different random phase shifts

$$\psi(x, y, 0) = -i\frac{1}{2} \sum_{3 \le k \le 10} \frac{e^{i\varphi_k} e^{ikx}}{k^p} + \text{c.c.}, \tag{5.89}$$

yet with the first two components set to zero. Hence the initial conditions represent a small $O(0.1)$ perturbation of the background mean flow. We set $\beta = 2$, $V(0) = 1$, and observe the evolution of $V(t)$ in the right frame of Figure 5.13, as it displays again erratic transitions between blocked and unblocked states. The initial and final total velocity fields are depicted in Figure 5.14, and they confirm the initially zonal and subsequent completely blocked flow configurations which occur for erratic intervals of time.

5.5.4 Symmetry breaking perturbations and topographic blocking events

Throughout this chapter we have restricted ourselves to the special exact solutions from the *Layered topographic equations*. Thereby we eliminated the non-linear vortical interactions in the flow, and effectively isolated the non-linear topographic interaction with the mean. Hence, none of the flows studied so far contained any vortices. Therefore, the relevance of topographic stress as a mechanism for topographic blocking events remains to be addressed within the full non-linear framework. In this section we present some preliminary numerical experiments in the full non-linear context of the equations in (5.1)–(5.3).

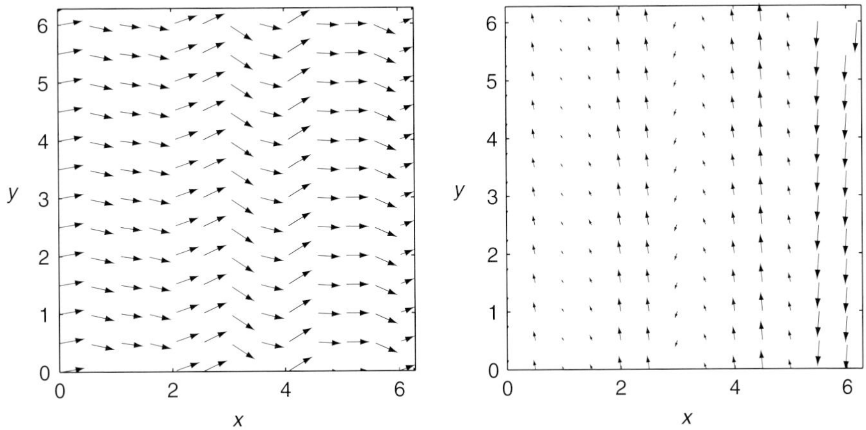

Figure 5.14 The initial and final total velocity fields, $\nabla^{\perp}\Psi(x, y, t)$, are shown for the random multi-mode topography in (5.85), the initial conditions in (5.89), and $U_0 = 1$, with $\beta = 2$: left, unblocked zonal flow at $t = 0$; right, blocked flow at $t = 1000$.

To avoid adding artificial smoothing or viscosity and instead remain within the realm of "exact" energy-conserving numerical schemes, we consider the truncated dynamics equations, which are obtained by projecting the equations in (5.1)–(5.3) on a subspace spanned by a finite number of Fourier modes

$$\psi(x, y, t) = \sum_{0<|\vec{k}|^2\leq\Lambda_k} \psi_{\vec{k}}(t)e^{i\vec{k}\cdot\vec{x}}.$$

We emphasize that here $\vec{k}$ is not restricted only to layered modes. In general we cannot expect that truncating conserved quantities will yield conserved quantities of the truncated equations. However, it is remarkable that, for the quadratic conserved quantities, the energy and the enstrophy, their truncations, E_{Λ_k} and $\mathcal{E}_{\Lambda_k}$, yield conserved quantities for the truncated equations. Therefore, the full non-linear system in (5.1)–(5.3), with the Fourier components ψ_k restricted to the modes $0 < |\vec{k}|^2 \leq \Lambda_k$, conserves both the truncated energy and the truncated enstrophy

$$E_{\Lambda_k} = \frac{V^2}{2} + \frac{1}{2} \sum_{0<|\vec{k}|^2\leq\Lambda_k} |\vec{k}|^2|\psi_{\vec{k}}|^2,$$

$$\mathcal{E}_{\Lambda_k} = \beta V + \frac{1}{2} \sum_{0<|\vec{k}|^2<\Lambda_k} |h_{\vec{k}} - |\vec{k}|^2\psi_{\vec{k}}|^2.$$

The numerical solution of the truncated dynamics equations is computed using a standard spectral method with fourth-order Runge–Kutta time integration.

Following Subsection 5.5.2, we consider the two-mode topography (5.87) and the initial conditions (5.88) with $\gamma = 0.1$. Thus $\beta = 5\sqrt{0.1}$ and we set $V_0 = 3\sqrt{0.1}$, which corresponds to the situation depicted in the lower frame of Figure 5.11. The mean, initially at $V' = 3$, tends to linger in that zonal state before proceeding towards the lower resonant zone, which corresponds to $\vec{k} = (2, 0)$. Since the layered solutions used in Subsection 5.5.2 are exact solutions of the full non-linear system, we need to perturb the initial flow configuration to create non-linear vortical interactions in the flow. To do so, we slightly perturb the initial conditions by adding the two vortical modes, $\vec{q} = (0, 1)$ and $\vec{p} = (2, 1)$, which will create a non-linear resonant triad with the mode $\vec{k} = (2, 0)$. These two modes satisfy the criteria for unstable triad resonance (Vallis, 1985)

$$\vec{q} + \vec{k} = \vec{p}, \tag{5.90}$$

$$|\vec{q}|^2 < |\vec{k}|^2 < |\vec{p}|^2. \tag{5.91}$$

Thus we set the small-scale initial conditions to

$$\begin{aligned} \psi(x, y, 0) = &-1.9\cos x - 4.5\cos 2x + 0.005[\sin(y + 0.24) \\ &+ \sin(2x + y + 0.97)], \end{aligned} \tag{5.92}$$

where the terms in square brackets correspond to the modes $\vec{q}$ and $\vec{p}$, respectively, with random phase shifts.

We begin with a low-dimensional system truncated at $\Lambda_k = 5$, and observe the evolution of the flow. In the top of Figure 5.15 the total stream functions, Ψ, of the initial and final flow configurations are shown. In the lower left frame we show the amount of energy present in the mean and the layered modes as a fraction of the total energy, whereas in the lower right frame we show the evolution of $V(t)$. Up to $t \simeq 50$, the dynamics follows closely the behavior predicted by the study in Subsection 5.5.2: initially most of the energy resides in the mean and the layered modes, while $V(t)$ oscillates about its initial value. As V approaches the second resonance zone, $V \simeq 0.4$, the non-linear triad interaction strongly amplifies the vortical modes, which extract up to 90 % of the total energy. From that point on, the dynamics diverge from the unperturbed solution; the average mean flow is clearly westward but for a few erratic bursts into the lower (eastward) resonance zone. Still, the North–South direction of the total stream function at final time clearly corresponds to a blocked flow situation.

Finally, we repeat the same numerical experiment but set $\Lambda_k = 16$ to include more modes in the truncated dynamics system. The initial and final total stream functions, Ψ, are depicted in the upper two frames in Figure 5.16, while the evolution of $V(t)$ is shown in the lower right frame. Again up to $t \simeq 50$, most

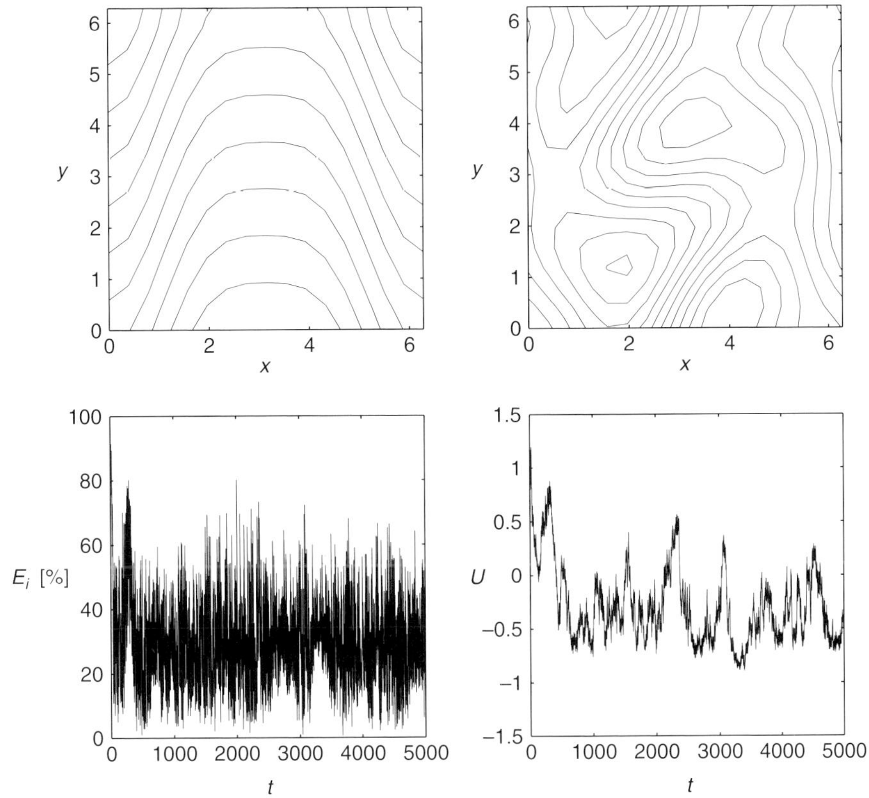

Figure 5.15 Non-linear truncated dynamics are shown for the two-mode topography in (5.87) and the initial conditions in (5.92), with $\Lambda_k = 5$. Top: the total stream function at $t = 0$ (left) and $t = 5000$ (right); bottom left, the energy contained in the mean and the layered modes as a fraction of the total energy (in %); bottom right, the mean $U(t)$.

of the energy resides in the layered modes and the mean as V slowly proceeds towards the resonant value $V \simeq 0.4$. At this stage up to 60% of the energy is suddenly transferred into the non-linear vortical modes due to the resonant triad, and the dynamics diverge from the un perturbed case. At later times the mean completely reverses and lingers about an average value slightly below -1, where the flow configuration typically consists of a westward jet meandering between arrays of vortices. The large-scale westward component is consistent with the theoretical prediction for the statistically most probable flow configuration of barotropic flow on a beta-plane (Carnevale and Frederiksen, 1987). From the lower left frame in Figure 5.16, we note that the energy exchange between the layered and the vortical (non-layered) modes, with about 50% of the energy

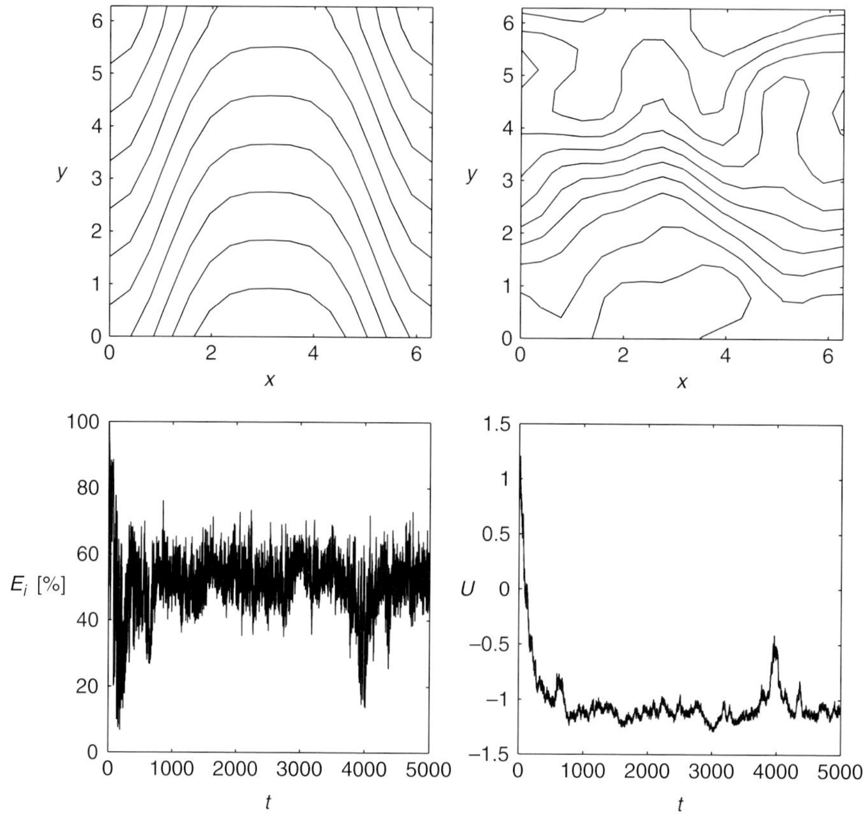

Figure 5.16 Non-linear truncated dynamics are shown for the two-mode topography in (5.87) and the initial conditions in (5.92), with $\Lambda_k = 16$. Top: the total stream function at $t = 0$ (left) and $t = 5000$ (right); bottom left, the energy contained in the mean and the layered modes as a fraction of the total energy (in %); bottom right, the mean $U(t)$.

contained in each of one them, is not quite as violent as in the previous case with less resolution. For the larger number of Fourier modes in the second example, the resonant instability mechanism in (5.90) and (5.91) is less prominent as more energy is allowed to feed back into the layered configuration through other triad interactions.

References

Allen, J. S., Samelson, R. M., and Newberger, P. A. (1991), Chaos in a model of forced quasigeostrophic flow over topography: an application of Melnikov's method. *J. Fluid Mech.* **226**, 511–547.

Arnold, V. I. (1989), *Mathematical Methods of Classical Mechanics*, 2nd edn. New York: Springer-Verlag.

Arnold, V. I., Kozlov, V. V., and Neishtadt, A. I. (1997), *Mathematical Aspects of Classical and Celestial Mechanics*, 2nd. edn. New York: Springer-Verlag.

Carnevale, G. and Frederiksen, J. (1987), Nonlinear stability and statistical mechanics of flow over topography. *J. Fluid Mech.* **175**, 157–181.

Charney, J. G. and DeVore, J. G. (1979), Multiple flow equilibria in the atmosphere and blocking. *J. Atmos. Sci.* **36**, 1205–1216.

Chirikov, B. V. (1979), A universal instability of many dimensional oscillator systems. *Phys. Rep.* **52**, 263–379.

Denzler, J. (1987), Mather sets for plane Hamiltonian systems. *J. Appl. Math. Phys. (ZAMP)* **38**, 791–812.

Edy, M. and Holloway, G. (1994), Sensitivity of a large-scale ocean model to a parametrization of topographic stress. *J. Phys. Oceanogr* **24**, 2577–1046.

Embid, P. and Majda, A. J. (1996), Averaging over fast gravity waves for geophysical flows with arbitrary potential vorticity. *Comm. Partial Diff. Eqs.* **21**, 619–658.

Escande, D. F. (1985), Stochasticity in classical hamiltonian systems: universal aspects. *Phys. Reports* **121**(3 & 4), 165–261.

Grote, M. A., Majda, A. J., and Ragazzo, C. G. (1999), Dynamic mean flow and small-scale interaction through topographic stress. *J. Nonlinear Sci.* **9**, 89–130.

Guckenheimer, J. and Holmes, P. (1990), *Nonlinear Oscillations, Dynamical Systems, and Bifurcations of Vector Fields*. New York: Springer-Verlag.

Hart, J. E. (1979), Barotropic quasi-geostrophic flow over anisotropic mountains. *J. Atmos. Sci.* **36**, 1736–1746.

Holloway, G. (1987), Systematic forcing of large-scale geophysical flows by eddy-topography interaction. *J. Fluid Mech.* **184**, 463–476.

Kaper, T. J. and Kovacic, G. (1994), Geometric criterion for adiabatic chaos. *J. Math. Phys.* **35**, 1202–1217.

Mather, J. N. (1993), Variational construction of connecting orbits. *Ann. Inst. Four.* **43**, 1349–1386.

McLaughlin, D. and Overman, E. (1995), Whiskered tori for integrable pde's: chaotic behavior in near integrable pde's. *Surveys in Appl. Math.* **1**, 83–200.

Meiss, J. D. (1992), Symplectic maps, variational principles, and transport. *Rev. of Modern Phys.* **64**, 795–848.

Menyuk, C. R. (1985), Particle Motion in the field of a modulated wave. *Phys. Rev. A* **31**, 3282–3290.

Moser, J. (1973), *Stable and Random Motion in Dynamical Systems*. Princeton, NJ: Princeton University Press.

Pedlosky, J. (1987), *Geophysical Fluid Dynamics*, 2nd ed. New York: Springer-Verlag.

Samelson, R. M. and Allen, J. S. (1987), Quasi-geostrophic topographically generated mean flow over the continental margin. *J. Phys. Oceanogr* **17**, 2043–2064.

Vallis, G. K. (1985), Instability and flow over topography. *Geophys. Astrophys. Fluid Dynamics* **34**, 1–38.

Weeks, E. R., Tian, Y., Urbach, J. S., Ide, K., Swinney, H. L., and Ghil, M. (1997), Transitions between blocked zonal flows in a rotating annulus with topography. *Science*, **278**, 1598–1601.

Zaslavsky, G. M. (1994), The width of the exponentially narrow stochastic layers. *Chaos* **4**, 589–591.

Zaslavsky, G. M. and Filonenko, N. N. (1968), Stochastic instability of trapped particles and conditions of applicability of the quasi-linear approximation. *Sov. Phys. JETP* **25**, 851–857.

Zaslavsky, G. M., Sagdeev, R. Z., Usikov, D. A., and Chernikov, A. A. (1991), *Weak Chaos and Quasi-Regular Patterns*. Cambridge: Cambridge University Press.

Appendix 1

Here we outline a standard Melnikov integral calculation, which leads to (5.82). First, we define the variable

$$z = kV - \frac{\beta}{k}.$$

In terms of z equations (5.78) and (5.79) reduce to

$$\begin{aligned}
\dot{z} &= -2|h_k||\tilde{q}_k|\sin(\Lambda') - 2\frac{|h_l||\tilde{q}_l|k}{l}\sin(c_1\Lambda' + c_2 t + c_3),\\
\dot{\Lambda}' &= z.
\end{aligned}$$

For $|h_l| = 0$ these equations have the following homoclinic solution to the equilibrium $(z, \Lambda') = (0, \pm\pi)$

$$\Lambda'(t) = 4\arctan\left[\exp\left(\sqrt{2|h_k||\tilde{q}_k|}\right)t\right] - \pi, \quad z(t) = \dot{\Lambda}'.$$

In this case the Melnikov integral is given by (Guckenheimer and Holmes, 1990)

$$M(t_0) = 2\frac{|\tilde{q}_l|k}{l}\int_{-\infty}^{+\infty}\dot{\Lambda}'(t)\sin(c_1\Lambda'(t) + c_2 t + c_3 + t_0)\,dt.$$

Since $\Lambda'(t)$ is odd for this homoclinic solution, we conclude that $M(-c_3) = 0$. Therefore

$$\begin{aligned}
\dot{M}(-c_3) &= 2\frac{|\tilde{q}_l|k}{l}\int_{-\infty}^{+\infty}\dot{\Lambda}'(t)\cos(c_1\Lambda'(t) + c_2 t)\,dt,\\
&= 2\frac{|\tilde{q}_l|k}{l}\int_{-\infty}^{+\infty}\dot{\Lambda}'(t)\exp(ic_1\Lambda'(t))\exp(ic_2 t)\,dt.
\end{aligned}$$

To simplify the algebra we restrict ourselves to the case

$$c_1 = \frac{l}{k} = \frac{n}{2}, \quad n = 1, 2, \ldots$$

From the identity

$$\arctan(u) = \frac{1}{2i}\ln\frac{1+iu}{1-iu}, \quad |u| < 1,$$

and the definition of $\dot{\Lambda}'(t)$, we write the above integral as

$$\dot{M}(-c_3) = 8\alpha\frac{|\tilde{q}_l|}{l}i^n\int_{-\infty}^{+\infty}\exp(\alpha t)\frac{[-i+\exp(\alpha t)]^{n-1}}{[i+\exp(\alpha t)]^{n+1}}\exp(ic_2 t)\,dt,$$

where $\alpha = \sqrt{2|h_k||\tilde{q}_k|}$. By extending the functions in this integral to the complex plane, integrating over an "infinite rectangle" of sides $\mathrm{Im}(t) = 0$ and $\mathrm{Im}(t) = 2\pi/\alpha$, and using residues, we obtain

$$\dot{M}(-c_3) = 8\alpha \frac{|\tilde{q}_l|k}{l} i^n \frac{2\pi i}{1 - \exp(-2\pi c_2/\alpha)} R. \tag{5.93}$$

Here R is the residue of

$$\exp(\alpha t) \frac{[-i + \exp(\alpha t)]^{n-1}}{[i + \exp(\alpha t)]^{n+1}} \exp(i c_2 t)$$

at the pole $t = 3\pi i/(2\alpha)$, whose order increases with n. We do not have an expression for R for a generic n. For instance, for $n = 1$

$$R_1 = -\frac{c_2}{\alpha^2} \exp[-c_2 3\pi/(2\alpha)], \tag{5.94}$$

and for $n = 4$

$$R_4 = -i \left[-\frac{2c_2^2}{3\alpha^3} + \frac{c_2^4}{3\alpha^5} \right] \exp[-c_2 3\pi/(2\alpha)].$$

Appendix 2

We briefly outline the fast wave averaging procedure to compute the asymptotic behavior in the limit $\beta \to \infty$. The rigorous validity of this method within a more general framework can be found in Embid and Majda (1996). First, we consider the general system with arbitrary topography (5.1)–(5.3) and expand the periodic small variables ψ and h in Fourier series

$$\psi(x, y, t) = \sum_{\vec{k} \neq 0} \psi_{\vec{k}}(t)\, e^{i\vec{k}\cdot\vec{x}}, \quad h(x, y) = \sum_{\vec{k} \neq 0} h_{\vec{k}}\, e^{i\vec{k}\cdot\vec{x}}, \quad \vec{x} = (x, y).$$

Next, we set $\varepsilon = \beta^{-1}$, and rewrite (5.1), (5.3) componentwise as

$$\dot{\psi}_{\vec{k}} = -\frac{1}{|\vec{k}|^2} \sum_{\vec{k}_1 + \vec{k}_2 = \vec{k}} \vec{k}_1^{\perp} \cdot \vec{k}_2 \psi_{\vec{k}_1} (-|\vec{k}_2|^2 \psi_{\vec{k}_2} + h_{k_2}) \tag{5.95}$$

$$-Vik_x \left(\psi_{\vec{k}} + \frac{h_{\vec{k}}}{|\vec{k}|^2} \right) + \frac{i}{\varepsilon} \frac{k_x}{|\vec{k}|^2} \psi_{\vec{k}}, \tag{5.96}$$

$$\dot{V} = \sum_{\vec{k} \neq 0} ik_x \psi_{\vec{k}} h_{-\vec{k}}, \qquad \vec{k} = (k_x, k_y).$$

We now wish to study the formal asymptotic limit $\varepsilon \to 0(\beta \to \infty)$. The leading order term, $O(\varepsilon^{-1})$, suggests to separate the fast time scale, t/ε, from the slow time scale t, and thus we write

$$\psi_{\vec{k}}(t) = e^{i\omega(\vec{k})t/\varepsilon} A_{\vec{k}}(t), \quad \omega(\vec{k}) = \frac{k_x}{|\vec{k}|^2}.$$

The next order ($O(1)$) terms yield the:

Reduced dynamical equations to leading order

$$\dot{A}_{\vec{k}} = \frac{1}{|\vec{k}|^2} \sum_{\vec{k}_1+\vec{k}_2=\vec{k},\, \omega(\vec{k}_1)+\omega(\vec{k}_2)=\omega(\vec{k})} \vec{k}_1^{\perp} \cdot \vec{k}_2 |\vec{k}_2|^2 A_{\vec{k}_1} A_{\vec{k}_2} \tag{5.97}$$

$$-\frac{1}{|\vec{k}|^2} \sum_{\vec{k}_1+\vec{k}_2=\vec{k},\, \omega(\vec{k}_1)=\omega(\vec{k})} \vec{k}_1^{\perp} \cdot \vec{k}_2 A_{\vec{k}_1} h_{\vec{k}_2} - Uik_x A_{\vec{k}}, \tag{5.98}$$

$$\dot{V} = 0. \tag{5.99}$$

We note the presence of resonant triads for the Rossby waves

$$\vec{k}_1 + \vec{k}_2 = \vec{k}, \quad \omega(\vec{k}_1) + \omega(\vec{k}_2) = \omega(\vec{k}),$$

and waves scattered by topography

$$\vec{k}_1 + \vec{k}_2 = \vec{k}, \quad \omega(\vec{k}_1) = \omega(\vec{k}).$$

However the mean flow topographic stress interaction is completely suppressed for β large enough as U remains constant in time. In the special case of layered solutions of the equations in (5.7), (5.8), the non-linear terms vanish and we obtain the:

Reduced dynamics with layered topography

$$\dot{A}_{\vec{k}} = -Vik_x A_{\vec{k}}, \tag{5.100}$$

$$\dot{U} = 0. \tag{5.101}$$

Therefore, to leading order, the solutions in the asymptotic limit $\beta \to \infty$ are

$$\psi(x, y, t) = \sum_{k\neq 0} \psi_k^0 e^{i\omega(k\vec{l})t/\varepsilon - i\vec{l}_x V_0 t} e^{ik\vec{l}\cdot\vec{x}}, \tag{5.102}$$

$$V(t) = V_0. \tag{5.103}$$

For large β the solution behaves like an array of independent Rossby waves and coincides with that obtained without topography in Subsection 5.3.1, while the mean flow does not change in time as observed in the numerical experiments from Subsection 5.5.1 with $\beta \gg 1$.

6

Introduction to information theory and empirical statistical theory

6.1 Introduction

In this chapter we will lay down the foundations for statistical theories of geophysical flows and predictability that will be developed in the following chapters. The mathematical underpinning for these applications is a systematic use of information theory following Shannon (1948) Shannon and Weaver (1949) and Jaynes (1957). We develop such ideas beginning with elementary examples in this chapter. The motivation for the statistical studies of barotropic quasi-geostrophic flows lies in the generic occurrence of coherent large-scale flow patterns in physical observations and numerical simulations of flows that are approximately two dimensional. Examples of these flows are large isolated eddies in the atmosphere (a well-known example is the great red spot in Jupiter's atmosphere, see Chapter 13), and the discovery of mesoscale eddies in the ocean. Large-scale organized flow patterns emerge under a wide variety of initial conditions of the flow and topography. The robustness of these patterns seems to indicate that these large-scale coherent flows do not depend on the fine details in the dynamics of the flow or the topography. In addition, these flow patterns persist for a long time, and are essentially steady in nature. Therefore, it is plausible that an explanation for the observed two-dimensional coherent patterns can be found with considerations from equilibrium statistical mechanics, where we are interested in the large-scale features of the flow rather than all the fine details, and where only a few bulk properties of the flow, such as averaged conserved quantities like the energy and enstrophy enter the analysis.

The objective of the statistical theories is the predictions of the most probable steady state that will develop in the flow under the constraints imposed by those bulk properties that are retained in the theory. Within the context of geophysical fluids, all the statistical theories considered here will be based on the barotropic quasi-geostrophic equations without dissipation and forcing

$$\frac{\partial q}{\partial t} + \nabla^{\perp}\psi \cdot \nabla q = 0, \quad \frac{dV}{dt}(t) = -\fint \frac{\partial h}{\partial x}\psi', \qquad (6.1)$$
$$q = \Delta\psi + h + \beta y \quad \psi = -V(t)y + \psi'.$$

In the previous chapters, we studied these equations extensively. Among other things, we know that the above equations possess conserved quantities, such as the energy and the generalized enstrophy, which act as constraints on the flow. Furthermore, we studied special stable steady state flows, where the potential vorticity q and the stream function ψ were functionally related

$$\psi = g(q).$$

We also know that these stable steady states satisfy the generalized selective decay principle, whereby the steady state is a minimizer of the generalized enstrophy, subject to the constraint of fixed energy. In later chapters we will see that these stable steady state flows also arise as the most probable states as predicted by some statistical mechanics theories for geophysical flows.

There are many possible ways to formulate a statistical mechanics theory for the most probable state of the large-scale flow; for example, as shown in subsequent chapters, different theories may result depending on the number of conserved quantities imposed as constraints in the theory. However, any statistical formulation must include a way to select the most probable state of the large-scale flow. Here we will adapt the information-theoretical approach developed by Shannon (1948) Shannon and Weaver (1949) and Jaynes (1957). According to the information theory point of view, this probability distribution can be selected in a natural manner as the least biased probability measure that is consistent with the given external constraints imposed in the theory. The notion of bias is itself quantified in an essentially unique way in terms of a functional called the information-theory entropy, and also known as the Shannon entropy as we show below.

6.2 Information theory and Shannon's entropy

The discussion of the statistical theories for large-scale coherent states and predictability presented here utilizes ideas from the information theory developed by Shannon (1948) Shannon and Weaver (1949), and adapted to statistical mechanics by Jaynes (1957). In this alternative conception of statistical mechanics, we do not start with a given probability measure and then do statistical measurements. Instead, we start with the information given by the measurements, and ask for the probability measure that has the least bias (most uncertainty) and is consistent with the measurements. The key fact in the determination of such probability distribution rests on the existence of an essentially unique quantity that measures the amount of uncertainty of a probability distribution; this quantity is the Shannon entropy. Next we define the Shannon entropy for discrete and finite probability measures.

Definition 6.1 *Let p be a finite, discrete probability measure on the sample space $\mathcal{A} = \{a_1, \ldots, a_n\}$*

$$p = \sum_{i=1}^{n} p_i \delta_{a_i}, \quad p_i \geq 0, \quad \sum_{i=1}^{n} p_i = 1. \tag{6.2}$$

The Shannon entropy $\mathcal{S}(p)$ of the probability p is defined as

$$\mathcal{S}(p) = \mathcal{S}(p_1, \ldots, p_n) = -\sum_{i=1}^{n} p_i \ln p_i. \tag{6.3}$$

Here and elsewhere, δ_{a_i} denotes the delta function at a point a_i.

The functional $\mathcal{S}$ is called the information-theoretic entropy because of the formal resemblance of the formula for $\mathcal{S}(p)$ in equation (6.3) to the expression for the entropy of the canonical ensemble in statistical mechanics and because Shannon utilized $\mathcal{S}(p)$ to measure information. It is worthwhile here to briefly recall Shannon's intuition from the theory of communication for the reason that (6.3) measures lack of information.

The simplest such problem involves representing a "word" in a message as a sequence of binary digits with length n, i.e., we need n-digits with length n to characterize it. The set $\mathcal{A}_{2^n}$ of all words of length n has $2^n = N$ elements and, clearly, the amount of information needed to characterize one element is $n = \log_2 N$. Continuing this type of reasoning, it follows that the amount of information needed to characterize an element of any set, $\mathcal{A}_N$, is $\log_2 N$ for general N. Now consider the situation of a set $\mathcal{A} = \mathcal{A}_{N_1} \cup \cdots \cup \mathcal{A}_{N_k}$, where the sets $\mathcal{A}_{N_i}$ are pairwise disjoint from each other with $\mathcal{A}_{N_i}$ having N_i total elements. Set p_i to be given by $p_i = N_i/N$, where $N = \sum N_i$. If we know that an element of $\mathcal{A}$ belongs to some $\mathcal{A}_{N_i}$, then we need $\log_2 N_i$ additional information to determine it completely. Thus, the average amount of information we need to determine an element, provided that we already know the $\mathcal{A}_{N_i}$ to which it belongs, is given by

$$\sum_i \frac{N_i}{N} \log_2 N_i = \sum p_i \log_2 p_i + \log_2 N.$$

Recall from our discussion above that $\log_2 N$ is the information that we need to determine an element given the set $\mathcal{A}$ if we do not know to which $\mathcal{A}_{N_i}$ a given element belongs. Thus, the corresponding average lack of information is

$$-\sum p_i \log_2 p_i$$

and we arrive at (6.3) as a measure of lack of information.

Next we show that the Shannon entropy is essentially unique. For this purpose we first introduce the space $\mathcal{PM}_n(\mathcal{A})$ of discrete probability measures on the sample space $\mathcal{A}$

$$\mathcal{PM}_n(\mathcal{A}) = \left\{ p = \sum_{i=1}^{n} p_i \delta_{a_i}, \quad p_i \geq 0, \quad \sum_{i=1}^{n} p_i = 1 \right\}. \tag{6.4}$$

Proposition 6.1 *(Jaynes, 1957) Let H_n be a function defined on the space of discrete probability measures $\mathcal{PM}_n$, and satisfying the following properties:*

1. *$H_n(p_1, \ldots, p_n)$ is a continuous function.*
2. *$A(n) = H_n(1/n, \ldots, 1/n)$ is monotonic increasing in n, i.e. H_n increases with increasing uncertainty.*
3. *Composition law: If the sample space $\mathcal{A} = \{a_1, \ldots, a_n\}$ is divided into two subsets $\mathcal{A}_1 = \{a_1, \ldots, a_k\}$ and $\mathcal{A}_2 = \{a_{k+1}, \ldots, a_n\}$ with probabilities $w_1 = p_1 + \cdots + p_k$, $w_2 = p_{k+1} + \cdots + p_n$, and conditional probabilities $(p_1/w_1, \ldots, p_k/w_1)$, and $(p_{k+1}/w_2, \ldots, p_n/w_2)$, then the amount of uncertainty with the information split in this way is the same as it was originally*

$$H_n(p_1, \ldots, p_n) = H_2(w_1, w_2) + w_1 H_k(p_1/w_1, \ldots, p_k/w_1) + w_2 H_{n-k}(p_{k+1}/w_2, \ldots, p_n/w_2).$$

Then H_n is a positive multiple of the Shannon entropy

$$H_n(p_1, \ldots, p_n) = K\mathcal{S}(p_1, \ldots, p_n) = -K \sum_{i=1}^{n} p_i \ln p_i \tag{6.5}$$

with $K > 0$.

Proof: Since $H_n(q_1, \ldots, q_n)$ is a continuous function, it is enough to prove that equation (6.5) holds for rational values of $q_1, \ldots, q_n$. Clearly, any set of rational numbers $q_1, \ldots, q_n$, with $0 < q_i \leq 1, i = 1, \ldots, n$, and $\sum_{i=1}^{n} q_i = 1$ can be written as $q_i = \nu_i / N$, where $N = \sum_{i=1}^{n} \nu_i$. Consider now the sample space $\mathcal{A} = \{a_1, \ldots, a_N\}$ with probabilities $\{p_1, \ldots, p_N\}$. We partition $\mathcal{A}$ into n subsets $\mathcal{A}_i = \{a_{k_{i-1}+1}, \ldots, a_{k_i}\}, i = 1, \ldots, n$, where $k_0 = 0$ and $k_i = k_{i-1} + \nu_i, i = 1, \ldots, n$. We also denote by w_i the probability associated with $\mathcal{A}_i$, $w_i = p_{k_{i-1}+1} + \ldots + p_{k_i}$. From property 3 it follows that

$$H_N(p_1, \ldots, p_N) = H_n(w_1, \ldots, w_n) + \sum_{i=1}^{n} w_i H_{\nu_i}(p_{k_{i-1}+1}/w_i, \ldots, p_{k_i}/w_i). \quad (6.6)$$

In particular, if we let $p_i = 1/N, i = 1, \ldots, N$, then $w_i = n_i/n = q_i$, and equation (6.6) yields

$$H_N(1/N, \ldots, 1/N) = H_n(q_1, \ldots, q_n) + \sum_{i=1}^{n} q_i H_{\nu_i}(1/\nu_i, \ldots, 1/\nu_i). \quad (6.7)$$

Utilizing property 2, and recalling that $A(n) = H_n(1/n, \ldots, 1/n)$, then equation (6.7) reduces to

$$A(N) = H_n(q_1, \ldots, q_n) + \sum_{i=1}^{n} q_i A(\nu_i). \quad (6.8)$$

In particular, let us set all $\nu_i = \nu$, then $N = n\nu$ and $q_i = \nu_i/N = 1/N$ for all i. In this case equation (6.8) reduces to

$$A(n\nu) = A(n) + A(\nu). \quad (6.9)$$

However, it is a well known fact that the only continuous function $A(n)$ satisfying this condition is

$$A(n) = K \ln n, \quad (6.10)$$

where the constant K is chosen to be positive so that $A(n)$ is monotonically increasing, as required by property 2. Finally, inserting this equation back into equation (6.8) and solving for $H_n(q_1, \ldots, q_n)$ yields

$$H_n(q_1, \ldots, q_n) = K \ln n - K \sum_{i=1}^{n} q_i \ln n_i = -K \sum_{i=1}^{n} q_i \ln q_i. \quad (6.11)$$

This concludes the proof of the proposition.

With a given probability measure $p \in \mathcal{PM}(\mathcal{A})$, various statistical measurements can be made with respect to this probability measure p. We recall that the expected value, or statistical measurement, of f with respect to p is given by

$$\langle f \rangle_p = \sum_{i=1}^{n} f(a_i) p_i. \quad (6.12)$$

The main principle of the subjective probability theory of Jaynes is to seek the least biased probability distribution consistent with the constraints imposed by the information as given; for example, by the statistical measurements of certain functions $f_j, j = 1, \ldots, r$, as in equation (6.12). In terms of the Shannon entropy this principle is known as the maximum entropy principle.

Definition 6.2 (Empirical maximum entropy principle) *Given a set of constraints $\mathcal{C}$ defined by*

$$\mathcal{C} = \{p \in \mathcal{PM}(\mathcal{A}) | \langle f_j \rangle_p = F_j, 1 \leq j \leq r\}, \tag{6.13}$$

the least biased probability distribution $p^* \in \mathcal{C}$ is given by maximizing the Shannon entropy $\mathcal{S}(p)$ subject to the constraints imposed by the statistical measurements of $f_j, j = 1, \ldots, r$ given by $\mathcal{C}$

$$\max_{p \in \mathcal{C}} \mathcal{S}(p) = \mathcal{S}(p^*), \quad p^* \in \mathcal{C}. \tag{6.14}$$

Next we consider several examples to illustrate the maximum entropy principle. Here in this chapter and elsewhere, we will proceed formally in seeking the maximum for (6.14) rather than rigorously proving that such a maximum exists (see Lasota and Mackey, 1994, for this type of rigorous argument.

Example 1: Find the least biased probability distribution p on $\mathcal{A} = \{a_1, \ldots, a_n\}$ with no additional constraints.

Since in this example there is no additional information available besides the fact that p is a probability measure, we expect that the least biased probability measure is going to be the uniform measure which assigns the same probability to every point in the sample space $\mathcal{A}$. We will verify this expectation by maximizing the Shannon entropy $\mathcal{S}(p_1, \ldots, p_n) = -\sum_{i=1}^{n} p_i \ln p_i$ subject only to the constraints that $p_i \geq 0$ and that $\sum_{i=1}^{n} p_i = 1$. By the Lagrange multiplier rule there is a multiplier λ so that the minimum $p = p^*$ satisfies

$$-\nabla_p \mathcal{S} + \lambda \nabla_p \left(\sum_{i=1}^{n} p_i \right) \bigg|_{p=p^*} = 0, \tag{6.15}$$

and subject to the constraint of being a probability measure. Componentwise, equation (6.15) yields n equations that must be satisfied by $p = p^*$

$$\ln p_i^* + 1 + \lambda = 0, i = 1, \ldots, n,$$

and this implies that all the probabilities $p_i^*, i = 1, \ldots, n$, are equal. Since the sum of the probabilities p_i^* is one, we conclude that

$$p_i^* = 1/n, \tag{6.16}$$

and conclude that the least biased measure is the uniform measure.

Example 2: Find the least biased probability measure that is consistent with a finite number $r \leq n-1$ of statistical measurements F_j of given functions f_j, $j = 1, \dots, r$

$$F_j = \langle f_j \rangle_p = \sum_{i=1}^{n} f_j(a_i) p_i, \quad j = 1, \dots, r.$$

In this example we want to maximize the Shannon entropy $\mathcal{S}(p_1, \dots, p_n) = -\sum_{i=1}^{n} p_i \ln p_i$ subject to the $r+1$ constraints

$$F_j = \langle f_j \rangle_p = \sum_{i=1}^{n} f_j(a_i) p_i, \quad j = 1, \dots, r, \quad \sum_{i=1}^{n} p_i = 1. \tag{6.17}$$

Notice that we restricted the number of additional constraints to be $r \leq n-1$, so that the system of algebraic equations (6.17) is not over-determined. Once more, the Lagrange multiplier rule asserts the existence of $r+1$ multipliers λ_0 and λ_j, $j = 1, \dots, r$ such that

$$-\nabla_p \mathcal{S} + \sum_{j=1}^{r} \lambda_j \nabla_p \langle f_j \rangle_p + \lambda_0 \nabla_p \left(\sum_{i=1}^{n} p_i \right) \Bigg|_{p^*} = 0. \tag{6.18}$$

Componentwise equation (6.18) yields a system of n equations for the unknowns p_i^*

$$\ln p_i^* = -\sum_{j=1}^{r} \lambda_j f_j(a_i) - (\lambda_0 + 1), \quad i = 1, \dots, n,$$

and solving for p_i^* we obtain

$$p_i^* = \exp\left(-\sum_{j=1}^{r} \lambda_j f_j(a_i) - (\lambda_0 + 1) \right). \tag{6.19}$$

To eliminate the multiplier λ_0, we utilize the constraint that the sum of all the probabilities p_i^* is 1. This simplifies the formula for p_i^* in equation (6.19) to

$$p_i^* = \frac{\exp\left(-\sum_{j=1}^{r} \lambda_j f_j(a_i)\right)}{\sum_{i=1}^{n} \exp\left(-\sum_{j=1}^{r} \lambda_j f_j(a_i)\right)}. \tag{6.20}$$

The other constraints $\lambda_i, i = 1, \dots, r$ are obtained by solving for the remaining constraint equations for the measurements $\langle f_j \rangle_p$ in equation (6.17). Interestingly, if we define the partition function $\mathcal{Z}(\vec{\lambda})$ by

$$\mathcal{Z}(\vec{\lambda}) = \sum_{i=1}^{n} \exp\left(-\sum_{j=1}^{r} \lambda_j f_j(a_i) \right), \tag{6.21}$$

then $\mathcal{Z}(\vec{\lambda})$ satisfies

$$-\frac{\partial}{\partial \lambda_j} \ln \mathcal{Z}(\vec{\lambda}) = \langle f_j \rangle_{p^*}. \tag{6.22}$$

In this fashion we have recovered the partition function $\mathcal{Z}(\vec{\lambda})$ of statistical mechanics, utilizing only the maximum entropy principle from information theory. See the book by Thompson (1972) for an excellent elementary introduction to mathematical statistical mechanics.

6.3 Most probable states with prior distribution

In the previous section we discussed the maximum entropy principle for the case where the only additional constraints are given in terms of a finite number of statistical measurements. However, there are many physically interesting cases where additional information cannot be formulated in terms of statistical measurement constraints. One such case occurs when there is an additional external bias.

To illustrate the presence of an external bias we consider the following example. Let us assume that the sample space $\mathcal{A}$ is the union of two disjoint subsets, $\mathcal{A}_1 = \{1, 2, 3\}$, with three sample points, and $\mathcal{A}_2 = \{4, 5\}$, with two sample points. In addition, assume that we know that these two sample subspaces are unrelated in the sense that any information we know about either set does not give additional information about the other set; this constraint constitutes an additional external bias. Now we ask for the least biased probability measure p consistent with the given constraints. The obvious guess for p is the uniform distribution, with $p_j = 1/5$ for $j = 1, \dots, 5$. However this answer is unlikely to give the least biased distribution; since $\mathcal{A}_1$ and $\mathcal{A}_2$ are unrelated, the maximum entropy principle applied to the probability distributions on $\mathcal{A}_1$ yields the uniform distribution $p_j = 1/3$, for $j = 1, 2, 3$, but, when it is applied to the probability distributions on $\mathcal{A}_2$, it yields the uniform distribution $p_j = 1/2$, for $j = 4, 5$. The obvious guess will not work because it ignores the additional constraint imposed on the sample space. These considerations are extended in the following example.

Example 3: Let $\mathcal{A} = \mathcal{A}_1 \cup \mathcal{A}_2$, where $\mathcal{A}_1, \mathcal{A}_2$ represent disjoint sample spaces which are completely unrelated

$$\mathcal{A}_1 = \{a_1, \dots, a_l\}, \quad \mathcal{A}_2 = \{a_{l+1}, \dots, a_n\}.$$

By applying the maximum entropy principle to each set, we get the least biased measure for each set, i.e.

$$p_0^{(1)} = \sum_{j=1}^{l} \frac{1}{l} \delta_{a_j}$$

is the natural measure on $\mathcal{A}_1$ and

$$p_0^{(2)} = \sum_{j=1}^{n-l} \frac{1}{n-l} \delta_{a_{l+j}}$$

is the natural measure on $\mathcal{A}_2$

If we know that statistical measurements on each set are equally important, then the least biased probability measure for the whole set should be

$$p_0 = \frac{1}{2} \sum_{j=1}^{l} \frac{1}{l} \delta_{a_j} + \frac{1}{2} \sum_{j=1}^{n-l} \frac{1}{n-l} \delta_{a_{l+j}},$$

or, more generally, possibly two disjoint sets may each have a different importance. In this case, we have a weighted probability measure.

$$p_0 = \gamma p_0^{(1)} + (1-\gamma) p_0^{(2)}, \quad 0 < \gamma < 1.$$

Situations where external bias considerations are applicable do occur naturally in the context of geophysical fluids, as in the case of flows with additional symmetries. One example of flows with additional symmetries is given by the array of Taylor vortices introduced in Chapter 1 with stream-function $\psi = \sin x \sin y$. The streamlines of this flow are depicted in Figure 6.1 over one period cell. Now consider a passive tracer, T, which satisfies

$$\frac{\partial T}{\partial t} + \nabla^{\perp} \psi \cdot \nabla T = 0.$$

Because of the additional symmetries, the period cell is subdivided into four smaller period cells; since the streamlines determine the flow, it is clear that the tracer behavior in each sub-cell is independent from the other sub-cells, and whatever information we have about a given sub-cell is unrelated to the other neighboring cells. In particular, if a tracer initially is confined to one sub-cell, measurements should be performed reflecting this external bias.

The above discussion makes it clear that the maximum entropy principle needs to be extended to accommodate the existence of additional external bias. Suppose we already have the external bias, i.e. we know the weighted importance (or probability) of each sample point, which is given as

$$p_0 = \sum_{i=1}^{n} p_i^0 \delta_{a_i}. \tag{6.23}$$

Also we assume that we know r additional constraints involving other measurements, i.e.

$$F_j = \langle f_j \rangle_p, \quad j = 1, \ldots, r. \tag{6.24}$$

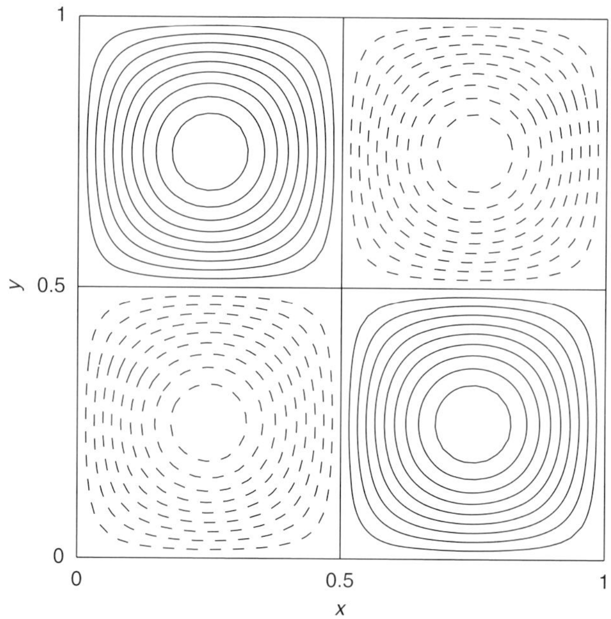

Figure 6.1 Streamlines of Taylor vortices with stream function $\psi = \sin x \sin y$. Counter-clockwise rotation (—); clockwise rotation (- -). x and y are both normalized by 2π.

The goal now is to select the least biased probability distribution p^* consistent with the given measurements while retaining the external bias expressed by p_0. In order to accomplish this goal, we modify the definition of Shannon entropy so that it can take into account the external bias. For this purpose we define the relative Shannon entropy $\mathcal{S}(p, p_0)$ by

$$\mathcal{S}(p, p_0) = -\sum_{i=1}^{n} p_i \ln\left(\frac{p_i}{p_i^0}\right), \tag{6.25}$$

and then we reformulate the maximum entropy principle in terms of the relative entropy.

Definition 6.3 (Maximum relative entropy principle) *The least biased probability measure p^*, given all the constraints and external bias, is the one that satisfies*

$$\max_{p \in \mathcal{C}} \mathcal{S}(p, p_0) = \mathcal{S}(p^*, p_0). \tag{6.26}$$

The above definitions of relative entropy and the maximum relative entropy principle represent reasonable extensions of the previous notions of entropy and the maximum entropy principle, provided that they satisfy the following two conditions. First, if there is no real external bias, i.e. if p_0 is given by the uniform distribution

$$p_0 = \sum_{i=1}^{n} \frac{1}{n} \delta_{a_i},$$

then the maximum relative entropy principle should yield the same probability distribution p^* as the maximum entropy principle from the previous section. Indeed, if $p_0 = \sum_{i=1}^{n} \frac{1}{n} \delta_{a_i}$, then the relative entropy $\mathcal{S}(p, p_0)$ differs from the entropy $\mathcal{S}(p)$ by a constant, $\mathcal{S}(p, p_0) = \mathcal{S}(p) - \ln n$. Therefore p^* maximizes both the entropy and the relative entropy.

The second condition that must be satisfied is the following. If the only constraint on the system is given by the external bias p_0, then the probability distribution predicted by the maximum relative entropy principle must be the external bias itself, $p^* = p_0$. Indeed, in this case the constraint $\mathcal{C}$ is given by $\mathcal{C} = \{p | p \in \mathcal{PM}(\mathcal{A})\}$. Thus, the least biased measure p^*, restricted only by the external condition, should be the original bias p_0 such that

$$\max_{p \in \mathcal{C}} \mathcal{S}(p, p_0) = \mathcal{S}(p_0, p_0) \equiv 0, \tag{6.27}$$

and subject to the condition that $\sum_{i=1}^{n} p_i = 1$. According to the Lagrange multiplier rule

$$-\nabla_p \mathcal{S}|_{p=p^*} + \lambda \nabla_p \left(\sum_{i=1}^{n} p_i \right) |_{p=p^*} = 0, \tag{6.28}$$

and componentwise it yields

$$\ln \left(\frac{p_i^*}{p_i^0} \right) + (1 + \lambda) = 0, \quad i = 1, \ldots, n. \tag{6.29}$$

This in turn implies that the ratio p_i/p_i^0 is a constant independent of i, and, since both p^* and p_0 are probability distributions, we conclude that $p^* = p_0$, as wanted.

Next we consider the problem of determining the least bias probability distribution p consistent with a finite number of statistical measurement constraints, together with an external bias constraint.

Example 4: Find the least biased probability measure $p = p^*$ on $\mathcal{A} = \{a_1, \ldots, a_n\}$ that is consistent with a finite number $r \leq n - 1$ of statistical measurements F_j of given functions f_j, $j = 1, \ldots, r$

$$F_j = \langle f_j \rangle_p = \sum_{i=1}^{n} f_j(a_i) p_i, \; j = 1, \ldots, r,$$

and the external bias given by p_0. In this case we maximize the relative Shannon entropy $\mathcal{S}(p, p_0)$ subject to $r+1$ constraints

$$F_j = \langle f_j \rangle_p = \sum_{i=1}^{n} f_j(a_i) p_i, \; j = 1, \ldots, r, \quad \sum_{i=1}^{n} p_i = 1. \tag{6.30}$$

In this case the Lagrange multiplier rule yields

$$-\nabla_p \mathcal{S}(p, p_0) + \sum_{j=1}^{r} \lambda_j \nabla_p \langle f_j \rangle_p + \lambda_0 \nabla_p \left(\sum_{i=1}^{n} p_i \right) \Bigg|_{p=p^*} = 0, \tag{6.31}$$

or componentwise

$$\ln \left(\frac{p_i^*}{p_i^0} \right) = -\sum_{j=1}^{r} \lambda_j f_j(a_i) - (\lambda_0 + 1), \quad 1 \leq i \leq n. \tag{6.32}$$

Elimination of λ_0 with the help of the normalization constraint $\sum_{i=1}^{n} p_i = 1$ yields the least biased probability measure

$$p_i^* = \frac{\exp\left(-\sum_{j=1}^{r} \lambda_j f_j(a_i)\right) p_i^0}{\sum_{i=1}^{n} \exp\left(-\sum_{j=1}^{r} \lambda_j f_j(a_i)\right) p_i^0}, \quad i = 1, \ldots, n. \tag{6.33}$$

In analogy to Example 6.2, we define the relative partition function $\mathcal{Z}(\vec{\lambda}, p_0)$ by

$$\mathcal{Z}(\vec{\lambda}, p_0) = \ln \left(\sum_{i=1}^{n} \exp \left(-\sum_{j=1}^{r} \lambda_j f_j(a_i) \right) p_i^0 \right) \tag{6.34}$$

and then it is straightforward to verify that the relative partition function satisfies

$$-\frac{\partial \mathcal{Z}(\vec{\lambda}, p_0)}{\partial \lambda_j} = \langle f_j \rangle_{p^*}. \tag{6.35}$$

6.4 Entropy for continuous measures on the line

6.4.1 Continuous measure on the line

A continuous probability density, $\rho(\lambda)$, on the line satisfies the two requirements

$$\rho(\lambda) \geq 0, \quad \int_{\mathcal{R}^1} \rho(\lambda) d\lambda = 1. \tag{6.36}$$

Such probability densities often arise as the probability distribution function of some random variable, q, so that

$$\text{Prob}\{\alpha \leq q < \beta\} = \int_{\alpha}^{\beta} \rho(\lambda) d\lambda. \tag{6.37}$$

The book of Lamperti (1966) is an elementary reference for the material presented next.

Let $\langle \cdot \rangle$ denote the expected value of a random variable over the probability space, where the random variable, q, is defined. Then, with the density $\rho(\lambda)$ in (6.37)

$$\langle F(q) \rangle = \int_{\mathcal{R}^1} F(\lambda)\rho(\lambda) d\lambda. \tag{6.38}$$

The mean, $\bar{q}$, and variance, σ^2, of a random variable are defined respectively by

$$\begin{aligned} \bar{q} &\equiv \langle q \rangle = \int_{\mathcal{R}^1} \lambda \rho(\lambda)\, d\lambda, \\ \sigma^2 &\equiv \langle (q-\bar{q})^2 \rangle = \int_{\mathcal{R}^1} (\lambda - \bar{q})^2 \rho(\lambda)\, d\lambda. \end{aligned} \tag{6.39}$$

Perhaps the most well-known and important examples are Gaussian densities, which are uniquely determined by their mean and variance as follows:

Definition 6.4 (Gaussian distribution) *A Gaussian distribution with given mean $\bar{\lambda}$ and variance σ has the probability density function*

$$\rho_{\bar{\lambda},\sigma}(\lambda) \equiv \frac{1}{\sqrt{2\pi}\sigma} e^{-\frac{(\lambda-\bar{\lambda})^2}{2\sigma^2}} \tag{6.40}$$

so that

$$\bar{\lambda} = \int \lambda \rho_{\bar{\lambda},\sigma}(\lambda) d\lambda, \quad \sigma^2 = \int (\lambda - \bar{\lambda})^2 \rho_{\bar{\lambda},\sigma}(\lambda) d\lambda. \tag{6.41}$$

6.4.2 Entropy and maximum entropy principle

Next we want to generalize the concept of Shannon entropy and relative entropy to a continuous probability density on the line. The Shannon entropy discussed in Section 6.2 readily extends to countable discrete probability measures

$$\mathcal{S}(\{p_i\}) = -\sum_i p_i \ln p_i, \quad p_i \geq 0, \quad \sum_i p_i = 1. \tag{6.42}$$

A natural generalization of this formula to the case with a continuous probability density function $\rho(\lambda)$ is to replace the summation by integration

$$\mathcal{S}(\rho) = -\int_{\mathcal{R}^1} \rho \ln(\rho)\, d\lambda. \tag{6.43}$$

Similarly the relative entropy of a probability density ρ relative to the probability ρ_0 is defined by

$$\mathcal{S}(\rho, \rho_0) = -\int_{\mathcal{R}^1} \rho \ln(\rho/\rho_0)\, d\lambda. \tag{6.44}$$

Here and below we use the minus sign for the relative entropy in order to formulate the maximum entropy principle with an external bias defined through ρ_0, which generalizes the maximum entropy principle without an external bias. The usual convention for the relative entropy has the plus sign in (6.44) so that an asymmetric distance function on probability measures is defined in this fashion. We discuss this briefly at the end of this section. We will also switch to this point of view when we discuss predictability in subsequent chapters.

Given some set of constraints, $\mathcal{C}$, on the space of probability densities, we define the probability density with the least bias (least information) for further measurements, given the constraints in $\mathcal{C}$ through the maximum entropy principle: *Find the probability density $\rho^*(\lambda) \in \mathcal{C}$, so that*

$$\mathcal{S}(\rho^*) = \max_{\rho \in \mathcal{C}} \mathcal{S}(\rho). \tag{6.45}$$

Of course there is a similar principle for relative entropy where ρ_0 represents the probability density measuring external bias:
Given the probability density ρ_0 measuring the external bias, find the probability density $\rho^*(\lambda) \in \mathcal{C}$, so that

$$\mathcal{S}(\rho^*, \rho_0) = \max_{\rho \in \mathcal{C}} \mathcal{S}(\rho, \rho_0). \tag{6.46}$$

To illustrate the maximum entropy principle in (6.45), we show that the Gaussian distributions in (6.40) are the probability densities with the least bias, given constraints $\mathcal{C}$ defined by the first and second moments. This fact supports the idea well known from the central limit theorem that Gaussian densities are the most universal distributions with given first and second moments. In other words, let the constraint set $\mathcal{C}$ be defined by

$$\begin{aligned} &\rho(\lambda) \geq 0, \quad \int_{\mathcal{R}^1} \rho(\lambda)\, d\lambda = 1, \\ &\bar{\lambda} = \int_{\mathcal{R}^1} \lambda \rho(\lambda)\, d\lambda, \quad \sigma^2 = \int_{\mathcal{R}^1} (\lambda - \bar{\lambda})^2 \rho(\lambda)\, d\lambda. \end{aligned} \tag{6.47}$$

We want to find the probability measure $\rho^*(\lambda)$ satisfying (6.45). As in Section 1.5, the variational derivative of the entropy is given by

$$\frac{\delta \mathcal{S}}{\delta \rho} = -(1 + \ln \rho). \tag{6.48}$$

The first and second moment constraints in (6.47) are linear functionals of the density ρ so that it is easy to calculate

$$\frac{\delta \bar{\lambda}}{\delta \rho} = \lambda, \quad \frac{\delta \sigma^2}{\delta \rho} = (\lambda - \bar{\lambda})^2. \tag{6.49}$$

From the Lagrange multiplier principle we have, at the entropy maximum

$$-\frac{\delta \mathcal{S}}{\delta \rho}|_{\rho=\rho^*} = -\mu_0 - \mu_1 \frac{\delta \bar{\lambda}}{\delta \rho}|_{\rho=\rho^*} - \mu_2 \frac{\delta \sigma^2}{\delta \rho}|_{\rho=\rho^*}, \tag{6.50}$$

where μ_0, μ_1, μ_2 are the Lagrange multipliers for the constraints in (6.47). With (6.48), (6.49), the equation in (6.50) becomes

$$\ln \rho^* = (-\mu_0 + 1) - \mu_1 \lambda - \mu_2 (\lambda - \bar{\lambda})^2. \tag{6.51}$$

However, (6.51) defines ρ^* as a Gaussian probability density with mean $\bar{\lambda}$ and variance σ^2 so that $\rho^* = \rho_{\bar{\lambda},\sigma}(\lambda)$, the Gaussian density defined in (6.40). Below in Subsection 6.4.4, we give a rigorous direct proof that the Gaussian density is the maximum of (6.45) with the constraints in (6.47).

6.4.3 Coarse graining and loss of information

Coarse graining is the terminology utilized in statistical physics (see Thompson, 1972) for processes that involve averaging over small scales. Intuitively, coarse graining means that information is lost so that the entropy, as a quantitative measure of lack of information, should increase. Here we establish this fact rigorously for probability measures on the line with a simple natural coarse-graining procedure.

Let $\lambda_i < \lambda_{i+1}$ for integers i, $-\infty < i < \infty$ define a countable partition of the line with equi-distance spacing $\Delta\lambda$ and given a probability measure, $\rho(\lambda)$, define the discrete coarse-grained probability measure p by

$$p = \sum_i \rho_i \delta_{\lambda_i}, \tag{6.52}$$

where the ρ_i are determined by the "coarse-grained" average of $\rho(\lambda)$

$$\rho_i = \int_{\lambda_i}^{\lambda_{i+1}} \rho(\lambda) d\lambda. \tag{6.53}$$

We expect that the coarse-grained p contains less information than the original density $\rho(\lambda)$. This intuition is confirmed quantitatively by the following general inequality establishing the strict increase of entropy through coarse graining for $\Delta\lambda \leq 1$

$$\mathcal{S}(p) \geq \mathcal{S}(\rho) + \ln((\Delta\lambda)^{-1}), \tag{6.54}$$

where

$$\mathcal{S}(p) = -\sum \rho_i \ln \rho_i$$

and $\mathcal{S}(\rho)$ is given in (6.43).

To prove (6.54) we use the elementary inequality

$$x(\ln(x) - \ln(y)) \geq x - y. \tag{6.55}$$

Assuming this inequality for the moment, we rewrite it in the more useful form

$$x \ln(x) \geq x \ln(y) + x - y. \tag{6.56}$$

Next, set $y = \rho_i / \Delta\lambda$ and $x = \rho(\lambda)$ and integrate (6.56) over the interval $(\lambda_i, \lambda_{i+1})$ to get

$$\int_{\lambda_i}^{\lambda_{i+1}} \rho(\lambda) \ln \rho(\lambda) d\lambda \geq \rho_i \ln \rho_i - \rho_i \ln(\Delta\lambda). \tag{6.57}$$

Summing the inequalities in (6.57) over i clearly implies (6.54) as claimed above. To verify the elementary inequality in (6.55), recall that the mean value theorem guarantees that

$$\begin{aligned} \frac{\ln(x) - \ln(y)}{x - y} &\geq \frac{1}{x}, \quad x > y, \\ \frac{\ln(x) - \ln(y)}{x - y} &\leq \frac{1}{x}, \quad x < y, \end{aligned} \tag{6.58}$$

and both of the inequalities in (6.58) guarantee (6.55).

6.4.4 Relative entropy as a "distance" function

We now introduce the functional $\mathcal{P}(p, \pi_0) = -\mathcal{S}(p, \pi_0)$. It is called the relative entropy "distance" function from the probability measure p to the probability measure π_0 where

$$\mathcal{P}(p, \pi_0) = \int_{\mathcal{R}^1} p \ln\left(\frac{p}{\pi_0}\right) = -\mathcal{S}(p, \pi_0). \tag{6.59}$$

The name "distance" is partially justified by the following claim

$$\mathcal{P}(p, \pi_0) \geq 0 \quad \text{with equality only for } p = \pi_0. \tag{6.60}$$

For any measure p with $\mathcal{P}(p, \pi_0) < \infty$, necessarily there is a function $\rho(\lambda)$ such that

$$p = \rho\pi_0 \quad \text{with} \int_{\mathcal{R}^1} \rho(\lambda)\pi_0(\lambda) d\lambda = 1, \text{ and } \rho \geq 0, \tag{6.61}$$

and (6.59) becomes

$$\mathcal{P}(p, \pi_0) = \int_{\mathcal{R}^1} \rho(\lambda) \ln(\rho) \pi_0(\lambda) d\lambda. \tag{6.62}$$

The identity in (6.62) gives us a useful alternative characterization for relative entropy in a form where the formula for relative entropy resembles that for entropy with the weight π_0. Next, we use Jensen's inequality (see Rudin (1974) for a more general statement and proof) to prove (6.60). In the special set-up we have here, this inequality states that for any probability density π_0, convex function φ, and arbitrary function $f(\lambda)$

$$\varphi(\int_{\mathcal{R}^1} f(\lambda)\pi_0(\lambda)d\lambda) \leq \int_{\mathcal{R}^1} \varphi(f(\lambda))\pi_0(\lambda)d\lambda, \tag{6.63}$$

i.e. with the notation from (6.38)

$$\varphi(\langle f\rangle_{\pi_0}) \leq \langle \varphi(f)\rangle_{\pi_0} \tag{6.64}$$

for any convex function φ, and probability density π_0. Recall that a smooth function φ is convex if $\varphi'' \geq 0$. To establish the claim in (6.60), we utilize the alternative representation for relative entropy in (6.62) and Jensen's inequality in (6.64) for the strictly convex function $\varphi(x) = x\ln(x)$ with the choice $f = \rho$. This yields

$$\mathcal{P}(p, \pi_0) = \int_{\mathcal{R}^1} \rho(\lambda)\ln(\rho)\pi_0(\lambda)d\lambda \geq x\ln(x)|_{x=\int \rho\pi_0\, d\lambda} = 0 \tag{6.65}$$

since $\int \rho\pi_0 d\lambda = 1$ from (6.61). It is easy to show that equality happens in (6.65) only for $\rho \equiv 1$ because $x\ln(x)$ is strictly convex so that (6.60) is now clear. It is worth pointing out here that $\mathcal{P}(p, \pi_0)$ is not a distance function since it is not symmetric. In other words, the distance from p to π_0 is not the same as the distance from π_0 to p, using $\mathcal{P}$ as the distance function in general.

Now we return to the problem discussed at the end of Subsection 6.4.2 and use the relative entropy to supply a simple direct proof that the Gaussian distribution, $\rho_{\bar{\lambda},\sigma}(\lambda)$ in (6.40) maximizes the entropy among all probability distributions, ρ, with the same mean and second moment as in (6.47). For such a probability distribution ρ satisfying (6.47) with $\rho \neq \rho_{\bar{\lambda},\sigma}$, (6.60) guarantees

$$\begin{aligned}
0 &\leq \mathcal{P}(\rho, \rho_{\bar{\lambda},\sigma}(\lambda)) \\
&= \int \rho\ln\rho - \int \rho\ln\rho_{\bar{\lambda},\sigma}(\lambda) \\
&= \int \rho\ln\rho - \int \rho\ln\left(\frac{\sigma}{\sqrt{2\pi}}\right) - \int \rho\frac{(\lambda-\bar{\lambda})^2}{\sigma^2} \\
&= \int \rho\ln\rho - \int \rho_{\bar{\lambda},\sigma}\ln\left(\frac{\sigma}{\sqrt{2\pi}}\right) - \int \rho_{\bar{\lambda},\sigma}\frac{(\lambda-\bar{\lambda})^2}{\sigma^2} \\
&= \mathcal{S}(\rho_{\bar{\lambda},\sigma}) - \mathcal{S}(\rho)
\end{aligned} \tag{6.66}$$

so that $\mathcal{S}(\rho_{\bar{\lambda},\sigma}) > \mathcal{S}(\rho)$ as required. The crucial third inequality in (6.66) is true because, according to (6.47), ρ and $\rho_{\bar{\lambda},\sigma}$ have the same first and second moments.

6.4.5 Information theory and the finite-moment problem for probability measures

Given a probability density, $p(\lambda)$, generated by some empirical observational procedure, it is often interesting to quantify the information contained in the moments of $p(\lambda)$

$$\bar{\lambda} = M_1 = \int \lambda p(\lambda)\, d\lambda; \quad M_j = \int (\lambda - \bar{\lambda})^j p(\lambda)\, d\lambda, \quad 2 \le j \le 2L, \tag{6.67}$$

in addition to $\int p(\lambda)\, d\lambda = 1$.

Thus given only the information in (6.67) we are interested in the probability distribution, $p^*_{2L}(\lambda)$ with the least bias given these moment constraints, i.e. the one that maximizes the entropy

$$\mathcal{S}(p^*_{2L}) = \max_{p \in PM_{2L}} \mathcal{S}(p). \tag{6.68}$$

In (6.68), PM_{2L} is the family of probability measures which satisfies the moment constraints in (6.67). A straightforward generalization of the calculations in (6.47)–(6.51) yields formally that

$$p^*_{2L} = \exp\left(\sum_{m=0}^{2L} \alpha_m (\lambda - \bar{\lambda})^m \right), \tag{6.69}$$

where α_m are the appropriate Lagrange multipliers. Notice that we need even powers and $\alpha_{2L} < 0$ to guarantee a finite probability measure, The case with $L = 1$ generates the Gaussian probability density, $p^*_2 = p^*_G$, already discussed in Subsections 6.4.2 and 6.4.4. Also, since adding more moments increases information, it follows that

$$\mathcal{S}(p^*_{2L_2}) \le \mathcal{S}(p^*_{2L_1}), \text{ for } L_1 \le L_2. \tag{6.70}$$

Next, we use the relative entropy introduced in (6.44) to quantify the information loss in utilizing only the $2L$ moments of p. As in (6.66) above, we calculate

$$\begin{aligned} 0 \le \mathcal{P}(p, p^*_{2L}) &= \int p \ln p - \int p \ln p^*_{2L} \\ &= \int p \ln p - \int p^*_{2L} \ln p^*_{2L} \\ &= \mathcal{S}(p^*_{2L}) - \mathcal{S}(p). \end{aligned} \tag{6.71}$$

With the form for p^*_{2L} in (6.69) it follows that $\ln p^*_{2L}$ is a sum of the $2L$ moment constraints in (6.67) so that automatically

$$\int p \ln p^*_{2L} = \int p^*_{2L} \ln p^*_{2L} \tag{6.72}$$

and this is the key step in the identity in (6.71). From (7.71) we immediately have, for $L_1 < L_2$

$$\mathcal{P}(p, p^*_{2L_1}) = \mathcal{P}(p, p^*_{2L_2}) + \mathcal{P}(p^*_{2L_2}, p^*_{2L_1}). \tag{6.73}$$

We can view measuring fewer moments of p as a coarse-grained measurement of p and *(6.73) precisely measures the information loss in this coarse-graining procedure* through the term

$$\mathcal{P}(p^*_{2L_2}, p^*_{2L_1}) = \mathcal{S}(p^*_{2L_1}) - \mathcal{S}(p^*_{2L_2}) \tag{6.74}$$

and also provides a quantitative estimate for the information inequalities in (6.70).

An important practical question is when is a probability distribution significantly non-Gaussian, i.e. how much additional information is contained in the third and forth moments of a probability distribution beyond the Gaussian estimate, p^*_G, using only the first and second moments. Thus it is interesting to compare $\mathcal{P}(p^*_4, p^*_G)$ for varying values of the third and forth moments in (6.67) with fixed values for the mean and second moment. The third and fourth moments of p^*_4 are characterized by the skewness and flatness

$$\text{Skew} = \frac{\int(\lambda-\bar{\lambda})^3 p^*_4}{\left(\int(\lambda-\bar{\lambda})^2 p^*_4\right)^{3/2}}, \qquad \text{Flat} = \frac{\int(\lambda-\bar{\lambda})^4 p^*_4}{\left(\int(\lambda-\bar{\lambda})^2 p^*_4\right)^{2}}. \tag{6.75}$$

In Figure 6.2 we graph the probability distributions, p^*_4 and p^*_G for zero mean, variance one, and varying skewness and flatness. The bimodal and skewed behavior of the probability distribution p^*_4 compared with the Gaussian is clearly evident. Thus, p^*_4 can capture important significant non-Gaussian features of a probability distribution. Table 6.1 presents the quantitative values of the information loss $\mathcal{P}(p_4, p_G)$, demonstrating the same trends from Figure 6.2.

Quantitative estimates for non-Gaussian features such as bimodality are extremely important in practical ensemble predictions. See Chapter 15 and the references listed there. Section 15.4 contains a practical demonstration. In particular, the simple ideas put forward here can be generalized in a computationally feasible practical fashion to ensemble predictions in many variables. The interested reader can also consult Mead and Papanicolaou (1984) for one-dimensional examples and Abramov and Majda (2004), Abramov *et al.* (2005) for the practical estimation of highly non-Gaussian two-dimensional probability distributions generated from observational data using the four moment estimator p^*_4.

6.5 Maximum entropy principle for continuous fields

Here we extend the concepts of the Shannon entropy and the maximum entropy principle to continuous random fields. In other words, we regard the potential

Table 6.1 *The information loss $\mathcal{P}(p_4, p_G)$ for various skewness and flatness of p_4, both p_4 and p_G have zero mean and unit variance. The utility monotonically increases with increasing skewness and/or decreasing flatness*

$\mathcal{P}(p_4, p_G)$	$Skew_{p_4} = 0$	$Skew_{p_4} = 0.3$	$Skew_{p_4} = 0.5$
$Flat_{p_4} = 1.5$	0.2772	0.3921	0.6921
$Flat_{p_4} = 2$	$6.015 \cdot 10^{-2}$	$9.879 \cdot 10^{-2}$	0.1885
$Flat_{p_4} = 2.8$	$1.016 \cdot 10^{-3}$	$1.226 \cdot 10^{-2}$	$3.842 \cdot 10^{-2}$

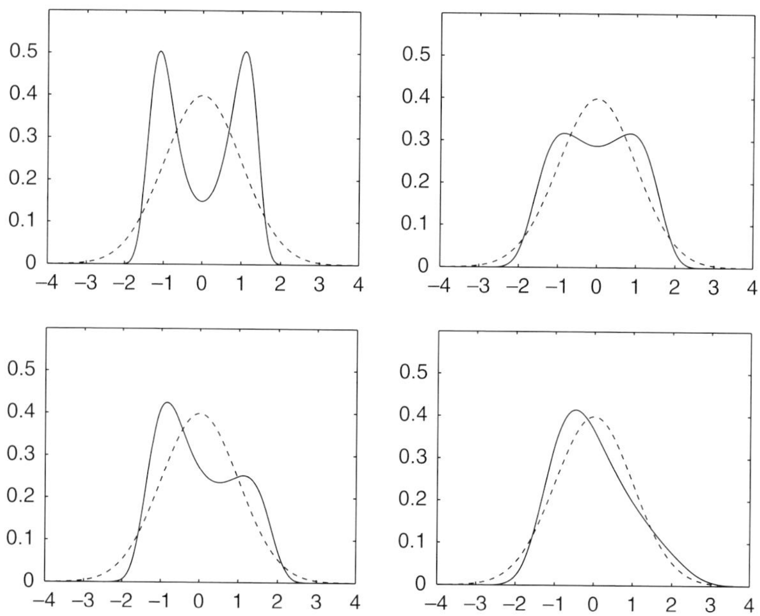

Figure 6.2 The probability density functions p_4^* from (6.69) with $L = 2$, zero mean, unit variance, various skewness and flatness (solid lines) versus Gaussians (dashed lines). Upper left – zero skewness, flatness 1.5; upper right – zero skewness, flatness 2; lower left – skewness 0.3, flatness 2; lower right – skewness 0.5, flatness 2.8.

vorticity distribution as involving small-scale random fluctuations, which can only be determined statistically. Thus, for each $\vec{x} = (x, y)$ in the domain of interest, we assume that there is a probability density $\rho(\vec{x}, \lambda)$ defined on the real line and satisfying (6.36) so that

$$\text{Prob}\,\{\alpha \le q(\vec{x}) < \beta\} = \int_\alpha^\beta \rho(\vec{x}, \lambda)\, d\lambda. \tag{6.76}$$

How do such small-scale random fluctuations of potential vorticity affect the large-scale feature of the flow? We discuss this later in the chapter in a simple context through the maximum entropy principle. First, we need to quantify the statistical information in (6.76) more precisely. The conditions in (6.76) encode only *one-point statistical* information for the potential vorticity but ignore the statistical information from correlations of different points $\vec{x}_1, \vec{x}_2$ etc. We have the following:

Definition 6.5 (One-point statistics for potential vorticity) *The one-point statistics for the potential vorticity q consist of a probability density $\rho(\vec{x}, \lambda)$ on $\Omega \times \mathcal{R}^1$ associated to the potential vorticity, which satisfies the following properties:*

1. ρ is a probability density on $\Omega \times \mathcal{R}^1$

$$\rho(\vec{x}, \lambda) \geq 0, \quad ⨍_{\Omega} \int_{\mathcal{R}^1} \rho(\vec{x}, \lambda)\, d\lambda\, d\vec{x} = 1. \tag{6.77}$$

2. For almost all $\vec{x} \in \Omega$, $\rho(\vec{x}, \lambda)$ is a probability density on $\mathcal{R}^1$

$$\int_{\mathcal{R}^1} \rho(\vec{x}, \lambda)\, d\lambda = 1. \tag{6.78}$$

3. For almost all $\vec{x} \in \Omega$, $\rho(\vec{x}, \lambda)$ is the probability density associated with the potential vorticity $q(\vec{x})$ at the location $\vec{x}$

$$\int_{q_-}^{q_+} \rho(\vec{x}, \lambda)\, d\lambda = \text{Prob}\, \{q_- \leq q(\vec{x}) < q_+\}. \tag{6.79}$$

In terms of the one-point statistics, at any specified location $\vec{x}$, the measurement of any function $F(\vec{x}, q)$ of the potential vorticity q is given by the expected value of $F(\vec{x}, \cdot)$ relative to the measure ρ, in the same fashion as it was defined in Section 6.4

$$\overline{F}_{\rho}(\vec{x}) \equiv \langle F(\vec{x}, q) \rangle_{\rho} = \int_{\mathcal{R}^1} F(\vec{x}, \lambda) \rho(\vec{x}, \lambda)\, d\lambda, \tag{6.80}$$

and the overall expected value is

$$< F(q) >_{\rho} = ⨍_{\Omega} \langle F(\vec{x}, q) \rangle_{\rho} d\vec{x}. \tag{6.81}$$

Here Ω is either a periodic domain or a channel domain. Next we generalize the definition of prior distribution and relative Shannon entropy from Section 6.4 to the present setting.

Definition 6.6 (Prior distribution) *A prior distribution $\Pi_0(\vec{x}, \lambda)$ is a probability density on $\Omega \times R^1$ which satisfies:*

1. Π_0 is a probability density on $\Omega \times R^1$

$$\Pi_0(\vec{x}, \lambda) \geq 0, \quad ⨍_{\Omega} \int_{\mathcal{R}^1} \Pi_0(\vec{x}, \lambda)\, d\lambda\, d\vec{x} = 1, \tag{6.82}$$

2. *For almost all* $\vec{x} \in \Omega$, $\Pi_0(\vec{x}, \lambda)$ *yields a probability measure in* $\mathcal{R}^1$

$$\int_{\mathcal{R}^1} \Pi_0(\vec{x}, \lambda)\, d\lambda = 1. \tag{6.83}$$

For this type of prior distribution, the relative entropy from (6.44) generalizes straightforwardly to the following:

Definition 6.7 (Relative entropy) *The relative entropy* $\mathcal{S}(\rho, \Pi_0(\vec{x}, \lambda))$ *of a probability measure* $\rho(\vec{x}, \lambda)$ *on* $\Omega \times R^1$ *is defined by*

$$\mathcal{S}(\rho, \Pi_0) \equiv -\fint_{\Omega} \int_{\mathcal{R}^1} \rho(\vec{x}, \lambda) \ln\left(\frac{\rho(\vec{x}, \lambda)}{\Pi_0(\vec{x}, \lambda))}\right) d\lambda\, d\vec{x}. \tag{6.84}$$

The calculation of the most probable state is then based on the maximum entropy principle: the most probable state ρ^* for a given prior distribution Π_0 and subject to a set of constraints $\mathcal{C}$ is obtained by maximizing the relative entropy $\mathcal{S}(\rho, \Pi_0)$ among all the probability distributions ρ satisfying the constraints in $\mathcal{C}$, i.e.

$$\mathcal{S}(\rho^*, \Pi_0) = \max_{\rho \in \mathcal{C}} \mathcal{S}(\rho, \Pi_0). \tag{6.85}$$

In particular, if there are no constraints, it is easy to check that the most probable state ρ^* is exactly the prior distribution $\Pi_0(\vec{x}, \lambda)$.

6.6 An application of the maximum entropy principle to geophysical flows with topography

Here we illustrate how the maximum entropy principle for continuous fields from (6.85) can be utilized to make interesting statistical predictions for large-scale geophysical flows with topography. This is a major topic in subsequent chapters of the book. Thus, we have in mind the underlying dynamics

$$\frac{\partial q}{\partial t} + \nabla^{\perp}\psi \cdot \nabla q = 0, \quad q = \Delta\psi + h \tag{6.86}$$

in periodic geometry.

6.6.1 *The Prior distribution*

What would be the simplest model for the small-scale random fluctuations of the potential vorticity? Here we model the small-scale potential vorticity fluctuation by the probability measure with the least bias with a given variance and zero mean. In Section 6.4, we showed that the least biased measure of this type is given by the Gaussian density. Thus, we assume that the prior distribution, $\Pi_0(\vec{x}, \lambda)$, in the maximum entropy principle is given by a Gaussian without explicit $\vec{x}$-dependence

$$\Pi_0(\lambda) = \frac{\sqrt{\alpha}}{\sqrt{2\pi}} \exp\left(-\frac{\alpha}{2}\lambda^2\right), \tag{6.87}$$

where $\alpha = \frac{1}{\sigma^2}$ with σ^2 the variance. This is the simplest conceivable model for the one-point statistics of the small-scale fluctuations of potential vorticity. Clearly in this model, the statistical fluctuations of the enstrophy at each point are determined by (6.80) as given by

$$\frac{1}{2}\langle q^2(\vec{x})\rangle = \frac{1}{2}\sigma^2 = \frac{1}{2}\alpha^{-1}. \tag{6.88}$$

6.6.2 Constraints on the potential vorticity distribution

Given the prior probability distribution, $\Pi_0(\lambda)$, in (6.87), we would like to determine the most probable distribution which emerges from suitable constraints and the maximum entropy principle. From (6.77) and (6.78), the constraint set, $\mathcal{C}$, should include the requirement

$$\int_{\mathcal{R}^1} \rho(\vec{x}, \lambda)\, d\lambda = 1, \text{ for all } \vec{x}. \tag{6.89}$$

While the Gaussian prior distribution encodes the small-scale fluctuations, the kinetic energy incorporates large-scale features of the flow. How do we define the kinetic energy for a probability density $\rho(\vec{x}, \lambda)$ satisfying definition 6.5? For a stream function ψ, the energy is given by

$$E = -\frac{1}{2}\fint_\Omega \psi(q-h), \qquad q = \Delta\psi + h. \tag{6.90}$$

For a candidate probability density, $\rho(\vec{x}, \lambda)$, we define the mean potential vorticity $\bar{q}(\vec{x})$ by using equation (6.80)

$$\bar{q}(\vec{x}) = \int_{\mathcal{R}^1} \lambda\rho(\vec{x}, \lambda)\, d\lambda, \tag{6.91}$$

and the associated stream function $\bar{\psi}(\vec{x})$ is then determined by

$$\Delta\bar{\psi}(\vec{x}) + h = \bar{q}(\vec{x}). \tag{6.92}$$

With equations (6.90)–(6.92), we define the mean *kinetic energy* for a probability density $\rho(\vec{x}, \lambda)$ by

$$E(\rho) = -\frac{1}{2}\fint \bar{\psi}(\bar{q} - h). \tag{6.93}$$

Thus, the constraint set $\mathcal{C}$ is defined by the requirements in equations (6.89) and equation (6.93). Intuitively, the mean energy measures the large-scale features of the flow, while the enstrophy measures the small-scale fluctuation of the flow as we saw earlier in the analysis in Chapters 2 and 3.

6.6.3 Statistical predictions of the maximum entropy principle

Next, we calculate explicitly the most probable state, $\rho^*(\vec{x}, \lambda)$, which satisfies

$$\mathcal{S}(\rho^*, \Pi_0) = \max_{\rho \in \mathcal{C}} \mathcal{S}(\rho, \Pi_0), \tag{6.94}$$

with Π_0, the Gaussian measure in (6.87), and the constraints, $\mathcal{C}$, defined by

$$\int_{\mathcal{R}^1} \rho(\vec{x}, \lambda)\, d\lambda = 1, \text{for all } \vec{x} \in \Omega, \quad E(\rho) = E_0. \tag{6.95}$$

Recall from Section 1.5 that the variational derivative of the kinetic energy with respect to the potential vorticity is given by

$$\frac{\delta E}{\delta \bar{q}} = -\bar{\psi}.$$

On the other hand, notice that the mean potential vorticity as defined in (6.91) is the first (linear) moment of the probability density function ρ, thus we have

$$\frac{\delta \bar{q}}{\delta \rho} = \lambda.$$

Hence, we deduce from the chain rule for variational derivatives that

$$\frac{\delta E}{\delta \rho} = -\bar{\psi}(\vec{x})\lambda. \tag{6.96}$$

As in equation (6.48), it is also straightforward to show that the variational derivative of the relative Shannon entropy $\mathcal{S}(\rho, \Pi_0)$ from (6.84) is given by

$$\frac{\delta \mathcal{S}}{\delta \rho} = -1 - \ln \rho + \ln \Pi_0. \tag{6.97}$$

Now with the help of the Lagrange multiplier rule, we deduce that the most probable probability density ρ^* must satisfy, according to the maximum entropy principle in (6.94)

$$\left.\frac{\delta \mathcal{S}}{\delta \rho}\right|_{\rho=\rho^*} = \left.\left(\theta \frac{\delta E}{\delta \rho} + \tilde{\gamma}(\vec{x})\right)\right|_{\rho=\rho^*}, \tag{6.98}$$

where θ is the Lagrange multiplier for the energy, and $\tilde{\gamma}(\vec{x})$ is the Lagrange multiplier for (6.89). Equation (6.98) can be rewritten as, thanks to equation (6.87), (6.96), and (6.97)

$$\ln(\rho^*) = \theta\lambda\bar{\psi}^* - (\tilde{\gamma}(\vec{x}) + 1 - \ln\Pi_0),$$

i.e.

$$\rho^*(\vec{x}, \lambda) = \frac{\sqrt{\alpha}}{\sqrt{2\pi}} \exp\left(-\frac{\alpha}{2}(\lambda - \frac{\theta}{\alpha}\bar{\psi}^*)^2\right). \tag{6.99}$$

In other words, the most probable measure ρ^* is a Gaussian measure for each $\vec{x}$. In this case, at each point $\vec{x}$, the first and second moments of ρ^* are given by

$$\begin{aligned} \int \lambda\rho^*(\vec{x}, \lambda)\, d\lambda &= \frac{\theta}{\alpha}\bar{\psi}^*(\vec{x}), \\ \int \lambda^2\rho^*(\vec{x}, \lambda)\, d\lambda &= \left(\frac{\theta}{\alpha}\right)^2 (\bar{\psi}^*(\vec{x}))^2 + \alpha^{-1}. \end{aligned} \tag{6.100}$$

Next we use (6.100) to determine an explicit equation for $\bar{\psi}^*(\vec{x})$. Recall that the mean field $\bar{\psi}^*(\vec{x})$ is determined by equation (6.92) and the mean potential vorticity is determined by (6.91). Combining these equations and (6.100) we arrive at the following **mean field equation**

$$\Delta\bar{\psi}^* + h = \frac{\theta}{\alpha}\bar{\psi}^* = \mu\bar{\psi}^*, \quad \text{with } \mu = \frac{\theta}{\alpha}, \alpha > 0. \tag{6.101}$$

From the second equation in (6.100), the statistical fluctuations about the mean of the potential vorticity field remain Gaussian with variance α^{-1}. The mean field equation in (6.101) is the same as encountered earlier in Chapter 1, which corresponds to special steady states with linear q–ψ relation, and whose stability was analyzed in Chapter 4. Thus, the maximum principle predicts interesting large-scale steady states in this simplest situation.

6.6.4 Determination of the multipliers and geophysical effect

Our next task is to determine the parameter μ with the given energy constraint $E(\rho^*) = E_0$. We will see that topography plays a very important role in determining the behavior of the mean states. Interesting enough, the most probable one-point statistics predicted will be those whose mean fields are selective decay states (enstrophy minimizer with given energy) as was discussed in Section 4.5.

We observe, thanks to (6.99), that the most probable distribution is determined by the mean field which is further determined by the mean field equation (6.101) together with the energy constraint. However, a closer look into the mean field equation indicates that solutions are not necessarily unique with a given energy level. Without loss of generality we assume that the smallest value in the spectrum of the topography is Λ, i.e.

$$h = \sum_{|\vec{k}|^2 \geq \Lambda} e^{i\vec{x}\cdot\vec{k}} \hat{h}_{\vec{k}}, \tag{6.102}$$

$$\hat{h}_{\vec{k}} \neq 0, \text{ for some } \vec{k} \text{ with } |\vec{k}|^2 = \Lambda.$$

The solutions to the mean field equation (6.101) can then be written explicitly as

$$\psi_\mu = \begin{cases} \sum\limits_{|\vec{k}|^2 \geq \Lambda} \frac{\hat{h}_{\vec{k}}}{|\vec{k}|^2+\mu} e^{i\vec{x}\cdot\vec{k}}, & \text{if } \mu \neq -|\vec{k}|^2, \\ \sum\limits_{|\vec{k}|^2 \geq \Lambda, |\vec{k}| \neq |\vec{k}_0|} \frac{\hat{h}_{\vec{k}}}{|\vec{k}|^2+\mu} e^{i\vec{x}\cdot\vec{k}} + \sum\limits_{|\vec{k}|=|\vec{k}_0|} c_{\vec{k}} e^{i\vec{x}\cdot\vec{k}}, & \text{if } \mu = -|\vec{k}_0|^2 \notin \Sigma(h), \end{cases} \tag{6.103}$$

where $c_{\vec{k}}$ are arbitrary constants satisfying the reality condition and $\Sigma(h)$ is the spectrum of the topography defined as

$$\Sigma(h) = \left\{ |\vec{k}|^2 | \hat{h}_{\vec{k}} \neq 0 \right\}. \tag{6.104}$$

The corresponding energy $E = E_\mu$ and enstrophy are given by:

- Case $\mu \neq -|\vec{k}|^2$, for all $\vec{k}$

$$E(q_\mu) = \frac{1}{2} \sum_{|\vec{k}|^2 \geq \Lambda} \frac{|\vec{k}|^2 |\hat{h}_{\vec{k}}|^2}{(|\vec{k}|^2+\mu)^2}, \quad \mathcal{E}(q_\mu) = \frac{1}{2} \sum_{|\vec{k}|^2 \geq \Lambda} \frac{|\vec{k}|^4 |\hat{h}_{\vec{k}}|^2}{(|\vec{k}|^2+\mu)^2}. \tag{6.105}$$

- Case $\mu = -|\vec{k}_0|^2 \notin \Sigma(h)$

$$\begin{aligned} E(q_\mu) &= \frac{1}{2} \sum_{|\vec{k}|^2 \geq \Lambda, |\vec{k}| \neq |\vec{k}_0|} \frac{|\vec{k}|^2 |\hat{h}_{\vec{k}}|^2}{(|\vec{k}|^2+\mu)^2} + \frac{1}{2} \sum_{|\vec{k}|=|\vec{k}_0|} |c_{\vec{k}}|^2 |\vec{k}_0|^2, \\ \mathcal{E}(q_\mu) &= \frac{1}{2} \sum_{|\vec{k}|^2 \geq \Lambda, |\vec{k}| \neq |\vec{k}_0|} \frac{|\vec{k}|^4 |\hat{h}_{\vec{k}}|^2}{(|\vec{k}|^2+\mu)^2} + \frac{1}{2} \sum_{|\vec{k}|=|\vec{k}_0|} |c_{\vec{k}}|^2 |\vec{k}_0|^4. \end{aligned} \tag{6.106}$$

It is then clear, see for instance Figure 1.5 for a plot of energy versus μ for the case of (6.105), that for a given energy level E_0, which is high enough, there could be more than one solution to the mean field equation with given energy.

The natural question then is which solution corresponds to the most probable state. In order to answer this question, we compute the entropy of the candidates of most probable one-point statistics ρ^*. We have, thanks to (6.84), (6.87), (6.91), (6.99), and (6.100)

$$\begin{aligned}
\mathcal{S}(\rho^*, \Pi_0) &= -\fint \int \frac{\sqrt{\alpha}}{\sqrt{2\pi}} \exp\left(-\frac{\alpha}{2}\left(\lambda - \frac{\theta}{\alpha}\bar{\psi}^*\right)^2\right)\left(-\frac{\alpha}{2}\left(\lambda - \frac{\theta}{\alpha}\bar{\psi}^*\right)^2 + \frac{\alpha}{2}\lambda^2\right) \\
&= -\fint \int \frac{\sqrt{\alpha}}{\sqrt{2\pi}} \exp\left(-\frac{\alpha}{2}\left(\lambda - \frac{\theta}{\alpha}\bar{\psi}^*\right)^2\right) \\
&\quad \times \left(\theta\bar{\psi}^*\left(\lambda - \frac{\theta}{\alpha}\bar{\psi}^*\right) + \frac{\theta^2}{2\alpha}(\bar{\psi}^*)^2\right) \\
&= -\frac{\theta^2}{2\alpha}\fint (\bar{\psi}^*)^2 \\
&= -\alpha\mathcal{E}(\overline{q}^*).
\end{aligned} \tag{6.107}$$

This leads to the following relationship between most probable states and selective decay states.

Proposition 6.2 *A probability density function ρ^* given by (6.99) is a most probable measure if and only if its mean field $\overline{q}^*$ is a selective decay state, i.e. enstrophy minimizer with given energy.*

Proof: Suppose that ρ^* is a most probable measure. Then $\overline{q}^*$ minimizes enstrophy among all solutions of the mean field equation (6.101) with given energy E_0, thanks to (6.107). We also observe that the mean field equation (6.101) is exactly the Euler–Lagrange equation for the enstrophy minimization problem with given energy according to Section 4.5. Thus $\overline{q}^*$ minimizes enstrophy among all possible candidates of minimizers of enstrophy with given energy. Hence we conclude that $\overline{q}^*$ is a global minimizer of the enstrophy with given energy, i.e. a selective decay state.

On the other hand, if $\overline{q}^*$ is a selective decay state, then $\overline{q}^*$ minimizes the enstrophy $\mathcal{E}(\overline{q}^*)$ among all solutions to the mean field equation (6.101) with given energy E_0. This further implies, thanks to (6.107), the corresponding probability density function ρ^* maximizes the entropy and hence ρ^* is a most probable measure.

This completes the proof of the proposition.

Thanks to Proposition 6.2, the problem of determining the most probable state is equivalent to determining selective decay states. This, together with the explicit form of solutions (6.103) to the mean field equation, together with the characterization of the selective decay states in Section 4.5, enables us to determine the Lagrange multiplier μ.

We first observe, thanks to (6.105), that the energy is a monotone decreasing function of μ in the parameter region $\mu > -\Lambda$. Moreover we have:

- $E(q_\mu) \to \infty$ as $\mu \to -\Lambda$,
- $E(q_\mu) \to 0$ as $\mu \to \infty$.

According to Proposition 4.2, a solution to the mean field equation (6.101) is a selective decay state if and only if $\mu \geq -1$, and it is the unique selective decay state which is nonlinear stable if $\mu > -1$. Thus, depending on the spectrum of the topography, there are two cases:

- Generic topography with $\Lambda = 1$.
  In this case, for each energy level E_0, the mean field equation (6.101) has only one solution ψ_μ given by the first case of (6.103) with $\mu = \mu(E_0) \in (-1, \infty)$. This is the unique mean field corresponding to the most probable state and is non-linearly stable.
- Topography with degenerate spectrum, i.e. $\Lambda > 1$.

  Subcase of low energy: $E_0 < E_1 = \frac{1}{2} \sum_{|\vec{k}|^2 \geq \Lambda} \frac{|\vec{k}|^2 |\hat{h}_{\vec{k}}|^2}{(|\vec{k}|^2 - 1)^2}$.
  In this subcase case, for each energy level E_0, the mean field equation (6.101) has only one solution ψ_μ given by the first case of (6.103) with $\mu = \mu(E_0) \in (-1, \infty)$ satisfying the energy constraint. This is the unique mean field corresponding to the most probable state and is non-linear stable.

  Subcase of high energy: $E_0 > E_1$.
  In this subcase, for each energy level E_0, there are infinitely many solutions to the mean field equation (6.101) given by the second case of (6.103) with $\mu = -1$ and the $c_{\vec{k}}$s satisfying the energy constraint. These solutions are selective decay states, thanks to Section 4.5, and hence correspond to most probable states according to Proposition 6.2. These states are stable in the restricted sense. See Subsection 2.2.1. However, they are not necessarily non-linearly stable.

Without geophysical effects so that $h \equiv 0$, the above theory still applies but predicts the trivial mean field equation

$$\Delta \bar{\psi}^* = \mu \bar{\psi}^*, \tag{6.108}$$

i.e. μ must be an eigenvalue of the Laplacian. Hence the mean field corresponding to most probable states must be the eigenfunction corresponding to the first eigenvalue, i.e. $\mu = -1$. Thus, the theory makes a very interesting and different prediction when geophysical effects are included, as the reader can surmise from our discussion of these simple exact solutions with and without topography from Chapter 1.

6.7 Application of the maximum entropy principle to geophysical flows with topography and mean flow

In this section, we apply the maximum entropy principle to geophysical flows with topography and large-scale mean velocity

$$\begin{aligned}
&\frac{\partial q}{\partial t} + \nabla^{\perp}\psi \cdot \nabla q = 0, \\
&\frac{dV}{dt} = -\int \frac{\partial h}{\partial x}\psi', \\
&q = \Delta\psi + h + \beta y = q' + \beta y, \quad \psi = -V(t)y + \psi',
\end{aligned} \tag{6.109}$$

with periodic geometry, where q is the potential vorticity, ψ is the stream function, V is the large-scale mean velocity, h is the topography, and β is the beta-plane constant, q' is the small-scale potential vorticity, and ψ' is the small-scale stream function, with $q' = \Delta\psi' + h = \omega + h$, where $\omega = \Delta\psi'$ is the relative vorticity.

Instead of assuming a prior distribution describing the small-scale random fluctuation of the potential vorticity, we proceed here with the maximum entropy principle without external bias and treat the two conserved quantities, i.e. the energy and enstrophy (see Section 1.3)

$$E(t) = \frac{1}{2}V^2(t) + \frac{1}{2}\int |\nabla^{\perp}\psi'|^2, \quad \mathcal{E}(t) = \beta V(t) + \frac{1}{2}\int |q'|^2 \tag{6.110}$$

as constraints on the one-point statistics. Surprising enough, the statistical prediction is very similar.

6.7.1 One-point statistics for potential vorticity and large-scale mean velocity and Shannon entropy

Definition 6.8 (One-point statistics for potential vorticity and large-scale mean velocity) *The one-point statistics for the potential vorticity q and large-scale mean velocity V consist of a probability density $(\rho(\vec{x},\lambda), \rho_0(v))$ on $\Omega \times \mathcal{R}^1 \times \mathcal{R}^1$ associated with the potential vorticity and large-scale mean velocity, which satisfies the following properties:*

1. *$\rho(\vec{x},\lambda)$ is a probability density function on $\Omega \times \mathcal{R}^1$. Moreover, for almost all $\vec{x} \in \Omega$, $\rho(\vec{x},\lambda)$ is a probability density function on $\mathcal{R}^1$, and ρ_0 is a probability density function on $\mathcal{R}^1$*

$$\rho(\vec{x},\lambda) \geq 0, \quad \int_{\mathcal{R}^1} \rho(\vec{x},\lambda)\, d\lambda = 1, \tag{6.111}$$

$$\rho_0(v) \geq 0, \quad \int_{\mathcal{R}^1} \rho_0(v)\, dv = 1. \tag{6.112}$$

2. *For almost all $\vec{x} \in \Omega$, $\rho(\vec{x}, \lambda)$ is a probability density on $\mathcal{R}^1$ and it is the probability density associated to the small-scale potential vorticity $q'(\vec{x})$ at the location $\vec{x}$*

$$\int_{q_-}^{q_+} \rho(\vec{x}, \lambda)\, d\lambda = \text{Prob}\{q_- \le q'(\vec{x}) < q_+\}. \tag{6.113}$$

3. *ρ_0 is the probability density function on $\mathcal{R}^1$ associated to the large-scale mean velocity V*

$$\int_{V_-}^{V_+} \rho_0(v)\, dv = \text{Prob}\{V_- \le V < V_+\}. \tag{6.114}$$

With these one-point statistics, the average of functions of the potential vorticity and large-scale mean velocity, $F(\vec{x}, q, V)$, can be defined analogous to (6.80) and (6.81) with ρ replaced by (ρ, ρ_0). The **Shannon entropy** is then defined as

$$\mathcal{S}((\rho, \rho_0)) = -\int_{\mathcal{R}^1} ⨍_{\Omega} \rho(\vec{x}, \lambda) \ln \rho\, d\vec{x}\, d\lambda - \int_{\mathcal{R}^1} \rho_0(v) \ln \rho_0(v)\, dv. \tag{6.115}$$

One word of caution on the notation. Here $\mathcal{S}((\rho, \rho_0))$ represents entropy, while $\mathcal{S}(\rho, \rho_0)$ from Section 6.5 represents relative entropy.

6.7.2 The constraints on the one-point statistics

Similar to the arguments leading to (6.89)–(6.93), we can formulate the constraints on the one-point statistics in the following way

$$\mathcal{C} = \mathcal{C}^{(0)} \cap \mathcal{C}^{(1)} \cap \mathcal{C}^{(\mathcal{E})} \cap \mathcal{C}^{(E)}, \tag{6.116}$$

with

$$\mathcal{C}^{(0)} = \left\{\rho_0 \,|\, \int_{\mathcal{R}^1} \rho_0(v)\, dv = 1\right\} \tag{6.117}$$

$$\mathcal{C}^{(1)} = \left\{\rho \,|\, \int_{\mathcal{R}^1} \rho(\vec{x}, \lambda)\, d\lambda = 1 \text{ for each } \vec{x}\right\} \tag{6.118}$$

$$\mathcal{C}^{(\mathcal{E})} = \left\{\rho, \rho_0 \,|\, \mathcal{E}_0 = \mathcal{E}(\rho, \rho_0) = \frac{1}{2} ⨍_{\Omega} \int_{\mathcal{R}^1} \lambda^2 \rho(\vec{x}, \lambda)\, d\lambda\, d\vec{x}\right. \tag{6.119}$$

$$\left. + \beta \int_{\mathcal{R}^1} v \rho_0(v)\, dv\right\} \tag{6.120}$$

$$\mathcal{C}^{(E)} = \left\{\rho, \rho_0 \,|\, E_0 = E(\rho, \rho_0) = -\frac{1}{2} ⨍_{\Omega} \overline{\psi'}(\overline{q'} - h)\, d\vec{x} + \frac{1}{2}\overline{V^2}\right\} \tag{6.121}$$

and

$$\overline{q'}(\vec{x}) = \int_{\mathcal{R}^1} \lambda \rho(\vec{x}, \lambda)\, d\lambda, \quad \Delta\overline{\psi'} + h = \overline{q'}(\vec{x}), \quad \overline{V^2} = \int_{\mathcal{R}^1} v^2 \rho_0(v)\, dv. \tag{6.122}$$

The constraint $\mathcal{C}^{(0)}$ says that ρ_0 is a probability measure, and the constraint $\mathcal{C}^{(1)}$ says that $\rho(\vec{x},\cdot)$ is a probability density for all $\vec{x}\in\Omega$. The constraint $\mathcal{C}^{(E)}$ corresponds to conservation of energy and the constraint $\mathcal{C}^{(\mathcal{E})}$ corresponds to conservation of enstrophy.

It is easy to check that when the associated probability measures are Dirac delta measures, the averaged energy and enstrophy defined above in (6.119)–(6.120) reduce to the original definition given in (6.110). This partially justifies the definition of such averaged energy and enstrophy.

6.7.3 Maximum entropy principle and statistical prediction

We now invoke the maximum entropy principle which dictates that the most probable probability density (ρ^*,ρ_0^*) corresponding to one-point statistics for flows satisfying the constraints of conservation of energy $E=E_0$ and conservation of enstrophy $\mathcal{E}=\mathcal{E}_0$, is given by the one that maximizes the Shannon entropy

$$\mathcal{S}((\rho^*,\rho_0^*)) = \max_{(\rho,\rho_0)\in\mathcal{C}} \mathcal{S}((\rho,\rho_0)), \tag{6.123}$$

where the constraint set $\mathcal{C}$ consists of the constraints described in (6.116)–(6.121).

To calculate the most probable state, we utilize the Lagrange multiplier method as usual, and we deduce that (ρ^*,ρ_0^*) must satisfy

$$\left.\frac{\delta\mathcal{S}}{\delta\rho}\right|_{\rho=\rho^*,\rho_0=\rho_0^*} = \left.\left(\theta\frac{\delta E}{\delta\rho}+\alpha\frac{\delta\mathcal{E}}{\delta\rho}+\tilde{\gamma}(\vec{x})\right)\right|_{\rho=\rho^*,\rho_0=\rho_0^*}, \tag{6.124}$$

$$\left.\frac{\delta\mathcal{S}}{\delta\rho_0}\right|_{\rho=\rho^*,\rho_0=\rho_0^*} = \left.\left(\theta\frac{\delta E}{\delta\rho_0}+\alpha\frac{\delta\mathcal{E}}{\delta\rho_0}+\gamma\right)\right|_{\rho=\rho^*,\rho_0=\rho_0^*}, \tag{6.125}$$

where θ is the Lagrange multiplier for the energy constraint, α is the Lagrange multiplier for the enstrophy constraint, $\tilde{\gamma}(\vec{x})$ is the Lagrange multiplier for $\mathcal{C}^{(1)}$, and γ is the Lagrange multiplier for $\mathcal{C}^{(0)}$.

Simple calculations similar to those in Section 1.5 and (6.96)–(6.97) yield the following variational derivatives

$$\begin{aligned}
\frac{\delta E}{\delta\rho} &= -\overline{\psi}'(\vec{x})\lambda, \quad & \frac{\delta E}{\delta\rho_0} &= \frac{1}{2}v^2,\\
\frac{\delta\mathcal{E}}{\delta\rho} &= \frac{1}{2}\lambda^2, \quad & \frac{\delta\mathcal{E}}{\delta\rho_0} &= \beta v,\\
\frac{\delta\mathcal{S}}{\delta\rho} &= -(1+\ln\rho), \quad & \frac{\delta\mathcal{S}}{\delta\rho_0} &= -(1+\ln\rho_0).
\end{aligned}$$

Hence (ρ^*, ρ_0^*) must satisfy

$$\ln(\rho^*) = \theta\lambda\overline{\psi'}^* - \left(\frac{\alpha}{2}\lambda^2 + \tilde{\gamma}(\vec{x}) + 1\right),$$

$$\ln(\rho_0^*) = -\frac{\theta}{2}v^2 - \alpha\beta v - \gamma - 1,$$

i.e.

$$\rho^*(\vec{x}, \lambda) = \frac{\sqrt{\alpha}}{\sqrt{2\pi}} \exp\left(-\frac{\alpha}{2}\left(\lambda - \frac{\theta}{\alpha}\overline{\psi'}^*\right)^2\right), \tag{6.126}$$

$$\rho_0^*(v) = \frac{\sqrt{\theta}}{\sqrt{2\pi}} \exp\left(-\frac{\theta}{2}\left(v + \frac{\alpha\beta}{\theta}\right)^2\right). \tag{6.127}$$

In other words, the most probable measures ρ^* and ρ_0^* are Gaussian measures for each $\vec{x}$. In this case, at each point $\vec{x}$ the first and second moments of ρ^* and the mean of ρ_0^* are given by

$$\int_{\mathcal{R}^1} \lambda\rho^*(\vec{x}, \lambda)\, d\lambda = \frac{\theta}{\alpha}\overline{\psi'}^*(\vec{x}), \tag{6.128}$$

$$\int_{\mathcal{R}^1} \lambda^2\rho^*(\vec{x}, \lambda)\, d\lambda = \left(\frac{\theta}{\alpha}\right)^2 (\overline{\psi'}^*(\vec{x}))^2 + \alpha^{-1}, \tag{6.129}$$

$$\int_{\mathcal{R}^1} v\rho_0^*(v)\, dv = -\frac{\alpha\beta}{\theta}. \tag{6.130}$$

Recall that the mean fields $\overline{\psi'}^*$ and $\overline{V}^*$ are given by

$$\overline{q'}^*(\vec{x}) = \int_{\mathcal{R}^1} \lambda\rho^*(\vec{x}, \lambda)\, d\lambda, \tag{6.131}$$

$$\overline{V}^* = \int_{\mathcal{R}^1} v\rho_0^*(v)\, dv, \tag{6.132}$$

and hence, after combining (6.127), (6.129), (6.130), and (6.131), we deduce the following **mean field equations**

$$\Delta\overline{\psi'}^* + h = \overline{q'}^*(\vec{x}) = \frac{\theta}{\alpha}\overline{\psi'}^* = \mu\overline{\psi'}^*, \tag{6.133}$$

$$\overline{V}^* = \int v\rho_0^*(v)\, dv = -\frac{\alpha\beta}{\theta} = -\frac{\beta}{\mu}, \tag{6.134}$$

with $\mu = \frac{\theta}{\alpha}$, $\alpha > 0$.

We observe that the mean field equations (6.133) and (6.134) are the same ones encountered earlier in Chapter 1, corresponding to special steady states with a linear q–ψ relation, and whose stability was analyzed in Section 4.2.

6.7.4 Determination of the multipliers and geophysical effects

Our next task is to determine the parameter μ with the given energy constraint $E(\rho^*, \rho_0^*) = E_0$ and the enstrophy constraint $\mathcal{E}(\rho^*, \rho_0^*) = \mathcal{E}_0$. We will see that geophysical effects like topography and beta-plane play very important roles in determining the behavior of the mean states.

Case 1: Non-trivial large-scale mean flow and topography

We first observe, thanks to equations (6.125) and (6.126), that $\alpha > 0$, $\theta > 0$. Hence

$$\mu = \frac{\theta}{\alpha} > 0. \tag{6.135}$$

In this case the mean field equations (6.132)–(6.133) possess a unique solution for each $\mu > 0$, which is given by

$$V_\mu = \overline{V}^* = -\frac{\beta}{\mu}, \tag{6.136}$$

$$\psi'_\mu = \overline{\psi'}^* = \sum_{|\vec{k}|\neq 0} \frac{\hat{h}_{\vec{k}}}{|\vec{k}|^2 + \mu} \exp(i\vec{k}\cdot\vec{x}). \tag{6.137}$$

In this case the total energy is given in terms of Fourier coefficients

$$E(q'_\mu, V_\mu) = \frac{1}{2} \sum_{|\vec{k}|\neq 0} \frac{|\vec{k}|^2 |\hat{h}_{\vec{k}}|^2}{(|\vec{k}|^2 + \mu)^2} + \frac{1}{2}\left(\frac{\beta}{\mu}\right)^2. \tag{6.138}$$

The total energy is a monotone decreasing function of μ in the parameter region $\mu > 0$. Moreover we have

$$E(q'_\mu, V_\mu) \to \infty \qquad \text{as } \mu \to 0,$$
$$E(q'_\mu, V_\mu) \to 0 \qquad \text{as } \mu \to \infty.$$

Thus, $0 < \mu < \infty$ *is uniquely determined by the energy* E, i.e. $\mu = \mu(E)$. Once μ is determined by the energy constraint, the mean field is then uniquely determined by (6.135)–(6.136). These states correspond to nonlinearly stable steady states with westward mean flow as previously discussed in Section 4.2.

In general, the total enstrophy of the mean field is always less than $\mathcal{E}_0 = \mathcal{E}((\rho^*, \rho_0^*))$. This is a consequence of a more general fact, indicating that the total enstrophy associated with any ρ, ρ_0 is equal to, or greater than, the total enstrophy associated with the pointwise mean of ρ, ρ_0, i.e.

$$\mathcal{E}(\overline{q'}, \overline{V}) \leq \mathcal{E}(\rho, \rho_0), \tag{6.139}$$

with

$$\overline{q'}(x) = \int \lambda\rho(\vec{x}, \lambda)\, d\lambda, \quad \overline{V} = \int v\rho_0(v)\, dv.$$

Indeed, we notice that $\lambda\rho = \lambda\rho^{\frac{1}{2}}\rho^{\frac{1}{2}}$ and use the Cauchy–Schwarz inequality $((\int fg)^2 \leq \int f^2 \int g^2)$ to get

$$\begin{aligned}
\mathcal{E}(\overline{q'}, \overline{V}) &= \frac{1}{2} \fint \overline{q'}^2(\vec{x})\, d\vec{x} + \frac{1}{2}\beta\overline{V} \\
&= \frac{1}{2} \fint \left(\int \lambda\rho(\vec{x}, \lambda)\, d\lambda \right)^2 d\vec{x} + \frac{1}{2}\beta \int v\rho_0(v)\, dv \\
&\leq \frac{1}{2} \fint \int \lambda^2 \rho(\vec{x}, \lambda)\, d\lambda \int \rho(\vec{x}, \lambda)\, d\lambda\, d\vec{x} + \frac{1}{2}\beta \int v\rho_0(v)\, dv \\
&= \frac{1}{2} \fint \int \lambda^2 \rho(\vec{x}, \lambda)\, d\lambda\, d\vec{x} + \frac{1}{2}\beta \int v\rho_0(v)\, dv \\
&= \mathcal{E}(\rho, \rho_0).
\end{aligned} \tag{6.140}$$

Moreover, thanks to (6.110), (6.119), and (6.128)–(6.133)

$$\begin{aligned}
\mathcal{E}_0 &= \mathcal{E}((\rho^*, \rho_0^*)) \\
&= \frac{1}{2}\alpha^{-1} + \frac{\mu^2}{2} \fint (\overline{\psi'}^*(\vec{x}))^2 d\vec{x} + \beta\overline{V}^* \\
&= \frac{1}{2}\alpha^{-1} + \mathcal{E}(\overline{q'}^*, \overline{V}^*) \\
&= \frac{1}{2}\alpha^{-1} + \mathcal{E}_*.
\end{aligned} \tag{6.141}$$

Thus we can calculate explicitly that

$$\alpha = (2(\mathcal{E}_0 - \mathcal{E}_*))^{-1}, \tag{6.142}$$

and hence we may recover the Lagrange multiplier for the energy as

$$\theta = \alpha\mu = \frac{\mu}{2(\mathcal{E}_0 - \mathcal{E}_*)}. \tag{6.143}$$

To summarize, in this case with non-trivial large-scale mean velocity and topography, the statistical theory predicts the unique most probable one point statistics and mean field. The one-point statistics are Gaussians. The mean field is uniquely determined by the energy level which satisfies a linear $q - \psi$ relation. These mean states correspond to westward mean flows ($\mu > 0$) and are nonlinearly stable due to the stability results in Chapter 4. We may also verify, via a straightforward calculation, that the most probable one-point statistics are those whose mean fields are the enstrophy minimizer with given energy level, i.e. selective decay states.

Case 2: No large-scale mean flow and non-trivial topography

Next we consider the important special case where there is no mean flow ($V = 0$) but there is a non-trivial topography ($h \neq 0$). In this case there is no beta-plane effect ($\beta = 0$), and therefore the only geophysical effect is due to the topography h. This is the same situation as in Section 6.6. However we are taking a slightly different approach here.

In this case the mean field equations reduce to a single equation

$$\Delta\overline{\psi}^* + h = \mu\overline{\psi}^*, \tag{6.144}$$

where the solutions are characterized by formula (6.103).

Again we face the problem of non-uniqueness of solutions to the mean field equation. Once again we compute the entropy associated with all candidates of most probable states to determine which solution corresponds to the most probable state(s). Thanks to (6.125), (6.141), and (6.143), we have

$$\begin{aligned}
\mathcal{S}(\rho^*) &= -⨍\int \frac{\sqrt{\alpha}}{\sqrt{2\pi}} \exp\left(-\frac{\alpha}{2}\left(\lambda - \frac{\theta}{\alpha}\bar{\psi}^*\right)^2\right) \\
&\quad \times \left(-\frac{\alpha}{2}\left(\lambda - \frac{\theta}{\alpha}\bar{\psi}^*\right)^2 + \ln\left(\sqrt{\frac{\alpha}{2\pi}}\right)\right) \\
&= \frac{1}{2} - \ln\left(\sqrt{\frac{\alpha}{2\pi}}\right) \\
&= \frac{1}{2} + 2\ln\left(\sqrt{\pi(\mathcal{E}_0 - \mathcal{E}_*)}\right).
\end{aligned} \tag{6.145}$$

Thus the entropy is maximized when the enstrophy of the mean field is minimized. In other words, the most probable one-point statistics are exactly those whose mean fields are selective decay states. Hence we are in the same situation as in Section 6.6.4, and the parameter μ is determined in exactly the same fashion. Thus, for generic topography with $\Lambda = 1$ in (6.102), $\mu = \mu(E_0) \in (-1, \infty)$ is uniquely determined by the energy, since the energy given by (6.105) is monotonic in μ and covers all positive values for $\mu \in (-1, \infty)$. The corresponding mean field given by the first case of (6.103) is non-linearly stable. For the case of topography with degenerate spectrum, i.e. $\Lambda > 1$ in (6.102), $\mu = \mu(E_0) \in (-1, \infty)$ is uniquely determined by the energy in the low energy case ($E_0 < E_1$) with a unique associated mean field which is non-linearly stable; $\mu = -1$ in the high energy case ($E_0 > E_1$) with infinitely many associated mean field solutions which are stable in the restricted sense.

Case 3: No large-scale mean flow or topography

We consider the special case of no geophysical effects, or $\beta = h = V = 0$. In other words we consider the special case of the energy enstrophy statistical theory for the two-dimensional Euler equations. In this case the mean field equation reduces to

$$\Delta\overline{\psi}^{*} = \mu\overline{\psi}^{*},$$

and it has a non-trivial solution only if $-\mu$ is one of the eigenvalues of the Laplacian operator on the torus $T^2 = \Omega$. Again the selective decay states are the most probable mean fields. Thus the statistical theory predicts the ground energy level states as the most probable mean fields.

This picture is very different from the ones with geophysical effects. We also notice that the mean states predicted by this statistical theory for the Euler equations are only marginally stable in this case. See the discussions in Chapters 2 and 4.

References

Abramov, R. and Majda, A. (2004), Quantifying uncertainty for non-Gaussian ensembles in complex systems. *SIAM J. Sci. Stat. Comp.* **26**(2), 411–447.

Abramov, R. Majda, A., and Kleeman, R. (2005), Information theory and predictability for low frequency variability. *J. Atmos. Sci.* **62**(1), 65–87.

Jaynes, E. (1957), Information theory and statistical mechanics. *Physical Review*, **106**, 620–630.

Lamperti, J. (1966), *Probability*. New York: Benjamin Inc.

Lasota, A. and Mackey, M. (1994), Chaos, fractals, and noise. *Applied Math. Sciences* **97**, New York: Springer-Verlag.

Majda, A., and Bertozzi, A. (2001), *Vorticity and Incompressible Flow*, Cambridge: Cambridge University Press.

Mead, L. and Papanicolaou, N. (1984), Maximum entropy in the problem of moments. *J. Math. Phys.* **25**, 2404–2417.

Rudin, W., 1974, *Real and Complex Analysis*, 2nd edn. New York: McGraw-Hill.

Shannon, C. (1948), The mathematical theory of communication, *Bell Sys. Tech. Journal*.

Shannon, C. and Weaver, W. (1949), The Mathematical Theory of Communication, Urbana, Ill: The University of Illinois Press.

Thompson, C. (1972), Mathematical Statistical Mechanics, Princeton, NJ: Princeton University Press.

7

Equilibrium statistical mechanics for systems of ordinary differential equations

7.1 Introduction

In this chapter we introduce theories of equilibrium statistical mechanics for large non-linear systems of ordinary differential equations (ODEs)

$$\frac{d\vec{X}}{dt} = \vec{F}(\vec{X}), \vec{X} \in \mathcal{R}^N, \ \vec{F} = (F_1, \ldots, F_N), \quad N \gg 1 \tag{7.1}$$
$$\vec{X}|_{t=0} = \vec{X}_0,$$

and their applications. Such large non-linear systems of ODEs may arise as the governing equations of particle systems such as the point vortex models in Chapter 9 or (Fourier mode) truncations of quasi-geostrophic equations as discussed in Chapter 8. A common feature of these systems is that the dynamics are highly chaotic and thus a single trajectory is highly unpredictable beyond a certain time due to sensitive dependence on data. Alternatively, observations as well as physical and numerical experiments often indicate the existence of coherent patterns out of large ensembles of trajectories. Hence it makes sense to study the ensemble behavior of trajectories instead of a single trajectory. This is reminiscent of atmosphere/ocean dynamics, where the prediction of one trajectory, namely the weather, is basically impossible beyond a certain time due to chaotic dynamics; while the prediction of ensembles of trajectories, corresponding to certain features of the climate, is relatively easier. We will give examples of these issues in Chapter 15.

An indispensable tool for the study of ensembles of solutions to the large systems of ODEs in (7.1) in a quantitative fashion is probability measures on the phase space ($\mathcal{R}^N$ in this case), or *statistical solutions*. The flow map naturally carries an initial probability density function $p_0(\vec{X})$ to a probability density function $p(\vec{X}, t)$ on the phase space at any future time as long as the flow exists and remains sufficiently smooth. A condition on the ODEs that ensures that $p(\vec{X}, t)$ is the simple pull-back of initial pdf $p_0(\vec{X})$ is the so-called Liouville property (7.2). The Liouville property is the first key ingredient for doing statistical mechanics for ODEs. The Liouville property is equivalent to saying that the flow map is volume preserving in the phase space. Thus, if an ensemble of initial data in the phase space is stretched in some direction under the flow map, it must be compensated for by squeezing in some other direction under the flow map and vice versa. This kind of stretching/squeezing together with the bending/twisting (mixing) implied by the non-linearity is a key mechanism in the complex/chaotic behavior of the dynamical system and in the tendency towards statistical equilibrium for large ensembles of trajectories. In the framework of equilibrium statistical mechanics, the evolution of the probability measures on phase space enables us to define the associated information-theoretic entropy which sets the stage for the application of the maximum entropy principle. In order to ensure non-trivial prediction, we need the second important ingredient for doing statistical mechanics, i.e. the existence of conserved quantities which translate into various constraints on the probability distribution at all times, thanks to the Liouville property. With the Liouville property and conserved quantities, we may invoke the maximum entropy principle to obtain the most probable states (probability distributions), or the Gibbs measure, on the phase space. The Gibbs measure turns out to be an invariant measure of the flow maps, or a stationary statistical solution. The ensemble behavior such as the mean field equations can then be deduced by taking appropriate averages with respect to the most probable probability distribution. Although more complex, this approach is more appealing than the empirical statistical approach that we introduced in Chapter 6, in the sense that we have incorporated some dynamics of the ODE system together with conserved quantities into the process of determining the most probable states. This approach also allows for more complete predictions beyond the mean state, such as statistical predictions for fluctuations about the mean. Of course the Liouville property and the existence of conserved quantities themselves are not sufficient to ensure the tendency towards the statistical equilibrium state (most probable state). Here and elsewhere we implicitly assume that the dynamics of the system is chaotic and instability is abundant so that there is some sort of ergodicity of the system, see for instance (7.32). The rigorous mathematical underpinning of the exact conditions, ensuring the convergence towards statistical equilibrium, is beyond the scope of this book and is a major unsolved mathematical problem. However, we will use elementary numerical experiments to test and confirm these predictions.

The chapter is organized as follows. In Section 7.2, we give an introduction to the general theory of equilibrium statistical mechanics for systems of ODEs satisfying the Liouville property; in Section 7.3, we apply the theory to ODEs that are the Galerkin truncation of the inviscid Burgers–Hopf equation. This is an appealing toy model for low-frequency variability of the atmosphere. Application to truncated quasi-geostrophic equations will be introduced in Chapter 8 as an integral part of the complete statistical mechanics applied to quasi-geostrophic equations, while another different application to point vortices is presented in Chapter 9. In Section 7.4, we introduce and study the Lorenz 96 "toy" model for atmospheric dynamics. In general, this model has dissipation and forcing, and, as explained, has an analogue of the weather wave packets of the actual atmosphere. When the damping and forcing vanish, this model is another system satisfying the behavior of equilibrium statistical mechanics. Quantitative comparison between the predictions of equilibrium statistical mechanics and the damped forced Lorenz 96 model are presented in detail.

7.2 Introduction to statistical mechanics for ODEs

In this section we introduce a general framework for conducting equilibrium statistical mechanics for large systems of ODEs. As was discussed above, there are two important ingredients for the application of equilibrium statistical mechanics to systems of ODEs; namely, the Liouville property, which ensures the applicability of the equilibrium statistical theory, and the existence of normalizable conserved quantities so that non-trivial most probable probability measures for doing further measurements (Gibbs measures) can emerge out of a suitable maximum entropy principle.

7.2.1 The Liouville property

We first introduce the Liouville property.

Definition 7.1 (Liouville property) *A vector field $\vec{F}(\vec{X})$ is said to satisfy the Liouville property if it is divergence free, i.e.*

$$div_{\vec{X}}\vec{F} = \sum_{j=1}^{N} \frac{\partial F_j}{\partial X_j} = 0. \tag{7.2}$$

An important consequence of the Liouville property is that it implies the flow map associated with (7.1) is volume preserving or measure preserving on the phase space. For this purpose we define the flow map $\{\Phi^t(\vec{X})\}_{t\geq 0}$, $\Phi^t : \mathcal{R}^N \mapsto \mathcal{R}^N$ associated with the finite-dimensional ODE system (7.1) by

$$\frac{d}{dt}\Phi^t(\vec{X}) = \vec{F}(\Phi^t(\vec{X})), \qquad \Phi^t(\vec{X})|_{t=0} = \vec{X}_0. \tag{7.3}$$

The divergence free nature of the vector field $\vec{F}$ gives an important property to the flow map $\Phi^t(\vec{X})$, i.e. the flow map $\Phi^t(\vec{X})$ is volume preserving or measure preserving on the phase space. This property can be shown exactly in the same way as we prove that the flow map of an incompressible fluid is volume preserving.

Proposition 7.1 $\Phi^t(\vec{X})$ *is volume preserving or measure preserving on the phase space, i.e.*

$$J(t) \stackrel{def}{=} \det\left(\nabla_{\vec{X}} \Phi^t(\vec{X})\right) \equiv 1, \tag{7.4}$$

for all time $t \geq 0$.

Proof: The proof follows from the calculus identity (cf. Majda and Bertozzi, 2001 or Chorin and Marsden, 1993)

$$\frac{dJ(t)}{dt} = \operatorname{div} \vec{F}(\Phi^t(\vec{X}))J(t), \quad J(0) = 1. \tag{7.5}$$

Since $\operatorname{div} \vec{F} = 0$, we get $J(t) = J(0) = 1$. This completes the proof of the proposition.

7.2.2 Evolution of probability measures and the Liouville equation

Since we are interested in the behavior of an ensemble of solutions, it is natural to consider probability measures on the phase space $\mathcal{R}^N$. Given an initial ($t = 0$) probability density function $p_0(\vec{X})$ (so that the associated probability measure is absolutely continuous with respect to the Lebesgue measure on $\mathcal{R}^N$), for any future time $t > 0$ we may define a probability density $p(\vec{X}, t)$ by the pull-back of the initial probability density $p_0(\vec{X})$ with the flow map $\Phi^t(\vec{X})$, i.e.

$$p(\vec{X}, t) = p_0\left((\Phi^t)^{-1}(\vec{X})\right). \tag{7.6}$$

Utilizing the measure preserving property of the flow map $\Phi^t(\vec{X})$ we may verify that this definition indeed yields a probability measure. In fact, we can prove that the one parameter family of probability measures $p(t)$ satisfies the Liouville equation, or, equivalently, that the probability density functions are transported by the vector field $\vec{F}$.

Proposition 7.2 $p(\vec{X}, t) = p_0\left((\Phi^t)^{-1}(\vec{X})\right)$ *is transported by the vector field* $\vec{F}$, *i.e. it satisfies the Liouville's equation*

$$\frac{\partial p}{\partial t} + \vec{F} \cdot \nabla_{\vec{X}} p = 0, \tag{7.7}$$

and hence $p(\vec{X}, t)$ *is a probability density function for all time.*

Proof: From the transport theorem (cf. Majda and Bertozzi, 2001 or Chorin and Marsden, 1993), we know that for any function $f(\vec{X}, t)$ in phase space

$$\frac{\partial}{\partial t}\int_{\Phi^t(\Omega)} f(\vec{X}, t) d\vec{X} = \int_{\Phi^t(\Omega)} \left(\frac{\partial f}{\partial t} + \operatorname{div}(f\vec{F})\right) d\vec{X}, \tag{7.8}$$

and since $\vec{F}(\vec{X}, t)$ is divergence free, this equation reduces to

$$\frac{\partial}{\partial t}\int_{\Phi^t(\Omega)} f(\vec{X}, t) d\vec{X} = \int_{\Phi^t(\Omega)} \left(\frac{\partial f}{\partial t} + \vec{F}\cdot\nabla_{\vec{X}} f\right) d\vec{X}. \tag{7.9}$$

Next we let $f(\vec{X}, t) = p(\vec{X}, t)$ in the last equation. From Definition 7.6 on the pull-back probability density function p, we have

$$\begin{aligned}\int_{\Phi^t(\Omega)} p(\vec{X}, t) d\vec{X} &= \int_{\Phi^t(\Omega)} p_0\left((\Phi^t)^{-1}(\vec{X})\right) d\vec{X} \\ &= \int_{\Omega} p_0(\vec{Y}) \det\left(\nabla_{\vec{Y}}\Phi^t(\vec{Y})\right) d\vec{Y} = \int_{\Omega} p_0(\vec{Y}) d\vec{Y}, \end{aligned} \tag{7.10}$$

where we made the change of variables $\vec{X} = \Phi^t(\vec{Y})$ in the equation, and utilized the fact that the Jacobian determinant of the flow map is identically equal to one. This implies that the left-hand side of (7.9) is identically zero, and thus

$$0 = \frac{\partial}{\partial t}\int_{\Phi^t(\Omega)} p(\vec{X}, t) d\vec{X} = \int_{\Phi^t(\Omega)} \left(\frac{\partial p}{\partial t} + \vec{F}\cdot\nabla_{\vec{X}} p\right) d\vec{X}. \tag{7.11}$$

Since the region Ω in phase space is arbitrary, we conclude that pointwise

$$\frac{\partial p}{\partial t} + \vec{F}\cdot\nabla_{\vec{X}} p = 0.$$

This proves Liouville's theorem.

To see that $p(t)$ is a probability density function, we notice that $p(\vec{X}, t)$ is non-negative by equation (7.6) and the assumption that p_0 is a probability density function. Next we set $\Omega = \mathcal{R}^N$ in (7.10) and we have

$$\int_{\mathcal{R}^N} p(\vec{X}, t) d\vec{X} = \int_{\mathcal{R}^N} p_0(\vec{X}) d\vec{X} = 1 \tag{7.12}$$

and therefore $p(\vec{X}, t)$ defines a probability density for $t > 0$. This concludes the proof of the proposition.

Notice that the Liouville equation (7.7) is a linear equation in p; it then follows that any function of p also satisfies the Liouville equation. Thus we have:

Corollary 7.1 *Let $G(p)$ be any function of the probability density p. Then*

$$\frac{\partial G(p)}{\partial t} + \vec{F}\cdot\nabla_{\vec{X}} G(p) = 0, \tag{7.13}$$

i.e. $G(p)$ satisfies Liouville's equation. This further implies that

$$\frac{d}{dt}\int_{\mathcal{R}^N} G(p(\vec{X},t)) = 0. \tag{7.14}$$

Proof: It follows directly from the chain rule and the last proposition

$$\frac{\partial G(p)}{\partial t} + \vec{F}\cdot\nabla_{\vec{X}} G(p) = G'(p)\left(\frac{\partial p}{\partial t} + \vec{F}\cdot\nabla_{\vec{X}} p\right) = 0. \tag{7.15}$$

Next, we integrate (7.15) over $\mathcal{R}^N$. Since $\vec{F}$ is divergence free, we have

$$\begin{aligned}\frac{d}{dt}\int_{\mathcal{R}^N} G(p(\vec{X},t)) &= \int_{\mathcal{R}^N} \frac{\partial G(p)}{\partial t} = -\int_{\mathcal{R}^N} \vec{F}\cdot\nabla_{\vec{X}} G(p)\\ &= -\int_{\mathcal{R}^N} \operatorname{div}\left(G(p)\vec{F}\right) = 0,\end{aligned} \tag{7.16}$$

and this proves the corollary. Of course, in our formal proof above, we did not worry about the precise behavior of p for large values in $\mathcal{R}^N$ when we integrated by parts.

7.2.3 Conserved quantities and their ensemble averages

We now introduce the second ingredient for doing equilibrium statistical mechanics for ODEs. We assume that the system (7.1) possesses L conserved quantities $E_l(\vec{X}(t))$, i.e.

$$E_l(\vec{X}(t)) = E_l(\vec{X}_0), \quad 1 \le l \le L. \tag{7.17}$$

These conserved quantities could be the truncated energy, enstrophy, or higher moments for the truncated quasi-geostrophic equations, or the truncated energy and linear momentum for the truncated Burgers–Hopf equation, or the Hamiltonian and angular momentum for the point-vortex system etc. The ensemble average of these conserved quantities with respect to a probability density function p is defined as

$$\overline{E}_l = \langle E_l\rangle_p \equiv \int_{\mathcal{R}^N} E_l(\vec{X})p(\vec{X})d\vec{X}, \quad 1 \le l \le L. \tag{7.18}$$

We naturally expect that these ensemble averages are conserved in time. Indeed, we have the following result:

Proposition 7.3

$$\langle E_l\rangle_{p(t)} = \langle E_l\rangle_{p_0}, \textit{for all } t. \tag{7.19}$$

Proof: The proof is a simple application of the Liouville property and a change of variable.

$$\begin{aligned}\langle E_l \rangle_{p(t)} &= \int_{\mathcal{R}^N} E_l(\vec{X}) p(\vec{X}, t) d\vec{X} = \int_{\mathcal{R}^N} E_l(\vec{X}) p_0((\Phi^t)^{-1}(\vec{X})) d\vec{X} \\ &= \int_{\mathcal{R}^N} E_l(\Phi^t(\vec{Y})) p_0(\vec{Y}) d\vec{Y} = \int_{\mathcal{R}^N} E_l(\vec{Y}) p_0(\vec{Y}) d\vec{Y} \\ &= \langle E_l \rangle_{p_0},\end{aligned}$$

where we have performed the change of variable $\vec{Y} = (\Phi^t)^{-1}(\vec{X})$, utilized the Liouville property (7.2) which implies that this change of variable is volume preserving as proved in Proposition 7.1, and the assumption that E_l is conserved in time.

7.2.4 Shannon entropy and the maximum entropy principle

Once we have a probability density function p on $\mathcal{R}^N$, we may naturally generalize the definition of information theoretic entropy on the real line from Section 6.4 to this case of an arbitrary Euclidean space $\mathcal{R}^N$. We define *the Shannon entropy* $\mathcal{S}$ for the probability density function $p(\vec{X})$ on $\mathcal{R}^N$ that is absolutely continuous with respect to the Lebesgue measure by

$$\mathcal{S}(p) = -\int_{\mathcal{R}^N} p(\vec{X}) \ln p(\vec{X}) d\vec{X}. \tag{7.20}$$

Note that the Shannon entropy is exactly identical with the Boltzmann entropy in the statistical mechanics of gas particles. Furthermore, by setting $G(p) = -p \ln p$ in Corollary 7.1, we can see that the Shannon entropy $\mathcal{S}$ is conserved in time

$$\mathcal{S}(p(t)) = \mathcal{S}(p_0). \tag{7.21}$$

The question now is how to pick our probability measure p with the least bias for doing future measurements. We invoke the maximum entropy principle and claim that the probability density function p with the least bias for conducting further measurements should be the one that maximizes the Shannon entropy (7.20), subject to the constraint set of measurements (conserved quantities) as defined through (7.18). More precisely, let

$$\mathcal{C} = \left\{ p(\vec{X}) \geq 0, \int_{\mathcal{R}^N} p(\vec{X}) d\vec{X} = 1, \langle E_l \rangle_p = \overline{E}_l, 1 \leq l \leq L \right\}. \tag{7.22}$$

The maximum entropy principle predicts that the most probable probability density function $p^* \in \mathcal{C}$ is the one that satisfies

$$\mathcal{S}(p^*) = \max_{p \in \mathcal{C}} \mathcal{S}(p). \tag{7.23}$$

7.2.5 The most probable state and Gibbs measure

In order to compute the most probable probability density function, we invoke the Lagrange multiplier method. Recall from Chapter 6 that the variational derivative of the entropy with respect to the probability density function is calculated as

$$\frac{\delta \mathcal{S}(p)}{\delta p} = -(1+\ln p), \tag{7.24}$$

and the variational derivatives of the constraints in (7.22) are calculated easily as

$$\frac{\delta \overline{E}_l}{\delta p} = E_l(\vec{X}), \tag{7.25}$$

since the constraints are linear in p. Thus the Lagrange multiplier method predicts that the most probable state must satisfy

$$-(1+\ln p^*) = \theta_0 + \sum_{l=1}^{L} \theta_l E_l(\vec{X}), \tag{7.26}$$

where θ_0 is the Lagrange multiplier for the constraint that p must be a probability density function, and θ_l is the Lagrange multiplier for $\overline{E}_l$ for each l respectively. Equation (7.26) can be rewritten as

$$p^*(\vec{X}) = c \exp\left(-\sum_{l=1}^{L} \theta_l E_l(\vec{X})\right),$$

or equivalently, we may write the most probable probability density function in the form analogous to the *Gibbs measure* in statistical mechanics for gas particle systems

$$p^*(\vec{X}) = \mathcal{G}_{\vec{\theta}}(\vec{X}) = C^{-1} \exp\left(-\sum_{l=1}^{L} \theta_l E_l(\vec{X})\right), \tag{7.27}$$

provided that the constraints are normalizable, i.e.

$$C = \int_{\mathcal{R}^N} \exp\left(-\sum_{l=1}^{L} \theta_l E_l(\vec{X})\right) d\vec{X} < \infty \tag{7.28}$$

and the θ_ls are the Lagrange multipliers so that $\mathcal{G}_{\vec{\theta}}$ satisfies (7.18).

Recall that each probability density function is transported by the vector field $\vec{F}$, thus satisfying the Liouville equation. Also, the entropy is conserved in time. Thus we expect the Gibbs measure to solve the steady state Liouville equation, since the Gibbs measure maximizes the entropy within the constraint set of (7.22). Indeed, this is a special case of the following general fact:

Proposition 7.4 *Let E_j, $1 \leq j \leq J$ be conserved quantities of the ODE (7.1). For any smooth function $G(E_1, \ldots, E_J)$, $G(E_1, \ldots, E_J)$ is a steady state solution to the Liouville equation. In particular, the most probable probability density function given by (7.27) is a steady state solution to the Liouville equation, i.e.*

$$\vec{F} \cdot \nabla_{\vec{X}} \mathcal{G}_{\vec{\theta}} = 0. \tag{7.29}$$

Proof: Since $E_j(\vec{X}(t))$ is conserved in time, we have

$$0 = \frac{d}{dt} E_j(\vec{X}(t)) = \frac{\partial E_j}{\partial t} + \vec{F} \cdot \nabla_{\vec{X}} E_j = \vec{F} \cdot \nabla_{\vec{X}} E_j, \text{ for all } j.$$

Thus

$$\vec{F} \cdot \nabla_{\vec{X}} G(E_1, \ldots, E_J) = \sum_{j=1}^{J} \vec{F} \cdot \nabla_{\vec{X}} E_j \frac{\partial G}{\partial E_j} = 0. \tag{7.30}$$

Finally setting $G = \mathcal{G}_{\vec{\theta}}$ in (7.30) we deduce (7.29). This completes the proof of the proposition.

An important consequence of a probability density function satisfying the steady state Liouville equation is that the associated probability measure is an *invariant probability measure* on the phase space.

Definition 7.2 (Invariant measure) *A measure μ on $\mathcal{R}^N$ is said to be invariant under the flow map Φ^t if*

$$\mu((\Phi^t)^{-1}(\Omega)) = \mu(\Omega) \tag{7.31}$$

for all (measurable) set $\Omega \subset \mathcal{R}^N$ and all time t.

An easy consequence of this definition and proposition 7.4 is

Corollary 7.2 *The Gibbs measure $\mathcal{G}_{\vec{\theta}}$ is an invariant measure of the dynamical system (7.1), i.e. it is a stationary statistical solution to (7.1).*

Proof: The proof is a straightforward calculation. For any (measurable) set Ω, we have

$$\begin{aligned}
\frac{d}{dt} \int_{(\Phi^t)^{-1}(\Omega)} \mathcal{G}_{\vec{\theta}}(\vec{X}) d\vec{X} &= \frac{d}{dt} \int_{\Omega} \mathcal{G}_{\vec{\theta}}(\Phi^t(\vec{Y})) d\vec{Y} \\
&= \int_{\Omega} \vec{F} \cdot \nabla_{\vec{X}} \mathcal{G}_{\vec{\theta}}(\vec{X})|_{\vec{X}=\Phi^t(\vec{Y})} \, d\vec{Y} \\
&= 0,
\end{aligned}$$

where in the first equality we have performed the change of variable $\vec{X} = \Phi^t(\vec{Y})$ and utilized the volume-preserving property of the flow Φ^t as well as the fact that Φ^t is the flow generated by the vector field $\vec{F}$. From this we conclude that $\int_{(\Phi^t)^{-1}(\Omega)} \mathcal{G}_{\vec{\theta}}(\vec{X})d\vec{X}$ is independent of time, and therefore

$$\int_{(\Phi^t)^{-1}(\Omega)} \mathcal{G}_{\vec{\theta}}(\vec{X})d\vec{X} = \int_{\Omega} \mathcal{G}_{\vec{\theta}}(\vec{X})d\vec{X},$$

which implies that the Gibbs measure is an invariant probability measure of Φ^t. This ends the proof of the corollary.

7.2.6 Ergodicity and time averaging

The equilibrium statistical mechanics formulation gives us clear and simple predictions of behavior of ensemble averages of solutions to the ODE system (7.1). The natural question then is if such a theory is valid. In other words, we want to know if a typical ensemble of solutions does behave as predicted by the statistical theory. Taking the ensemble average of exact solutions of (7.1) analytically is an extremely challenging problem. A practical approach is to utilize numerical simulations. The obvious direct numerical approach is to simulate a very large ensemble of trajectories with initial distribution satisfying the Gibbs measure and then take their average at a later (large) time. This is usually very costly in realistic systems for the atmosphere, and ocean and such ensemble predictions will be studied in Chapter 15. Instead, we usually assume that the Gibbs measure is *ergodic* so that the spatial average is equivalent to the time average, i.e.

$$\langle g(\vec{X})\rangle = \int_{\mathcal{R}^N} g(\vec{X})\mathcal{G}_{\vec{\theta}}(\vec{X})d\vec{X} = \lim_{T\to\infty}\frac{1}{T}\int_{T_0}^{T_0+T} g(\vec{X}(t))\,dt \tag{7.32}$$

for almost all trajectories $\vec{X}(t)$. Such an ergodicity assumption allows us to verify the statistical prediction utilizing the long time average of one "typical" trajectory.

7.2.7 A simple example violating the Liouville property

Clearly, the Liouville property in (7.2) for the vector field $\vec{F}$ is central for the above theory. In order to appreciate the importance of this assumption, we present a dynamical system, which has a family of conserved quantities, but violates the Liouville property (7.2) and has no invariant measure that is absolutely continuous with respect to the Lebesgue measure.

Let $\vec{X} = (X_1, X_2)^T$ and $\vec{F}^T = X_1\vec{X}^{\perp} = (-X_1X_2, X_1^2)$, i.e. consider the following dynamical system

$$\frac{dX_1}{dt} = -X_1X_2, \quad \frac{dX_2}{dt} = X_1^2.$$

The system has the family of conserved quantities

$$E(\vec{X}\cdot\vec{X})$$

for an arbitrary smooth function, $E(s)$. Also, the vector field $\vec{F}$ violates the Liouville property since

$$\mathrm{div}_{\vec{X}}\vec{F} = -X_2 \neq 0.$$

Statistical solutions of this equation satisfy the Fokker–Planck equation

$$\frac{\partial p_t}{\partial t} + \mathrm{div}_{\vec{X}}(\vec{F}p_t) = 0,$$

and invariant measures like the Gibbs measure, which are functions of the conserved quantities, necessarily are steady solutions of this equations, i.e.

$$\mathrm{div}_{\vec{X}}(\vec{F}p_t) = 0.$$

Now it is straightforward to verify that for any smooth conserved quantity $E(\vec{X}\cdot\vec{X})$

$$\mathrm{div}_{\vec{X}}(\vec{F}E) \neq 0$$

so that none of the non-trivial functions of the conserved quantities defines an invariant measure.

In fact, it is easy to see that all solutions converge to the vertical axis $(X_1 = 0)$ and thus all invariant measures must be supported on the vertical axis, and hence not absolutely continuous with respect to the Lebesgue measure. Indeed, we can show that all invariant measures for this dynamical system must be given in the form of

$$\delta(X_1)\times\rho(X_2), \quad \text{for any } \rho \in PM(\mathcal{R})$$

and thus it solves the Fokker–Planck equation in the sense of distribution and is not absolutely continuous with respect to the Lebesgue measure.

7.3 Statistical mechanics for the truncated Burgers–Hopf equations

In this section we apply the general theory of equilibrium statistical mechanics for ODEs that we introduced in Section 7.2 to the truncated Burgers–Hopf equations.

This is a simple model exhibiting many qualitative features of the atmosphere, such as long time correlation for the large-scale structures versus the short time correlation for the small-scale structures. Since atmosphere/ocean dynamics is extremely complicated, it makes sense to study simplified models that capture some specific features of the atmosphere/ocean dynamics. This is one feature of the approach in this book.

7.3.1 The truncated Burgers–Hopf systems and their conserved quantities

We first introduce the ODE systems that we want to investigate.

The inviscid Burgers–Hopf equation

$$\frac{\partial}{\partial t}u + \frac{1}{2}\left(u^2\right)_x = 0, \tag{7.33}$$

is one of the extensively studied models in applied mathematics. One of the prominent features of this equation is the formation and propagation of shocks (discontinuity) from smooth initial data. Here we consider spectral truncations of the Burgers–Hopf equation (7.33) which conserve energy. Such a numerical approximation would be disastrous if we are interested in the dynamics of (7.33), since shock waves form and dissipate energy. However, our interest here is models that capture certain salient features of the atmosphere/ocean dynamics, and not that of (7.33). Suppose the Burgers–Hopf equation is equipped with periodic boundary conditions so that Fourier series may be applied. Let $P_\Lambda f = f_\Lambda$ denote the finite Fourier series truncation of f, i.e.

$$P_\Lambda f = f_\Lambda = \sum_{|k|\le\Lambda} \hat{f}_k e^{ikx}. \tag{7.34}$$

Here and elsewhere in this section it is tacitly assumed that f is 2π-periodic and real valued so that the complex Fourier coefficients $\hat{f}_k$ satisfy $\hat{f}_{-k} = \hat{f}_k^*$, where $*$ denotes the complex conjugation in this context. The positive integer, Λ, from (7.34) defines the number of complex-valued degrees of freedom in the approximation. With these preliminaries, the model introduced and studied here is the approximation to (7.33)

$$(u_\Lambda)_t + \frac{1}{2}P_\Lambda(u_\Lambda^2)_x = 0. \tag{7.35}$$

This is the Fourier–Galerkin truncated approximation to (7.33). With the expansion

$$u_\Lambda(t) = \sum_{|k|\le\Lambda} \hat{u}_k(t)e^{ikx}, \qquad \hat{u}_{-k} = \hat{u}_k^* \tag{7.36}$$

the equations in (7.35) can be written equivalently as the following system of non-linear ordinary differential equations for the complex amplitudes $\hat{u}_k(t)$ with $|k| \le \Lambda$

$$\frac{d}{dt}\hat{u}_k = -\frac{ik}{2}\sum_{k=p+q,\,|p|,\,|q|\le\Lambda} \hat{u}_p\hat{u}_q. \tag{7.37}$$

It is elementary to show that solutions of the equations in either (7.35) or (7.37) conserve both momentum and energy, i.e.

$$M = \fint u_\Lambda(t) = \hat{u}_0 \tag{7.38}$$

and

$$E = \frac{1}{2} \fint u_\Lambda^2 dx = \frac{1}{2} |\hat{u}_0|^2 + \sum_{k=1}^{\Lambda} |\hat{u}_k|^2 \tag{7.39}$$

are constant in time for solutions of (7.35) or (7.37). The proof for conservation of energy is as follows

$$\frac{\partial}{\partial t} \int u_\Lambda^2 = -\frac{1}{2} \int u_\Lambda P_\Lambda \left(u_\Lambda^2\right)_x = -\frac{1}{2} \int u_\Lambda \left(u_\Lambda^2\right)_x = -\frac{1}{3} \int \left(u_\Lambda^3\right)_x = 0.$$

The momentum constraint in (7.38) is associated with trivial dynamical behavior and thus after a Galilean transformation, we have $M = 0$, so that $\hat{u}_0(t) \equiv 0$ in the formula for the energy, E, in (7.39). Also, all of the sums in (7.37) involve only k with $1 \leq k \leq \Lambda$. In addition to the conserved quantities in (7.38) and (7.39), it is easy to check that the third moment

$$H = \int u_\Lambda^3 dx \tag{7.40}$$

is also conserved. Indeed

$$\begin{aligned} \frac{d}{dt} \int u_\Lambda^3 \, dx &= 3 \int u_\Lambda^2 (u_\Lambda)_t \, dx = -\frac{3}{2} \int u_\Lambda^2 P_\Lambda \left(u_\Lambda^2\right)_x \\ &= -\frac{3}{2} \int P_\Lambda (u_\Lambda^2) P_\Lambda \left(u_\Lambda^2\right)_x = 0. \end{aligned}$$

The quantity, H, is actually the Hamiltonian for the truncated Burgers–Hopf system (see Abramov *et al.*, 2003).

It is well known that non-trivial smooth solutions of (7.33) develop discontinuities in finite time and thus exhibit a transfer of energy from large scales to small scales. For functions with $\hat{u}_k(t)$ identically zero instantaneously for $k > \frac{\Lambda}{2}$, the approximation in (7.35) or (7.37) represents this energy transfer exactly; however, once this transfer develops in a general solution of (7.35) or (7.37), the conservation of energy constraint rapidly redistributes the energy in the smaller scales to the larger-scale modes. This effect is responsible for the intuitive fact that the small-scale modes of the system should decorrelate more rapidly than the large-scale modes. We discuss this important effect quantitatively later in this section.

7.3.2 The Liouville property

In order to apply the equilibrium statistical mechanics procedure that we introduced in Section 7.2, we need to verify that the ODE system (7.35) or (7.37) satisfies the Liouville property. For this purpose, we need to write it in canonical form as in (7.1). Let $\hat{u}_k$, $1 \leq k \leq \Lambda$ be the defining modes for the equations in (7.37), and let

$$\hat{u}_k = a_k + ib_k, \quad 1 \leq k \leq \Lambda \tag{7.41}$$

with a_k and b_k being the real and imaginary part of $\hat{u}_k$ respectively. Thus we have $N = 2\Lambda$ number of unknowns. Let

$$\vec{X} = (a_1, b_1, \ldots, a_\Lambda, b_\Lambda), \quad X_{2k-1} = a_k, X_{2k} = b_k, \quad 1 \leq k \leq \Lambda. \tag{7.42}$$

Equation (7.37) can then be written in a compact form

$$\frac{d\vec{X}}{dt} = \vec{F}(\vec{X}), \quad \vec{X}(0) = \vec{X}_0.$$

with the vector field F satisfying the property

$$\begin{aligned} F_{2k-1}(\vec{X}) &= k(X_{2k}X_{4k-1} - X_{2k-1}X_{4k}) \\ &\quad + \tilde{F}_{2k-1}(X_1, \ldots, X_{2k-2}, X_{2k+1}, \ldots, X_N) \\ F_{2k}(\vec{X}) &= k(X_{2k-1}X_{4k-1} + X_{2k}X_{4k}) \\ &\quad + \tilde{F}_{2k}(X_1, \ldots, X_{2k-2}, X_{2k+1}, \ldots, X_N), \end{aligned} \tag{7.43}$$

where $X_k = 0$ if $k > N$. It is then straightforward to check that the vector field $\vec{F} = (F_1, \ldots, F_N)$ is divergence free and hence (7.35) and (7.37) satisfy the Liouville property.

7.3.3 The Gibbs measure and the prediction of equipartition of energy

First, assume that the momentum, $M_0 = 0$, and that the energy E in (7.39) is the relevant statistical quantity for further measurements. Combining the Liouville property together with the conserved energy and the general framework from Section 7.2 we deduce from (7.27) that the most probable distribution, or the canonical Gibbs measure, is given by

$$\mathcal{G}_\theta = C_\theta^{-1} \exp\left(-\frac{\theta}{2}\sum_{k=1}^{\Lambda} |\hat{u}_k|^2\right), \qquad \theta > 0. \tag{7.44}$$

This is a product of identical Gaussian measures. For a given value of the mean energy $\overline{E}$, the Lagrange multiplier θ is given by, thanks to (7.39)

$$\theta = \frac{\Lambda}{\overline{E}}, \quad \text{and } \operatorname{Var}\{a_k\} = \operatorname{Var}\{b_k\} = \frac{1}{\theta} = \frac{\overline{E}}{\Lambda} \tag{7.45}$$

where Var denotes the variance. The canonical Gibbs ensemble predicts a zero mean state and a spectrum with *equi-partition of energy* in all modes according to (7.44).

Next we ask whether the same procedure would work for the conserved quantity given by $H = \int u_\Lambda^3 \, dx$ from (7.40). This is an odd function of u_Λ and grows cubically at infinity so that the associated Gibbs measure is infinite. Thus, this conserved quantity provides an example which is not statistically normalizable so that the intergal in (7.28) is infinite even though it is the Hamiltonian for this system. We discuss the statistical relevance of this conserved quantity for the dynamics briefly later in this book (see Abramov *et al.*, 2003).

7.3.4 Numerical evidence of the validity of the statistical theory

We show below that the statistical prediction (7.44) of equi-partition of energy in all modes is satisfied with surprising accuracy for Λ of moderate size. For all the numerical results reported here (Majda and Timofeyev, 2000, 2002), a pseudo-spectral method of spatial integration combined with fourth-order Runge–Kutta time stepping is utilized for (7.37). We can anticipate from (7.44), which predicts energy equi-partition, that in dynamic simulations both the high and low spatial wave numbers are equally important, so increased spatial resolution in the pseudo-spectral algorithm is needed. This is achieved by increasing the length of the array containing the discrete Fourier coefficients by adding zeros for wave numbers k with $\Lambda < k \leq \Lambda_*$ and then performing the discrete Fourier transform to the x-space on this bigger array for pseudo-spectral computations. The choice $\Lambda_* \geq 4\Lambda$ works well in practice and is utilized below. The typical time step for the Runge–Kutta time integrator is $\Delta t = 2 \times 10^{-4}$, with the necessity to use smaller time steps for larger values of the total energy $\overline{E}$ (smaller θ and/or larger Λ). In all simulations presented below the energy computed from (7.39) is conserved within $10^{-4}\%$, a relative error of 10^{-6}.

All statistical quantities are computed as time averages, i.e. we assume that the Gibbs measure (7.44) is ergodic and hence (7.32) applies. In particular, the energy in the kth mode is computed by

$$\langle |\hat{u}_k|^2 \rangle = \frac{1}{T} \int_{T_0}^{T+T_0} |\hat{u}_k(t)|^2 dt. \tag{7.46}$$

In the simulation presented here, the initial averaging value, $T_0 = 100$, and averaging window, $T = 5000$, were utilized. This is a severe test because only the micro-canonical statistics for an individual solution of (7.37) are utilized rather than a Monte-Carlo average over many random initial data as given by the canonical Gibbs ensemble in (7.44).

The initial data are selected at random with Fourier coefficients u_k, $1 \leq k < \Lambda - 15$ sampled from a Gaussian distribution with mean zero and variance $0.8 \times \theta^{-1}$. The tail Fourier coefficients with $\Lambda - 15 \leq k \leq \Lambda$ are initialized with random phases and equal amplitudes

$$|u_k|^2 = \frac{1}{15}\left(\overline{E} - \sum_{j=1}^{\Lambda-16} |u_j|^2\right), \qquad \Lambda - 15 \leq k \leq \Lambda$$

to satisfy the energy constraint given by $\overline{E}$.

A wide range of simulations of (7.37) have been performed in several parameter regimes. Here we discuss the case $\theta = 10$ and $\Lambda = 100$. Truncations varying both Λ and θ (Majda and Timofeyev, 2000, 2002) exhibit qualitatively similar behavior and we focus our attention on the numerical simulations with $\theta = 10$, $\Lambda = 100$. Many more results of this type can be found in the two papers mentioned above.

The energy spectrum for the real and imaginary parts of the Fourier modes is presented in the top and the bottom parts of Figure 7.1, respectively. The straight lines in Figure 7.1 correspond to the theoretically predicted value

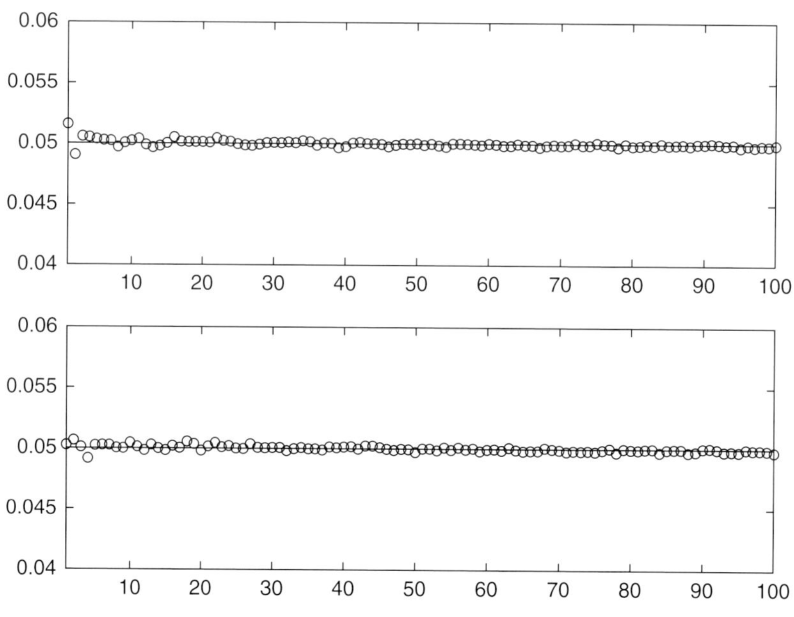

Figure 7.1 Galerkin truncation with $\theta = 10$, $\Lambda = 100$. Energy spectrum; circles – Numerics, solid line – analytical predictions. Note the small vertical scale consistent with errors of at most a few percent.

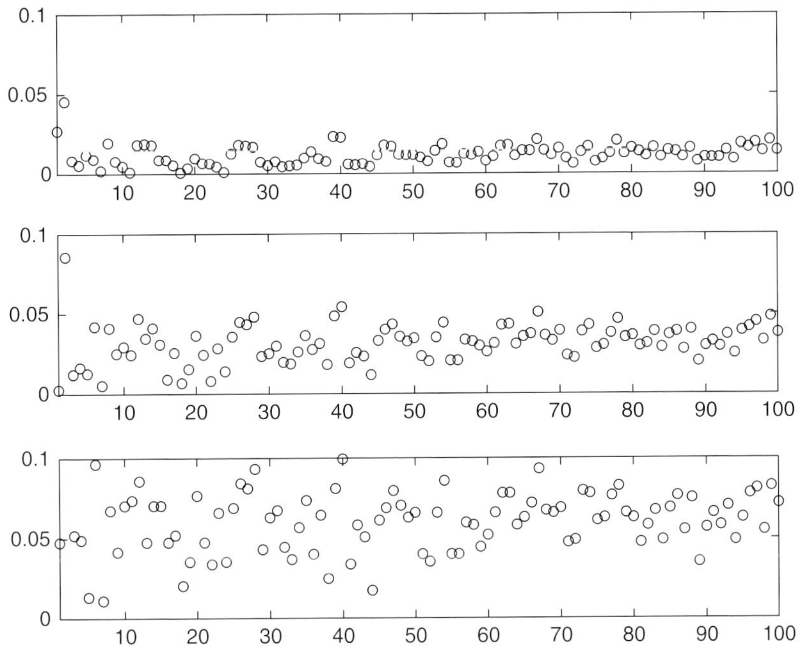

Figure 7.2 Galerkin truncation with $\theta = 10$, $\Lambda = 100$. Relative errors for numerical and analytical estimates for 4th, 6th, and 8th moments; vertical axis = 10% relative error. Note the small vertical scale consistent with the small errors.

$\mathrm{Var}\{\mathrm{Re}\hat{u}_k\} = \mathrm{Var}\{\mathrm{Im}\hat{u}_k\} = 0.05$ with $\theta = 10$. Clearly, there is statistical equi-partition of energy for times t with $t \geq 100$ in the solution of (7.37); the relative errors mostly occur at large scales and do not exceed 3.5%.

The equi-partition of the energy predicted by the theory in (7.44) is very robust. The statistical predictions of (7.44) go beyond the energy spectrum and also predict Gaussian behavior for the higher moments. Here we observe in Figure 7.2 strong numerical evidence supporting the Gaussian nature of the dynamics in (7.44) with the large variance, $\theta = 10$. Recall that for the Gaussian distribution in (7.44), the following relationships between second and higher moments hold

$$\begin{aligned}
\langle |\mathrm{Re}\hat{u}_k|^4 \rangle &= 3\langle |\mathrm{Re}\hat{u}_k|^2 \rangle^2 = \frac{3}{4}\theta^{-2} \\
\langle |\mathrm{Re}\hat{u}_k|^6 \rangle &= 15\langle |\mathrm{Re}\hat{u}_k|^2 \rangle^3 = \frac{15}{8}\theta^{-3} \\
\langle |\mathrm{Re}\hat{u}_k|^8 \rangle &= 105\langle |\mathrm{Re}\hat{u}_k|^2 \rangle^4 = \frac{105}{16}\theta^{-4}.
\end{aligned} \tag{7.47}$$

To check the Gaussianity of the dynamics we compute the relative error between the analytical predictions in (7.47) and numerical estimates given by

$$\begin{aligned}
\text{Rel. Err.}\{\text{4th Moment}\} &= |\langle|\text{Re}\hat{u}_k|^4\rangle - \frac{3}{4}\theta^{-2}| \times \frac{4}{3}\theta^2 \\
\text{Rel. Err.}\{\text{6th Moment}\} &= |\langle|\text{Re}\hat{u}_k|^6\rangle - \frac{15}{8}\theta^{-3}| \times \frac{8}{15}\theta^3 \\
\text{Rel. Err.}\{\text{8th Moment}\} &= |\langle|\text{Re}\hat{u}_k|^8\rangle - \frac{105}{16}\theta^{-4}| \times \frac{16}{105}\theta^4.
\end{aligned} \tag{7.48}$$

In Figure 7.2, the relative error in (7.48) for the fourth, sixth, and eighth moments is computed as a function of the wave number from the simulations of (7.37) with $\theta = 10$ and $\Lambda = 100$. Relative errors in the fourth moment prediction (the top part of Figure 7.2) are less than 2% for almost all wave numbers and never exceed 3.5%, with the largest errors for the low wave numbers. For the sixth moments (the middle part of Figure 7.2) the relative errors are less than 5% for most of the wave numbers and do not exceed 9% overall. For the eighth moment (the bottom part of Figure 7.2) the relative errors are about 7–8% for most of the Fourier modes. Thus, the higher-order statistics agree with the predictions of the invariant measure in (7.44), and the Gaussianity of the dynamics of the equations in (7.37) is confirmed with surprising accuracy.

7.3.5 Truncated Burgers–Hopf equation as a model with statistical features in common with atmosphere

In this section, we show that the truncated Burgers–Hopf equation is a simple model with many degrees of freedom with many statistical properties similar to those occurring in dynamical systems relevant to the atmosphere (Majda and Timofeyev, 2000, 2002). These properties include long time correlated large-scale modes of low frequency variability and short time correlated "weather modes" at small scales (see Abramov *et al.*, 2005 and references therein). A simple correlation scaling theory for the model is developed below and is confirmed to be valid over a wide range of scales by simple numerical experiments (Majda and Timofeyev, 2000, 2002, Abramov *et al.*, 2003). Subsequently in Chapter 15 of this book, this model is utilized to test recent theories of predictability utilizing relative entropy.

A scaling theory for temporal correlations

It is a simple matter to present a scaling theory which predicts that the temporal correlation times of the large-scale modes are longer than those for the small-scale modes. Recall from (7.45) that the statistical predictions for the energy per mode is $\overline{E}/\Lambda = \theta^{-1}$; since $\overline{E}/\Lambda$ has units length^2/time^2, and the wave number k has units length^{-1}, the predicted eddy turnover time for the kth mode is given by

$$\left(\frac{\Lambda}{\overline{E}}\right)^{1/2} \frac{1}{k} = \frac{\sqrt{\theta}}{k}, \qquad 1 \le k \le \Lambda.$$

If the physical assumption is made that the kth mode decorrelates on a time scale proportional to the eddy turnover time with a universal constant of proportionality, C_0, a simple plausible scaling theory for the dynamics of the equation in (7.35) or (7.37) emerges and predicts that the correlation time for the kth mode, T_k, is given by

$$T_k = \frac{C_0\sqrt{\theta}}{k}, \qquad 1 \le k \le \Lambda. \tag{7.49}$$

Thus, the scaling theory implied by (7.49) shows that the larger-scale modes in the system should have longer correlation times than the smaller-scale modes. This basic qualitative fact is always confirmed in the numerical simulations for the Fourier Galerkin truncation with $\Lambda \ge 5$ (Majda and Timofeyev, 2000, 2002; Abramov *et al.*, 2003).

Next we present one example confirming this numerical behavior.

Numerical evidence for the correlation scaling theory

For the numerical results reported here, the correlation function of the kth mode, $X_k = \text{Re } \hat{u}_k$, is computed from time averaging as in (7.46), i.e. for any time τ, the correlation function $C_k(\tau)$ is computed by

$$C_k(\tau) = \frac{1}{T}\int_{T_0}^{T_0+T} X_k(t+\tau)X_k(t)dt, \tag{7.50}$$

with T picked sufficiently large. We also define the correlation time for the kth mode through

$$T_k = \int_0^\infty |C_k(\tau)|d\tau. \tag{7.51}$$

Here results are presented for the same case, $\theta = 10$, $\Lambda = 100$, studied earlier to confirm equipartition of energy. The time correlations of Re $\hat{u}_k$ for $k = 1, 2, 3, 10, 15, 20$ are presented in Figure 7.3 illustrating the wide range of time scales present in the system. The elementary scaling theory for correlation times developed in (7.49) is compared with the numerically computed correlation times in Figure 7.4. There the scaling formula in (7.49) is multiplied by a normalization constant, C_0, to exactly match the correlation time for the mode with $k = 1$. As shown in Figure 7.4, the simple theory proposed in (7.49) is an

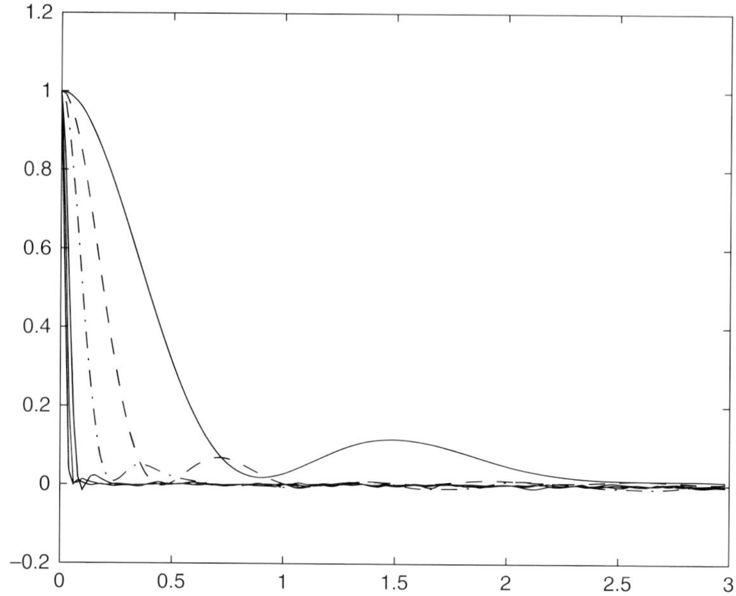

Figure 7.3 Galerkin truncation with $\beta = 10$, $\Lambda = 100$. Correlation functions for modes $k = 1, 2, 3, 10, 15, 20$.

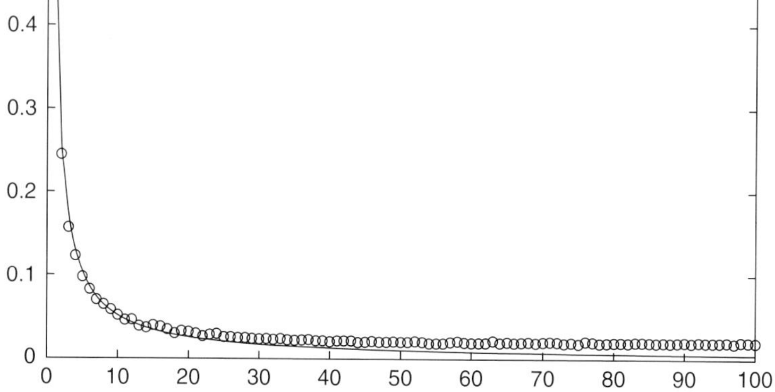

Figure 7.4 Galerkin truncation with $\beta = 10$, $\Lambda = 100$. Correlation times v. wave number k. Circles – DNS, solid line – predictions of the scaling theory.

excellent fit for the large-scale wave numbers, k, with $k \leq 15$, which exhibit the largest range of scaling behavior for the temporal correlations. These results are very robust with even better agreement in many diverse circumstances (Majda and Timofeyev, 2000, 2002; Abramov *et al.*, 2003).

7.4 The Lorenz 96 model

Here we discuss the Lorenz 96 model (L96) as a simple toy model with the same statistical features of the weather wave packets of the atmosphere. We also carefully compare the behavior of L96 with the undamped unforced IL96 model regarding its climatology.

The L96 model is given by

$$\frac{du_j}{dt} = (u_{j+1} - u_{j-2})u_{j-1} - u_j + F, \quad j = 0 \dots J \tag{7.52}$$

with periodic boundary conditions, $u_0 = u_J$. It is easy to see that (7.52) is consistent (a finite difference discretization) with the following damped forced Burgers–Hopf equation

$$u_t - uu_x = -u + F,$$

with grid size $\Delta x = 1/3$.

7.4.1 Geophysical properties of the Lorenz 96 model

As mentioned before, the L96 model possesses some general properties of geophysical models, namely energy-preserving advection, damping, and forcing. The term $-u_j$ in (7.52) represents damping (with a unit time scale of five days), while F represents constant "solar forcing." The model in (7.52) is designed to mimic midlatitude weather and climate behavior, so periodic boundary conditions are appropriate. The unit spatial scale between discrete nodes is regarded as a non-dimensional midlatitude Rossby radius $\approx$800 km; the discrete system size is set to be $J = 40$ nodes, so the discrete system with 40 modes corresponds to a midlatitude belt of roughly 30 000 km.

In midlatitude weather systems, the main "weather waves," the Rossby waves, have westward (toward negative x) phase velocity, but from our own anecdotal experience, weather systems collectively move eastward (toward positive x) with unstable behavior. The models in (7.52) have analogous behavior. To see this, note that $u_j = F$ defines a steady state solution of (7.52). We carry out this linear analysis next and confirm the above intuition.

Linear stability of simple steady states

With the steady state solution, $u_j = F$, we write (7.52) in terms of perturbations around the steady state: taking

$$v_j = F + u_j,$$

the L96 model is written in new variables as

$$\frac{dv_j}{dt} = F(v_{j+1} - v_{j-2}) + (v_{j+1} - v_{j-2})v_{j-1} - v_j. \tag{7.53}$$

To check linear stability we linearize (7.53) and obtain

$$\frac{dv_j}{dt} = F(v_{j+1} - v_{j-2}) - v_j. \tag{7.54}$$

Proceeding with linear analysis of the Fourier modes, we introduce the discrete Fourier transform

$$\hat{v}_k = \frac{1}{J}\sum_{j=0}^{J-1} v_j e^{-2\pi ikj/J}.$$

In terms of Fourier coefficients the equation (7.54) becomes

$$\frac{d\hat{v}_k}{dt} = a_k \hat{v}_k, \tag{7.55}$$

where

$$a_k = [(e^{2\pi ik/J} - e^{-4\pi ik/J})F - 1].$$

The real part of a_k is

$$\Re(a_k) = [\cos(2\pi k/J) - \cos(4\pi k/J)]F - 1,$$

and the linear instability occurs when $\Re(a_k) > 0$, or

$$\cos(2\pi k/J) - \cos(4\pi k/J) > \frac{1}{F}. \tag{7.56}$$

The next section offers detailed information about geophysical properties of the model based on these linearly unstable waves. See Table 7.1 for a list of bands of linearly stable and unstable waves.

Rossby waves

As mentioned earlier, the L96 model has the analogues of Rossby waves for geophysical flows. We again write the linearized model (7.54)

$$\frac{dv_j}{dt} = F(v_{j+1} - v_{j-2}) - v_j.$$

Table 7.1 *Bands of linearly stable and unstable waves for different values of F and ū, J = 40*

F	$\bar{u}$	k_{begin}	k_{end}
5	1.63668	5	11
6	2.0109	4	12
8	2.34299	4	12
16	3.08713	3	12
32	3.89012	3	12
∞	∞	0	13

Assuming v_j to be of the form

$$v_j = \sum_k A_k \exp\left(\frac{2\pi i k j}{J} - i\omega(k)t\right), \tag{7.57}$$

we substitute (7.57) into (7.54) and obtain the dispersion relation $\omega(k)$

$$\omega(k) = i\left[\left(e^{\frac{2\pi i k}{J}} - e^{-\frac{4\pi i k}{J}}\right)F - 1\right]. \tag{7.58}$$

The imaginary part of $\omega(k)$ determines linear stability/instability of the wave, which has already been shown in the previous section. Denoting $\tilde{\omega}(k) \equiv \Re(\omega(k))$, for $\tilde{\omega}(k)$ we write

$$\tilde{\omega}(k) = -\left[\sin\left(\frac{2\pi k}{J}\right) + \sin\left(\frac{4\pi k}{J}\right)\right]F. \tag{7.59}$$

The equation in (7.59) determines the phase and group velocities of the wave packet in (7.57). The phase velocity (velocity of a single wave) of (7.57) is defined as

$$V(k) = \frac{\tilde{\omega}(k)}{k} = -\left[\sin\left(\frac{2\pi k}{J}\right) + \sin\left(\frac{4\pi k}{J}\right)\right]\frac{F}{k}, \tag{7.60}$$

while the group velocity (velocity of a packet of waves) is defined as

$$V_{gr}(k) = \frac{d\tilde{\omega}(k)}{dk} = -\frac{2\pi F}{J}\left[\cos\left(\frac{2\pi k}{J}\right) + \cos\left(\frac{4\pi k}{J}\right)\right]. \tag{7.61}$$

We denote the direction of positive Vs as "eastward," and the direction of negative Vs as "westward." The most interesting waves are those which have "westward" phase velocities like Rossby waves, and "eastward" group velocities like actual weather systems as discussed earlier in Chapter 1. From (7.60), we find that:

- Waves with $\left(\frac{1}{3} < \frac{k}{J} < \frac{1}{2}\right) \bigcup \left(\frac{2}{3} < \frac{k}{J} < 1\right)$ have "eastward" phase velocity.
- Waves with $\left(0 < \frac{k}{J} < \frac{1}{3}\right) \bigcup \left(\frac{1}{2} < \frac{k}{J} < \frac{2}{3}\right)$ have "westward" phase velocity.

Looking at (7.61) we observe that:

- Waves with $\left(0.1489 < \frac{k}{J} < 0.4096\right) \bigcup \left(0.5904 < \frac{k}{J} < 0.8511\right)$ have "eastward" group velocity.
- The rest of the waves has "westward" group velocity.

Combining the two results we obtain

$$\left(0.1489 < \frac{k}{J} < \frac{1}{3}\right) \bigcup \left(0.5904 < \frac{k}{J} < \frac{2}{3}\right). \tag{7.62}$$

The wave numbers in (7.62) have "westward" phase and "eastward" group velocities.

Linearly unstable waves play an important role in the dynamics of the L96 model. The linear instability criterion is already given in (7.62). Figures 7.5–7.9 display phase and group velocities and stable and unstable wave numbers for $J = 40$, and $F = 5$, 6, 8, 16, and 32. Note that the wave numbers of interest (which are linearly unstable and have phase and group velocities of opposite signs) generally lie between wave numbers $k = 5$ and $k = 12$, and these correspond to the reasonable wave lengths of baroclinic instability in the atmosphere (Pedlosky, 1987).

Below we call the solutions of (7.52) without damping and forcing the undamped unforced IL96 model and denote it by $F = 0$. In Figure 7.10 we demonstrate the propagation of typical solutions of the L96 model in time with

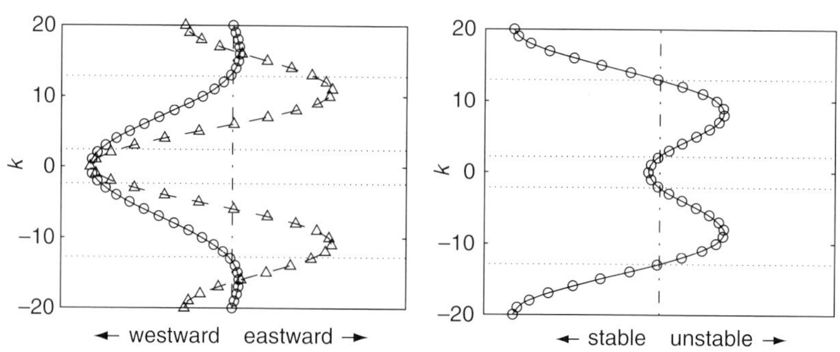

Figure 7.5 The phase/group velocities (left) and linearly stable/unstable wave numbers (right) for $F = 5$ dynamical regime. Left plot: solid line with circles – phase velocity, dashed line with triangles – group velocity, vertical dash–dotted line marks zero velocity, horizontal dotted lines mark boundaries between stable and unstable regions. Right plot: solid line with circles is $[\cos(2\pi k/J) - \cos(4\pi k/J) - (1/F)]$, vertical dash–dotted line marks zero (neither stability nor instability), horizontal dotted lines mark boundaries between stable and unstable regions.

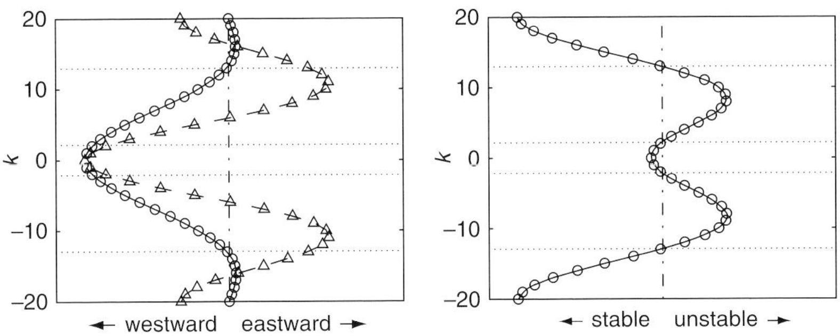

Figure 7.6 The phase/group velocities (left) and linearly stable/unstable wave numbers (right) for $F = 6$ dynamical regime. Left plot: solid line with circles – phase velocity, dashed line with triangles – group velocity, vertical dash–dotted line marks zero velocity, horizontal dotted lines mark boundaries between stable and unstable regions. Right plot: solid line with circles is $[\cos(2\pi k/J) - \cos(4\pi k/J) - (1/F)]$, vertical dash–dotted line marks zero (neither stability nor instability), horizontal dotted lines mark boundaries between stable and unstable regions.

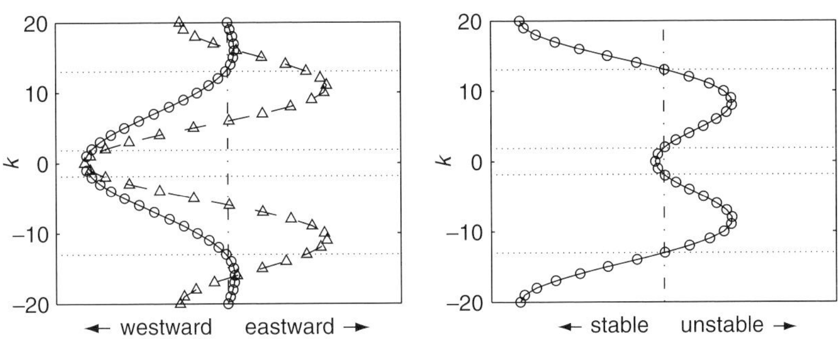

Figure 7.7 The phase/group velocities (left) and linearly stable/unstable wave numbers (right) for $F = 8$ dynamical regime. Left plot: solid line with circles – phase velocity, dashed line with triangles – group velocity, vertical dash–dotted line marks zero velocity, horizontal dotted lines mark boundaries between stable and unstable regions. Right plot: solid line with circles is $[\cos(2\pi k/J) - \cos(4\pi k/J) - (1/F)]$, vertical dash–dotted line marks zero (neither stability nor instability), horizontal dotted lines mark boundaries between stable and unstable regions.

constant forcing, set to $F = 6$, $F = 8$, and $F = 0$, which emerge from numerical simulations at long times. These are typical solutions on the "attractor." We can observe that solutions for $F = 6$ and $F = 8$ definitely have patterns with opposite phase and group velocities, whereas the regime $F = 0$ is completely chaotic.

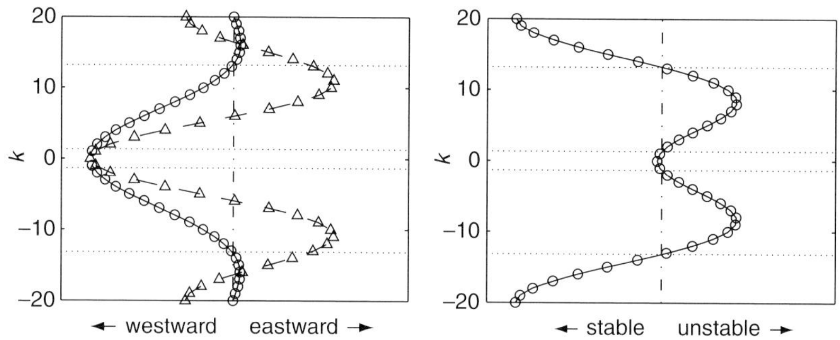

Figure 7.8 The phase/group velocities (left) and linearly stable/unstable wave numbers (right) for $F = 16$ dynamical regime. Left plot: solid line with circles – phase velocity, dashed line with triangles – group velocity, vertical dash–dotted line marks zero velocity, horizontal dotted lines mark boundaries between stable and unstable regions. Right plot: solid line with circles is $[\cos(2\pi k/J) - \cos(4\pi k/J) - (1/F)]$, vertical dash–dotted line marks zero (neither stability nor instability), horizontal dotted lines mark boundaries between stable and unstable regions.

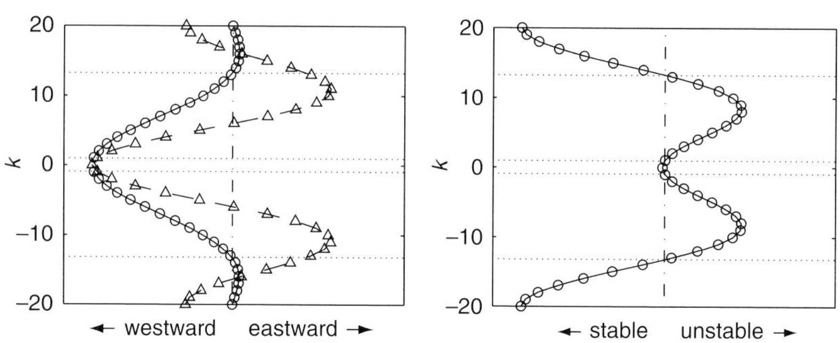

Figure 7.9 The phase/group velocities (left) and linearly stable/unstable wave numbers (right) for $F = 32$ dynamical regime. Left plot: solid line with circles – phase velocity, dashed line with triangles – group velocity, vertical dash–dotted line marks zero velocity, horizontal dotted lines mark boundaries between stable and unstable regions. Right plot: solid line with circles is $[\cos(2\pi k/J) - \cos(4\pi k/J) - (1/F)]$, vertical dash–dotted line marks zero (neither stability nor instability), horizontal dotted lines mark boundaries between stable and unstable regions.

7.4.2 Equilibrium statistical theory for the undamped unforced L-96 model

We consider the L96 model without dissipation and forcing which is given by

$$\frac{du_j}{dt} = (u_{j+1} - u_{j-2})u_{j-1}, \; j = 0, \ldots, J-1 \tag{7.63}$$

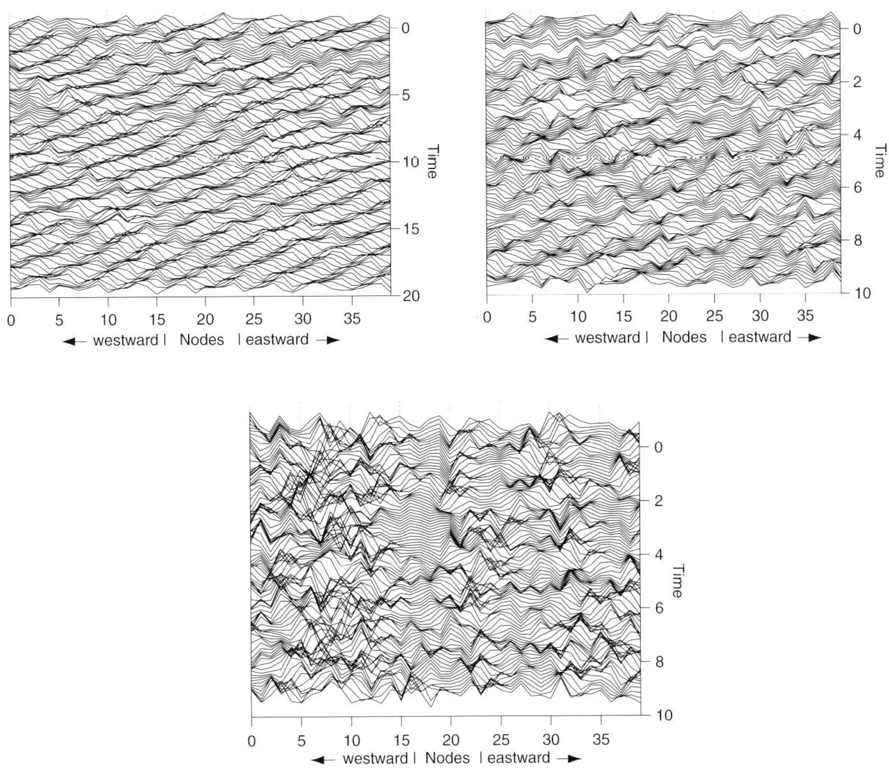

Figure 7.10 The snapshots of a numerical solution for the dynamical regimes $F = 6$ (upper left), $F = 8$ (upper right) and $F = 0$ (center).

with periodic boundary conditions. Below (7.64) we have already shown that this system conserves the energy

$$E = \frac{1}{2}\sum_j u_j^2 \tag{7.64}$$

and in general does not conserve linear momentum. This system of ODEs also trivially has the strong Liouville property, since with $\vec{u} = (u_j)$ and $\vec{F} = (F_j)$ with $F_j = (u_{j+1} - u_{j-2})u_{j-1}$, clearly $\frac{\partial F_j}{\partial u_j} = 0$. Thus, the general theory of equilibrium statistical mechanics from Section 7.2 yields the equilibrium statistical invariant Gaussian–Gibbs measure

$$p^* = \mathcal{Z}^{-1}\exp\left(-\frac{\theta}{2}\sum_j u_j^2\right) = \mathcal{Z}^{-1}\exp(-\theta E) \tag{7.65}$$

with θ the Lagrange multiplier matching the mean energy. As in our discussion in 7.3 for the TBH model, this measure predicts

$$\begin{gathered}\text{equipartition of energy among the modes, } u_j, \\ \text{in long time average of solutions of (7.63).}\end{gathered} \tag{7.66}$$

Even though this system is another discrete approximation to the Burgers–Hopf equation, $u_t + uu_x = 0$, this model can exhibit a very different decay of correlation times with wave number than we predicted and established for the TBH model in Section 7.3 above. Through numerical experiments in the next section, we verify the prediction in (7.66) with very high accuracy for IL96 (7.63) and also show the interesting result that the correlation functions have essentially the same correlation time at all wave numbers in Fourier space in contrast to the TBH model. See Majda and Timofeyev (2002) for the statistical behavior of other finite-difference approximations for the Burgers–Hopf equation, which conserve energy and also linear momentum. In the next section, we also study the statistical properties of the damped forced L96 model and compare and contrast these with those for the undamped unforced IL96 model.

7.4.3 *Statistical properties of the damped forced and undamped unforced L96 models*

Here we verify the equilibrium statistical predictions of the last section and compare and contrast them with those for the L96 equation as we vary the forcing, F. It is interesting in many physical systems to understand how the statistical properties of a damped forced chaotic system like L96 are related to the predictions through equilibrium statistical mechanics of the corresponding undamped unforced equations, IL96. This comparison is what we will develop below through some simple theory and numerical experiments. We will find that many statistical features of L96 are predicted and captured by IL96 for moderately large values of F like $F = 8, 16, 32$. While more details are presented below, first we need to provide some intuition into these developments. First, as in Section 7.3, statistics are gathered for solutions of L96 from (7.52) or IL96 from long time averages. Thus, the statistics of a functional, $\mathcal{F}(\vec{u})$, of the variables, u_j, is given by an empirical long time average

$$\overline{\mathcal{F}(\vec{u})} = \frac{1}{T} \int_{T_0}^{T_0+T} \mathcal{F}(\vec{u}(t))dt$$

with T_0 some initial start-up time for the numerical solution to adjust to the attractor and T sufficiently large.

Table 7.2 *The values of energy of the mean $E(\bar{u})$ and the energy of perturbations E_p for different dynamical regimes*

F	$\bar{u}$	$E(\bar{u})$	E_p	$\bar{u}/\sqrt{E_p}$
0	0.004	0.0003	0.998634	$0.00113 \cdot 10^{-2}$
5	1.63668	53.5745	110.091	$0.156 \cdot 10^{-2}$
6	2.0109	80.8742	160.425	$0.1588 \cdot 10^{-2}$
8	2.34299	109.792	265.028	$0.1439 \cdot 10^{-2}$
16	3.08713	190.607	797.126	$0.1093 \cdot 10^{-2}$
32	3.89012	302.661	2186.68	$0.08319 \cdot 10^{-2}$

The first statistical difference between the damped unforced IL96 and the L96 model is that the simplest statistic, the ensemble mean, has a completely different character in these two models. In particular, $\bar{u}_j = 0$ for IL96 according to Subsection 7.4.2, while $\bar{u}_j = \bar{u} \neq 0$ for L96, where for a given value of forcing, F, $\bar{u}$ is determined by a subtle balance of dissipation, forcing, and non-linearity.

We note that the L96 model in (7.52) is translation invariant, so the statistical ensemble mean, at any given coordinate value is necessarily the same for all j and denoted by $\bar{u}$ with $\bar{u}$ a non-linear function of F. This statistical mean value $\bar{u}$ is called the climatology below by analogy with realistic geophysical systems. In Table 7.2 below, we tabulate $\bar{u}$ and we see that $\bar{u}$ from the simulations is nearly zero, i.e. 10^{-3} for IL96 as predicted from Subsection 7.4.2, while $\bar{u}$ is clearly a non-linear function of F for $F = 5, 6, 8, 16, 32$, which does not coincide with the trivial steady state $u_j = F$ utilized in the discussion in Subsection 7.4.1. The energy of this mean state, $E(\bar{u})$, is given by

$$E(\bar{u}) = \frac{J}{2}\bar{u}^2. \tag{7.67}$$

The second bulk statistical quantity of interest for L96 is the strength of fluctuations about the given climatology; this is measured by the average variance in energy fluctuations about the mean

$$E_p = \frac{1}{2}\sum_{j=0}^{J} \overline{(u_j - \bar{u})^2}. \tag{7.68}$$

Once again for the undamped unforced IL96, there is the precise equilibrium statistical prediction from (7.64) and (7.66), $E_p = E$, the prescribed energy of fluctuations; on the other hand, $E_p(F)$ is a non-linear function of the forcing F, which involves subtle balances of forcing, dissipation, and non-linearity, with the values of $E_p(F)$ plotted in Table 7.2. It is clear from Table 7.2 that the energy of the fluctuations dominates the energy of the mean as F increases. We contend that as F increases the statistics of the L96 model are in many respects converging to those of IL96 established in Subsection 7.4.3. To facilitate such a comparison it is natural to rewrite the L96 model in re-scaled variables using $\bar{u}$ and E_p so that equilibrium statistical mechanics predictions comparisons can be made for a fixed energy.

Rescaling the damped forced L96 model

The simple idea of the rescaling $u_j \to \tilde{u}_j$, $t \to \tilde{t}$ is to set the mean state to zero in new coordinates, and the energy of perturbations to 1, while retaining the same form of the non-linear term in the right-hand side of the L96 model in (7.52). The corresponding coordinate-time rescaling is given by

$$u_j = E_p^{1/2}\tilde{u}_j + \bar{u}, \quad t = E_p^{-1/2}\tilde{t}. \tag{7.69}$$

The L96 model in (7.52), written through the coordinates $\tilde{u}_j$ and $\tilde{t}$, becomes

$$\frac{d\tilde{u}_j}{d\tilde{t}} = (\tilde{u}_{j+1} - \tilde{u}_{j-2})\tilde{u}_{j-1} + E_p^{-1/2}[(\tilde{u}_{j+1} - \tilde{u}_{j-2})\bar{u} - \tilde{u}_j] + E_p^{-1}(F - \bar{u}). \tag{7.70}$$

Obviously, if the forcing F increases, both the mean state $\bar{u}$ and the energy of perturbations E_p will also increase as shown in Table 7.1. Thus, if the limits

$$E_p^{-1/2}\bar{u} \to 0 \quad \text{and} \quad E_p^{-1}(F - \bar{u}) \to 0 \quad \text{as} \quad F \to \infty \tag{7.71}$$

are valid, then the rescaled dynamics in (7.70) should approach the dynamics of the L96 model with no damping and forcing with energy $E_p = 1$.

Linear stability of the mean state

The motion of the rescaled system in (7.70) is centered, by construction, around its statistical mean state $\bar{u}$. Thus, it is important to check whether its solution is linearly stable or unstable for small $\tilde{u}_j$. Following (7.54), the rescaled linearized system is written in the Fourier space as

$$\dot{\hat{u}}_k = \omega(k)\hat{u}_k + E_p^{-1}(F - \bar{u}), \quad \omega(k) = E_p^{-1/2}\left[\bar{u}(e^{2\pi ik/J} - e^{-4\pi ik/J}) - 1\right], \tag{7.72}$$

where J denotes the number of gridpoints in the model. The linear stability criterion is

$$\begin{aligned} \Re\omega(k) &= E_p^{-1/2}\left[\bar{u}\left(\cos(2\pi k/J) - \cos(4\pi k/J)\right) - 1\right], \\ \Re\omega(k) &< 0 \qquad \text{stable}, \\ \Re\omega(k) &> 0 \qquad \text{unstable}. \end{aligned} \tag{7.73}$$

Solving the equation $\Re\omega(k) = 0$, we find out that the band of unstable waves emerges when

$$\bar{u} > \frac{8}{9}$$

begins at

$$k_{begin} = \frac{J}{2\pi} \arccos\left(\frac{1 + 3\sqrt{1 - \frac{8}{9}\bar{u}^{-1}}}{4}\right) \tag{7.74}$$

and ends at

$$k_{end} = \frac{J}{2\pi} \arccos\left(\frac{1 - 3\sqrt{1 - \frac{8}{9}\bar{u}^{-1}}}{4}\right). \tag{7.75}$$

In particular, in the limiting case with $F \to \infty$ and, therefore, $\bar{u} \to \infty$

$$k_{begin} = 0, \quad k_{end} = \frac{J}{3}. \tag{7.76}$$

The boundaries of the band of stable and unstable waves are shown in the Table 7.1 for different F and $\bar{u}$ for the number of gridpoints $J = 40$. Note that even as $F \to \infty$, the wave numbers beyond $k = 13$ remain linearly stable in the formal limit.

The bulk behavior of the rescaled problem

Numerical solutions of both the damped forced L96 model in (7.52) and the IL96 model without forcing and damping in (7.63) are found by numerical time integration via the fourth-order Adams–Bashforth multistep method with the time step $\Delta t = 1/64$. All statistical experiments are carried out by a long-time integration of the two models on a time averaging window of $T = 100\,000$ units with the initial spin-up time of $T_0 = 1000$ time units. The computations are performed for the forced damped regimes $F = 5, 6, 8, 16, 32$ and the regime without forcing and damping for IL96 (sometimes denoted below as the $F = 0$ regime).

The decay of the quantity $E_p^{-1/2}\bar{u}$ in (7.71) is equivalent to the decay of the ratio between the energy of the mean state $E(\bar{u})$ and the mean energy of perturbations E_p, and we choose to show the ratio of energies instead. Magnitudes of forcing F and corresponding values of $E(\bar{u})$ and E_p are shown in Table 7.2. It can be seen that E_p grows faster than $E(\bar{u})$ as F increases. We can also observe that despite the fact that the band of linearly unstable waves grows as energy increases, the magnitude of the linear amplification factor $\Re\omega(k)$ from (7.72) decays to zero at the same time, which is shown in Figure 7.11. This means that, as $F \to \infty$, linearly unstable waves become almost neutrally stable.

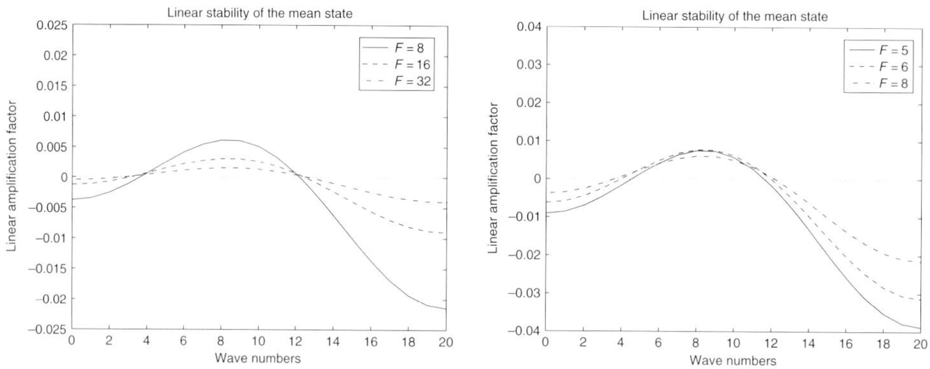

Figure 7.11 The linear amplification factor $\Re\omega(k)$ of the linearization about the mean state $\bar{u}$ for the regimes $F = 5, 6, 8, 16, 32$.

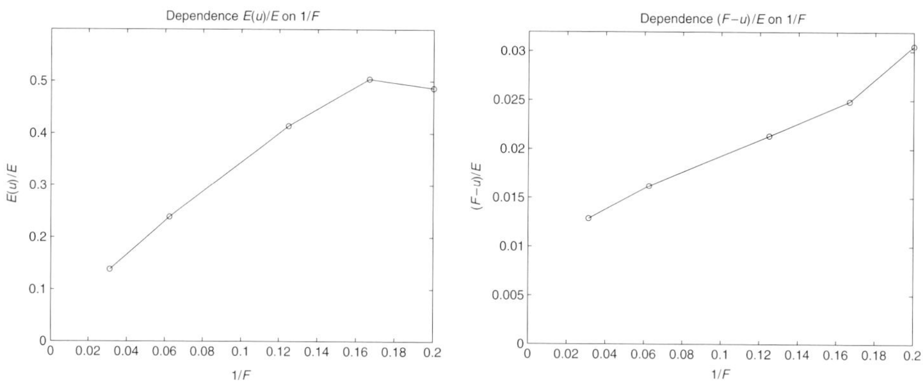

Figure 7.12 Left: the dependence of $E(\bar{u})/E_p$ on $1/F$. Right: the dependence of $(F - \bar{u})/E_p$ on $1/F$.

In Figure 7.12 we demonstrate the experimental dependence of energy and mean state ratios from (7.71) on the forcing F. The left plot of Figure 7.12 demonstrates how the ratio $E(\bar{u})/E_p$ depends on the forcing F (reciprocal scale $1/F$ is shown along the horizontal axis). All three results for the dynamical regimes $F = 8$, $F = 16$, and $F = 32$ lie approximately on a straight line which goes through the origin. Though it cannot be established convincingly with only three data points, the assumption that the ratio of energy of the mean state and the mean energy of the perturbations goes to zero, as the forcing F increases to infinity, seems to be valid. The right plot of Figure 7.12, which demonstrates the dependence of the ratio $(F - \bar{u})/E_p$ on the forcing F, does not, however, show the convergence to zero as F increases, but tends toward a finite value in this range of F.

The climatology of different forcing regimes in rescaled coordinates

Earlier we made an assumption that with increasing forcing F in the damped forced L96 model, the rescaled dynamics in (7.70) should approach the dynamics of the undamped unforced IL96 model. If that assumption holds, we should be able to observe the statistical similarities in the climatology of the L96 model in the rescaled space–time coordinates. The rescaled variance spectrum of the L96 model is shown in Figure 7.13 for the regimes $F = 0, 5, 6, 8, 16$, and 32. The variance spectrum is defined as a set of time averaged variances for the Fourier coefficients

$$\mathrm{Var}(\hat{u}_k) = \overline{\left|\hat{u}_k(t) - \bar{\hat{u}}_k\right|^2}, \tag{7.77}$$

where the overbar denotes the time average, such that $\bar{\hat{u}}_k$ is the time average of $\hat{u}_k(t)$. In practice the time-averaging window T is finite but large (10^5 time units in our simulations). For $F = 0$ the spectrum is flat as expected, and strongly confirms the equipartition of energy for IL96. The rescaled spectra for $F \neq 0$ are non-uniform, with a hump where the band of linearly unstable wave modes is located. However, we can see that in the part of the spectrum for the wave numbers $k = 0 \ldots 10$ there is a systematic trend: the spectrum for $F \neq 0$ in that part tends to approach the undamped unforced flat spectrum as F increases. The same trend, however, does not hold for the range of wave numbers $k = 13 \ldots 20$, where the spectrum $F \neq 0$ does not systematically approach the inviscid case. The possible reason could be that these wave numbers remain linearly stable even when $F \to \infty$, according to Table 7.1 and our earlier discussion.

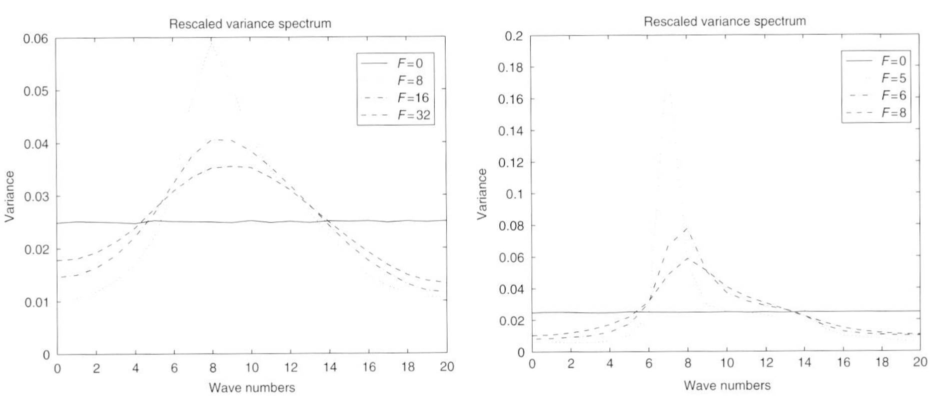

Figure 7.13 The variance spectrum for the Lorenz 96 model in different dynamical regimes. Left plot: solid line – $F = 0$, dotted line – $F = 8$, dot–dashed line – $F = 16$, dashed line – $F = 32$. Right plot: solid line – $F = 0$, dotted line – $F = 5$, dot–dashed line – $F = 6$, dashed line – $F = 8$.

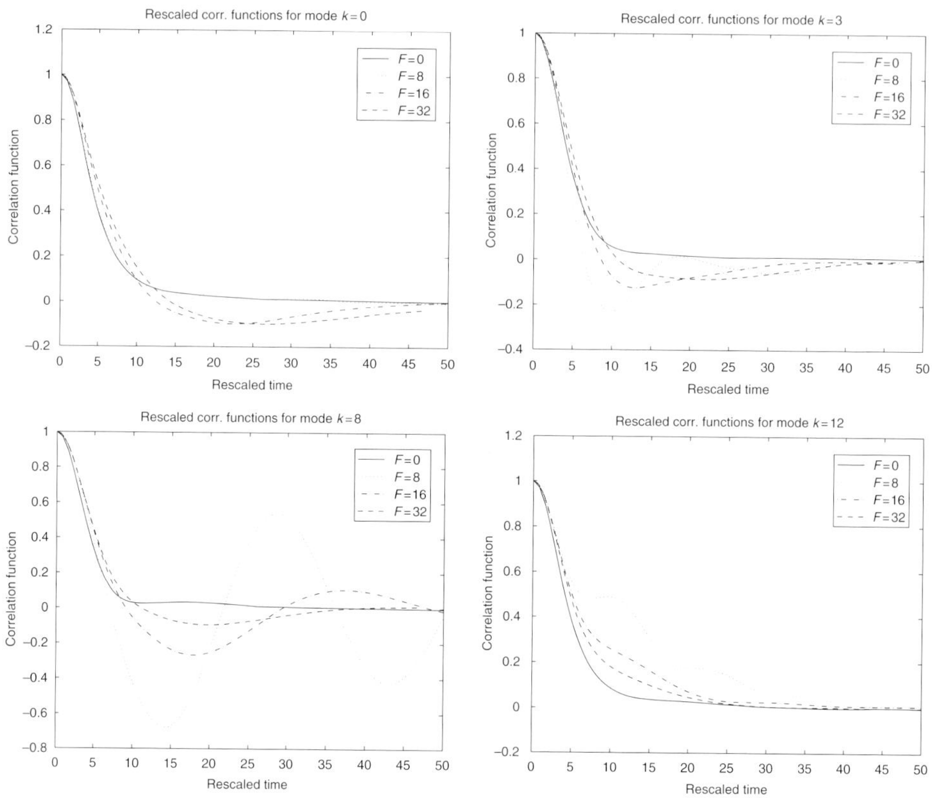

Figure 7.14 The time autocorrelation functions for the wave numbers $k = 0, 3, 8, 12$, Lorenz 96 model in different dynamical regimes. Solid line – $F = 0$, dotted line – $F = 8$, dot–dashed line – $F = 16$, dashed line – $F = 32$.

The time autocorrelation functions are shown in Figures 7.14 and 7.15 for the wave numbers $k = 0, 8, 12$ for the dynamical regimes $F = 0, 5, 6, 8, 16$, and 32. Recall the general formula for the autocorrelation function of a variable $a(t)$ is written as

$$\mathrm{Corr}(\tau) = \frac{\overline{(a(t)-\bar{a})(a(t+\tau)-\bar{a})}}{\overline{(a(t)-\bar{a})^2}}, \tag{7.78}$$

where the overbar denotes the time average, such that $\bar{a}$ is the time average of $a(t)$. In practice the time-averaging window T is finite but large (10^5 time units in our simulations). The autocorrelation function in (7.78) measures mixing in a dynamical system, and in some sense can be regarded as an indicator of how fast a variable "forgets" about its behavior in the past. Each plot on Figures 7.14 and 7.15 reveals the following fact: first there is a stage until approximately time $t = 5-10$ (depending on the wave number shown) where there are no significant differences in the time autocorrelations between different wave numbers with various stability

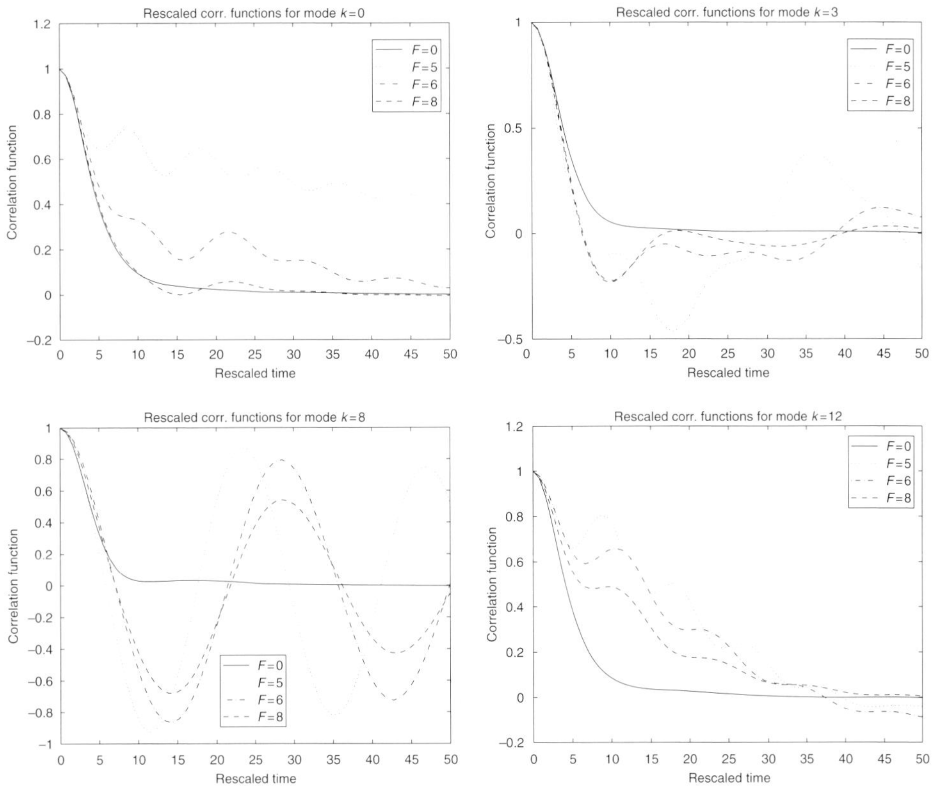

Figure 7.15 The time autocorrelation functions for the wave numbers $k = 0, 3, 8, 12$, Lorenz 96 model in different dynamical regimes. Solid line – $F = 0$, dotted line – $F = 5$, dot–dashed line – $F = 6$, dashed line – $F = 8$.

characteristics. After that time, however, the time autocorrelation functions for different dynamical regimes diverge from each other. The remarkable observation here is that, while the time correlations for the linearly unstable wave number $k = 8$ and high values of F like $F = 16$ and $F = 32$ are close to non-oscillatory $F = 0$, which decays fast (Figure 7.14), the correlation functions for $F = 5, 6, 8$ are highly oscillatory and exhibit rather weak decay in time (Figure 7.15). There is a systematic trend: the larger the forcing is, the closer the corresponding correlation function is to the case with no forcing and no damping. The observed results on Figures 7.14 and 7.15 generally confirm the assumption that for large values of the forcing F the rescaled dynamics in (7.70) roughly approach that of the undamped unforced IL96 model.

As discussed earlier for the TBH model, the correlation time is defined as a typical time of the decay of the autocorrelation function. Usually it is given through the time integral of the autocorrelation function. Here we look at two measures of the correlation time: one is the integral of the autocorrelation function

$$T = \int_0^\infty \mathrm{Corr}(\tau)d\tau, \tag{7.79}$$

and the other is the integral of the absolute value of the autocorrelation function

$$T_{\mathrm{abs}} = \int_0^\infty |\mathrm{Corr}(\tau)|\, d\tau. \tag{7.80}$$

The first measure allows for oscillatory cancellation in the correlations, while the second does not. The two types of correlation time spectra are presented in Figure 7.16: the correlation times on the left plots of Figure 7.16 are obtained by

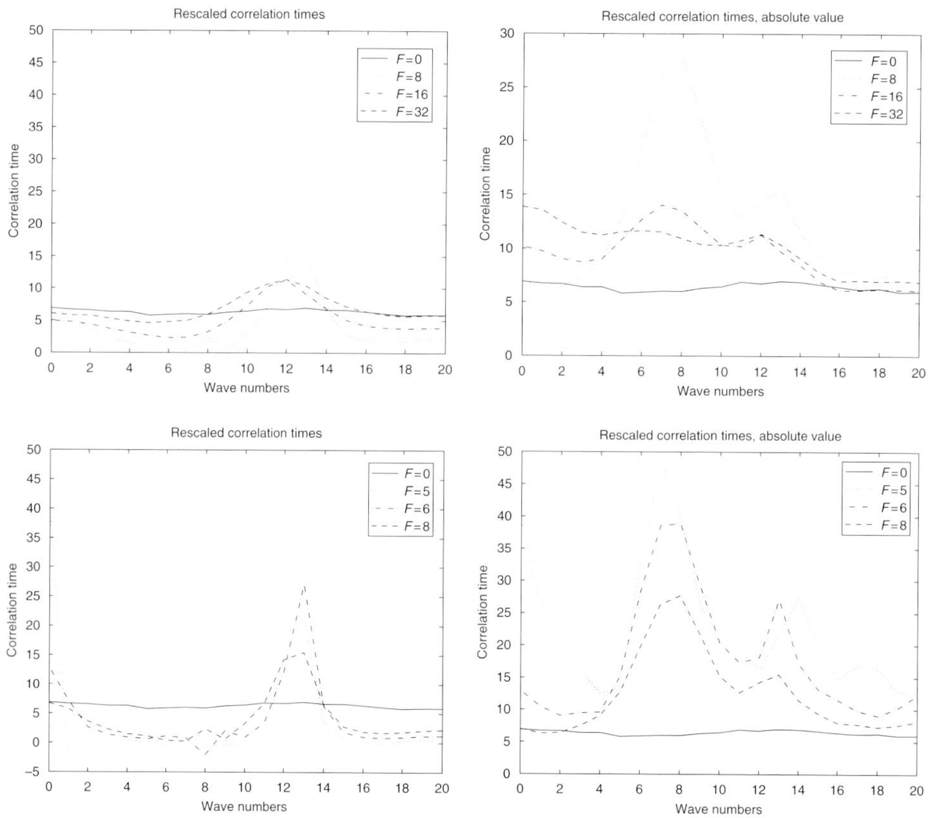

Figure 7.16 The spectra of correlation times (integrals of correlation functions) for the Lorenz 96 model in different dynamical regimes. Left plots: the integrals are taken with the time correlation functions directly. Right plots: the integrals are taken with the absolute values of the correlation functions. Upper plots: solid line – $F=0$, dotted line – $F=8$, dot–dashed line – $F=16$, dashed line – $F=32$. Lower plots: solid line – $F=0$, dotted line – $F=5$, dot–dashed line – $F=6$, dashed line – $F=8$.

integrating the time autocorrelation functions, whereas the correlation times on the right plots of Figure 7.16 are obtained by integrating the absolute value of the corresponding correlation function. The correlation time spectra for the $F \neq 0$ regimes using the measure in (7.79) tend to approach the unforced undamped case as F increases. The highly oscillatory correlation functions for linearly unstable wave numbers in the regimes $F = 5, 6, 8$ produce completely different correlations T and T_{abs}; while T are even smaller than those for the IL96 model due to cancellation of positive and negative parts of correlation functions, T_{abs} are huge, differing almost by an order of magnitude. Also note that Figure 7.16 for $F = 0$, i.e. IL96, shows that all Fourier modes decay at the same rate in this conservative model, in contrast to what we established in Section 7.3 for the TBH model.

References

Abramov, R., Kovacic, G., and Majda, A. (2003), Hamiltonian structure and statistically relevant conserved quantities for the truncated Burgers–Hopf equation. *Comm. Pure Appl. Math.* **56**(1), 1–46.

Abramov, R., Majda, A., and Kleeman, R. (2005), Information theory and predictability for low frequency varaibility. *J. Atmos. Sci.* **62**(1), 65–87.

Chorin, A., and Marsden, J. (1993), *A Mathematical Introduction to Fluid Mechanics*. 3rd edn, New York: Springer- Verlag.

Kraichnan, R. H. (1975), Statistical dynamics of two-dimensional flow. *J. Fluid Mech.* **67**, 155–175.

Lorenz, E. (1996), *Predictability: a problem partly solved in Proceedings of Seminar on Predictability*, ECMWF.

Lorenz, E. and Emanuel, K. (1998), Optimal sites for supplementary weather observations. *J. Atmos. Sci.* **55**, 399–414.

Majda, A., and Bertozzi, A. (2001), *Vorticity and Incompressible Flow*, Cambridge: Cambridge University Press.

Majda, A. J. and Timofeyev, I. (2000), Remarkable statistical behavior for truncated Burgers–Hopf dynamics. *Proc. Nat. Acad. Sci.* **97**(23), 12413–12417.

Majda, A. J. and Timofeyev, I. (2002), Statistical mechanics for truncations of the Burgers–Hopf equations: a model for intrinsic stochastic behavior with scaling. Based on a Leonardo da Vinci Lecture of A. Majda in Milan, June 2000, *Milan J. Math.* **70**, 39–96.

Pedlosky, J. (1987), *Geophysical Fluid Dynamics*, 2nd edn. New York: Springer-Verlag.

8

Statistical mechanics for the truncated quasi-geostrophic equations

8.1 Introduction

In this chapter we set-up statistical mechanics and apply the theory from Chapter 7 to suitable truncations of the barotropic quasi-geostrophic equations without damping and forcing

$$\frac{\partial q}{\partial t} + J(\psi, q) = 0, \quad \frac{dV}{dt}(t) = -\int \frac{\partial h}{\partial x}\psi', \tag{8.1}$$

where

$$\begin{aligned} q &= \Delta\psi' + h + \beta y, \quad \omega = \Delta\psi', \\ \psi &= -V(t)y + \psi', \quad q' = \Delta\psi' + h, \\ \vec{v} &= \nabla^{\perp}\psi = \begin{pmatrix} V(t) \\ 0 \end{pmatrix} + \nabla^{\perp}\psi' \end{aligned}$$

and $\Omega = [0, 2\pi] \times [0, 2\pi]$ is the region occupied by the fluid.

Recall that, in the empirical theory introduced in Chapter 6, we postulated empirical one-point statistics which have no direct relationship with the dynamic equations (8.1). In this chapter we build an alternative theory, involving complete statistical mechanics, based on the work of Kraichnan (1975) for ideal fluid flow, and the work of Salmon, Holloway, and Hendershott (1976) for quasi-geostrophic flows. In this alternative approach, we will build a time-dependent transport theory, inherited from the dynamic equations (8.1), considering all the correlations of fluctuations and, finally, we derive the most probable state as well as its mean state.

This alternative approach has more traditional features, i.e. it has a keen analogy with the traditional statistical mechanics for the physical objects like ideal gas particles. In the traditional statistical mechanics for gas particles, finite particle systems are considered, and then the limiting procedure, called thermodynamic limit, is taken to reveal the limit behavior of the systems. The concepts like ensemble or ensemble average, which we did not use in the empirical theory in Chapter 6, are indispensable to developing the traditional theory. However, at the end of this chapter, we will see that the statistical mechanics actually gives basically the same prediction for the mean state as the empirical statistical theory from Chapter 6. This builds up our confidence in the empirical theory as an economic way of predicting statistical behavior.

Here we present the statistical theory for truncated quasi-geostrophic dynamics originally developed by Salmon *et al.* (1976) and Carnevale and Frederiksen (1987). This theory is carried out in several steps:

Step I: We make a finite-dimensional truncation of the barotropic quasi-geostrophic equations (8.1). This amounts to making a Galerkin approximation, where the equation is projected on to a finite-dimensional subspace. Since we are working with two-dimensional periodic boundary conditions, the truncation is readily accomplished with the standard Fourier basis.

Step II: We then observe that the finite-dimensional truncated equations have special properties:

1. The truncated equations conserve the truncated energy and enstrophy. However, higher moments of the truncated vorticity are no longer conserved.
2. The truncated equations satisfy the Liouville property. This implies that the truncated equations define an incompressible flow in phase space. Therefore the resulting flow map is measure preserving, and we can 'transport' measures with the flow.

Step III: We apply the equilibrium statistical mechanics theory for ODEs set up in Chapter 7 to the truncated system, i.e. find the probability measure, the Gibbs measure, that maximizes the Shannon entropy, subject to the constraints of fixed energy and enstrophy.

Step IV: We compute the mean state of the Gibbs measure of the finite-dimensional truncated equations and study the limit behavior of this mean state in the limit when the dimension $N \to \infty$. This is the continuum limit of the system.

The rest of the chapter is organized as follows. In Section 8.2 we introduce the spectrally truncated quasi-geostrophic equations, verify the Liouville property and the conservation of truncated energy and enstrophy. The non-linear stability of exact steady state solutions to the truncated system with linear potential vorticity–stream function relation will be discussed briefly as well, following the material in Chapter 4. In Section 8.3 we apply the equilibrium statistical mechanics theory for ODEs developed in Chapter 7 to the truncated quasi-geostrophic equations. The most probable measures, Gibbs measures, are Gaussians. Their mean states are exact steady state solutions to the truncated system with linear potential vorticity–stream function relation. These mean states are non-linearly stable in the presence of large-scale mean flow or generic topography. In Section 8.4

we provide some numerical evidence confirming the convergence of the time average of numerical solutions to the most probable mean state predicted by the equilibrium statistical theory for the truncated systems. As an alternative approach which emphasizes the fluctuation about the mean state, we apply equilibrium statistical mechanics theory to the fluctuations about the mean for the truncated systems in Section 8.5. Interesting enough, the statistical theory predicts equipartition of energy in the pseudo-energy coordinates. This resembles the case of truncated Burgers equations discussed in Section 7.3. Finally, in Section 8.6 we consider the limiting behavior, as the truncation eigenvalue Λ approaches infinity. We will see that geophysical effects play a very important role in this limiting procedure. Remarkably, the mean state predicted by the the complete statistical mechanics theory is the same kind as predicted by empirical statistical mechanics using two conserved quantities, namely energy and enstrophy as in Chapter 6 in the case of generic geophysical effects. Moreover we will see that these steady states predicted by the statistical theory will be non-linear stable in the presence of generic geophysical effects. They are also selective decay states as described in Section 4.5. In the case of no geophysical effects or the case with no large-scale mean flow but with topography with degenerate spectrum, the prediction of the complete statistical mechanics theory is slightly different from the prediction of the empirical statistical theory, in that the under-determined ground state modes in the empirical statistical theory become Gaussian random variables with mean zero and variance satisfying the appropriate energy constraint. All of these results provide strong supporting evidence for utilizing the empirical statistical theory with two conserved quantities when only the mean state is of interest.

8.2 The finite-dimensional truncated quasi-geostrophic equations

In this section we introduce certain truncated systems of (8.1). We also verify that these truncated systems satisfy the Liouville property and conserve the truncated energy and enstrophy. The non-linear stability of certain steady state solutions to the truncated equations will also be discussed.

8.2.1 The spectrally truncated quasi-geostrophic equations

We now introduce the spectrally truncated systems which approximate (8.1). The truncated dynamic equations are a finite-dimensional approximation of the barotropic quasi-geostrophic equations, which are obtained by projecting the equations in (8.1) onto a subspace involving only a finite number of Fourier modes. In another words, the truncated system is a Galerkin approximation of the barotropic quasi-geostrophic equations with the use of standard Fourier basis. To derive the truncated dynamics equations we proceed as follows. We first introduce the Fourier series expansions of the truncated small-scale stream function ψ'_Λ, the truncated vorticity ω_Λ, and the truncated topography h_Λ in terms of the basis $B_\Lambda = \left\{\exp\left(i\vec{k}\cdot\vec{x}\right) | 1 \le \left|\vec{k}\right|^2 \le \Lambda\right\}$

$$
\begin{aligned}
\psi'_\Lambda &\equiv \sum_{1\le|\vec{k}|^2\le\Lambda} \hat{\psi}_{\vec{k}}(t)e^{i\vec{x}\cdot\vec{k}} = -\sum_{1\le|\vec{k}|^2\le\Lambda} \frac{1}{|\vec{k}|^2}\hat{\omega}_{\vec{k}}(t)e^{i\vec{x}\cdot\vec{k}},\\
h_\Lambda &\equiv \sum_{1\le|\vec{k}|^2\le\Lambda} \hat{h}_{\vec{k}}(t)e^{i\vec{x}\cdot\vec{k}},\\
\omega_\Lambda &\equiv \sum_{1\le|\vec{k}|^2\le\Lambda} \hat{\omega}_{\vec{k}}(t)e^{i\vec{x}\cdot\vec{k}} = \sum_{1\le|\vec{k}|^2\le\Lambda} (-|\vec{k}|^2\hat{\psi}_{\vec{k}}(t))e^{i\vec{x}\cdot\vec{k}},
\end{aligned}
\tag{8.2}
$$

where the amplitudes $\hat{\psi}_{\vec{k}}$, $\hat{\omega}_{\vec{k}}$ and $\hat{h}_{\vec{k}}$ satisfy the reality conditions $\hat{\psi}_{-\vec{k}} = \hat{\psi}^*_{\vec{k}}$, $\hat{\omega}_{-\vec{k}} = \hat{\omega}^*_{\vec{k}}$, and $\hat{h}_{-\vec{k}} = \hat{h}^*_{\vec{k}}$ with $*$ denoting complex conjugation here. We also assume that the mean value $\hat{h}_0$ of the topography is zero, so that the solvability condition for the associated steady state equation is automatically satisfied. We also denote by P_Λ the orthogonal projection on to the finite-dimensional space $V_\Lambda = \text{span}\{B_\Lambda\}$. The truncated dynamical equations are obtained by projecting the barotropic quasi-geostrophic equations in (8.1) onto V_Λ

$$
\frac{\partial q'_\Lambda}{\partial t} + \beta\frac{\partial \psi'_\Lambda}{\partial x} + V\frac{\partial q'_\Lambda}{\partial x} + P_\Lambda\left(\nabla^\perp\psi'_\Lambda\cdot\nabla q'_\Lambda\right) = 0, \quad q'_\Lambda = \omega_\Lambda + h_\Lambda, \tag{8.3}
$$

$$
\frac{dV}{dt} - \int h_\Lambda \frac{\partial \psi'_\Lambda}{\partial x} = 0. \tag{8.4}
$$

Then, we get the finite-dimensional system of the ordinary differential equation (ODE), *the truncated dynamical equations*, for the Fourier coefficients with $1 \le |\vec{k}|^2 \le \Lambda$

$$
\frac{d\hat{\omega}_{\vec{k}}}{dt} - \frac{i\beta k_1}{|\vec{k}|^2}\hat{\omega}_{\vec{k}} + iVk_1(\hat{\omega}_{\vec{k}} + \hat{h}_{\vec{k}}) - \sum_{\substack{\vec{l}+\vec{m}=\vec{k},\\ |\vec{l}|^2\le\Lambda,|\vec{m}|^2\le\Lambda}} \frac{\vec{l}^\perp\cdot\vec{m}}{|\vec{l}|^2}\hat{\omega}_{\vec{l}}(\hat{\omega}_{\vec{m}} + \hat{h}_{\vec{m}}) = 0, \tag{8.5}
$$

$$
\frac{dV(t)}{dt} - i\sum_{1\le|\vec{k}|^2\le\Lambda} k_1\frac{\hat{h}_{-\vec{k}}\hat{\omega}_{\vec{k}}}{|\vec{k}|^2} = 0. \tag{8.6}
$$

8.2.2 Conserved quantities for the truncated system

According to Subsection 1.3.5, the quasi-geostrophic equations (8.1) possess two or more robust conserved quantities. An advantage of utilizing Fourier truncation is that the linear and quadratic conserved quantities survive the truncation. More precisely we have:

Proposition 8.1 *The truncated energy E_Λ and enstrophy $\mathcal{E}_\Lambda$ are conserved in the finite-dimensionally truncated dynamics, where*

$$E_\Lambda = \frac{1}{2}V^2 + \frac{1}{2}\fint |\nabla^\perp \psi'_\Lambda|^2 \, d\vec{x} = \frac{1}{2}V^2 + \frac{1}{2}\sum_{1\leq|\vec{k}|^2\leq\Lambda} |\vec{k}|^2|\hat{\psi}_{\vec{k}}|^2, \tag{8.7}$$

$$\mathcal{E}_\Lambda = \beta V + \frac{1}{2}\fint q'^2_\Lambda \, d\vec{x} = \beta V + \frac{1}{2}\sum_{1\leq|\vec{k}|^2\leq\Lambda} |-|\vec{k}|^2\hat{\psi}_{\vec{k}} + \hat{h}_{\vec{k}}|^2. \tag{8.8}$$

Proof: In order to show that the truncated total energy E_Λ is conserved in time it is equivalent to proving that the time derivative of it is identically zero for all time. For this purpose we take the time derivative of the truncated total energy E_Λ and utilize the truncated dynamic equations (8.3) and (8.4) and we repeatedly carry out integration by parts

$$\begin{aligned}
\frac{d}{dt}E_\Lambda &= V\frac{dV}{dt} - \fint \psi'_\Lambda \frac{\partial \omega_\Lambda}{\partial t} d\vec{x} \\
&= V\frac{dV}{dt} - \fint \psi'_\Lambda \frac{\partial q'_\Lambda}{\partial t} d\vec{x} \\
&= V\fint h_\Lambda \frac{\partial \psi'_\Lambda}{\partial x} + \beta \fint \psi'_\Lambda \frac{\partial \psi'_\Lambda}{\partial x} d\vec{x} + V\fint \psi'_\Lambda \frac{\partial q'_\Lambda}{\partial x} d\vec{x} \\
&\quad + \fint P_\Lambda \left(\nabla^\perp \psi'_\Lambda \cdot \nabla q'_\Lambda\right) \psi'_\Lambda \\
&= V\fint h_\Lambda \frac{\partial \psi'_\Lambda}{\partial x} + V\fint \psi'_\Lambda \frac{\partial h_\Lambda}{\partial x} d\vec{x} + \fint \left(\nabla^\perp \psi'_\Lambda \cdot \nabla q'_\Lambda\right) \psi'_\Lambda \\
&= 0,
\end{aligned}$$

The conservation of the truncated total enstrophy can be shown in a similar fashion. We end the proof of the proposition.

8.2.3 Non-linear stability of some exact solutions to the truncated system

It is easy to see that the truncated system possesses exact solutions having a linear $q_\Lambda - \psi_\Lambda$ relation similar to those introduced in Chapter 1 for the continuum equations

$$\overline{q}_\Lambda = \mu \overline{\psi}_\Lambda. \tag{8.9}$$

This condition is the same as

$$\Delta \overline{\psi}'_\Lambda + h_\Lambda = \mu \overline{\psi}'_\Lambda, \quad \overline{V} = -\frac{\beta}{\mu}. \tag{8.10}$$

The non-linear stability of these type of steady states can be studied using the same techniques as those in Section 4.2. More precisely, we consider a linear combination of the truncated energy and enstrophy to form a positive quadratic form for perturbations. Let $(\delta q_\Lambda, \delta V)$ be the perturbations, we then have, following Section 4.2 and utilizing (8.10)

$$\mu E_\Lambda(q_\Lambda, V) + \mathcal{E}_\Lambda(q_\Lambda, V) = \mu E_\Lambda(\overline{q}_\Lambda, \overline{V}) + \mathcal{E}_\Lambda(\overline{q}_\Lambda, \overline{V}) + \mathcal{W}_\mu(\delta q_\Lambda, \delta V), \quad (8.11)$$

where

$$\begin{aligned} \mathcal{W}_\mu(\delta q_\Lambda, \delta V) &= \frac{\mu}{2}(\delta V)^2 + \frac{1}{2}\sum_{1\le|\vec{k}|^2\le\Lambda}\left(1+\frac{\mu}{|\vec{k}|^2}\right)(\delta q_\Lambda)^2 \\ &= \frac{\mu}{2}(V-\overline{V})^2 + \frac{1}{2}\sum_{1\le|\vec{k}|^2\le\Lambda}|\vec{k}|^2(\mu+|\vec{k}|^2)(\hat{\psi}_{\vec{k}} - \overline{\hat{\psi}}_{\vec{k}})^2. \end{aligned} \quad (8.12)$$

The same argument as in Section 4.2 implies that the steady state solution $(\overline{q}_\Lambda, \overline{V})$ is non-linearly stable for $\mu > 0$ in general, or for $\mu > -1$ when $V \equiv 0$.

8.2.4 The Liouville property

Next we verify the Liouville property for the truncated equations (8.3) and (8.4). These equations can be written as a system of ODEs with real coefficients. Indeed, let

$$S = \left\{\vec{k}_1, \ldots, \vec{k}_M\right\} \quad (8.13)$$

be a defining set of modes for $\{1 \le |\vec{k}|^2 \le \Lambda\}$ satisfying

$$\vec{k} \in S \Rightarrow -\vec{k} \notin S, \quad S \cup (-S) = \{1 \le |\vec{k}|^2 \le \Lambda\}. \quad (8.14)$$

Let $N = 2M+1$ and define $\vec{X} \in \mathcal{R}^N$

$$\vec{X} \equiv \left(V, \mathrm{Re}\,\hat{\psi}_{\vec{k}_1}, \mathrm{Im}\,\hat{\psi}_{\vec{k}_1}, \ldots, \mathrm{Re}\,\hat{\psi}_{\vec{k}_M}, \mathrm{Im}\,\hat{\psi}_{\vec{k}_M}\right), \quad \vec{X} \in R^{2M+1} = \mathcal{R}^N, N \gg 1. \quad (8.15)$$

We then notice that each point $\vec{X}$ in a big space $\mathcal{R}^N$ can represent the entire state of the finite-dimensional system. With these notations, the truncated dynamic equations (8.5) and (8.6) can be written in a more compact form

$$\frac{d\vec{X}}{dt} = \vec{F}(\vec{X}), \quad \vec{X}|_{t=0} = \vec{X}_0, \quad (8.16)$$

with the vector field $\vec{F}$ satisfying the property

$$F_j(\vec{X}) = F_j(X_1, \ldots, X_{j-1}, X_{j+1}, \ldots, X_N), \tag{8.17}$$

i.e. F_j does not depend on X_j. This immediately implies the Liouville property. In fact, $\vec{F}$ satisfies the so-called detailed Liouville property, since it is satisfied locally in terms of the Fourier coefficients.

It is easy to see why (8.17) is true. It is obvious that F_1 is independent of $X_1 = V$. Observe that F_{2j} and F_{2j+1} correspond to $\hat{\psi}_{\vec{k}_j}$ in (8.5). It is obvious that the contributions from the linear terms in (8.5) are either independent of $\hat{\psi}_{\vec{k}_j}$ or cause a rotation of $X_{2j} = \mathrm{Re}\,\hat{\psi}_{\vec{k}_j}$ and $X_{2j+1} = \mathrm{Im}\,\hat{\psi}_{\vec{k}_j}$. As for the non-linear term in (8.5), the contribution from $\hat{\psi}_{\vec{k}_j}$ and $\hat{\psi}_{-\vec{k}_j}$ (the same as from X_{2j} and X_{2j+1}) is zero, since the restriction on the summation indices requires either $\vec{l} = 0$, $\vec{m} = \vec{k}_j$, or $\vec{l} = \vec{k}_j$, $\vec{m} = 0$, or $\vec{l} = -\vec{k}_j$, $\vec{m} = 2\vec{k}_j$, or $\vec{l} = 2\vec{k}_j$, $\vec{m} = -\vec{k}_j$. In either case we have $\vec{l}^{\perp} \cdot \vec{m} = 0$.

8.3 The statistical predictions for the truncated systems

We now have all the ingredients for the application of the equilibrium statistical theory introduced in Section 7.2. We base the theory on the truncated energy and enstrophy from (8.7) and (8.8), which are the two conserved quantities in this truncated system.

Let α and θ be the Lagrange multipliers for the enstrophy $\mathcal{E}_\Lambda$ and energy E_Λ respectively. Let

$$\mu = \frac{\theta}{\alpha}, \quad \text{if}\, \alpha \neq 0. \tag{8.18}$$

Thanks to the formula in (7.27), and (8.7)–(8.8), the Gibbs measure is then given by

$$\mathcal{G}_{\alpha,\theta} = c \exp\left(-\alpha\left(\beta V + \frac{1}{2}\sum_{1\le|\vec{k}|^2\le\Lambda} |-|\vec{k}|^2\hat{\psi}_{\vec{k}} + \hat{h}_{\vec{k}}|^2\right)\right.$$
$$\left.-\theta\left(\frac{1}{2}V^2 + \frac{1}{2}\sum_{1\le|\vec{k}|^2\le\Lambda} |\vec{k}|^2|\hat{\psi}_{\vec{k}}|^2\right)\right), \tag{8.19}$$

where α and θ are determined by the average enstrophy and energy constraints. Due to the resemblance of the most probable distribution (8.19) to the thermal equilibrium ensemble, θ is sometimes referred to as the inverse temperature and α as the thermodynamic potential.

In order to guarantee that the Gibbs measure is a probability measure we need to ensure that the coefficients of the quadratic terms are negative, i.e.

$$\alpha|\vec{k}|^4+\theta|\vec{k}|^2>0, \text{ for all } \vec{k} \text{ satisfying } |\vec{k}|^2\leq\Lambda, \text{ and}$$
$$\theta>0, \text{ if } V\neq 0.$$

This implies either

$$\alpha>0, \quad \mu>0, \tag{8.20}$$

or

$$V\equiv 0, \quad \alpha>0. \quad \mu>-1, \tag{8.21}$$

or

$$\alpha<0, \quad \mu<-\Lambda, \quad \theta>0. \tag{8.22}$$

Obviously, case (8.22) is a spurious condition due to the truncation only and hence is not physically relevant.

Under the realizability condition in (8.20), (8.21), we may introduce

$$\overline{V}=-\frac{\beta}{\mu}, \quad \overline{\hat{\psi}}_{\vec{k}}=\frac{\hat{h}_{\vec{k}}}{\mu+|\vec{k}|^2}, \tag{8.23}$$

and

$$\overline{\psi}'_\Lambda(\vec{x},t)=\sum_{1\leq|\vec{k}|^2\leq\Lambda}\overline{\hat{\psi}}_{\vec{k}}e^{i\vec{x}\cdot\vec{k}}. \tag{8.24}$$

We then observe that $(\overline{V},\overline{\psi}'_\Lambda)$ satisfies equation (8.10) and hence it is non-linearly stable under the realizability condition (8.20). If $V\equiv 0$, then $\overline{\psi}'_\Lambda$ is a solution to the first equation in (8.10) and it is non-linearly stable if the realizability condition (8.21) is satisfied.

We may rewrite the Gibbs measure, thanks to (8.11) and (8.12), as

$$\begin{aligned}\mathcal{G}_{\alpha,\mu}&=c\exp(-(\alpha\mathcal{E}_\Lambda+\alpha\mu E_\Lambda))\\&=c\exp(-\alpha(\mathcal{E}_\Lambda+\mu E_\Lambda))\\&=c_{\alpha,\mu}\exp(-\alpha\mathcal{W}_\mu(\delta q,\delta V))\\&=c_{\alpha,\mu}\exp\left(-\alpha\left(\frac{\mu}{2}(V-\overline{V})^2\right.\right.\\&\qquad\left.\left.+\frac{1}{2}\sum_{1\leq|\vec{k}|^2\leq\Lambda}|\vec{k}|^2(|\vec{k}|^2+\mu)(\hat{\psi}_{\vec{k}}-\overline{\hat{\psi}}_{\vec{k}})^2\right)\right),\end{aligned}\tag{8.25}$$

or equivalently

$$\mathcal{G}_{\alpha,\mu}(\vec{X}) = \prod_{j=1}^{N} \mathcal{G}^{j}_{\alpha,\mu}(X_j). \tag{8.26}$$

Thus, the invariant Gibbs measure for the dynamics is a product of Gaussian measures with a non-zero mean. Moreover, $(\overline{V}, \overline{\psi'}_\Lambda)$ is exactly the ensemble average, or mean state, of (V, ψ'_Λ) with respect to the Gibbs measure

$$\langle \vec{X} \rangle = \int_{\mathcal{R}^N} \vec{X} \mathcal{G}_{\alpha,\mu}(\vec{X}) d\vec{X} = \left(\overline{V}, \overline{\hat{\psi}}_{\vec{k}_1}, \ldots, \overline{\hat{\psi}}_{\vec{k}_M}\right). \tag{8.27}$$

This implies, assuming ergodicity for the Gibbs measure as described in Chapter 7, the following remarkable prediction

$$\begin{aligned} \lim_{T\to\infty} \frac{1}{T} \int_{T_0}^{T_0+T} \psi'_\Lambda(\vec{x}, t) dt &= \overline{\psi}'_\Lambda, \\ \lim_{T\to\infty} \frac{1}{T} \int_{T_0}^{T_0+T} V(t) dt &= \overline{V} = -\frac{\beta}{\mu}. \end{aligned} \tag{8.28}$$

Hence the statistical theory predicts that the time average of the solutions to the truncated equations converge to the non-linearly stable exact steady state solutions to the truncated equations. In other words, a specific coherent large-scale mean flow will emerge from the dynamics of (8.3), (8.4) with generic initial data after computing a long time average.

8.4 Numerical evidence supporting the statistical prediction

The equilibrium statistical mechanics theory provides us with a striking prediction regarding the long time behavior of solutions to the spectrally truncated systems. Namely, the long time average of the solution should converge to the most probable mean state for generic initial data. We naturally ask if such a prediction is valid. Instead of trying to prove the ergodicity of the Gibbs measure, an impossible theoretical task at the present stage of our knowledge, we utilize careful numerical experiments as we did earlier in Chapter 7. Here we report some numerical evidence supporting the convergence of the time averages of numerical solutions to the most probable mean state predicted by the equilibrium statistical mechanics theory in Section 8.3 for the spectrally truncated barotropic quasi-geostrophic equations on a periodic box with topography but no large-scale mean flow. In other words, we provide numerical evidence for the validity of equation (8.28).

As usual, the time average of a function $f(t)$ over an averaging window T is defined as

$$\langle f\rangle(T) = \frac{1}{T}\int_{T_0}^{T_0+T} f(t)dt \tag{8.29}$$

as a function of T, where T_0 is a time when averaging begins. In our case we choose $T_0 = 1000$ and T goes up to 5000.

In order to make the the visualization easy, we choose the following layered topography

$$h = 0.2\cos(x) + 0.4\cos(2x) \tag{8.30}$$

so that the most probable mean state predicted by the statistical theory is also layered, i.e. function of x only. The initial conditions are generated using a generic random initial data with energy $E = 7$ and potential enstrophy $\mathcal{E} = 20$. The parameter μ is determined at the end of the calculation numerically, and is approximately -0.9029. The numerical simulation was performed on the 11×11 grid of Fourier coefficients; however, the results presented here are the stream functions and flows (velocity fields) on the 32×32 grid in physical space after interpolation of these results.

The convergence is measured in terms of the relative L^2 error, which is defined as

$$L_2[\mathbf{f}, \mathbf{g}] = \frac{\|\mathbf{f} - \mathbf{g}\|_0}{\|\mathbf{f}\|_0} \tag{8.31}$$

for two functions $\mathbf{f}$ and $\mathbf{g}$.

According to the above definition, $L_2[\bar{\psi}_\Lambda, \langle\psi_\Lambda\rangle(T)]$ denotes the relative L_2-error for the stream function, and $L_2[\nabla^\perp\bar{\psi}_\Lambda, \langle\nabla^\perp\psi_\Lambda\rangle(T)]$ denotes the relative L_2-error for the velocity (flow).

T	$L_2[\bar{\psi}_\Lambda, \langle\psi_\Lambda\rangle(T)]$	$L_2[\nabla^\perp\bar{\psi}_\Lambda, \langle\nabla^\perp\psi_\Lambda\rangle(T)]$
10	1.404	1.415
20	1.318	1.342
50	1.109	1.116
70	0.8953	0.9008
100	0.4333	0.4473
200	0.2451	0.2604
500	0.2416	0.2496
1000	$8.648 \cdot 10^{-2}$	0.1017
5000	$6.902 \cdot 10^{-2}$	$9.184 \cdot 10^{-2}$

(8.32)

Table (8.32) lists the relative errors against the averaging window size T. As we can see, the relative L_2-errors for both stream function and velocity decrease as the

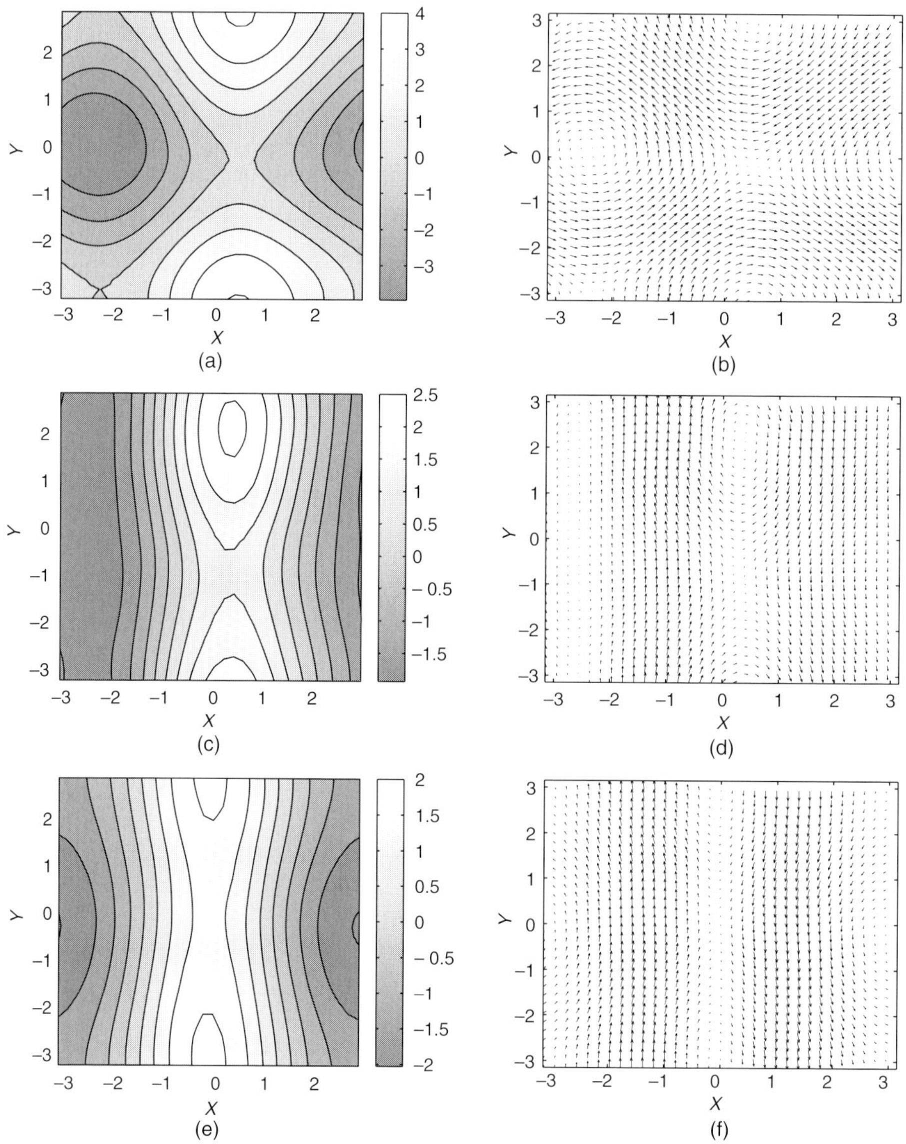

Figure 8.1 (a) time average of stream function with $T = 50$, (b) time average of velocity with $T = 50$, (c) time average of stream function with $T = 100$, (d) time average of velocity with $T = 100$, (e) time average of stream function with $T = 500$, (f) time average of velocity with $T = 500$.

averaging time increases. For a time average window of 5000, the relative error is less than 10%. Figure 8.1 depicts the time average of the numerical solutions over windows of size 50, 100, and 500. Figure 8.2 shows the time average of the numerical solutions over a window of 5000 versus the most probable mean

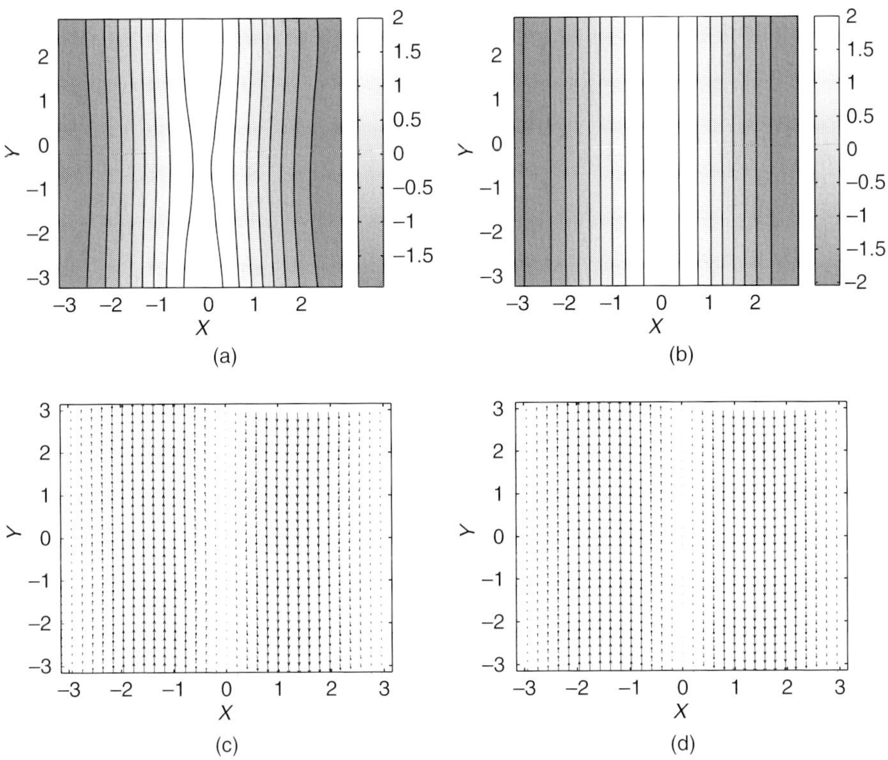

Figure 8.2 (a) time average of stream function with $T = 5000$, (b) stream function of the most probable mean state, (c) time average of velocity field with $T = 5000$, (d) velocity field of the most probable mean state.

state predicted by the equilibrium statistical theory. It is visually clear that the running time average is close to the predicted most probable mean state for a window size of 5000, but far from close for a window of size 50 or 100, with intermediate behavior for a window of size 500. This indicates that averaging over a sufficiently large window of time is essential for observing the predicted results in (8.28). The specific examples for layered topography in Chapter 5 show that such results are indeed generic in a statistical sense, i.e. not every individual solution satisfies (8.28).

8.5 The pseudo-energy and equilibrium statistical mechanics for fluctuations about the mean

In many problems in geophysical fluids, we can assume that the time averaged mean state, the climate, is known with reasonable accuracy and that the statistical fluctuations about the mean state are the quantities of interest. This point of view is an important one for stochastic modeling. See for instance, Majda, Timofeyev, and Vanden Eijnden (2001). Here we show how dynamical equations for fluctuations and equilibrium statistical mechanics can be applied directly for the truncated quasi-geostrophic equations treated earlier in this chapter. For simplicity in exposition, we assume that there is no large-scale mean flow, and the only geophysical effect is topography, so that the equations in (8.3)–(8.4) become

$$\frac{\partial q_\Lambda}{\partial t} + P_\Lambda(\nabla^\perp \psi_\Lambda \cdot \nabla q_\Lambda) = 0. \tag{8.33}$$

We consider the exact steady solutions from (8.9) or (8.10)

$$\overline{q}_\Lambda = \mu \overline{\psi}_\Lambda, \quad \overline{q}_\Lambda = \Delta \overline{\psi}_\Lambda + h_\Lambda \tag{8.34}$$

and consider perturbations about this mean state, i.e.

$$\psi_\Lambda = \overline{\psi}_\Lambda + \tilde{\psi}, \quad q_\Lambda = \overline{q}_\Lambda + \tilde{\omega}, \quad \Delta\tilde{\psi} = \tilde{\omega}. \tag{8.35}$$

Substituting (8.35) into (8.33) yields the equations

$$\begin{aligned}\frac{\partial \tilde{\omega}}{\partial t} &+ P_\Lambda(\nabla^\perp \overline{\psi}_\Lambda \cdot \nabla \overline{q}_\Lambda) + P_\Lambda(\nabla^\perp \tilde{\psi} \cdot \nabla \overline{q}_\Lambda) + P_\Lambda(\nabla^\perp \overline{\psi}_\Lambda \cdot \nabla \tilde{\omega}) \\ &+ P_\Lambda(\nabla^\perp \tilde{\psi} \cdot \nabla \tilde{\omega}) = 0.\end{aligned} \tag{8.36}$$

Utilizing (8.34) we have

$$\begin{gathered} P_\Lambda(\nabla^\perp \overline{\psi}_\Lambda \cdot \nabla \overline{q}_\Lambda) = 0, \\ P_\Lambda(\nabla^\perp \tilde{\psi} \cdot \nabla \overline{q}) = \mu P_\Lambda(\nabla^\perp \tilde{\psi} \cdot \nabla \overline{\psi}) = -\mu P_\Lambda(\nabla^\perp \overline{\psi} \cdot \nabla \tilde{\psi}). \end{gathered}$$

Hence (8.36) becomes the **equations for perturbations**

$$\begin{gathered} \frac{\partial \tilde{\omega}}{\partial t} + P_\Lambda(\nabla^\perp \overline{\psi}_\Lambda \cdot \nabla(\tilde{\omega} - \mu\tilde{\psi})) + P_\Lambda(\nabla^\perp \tilde{\psi} \cdot \nabla \tilde{\omega}) = 0, \\ \Delta \tilde{\psi} = \tilde{\omega}. \end{gathered} \tag{8.37}$$

The equations for perturbations involve the familiar truncated non-linear terms of ordinary fluid flow

$$P_\Lambda(\nabla^\perp \tilde{\psi} \cdot \nabla \tilde{\omega}) \tag{8.38}$$

and the linear operator reflecting the mean flow $\overline{\psi}_\Lambda$

$$P_\Lambda(\nabla^\perp \overline{\psi}_\Lambda \cdot \nabla(\tilde{\omega} - \mu\tilde{\psi})). \tag{8.39}$$

Next, we would like to set-up a statistical theory for fluctuations directly from the dynamics in (8.37). According to Chapter 7, we need to find the conserved quantities for (8.37) and also verify the Liouville property for the finite-dimensional system of ODEs. From Section 8.2, we know that the standard non-linear terms in

(8.38) conserve the energy and enstrophy; however, the linear operator in (8.39) conserves neither the energy nor the enstrophy. Nevertheless, the non-linear stability analysis from Section 4.2 and equation (8.12) suggests that there exists a single conserved quantity, $\tilde{E}$, for the dynamics in (8.37), involving fluctuations about the mean state given by the *pseudo-energy*

$$\tilde{E} = \frac{1}{2} \fint (\tilde{\omega}^2 - \mu \tilde{\psi} \tilde{\omega}) = \mathcal{E}_\Lambda + \mu E_\Lambda, \tag{8.40}$$

which is an appropriate linear combination of energy and enstrophy. To check the conservation of the pseudo-energy directly for the dynamic equations in (8.37), we focus only on the contribution arising from the linear terms in (8.39), since the non-linear terms in (8.38) automatically conserve both energy and enstrophy separately. We calculate from (8.37) and (8.39) that

$$\begin{aligned}
\frac{d\tilde{E}}{dt} &= \fint (\tilde{\omega} - \mu \tilde{\psi}) \frac{\partial \tilde{\omega}}{\partial t} \\
&= -\fint (\tilde{\omega} - \mu \tilde{\psi}) \nabla^\perp \overline{\psi}_\Lambda \cdot \nabla (\tilde{\omega} - \mu \tilde{\psi}) \\
&= -\fint \nabla^\perp \overline{\psi}_\Lambda \cdot \nabla \left(\frac{1}{2} (\tilde{\omega} - \mu \tilde{\psi})^2 \right) \\
&= 0,
\end{aligned} \tag{8.41}$$

which verifies the conservation of the pseudo-energy.

From Section 4.2 and Subsection 8.2.3, the quadratic form defining the pseudo-energy in (8.40) is given by

$$\tilde{E} = \frac{1}{2} \sum_{|\vec{k}|^2 \leq \Lambda} \left(1 + \frac{\mu}{|\vec{k}|^2} \right) |\tilde{\omega}_{\vec{k}}|^2 \tag{8.42}$$

and is positive definite only for $\mu > -1$ for arbitrary truncations Λ. In this case, we introduce the pseudo-energy variables, $p_{\vec{k}}$, defined by

$$p_{\vec{k}} = \left(1 + \frac{\mu}{|\vec{k}|^2} \right)^{1/2} \tilde{\omega}_{\vec{k}}. \tag{8.43}$$

Next we sketch how to set up equilibrium statistical mechanics directly for the fluctuation described by the dynamic equations (8.37).

With (8.43) we use the natural coordinates

$$\vec{X} = \left(\text{Re } p_{\vec{k}_1}, \text{Im } p_{\vec{k}_1}, \ldots, \text{Re } p_{\vec{k}_M}, \text{Im } p_{\vec{k}_M} \right) \tag{8.44}$$

over a suitably defining set S as in (8.13)–(8.14). It is a straightforward exercise for the reader to check that the Liouville property is satisfied directly for the

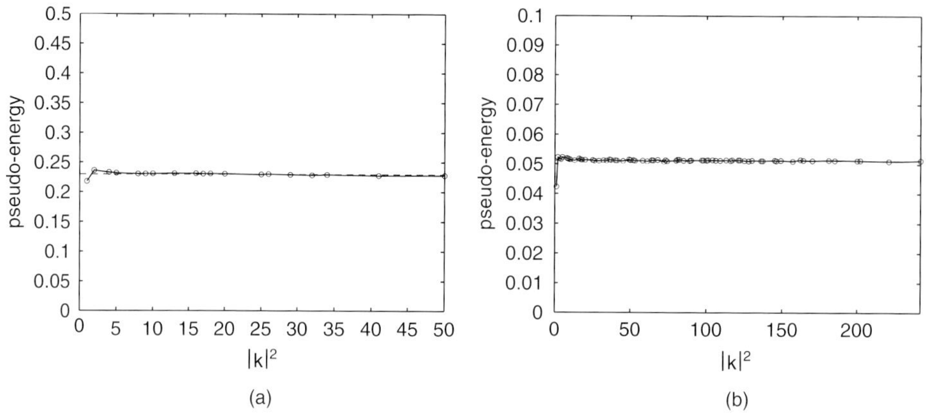

Figure 8.3 Pseudo-energy spectrum. Solid line with circles – numerical result, dashed line – analytical prediction. (a) 11×11 case: $\alpha = 4.355$, $\mu = -0.9029$, (b) 23×23 case: $\alpha = 19.52$, $\mu = -0.9545$.

dynamics (8.37) with the variables in (8.43) with the same structure as (8.16) and (8.17). Thus the equilibrium statistical theory developed in Chapter 7 applies directly to the fluctuations with the equilibrium statistical Gibbs measure given by

$$\mathcal{G}_\alpha = C_\alpha e^{-\alpha \tilde{E}} = C_\alpha e^{-\frac{\alpha}{2}\sum |p_{\vec{k}}|^2}. \tag{8.45}$$

This Gibbs measure for fluctuations has the same structure in the pseudo-energy variables as the Gibbs measure for the truncated Burgers equations in Section 7.3 and leads to predictions of equi-partition of pseudo-energy for fluctuations about the mean state as in Chapter 7 for the truncated Burgers equation. Clearly, the same result would have been obtained directly from the approach in Sections 8.2 and 8.3 by merely diagonalizing the quadratic form in (8.25); however, the dynamic statistical theory for fluctuations developed here has independent interest and provides an alternative point of view which we emphasize here. Figure 8.3 contains numerical results confirming the equi-partition of pseudo-energy predicted by (8.45). The numerical simulation is carried out in exactly the same way as described in Section 8.4 with the same layered topography given by (8.30). The pseudo-energy in each mode is calculated using time average just as in Subsection 7.3.4. It is visually clear that the equi-partition of energy holds very well.

8.6 The continuum limit

Recall that we are interested in the dynamics of the quasi-geostrophic equations (8.1). It is only in the limit of $\Lambda \to \infty$ that we recover the solution to (8.1) from solutions to the truncated equations (8.3), (8.4). Thus we are naturally interested in the asymptotic behavior of the invariant Gibbs measures as well as their mean states from Section 8.3 at large Λ. We are also interested in taking the continuum limit as Λ approaches infinity.

It is apparent that all predictions of the truncated system must depend on the truncation eigenvalue Λ, thus we occasionally append a suffix Λ in order to distinguish the dependence on Λ and avoid confusion.

There is a subtle issue when we take the limit $\Lambda \to \infty$, i.e. how do we pick $\alpha = \alpha_\Lambda$, $\mu = \mu_\Lambda$ satisfying the energy and the enstrophy constraints as $\Lambda \to \infty$.

We start by recalling two observations. First, the ensemble average of the truncated energy and enstrophy need not satisfy the exact energy/enstrophy constraint. It is in the limit of the cut-off wave number approaching infinity, i.e. $\Lambda \to \infty$, that these two constraints have to be satisfied, i.e.

$$\lim_{\Lambda\to\infty} \langle E_\Lambda \rangle = E_0, \qquad \lim_{\Lambda\to\infty} \langle \mathcal{E}_\Lambda \rangle = \mathcal{E}_0. \tag{8.46}$$

Second, there is a constraint on the energy E_0 and enstrophy $\mathcal{E}_0$ imposed. Apparently $\mathcal{E}_0$ must be at least the minimum enstrophy associated with the given energy level, i.e.

$$\mathcal{E}_0 \geq \min_{E(\psi)=E_0} \mathcal{E}(\psi) = \mathcal{E}_*(E_0). \tag{8.47}$$

In the case of equality, any state satisfying the energy/enstrophy constraint must be an enstrophy-minimizing (selective decay) state. These enstrophy-minimizing states are completely characterized in Section 4.5 and there is not much randomness present. Thus, we assume that the given enstrophy level $\mathcal{E}_0$ is strictly more than the minimum enstrophy associated with the given energy level E_0, i.e

$$\mathcal{E}_0 > \mathcal{E}_*(E_0) = \min_{E(\psi)=E_0} \mathcal{E}(\psi). \tag{8.48}$$

Since the constraints are given in terms of the ensemble average of the enstrophy and energy with respect to the Gibbs measure (8.25), which is a Gaussian, we explicitly calculate that

$$\begin{aligned} \int V^2 \mathcal{G}_{\alpha,\mu} &= (\alpha\mu)^{-1} + \frac{\beta^2}{\mu^2}, \\ \int |\hat{\psi}_{\vec{k}}|^2 \mathcal{G}_{\alpha,\mu} &= (\alpha|\vec{k}|^2(\mu + |\vec{k}|^2))^{-1} + |\overline{\hat{\psi}_{\vec{k}}}|^2. \end{aligned} \tag{8.49}$$

We then observe that the ensemble average of the energy and the enstrophy naturally separate into two parts, a mean part that corresponds to the mean state and a fluctuation part

$$\langle E_\Lambda \rangle = \langle E_\Lambda \rangle_{\mathcal{G}} = \overline{E}_\Lambda + E'_\Lambda,$$
$$\overline{E}_\Lambda = \frac{1}{2}\left(\frac{\beta^2}{\mu^2} + \sum_{1\le|\vec{k}|^2\le\Lambda} \frac{|\vec{k}|^2|\hat{h}_{\vec{k}}|^2}{(\mu+|\vec{k}|^2)^2}\right), \tag{8.50}$$
$$E'_\Lambda = \frac{\alpha^{-1}}{2}\left(\mu^{-1} + \sum_{1\le|\vec{k}|^2\le\Lambda} \frac{1}{\mu+|\vec{k}|^2}\right),$$

while for the enstrophy

$$\langle \mathcal{E}_\Lambda \rangle = \langle \mathcal{E}_\Lambda \rangle_{\mathcal{G}} = \overline{\mathcal{E}}_\Lambda + \mathcal{E}'_\Lambda,$$
$$\overline{\mathcal{E}}_\Lambda = -\frac{\beta^2}{\mu} + \frac{1}{2}\sum_{1\le|\vec{k}|^2\le\Lambda} \frac{\mu^2|\hat{h}_{\vec{k}}|^2}{(\mu+|\vec{k}|^2)^2}, \quad \mathcal{E}'_\Lambda = \frac{\alpha^{-1}}{2}\sum_{1\le|\vec{k}|^2\le\Lambda} \frac{|\vec{k}|^2}{\mu+|\vec{k}|^2}. \tag{8.51}$$

For the case without large-scale mean flow, i.e $V \equiv 0$, $\beta = 0$, we have

$$\overline{E}_\Lambda = \frac{1}{2}\sum_{1\le|\vec{k}|^2\le\Lambda} \frac{|\vec{k}|^2|\hat{h}_{\vec{k}}|^2}{(\mu+|\vec{k}|^2)^2}, \quad E'_\Lambda = \frac{\alpha^{-1}}{2}\sum_{1\le|\vec{k}|^2\le\Lambda} \frac{1}{\mu+|\vec{k}|^2}, \tag{8.52}$$

and

$$\overline{\mathcal{E}}_\Lambda = \frac{1}{2}\sum_{1\le|\vec{k}|^2\le\Lambda} \frac{\mu^2|\hat{h}_{\vec{k}}|^2}{(\mu+|\vec{k}|^2)^2}, \quad \mathcal{E}'_\Lambda = \frac{\alpha^{-1}}{2}\sum_{1\le|\vec{k}|^2\le\Lambda} \frac{|\vec{k}|^2}{\mu+|\vec{k}|^2}. \tag{8.53}$$

Observe that the fluctuation part of the energy is isotropic and independent of the mean state. We may estimate this part of the energy by replacing the summation by integration for large Λ

$$\begin{aligned}
E'_\Lambda &= \frac{\alpha^{-1}}{2}\sum_{1\le|\vec{k}|^2\le\Lambda} \frac{1}{\mu+|\vec{k}|^2} + \frac{(\alpha\mu)^{-1}}{2} \\
&\cong \frac{\alpha^{-1}}{2}2\pi\int_1^{\sqrt{\Lambda}} \frac{|\vec{k}|}{\mu+|\vec{k}|^2}d|\vec{k}| + \frac{(\alpha\mu)^{-1}}{2} \\
&= \frac{\alpha^{-1}}{2}\pi\ln(\mu+|\vec{k}|^2)\Big|_1^{\sqrt{\Lambda}} + \frac{(\alpha\mu)^{-1}}{2} \\
&= \frac{\alpha^{-1}}{2}\pi\ln\left(\frac{\mu+\Lambda}{\mu+1}\right) + \frac{(\alpha\mu)^{-1}}{2}.
\end{aligned} \tag{8.54}$$

Likewise, the fluctuation part of the total enstrophy can be estimated as

$$\begin{aligned}\mathcal{E}'_\Lambda &= \frac{1}{2\alpha}\sum_{1\le|\vec{k}|^2\le\Lambda}\frac{|\vec{k}|^2}{\mu+|\vec{k}|^2}\\ &\cong \frac{1}{2\alpha}2\pi\int_1^{\sqrt{\Lambda}}\frac{|\vec{k}|^3}{\mu+|\vec{k}|^2}d|\vec{k}|\\ &= \frac{\pi}{\alpha}\int_1^{\sqrt{\Lambda}}\frac{|\vec{k}|(\mu+|\vec{k}|^2)-\mu|\vec{k}|}{\mu+|\vec{k}|^2}d|\vec{k}|\\ &= \frac{\pi}{2\alpha}\left(\Lambda-1-\mu\ln\left(\frac{\mu+\Lambda}{\mu+1}\right)\right).\end{aligned} \tag{8.55}$$

Immediately, we can see that, if the parameters α and μ are bounded independent of Λ, the fluctuation part of the energy diverges to infinite as Λ approaches infinity, which contradicts the energy constraint. Thus, one of the parameters α and μ must approach infinity. In the Appendix, we will show that the parameter μ must remain bounded in the continuum limit. This is of course expected since large μ corresponds to small geophysical influence, which contradicts physical reality. In what follows, we consider the case that μ is bounded and hence α approaches infinity. This means that the variance of the Gaussian measures for individual wave numbers $|\vec{k}|$ in (8.25) tends to zero, i.e. the fluctuations are suppressed at all wave numbers. For the case without large-scale mean flow, the fluctuations are suppressed away from the ground energy shell, since μ could approach -1 in this case (see the realizability condition (8.21)). We will give more details in Subsection 8.6.2.

As we shall demonstrate below, the asymptotic behavior of the Gibbs measure and its mean state depend heavily on the geophysical effects. We will elaborate in four cases below: an easy case with large-scale mean flow; a case without large-scale mean, but with non-trivial generic topography; a case without geophysical effects; and a case without large-scale mean, but with non-trivial topography having a degenerate spectrum.

8.6.1 The case with a large-scale mean flow

This is the simplest case.

By the realizability condition (8.20) and the explicit formula for the ensemble energy given in (8.50), we observe that $E'_\Lambda > 0$ which further implies that

$$\frac{\beta^2}{2\mu_\Lambda^2} \le \overline{E}_\Lambda \le E_0 \tag{8.56}$$

asymptotically for large Λ. This leads to the following asymptotic lower bound on μ_Λ

$$\mu_\Lambda \geq \frac{\beta}{\sqrt{2E_0}}. \tag{8.57}$$

Substituting (8.57) into the second equation in (8.51), we see that

$$\overline{\mathcal{E}}_\Lambda \geq -\frac{\beta^2}{\mu_\Lambda} \geq -\beta\sqrt{2E_0}, \tag{8.58}$$

asymptotically for large Λ. Thus, we arrive at an asymptotic bound on the fluctuation part of the ensemble enstrophy

$$\mathcal{E}'_\Lambda \leq \mathcal{E}_0 - \overline{\mathcal{E}}_\Lambda \leq \mathcal{E}_0 + \beta\sqrt{2E_0}. \tag{8.59}$$

Now that μ_Λ is bounded above and is positive by the realizability condition (8.20), hence we deduce

$$\lim_{\Lambda\to\infty} \frac{\ln \frac{\mu_\Lambda+\Lambda}{\mu_\Lambda+1}}{\Lambda} = \lim_{\Lambda\to\infty} \frac{\ln \Lambda}{\Lambda} = 0. \tag{8.60}$$

Therefore, combining (8.60), (8.55), and (8.59), we have

$$\alpha_\Lambda \geq \frac{\pi\Lambda}{2(\mathcal{E}_0 + \beta\sqrt{2E_0})}, \tag{8.61}$$

asymptotically for large Λ. Utilizing (8.61), (8.57) and the boundedness of μ in (8.54) we deduce

$$E'_\Lambda \to 0, \quad \text{as } \Lambda \to \infty. \tag{8.62}$$

Thus, we conclude that *all energy must reside in the mean state asymptotically for large* Λ. This leads to the following choice of parameters and asymptotic behavior of the parameters as well as the mean states.

Step 1: (The choice of $\mu = \mu_\Lambda$) For the given energy E_0 and Λ, we can find a unique $\mu = \mu_\Lambda > 0$ such that

$$\overline{E}_\Lambda = E(\overline{\psi'}_{\mu_\Lambda}, \overline{V}_{\mu_\Lambda}) = E_0. \tag{8.63}$$

Then $\overline{\psi'}_{\mu_\Lambda}, \overline{V}_{\mu_\Lambda}$ satisfy

$$\Delta\overline{\psi'}_{\mu_\Lambda} + h_\Lambda = \mu_\Lambda \overline{\psi'}_{\mu_\Lambda}, \overline{V}_{\mu_\Lambda} = -\frac{\beta}{\mu_\Lambda}. \tag{8.64}$$

It is easy to see that μ_Λ is a non-decreasing function of Λ, since for fixed μ the energy associated with the mean state $\overline{E}_\Lambda$ given by (8.50) is a monotonic-increasing function in Λ because more terms are added. Moreover, it is easy to check that there exists $\mu(=\mu_\infty) > 0$ such that

$$\lim_{\Lambda\to\infty} \mu_\Lambda = \mu \tag{8.65}$$

in a monotonic-decreasing fashion. We observe that we have positive temperature for all energy levels in this case.

Step 2: (The limit of the mean states) The mean states also converge. It is easy to check, thanks to (8.23), (8.24), and (8.65), that

$$\overline{\psi'}_{\mu_\Lambda} \to \overline{\psi'}_\mu = \sum \frac{\hat{h}_{\vec{k}}}{\mu + |\vec{k}|^2} e^{i\vec{k}\cdot\vec{x}}, \quad \overline{V}_{\mu_\Lambda} \to \overline{V}_\mu = -\frac{\beta}{\mu}, \tag{8.66}$$

where $\mu = \mu_\infty > 0$ is the unique μ in $(0, \infty)$ such that the energy constraint is met by the limit mean state $(\overline{\psi'}_\mu, \overline{V}_\mu)$, i.e.

$$E(\overline{\psi'}_\mu, \overline{V}_\mu) = E_0. \tag{8.67}$$

We also observe that the limit mean state satisfies the limit mean field equation, thanks to (8.66)–(8.67)

$$\Delta\overline{\psi'}_\mu + h = \mu\overline{\psi'}_\mu, \quad \overline{V}_\mu = -\frac{\beta}{\mu}, \tag{8.68}$$

and therefore is non-linearly stable according to Section 4.5.

Step 3: (The choice of $\alpha = \alpha_\Lambda$ and the enstrophy constraint) We know that for the given energy E_0 there exists the minimal enstrophy, $\mathcal{E}_*(E_0)$, which is equal to $\mathcal{E}(\overline{\psi'}_\mu, \overline{V}_\mu)$ since $(\overline{\psi'}_\mu, \overline{V}_\mu)$ is the enstrophy-minimizing state (or selective decay state) with the topography h and β, according to Section 4.5. On the other hand, thanks to (8.66)–(8.68)

$$\lim_{\Lambda\to\infty} \overline{\mathcal{E}}_\Lambda = \mathcal{E}(\overline{\psi'}_\mu, \overline{V}_\mu) = \mathcal{E}_*(E_0). \tag{8.69}$$

Combining (8.48) and (8.69) we have, for large Λ

$$\mathcal{E}_0 > \overline{\mathcal{E}}_\Lambda = \mathcal{E}(\overline{\psi'}_{\mu_\Lambda}, \overline{V}_{\mu_\Lambda}). \tag{8.70}$$

We now pick α_Λ so that the enstrophy constraint is satisfied for all truncations, i.e.

$$\mathcal{E}_0 = \langle \mathcal{E}_\Lambda \rangle = \overline{\mathcal{E}}_\Lambda + \mathcal{E}'_\Lambda. \tag{8.71}$$

From (8.55), this amounts to requiring

$$\alpha_\Lambda \cong \frac{\frac{\pi}{2}\Lambda}{\mathcal{E}_0 - \overline{\mathcal{E}}_\Lambda} \tag{8.72}$$

for $\Lambda \gg 1$. Note that $\alpha \to \infty$ as $\Lambda_\Lambda \to \infty$. This limit is a *non-extensive thermodynamic limit* because the energy of individual fluctuation is not constant.

Step 4: (The energy constraint) It is easy to see that the energy fluctuation goes to zero as $\Lambda \to \infty$. More precisely, thanks to (8.72) and (8.54)

$$E'_\Lambda \cong \frac{\pi \ln \Lambda}{2\Lambda} \to 0. \tag{8.73}$$

Combining (8.73) with (8.63) and (8.50) we conclude that the energy constraint is satisfied in the sense of (8.46).

8.6.2 The case without large-scale mean flow but with generic topography

Here we assume no large-scale mean flow but generic topography in the sense that

$$V \equiv 0, \quad \beta = 0, \quad P_{\Lambda_1}(h) \neq 0, \tag{8.74}$$

where P_{Λ_1} is the projection on to the ground state modes (first energy shell). In this case, the mean and fluctuation part of the ensemble average of the energy and enstrophy are given in (8.52) and (8.53).

It is easy to see that

$$E_0 \geq \overline{E}_\Lambda \geq \frac{1}{2} \sum_{|\vec{k}|^2=1} \frac{|\vec{k}|^2 |\hat{h}_{\vec{k}}|^2}{(\mu_\Lambda + |\vec{k}|^2)^2} = \frac{1}{2} \frac{\|P_1(h)\|_0^2}{(\mu_\Lambda + 1)^2}, \tag{8.75}$$

asymptotically for large Λ. Thus

$$\mu_\Lambda \geq \frac{\|P_1(h)\|_0}{\sqrt{2E_0}} - 1 > -1, \tag{8.76}$$

asymptotically for large Λ. Therefore (8.60) holds in this case of no large-scale mean flow as well. Hence, after combining (8.60), (8.53), and (8.55), we deduce

$$\alpha_\Lambda \geq \frac{\pi\Lambda}{2\mathcal{E}'_\Lambda} \geq \frac{\pi\Lambda}{2\mathcal{E}_\Lambda}, \tag{8.77}$$

asymptotically for large Λ. This further implies that (8.62) is true in this case of no large-scale mean flow but with generic topography. We may then conclude that *all energy must reside in the mean state asymptotically for large* Λ. Therefore, we have the following choice of parameters as well as the asymptotic behavior of the mean states.

Step 1: For given energy level E_0 and Λ, we can pick a unique $\mu = \mu_\Lambda \in (-1, \infty)$ such that

$$\overline{E}_\Lambda = \frac{1}{2} \sum_{|\vec{k}|^2 \leq \Lambda} \frac{|\vec{k}|^2 |\hat{h}_{\vec{k}}|^2}{(\mu + |\vec{k}|^2)^2} = E(\overline{\psi}_{\Lambda,\mu}) = E_0. \tag{8.78}$$

This is possible since $\overline{E}_\Lambda$ approaches infinity as μ approaches $-\Lambda_1 = -1$ and approaches zero as μ approaches ∞; also the energy of the mean state is a monotonic function in μ for fixed Λ.

Indeed, we can verify that $\mu = \mu_\Lambda$ is a non-decreasing function of Λ (since more terms are added as the wave number increases) and there is a limit $\mu = \mu_\infty$ such that

$$\lim_{\Lambda \to \infty} \mu_\Lambda = \mu (= \mu_\infty) > -1, \quad E(\overline{\psi}_\mu) = E_0. \tag{8.79}$$

We observe that for low energy ($E_0 < E(\overline{\psi}_0)$) we have positive temperature, while for high energy ($E_0 > E(\overline{\psi}_0)$) we have negative temperature in this case.

Step 2: (The limit of the mean states) The mean states also converge. It is easy to check that

$$\overline{\psi}_{\mu_\Lambda} \to \overline{\psi}_\mu = \sum \frac{\hat{h}_{\vec{k}}}{\mu + |\vec{k}|^2} e^{i\vec{k}\cdot\vec{x}}, \tag{8.80}$$

and $\mu = \mu_\infty > -1$ is the unique μ in $(-1, \infty)$ such that the energy constraint is met by the limit mean state $\overline{\psi}_\mu$, i.e.

$$E(\overline{\psi}_\mu) = E_0, \tag{8.81}$$

and the limit mean state satisfies the limit mean field equation

$$\Delta \overline{\psi}_\mu + h = \mu \overline{\psi}_\mu. \tag{8.82}$$

Hence the limit mean state is non-linearly stable according to Section 4.5.

Step 3: (The choice of $\alpha = \alpha_\Lambda$ and the enstrophy constraint) We know that for the given energy E_0 there exists the minimal enstrophy, $\mathcal{E}_*(E_0)$ which is equal to $\mathcal{E}(\overline{\psi}_\mu)$ since $\overline{\psi}_\mu$ is the enstrophy-minimizing state (or selective decay state) with the topography h according to Section 4.5. On the other hand, thanks to (8.79)–(8.80)

$$\lim_{\Lambda \to \infty} \overline{\mathcal{E}}_\Lambda = \mathcal{E}(\overline{\psi}_\mu) = \mathcal{E}_*(E_0). \tag{8.83}$$

From (8.48) and (8.83) we deduce, for large Λ

$$\mathcal{E}_0 > \overline{\mathcal{E}}_\Lambda = \mathcal{E}(\overline{\psi}_{\mu_\Lambda}). \tag{8.84}$$

We now pick α_Λ so that the enstrophy constraint is satisfied for all truncation, i.e

$$\mathcal{E}_0 = \langle \mathcal{E}_\Lambda \rangle = \overline{\mathcal{E}}_\Lambda + \mathcal{E}'_\Lambda. \tag{8.85}$$

Thanks to (8.55), that amounts to

$$\alpha_\Lambda \cong \frac{\pi\Lambda}{a\mathcal{E}'_\Lambda} = \frac{\pi\Lambda}{2(\mathcal{E}_0 - \overline{\mathcal{E}}_\Lambda)} \cong \frac{\pi\Lambda}{2(\mathcal{E}_0 - \mathcal{E}_*(E_0))} \tag{8.86}$$

for $\Lambda \gg 1$. Note that $\alpha \to \infty$ as $\Lambda_\Lambda \to \infty$.

Step 4: (The energy constraint) It is easy to check that the energy fluctuation goes to zero as $\Lambda \to \infty$. More precisely, thanks to (8.86), (8.76), and (8.54)

$$E'_\Lambda \cong \frac{\pi \ln \Lambda}{\Lambda} \to 0. \tag{8.87}$$

Combining (8.87) with (8.78) and (8.50) we conclude that the energy constraint is satisfied in the sense of (8.46).

8.6.3 The case with no geophysical effects

Here we assume no geophysical effects

$$\beta = 0, \quad V \equiv 0, \quad h = 0, \tag{8.88}$$

i.e, we have two-dimensional Euler flow. This is the case treated by Kraichnan (1975). Observe that the mean state is identically zero; therefore, all energy must reside in the fluctuation part. Recall that the parameter $\mu = \mu_\Lambda$ is bounded below by -1 according to the realizability condition (8.21), and -1 is a singular point to $E_\Lambda = E'_\Lambda$. Hence we naturally decompose the fluctuation energy, E'_Λ, into the ground energy shell part and high energy shell part as

$$\begin{aligned} E'_\Lambda &= \frac{\alpha^{-1}}{2} \sum_{|\vec{k}|=1} \frac{1}{\mu + |\vec{k}|^2} + \frac{\alpha^{-1}}{2} \sum_{1<|\vec{k}|^2 \le \Lambda} \frac{1}{\mu + |\vec{k}|^2} \\ &= E'_{\Lambda,1} + E'_{\Lambda,2}. \end{aligned} \tag{8.89}$$

A similarly decomposition holds for the enstrophy

$$\begin{aligned} \mathcal{E}'_\Lambda &= \frac{\alpha^{-1}}{2} \sum_{|\vec{k}|=1} \frac{|\vec{k}|^2}{\mu + |\vec{k}|^2} + \frac{\alpha^{-1}}{2} \sum_{1<|\vec{k}|^2 \le \Lambda} \frac{|\vec{k}|^2}{\mu + |\vec{k}|^2} \\ &= \mathcal{E}'_{\Lambda,1} + \mathcal{E}'_{\Lambda,2}. \end{aligned} \tag{8.90}$$

Calculations similar to (8.54) and (8.55) yield

$$
\begin{aligned}
E'_{\Lambda,2} &\cong \frac{\alpha^{-1}}{2}\pi \ln\left(\frac{\mu+\Lambda}{\mu+2}\right), \\
\mathcal{E}'_{\Lambda,2} &\cong \frac{\pi}{2\alpha}\left(\Lambda-2-\mu\ln\left(\frac{\mu+\Lambda}{\mu+2}\right)\right)
\end{aligned}
\tag{8.91}
$$

for large Λ. Notice that

$$
\lim_{\Lambda\to\infty}\frac{\ln\frac{\mu_\Lambda+\Lambda}{\mu_\Lambda+2}}{\Lambda}=0, \tag{8.92}
$$

since μ is bounded above and bounded below by -1. Therefore, according to the second equation in (8.91) and (8.92)

$$
\mathcal{E}'_{\Lambda,2}\cong\frac{\pi\Lambda}{2\alpha_\Lambda}, \tag{8.93}
$$

which further implies

$$
\alpha_\Lambda\cong\frac{\pi\Lambda}{2\mathcal{E}'_{\Lambda,2}}\geq\frac{\pi\Lambda}{2\mathcal{E}_0}. \tag{8.94}
$$

It follows then, thanks to the first equation in (8.91)

$$
\lim_{\Lambda\to\infty}E'_{\Lambda,2}\leq\lim_{\Lambda\to\infty}\frac{\mathcal{E}_0}{\Lambda}\ln\left(\frac{\mu_\Lambda+\Lambda}{\mu_\Lambda+2}\right)=0. \tag{8.95}
$$

Henceforth, we conclude that *all energy must reside in the fluctuations on the ground energy shell asymptotically for large* Λ. Consequently, we have the following choice of parameters as well as their asymptotic behavior.

Step 1: (The choice of μ_Λ and α_Λ) For a given high energy level E_0, we choose μ_Λ and α_Λ such that

$$
E_0=E'_{\Lambda,1}=\frac{2}{\alpha_\Lambda(\mu_\Lambda+1)}. \tag{8.96}
$$

This does not completely determine μ_Λ and α_Λ.

Notice that E_0 is the energy associated with

$$
\tilde{\psi}_\Lambda=\sum_{|\vec{k}|=1}\frac{1}{\sqrt{\alpha_\Lambda(\mu_\Lambda+1)}}e^{i\vec{k}\cdot\vec{x}}. \tag{8.97}
$$

It is clear, according to Section 4.5, that $\tilde{\psi}_\Lambda$ is a selective decay state (enstrophy-minimizing state). Thus we have, thanks to (8.48)

$$
\mathcal{E}_0>\mathcal{E}_*(E_0)=\mathcal{E}(\tilde{\psi}_\Lambda)=\sum_{|\vec{k}|=1}\frac{2}{\alpha_\Lambda(\mu_\Lambda+1)}=\mathcal{E}'_{\Lambda,1}. \tag{8.98}
$$

Again we choose $\alpha_\Lambda, \mu_\Lambda$ so that the enstrophy constraint is satisfied at all truncations, i.e.

$$\mathcal{E}_0 = \mathcal{E}'_{\Lambda,1} + \mathcal{E}'_{\Lambda,2} = \mathcal{E}_*(E_0) + \mathcal{E}'_{\Lambda,2}. \tag{8.99}$$

Combining (8.99) and (8.93) and substituting the result into (8.96) we have

$$\begin{aligned} \alpha_\Lambda &\cong \frac{\pi\Lambda}{2(\mathcal{E}_0 - \mathcal{E}'_{\Lambda,1})} = \frac{\pi\Lambda}{2(\mathcal{E}_0 - \mathcal{E}_*(E_0))} = \frac{\pi\Lambda}{2(\mathcal{E}_0 - \mathcal{E}(\tilde{\psi}_\Lambda))} \\ \mu_\Lambda &= \frac{2}{\alpha_\Lambda E_0} - 1 \cong \frac{4(\mathcal{E}_0 - \mathcal{E}_*(E_0))}{\pi\Lambda E_0} - 1 \to -1. \end{aligned} \tag{8.100}$$

We observe that we have negative temperature regardless of energy level in this case.

Step 2: (The energy and enstrophy constraints) We can now verify that the energy and enstrophy constraints are satisfied in the sense of (8.46). Indeed, for the truncated energy we have, thanks to (8.96), (8.91), (8.92), and (8.100)

$$\begin{aligned} \lim_{\Lambda\to\infty} \langle E_\Lambda \rangle &= \lim_{\Lambda\to\infty} (E'_{\Lambda,1} + E'_{\Lambda,2}) \\ &= \lim_{\Lambda\to\infty} \left(E_0 + \frac{\pi \ln(\frac{\mu_\Lambda+\Lambda}{\mu_\Lambda+2})}{2\alpha_\Lambda} \right) \\ &= E_0. \end{aligned} \tag{8.101}$$

The enstrophy constraint is satisfied at all truncation levels by (8.99).

Step 3: (The limit of the Gibbs measure) We also observe that

$$\alpha_\Lambda |\vec{k}|^2 (|\vec{k}|^2 + \mu_\Lambda) \to \infty \quad \text{as } \Lambda \to \infty \quad \text{for } |\vec{k}|^2 \geq 2. \tag{8.102}$$

Thus the Gaussians degenerate into Dirac delta measure centered at 0 for high modes ($|\vec{k}|^2 \geq 2$). Fluctuation persists on the ground energy shell. Indeed, for $|\vec{k}|^2 = 1$

$$\alpha_\Lambda |\vec{k}|^2 (|\vec{k}|^2 + \mu_\Lambda) = \alpha_\Lambda (1 + \mu_\Lambda) = \frac{2}{E_0} \tag{8.103}$$

Hence the measures for the ground modes converge to Gaussians with mean zero and variance $\sqrt{\frac{E_0}{2}}$.

This also agrees with the empirical theory, except the under-determined ground modes with energy E_0 in the empirical theory are replaced by Gaussians with zero mean and variance $\sqrt{\frac{E_0}{2}}$.

8.6.4 The case with no large-scale mean flow but with topography having degenerate spectrum

Here we assume no large-scale mean flow but with non-trivial topography h having degenerate spectrum, i.e.

$$V = 0, \quad \beta = 0, \quad P_{\Lambda_1}(h) = 0, \tag{8.104}$$

where P_{Λ_1} is the projection on to the ground state modes (first energy shell).

In this case, the limiting procedure depends on whether the prescribed energy level E_0 is low or high when compared with the maximum energy that could reside in an enstrophy-minimizing mean state, i.e.

$$E_{-1} = \frac{1}{2} \sum_{|\vec{k}|^2 > 1} \frac{|\vec{k}|^2 |\hat{h}_{\vec{k}}|^2}{(-1 + |\vec{k}|^2)^2} = E(\overline{\psi}_{-1}). \tag{8.105}$$

Low energy subcase

In this subcase, we assume the imposed energy level E_0 is low in the sense that

$$E_0 < E_{-1}. \tag{8.106}$$

We then have

$$\overline{E}_\Lambda = \frac{1}{2} \sum_{2 \le |\vec{k}|^2 \le \Lambda} \frac{|\vec{k}|^2 |\hat{h}_{\vec{k}}|^2}{(\mu_\Lambda + |\vec{k}|^2)^2} \le E_0 < E_{-1}. \tag{8.107}$$

Observe that

$$E_\mu = \frac{1}{2} \sum_{2 \le |\vec{k}|^2} \frac{|\vec{k}|^2 |\hat{h}_{\vec{k}}|^2}{(\mu_\Lambda + |\vec{k}|^2)^2} \tag{8.108}$$

is a strictly monotoic decreasing function in μ; it approaches E_{-1} as μ approaches -1, and it approaches 0 as μ approaches infinity. Thus, when combined with (8.106), there exists a unique $\mu = \mu(E_0) \in (-1, \infty)$ such that

$$E_{\mu(E_0)} = E_0. \tag{8.109}$$

Notice that

$$\overline{E}_\Lambda \cong E_\mu, \quad \text{if } \mu_\Lambda \cong \mu, \tag{8.110}$$

asymptotically for large Λ. This implies

$$\mu_\Lambda \ge \mu(E_0) > -1 \tag{8.111}$$

asymptotically for large Λ. Indeed, if (8.111) were violated, we would have

$$E_0 \ge \overline{E}_{\Lambda,\mu_\Lambda} > \overline{E}_{\Lambda,\mu(E_0)} \cong E_\mu = E_0, \tag{8.112}$$

which is a contradiction. From (8.111) we deduce that (8.60) holds, which further implies, when combined with (8.55)

$$\frac{\pi\Lambda}{2\alpha_\Lambda} \cong \mathcal{E}'_\Lambda \leq \mathcal{E}_\Lambda \leq \mathcal{E}_0. \tag{8.113}$$

Thus (8.77) holds, which further implies that (8.62) holds in this subcase of low energy within the case of no large-scale mean flow and topography with degenerate spectrum. Henceforth, *all energy must reside in the mean part asymptotically for large* Λ. The detailed choice of the parameters as well as the asymptotic behavior of the parameters and mean states are the same as in the case with generic topography as long as the low energy assumption (8.106) is satisfied.

High energy subcase

In this subcase, we assume the imposed energy level E_0 is high in the sense that

$$E_0 > E_{-1}. \tag{8.114}$$

Then the same procedure for the case with generic topography does not work any more since the maximum amount of energy that an enstrophy-minimizing mean state could absorb is E_{-1}. This implies that part of the energy has to reside permanently in the fluctuation part. This is a mixture of the case with generic topography (Subsection 8.6.2) and the case with no geophysical effects (Subsection 8.6.3). Utilizing the decomposition of the fluctuation part of the ensemble energy and enstrophy in (8.89) and (8.90), as well as arguments similar to the arguments leading to (8.95), we may deduce that (8.95) holds in this case as well. We leave the detail to the interested reader. In conclusion, *the energy must reside in the mean state and the fluctuation on the ground energy shell asymptotically for large* Λ. This leads to the following choice of the parameters:

Step 1: (The choice of μ_Λ and α_Λ) For a given high energy level E_0, we choose μ_Λ and α_Λ such that

$$E_0 = E_{-1} + E'_{\Lambda,1} = E_{-1} + \frac{2}{\alpha_\Lambda(\mu_\Lambda + 1)}. \tag{8.115}$$

This does not completely determine μ_Λ and α_Λ. Intuitively $E_{-1} + E'_{\Lambda,1}$ approximates $\overline{E}_\Lambda + E'_{\Lambda,1}$.

Notice that E_0 is the energy associated with

$$\begin{aligned}\tilde{\psi}_\Lambda &= \overline{\psi}_{-1} + \sum_{|\vec{k}|=1} \frac{1}{\sqrt{\alpha_\Lambda(\mu_\Lambda+1)}} e^{i\vec{k}\cdot\vec{x}} \\ &= \sum_{|\vec{k}|^2>1} \frac{\hat{h}_{\vec{k}}}{-1+|\vec{k}|^2} + \sum_{|\vec{k}|=1} \frac{1}{\sqrt{\alpha_\Lambda(\mu_\Lambda+1)}} e^{i\vec{k}\cdot\vec{x}}.\end{aligned} \tag{8.116}$$

It is clear, thanks to Section 4.5, that $\tilde{\psi}_\Lambda$ is a selective decay state (enstrophy-minimizing state), since it satisfies the equation

$$\Delta\tilde{\psi}_\Lambda + h = -\tilde{\psi}_\Lambda. \tag{8.117}$$

Thus we have

$$\begin{aligned}\mathcal{E}_0 > \mathcal{E}_*(E_0) &= \mathcal{E}(\tilde{\psi}_\Lambda)\\ &= \frac{1}{2}\sum_{|\vec{k}|^2>1}\frac{|\hat{h}_{\vec{k}}|^2}{(-1+|\vec{k}|^2)^2} + \sum_{|\vec{k}|=1}\frac{2}{\alpha_\Lambda(\mu_\Lambda+1)}\\ &= \mathcal{E}_{-1} + \mathcal{E}'_{\Lambda,1}.\end{aligned} \tag{8.118}$$

We will see that $\mathcal{E}_{-1} + \mathcal{E}'_{\Lambda,1}$ approximates $\overline{\mathcal{E}}_\Lambda + \mathcal{E}'_{\Lambda,1}$.

Again we choose $\alpha_\Lambda, \mu_\Lambda$ so that the enstrophy constraint is satisfied approximate, i.e.

$$\mathcal{E}_0 = \mathcal{E}_{-1} + \mathcal{E}'_{\Lambda,1} + \mathcal{E}'_{\Lambda,2} = \mathcal{E}(\overline{\psi}_{-1}) + \mathcal{E}'_{\Lambda,1} + \mathcal{E}'_{\Lambda,2}. \tag{8.119}$$

The second equation in (8.91) and (8.92) suggests the following choice of α_Λ

$$\alpha_\Lambda \cong \frac{\pi\Lambda}{2\mathcal{E}'_{\Lambda,2}} = \frac{\pi\Lambda}{2(\mathcal{E}_0 - \mathcal{E}_*(E_0))} = \frac{\pi\Lambda}{2(\mathcal{E}_0 - \mathcal{E}(\tilde{\psi}_\Lambda))}, \tag{8.120}$$

which further implies, after substituting (8.120) into (8.115)

$$\mu_\Lambda = \frac{2}{\alpha_\Lambda(E_0 - E_{-1})} - 1 \cong \frac{4(\mathcal{E}_0 - \mathcal{E}_*(E_0))}{\pi\Lambda(E_0 - E_{-1})} - 1 \to -1. \tag{8.121}$$

Step 2: (The energy and enstrophy constraints) We can now verify that the energy and enstrophy constraints are satisfied in the sense of (8.46). Indeed, for the truncated energy we have, thanks to (8.91), (8.115)–(8.121)

$$\begin{aligned}\lim_{\Lambda\to\infty}\langle E_\Lambda\rangle &= \lim_{\Lambda\to\infty}(\overline{E}_\Lambda + E'_{\Lambda,1} + E'_{\Lambda,2})\\ &= \lim_{\Lambda\to\infty}\left(\frac{1}{2}\sum_{1<|\vec{k}|^2\le\Lambda}\frac{|\vec{k}|^2|\hat{h}_{\vec{k}}|^2}{(\mu_\Lambda+|\vec{k}|^2)^2} + \frac{\alpha_\Lambda^{-1}}{2}\sum_{|\vec{k}|=1}\frac{1}{\mu_\Lambda+|\vec{k}|^2}\right)\\ &\quad + \lim_{\Lambda\to\infty}\frac{\pi\ln(\frac{\mu_\Lambda+\Lambda}{\mu_\Lambda+2})}{2\alpha_\Lambda}\\ &= \lim_{\Lambda\to\infty}\left(\frac{1}{2}\sum_{1<|\vec{k}|^2\le\Lambda}\frac{|\vec{k}|^2|\hat{h}_{\vec{k}}|^2}{(-1+|\vec{k}|^2)^2} + \frac{2}{\alpha_\Lambda(\mu_\Lambda+1)}\right)\\ &= E(\tilde{\psi}_\Lambda)\\ &= E_0.\end{aligned} \tag{8.122}$$

As for the enstrophy, we have

$$\begin{aligned}
\lim_{\Lambda\to\infty}\langle\mathcal{E}_\Lambda\rangle &= \lim_{\Lambda\to\infty}(\overline{\mathcal{E}}_\Lambda+\mathcal{E}'_{\Lambda,1}+\mathcal{E}'_{\Lambda,2})\\
&= \lim_{\Lambda\to\infty}\left(\frac{1}{2}\sum_{1<|\vec{k}|^2\le\Lambda}\frac{|\vec{k}|^4|\hat{h}_{\vec{k}}|^2}{(\mu_\Lambda+|\vec{k}|^2)^2}+\mathcal{E}'_{\Lambda,1}+\mathcal{E}'_{\Lambda,2}\right)\\
&= \lim_{\Lambda\to\infty}\left(\frac{1}{2}\sum_{1<|\vec{k}|^2\le\Lambda}\frac{|\vec{k}|^4|\hat{h}_{\vec{k}}|^2}{(-1+|\vec{k}|^2)^2}+\mathcal{E}'_{\Lambda,1}+\mathcal{E}'_{\Lambda,2}\right)\\
&= \lim_{\Lambda\to\infty}(\mathcal{E}_{-1}+\mathcal{E}'_{\Lambda,1}+\mathcal{E}'_{\Lambda,2})\\
&= \mathcal{E}_0.
\end{aligned}\tag{8.123}$$

Step 3: (The limit of the mean states) The mean states converge as well. More precisely

$$\lim_{\Lambda\to\infty}\overline{\psi}_\Lambda=\overline{\psi}_{-1}=\sum_{|\vec{k}|^2>1}\frac{\hat{h}_{\vec{k}}}{-1+|\vec{k}|^2}e^{i\vec{k}\cdot\vec{x}}.\tag{8.124}$$

The limit mean state is stable in the restricted sense according to Section 4.5.

We also observe that

$$\alpha_\Lambda|\vec{k}|^2(|\vec{k}|^2+\mu_\Lambda)\to\infty\quad\text{as }\Lambda\to\infty\quad\text{for }|\vec{k}|^2\ge 2.\tag{8.125}$$

Thus the Gaussians degenerate into Dirac delta measure centered at $\hat{\overline{\psi}}_{\vec{k}}=\frac{\hat{h}_{\vec{k}}}{-1+|\vec{k}|^2}$ for high modes ($|\vec{k}|^2\ge 2$). Fluctuation persists on the ground energy shell. Indeed, for $|\vec{k}|^2=1$

$$\alpha_\Lambda|\vec{k}|^2(|\vec{k}|^2+\mu_\Lambda)=\alpha_\Lambda(1+\mu_\Lambda)=\frac{2}{E_0-E_{-1}}.\tag{8.126}$$

Hence the measures for the ground modes converge to Gaussian with mean zero and variance $\sqrt{\frac{E_0-E_{-1}}{2}}$.

This also agrees with the empirical theory except the under-determined ground modes with energy E_0-E_{-1} in the empirical theory are replaced by Gaussian with zero mean and variance $\sqrt{\frac{E_0-E_{-1}}{2}}$.

Combining the two subcases, we conclude that we have positive temperature for low energy ($E_0<E(\overline{\psi}_0)$) and negative temperature for high energy ($E_0>E(\overline{\psi}_0)$).

We may then conclude that the complete statistical mechanics theory predicts that the most probable mean state is the unique selective decay state with given energy E_0 in the case with large-scale mean flow or no large-scale mean flow but generic topography. The prediction is the same as those from the empirical theories using the energy and enstrophy constraints described in Chapter 6.

Now we see why the geophysical flows are more suitable for applying statistical theory than the ordinary 2-D Euler flows, which has no beta-plane effect, no topography. In the absence of those geophysical effects, the energy is distributed only over the fluctuating part, and this makes the limiting procedure a very different one where fluctuations in the large-scale dominate.

8.7 The role of statistically relevant and irrelevant conserved quantities

The theory of statistical mechanics for spectrally truncated quasi-geostrophic flow is simultaneously very rich yet unsatisfactory as regards the statistical description of solutions of the continuum equations in (8.1) because the equations (8.1) can preserve many more conserved quantities given by the generalized enstrophies (see Chapter 1) than the two conserved quantities in (8.7) and (8.8) for the spectrally truncated dynamics. In fact, the continuum equations (8.1) have exactly these two robust conserved quantities only for the special set-up with non-zero large-scale mean flow $V \neq 0$ with $\beta \neq 0$ and periodic geometry. Otherwise, there are infinitely many more robust conserved quantities for the continuum equations (see Chapter 1) ignored by the spectrally truncation procedure, which potentially have important statistical significance. A natural question arises regarding the statistical significance of these additional conserved quantities. We study these issues subsequently in Chapter 14 of this book by utilizing suitable more complex discrete approximations of (8.1) that have many more additional conserved quantities beyond (8.7) and (8.8). Also included in our discussion in that chapter is the Galerkin truncated Burgers–Hopf model discussed in Section 7.3 as a simple model for these issues. Recall that these equations conserve the Hamiltonian defined by the integral of the third power in addition to the momentum and energy, so that the statistical significance of the Hamiltonian provides an important model problem for these issues.

In the next chapter, we generalize the empirical statistical theory described in Chapter 6 to include more non-linear information as well as the use of additional conserved quantities. The relevance of these non-linear theories both for non-linear dynamics and observations will be established in suitable regimes in subsequent chapters.

References

Carnevale, G. F. and Frederiksen, J. S. (1987), Nonlinear stability and statistical mechanics of flow over topography. *J. Fluid Mech.* **175**, 157–181.

Kraichnan, R. H. (1975), Statistical dynamics of two-dimensional flow. *J. Fluid Mech.* **67**, 155–175.

Majda, A., Timofeyev, I. and Vanden Eijnden, E. (2001), A mathematical framework for stochastic climate models. *Comm. Pure and Appl. Math.* **54**, 891–974.

Salmon, R. (1998), *Lectures on Geophysical Fluid Dynamics*. Oxford: Oxford University Press.

Salmon, R., Holloway, G. and Hendershott, M. (1976), The equilibrium statistical mechanics of simple quasi-geostrophic models. *J. Fluid Mech.* **75**, 691–703.

Appendix 1

The purpose of this Appendix is to give a sketch of the proof that the quotient of the Lagrange multiplier θ for the truncated energy and the Lagrange multiplier α for the truncated enstrophy, i.e. $\frac{\theta}{\alpha} = \mu = \mu_\Lambda$, in the Gibbs measure (8.25) must be bounded from above independent of Λ if the energy and enstrophy constraints hold asymptotically for the Gibbs measure in the sense of (8.46).

In what follows we assume all limits exist. Otherwise we just go through a subsequence argument.

Indeed, according to (8.50) and (8.51), if $\mu_\Lambda \to \infty$, then all energy will be concentrated in fluctuations in high modes which shift to an infinite wave number. This is intuitively not consistent with the emergence of large-scale coherent structures. We may justify our intuition through the following by way of contradiction argument (BWOC).

Now suppose that

$$\mu_\Lambda \to \infty \tag{8.127}$$

we then have, according to (8.50) and (8.51)

$$\begin{aligned} \lim_{\Lambda\to\infty} \overline{E}_\Lambda &= 0, \\ \lim_{\Lambda\to\infty} \overline{\mathcal{E}}_\Lambda &= \frac{1}{2}\|h\|^2_{L^2} \geq 0. \end{aligned} \tag{8.128}$$

Thus, according to (8.46), (8.50)–(8.51), and (8.128)

$$\frac{\mathcal{E}_0}{E_0} \cong \frac{\langle \mathcal{E}_\Lambda \rangle}{\langle E_\Lambda \rangle} \cong \frac{\frac{1}{2}\|h\|^2_{L^2} + \mathcal{E}'_\Lambda}{E'_\Lambda} \geq \frac{\sum_{1\leq|\vec{k}|^2\leq\Lambda} \frac{|\vec{k}|^2}{\mu_\Lambda + |\vec{k}|^2}}{\sum_{1\leq|\vec{k}|^2\leq\Lambda} \frac{1}{\mu_\Lambda+|\vec{k}|^2} + \frac{1}{\mu_\Lambda}}. \tag{8.129}$$

We thus observe that the generalized Dirichlet quotient (enstrophy over energy) depends only on μ_Λ asymptotically under the assumption (8.127).

It is easy to see that the fluctuation part of the energy and enstrophy in each mode approaches zero as μ_Λ approaches infinity. Therefore, only the contribution from the high modes are relatively important in the last expression in (8.129). This means we may neglect the lower modes below a wave number $\sqrt{\Lambda_j}$ for any fixed Λ_j. This leads to a lower bound of Λ_j to (8.129) which further leads to a contradiction since j is arbitrary. On the other hand, the contribution from the high modes in (8.129) could be small as well. What we need to show is that the low modes are relatively smaller than the high modes. For this purpose, let us fix a j and consider the following decomposition of the denominator of the last expression in (8.129) into the low modes ($|\vec{k}|^2 < \Lambda_j$) and the high modes ($|\vec{k}|^2 \geq \Lambda_j$) parts, namely

$$\begin{aligned}
&\sum_{1\leq|\vec{k}|^2\leq\Lambda} \frac{1}{\mu_\Lambda + |\vec{k}|^2} + \frac{1}{\mu_\Lambda} \\
&= \left\{ \frac{1}{\mu_\Lambda} + \sum_{1\leq|\vec{k}|^2<\Lambda_j} \frac{1}{\mu_\Lambda + |\vec{k}|^2} \right\} + \left\{ \sum_{\Lambda_j\leq|\vec{k}|^2\leq\Lambda} \frac{1}{\mu_\Lambda + |\vec{k}|^2} \right\} \\
&= I + II. \qquad (8.130)
\end{aligned}$$

Observe that for each fixed eigenvalue Λ_k, there are at most $2\pi\sqrt{\Lambda_k}$ number of modes corresponding to this eigenvalue. Consequently we have the following crude upper bound on the contribution from the low modes

$$I \leq \frac{1}{\mu_\Lambda}(2\pi\sqrt{\Lambda_j} \times j) = \frac{2j\pi\sqrt{\Lambda_j}}{\mu_\Lambda}. \qquad (8.131)$$

On the other hand, a calculation similar to (5.54) leads to the following estimate on the contribution from the high modes

$$II \cong \frac{\pi}{2} \ln\left(\frac{\mu_\Lambda + \Lambda}{\mu_\Lambda + \Lambda_j}\right). \qquad (8.132)$$

We shall demonstrate below that I is of lower order to II in all scenarios. Without loss of generality (by going through a subsequence if necessary) we assume that the limit $\frac{\mu_\Lambda}{\Lambda}$ exists.

Case 1: First we assume that the μ_Λ grows much faster than Λ, i.e.

$$\frac{\mu_\Lambda}{\Lambda} \to \infty. \qquad (8.133)$$

In this case

$$\ln\left(\frac{\mu_\Lambda + \Lambda}{\mu_\Lambda + \Lambda_j}\right) = \ln\left(1 + \frac{\Lambda - \Lambda_j}{\mu_\Lambda + \Lambda_j}\right) \cong \frac{\Lambda - \Lambda_j}{\mu_\Lambda + \Lambda_j}. \qquad (8.134)$$

Therefore, combining (8.134), (8.132), and (8.131), we have asymptotically

$$\frac{I}{II} \leq \frac{4j\sqrt{\Lambda_j}}{\Lambda - \Lambda_j} \frac{\mu_\Lambda + \Lambda_j}{\mu_\Lambda} \cong \frac{4j\sqrt{\Lambda_j}}{\Lambda - \Lambda_j} \to 0. \qquad (8.135)$$

Case 2: Suppose now that μ_Λ and Λ grow at the same rate, i.e.

$$\frac{\mu_\Lambda}{\Lambda} \to c \neq 0. \tag{8.136}$$

In this case

$$\ln\left(\frac{\mu_\Lambda + \Lambda}{\mu_\Lambda + \Lambda_j}\right) \cong \ln\left(\frac{c+1}{c}\right). \tag{8.137}$$

Therefore, combining (8.137), (8.132), and (8.131), we have

$$\frac{I}{II} \leq \frac{4j\sqrt{\Lambda_j}}{\mu_\Lambda \ln(\frac{c+1}{c})} \to 0. \tag{8.138}$$

Case 3: Lastly we consider the case when μ_Λ grows slower than Λ, i.e.

$$\frac{\mu_\Lambda}{\Lambda} \to 0. \tag{8.139}$$

In this case

$$\ln\left(\frac{\mu_\Lambda + \Lambda}{\mu_\Lambda + \Lambda_j}\right) \to \infty. \tag{8.140}$$

Therefore, combining (8.140), (8.132), and (8.131), we have

$$\frac{I}{II} \to 0. \tag{8.141}$$

In any case, we proved that (8.141) is true. Utilizing (8.141) in (8.129) we deduce

$$\frac{\mathcal{E}_0}{E_0} \geq \frac{\sum_{1\leq|\vec{k}|^2\leq\Lambda} \frac{|\vec{k}|^2}{\mu_\Lambda+|\vec{k}|^2}}{\sum_{\Lambda_j\leq|\vec{k}|^2\leq\Lambda} \frac{1}{\mu_\Lambda+|\vec{k}|^2}} \geq \frac{\sum_{\Lambda_j\leq|\vec{k}|^2\leq\Lambda} \frac{|\vec{k}|^2}{\mu_\Lambda+|\vec{k}|^2}}{\sum_{\Lambda_j\leq|\vec{k}|^2\leq\Lambda} \frac{1}{\mu_\Lambda+|\vec{k}|^2}} \geq \Lambda_j. \tag{8.142}$$

This is a contradiction since j is arbitrary. Of course this implies that (8.127) cannot be true, i.e. μ_Λ *must be bounded independent of* Λ.

A last comment is that the approximation of the summation by integration is a valid one in the sense that the error is of lower order.

9

Empirical statistical theories for most probable states

9.1 Introduction

In this chapter we continue the study of statistical theories for the most probable state of the barotropic quasi-geostrophic equations

$$\frac{\partial q}{\partial t} + \nabla^{\perp}\psi \cdot \nabla q = 0, \quad \frac{dV}{dt} = -\fint \frac{\partial h}{\partial x}\psi', \tag{9.1}$$

where $q = \Delta\psi + h + \beta y$ is the potential vorticity, $\psi = -V(t)y + \psi'$ is the stream function, V is the large-scale mean flow, h is the topography, β is the beta-plane constant, ψ' is the small-scale stream function, $\omega = \Delta\psi'$ is the relative vorticity, and $q' = \Delta\psi' + h = \omega + h$ is the small-scale potential vorticity. The bar in the integral sign indicates that the space integral has been normalized by the area of the domain Ω. Here we assume either periodic geometry, where all the functions are 2π-periodic, so that $\Omega = [0, 2\pi] \times [0, 2\pi]$ and the normalization constant is $|\Omega| = 4\pi^2$ or channel geometry ($\Omega = [0, 2\pi] \times [0, \pi]$) as described in Section 1.4. We will also consider the case of disk domain with radius R and centered at the origin, or the entire plane for the point-vortex theory without mean flow.

In Chapters 6 and 8 we discussed the classical statistical theory with two conserved quantities both from the non-traditional point of view of empirical statistical theory in Chapter 6 and complete statistical mechanics in Chapter 8. This theory is based on the existence of two conserved quantities of the quasi-geostrophic equations (9.1), namely the total energy, E, and the total enstrophy, $\mathcal{E}$, which are the only two robust conserved quantities in periodic geometry with $\beta = 0$. However as we demonstrated in Sections 1.3 and 1.4 the barotropic quasi-geostrophic equations (9.1) possess many more conserved quantities depending on the beta-effect, the geometry of the underlying domain, and the mean flow. For instance, in the case of no large-scale mean flow ($V \equiv 0$) and no beta-plane effect ($\beta = 0$), we know that all generalized enstrophy of the form $\int G(q)$ are conserved together with the energy for the periodic case. For the special choice of $G(q) = q$, we have the conservation of circulation, which is not necessarily zero in the channel geometry, although it must be zero for the periodic geometry. See Subsection 1.3.5 for a review of conserved quantities in the periodic case and Subsection 1.4.3 for a review of conserved quantities in the channel geometry case.

In principle, the existence of additional constraints in the form of conserved quantities raises the interesting possibility of formulating alternative statistical theories that take into account the additional constraints imposed on the quasi-geostrophic flow.

In this chapter we introduce several alternative equilibrium statistical theories for the barotropic quasi-geostrophic equations (9.1), which take into account other constraints besides the conservation of energy constraint. In Section 9.2, we present an empirical statistical theory in channel geometry that includes a general prior distribution for small-scale single-point fluctuation of potential vorticity, and a few constraints at large-scale imposed by conserved quantities. First, we consider the empirical statistical theory with a general prior distribution and two conserved quantities, the energy and the total circulation without the beta-effect and large-scale mean flow. The mean field equation predicted by the statistical theory is a non-linear elliptic equation in general. The solutions are exact steady state solutions to (9.1), provided that the prior distribution is independent of the spatial location. Next, we consider the empirical statistical theory with a general prior distribution and three conserved quantities, the energy, the total circulation, and the impulse I for flow in a channel domain with non-trivial beta-effect and latitudinal topography. The mean field equation is again a non-linear elliptic equation in general. However the solutions to the mean field equation are not necessarily exact steady state solutions to (9.1), unless we assume that the prior distribution is independent of the spatial location and we take a special linear combination of the energy and impulse as a single conserved quantity replacing the separate conservation of energy and impulse. In Section 9.3 both empirical and complete statistical theories for point vortices are developed. First we show how the famous mean field equations of point vortex theory (Joyce and Montgomery, 1974) can be recovered from empirical statistical theory with a few constraints together with a special prior distribution arising from ensembles of point vortices. In the second part of Section 9.3, we introduce the complete statistical mechanics theory for point vortices (Onsager, 1949). Interestingly enough, the complete statistical mechanics theory predicts the same mean field equations for point vortices that we derived in the previous subsection directly using empirical statistical theory for the entire plane case. Thus, as established in Chapters 6 and 8 for the energy–enstrophy statistical theory, the mean field theory for point vortices can be derived either from an empirical statistical theory or from complete statistical mechanics at a suitable limit. In Section 9.4 we present the formal empirical statistical theory with infinitely many constraints, which attempts to take into account the energy and all the generalized enstrophies introduced earlier. This formal statistical theory is based on the work of Miller (1990) and Robert (1991). Clearly the constraints in this theory are more stringent than in any of the few constraints empirical statistical theories studied earlier, because this statistical theory imposes constraints on all the moments of the potential vorticity q, rather than merely the first moment, the total circulation, or the second moment, the enstrophy. We will show that conservation of all functions $G(q)$ of the potential vorticity q is equivalent to conservation of areas of the level sets of the potential vorticity. This fact will allow us to introduce at every point $\vec{x}$ in physical space a probability measure for the potential vorticity q at that point. Once a suitable space of probability measures has been defined, a corresponding measure of the Shannon entropy can be introduced, and the machinery from information theory will again yield the most probable state formally by maximizing the Shannon entropy over the given probability measures and under the appropriate constraints in energy and otherwise. The mean $\bar{\psi}$ of the predicted most probable state satisfies a non-linear elliptic equation. The non-linear stability of the most probable states predicted by various statistical theories is developed in Section 9.5 as a simple application of Chapter 4. The strengths and weakness of the various statistical theories as well as their potential applicability are discussed extensively in Chapters 10, 11, 12, and 13 below.

9.2 Empirical statistical theories with a few constraints

In this section we present the statistical theory with a prior distribution and a few conserved quantities. We will focus on the channel geometry ($\Omega = [0, 2\pi] \times [0, \pi]$) introduced in Section 1.4, which we shall recall later in this section.

Recall that there are infinitely many conserved quantities in this channel geometry. In particular, we have the conservation of energy $E = \frac{1}{2}V^2 - \frac{1}{2}\fint \psi'\omega$, the circulation $\Gamma = \fint q$, and the conservation of mean flow impulse difference $V - I = V - \fint yq$. In what follows we will develop empirical statistical theory based on either two conserved quantities, i.e. energy and circulation or three conserved quantities, i.e. energy, circulation, and impulse, depending on the physical context.

In general we should incorporate prior information or bias on the distribution of the small-scale potential vorticity (see Chapter 6). Such bias is formulated in terms of a prior distribution $\Pi_0(\vec{x}, \lambda)$, which is a probability density function on

$\Omega \times \mathcal{R}^1$. The reader is referred to Definition 6.6 from Chapter 6 for more detail on such a prior distribution.

In the special cases that we consider here, the large-scale mean flow is either identically zero or a constant of motion. Thus we only need to postulate the one-point statistics $\rho(\vec{x}, \lambda)$ for the potential vorticity q. The reader is referred to Definition 6.5 from Chapter 6 for more detail on one-point statistics.

We then recall from Chapter 6 that the Shannon information-theoretical relative entropy $\mathcal{S}(\rho, \Pi_0)$ is defined as

$$\mathcal{S}(\rho, \Pi_0) \equiv -\fint \int \rho(\vec{x}, \lambda) \ln\left(\frac{\rho(\vec{x}, \lambda)}{\Pi_0(\vec{x}, \lambda)}\right) d\lambda\, d\vec{x}. \tag{9.2}$$

The most probable one-point statistics ρ^* is then derived via the maximum entropy principle, i.e. ρ^* is the one that maximizes the Shannon relative entropy (9.2) with given constraints and prior distribution Π_0. In particular, if there are no constraints, it is easy to check that the most probable state ρ^* is exactly the prior distribution $\Pi_0(\vec{x}, \lambda)$.

Next we apply the maximum entropy principle to various situations, where the flows satisfy specific symmetries and conserved quantities constraints.

9.2.1 The energy–circulation empirical theory with a general prior distribution

Here we consider the special case of channel flow without large-scale mean flow ($V \equiv 0$) and no beta-plane effect ($\beta = 0$). Thus, according to Section 1.4, the stream function takes the form

$$\psi = \psi' = \sum_{k\geq 1}\sum_{j\geq 0}(a_{jk}\cos(jx) + b_{jk}\sin(jx))\sin(ky). \tag{9.3}$$

Recall that the energy and circulation are conserved in this case. They can be represented in terms of the one-point statistics as

$$E(\rho) = -\frac{1}{2}\fint \overline{\psi}\overline{\omega} d\vec{x}, \quad \Gamma(\rho) = \fint \int \lambda\rho(\vec{x}, \lambda) d\lambda d\vec{x}, \tag{9.4}$$

where the overline represents average with respect to the one-point statistics, i.e.

$$\overline{q}(\vec{x}) = \int \lambda\rho(\vec{x}, \lambda) d\lambda, \quad \overline{\omega} = \overline{q} - h, \quad \Delta\overline{\psi} + h = \overline{q}. \tag{9.5}$$

The conservation of energy and circulation translate into constraints on the one-point statistics. The set of constraints $\mathcal{C}$ in this case is given by

$$\mathcal{C} = \mathcal{C}^{(0)} \cap \mathcal{C}^{(E)} \cap \mathcal{C}^{(\Gamma)}, \tag{9.6}$$

where

$$\mathcal{C}^{(0)} = \{\rho | M = \int \rho(\vec{x}, \lambda)\, d\lambda = 1 \text{ for each } \vec{x} \in \Omega\},$$

$$\mathcal{C}^{(E)} = \{\rho | E(\rho) = -\frac{1}{2} \fint \overline{\psi}(\overline{q} - h)\, d\vec{x} = E_0\}, \tag{9.7}$$

$$\mathcal{C}^{(\Gamma)} = \{\rho | \Gamma(\rho) = \fint \int \lambda \rho(\vec{x}, \lambda)\, d\lambda d\vec{x} = \Gamma_0\}.$$

According to the Lagrange multiplier method, there exist constants θ, γ, and $\tilde{\mu}$, so that, if ρ^* is the most probable state which maximizes the relative entropy $\mathcal{S}(\rho, \Pi_0)$ subject to the constraints in $\mathcal{C}$ determined by (9.6)–(9.7), then the following variational equation holds

$$\left(-\frac{\delta \mathcal{S}(\rho, \Pi_0)}{\delta \rho} + \theta \frac{\delta E}{\delta \rho} + \gamma \frac{\delta \Gamma}{\delta \rho} \right) \Bigg|_{\rho = \rho^*} = \tilde{\mu} \frac{\delta M}{\delta \rho} \Bigg|_{\rho = \rho^*}. \tag{9.8}$$

It is easy to calculate the variational derivatives, see for instance Sections 1.5 and 6.6, so that

$$\frac{\delta \mathcal{S}(\rho, \Pi_0)}{\delta \rho} = \ln\left(\frac{\rho}{\Pi_0}\right) + 1,$$

$$\frac{\delta E}{\delta \rho} = -\lambda \psi, \quad \frac{\delta \Gamma}{\delta \rho} = \lambda, \quad \frac{\delta M}{\delta \rho} = \delta_{\vec{x}} \otimes 1.$$

Introducing these variational derivatives back into the variational equation (9.8) yields

$$\ln\left(\frac{\rho^*}{\Pi_0}\right) + 1 = \theta \psi^* \lambda - \gamma \lambda - \tilde{\mu}(\vec{x}). \tag{9.9}$$

Utilizing the condition $\int \rho^*(\vec{x}, \lambda)\, d\lambda = 1$, we deduce that the most probable one-point statistics must take the form

$$\rho^*(\vec{x}, \lambda) = \frac{e^{(\theta \psi^* - \gamma)\lambda} \Pi_0(\vec{x}, \lambda)}{\int e^{(\theta \psi^* - \gamma)\lambda} \Pi_0(\vec{x}, \lambda)\, d\lambda}. \tag{9.10}$$

Next, we introduce the *partition function*

$$\mathcal{Z}(\psi, \vec{x}) = \int e^{(\psi - \gamma)\lambda} \Pi_0(\vec{x}, \lambda)\, d\lambda. \tag{9.11}$$

Notice that the partition function is simply the moment-generating function of $\Pi_0(\vec{x}, \lambda)$ evaluated at $\psi(\vec{x}) - \gamma$. With the help of the partition function we can simplify the expression for the mean field q^* of the potential vorticity

$$q^* = \int \lambda \rho^*(\vec{x}, \lambda)\, d\lambda = \frac{1}{\theta} \frac{\partial}{\partial \psi} \ln \mathcal{Z}(\theta \psi, \vec{x}) \Bigg|_{\psi = \psi^*}. \tag{9.12}$$

Thus we arrive at the mean field equation for the most probable state

$$\Delta\psi^* + h = \frac{1}{\theta}\left(\frac{\partial}{\partial\psi}\ln \mathcal{Z}(\theta\psi, \vec{x})\right)\Big|_{\psi=\psi^*}. \tag{9.13}$$

This mean field equation for q^* depends on the partition function $\mathcal{Z}$ and therefore also on the prior probability. If there is $\vec{x}$-dependence in Π_0, the mean state q^* is, in general, not a steady state solution of fluid equations (9.1). For the special class of prior distribution

$$\Pi_0(\vec{x}, \lambda) = \Pi_0(\lambda), \tag{9.14}$$

we observe from (9.11) that the $\vec{x}$ dependence of $\mathcal{Z}$ is through ψ only and hence $\mathcal{Z}$ is a function of the stream function ψ and the parameters θ, γ only. This further implies that the mean field equation (9.13) can be rewritten as

$$\Delta\psi^* + h = \mathcal{F}(\psi^*). \tag{9.15}$$

This is a non-linear elliptic equation in general, and the solutions to this mean field equation are exact steady state solutions to the barotropic quasi-geostrophic equations (9.1), since (9.15) implies that the potential vorticity is a function of the stream function. The stability of these solutions will be discussed later in this chapter, while their predicted structure is the topic of Chapter 11.

Next we specialize the partition function $\mathcal{Z}$, and therefore the mean field equation (9.13), or (9.15), for special cases of the prior distribution Π_0.

Example 1: Uniform prior distribution. If only the maximum value Q_+ and the minimum value Q_- of the small-scale potential vorticity fluctuation are available, then the natural prior distribution Π_0 is uniformly distributed over the range $[Q_-, Q_+]$, i.e.

$$\Pi_0(\vec{x}, \lambda) = \frac{1}{2Q} I_{\overline{Q}-Q}^{\overline{Q}+Q}(\lambda), \tag{9.16}$$

where I_a^b represents the indicator function over the interval $[a, b]$, i.e.

$$I_a^b = I_a^b(\lambda)$$

and $\overline{Q}$ and Q are the interval midpoint and half-width respectively

$$\overline{Q} = \frac{1}{2}(Q_+ + Q_-), \quad Q = \frac{1}{2}(Q_+ - Q_-). \tag{9.17}$$

A straightforward calculation leads to the following explicit form for the mean field equation

$$\Delta\psi^* + h = \overline{Q} + Q\left(\coth(Q(\theta\psi^* - \gamma)) - \frac{1}{Q(\theta\psi^* - \gamma)}\right), \tag{9.18}$$

which is the well-known Langevin equation. Some interesting solutions of this equation will be discussed in Chapter 11 below.

Example 2: Gaussian prior distribution. Another useful case is when the prior distribution Π_0 is Gaussian

$$\Pi_0(\lambda) = \frac{\sqrt{\alpha}}{\sqrt{2\pi}} e^{-\frac{\alpha\lambda^2}{2}}. \tag{9.19}$$

This is the case when the only information that we have is the first and second moments of the distribution of the potential vorticity as discussed in Section 6.4. In this case, the mean field equation (9.15) recovers the prediction of the classical statistical theory based on the conservation of the total energy and the total enstrophy. Indeed, a straightforward algebraic manipulation permits us to compute exactly the partition function $\mathcal{Z}(\psi, \vec{x})$ in this case

$$\begin{aligned}\mathcal{Z}(\psi, \vec{x}) &= \int \exp((\psi - \gamma)\lambda)\Pi_0(\lambda)\, d\lambda \\ &= \int \exp((\psi - \gamma)\lambda)\frac{\sqrt{\alpha}}{\sqrt{2\pi}} \exp\left(-\frac{\alpha\lambda^2}{2}\right) d\lambda \\ &= e^{\frac{(\psi-\gamma)^2}{2\alpha}}.\end{aligned} \tag{9.20}$$

Therefore

$$\frac{\partial \ln \mathcal{Z}(\theta\psi)}{\partial\psi} = \frac{\theta(\theta\psi - \gamma)}{\alpha},$$

and the mean field equation (9.15) for the most probable state ψ^* takes the form

$$\Delta\psi^* + h = \frac{\theta\psi^* - \gamma}{\alpha} \tag{9.21}$$

In the special case of $\gamma = 0$, the mean field equation (9.21) reduces to the equation predicted by the classical statistical theory based on the energy and the enstrophy constraint from Chapters 6 and 8.

Example 3: Gamma prior distribution. Recall from Chapter 6 that the Gaussian distribution is the simplest and least-biased prior distribution using the first and second moments alone. In practice, prior distributions for small-scale fluctuations of potential vorticity should also be skewed and even with long tails on one side to reflect a preponderance of either cyclones or anti-cyclones in small-scale potential vorticity fluctuations in a given physical context. Thus, it is interesting to build a family of prior distributions $\Pi_0(\lambda)$ measuring deviations from Gaussian distributions through skewness, the third moment of $\Pi_0(\lambda)$, and also with a higher probability of large amplitude events going in the same direction as this skewness. With the normalizations for the first and second moments

$$\int \lambda \Pi_0(\lambda)\, d\lambda = 0, \quad \int \lambda^2 \Pi_0(\lambda)\, d\lambda = 1,$$

the skewness is given by the third moment

$$\int \lambda^3 \Pi_0(\lambda)\, d\lambda = \mathrm{Skew}\Pi_0.$$

An attractive choice of a family of prior distributions fitting these requirements is the family of centered Γ-distributions

$$\Pi_0^\varepsilon(\lambda) = |\varepsilon|^{-1} R(\varepsilon^{-1}[\lambda + \varepsilon^{-1}]; \varepsilon^{-2}),$$

where

$$R(z, a) = \Gamma(a)^{-1} z^{a-1} e^{-z}, \text{ if } z \geq 0, \quad R(z, a) = 0, \text{ if } z < 0,$$

denotes the standard Γ-density. This family has a one-sided exponential tail and compact support on the other side, with the Γ-distribution shifted to have zero mean. By construction, $\Pi_0^\varepsilon(\lambda)$ satisfies the normalizations on the first and second moments and the skewness if given by

$$\mathrm{Skew}\Pi_0 = 2\varepsilon.$$

Since the Γ-distribution, $R(z, a)$, has the explicit generating or partition function, $\mathcal{Z}(\psi, a) = (1 - \psi)^{-a}$, it is not difficult to show that, in this case, the mean field equation (9.15) is given by

$$\Delta\psi^* + h = \frac{\theta\psi^*}{1 - \varepsilon\theta\psi^*},$$

so that we recover the Gaussian prediction for $\varepsilon = 0$. For simplicity, we have ignored the circulation in our discussion here.

We provide an a priori justification of this empirical theory in Chapter 14 through considering the statistical behavior of truncations of the quasi-geostrophic equations with many conserved quantities. In Chapter 13, we use this theory to successfully predict the location of Jupiter's Red Spot.

It is easy to observe that the energy–circulation empirical statistical theory works provided the conservation of energy and circulation hold, and is not necessarily restricted to a channel domain. For instance, the theory applies to quasi-geostrophic flow in a (simply connected) bounded domain where the conservation of energy and circulation still hold, or to the case of flow occupying the entire plane $\mathcal{R}^2$ with the fluids decaying at infinity, or the swimming pool geometry discussed in Section 1.3.3, equation (1.30).

9.2.2 The energy–circulation impulse theory with a general prior distribution

Next we consider the special case where the topography is a function of the latitude only, $h = h(y)$. This implies that the large-scale mean flow V is a constant of motion. In this case the impulse

$$I = \fint yq\, d\vec{x} \tag{9.22}$$

is a conserved quantity. See Section 1.4 for more details.

Again we need to postulate the one-point statistics for the potential vorticity only, since the large-scale mean flow is a constant.

The new set of constraints $\mathcal{C}$ includes the impulse I in addition to the total energy E and the total circulation Γ

$$\mathcal{C} = \mathcal{C}^{(0)} \cap \mathcal{C}^{(E)} \cap \mathcal{C}^{(\Gamma)} \cap \mathcal{C}^{(I)}, \tag{9.23}$$

where

$$\begin{aligned}
\mathcal{C}^{(0)} &= \{\rho | M = \int \rho(\vec{x}, \lambda)\, d\lambda = 1 \text{ for each} \vec{x} \in \Omega\},\\
\mathcal{C}^{(E)} &= \{\rho | E(\rho) = -\frac{1}{2} \fint \overline{\psi}' (\overline{q} - h - \beta y)\, d\vec{x} = E_0\}\\
\mathcal{C}^{(\Gamma)} &= \{\rho | \Gamma(\rho) = \fint \int \lambda \rho(\vec{x}, \lambda)\, d\lambda d\vec{x} = \Gamma_0\},\\
\mathcal{C}^{(I)} &= \{\rho | I(\rho) = \fint y\, \overline{q}(\vec{x})\, d\vec{x} = \fint \int y\lambda \rho(\vec{x}, \lambda)\, d\lambda d\vec{x} = I_0\}.
\end{aligned} \tag{9.24}$$

Again, the most probable state ρ^* maximizes the relative entropy $\mathcal{S}(\rho, \Pi_0)$, subject to the constraints in $\mathcal{C}$. According to the Lagrange multiplier method, there exist constants θ, γ, χ, and $\tilde{\mu}$, so that the most probable state ρ^* satisfies the variational equation

$$\left(-\frac{\delta \mathcal{S}(\rho, \Pi_0)}{\delta \rho} + \theta \frac{\delta E}{\delta \rho} + \gamma \frac{\delta \Gamma}{\delta \rho} + \chi \frac{\delta I}{\delta \rho}\right)\bigg|_{\rho=\rho^*} = \tilde{\mu} \frac{\delta M}{\delta \rho}\bigg|_{\rho=\rho^*}. \tag{9.25}$$

We have already calculated the variational derivatives of $\mathcal{S}$, E, Γ. The variational derivative of the impulse is easily calculated as

$$\frac{\delta I}{\delta \rho} = \lambda y. \tag{9.26}$$

Substituting all the variational derivatives back into equation (9.24) and neglecting the overline for the mean fields yields

$$\ln\left(\frac{\rho^*}{\Pi_0}\right)+1=\theta\psi'^*\lambda-\gamma\lambda-\chi\lambda y-\tilde{\mu}(\vec{x}). \tag{9.27}$$

From the normalization condition $\int \rho^*(\vec{x},\lambda)\,d\lambda=1$, we conclude that the most probable one-point statistics must have the form

$$\rho^*(\vec{x},\lambda)=\frac{e^{(\theta\psi'^*-\gamma-\chi y)\lambda}\Pi_0(\vec{x},\lambda)}{\int e^{(\theta\psi'^*-\gamma-\chi y)\lambda}\Pi_0(\vec{x},\lambda)\,d\lambda}. \tag{9.28}$$

Next we introduce the *partition function*

$$\mathcal{Z}(\psi',\vec{x})=\int e^{(\psi'-\gamma-\chi y)\lambda}\Pi_0(\vec{x},\lambda)\,d\lambda. \tag{9.29}$$

Again, we utilize the partition function to simplify the expression for the mean value q^* of the potential vorticity

$$q^*=\int \lambda\rho^*(\vec{x},\lambda)\,d\lambda=\frac{1}{\theta}\frac{\partial}{\partial\psi'}\ln\mathcal{Z}(\theta\psi',\vec{x})\bigg|_{\psi'=\psi'^*}. \tag{9.30}$$

Since the potential vorticity q in the quasi-geostrophic equations (9.1) satisfies $q=\Delta\psi'+h+\beta y$, we finally conclude that the most probable mean field ψ^* satisfies the non-linear mean field equation

$$\Delta\psi'^*+h+\beta y=\frac{1}{\theta}\left(\frac{\partial}{\partial\psi'}\ln\mathcal{Z}(\theta\psi',\vec{x})\right)\bigg|_{\psi'=\psi'^*}. \tag{9.31}$$

Recall that we are primarily interested in steady state solutions to the barotropic quasi-geostrophic equations (9.1) emerging out of statistical theory. We also recall that a steady state solution to (9.1) in our current situation of zonal topography must satisfy either $q=\mathcal{F}(\psi)$ or $\psi=\mathcal{G}(q)$ locally for some function $\mathcal{F}$ or $\mathcal{G}$. We then observe that the solutions to the mean field equation (9.31) are not necessarily steady state solutions of (9.1), since the necessary functional relation between the potential vorticity q and the stream function ψ is not obvious and may not exist. In order to ensure that solutions to (9.31) correspond to steady state solutions to (9.1), we assume that the prior distribution is independent of the spatial location, i.e. takes the form $\Pi_0(\lambda)$, and the Lagrange multipliers for the impulse and energy are related through the large-scale mean flow in the following manner

$$\chi=\theta V \tag{9.32}$$

with V being the constant large-scale mean flow. We then verify that the partition function is a function of the stream function ψ, and, consequently, the right-hand side of the mean field equation (9.31) is a function of ψ, i.e.

$$\begin{aligned}
\frac{1}{\theta}\frac{\partial}{\partial\psi'}\ln \mathcal{Z}(\theta\psi', \vec{x}) &= \frac{1}{\theta}\frac{\partial}{\partial\psi'}\left(\ln\int e^{\theta(\psi'-Vy)\lambda}e^{-\gamma\lambda}\Pi_0(\lambda)d\lambda\right)\\
&= \frac{\int \lambda e^{\theta(\psi'-Vy)\lambda}e^{-\gamma\lambda}\Pi_0(\lambda)\,d\lambda}{\int e^{\theta(\psi'-Vy)\lambda}e^{-\gamma\lambda}\Pi_0(\lambda)\,d\lambda}\\
&= \frac{\int \lambda e^{\theta\psi\lambda}e^{-\gamma\lambda}\Pi_0(\lambda)\,d\lambda}{\int e^{\theta\psi\lambda}e^{-\gamma\lambda}\Pi_0(\lambda)\,d\lambda}\\
&= \mathcal{F}(\psi)
\end{aligned} \tag{9.33}$$

for some function $\mathcal{F}$. Thus, with (9.31) and (9.33), the most probable mean stream function $\psi'^*(\vec{x})$ defines an exact steady state solution of the quasi-geostrophic equations. Notice that $\psi|_{y=0} \neq \psi|_{y=\pi}$ in general. This is consistent with the left-hand side of the mean field equation (9.31), whose value at the top and bottom of the channel are not equal in general.

We leave the case of general topography and large-scale mean flow as an exercise to the interested reader.

There is an interpretation of the assumption (9.32) in terms of conserved quantities. Instead of imposing the conservation of energy and impulse separately, (9.32) is equivalent to imposing a single conserved quantity, which is the following linear combination of the energy and impulse

$$\tilde{E} = E + VI = -\frac{1}{2}\fint \psi'\omega + V\fint y\omega = \fint\left(-\frac{1}{2}\psi' + Vy\right)\omega, \tag{9.34}$$

and this ensures that the prediction of the large-scale structures from the statistical theory yield exact solutions to the barotropic quasi-geostrophic equations (9.1). Such a view will be utilized in Chapter 11 for the predictions and comparisons of various statistical theories.

9.3 The mean field statistical theory for point vortices

In this section, we neglect all geophysical effects, i.e. $V \equiv 0$, $h \equiv 0$, $\beta = 0$, so that the quasi-geostrophic equations (9.1) reduce to the classical incompressible Euler's equations for two-dimensional homogeneous fluids

$$\frac{\partial\omega}{\partial t} + \nabla^{\perp}\psi\cdot\nabla\omega = 0, \quad \Delta\psi = \omega, \tag{9.35}$$

in the stream function–vorticity formulation. We are interested in the large-scale statistical behavior of solutions arising as limits of point vortices.

9.3.1 Derivation of the mean field point-vortex theory from an empirical statistical theory

Here we show how to derive the mean field equation of classical point-vortex theory for two-dimensional fluid flows from an empirical statistical theory with a suitable prior distribution and the two constraints of conservation of energy and circulation and its generalization to three constraints determined by the energy, circulation, and angular momentum for a disk domain. The domain under consideration is either a periodic box, or a channel, or a disk of radius R centered at the origin ($B_R(0)$) with zero boundary condition for the stream function ψ. Below, we recover the limiting case of the whole space $\mathcal{R}^2$ by letting R approach infinity in the disk case.

The constraints are the conservation of energy E, the circulation Γ in general, with the addition of the angular momentum A in the case of the disk $B_R(0)$

$$E = -\frac{1}{2}\fint \psi\omega, \quad \Gamma = \fint \omega, \quad A = \fint |\vec{x}|^2\omega. \tag{9.36}$$

For the disk geometry, i.e. $\Omega = B_R(0)$ with $\psi\big|_{|\vec{x}|=R} = 0$, the proof of the conservation of the energy and circulation is straightforward and similar to the channel case discussed in Section 1.4. As for the conservation of angular momentum A, we directly verify, utilizing the conservation of circulation and repeated integration by parts that

$$\begin{aligned}
\frac{dA}{dt} &= \frac{d}{dt}\fint |\vec{x}|^2\omega = \frac{d}{dt}\fint (|\vec{x}|^2 - R^2)\omega = \fint (|\vec{x}|^2 - R^2)\frac{\partial\omega}{\partial t} \\
&= -\fint (|\vec{x}|^2 - R^2)\nabla^{\perp}\psi\cdot\nabla\omega = 2\fint \vec{x}\cdot\nabla^{\perp}\psi\,\Delta\psi \\
&= 2\fint\left(-x\frac{\partial\psi}{\partial y} + y\frac{\partial\psi}{\partial x}\right)\Delta\psi \\
&= \frac{2}{\pi R^2}\int_{|\vec{x}|=R}\left(-x\frac{\partial\psi}{\partial y} + y\frac{\partial\psi}{\partial x}\right)\frac{\partial\psi}{\partial r} \\
&\quad -2\fint\left(-x\frac{\partial}{\partial y}\nabla\psi\cdot\nabla\psi + y\frac{\partial}{\partial x}\nabla\psi\cdot\nabla\psi\right) \\
&= -\fint \operatorname{div}(y|\nabla\psi|^2, -x|\nabla\psi|^2) \\
&= 0.
\end{aligned}$$

The energy, circulation, and angular momentum conservation translate into constraints on the one-point statistics ρ for the vorticity ω

$$\begin{aligned} E(\rho) &= -\frac{1}{2}⨍ \bar{\psi}\bar{\omega} = E_0, \\ \Gamma(\rho) &= ⨍ \int \lambda \rho(\vec{x}, \lambda)\, d\lambda\, d\vec{x} = \Gamma_0, \\ A(\rho) &= ⨍ \int |\vec{x}|^2 \lambda \rho(\vec{x}, \lambda)\, d\lambda\, d\vec{x} = A_0. \end{aligned} \tag{9.37}$$

However, we will drop the condition

$$\int \rho(\vec{x}, \lambda)\, d\lambda = 1 \quad \text{for each } \vec{x}.$$

Instead we will just assume

$$\tilde{M} = ⨍ \int \rho(\vec{x}, \lambda)\, d\lambda\, d\vec{x} = 1, \tag{9.38}$$

so that point vortices are allowed. By point vortex we mean a vorticity field of the form

$$\omega(\vec{x}) = \omega_0 \delta_{\vec{x}_0}, \tag{9.39}$$

where ω_0 is a fixed constant and $\vec{x}_0 = (x_0, y_0)$ is a fixed point in the domain under consideration. It is called point vortex, since the associated velocity field is given by, assuming for simplicity $\vec{x}_0 = (0, 0)$ and $\Omega = \mathcal{R}^2$

$$\vec{v}(\vec{x}) = \frac{1}{2\pi} \begin{pmatrix} -y \\ x \end{pmatrix} |\vec{x}|^{-2}, \tag{9.40}$$

after solving for the stream function and taking the perpendicular derivative. Hence the fluid is swirling around $\vec{x}_0 = (0, 0)$. See Subsection 9.3.2 below and Majda and Bertozzi (2001) for more details.

Next we want to address the question of what is the natural prior distribution for point vortices. Suppose the prior information that we have is that there are a large number of vortices with uniform strength (same ω_0) and location $\vec{x}_i$ drawn completely at random from the uniform measure on Ω. Thus we are considering a collection of N such vortices normalized to give a probability measure by

$$\sum_{i=1}^{N} \frac{1}{N} \left(\delta_{\omega_0}(\lambda) \otimes \delta_{\vec{x}_i} \right), \tag{9.41}$$

where the locations $\vec{x}_i$ are chosen so that for $\Omega_1 \subseteq \Omega$

$$\text{Prob}\{\vec{x}_i \in \Omega_1\} = \frac{|\Omega_1|}{|\Omega|}. \tag{9.42}$$

Then the law of large numbers guarantees that as $N \to \infty$

$$\sum_{i=1}^{N} \frac{1}{N} \left(\delta_{\omega_0}(\lambda) \otimes \delta_{\vec{x}_i}\right) \rightharpoonup \delta_{\omega_0}(\lambda) \frac{d\vec{x}}{|\Omega|}, \tag{9.43}$$

for almost all choices $\vec{x}_i$, where the convergence is weak in the sense of measure (Lamperti, 1966: Chapter 2). In other words, for all nice (continuous) functions, $f(\vec{x}, \lambda)$

$$\sum_{i=1}^{N} \frac{1}{N} f(\vec{x}_i, \omega_0) \to \fint f(\vec{x}, \omega_0)\, d\vec{x}$$

for almost every choice of $\vec{x}_i$. Thus, the most natural prior distribution associated with the limit of sums of N-point vortices with uniformly distributed locations and strength ω_0/N is given by

$$\Pi_0(\lambda) = \delta_{\omega_0}(\lambda) \frac{d\vec{x}}{|\Omega|}. \tag{9.44}$$

We can now proceed with the Lagrange multiplier method for relative entropy maximization with a given prior distribution. We already have the variational derivatives for the relative entropy, energy, and circulation constraints. As for the angular momentum constraint and constraint (9.38), their variational derivatives are calculated easily as

$$\frac{\delta \tilde{M}}{\delta \rho} = 1, \quad \frac{\delta A}{\delta \rho} = |\vec{x}|^2 \lambda. \tag{9.45}$$

Therefore, the maximum entropy principle dictates that the most probable one-point statistics ρ^* must satisfy

$$\ln\left(\frac{\rho^*}{\Pi_0}\right) + 1 = \theta \psi^* \lambda - \gamma \lambda - \alpha |\vec{x}|^2 \lambda - \tilde{\mu}, \tag{9.46}$$

where $\theta, \gamma, \alpha, \tilde{\mu}$ are the Lagrange multipliers for the energy, circulation, angular momentum, and the normalization condition $\tilde{M}$ respectively.

By applying the condition $\fint \int \rho(\vec{x}, \lambda) d\lambda d\vec{x} = 1$, we get

$$\rho^*(\vec{x}, \lambda) = \frac{e^{(\theta\psi^* - \gamma - \alpha|\vec{x}|^2)\lambda} \Pi_0(\vec{x}, \lambda)}{\fint \int e^{(\theta\psi^* - \gamma - \alpha|\vec{x}|^2)\lambda} \Pi_0(\vec{x}, \lambda)\, d\lambda d\vec{x}}. \tag{9.47}$$

Finally, since $q^* = \omega^* = \int \lambda \rho^*(\vec{x}, \lambda)\, d\lambda$, the **mean field equation** for the mean state ψ^* is given by

$$\Delta \psi^* = \omega^* = \frac{\int \lambda e^{(\theta\psi^* - \gamma - \alpha|\vec{x}|^2)\lambda} \Pi_0(\vec{x}, \lambda)\, d\lambda}{\fint \int e^{(\theta\psi^* - \gamma - \alpha|\vec{x}|^2)\lambda} \Pi_0(\vec{x}, \lambda)\, d\lambda\, d\vec{x}}. \tag{9.48}$$

Next we utilize the prior distribution $\Pi_0(\lambda)$ for statistical ensembles of point vortices given by (9.44) and we deduce the following explicit form of the mean field equation

- **Case with no angular momentum.** We then recover the mean field equation for empirical statistical point-vortex theory

$$\Delta\psi^* = \frac{\omega_0 e^{\theta\omega_0\psi^*}}{⨍ e^{\theta\omega_0\psi^*}\, d\vec{x}}. \tag{9.49}$$

Introducing the notation

$$\theta\omega_0 = -\beta, \quad \Gamma_0 = \omega_0|\Omega|, \tag{9.50}$$

where Γ_0 is the circulation and β is the negative temperature, we may rewrite (9.49) as

$$\Delta\psi^* = \frac{\Gamma_0 e^{-\beta\psi^*}}{\int e^{-\beta\psi^*}\, d\vec{x}}. \tag{9.51}$$

- **Case with angular momentum.** In this case we must have a disk $B_R(0)$ and the mean field equation takes the form

$$\Delta\psi^* = \frac{\Gamma_0 e^{-\beta\psi^* - \alpha|\vec{x}|^2}}{\int_{B_R(0)} e^{-\beta\psi^* - \alpha|\vec{x}|^2}\, d\vec{x}}. \tag{9.52}$$

Letting R approaching infinity, we deduce the mean field equation on the whole plane

$$\Delta\psi^* = \frac{\Gamma_0 e^{-\beta\psi^* - \alpha|\vec{x}|^2}}{\int_{\mathcal{R}^2} e^{-\beta\psi^* - \alpha|\vec{x}|^2}\, d\vec{x}}. \tag{9.53}$$

Next, we will sketch how the mean field equations for point vortices can be derived from a more complete equilibrium statistical theory.

9.3.2 Complete statistical mechanics for point vortices

Here we consider flow over the entire plane, i.e. $\Omega = \mathcal{R}^2$, and ignore geophysical effects. Our objective is to give a heuristic derivation of the complete statistical mechanics for point vortices in the plane which is the analogue of the complete statistical mechanics for the energy–enstrophy theory developed in Chapter 8. There are beautiful rigorous mathematical derivations of these equations independently due to Kiessling (1993) and Caglioti *et al.* (1992, 1995) as well as novel extension of these ideas to ensembles of nearly parallel vortex filaments (Majda and Bertozzi, 2001: Chapter 7; Lions and Majda, 2000). The lectures by Lions (1998) provide a mathematical survey of these rigorous theories. All of that material is beyond the focus of this book and the interested reader should consult these references.

The dynamics of point vortices in the plane

The equations for point vortices arise from the heuristic approximation that the vorticity is a superposition of δ-functions for all times

$$\omega(\vec{x}, t) = \sum_{j=1}^{N} \Gamma_j \delta(\vec{x} - \vec{X}_j(t)), \qquad \vec{X}_j(t)|_{t=0} = \vec{X}_j^0. \tag{9.54}$$

From the vorticity-stream form, the formal induced velocity associated with the vorticity in (9.54) is given by

$$\vec{v}(\vec{x}, t) = \frac{1}{2\pi} \sum_{j=1}^{N} \Gamma_j \frac{(\vec{x} - \vec{X}_j)^{\perp}}{|\vec{x} - \vec{X}_j|^2}, \tag{9.55}$$

with $\vec{x}^{\perp} = (-y, x)$. Ignoring the fact that the velocity of a point vortex is infinite at its center, a point vortex induces no motion at its center. Thus, using the particle trajectory equations

$$\frac{d\vec{X}_j}{dt} = \vec{v}(\vec{X}_j, t), \quad \vec{X}_j|_{t=0} = \vec{X}_j^0,$$

and the above formal fact we arrive at the

Point-vortex equations

$$\frac{d\vec{X}_j}{dt} = \frac{1}{2\pi} \sum_{k=1, k\neq j}^{N} \Gamma_k \frac{(\vec{X}_j - \vec{X}_k)^{\perp}}{|\vec{X}_j - \vec{X}_k|^2}, \qquad \vec{X}_j|_{t=0} = \vec{X}_j^0. \tag{9.56}$$

It is well known, see for instance Onsager (1949), that the equations for point vortices form a Hamiltonian system, i.e. the dynamical equations in (9.56) can be rewritten in the generalized Hamiltonian form

$$\Gamma_j \frac{d}{dt} \vec{X}_j = J \nabla_{\vec{X}_j} H, \tag{9.57}$$

where J is the usual skew-symmetric matrix

$$J = \begin{pmatrix} 0 & -1 \\ 1 & 0 \end{pmatrix} \tag{9.58}$$

For the motion of point vortices in (9.56), the reader can readily verify that the Hamiltonian is given by

$$H = \frac{1}{4\pi} \sum_{j \neq k} \Gamma_j \Gamma_k \ln |\vec{X}_j - \vec{X}_k|. \tag{9.59}$$

Liouville property

The Liouville property follows directly from the Hamiltonian formulation (9.57). Indeed, for

$$\mathbf{X} = (\vec{X}_1, \cdots, \vec{X}_N) = (X_1, Y_1, \cdots, X_N, Y_N) \in \mathcal{R}^{2N}, \tag{9.60}$$

the point-vortex equations (9.56) or (9.57) can be written as

$$\frac{d\mathbf{X}}{dt} = \vec{F}(\mathbf{X}), \tag{9.61}$$

with

$$F_{2j} = -\frac{1}{\Gamma_j} \frac{\partial H}{\partial Y_j}, \qquad F_{2j+1} = \frac{1}{\Gamma_j} \frac{\partial H}{\partial X_j}. \tag{9.62}$$

It is then obvious that (9.61) satisfies the local Liouville property which implies the Liouville property for the point-vortex system (9.56) or (9.57).

Conserved quantities

In general, according to Noether's theorem, symmetries in a Hamiltonian lead to conserved quantities. For instance, for the entire plane geometry, the translational symmetry

$$H(\vec{X}_1 + \vec{Y}, \vec{X}_2 + \vec{Y}, \ldots, \vec{X}_N + \vec{Y}) = H(\vec{X}_1, \vec{X}_2, \ldots, \vec{X}_N), \tag{9.63}$$

for all $\vec{Y}$ leads to the conserved quantity for the dynamics

$$\mathbf{M} = \sum_{j=1}^{N} \Gamma_j \vec{X}_j(t), \tag{9.64}$$

involving the linear momentum. Likewise, for disk or entire plane geometry, the rotational symmetry of the Hamiltonian

$$H(\mathcal{O}(\theta)\vec{X}_1, \mathcal{O}(\theta)\vec{X}_2, \ldots, \mathcal{O}(\theta)\vec{X}_N) = H(\vec{X}_1, \vec{X}_2, \ldots, \vec{X}_N), \tag{9.65}$$

for any rotation matrix $\mathcal{O}(\theta)$ in the plane yields the conserved quantity for the dynamics

$$A = \sum_{j=1}^{N} \Gamma_j |\vec{X}_j(t)|^2, \tag{9.66}$$

which is the angular momentum. We leave the verification of these conserved quantities as an exercise for the interested reader. However, we remark here that **M** and A are discrete versions of the continuous first two moments of vorticity $\int \vec{x}\omega d\vec{x}$ and $\int |\vec{x}|^2\omega d\vec{x}$ which are conserved quantities for the 2-D Euler equations in (9.35). We note that the Hamiltonian for the point-vortex system is formally a discrete approximation to minus the kinetic energy,

$$E = -\frac{1}{2}\int \psi\Delta\psi \cong -\frac{1}{4\pi}\sum_{j\neq k}\Gamma_j\Gamma_k \ln|\vec{X}_j - \vec{X}_k| = -H(\vec{X}_1, \vec{X}_2, \dots, \vec{X}_N), \quad (9.67)$$

except for the non-trivial difference that the self-interaction energy of a point vortex is actually infinite, so the approximation in (9.67) is only a heuristic renormalization with these infinities due to self-interaction removed.

The statistical mechanics of N identical point vortices

With the Liouville property and conserved quantities in hand, we are ready to apply the equilibrium statistical mechanics theory for ODE's developed in Chapter 7 to the point-vortex system (9.56) or (9.57). For simplicity in exposition we will ignore the conserved linear momentum, **M**, related to the impulse since we may consider disk domain for which the linear momentum is not conserved, and even for the entire plane case its effect can be removed ultimately through the translation of coordinates in this context. Thus the conserved quantities that we use here are the "energy," $-H_N$, in (9.59), and the angular momentum, $A = A_N$, in (9.66). We also assume, without loss of generality, that $\Gamma_j = 1$ for $1 \leq j \leq N$.

According to the equilibrium statistical mechanics theory introduced in Chapter 7, the most probable (least-biased) probability measure for doing further statistical measurements of the point-vortex system is given by the Gibbs measure

$$\mathcal{G}_N(\mathbf{x}) = \frac{e^{\theta H_N(\vec{x}_1,\dots,\vec{x}_N) - \alpha|\mathbf{x}|^2}}{Z(N)}, \quad (9.68)$$

with $\mathbf{x} = (\vec{x}_1, \dots, \vec{x}_N)$. Here, $Z(N)$ is the normalizing factor that guarantees that $\mathcal{G}_N$ is a probability density function so that $\int \mathcal{G}_N d\mathbf{x} = 1$. Thus, $Z(N)$ is given by

$$Z(N) = \int_{\mathcal{R}^{2N}} e^{\theta H_N(\vec{x}_1,\dots,\vec{x}_N) - \alpha|\mathbf{x}|^2} d\vec{x}_1 \cdots d\vec{x}_N, \quad (9.69)$$

where θ and α are the Lagrange multipliers associated with the given "energy" and angular momentum constraints. In order to ensure $Z(N) < +\infty$, for given parameter values, θ and α, in the Gibbs measure, we clearly need α to satisfy $\alpha > 0$.

A somewhat more subtle constraint on θ is needed to guarantee that $Z(N)$ is finite (see Caglioti *et al.*, 1992, for more detail). Substituting the explicit form of H_N from (9.59) into (9.68) yields

$$Z(N) = \int_{\mathcal{R}^{2N}} e^{\frac{\theta}{4\pi}\sum_{j\neq k}\ln|\vec{x}_j-\vec{x}_k|} e^{-\alpha|\mathbf{x}|^2} d\vec{x}_1\cdots d\vec{x}_N$$
$$= \int_{\mathcal{R}^{2N}} \prod_{j\neq k} |\vec{x}_j - \vec{x}_k|^{\frac{\theta}{4\pi}} e^{-\alpha|\mathbf{x}|^2} d\vec{x}_1\cdots d\vec{x}_N. \tag{9.70}$$

The above expression for $Z(N)$ is finite if and only if the following two-dimensional integral is finite

$$\int_{\mathcal{R}^2} |\vec{x}|^{\frac{N\theta}{4\pi}} e^{-\alpha|\vec{x}|^2} d\vec{x},$$

which is satisfied provided that $\frac{N\theta}{4\pi} > -2$, i.e.

$$\theta > -\frac{8\pi}{N}. \tag{9.71}$$

Otherwise, the stronger local singularity for $|\vec{x}| \cong 0$ in the integrand would make the integral infinite.

Finally, we remark that averaged measurement of a function $F(\vec{x}_1,\ldots,\vec{x}_N)$ is determined by the ensemble average

$$\langle F\rangle_{\mathcal{G}_N} = \int_{\mathcal{R}^{2N}} F(\vec{x}_1,\ldots,\vec{x}_N)\mathcal{G}_N(\vec{x}_1,\ldots,\vec{x}_N)d\vec{x}_1,\ldots,d\vec{x}_N, \tag{9.72}$$

where $\mathcal{G}_N$ is the Gibbs measure defined in (9.68).

The mean field limit equations as $N\to\infty$

The requirement in (9.71) suggests that we replace the inverse temperature, θ, by θ/N if we attempt to take the limit as $N\to\infty$. Since we have N identical vortices, and the Gibbs measure (9.68) is symmetric in $\vec{x}_1,\cdots,\vec{x}_N$, it is extremely natural to focus on the marginal distribution of a single vortex, $\rho_N(\vec{x}), \vec{x}\in\mathcal{R}^2$. From (9.68) this probability density is defined by

$$\rho_N(\vec{x}) = \frac{\int_{\mathcal{R}^{N-1}} e^{\frac{\theta}{N}H_N(\vec{x},\vec{x}_1,\ldots,\vec{x}_{N-1})-\alpha(|\vec{x}|^2+\sum|\vec{x}_j|^2)} d\vec{x}_1\ldots d\vec{x}_{N-1}}{Z(N)}. \tag{9.73}$$

In order to take the continuum limit, we make the following technical assumption regarding the asymptotic behavior of the marginal distribution (9.73)

$$\frac{1}{N}\sum_{i=1}^{N}\delta_{\vec{x}_i} \rightharpoonup \rho(\vec{x}),$$
$$\rho_N(\vec{x}) \rightharpoonup \rho(\vec{x}) \text{ as } N\to\infty, \tag{9.74}$$

where the locations of the N identical point vortices, $\vec{x}_i$, are drawn at random from the marginal Gibbs distribution in (9.73). Assumption (9.74) is a generalization of the law of large numbers, which can be verified a priori (see for instance Kiessling, 1993; Caglioti *et al.*, 1992, 1995). The meaning of the first condition in (9.74) is that for nice functions $f(\vec{y})$, as $N \to \infty$

$$\sum_{i=1}^{N} \frac{1}{N} f(\vec{x}_i) \to \int_{\mathcal{R}^2} f(\vec{y}) \rho(\vec{y}) d\vec{y}.$$

In particular, through an explicit computation we have

$$\frac{1}{N} H_N(\vec{x}, \vec{x}_1, \ldots, \vec{x}_{N-1}) = \frac{1}{N} H_{N-1}(\vec{x}_1, \ldots, \vec{x}_{N-1}) + \frac{1}{2\pi} \sum_{i=1}^{N-1} \frac{1}{N} \ln(|\vec{x} - \vec{x}_i|), \tag{9.75}$$

where from (9.74) the second sum is approximately

$$\frac{1}{2\pi} \sum_{i=1}^{N-1} \frac{1}{N} \ln(|\vec{x} - \vec{x}_i|) \to \frac{1}{2\pi} \int_{\mathcal{R}^2} \ln(|\vec{x} - \vec{y}|) \rho(\vec{y}) d\vec{y}, \tag{9.76}$$

as $N \to \infty$. Notice that

$$\psi(\vec{x}) = \frac{1}{2\pi} \int_{\mathcal{R}^2} \ln(|\vec{x} - \vec{y}|) \rho(\vec{y}) d\vec{y}, \tag{9.77}$$

is the stream function, ψ, satisfying

$$\Delta \psi = \rho. \tag{9.78}$$

With (9.75)–(9.78), it is a simple matter to formally derive the limiting mean field equation satisfied by ρ or equivalently, ψ. From (9.73) and (9.75)–(9.77), for $N >> 1$

$$\rho(\vec{x}) \cong \rho_N(\vec{x}) \cong \frac{e^{-\alpha|\vec{x}|^2} e^{-\theta\psi(\vec{x})} \tilde{Z}(N-1)}{Z(N)} \tag{9.79}$$

with

$$\tilde{Z}(N-1) = \int_{\mathcal{R}^{N-1}} e^{-\frac{\theta}{N} H_{N-1}(\vec{x}_1, \ldots, \vec{x}_{N-1})} e^{-\alpha \sum_{j=1}^{N-1} |\vec{x}_j|^2} d\vec{x}_1 \cdots d\vec{x}_{N-1}. \tag{9.80}$$

Since $\rho_N(\vec{x})$ and $\rho(\vec{x})$ are both probability densities, we necessarily have the relations that as $N \to \infty$

$$\frac{Z(N)}{\tilde{Z}(N-1)} \cong \int_{\mathcal{R}^2} e^{-\alpha|\vec{x}|^2 - \theta\psi(\vec{x})} d\vec{x}. \tag{9.81}$$

Combining (9.78)–(9.81), we deduce that the probability density for point vortices with unit strength as $N \to \infty$ must satisfy the:

Mean field equation

$$\Delta\psi = \frac{e^{-\alpha|\vec{x}|^2-\theta\psi(\vec{x})}}{\int_{\mathcal{R}^2} e^{-\alpha|\vec{x}|^2-\theta\psi(\vec{x})}\, d\vec{x}}. \tag{9.82}$$

This equation is clearly the same equation as (9.53) that we derived in Subsection 9.3.1 directly from a maximum entropy principle with a suitable prior distribution for ensembles of point vortices. For simplicity in exposition, we have normalized the total circulation to be unity in our discussion in this section.

It is interesting for the reader to compare the complete statistical mechanics derivation of the point-vortex mean field theory presented here with that for the energy–enstrophy theory developed in Chapter 8. Both theories rely on (very different) finite-dimensional dynamics in the approximation. The rigorous mathematical proofs of Kiessling (1993) and Caglioti *et al.* (1992, 1995) establish the key facts in (9.74)–(9.82) *a priori*. The behavior of the solutions of (9.82) for $\theta > -8\pi$ is studied in detail by Caglioti *et al.* (1992, 1995). In particular, as θ decreases to -8π, solutions tend to a point vortex. Also see our discussion in Chapters 10 and 11.

9.4 Empirical statistical theories with infinitely many constraints

9.4.1 Maximum entropy principle incorporating all generalized enstrophies

We present these theories in the simplest physical setup. We recall from Chapter 1 that for the barotropic quasi-geostrophic equations (9.1) without mean flow ($V \equiv 0$) and either without the beta-plane effect, i.e. $\beta = 0$, or with the channel symmetry (9.3) the following quantities are conserved:

- the energy, and
- the generalized enstrophies, $\int G(q(\vec{x}))\, d\vec{x}$ for any function $G(q)$.

Here we show how to attempt formally to incorporate all of these constraints through an empirical statistical theory. This approach was originally developed by Robert (1991) and is related to attempts by Miller (1990) and Miller *et al.*, (1992) to develop complete statistical theories with infinitely many invariants. This theory is critiqued both theoretically and from a practical view point in Chapters 10 and 11 below.

There are two issues that we need to address. First, how to incorporate the infinitely many constraints induced by the infinitely many conserved enstrophies and, second, how to pick the prior distribution.

The first issue we need to address is how to deal with the infinite number of constraints. A naive approach would be to impose infinitely many constraints on the one-point statistics ρ induced by the conservation of generalized enstrophies. This is practically impossible, since we are not able to invoke the Lagrange multiplier method with infinitely many constraints. Fortunately it turns out that there is a way to circumvent such a difficulty. We recall the following well-known fact (see for instance Rudin, 1982). For each uniformly bounded vorticity field $q(\vec{x})$, there exists an associated probability measure $\Pi_q(\lambda)$ on $\mathcal{R}^1$, called the distribution function of q (here $\Pi_q(\lambda)$ is the associated density function), such that

$$\begin{aligned} ⨍ 1_{\{\vec{x}|\alpha\le q(\vec{x})<\beta\}}\, d\vec{x} &= \int_\alpha^\beta \Pi_q(\lambda)\, d\lambda, \\ \int \Pi_q(\lambda)\, d\lambda &= 1, \\ ⨍ G(q)\, d\vec{x} &= \int G(\lambda)\,\Pi_q(\lambda)\, d\lambda, \quad \text{for arbitrary } G. \end{aligned} \tag{9.83}$$

Now let q_0 be the initial data for the following PDE

$$\frac{\partial q}{\partial t} + \vec{v}\cdot\nabla q = 0, \quad \operatorname{div}\vec{v} = 0. \tag{9.84}$$

Here the PDE is not necessarily the quasi-geostrophic equations, but any transport equation involving transport by an incompressible velocity field. Then, for the solution $q(\vec{x}, t)$ at any time, the equality

$$\int G(q(\vec{x}, t))\, d\vec{x} = \int G(q_0(\vec{x}))\, d\vec{x} \tag{9.85}$$

holds. Notice the above equality implies

$$\int G(\lambda)\,\Pi_q(\lambda)\, d\lambda = \int G(\lambda)\,\Pi_{q_0}(\lambda)\, d\lambda, \tag{9.86}$$

using the associated probability measure representation (9.83). Therefore, the conservation of arbitrary $⨍ G(q(\vec{x}))\, d\vec{x}$ implies

$$\Pi_{q(t)}(\lambda) = \Pi_{q_0}(\lambda) \tag{9.87}$$

for all times t, i.e. the distribution function of the potential vorticity is invariant in time. So conservation of all the conserved quantities in (9.85) in time is equivalent to studying an ensemble of functions of the potential vorticity with the same distribution function. Thus, if we are interested in predicting large-scale behavior from statistical (time) averages over dynamics which conserves all the generalized enstrophies, the natural prior distribution is given by the distribution function, $\Pi_{q_0}(\lambda)$, i.e.

$$\Pi_0(\lambda) = \Pi_{q_0}(\lambda). \tag{9.88}$$

This is the choice for the prior distribution which we utilize below in the maximum entropy principle.

Now suppose we have a probability density $\rho(\vec{x}, \lambda)$ representing the one-point statistics for the potential vorticity (see Definition 6.5 from Chapter 6 for more details). We need to find a way to represent the conservation of all generalized enstrophies, $⨍ G$, discussed earlier in (9.84)–(9.87)

$$⨍ G(q(\vec{x}))d\vec{x} = ⨍ G(q_0(\vec{x}))d\vec{x} = \int G(\lambda)\Pi_{q_0}(\lambda)d\lambda \tag{9.89}$$

for the one-point statistical measure $\rho(\vec{x}, \lambda)$.

The obvious way to achieve this is to recall from Chapter 6 that the pointwise and overall expected value of a non-linear function, $G(\lambda)$, with respect to the probability density, $\rho(\vec{x}, \lambda)$, is given by

$$\begin{aligned} \langle G\rangle_{\rho(\vec{x})} &= \int G(\lambda)\rho(\vec{x}, \lambda)\, d\lambda, \\ \langle G\rangle_{\rho} &= ⨍ \int G(\lambda)\rho(\vec{x}, \lambda)\, d\lambda d\vec{x}. \end{aligned} \tag{9.90}$$

Thus, combining (9.90) and (9.89), the conservation of generalized enstrophy for such a probability measure becomes the requirements

$$\int G(\lambda) ⨍ \rho(\vec{x}, \lambda)\, d\vec{x}\, d\lambda = \int G(\lambda)\, \Pi_{q_0}(\lambda)\, d\lambda \tag{9.91}$$

for all non-linear functions, $G(\lambda)$. From (9.91) we see that for the one-point statistical density $\rho(\vec{x}, \lambda)$ conservation of all the generalized enstrophies consistent with the prior distribution, $\Pi_{q_0}(\lambda)$, requires that

$$⨍ \rho(\vec{x}, \lambda)d\vec{x} = \Pi_{q_0}(\lambda), \text{ for any } \lambda \in \text{supp } \Pi_{q_0}. \tag{9.92}$$

A one-point statistic $\rho(\vec{x}, \lambda)$ satisfying (9.92) is called a *macrostate associated with the microstate* q_0. This terminology will be used in Chapter 11 for discussion of statistical sharpness of mean field predictions of various statistical theories.

Note that, in general, the requirements in (9.92) are an infinite number of constraints. Also, the interested reader can easily check that, in the special case that the probability measure, $\rho(\vec{x}, \lambda)$, is associated with a single function, i.e. $\rho(\vec{x}, \lambda) = \delta_{q(\vec{x})}(\lambda)$, the conditions in (9.92) reduce to the requirement that $\Pi_q(\lambda) = \Pi_{q_0}(\lambda)$, as we would expect.

Finally, we are ready to formulate the maximum entropy principle which encodes all of the generalized enstrophies. With the prior distribution, $\Pi_{q_0}(\lambda)$, we define the relative entropy with respect to Π_{q_0}, i.e.

$$\mathcal{S}(\rho, \Pi_{q_0}(\lambda)) = -\fint \int \ln\left(\frac{\rho(\vec{x}, \lambda)}{\Pi_{q_0}(\lambda)}\right) \rho(\vec{x}, \lambda)\, d\lambda\, d\vec{x}. \tag{9.93}$$

We now follow a familiar routine calculation to find the most probable states. Using the above properties, we may write the constraint set as

$$\begin{aligned} E(\rho) &= -\frac{1}{2}\fint \overline{\psi\omega} = E_0, \\ \int \rho(\vec{x}, \lambda)\, d\lambda &= 1 \quad \text{for each } \vec{x}, \\ \fint \rho(\vec{x}, \lambda)\, d\vec{x} &= \Pi_{q_0}(\lambda) \text{ for any } \lambda \in \operatorname{supp} \Pi_{q_0}, \end{aligned} \tag{9.94}$$

and the most probable state is the one that maximizes the entropy (9.93) subject to the constraints in (9.94).

9.4.2 The most probable state and the mean field equation

Next we invoke the Lagrange multiplier method. By the maximum entropy principle, the most probable state ρ^* is the one that maximizes the relative entropy (9.93) under the given constraints (9.94). Since

$$\frac{\delta}{\delta\rho}\left(\fint \rho(\vec{x}, \lambda)\, d\vec{x}\right) = \delta_\lambda, \tag{9.95}$$

the most probable state ρ^* satisfies

$$\frac{\delta}{\delta\rho}\mathcal{S}(\rho, \Pi_{q_0})\bigg|_{\rho=\rho^*} = \left(\theta\frac{\delta E}{\delta\rho} + \tilde{\gamma}(\lambda) + \tilde{\gamma}_2(\vec{x})\right)\bigg|_{\rho=\rho^*}, \tag{9.96}$$

where $\tilde{\gamma}(\lambda)$ is the Lagrange multiplier associated with the generalized enstrophy constraints in the third equation in (9.94). Hence, by a routine calculation of the variational derivatives, we obtain

$$\ln\left(\frac{\rho^*}{\Pi_{q_0}}\right) + 1 = \theta\psi^*(\vec{x})\lambda - \tilde{\gamma}(\lambda) - \tilde{\gamma}_2(\vec{x}). \tag{9.97}$$

By requiring $\int \rho^*(\vec{x}, \lambda)\, d\lambda = 1$, we get the most probable state

$$\rho^*(\vec{x}, \lambda) = \frac{e^{\theta\psi^*\lambda - \tilde{\gamma}(\lambda)}\Pi_{q_0}(\lambda)}{\int e^{\theta\psi^*\lambda - \tilde{\gamma}(\lambda)}\Pi_{q_0}(\lambda)\, d\lambda}. \tag{9.98}$$

For convenience, as in Chapter 6, we define the *partition function* as

$$\mathcal{Z}(\psi, \tilde{\gamma}(\lambda)) = \int e^{\psi\lambda - \tilde{\gamma}(\lambda)} \Pi_{q_0}(\lambda)\, d\lambda. \tag{9.99}$$

Since

$$q^* = \int \lambda \rho^*(\vec{x}, \lambda)\, d\lambda = \frac{1}{\theta} \frac{\partial \ln \mathcal{Z}(\theta\psi)}{\partial \psi}\bigg|_{\psi=\psi^*}, \tag{9.100}$$

we obtain that the mean field of the most probable state must satisfy the:

Mean field equation

$$\Delta\psi^* + h = \frac{1}{\theta}\frac{\partial \ln \mathcal{Z}(\theta\psi)}{\partial \psi}\bigg|_{\psi=\psi^*}. \tag{9.101}$$

Note that the multipliers θ, $\tilde{\gamma}(\lambda)$ are determined by the constraints

$$E(q^*(\theta, \tilde{\gamma}(\lambda))) = E_0,$$
$$⨍ \rho^*(\vec{x}, \lambda) d\vec{x} = \Pi_{q_0}(\lambda) \text{ for all } \lambda \in supp\, \Pi_{q_0},$$

and the last equations determine $\tilde{\gamma}(\lambda)$ from the requirements

$$⨍ \frac{e^{\theta\psi^*\lambda - \tilde{\gamma}(\lambda)}}{\int e^{\theta\psi^*\lambda - \tilde{\gamma}(\lambda)} \Pi_{q_0}(\lambda)\, d\lambda}\, d\vec{x} = 1 \text{ for all } \lambda \in supp\, \Pi_{q_0}. \tag{9.102}$$

The case with the beta-effect for the channel geometry can be treated similarly by incorporating the conserved quantity given by the impulse as in Subsection 9.2.2.

A final remark is that no complete statistical mechanics theory which incorporates the infinitely many conserved generalized enstrophies has been developed yet. The limitations in the attempts to do this will be discussed briefly in Chapter 11.

9.5 Non-linear stability for the most probable mean fields

We observe from Sections 9.2, 9.3, and 9.4 as well as Chapter 6, for the case without large-scale mean flow ($V = 0$) and conservation of impulse, that the most probable mean field q^* predicted by various statistical theories under the assumption that the prior distribution is independent of the spatial location $(\Pi_0(\vec{x}, \lambda) = \Pi_0(\lambda))$, satisfies the mean field equation

$$q^* = F(\psi^*), \quad F(\psi) = \frac{1}{\theta} \frac{\partial \ln \mathcal{Z}(\theta\psi)}{\partial \psi}, \tag{9.103}$$

where $\mathcal{Z}$ is the partition function defined by

$$\mathcal{Z}(\psi) = \int e^{\psi\lambda - \tilde{\gamma}(\lambda)} \Pi_0(\lambda)\, d\lambda. \tag{9.104}$$

The mean field of infinitely many constraints theory given by (9.101) can be recovered from (9.103)–(9.104), with $\Pi_{q_0}(\lambda)$ replacing $\Pi_0(\lambda)$ in (9.104), and the mean field for the energy circulation theory given by (9.13) can be recovered with $\gamma\lambda$ replacing $\tilde{\gamma}(\lambda)$ in (9.104). Equation (9.103) falls into the category of exact steady state solutions of the barotropic quasi-geostrophic equations (9.1) whose non-linear stability we studied in Section 4.3. Thus the answer to the question regarding the stability of the predicted most probable mean field reduces to the simple problem of verifying the appropriate assumption in Theorem 4.3. Hence we have:

Theorem 9.1 *The mean fields associated with the most probable state given by (9.103)–(9.104) are non-linearly stable, provided one of the following conditions holds:*

(A) either we have positive temperature, i.e. $\theta > 0$, and bounded support for the prior distribution Π_0; or

(B) we have negative temperature, i.e. $\theta < 0$, but the support of the prior distribution is bounded, i.e. there exists a constant a, such that $supp(\Pi_0) \subset [-a, a]$. Moreover we assume $-\theta a^2 < 2$.

Proof: As mentioned above, this is a simple application of the stability results for steady state solutions in Section 4.3.

The assumption of bounded support of the prior distribution as imposed in the negative temperature case (B) is sufficient, but not necessary. For instance, the energy theory with Gaussian prior distribution from Section 6.6 predicts a linear potential vorticity–stream function relation. The mean field is non-linearly stable provided $\mu > -1$ or equivalently, $\theta = \frac{\mu}{\alpha} > -\frac{1}{\alpha}$. Recall we assumed a Gaussian prior distribution with zero mean and variance $\frac{1}{\alpha}$. This prior distribution does not have compact(bounded) support.

We first need to verify that the solutions of (9.103) may be written in the form $\psi = G(q)$ for some appropriate function G.

It is easy to check that $\ln \mathcal{Z}$ from (9.104) is convex in ψ. Indeed

$$\frac{\partial \ln \mathcal{Z}}{\partial \psi} = \frac{\int \lambda \exp(\psi\lambda - \tilde{\gamma}(\lambda))\Pi_0(\lambda) d\lambda}{\int \exp(\psi\lambda - \tilde{\gamma}(\lambda))\Pi_0(\lambda) d\lambda}, \tag{9.105}$$

and hence

$$\frac{\partial^2 \ln \mathcal{Z}}{\partial \psi^2} = \frac{1}{\mathcal{Z}^2}\bigg(\int \exp(\psi\lambda - \tilde{\gamma}(\lambda))\Pi_0(\lambda)d\lambda \int \lambda^2 \exp(\psi\lambda - \tilde{\gamma}(\lambda))\Pi_0(\lambda)d\lambda$$
$$- \Big(\int \lambda \exp(\psi\lambda - \tilde{\gamma}(\lambda))\Pi_0(\lambda)d\lambda\Big)^2\bigg)$$
$$\geq 0 \tag{9.106}$$

by the Cauchy–Schwarz inequality $(\int f^2 \int g^2 - (\int fg)^2 \geq 0)$. In fact it is strictly convex i.e.

$$\frac{\partial^2 \ln \mathcal{Z}}{\partial \psi^2} > 0 \tag{9.107}$$

except for the trivial uninteresting case in which Π_{q_0} is a Dirac measure.

In order to apply the stability results from Section 1.4, we would like to define

$$G = F^{-1} \tag{9.108}$$

so that we have

$$\psi = G(q). \tag{9.109}$$

We need to check the invertability of F. Fortunately we have

$$F'(\psi^\star) = \frac{1}{\theta}\frac{\partial^2 \ln \mathcal{Z}(\theta\psi)}{\partial \psi^2}\bigg|_{\psi=\psi^\star} > 0. \tag{9.110}$$

Hence F is invertible and thus such a function G exists.

To get the stability in the case of (A), we notice that

$$G' = (F')^{-1} > 0 \tag{9.111}$$

and it is easy to observe that it is bounded away from 0 and ∞. Thus stability follows immediately.

To derive stability in the case of condition (B), we need to show, according to Section 4.3, that

$$G' < -\frac{1}{2}, \tag{9.112}$$

which is equivalent to

$$-F' = -\frac{1}{\theta}\frac{\partial^2 \ln \mathcal{Z}(\theta\psi)}{\partial^2 \psi}\bigg|_{\psi=\psi^\star} < 2. \tag{9.113}$$

Since

$$\frac{\partial^2 \ln \mathcal{Z}(\theta\psi)}{\partial \psi^2} = \frac{1}{\mathcal{Z}^2(\theta\psi)}\left(\mathcal{Z}(\theta\psi)\frac{\partial^2 \mathcal{Z}(\theta\psi)}{\partial \psi^2} - \left(\frac{\partial \mathcal{Z}(\theta\psi)}{\partial \psi}\right)^2\right) \leq \frac{1}{\mathcal{Z}(\theta\psi)}\frac{\partial^2 \mathcal{Z}(\theta\psi)}{\partial \psi^2}, \tag{9.114}$$

it suffices to verify

$$-\frac{1}{\theta \mathcal{Z}(\theta\psi)} \frac{\partial^2 \mathcal{Z}(\theta\psi)}{\partial \psi^2} < 2. \tag{9.115}$$

Notice

$$\frac{1}{\mathcal{Z}(\theta\psi)} \frac{\partial^2 \mathcal{Z}(\theta\psi)}{\partial \psi^2} = \frac{\int \theta^2 \lambda^2 \exp(\theta\lambda\psi - \tilde{\gamma}(\lambda))\Pi_0(\lambda)d\lambda}{\int \exp(\theta\lambda\psi - \tilde{\gamma}(\lambda))\Pi_0(\lambda)d\lambda} \leq \theta^2 a^2. \tag{9.116}$$

Hence the stability condition (assumption) is verified under the condition (B).

This completes the proof of the theorem.

The general case with large-scale mean flow and conservation of mean flow–impulse difference can be treated similarly and is left as an exercise to the interested reader.

Finally it is worthwhile to point out that the stability result here is not as sharp as that presented in Chapter 4. Better stability results may be available if we have more information on the prior distribution $\Pi_q(\lambda)$.

References

Caglioti, E., Lions, P. L., Marchioro, C., and Pulvirenti, M. (1992), A special class of stationary flows for two-dimensional Euler equations: a statistical mechanics description. *Commun. Math. Phys.* **143**, 501–525.

Caglioti, E., Lions, P. L., Marchioro, C., and Pulvirenti, M. (1995), A special class of stationary flows for two-dimensional Euler equations: a statistical mechanics description. *Commun. Math. Phys.* **174**, 229–260.

Joyce, G. and Montgomery, D. C. (1974), Negative temperature states for the two-dimensional guiding center plasma. *J. Plasma Phys.* **10**, 107–121.

Kiessling, M. K.-H. (1993), Statistical mechanics of classical particles with logarithmic interactions. *Commun. Pure Appl. Math.* **47**, 27–56.

Lamperti, J. (1966), *Probability*. New York: Benjamin Inc.

Lions, P.-L. (1998), On Euler Equations and Statistical Physics. Cattecha GaliPeiana, Pisa, 1997. Scuola Normale Supanore, Pisa.

Lions, P. L. and Majda, A. (2000), Equilibrium statistical theory for nearly parallel vortex filaments. *Comm. Pure Appl. Math.* **53**, 76–142.

Majda, A. and Bertozzi, A. (2001), *Vorticity and Incompressible Flow*. Cambridge: Cambridge University Press.

Miller, J. (1990), *Statistical mechanics of Euler equations in two dimensions. Phys. Rev. Lett.* **65**, 2137–2140.

Miller, J., Weichman, P. B., and Cross, M. C. (1992), Statistical mechanics, Euler's equation, and Jupiter's Red Spot. *Phys. Rev. A* **45**, 2328–2359.

Montgomery, D. and Joyce, G. (1974), Statistical mechanics of negative temperature states. *Phys. Fluids* **17**, 1139– 1145.

Onsager, L. (1949), Statistical hydrodynamics, *Nuovo Cimento* (9) 6, 279–287, (1949). Supplemento, no. 2(Convegno Internazionale di Meccanica Statistica).

Robert, R. (1991), A maximum entropy principle for two- dimensional Euler equations. *J. Stat. Phys.* **65**, 531– 553.

Rudin, W. K. (1987), *Real and Complex Analysis*, 3rd edn. McGraw-Hill.

10

Assessing the potential applicability of equilibrium statistical theories for geophysical flows: an overview

10.1 Introduction

In Chapters 8 and 9, we have introduced and discussed four different ways to utilize empirical statistical information to develop statistical theories for the large-scale features of geophysical flows. These theories differ in the way small-scale statistical information is encoded in a prior distribution and also in the number of conserved integrals of the non-linear potential vorticity for ideal fluid motion, which are regarded as statistically relevant; they also depend on suitable externally imposed conserved quantities for ideal fluid motion, such as the energy, circulation, and impulse.

Consider the simplest geophysical set-up

$$\frac{\partial q}{\partial t} + \nabla^{\perp}\psi \cdot \nabla q = \mathcal{D}\psi + \mathcal{F}, \qquad q = \Delta\psi + h, \quad \omega = \Delta\psi \tag{10.1}$$

in a periodic or channel geometry. The various equilibrium statistical theories ignore the dissipation and forcing in (10.1) and differ in the statistical importance which is assigned to various members in the infinite list of inviscid conserved quantities

$$\mathcal{G}(q) = \fint G(q)d\vec{x}, \quad \text{for any function } G(q). \tag{10.2}$$

The externally imposed conserved quantities for the unforced non-dissipative dynamics imposed by these statistical theories are

$$\begin{aligned} \text{Energy: } E &= -\frac{1}{2}\fint \psi\omega, \\ \text{Circulation: } \Gamma &= \fint q, \\ \text{Impulse: } I &= \fint yq \end{aligned} \tag{10.3}$$

for the appropriate underlying geometry, i.e. periodic flows, channel flows, basin etc.

These four statistical theories are the following: (10.4)

(1) The energy–enstrophy statistical theory (EEST) (Chapter 8 and Section 6.7).
(2) The point-vortex statistical theory (PVST) (Section 9.3).
(3) The empirical statistical theory with a prior distribution (ESTP) (Section 9.2 and Section 6.6).
(4) The empirical statistical theory with many constraints (ESTMC) attempting to include all the conserved quantities in (10.2) (Section 9.4).

Below, we often use the abbreviations in (10.4) as shorthand for the statistical theories. In periodic geometry, all four of the statistical theories in (10.4) impose energy as a large-scale constraint, but differ substantially in the fashion in which various members of the list in (10.2) are deemed as statistically relevant.

As established in Chapter 9, all of these statistical theories can be formulated to begin with an assumed prior probability distribution $\Pi_0(\lambda)$ for the one-point statistical fluctuations of potential vorticity at small scales, i.e.

$$\text{Prob}\{a \leq q(\vec{x}, t) < b\} = \int_a^b \Pi_0(\lambda) d\lambda \tag{10.5}$$

and predict that the stream function of the mean value of the coarse-grained large-scale fluctuations, $\overline{\psi}$, is an exact solution of the idealized dynamics in (10.1) without dissipation and forcing, i.e. $\overline{\psi}$ satisfies the mean field equation

$$\overline{q} = \Delta\overline{\psi} + h = F(\overline{\psi}, E, \Gamma). \tag{10.6}$$

Here the non-linear function F depends on both the nature of the statistical theory and the conserved quantities, E, Γ. Of course, some of the statistical theories such as EEST include much more information, such as the behavior of statistical fluctuation about this mean state, but all of them predict the coarse-grained probability density for potential vorticity fluctuations.

10.2 Basic issues regarding equilibrium statistical theories for geophysical flows

With the above brief summary of the developments in Chapters 8 and 9, various theoretical and practical issues naturally arise in assessing the potential applicability of the equilibrium statistical approaches in (10.4) to observed geophysical phenomena as well as to geophysical models like those in (10.1). Here is a list of such basic issues which we split roughly into theoretical and applied categories.

Some basic applied issues

(A-1) Geophysical phenomena: What type of geophysical phenomena are predicted at large scales by the solutions of the mean field equations in (10.6) for the various statistical theories in (10.4)?

(A-2) Coherent structures: Do the coherent structures predicted by these statistical theories correspond to observed geophysical structures such as the Great Red Spot of the Jupiter, Gulf Stream rings in the ocean, prototype atmospheric dipole blocking patterns etc? (See Flierl, 1987; Marcus, 1993; Dowling, 1995 for review articles on such coherent structures.)

(A-3) Rhines scale: Can transitions from large-scale zonal jets to vortices in a beta-plane channel (Rhines, 1975) be predicted automatically and quantitatively by an equilibrium statistical theory?

(A-4) Forcing and dissipation: Can the equilibrium statistical theories be utilized to approximate in a statistical sense the geophysical flows in (10.1) with realistic damping and driving? Which regimes of damping and driving are needed for this? How do these results depend on the nature of the statistical theories in (10.4)?

(A-5) Practical choice of a prior distribution: Given a practical geophysical context, how do we select the prior distribution, $\Pi_0(\lambda)$, and the number of constraints to be utilized for the given statistical theory based upon the available observational or numerical record?

Some basic theoretical issues

(T-1) Comparison of predictions of statistical theories: How do the predictions of the various equilibrium statistical theories in (10.4) compare and contrast with each other?

(T-2) Additional information in predictions of ESTMC versus other statistical theories: How much additional information is actually contained in the predictions of ESTMC at large scales as compared with the predictions of the simpler EEST, PVST, and ESTP?

(T-3) Complete statistical mechanics for ESTMC and ESTP: As established in Chapters 6, 8, and 9, the large-scale mean predictions of both the EEST and PVST can be regarded as derived from an empirical statistical theory; however, they also can be derived more rigorously from the complete equilibrium statistical mechanics (Chapter 8 and Section 9.3) of a suitable approximate model for (10.1) in an appropriate limit. Is there a similar complete statistical mechanics theory for ESTMC or ESTP?

(T-4) Theoretical derivation of prior distribution for ESTP: Can the prior distribution for ESTP be derived in a fundamental fashion from theoretical information?

(T-5) Selective decay versus statistical behavior: For the special case of (10.1) with $h = 0$, the rigorous predictions of selective decay from Chapter 3 coincide with the predictions in Chapter 8 of the EEST in the continuum limit without geophysical effects. For purely decaying flows for (10.1), do the PVST, ESTP, and ESTMC make improved statistical predictions beyond selective decay?

Of course, the applied issues (A-1) to (A-5) and the theoretical issues (T-1) to (T-5) are not mutually disjoint and in the next several chapters, we demonstrate how mathematical theory, numerical simulations, and actual observations of physical phenomena interact symbiotically to give insight into all of the basic issues mentioned above. The remainder of this chapter involves a brief review of some of the developments in addressing those issues and also an overview of the material in the next few chapters of this book.

10.3 The central role of equilibrium statistical theories with a judicious prior distribution and a few external constraints

In Chapter 11, we discuss the predictions and comparisons of equilibrium statistical theories, while in Chapter 12 we describe the role of equilibrium statistical theories and the dynamical modeling of suitable flows with forcing and dissipation. The basic theme of these two chapters is to present both theoretical and computational evidence that non-linear statistical theories such as ESTP, PVST, and EEST, which utilize a suitable prior distribution combined with only a few practical external constraints, such as the energy, circulation, and the impulse, are the robust theories for potential applications. As we have shown in Chapter 9, the large-scale mean field predictions of both the PVST and EEST theories can be derived as special cases of the ESTP with two very special choices of the prior distribution. Thus, the evidence accumulated in Chapters 11 and 12 has the theme: *the ESTP with a judicious choice for a prior distribution is the most appropriate statistical formulation for applications.* This claim is strongly supported by the application of ESTP presented in Chapter 13 to the observational record on Jupiter with a physically motivated prior distribution; the theory successfully predicts self-consistently the observations of both the Voyager and Galileo space missions. Thus, the issue in (A-5) is addressed in this specific applied context of Jupiter in Chapter 13. The numerical experiments in Chapter 14 provide further compelling evidence for the use of ESTP. Holloway and his coworkers have emphasized the role of the EEST for explaining and parameterizing certain oceanographic flows with topography (Holloway, 1986, 1992; Holloway and Eby, 1993; Merryfield and Holloway, 1996), while Frederiksen and Sawford (1980) has emphasized the role of EEST for the atmosphere (see Chapter 16). DiBattista and Majda (2000), DiBattista, Majda, and Marshall (2002) have utilized fully the non-linear ESTP with a suitable prior distribution measuring the intensity and duration of surface buoyancy forcing to parameterize the seasonal spread of the cold water mass during deep ocean convection.

In the latter part of Chapter 11, theoretical evidence is presented (Majda and Holen, 1997) for the central role of ESTP theories. This is done by beginning with the more complex ESTMC theory described in Section 9.4 and using elementary rigorous mathematical arguments which prove that *all of the convex non-linear constraints $\mathcal{G}(q)$ from (10.2) with $G''(q) \geq 0$ and $G''(q)$ not identically zero, necessarily are decreased in passing from the microstate prior distribution $\Pi_{q_0}(\lambda)$ in the ESTMC (see Section 9.4) to the macrostate, unless the microstate is extremely special, i.e. statistically sharp, while the robust conserved quantities energy, circulation, and impulse are exactly conserved.* Furthermore, it is shown there that the *mean field macrostates* of the EEST and PVST are statistically sharp, so that the ESTMC has no additional information beyond these statistical predictions. These simple mathematical theorems provide compelling theoretical evidence that ESTP with a *judicious prior distribution* and *a few large-scale constraints* are appropriate theories for applications. This fits nicely with potential practical applications where the information needed for measuring the infinitely many constraints in (10.2) at the microscales is usually not available. Further, quantitative comparison of EEST and ESTMC for flow with topography is presented at the end of Chapter 11. Thus, the material in the second part of Chapter 11 provides substantial insight into the theoretical issues in (T-1), (T-2). With this as background, an extremely important theoretical issue is the one in (T-4): Is there a theoretical derivation of a suitable judicious prior distribution for ESTP under appropriate circumstances? In an important paper, Turkington (1999) has made substantial progress on (T-4). He presents rigorous mathematical arguments supporting the intuitive fact that if only the extrema of the small-scale fluctuations of potential vorticity are known, the suitable small-scale prior distribution for ESTP is

$$\Pi_0(\vec{x}, \lambda) = \frac{1}{2Q} I_{\overline{Q}-Q}^{\overline{Q}+Q}(\lambda), \tag{10.7}$$

where I_a^b represents the indicator function over the interval $[a, b]$, i.e.

$$I_a^b(\lambda) = \begin{cases} 1 & a \leq \lambda < b, \\ 0 & \text{otherwise} \end{cases}$$

and $\overline{Q}$ and Q are the interval midpoint and half-width respectively, i.e.

$$\overline{Q} = \frac{1}{2}(Q_+ + Q_-), \quad Q = \frac{1}{2}(Q_+ - Q_-), \tag{10.8}$$

(see Section 9.2 above). Turkington also argues persuasively through mathematical theory that for a suitable dilute mixture of positive and negative potential vorticity, the natural prior probability distribution $\Pi_0(\lambda)$ for ESTP is given by

$$\Pi_0(\lambda) = \frac{\alpha}{2} e^{-\alpha|\lambda|}. \tag{10.9}$$

The prior distribution in (10.9) is associated with conservation of the total circulation magnitude, i.e.

$$A = \fint |q|. \tag{10.10}$$

The context for Turkington's theoretical arguments will be presented later in this chapter in our brief discussion below of the issue in (T-3).

With the explicit theoretically motivated prior distributions in (10.7) and (10.9), an important practical test is the nature and range of the geophysical phenomena predicted by the ESTP with these explicit prior distributions. At the beginning of Chapter 11, we show that these equilibrium statistical theories give genuine insight into the practical issues in (A-1), (A-2), (A-3) by demonstrating that the mean field predictions of ESTP with these prior distributions in a channel geometry have many of the qualitative and quantitative features required in (A-1), (A-2), (A-3).

10.4 The role of forcing and dissipation

As in (10.1), virtually all practical geophysical flows are subject to both forcing and dissipation. Both the nature of the forcing and the dissipation can vary from application to application. As in Chapter 2, there can be large-scale forcing of the atmosphere through radiative heating, or the ocean through surface wind stress, but also intense random small-scale forcing of the earth's atmosphere from convective storms, and the ocean from unresolved baroclinic instability processes on small length scales. As we indicate in Chapter 13, recent observational data suggest that the weather layer of Jupiter is subject to intense small-scale forcing from convective plumes. The main premise of all the equilibrium statistical theories listed in (10.4) is the use of exactly conserved quantities by the idealized dynamics without dissipation and forcing. Thus, the issue in (A-4) is an extremely important one for practical applications. The theoretical issue in (T-5) arises as the idealized special case, where the external forcing, $\mathcal{F}$, in (10.1) vanishes and all non-ideal fluid features arise through viscous damping. Chapter 12 is centered about these two issues. Recall from Chapter 3, that for decaying geophysical flows the ultimate long time behavior is given by the theory of selective decay. We begin by discussing the simpler situation of decaying flows.

The first numerical evidence for the significance of the mean states of equilibrium statistical theories for describing unforced decaying turbulent flows is in the important paper by Montgomery *et al.* (1992). These authors established that for intermediate long time intervals that are much longer than the initial interval of time with rapid decay of enstrophy but much shorter than the selective decay time, the numerical solution was very strongly correlated with the solution of the sinh–Poisson equation (Montgomery and Joyce, 1974) arising as the mean field prediction of the point-vortex statistical theory in periodic geometry. Furthermore, this degree of correlation was always significantly larger than that with the energy–enstrophy theory describing selective decay. In Chapter 12, we study additional decaying numerical simulations in two different ways to gain insight into the basic issues in (A-4) and (T-5). In one setting, explicit solutions of the sinh–Poisson equation for the PVST are utilized to build initial data involving suitable large amplitude random perturbations of these states in basin (swimming pool) geometry; numerical simulations establish the much stronger correlation with the PVST than the selective decay states of EEST, even throughout regimes with large dynamic changes over time in both the energy and circulation (Majda and Holen, 1997). Here we report on results which demonstrate that this decaying behavior can even be predicted a prior by crude closure dynamics (Grote and Majda, 1997). The above results apply to decaying 2-D fluid flows without geophysical effects. A simple study of decaying flows with geophysical effects involving a channel flow with beta-effect is also presented in Chapter 12. Here the initial data are taken as a mean field state arising from ESTP with the prior distribution from (10.7). The new feature is that the decaying solution undergoes a topological transition from a coherent vortex to a zonal shear flow as the solution decays. It is demonstrated in Chapter 12 that the mean field predictions of ESTP, with time-varying energy and circulation, accurately track the flow through this entire topological transition. In a classical study, Bretherton and Haidvogel (1976) have studied selective decay with topography as an alternative explanation for the success of EEST for flows with topography (Holloway, 1986).

The reader might validly question whether the success of the mean field predictions of the equilibrium statistical theories in describing decaying flows over long time intervals is just a direct consequence of the deterministic non-linear stability of these steady states for inviscid dynamics (see Section 4.5) coupled with small dissipation effects on large scales at large Reynolds numbers. Systematic numerical studies of geophysical flows in (10.1) with both forcing and damping can address this issue. For prototype oceanographic flows in a basin with steady large-scale forcing, it has been shown that the predictions of EEST can either succeed or fail dramatically in describing the ultimate steady state (Griffa and Salmon, 1989; Wang and Vallis, 1994), depending on the structure of the forcing. In Chapter 12, we study the solutions of (10.1) with random small-scale forcing and dissipation; as mentioned above, this is another situation of considerable geophysical interest. The numerical results presented there demonstrate significant regimes of strong small-scale random forcing with dissipation, where the flux of energy to large scales is sufficiently weak so that the equilibrium statistical theory quantitatively describes the evolving large-scale coherent state without resolving any of the complex statistical details of the inverse cascade process (DiBattista, Majda, and Grote, 2001). This fact is demonstrated a priori in a dramatic fashion in Chapter 12 through crude closure algorithms for dynamic evolution (Grote and Majda, 1997, 2000), which involve non-linear predictive equations for only two evolving parameters in the mean field state of an equilibrium statistical theory compared with the result of a resolved numerical simulation with $O(10^5)$ to $O(10^6)$ degrees of freedom. This success of the equilibrium theory in describing appropriate regimes of random small-scale forcing provides background physical motivation for applying such theories to Jupiter in Chapter 13, where the most important forcing seems to occur at small scales through random convective plumes.

Finally we complete our overview in this chapter with the following topic.

10.5 Is there a complete statistical mechanics theory for ESTMC and ESTP?

First we mention that there is a remarkable finite-dimensional discretization of the ideal fluid equations in (10.1) without forcing and dissipation, the sine-bracket truncation, which for a given $N \times N$ spatial discretization, automatically conserves both the energy and N-independent discrete analogues of the invariants in (10.2). This discrete truncation is a suitably modified version of the Fourier spectral method in Chapter 8 and is due independently to Zeitlin (1991) and Miller *et al.* (1992). These approximations are a very natural context to study the interesting issues of statistically relevant and irrelevant conserved quantities for geophysical flows. Chapter 14 of this book contains a study of these issues both for the sine-bracket truncated geophysical fluid equations and the truncated Burgers–Hopf equations (see Chapter 7) through a series of careful systematic numerical experiments. While it is an appealing open problem, no one has succeeded in deriving the appropriate continuum statistical theory for these approximations in the limit as $N \to \infty$. One obstacle to overcome in such a derivation is that the standard finite-dimensional Gibbs ensemble with these finite invariants formally diverges and is not renormalizable (see Chapter 7).

However, there are alternative interesting attempts (Miller, 1990; Miller *et al.*, 1992) to derive ESTMC from a formal equilibrium statistical mechanics lattice model, which does not respect the actual dynamics of ideal two-dimensional fluid flow, but nevertheless is a dynamic model with discrete analogues of the many invariants in (10.2). An application of formal statistical mechanics coarse-graining approximations to these lattice models in a mean field limit yields the empirical statistical theory described earlier in Section 9.4 (Miller *et al.*, 1992). Chorin (1996) and Turkington (1999) have made insightful criticisms of the shortcomings of such lattice models in describing the actual properties of ideal fluid flow.

In fact, the criticism of this work by Turkington (1999) naturally leads to the ESTP with the prior distribution singled out in (10.7) and (10.9). The basic criticism of Chorin and Turkington is easy to understand intuitively. The lattice model by definition has a smallest scale Δx; ignore geophysical effects and consider a prior probability distribution for vorticity in periodic geometry at each lattice site taking on only the two values, $-\lambda$ and λ. This is the probability density naturally associated with initial data consisting of two patches of vorticity with values, $-\lambda$ and λ and equal area (Majda and Bertozzi, 2001 Chapter 8). The coarse-graining procedure in Miller *et al.* (1992) involves building a probability measure on a coarse-grained lattice with spacing $Q\Delta x$, $Q \gg 1$ by averaging over all Q^2 terms from the original lattice with spacing Δx centered about the coarse-grained lattice site. For Q large enough, this original probability distribution with only two values is replaced approximately by the uniform probability density in (10.7) with $\overline{Q} = 0$, $Q = \lambda$. This procedure gives the heuristic reasoning for why the coarse graining of the lattice model in Miller *et al.* (1992) yielding the equations in Section 9.4 for a vortex patch prior distribution is not the correct mean field continuum limit and suggests instead the use of ESTP with the prior in (10.7) as the appropriate answer. The basic reason is that two-dimensional idealized fluid flow does not have a smallest scale, while a lattice model always has a smallest scale; if the prior probability distribution from the initial data, $\Pi_{q_0}(\lambda)$, from Section 9.4, is imposed on this smallest scale, detailed information on this prior distribution is necessarily lost on any coarse-grained scale.

Turkington (1999) suggests that the remedy for overcoming this defect in the lattice models is to regard the discrete lattice as providing an "observation window" in a suitably precise sense for the ideal Euler dynamics; in this fashion, the equalities in Section 9.4 associated with the convex conserved quantities in (10.2) become inequalities. He goes on to show that these inequalities are satisfied for the coarse-grained large-scale limit for vortex patch initial data, provided that the ESTP equations with the prior distribution in (10.7) are utilized (see Section 5.1 of Turkington, 1999). He also considers prior distributions for the initial data associated with two equal strengths, equal area vortex patches surrounded by irrotational flow, i.e. the probability density associated with the initial data is given by the sum of these point masses

$$V\delta_{-\lambda} + V\delta_{\lambda} + (1-2V)\delta_0. \tag{10.11}$$

In Section 5.3 of Turkington (1999), he uses the observation window approach in the dilute vortex limit, $V \to 0$, $\lambda \to \infty$ with normalized circulation, $\lambda V = 1$, to show that the coarse-grained large-scale limit satisfies the mean field equation associated with ESTP with the prior distribution in (10.9). In an interesting recent work, Boucher, Ellis, and Turkington (2000) have applied large deviation theory to the lattice model in the coarse-grained continuum limit to rigorously justify the mean field prediction by ESTP, with the prior distribution in (10.7) or (10.9) as the most probable state. Of course, when the ideal 2-D Euler equations are subject to damping and driving, there is no reason to believe a priori that the convex conserved quantities from (10.2) are automatically decreased in such a process.

References

Boucher, C., Ellis, R., and Turkington, B. (2000), Derivation of maximum entropy principles in two-dimensional turbulence via large deviations. *J. Stat. Phys.* **98**(5–6), 1235–1278.

Bretherton, F. B. and Frederiksen, J. (1987), Two dimensional turbulence above topography, *J. Fluid Mech.* **78**, 129–154.

Chorin, A., 1996, Partition functions and equilibrium measures in two dimensional and quasi-three-dimensional turbulence. *Phys. Fluids* **8**(10), 2656–2660.

DiBattista, M. T. and Majda, A. J. (2000), An equilibrium statistical theory for large-scale features of open-ocean convection. *J. Phys. Oceanogr.* **30**(6), 1325–1353.

DiBattista, M. T., Majda, A. J., and Grote, M. J. (2001), Meta-stability of equilibrium statistical structures for prototype geophysical flows with damping and driving. *Phys.* D 151(2–4), 271–304.

DiBattista, M. T., Majda, A. J., and Marshall, J. (2002), A statistical theory for the "patchiness" of open-ocean deep convection: the effect of preconditioning. *J. Phys. Oceanogr.* **32**, 599–626.

DiBattista, M. T., Majda, A. J., and Turkington, B. (1998), Prototype geophysical vortex structures via large-scale statistical theory. *Geophys. Astrophys. Fluid Dynam.* **89**(3–4), 235–283.

Dowling, T. E. (1995), Dynamics of Jovian atmospheres. *Ann. Rev. Fluid Mech.* **27**, 293–334.

Flierl, G. R. (1987), Isolated eddy models in geophysics. *Ann. Rev. Fluid Mech.* **19**, 493–530.

Frederiksen, J. S. and Sawford, B. L. (1980), Statistical dynamics of two-dimensional inviscid flow on a sphere. *J. Atmospheric Sci.* **37**(4), 717–732.

Griffa, A. and Salmon, R. (1989), Wind-driven ocean circulation and equilibrium statistical mechanics. *J. Marine Res.* **47**, 457–92.

Grote, M. J. and Majda, A. J. (1997), Crude closure dynamics through large scale statistical theories. *Phys. Fluids* **9**(11), 3431–3442.

Grote, M. J. and Majda, A. J. (2000), Crude closure for flow with topography through large-scale statistical theory. *Nonlinearity* **13**(3), 569–600.

Holloway, G. (1986), Eddies, waves, circulation and mixing: statistical geofluid mechanics. *Ann. Rev. Fluid Mech.* (18), 91–147.

Holloway, G. (1992), Representing topographic stress for large scale ocean models. *J. Phys. Oceanography* **22**, 1033–1046.

Holloway, G. and Eby, M. (1993), Hybrid Statistical Mechanics–Ocean Circulation for Efficient Large Scale Modelling. *Proc. of Ocean '93* (Victoria, BC, Canada) Vol. 1 (Piscataway, NJ: IEEE Order Dept), p. 323.

Joyce, G. and Montgomery, D. C. (1974), Negative temperature states for the two-dimensional guiding center plasma. *J. Plasma Phys.* **10**, 107–121.

Majda, A. and Bertozzi, A. (2001), *Vorticity and Incompressible Flow*. Cambridge: Cambridge University Press.

Majda, A. J. and Holen, M. (1997), Dissipation, topography, and statistical theories for large-scale coherent structure. *Comm. Pure Appl. Math.* **50**(12), 1183–1234.

Marcus, P. S. (1993), Jupiter's Great Red Spot and other vortices. *Ann. Rev. Astrophys.* **31**, 523–573.

Merryfield, W. J. and Holloway, G. (1997), Topographic stress parameterization in a quasi-geostrophic barotropic model. *J. Fluid Mech.* **341**, 1–18.

Miller, J. (1990), Statistical mechanics of Euler equations in two dimensions. *Phys. Rev. Lett.* **65**, 2137–2140.

Miller, J., Weichman, P. B., and Cross, M. C. (1992), Statistical mechanics, Euler's equation, and Jupiter's Red Spot. *Phys. Rev.* **A 45**, 2328–2359.

Montgomery, D. and Joyce, G. (1974), Statistical mechanics of negative temperature states. *Phys. Fluids* **17**, 1139–1145.

Montgomery, D., Matthaeus, W. H., Martinez, D., and Oughton, S. (1992), Relaxation in two dimensions and sinh–Poisson equation. *Phys. Fluids* **A4**, 3–6.

Rhines, P. (1975), Waves and turbulence on a beta plane. *J. Fluid Mech.* **69**, 417–443.

Turkington, B. (1999), Statistical equilibrium measures and coherent states in two-dimensional turbulence. *Comm. Pure Appl. Math.* **52**(7), 781–809.

Wang, J. and Vallis, G. (1994), Emergence of Fofonoff states in inviscid and viscous ocean circulation models. *J. Marine Res.* **52**, 83–127.

Zeitlin, V. (1991), Finite mode analogs of 2D ideal hydrodynamics: coadjoint orbits and local canonical structure. *Phys. D* **49**(3), 353–362.

11
Predictions and comparison of equilibrium statistical theories

11.1 Introduction

In the previous Chapters 6, 8, and 9, various statistical theories have been introduced to predict large-scale structures for geophysical flows. These statistical theories range from the simple energy–enstrophy theory (EEST) developed in Chapters 6 and 8, point vortex theory (PVST) from Section 9.3, empirical statistical theory with a prior distribution and a few external constraints (ESTP) introduced in Chapter 9 and recalled in Chapter 10, to the elaborate equilibrium statistical theory involving many constraints (ESTMC) developed in Section 9.4. In the presence of such a wide variety of theories, a central question is the applicability of these statistical theories to geophysical flows. As indicated in Chapter 10, the purpose of this chapter is to address several practical as well as theoretical issues pertinent to the potential applicability of various equilibrium statistical theories. In particular we will address three applied issues, (A-1)–(A-3), and two theoretical issues, (T-1)–(T-2) of Chapter 10. We will provide strong evidence, both numerical and analytical, in supporting the central role of equilibrium statistical theories with judicious prior distribution and few external constraints. In doing so, we also provide answers, either numerical or analytical or both, to the questions in (A-1)–(A-3) and (T-1)–(T-2), i.e. the type of geostrophic phenomena that are predicted at large scales by the solutions of the mean field equations (10.6) for various statistical theories (10.4); whether coherent structures predicted by various equilibrium statistical theories correspond to observed geophysical structures; if the transitions from large-scale zonal jets to vortices in a beta-plane channel (Rhines, 1975) can be predicted automatically and quantitatively by an equilibrium statistical theory; whether the predictions of the various equilibrium statistical theories in (10.4) compare and contrast with each other; and how much additional information is actually contained in the predictions of ESTMC at large scales as compared with the simpler EEST, PVST, and ESTP predictions.

The chapter is organized as follows. In the second section we first give a brief review of ESTP with two prior distributions from (10.7) and (10.9) introduced by Turkington (1999). The external constraints are the conservation of energy and circulation in a beta-plane channel with mean flow. The predictions of the statistical theory, which is a set of elliptic partial differential equations, are evaluated/solved using a carefully designed accurate numerical procedure for a wide range of parameter values. Numerical evidence suggests that the most probable states predicted by this theory range from zonal shear flows to monopole vortices and dipole vortices as the geophysical parameter, β, varies as well as the mean flow, V. We also establish that various coherent structures can maximize the entropy and be the most probable large-scale structures when consistent symmetries in the flow field are imposed, while other coherent structures actually maximize the empirical entropy when such symmetries are broken. The predictions of the statistical theory with and without geophysical effects are all considered. This partially addresses the applied issues (A-1) and (A-2) from Chapter 10. An interesting feature of the numerical results is its connection with the Rhines scale. Given the energy density, $E' = E/HL$, where H is the width and L is the length of the channel, and the geophysical beta-effect, β, Rhines (1975, 1979) introduced the scale, $L_R = E'^{\frac{1}{4}}/\beta^{\frac{1}{2}}$, as a fundamental length scale and predicted that zonal jets dominate for scales $L \gtrsim L_R$, while vortices dominate for scales $L \lesssim L_R$. Statistical theory yields a prediction for large-scale coherent structure consistent with predictions from the Rhines' scale (see Subsection 11.2.3). This addresses the applied issue (A-3) from Chapter 10. Most of the material in this section is due to DiBattista, Majda, and Turkington (1998). In the third section we consider the problem of statistical sharpness of statistical theories with a few constraints (such as EEST and PVST) in the more elaborate theory involving infinitely many constraints (ESTMC). We discover that the classical few-constraint statistical theories, involving energy–enstrophy principles (EEST) or point vortices (PVST), are statistically sharp with an infinite number of constraints (ESTMC). In other words, at the macrostates of the few-constraint theories, the many constraint theories provide no additional statistical information. This is established through a general link between statistical theories, generalized selective decay principles (see Section 4.5), and statistical sharpness. This provides an answer to the theoretical issue (T-2) from Chapter 10. In the fourth section we consider an asymptotic procedure with small potential vorticity. We show that the many constraint theories of flows (ESTMC) yield the simpler energy–enstrophy (EEST) macroscopic states at leading order with systematic higher-order corrections, involving a renormalized topography that includes higher moments of the microscopic potential vorticity distribution. This offers an answer to the theoretical issue (T-1) from

Chapter 10. The rigorous results in Sections 3 and 4 are mainly due to Majda and Holen (1997).

In Chapter 13, we will provide further information regarding the applied issue (A-2), where the first quantitatively accurate prediction of the Great Red Spot and White Ovals on Jupiter will be presented utilizing ESTP. Issues involving forcing and dissipation will be addressed in Chapter 12.

11.2 Predictions of the statistical theory with a judicious prior and a few external constraints for beta-plane channel flow

11.2.1 Statistical theory

Here we consider ideal barotropic flows on a beta-plane channel in the absence of topographic effect ($h \equiv 0$). In order to talk about Rhines' scale, here we consider a channel which is periodic in x with period L and extent $-H/2 < y < H/2$, although the numerical simulations are performed in the channel geometry with lengths, $L = 2\pi$, $\quad H = 2\pi$. The dynamics of the stream function or potential vorticity are governed by the usual barotropic quasi-geostrophic equations without damping and external forcing

$$
\begin{gathered}
\frac{\partial q}{\partial t} + \nabla^{\perp}\Psi \cdot \nabla q = 0, \\
q = \Delta\Psi + \beta y, \quad \Psi = -Vy + \psi
\end{gathered}
\tag{11.1}
$$

together with the boundary conditions that all quantities are periodic in x with period L and the small-scale stream function ψ vanishes at the walls ($y = \pm H/2$) of the channel. See Section 1.4 for more details. Notice here we use ψ and Ψ to denote the small-scale and total stream functions respectively, while we used ψ' and ψ to denote the small-scale and total stream functions respectively in previous chapters. The incompressible velocity is defined as

$$
\vec{v} = \nabla^{\perp}\Psi = \begin{pmatrix} V \\ 0 \end{pmatrix} + \nabla^{\perp}\psi. \tag{11.2}
$$

All of the solutions presented here involve a channel with unit aspect ratio, except for our discussion of the Rhines scale in Subsection 11.2.3.

As established in Section 1.4, the dynamic equations (11.1) conserve the total kinetic energy, total circulation, and impulse

$$
E = \frac{1}{2}\fint |\vec{v}|^2 d\vec{x}, \quad \Gamma = \fint q d\vec{x}, \quad I = \fint y\, q\, d\vec{x} \tag{11.3}
$$

as well as the infinite number of integrals (the generalized enstrophies) involving arbitrary functions of potential vorticity

$$\mathcal{G}(q) = \int G(q)d\vec{x} \tag{11.4}$$

for any function G. Included in this infinite list are two that take on special importance in the statistical theory utilized below: the extrema of q

$$Q_+ = \max q, \quad Q_- = \min q \tag{11.5}$$

and the mean absolute potential vorticity

$$A = \fint |q| d\vec{x}. \tag{11.6}$$

See (10.7) and (10.9) and the discussion in Section 10.5 for the importance of (11.5) and (11.6).

As was discussed in Subsection 9.2.2, we utilize the following modified "energy" and circulation as conserved quantities in our statistical theory below

$$\tilde{E} = \fint \left(-\frac{1}{2}\psi + Vy\right)(q - \beta y), \quad \Gamma = \fint q\, d\vec{x} \tag{11.7}$$

instead of the three conserved quantities given in (11.3). Such a choice of conserved quantities ensures that the large-scale mean field predicted by the statistical theory be exact steady state solutions to the barotropic quasi-geostrophic equations (11.1). For such a choice of modified "energy", the impulse is no longer a free parameter, and thus we may consider the correlation between the impulse and the large-scale mean for the most probable mean states below.

Here we consider the ESTP derived in Section 9.2 in empirical fashion with a given prior distribution, $\Pi_0(\lambda)$, and the constraints in (11.7). We utilize the two choices of prior distribution in Turkington (1999) from (10.7) and (10.9) with a derivation discussed briefly in Section 10.5. Thus, considering only the potential vorticity extrema Q_+ and Q_-, we use a prior distribution of the form

$$\Pi_0(\lambda) = \frac{1}{2Q} I_{\overline{Q}-Q}^{\overline{Q}+Q}(\lambda), \tag{11.8}$$

in which I_b^a represents the indicator function over the interval $[a, b]$, and $\overline{Q}$ and Q are the interval midpoint and half-width respectively, i.e.

$$\overline{Q} = \frac{1}{2}(Q_+ + Q_-), \quad Q = \frac{1}{2}(Q_+ - Q_-). \tag{11.9}$$

As mentioned already in Section 9.2, this prior distribution leads to a *Langevin equation* for the mean field potential vorticity.

Increasing the complexity only slightly, we consider, in addition to the potential vorticity extrema in (11.5), the constraint on the mean absolute potential vorticity in (11.6) in the *dilute PV limit*. See the discussion above equation (10.11) and Turkington (1999). This added constraint on the potential vorticity provides additional information on the fine-scale mixing, which is encoded in the prior distribution

$$\Pi_0(\lambda) = \frac{\alpha}{2} e^{-\alpha|\lambda|}, \tag{11.10}$$

in which α is a parameter with a value which ensures that the constraint on the mean absolute potential vorticity, A, in (11.6) is satisfied.

For the explicit prior distributions in (11.8) and (11.10) the mean field equation for the ESTP from Section 9.2 takes the explicit forms:

- Langevin theory

$$\overline{q} = \overline{Q} + Q\left(\coth Q\Theta - \frac{1}{Q\Theta}\right) \equiv \overline{Q} + QL[Q\Theta] \tag{11.11}$$

- Dilute PV theory

$$\overline{q} = \frac{2\Theta}{\alpha^2 - \Theta^2}, \tag{11.12}$$

where $\Theta = \theta(\overline{\psi} - Vy) - \gamma$, and $L[x] \equiv \coth[x] - 1/x$ is the Langevin function. We will describe later in this subsection an accurate iterative algorithm that simultaneously determines the Lagrange multipliers, θ and γ, and solves the mean field equations for the coarse-grained (mean field) potential vorticity field, $\overline{q}$.

As we remarked in Section 9.2, solutions of the mean field equations in (11.11) and (11.12) are exact steady solutions of the barotropic flow equations on a beta-plane in channel geometry. These solutions are traveling waves, moving with velocity V. These exact solutions are the large-scale coherent structures predicted by the statistical theory, given the imposed values of the conserved quantities.

Finally, we list the independent parameters that determine solutions in the statistical theories. Naturally, this must include the large-scale constraints for the "energy," $\tilde{E}$, and circulation, Γ, the eastward mean flow, $V > 0$, and geophysical beta-parameter. We have also introduced additional parameters $\overline{Q}$ and Q in (11.9), and α, whose values characterize the shape of the various prior distributions (11.8) and (11.10). Of these additional parameters, however, it is always possible to eliminate one due to a scaling invariance in the potential vorticity field. Thus, in the Langevin theory we set the vorticity range half-width $Q = 1$, and vary five parameters $\tilde{E}, \Gamma, V, \beta$, and $\overline{Q}$. In the dilute PV theory we set $\alpha = 2$ and vary the four parameters $\tilde{E}, \Gamma, V$, and β (here, $\overline{Q}$ is zero and Q is infinite). We shall see that various coherent structures of geophysical interest maximize entropy in distinct ranges of these parameters.

Numerical techniques

To compute the maximum entropy solutions described above, we adapt a rapid and accurate algorithm developed by Turkington and Whitaker (1996). The main idea is to iterate over a convergent set of simpler problems by replacing the "energy" constraint, $\mathcal{C}^{(1)}$, with a series of linearized inequalities. In the limit, we recover the original "energy" equality. Thus, in the kth step of the algorithm the mean field potential vorticity, $\overline{q}^k$, is defined by the revised constraints, $\mathcal{C}^{(1')}$ and $\mathcal{C}^{(2')}$ (we omit the probability constraint corresponding to $\mathcal{C}^{(3)}$ for simplicity)

$$\mathcal{C}^{(1')} = \Big\{\rho^k \mid \tilde{E}(\rho^k) = \fint \left(-\frac{1}{2}\overline{\psi}^{k-1} + Vy\right)(\overline{q}^k - \beta y)\, d\vec{x}$$

$$= \tilde{E} + \tilde{E}^{k-1} - VI^{k-1} > \tilde{E}\Big\}$$

$$\mathcal{C}^{(2')} = \left\{\rho^k \mid \Gamma(\rho^k) = \fint \overline{q}^k\, d\vec{x} = \Gamma\right\}, \tag{11.13}$$

in which the "energy" integral $\mathcal{C}^{(1')}$ is linearized about the stream function $\overline{\psi}^{k-1}$, which is calculated in the previous step.

This problem is solved by a standard two-part process, which simultaneously constructs the mean field potential vorticity, $\overline{q}$, and determines the unknown Lagrange multipliers, θ and γ. First, upon completion of the $(k-1)$th step, the stream function, $\overline{\psi}^{k-1}$, "energy", $\tilde{E}^{k-1}$, and impulse, I^{k-1}, are known. From these quantities and the appropriate mean field equation (11.11) and (11.12), we form a matrix equation from the linear constraints, $\mathcal{C}^{(1')}$ and $\mathcal{C}^{(2')}$, whose solution, which is calculated by Newton's method, determines the updated values of the Lagrange multipliers θ^k and γ^k. Then, using the new Lagrange multipliers, we calculate the updated vorticity field $\overline{q}^k$. Finally, by inverting the Laplacian operator, we calculate the stream function, $\overline{\psi}^k$, the "energy," $\tilde{E}^k$, and the impulse, I^k. We repeat this process until the difference between the calculated and target "energies" is less than 10^{-8}.

Turkington and Whitaker (1996) show that the potential vorticity field $\overline{q}^k$ approaches the true entropy maximizer as the "energy" difference, $\tilde{E} - \tilde{E}^k$, converges to zero; more details of both the algorithm and its properties may be found in their paper.

Role of symmetries

Solutions to the mean field equations (11.11), and (11.12) are frequently symmetric with respect to the channel domain. Typically, solutions exhibiting different symmetries lie upon distinct branches that emerge from a common bifurcation point. We can select solutions exhibiting particular symmetries by introducing simple restrictions in our algorithm. This allows us to calculate classes of solutions that may not maximize entropy overall, but do so within a certain symmetry class. Furthermore, in a damped and driven flow we may observe these classes of solutions when either the domain or the forcing respects an underlying symmetry (see Chapter 10).

There are four types of symmetry groups that the potential vorticity, $q(x, y)$, typically possess:

$$\text{streamwise translational symmetry:} \quad \overline{q}(x, y) = \overline{q}(y) \tag{11.14}$$

$$\text{spanwise symmetry:} \quad \overline{q}(x, y) = \overline{q}(x, -y) \tag{11.15}$$

$$\text{spanwise antisymmetry:} \quad \overline{q}(x, y) = -\overline{q}(x, -y) \tag{11.16}$$

$$\text{diagonal antisymmetry:} \quad \overline{q}(x, y) = -\overline{q}(-x, -y). \tag{11.17}$$

Here, the origin of the coordinate system is placed in the center of the periodic channel. We take care in the subsequent discussion to present the different classes of solutions that solve the mean field equations (11.11) and (11.12), and the symmetry restrictions under which they maximize entropy.

11.2.2 Coherent geophysical vortices (monopoles and dipoles)

Basic solutions with no geophysical effects ($V = \beta = 0$)

We now present the prototype large-scale structures that emerge as the most probable states in the Langevin and dilute PV statistical theories according to the numerical results reported by DiBattista, Majda, and Turkington (1998).

The numerical procedure of constructing the most probable states involves solving the mean field equations (11.11) or (11.12), by the iterative algorithm described in the previous subsection, for a given set of parameter values. In this section we concentrate on the basic structures that satisfy these equations with vanishing mean flow and beta-parameter, i.e. with $V = \beta = 0$. In this case, the "energy", $\tilde{E}$, defined in (11.7) coincides with the total kinetic energy, E.

In fact, these basic solutions span the complete range of structures that arise within these statistical theories. Langevin theory produces shear flows and, at sufficiently high values of energy, E, circulation, Γ, and $\overline{Q}$, monopole vortices as the most probable states. The solutions that arise from dilute PV theory, however, span a larger range that encompasses shear flows, monopole vortices, vortex streets, and vortex dipoles, each of which appears spontaneously at sufficiently high energies.

Langevin theory

Recall that solutions to the Langevin mean field equation with $\overline{Q} = 0$ are determined solely by their energy, and circulation, Γ.

For a fixed non-zero value of E, circulation, the entropy-maximizing mean field state is either a spanwise symmetric shear flow at low energies or a monopolar vortex at high energies.

The monopole vortex, bifurcates from the spanwise-symmetric shear flows with positive circulation at high energies, provided that $\overline{Q} \approx Q$ is not zero and the potential vorticity is nearly all single signed. A plot of the energies and entropies of solutions in which the circulation $\Gamma = 1.0$ and $\overline{Q} = Q = 1.0$ is shown in Figure 11.1. The bifurcation point in this energy–entropy diagram is near the value, $E = 0.071$, and clearly shows the emergence of the coherent vortex from the underlying shear flow. Monopole vortices are the most probable states at higher energies, whereas shear flows, which exist throughout the entire range of energies, are maximum entropy solutions only at lower energies. In Figure 11.2 we show an example of a monopole solution for the energy, $E = 0.115$, and the circulation, Γ, and $\overline{Q}$ given above. This vortex therefore lies near the end of the upper branch shown in Figure 11.1, with an energy near the maximum possible value. The shape is near a top-hat, with a central patch of nearly uniform potential vorticity surrounded by a near-neutral fluid. The region of parameter space in which the monopole vortices exist is a complex function of E, Γ, and $\overline{Q}$. Indeed, for even higher values of circulation than shown in the entropy–energy diagram in Figure 11.1, the very high energy monopole vortices elongate and may even revert back to shear flows, being no longer able to support their compact shape at large aspect ratios.

This completes the survey of structures that maximize entropy in the Langevin theory for vanishing mean flow and beta-parameter, a set which comprises only shear flows and monopole vortices. As a final note, however, we point out that the low energy solutions in Figure 11.1 have positive slope and thus positive temperature $1/\theta$; the potential vorticity of solutions in this regime is largely accumulated near the sides of the channel.

Dilute PV Limit

We turn now to the basic solutions in the dilute PV theory, which span a much richer assortment of coherent structures. Here we consider the situation in which the mean flow, V, the beta-parameter, and the circulation, Γ, all vanish. Thus the "energy" defined in (11.7) is the total kinetic energy. At low energies the entropy-maximizing solutions are shear flows, as they are in the Langevin theory. However, at higher energies coherent structures such as monopole vortices, vortex streets, and vortex pairs systematically emerge from the underlying shear flow solutions.

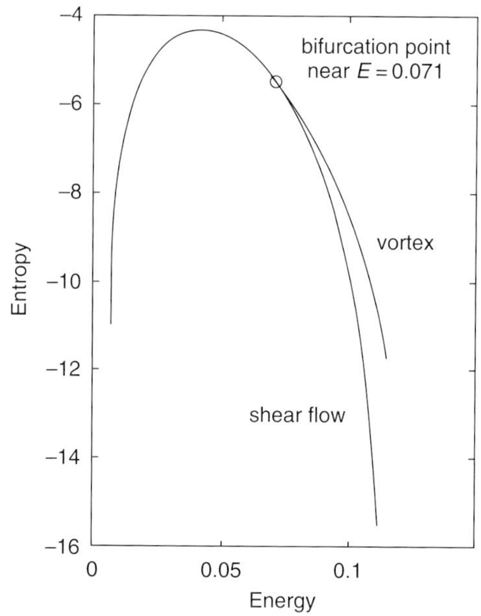

Figure 11.1 Entropies and energies of maximum entropy solutions in Langevin theory with circulation $\Gamma = 1.0$, and $\overline{Q} = Q = 1.0$, $V = \beta = 0$.

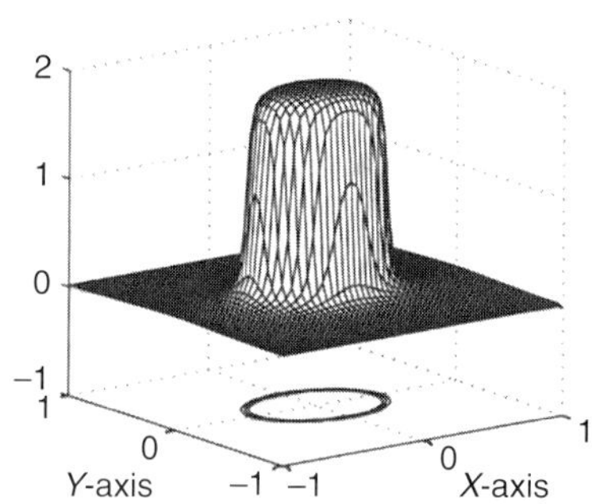

Figure 11.2 Monopolar vortex maximum entropy solution in Langevin theory with energy $E = 0.115$, circulation $\Gamma = 1.0$, and $\overline{Q} = Q = 1.0$, $V = \beta = 0$.

All solutions, both shear flows and vortex structures, are shown in the entropy–energy diagram Figure 11.3. The long, dotted curve sweeping the length of the graph indicates the position of the shear flow solution; notice that for low energies these solutions are the entropy maximizers. There are two bifurcation points along this curve: the first, near the point $E = 0.032$, indicates the energy at which two higher-entropy branches simultaneously split from the shear flow solutions. Monopole vortices lie on the upper branch, while vortex streets lie on the lower branch. The second bifurcation point, near the point $E = 0.077$, marks

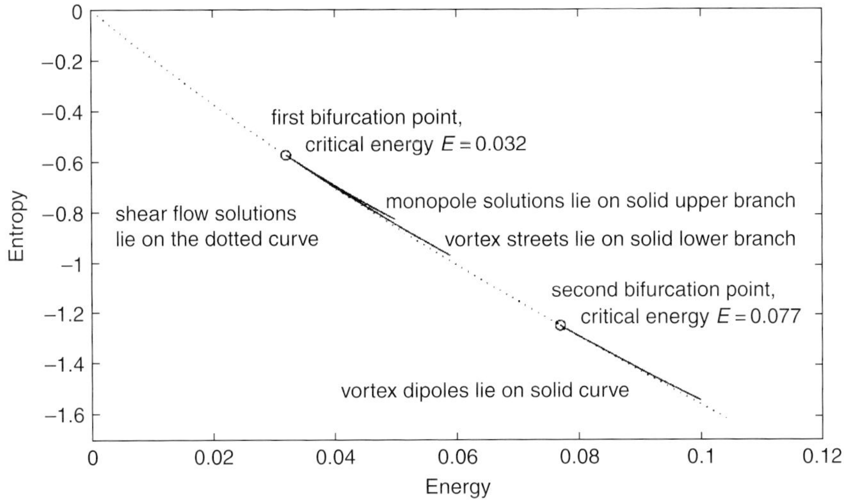

Figure 11.3 Entropies and energies of maximum entropy solutions in dilute PV theory, with $\alpha = 2$, $V = \beta = \Gamma = 0$.

the energy above which vortex pairs exist. Whereas in the Langevin theory vortex structures appear as most probable states only for cases in which the vorticity is predominantly single signed, the vortex structures in the dilute PV limit can appear in a fluid comprised of both positive and negative vorticity.

Each of these classes of solutions exhibit one of the symmetries discussed in the previous subsection; the monopole vortices are spanwise symmetric, as in (11.15), the vortex dipoles spanwise antisymmetric, as in (11.16), and the vortex streets diagonal antisymmetric, as in (11.17). Each of the branches shown in Figure 11.3, therefore, represent the maximum entropy solutions for periodic channel flow with vanishing mean flow and beta-parameter under the appropriate symmetry restrictions.

In Figures 11.4(a), (b) we show examples of the solutions with vortex structures represented in the entropy–energy diagram of Figure 11.3.

The vortex structures, two of which are monopole vortices and two of which are vortex streets, lie on branches originating from the first bifurcation point near $E = 0.032$ and are shown in Figure 11.4(a). The vortex structures do *not* all share the same entropy; the monopoles, which lie on the left half of the diagram, possess higher entropy than the vortex streets, which lie on the right half. At higher energies these vortex solutions grow without bound, sharpening into tall, narrow spikes as the surrounding vorticity field flattens to a constant level. At some point the resolution of our discrete mesh is insufficient to capture these peaked solutions; these values are the points at which the solution curves terminate in the entropy–energy diagram of Figure 11.3.

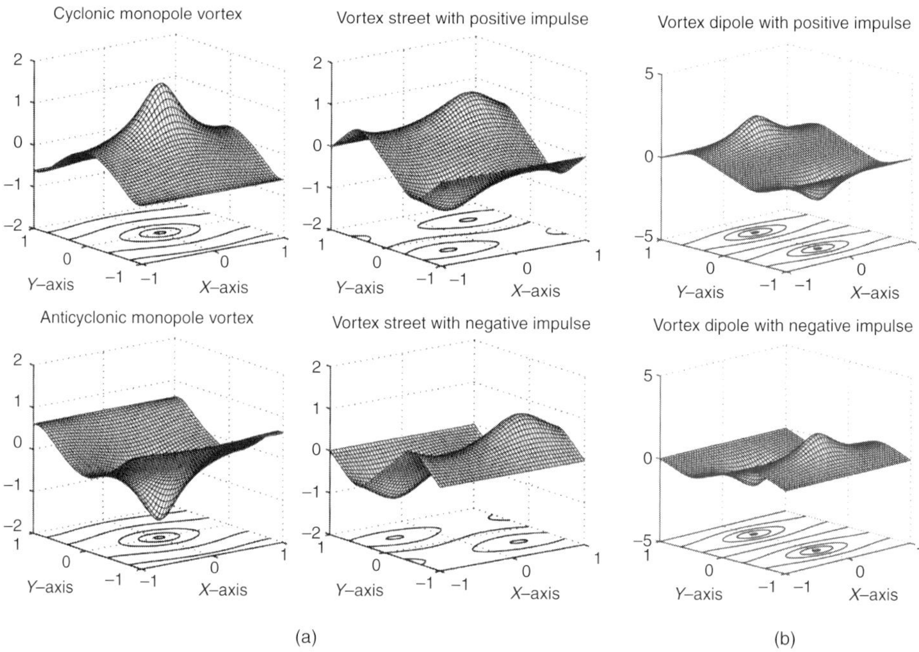

Figure 11.4 Monopoles, vortex streets, and dipoles predicted by dilute PV. (a) Plot of the potential vorticity with energy near the bifurcation point $E = 0.032$. Left column: monopole vortices on the upper branch with higher entropy; right column: vortex streets on the lower branch with lower entropy; (b) plot of the potential vorticity with energy near the second bifurcation point $E = 0.077$. Vortex dipoles.

The high-energy vortex structures, which are vortex dipoles, that lie on the branch originating from the second bifurcation point near $E = 0.077$ are shown in Figure 11.4(b). These solutions spring from high-energy positive-impulse shear flows (positive-impulse vortex pair) and negative-impulse shear flows (negative-impulse vortex pair). No similar structure appears to bifurcate from the zero-impulse symmetric shear flows at this point. Note that although these vortex dipoles are paired configurations they do not travel, since the mean flow speed is zero. The pair therefore is strongly affected by the boundaries and each of the vortices appears midway between the edge and centerline of the channel. Like the monopoles and vortex streets above, the vortices in the pair grow narrow and more peaked at higher energies.

Geophysical solutions ($V \neq 0$, $\beta \neq 0$)

We now consider the case with geophysical effects. We observe that, in both the Langevin and dilute PV theories, no additional structures appear to maximize entropy in the geophysical regime beyond those presented in the previous subsection. However, in both statistical theories an eastward mean flow raises the entropy of the positive-impulse solutions and lowers the entropy of the those with negative impulse (these are equal without mean flow), while a positive beta-parameter causes the reverse effect. Systematic changes in these parameters yields systematic alterations of the shapes of the mean field potential vorticity surfaces and movement of the bifurcation points in the enlarged parameter space. The predicted coherent structures in the dilute PV theory are typically either dipolar-vortex streets or shear flows, depending on the mean flow, beta-parameter, and "energy".

For more detailed discussions beyond that given below, we refer to the original article of DiBattista, Majda, and Turkington (1998).

Langevin theory

Here we systematically increase the mean flow while keeping β and the "energy" fixed. We consider a regime like the one which produced the monopole vortex in Figure 11.2 and systematically increase the mean flow from zero to positive values; as mentioned above, we will observe an increasing impulse in the mean field state as V increases. Figure 11.5 represents a series of six plots of the potential vorticity for the maximum entropy mean field state. The "energy" and circulation are fixed at the values $\tilde{E} = 0.03$ and $\Gamma = 1.0$ respectively, with $\beta = 0.3$ and a variable mean flow ranging from $V = 0.0$ to $V = 0.12$. Also, here $\overline{Q} = 0.9$ and $\underline{Q} = 1.0$ so that nearly all the potential vorticity is positive. Of these six solutions, two are shear flows with a radically different structure, while four are monopole vortices; clearly, the point at which the vortex emerges from the shear flows is a complex function of all these parameters. The shear flow solutions shown in the upper left corner of Figure 11.5 has positive beta-parameter but no mean flow and forms a single ridge with negative impulse located in the lower half of the channel. As the mean flow increases, the ridge moves northward and a coherent vortex forms. Near the critical ratio $\beta/V \sim 10$, the vortex is roughly centered in the channel and crosses into the upper half of the channel as V is increased further. At larger values of V, the vortex approaches the channel edge and instead becomes a strong shear flow with positive impulse, hugging the northern boundary.

Dilute PV theory

We have shown earlier in this section that the dilute potential vortex theory (see Figure 11.4(a)) can have dipole vortices as mean field statistical predictions when there are no geophysical effects or mean flow. In this situation, due to symmetry, both positive and negative impulse solutions are the most probable predicted state. With the geophysical effect of β included and an eastward mean flow, $V > 0$,

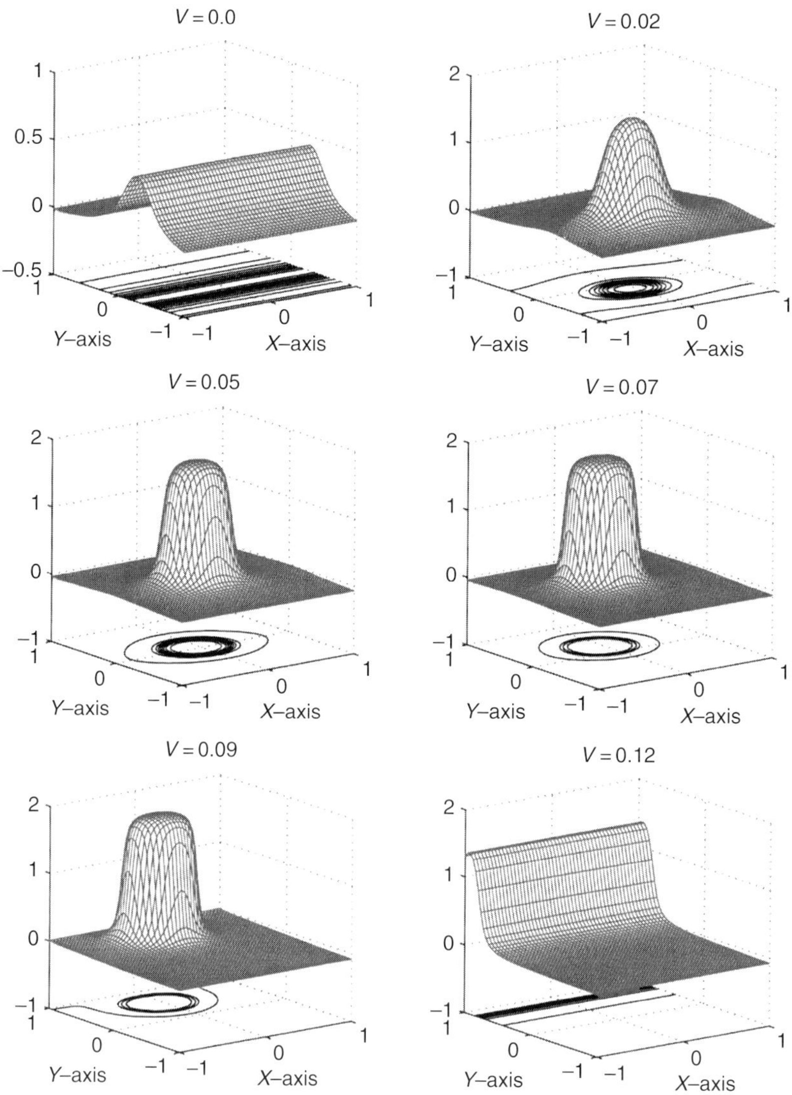

Figure 11.5 Plot of potential vorticity of maximum entropy solutions in Langevin theory, with "energy" $\tilde{E} = 0.03$, circulation $\Gamma = 1.0$, $\beta = 0.3$, $\overline{Q} = 0.9$ and $Q = 1.0$, a variable mean flow ranging from $V = 0.0$ to $V = 0.12$.

it is interesting to obtain most probable mean field states with dipole structure with negative impulse so that the anti-cyclones (high pressure region) sit above the cyclones (low pressure region) and produce simple prototype models for atmospheric blocking (Flierl, 1987). In DiBattista *et al.* (1998), it is established that there is a transition ratio $\beta/V = 10$ so that the most probable states of the dilute potential vorticity theory have positive impulse (cyclones above anti-cyclones) for

$\beta/V \leq 10$ and negative impulse (anti-cyclones above cyclones and thus, prototype blocking patterns) for $\beta/V \geq 10$. Next, we illustrate this behavior below.

First, we demonstrate the positive impulse solutions with $\beta/V \leq 10$. We fix $V = 0.06$ and $\beta = 0.3$ so that $\beta/V = 5$ and compute the mean field statistical states as the "energy" varies from $\tilde{E} = 0.11$ to $\tilde{E} = 0.14$ in equal increments. The corresponding potential vorticity plots are shown in Figure 11.6 with the clear emergence of a vortex dipole pair with positive impulse as the "energy" increases. On the other hand, we decrease the mean flow V to $V = 0.02$ and keep $\beta = 0.3$ so that $\beta/V = 15$ and we vary the "energy" in equal increments from $\tilde{E} = 0.05$ to $\tilde{E} = 0.08$. Potential vorticity plots of the most probable states are presented in Figure 11.7 and show the emergence of dipole pairs with negative impulse as the "energy" increases. These are prototype omega blocking patterns produced by the statistical theory. More conventional symmetric blocking patterns with lower entropy and negative impulse like those in Figure 11.4(b) can also be produced from the statistical theory by imposing additional symmetries in the solution as we have discussed earlier (see DiBattista *et al.*, 1998).

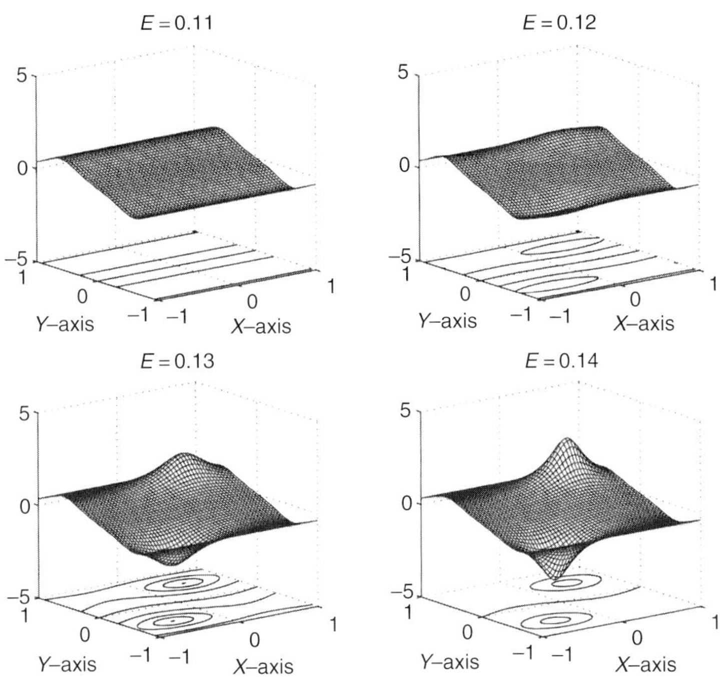

Figure 11.6 Plot of potential vorticity of maximum entropy solutions in dilute PV theory, with $\alpha = 2$, $V = 0.06$, $\beta = 0.3 (\beta/V = 5)$. Observe the emergence of a vortex dipole pair with positive impulse as the "energy" increases.

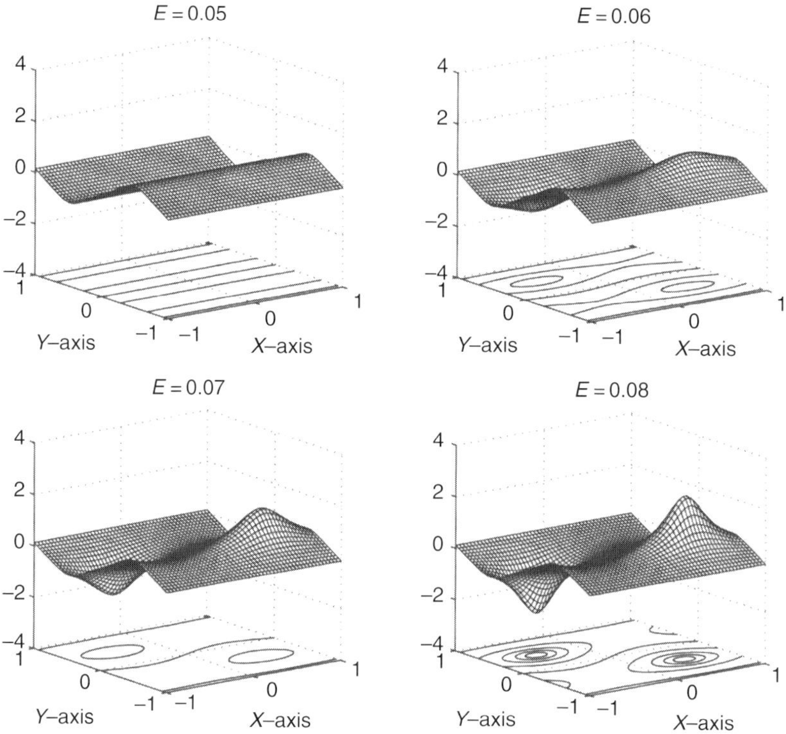

Figure 11.7 Plot of potential vorticity of maximum entropy solutions in dilute PV theory, with $\alpha = 2$, $V = 0.02$, $\beta = 0.3(\beta/V = 15)$. Observe the emergence of a vortex dipole pairs with negative impulse as the "energy" increases.

11.2.3 Statistical predictions of generalized Rhines' scale

Recall that the energy density, E', has the units $[E'] = \frac{L^2}{T^2}$, while β has the units, $[\beta] = L^{-1}T^{-1}$. Thus, it is apparent that the only length scale that we could derive out of the energy density, E', in the large-scale flow and the geophysical beta-effect, β, is

$$L_R = \frac{E'^{\frac{1}{4}}}{\beta^{\frac{1}{2}}}, \tag{11.18}$$

which was first introduced by Rhines (1975, 1979) as one of the fundamental scales for geophysical flows on a beta-plane. He also argued that zonal jets dominate for scales $L \gtrsim L_R$ while vortices dominate for scales $L \lesssim L_R$. Rhines applied this length scale to simulations of decaying two-dimensional turbulence, showing that the transition between a vortex-dominated flow and a wavelike propagation region is well-predicted by this measure. We show in this subsection that a similar universal length scale predicts the point at which vortices, both

Langevin monopoles and dilute PV vortex streets, emerge as maximum entropy states of a statistical theory from the underlying shear flows. Thus, the equilibrium statistical theory automatically predicts a generalized Rhines' scale in the nature of the coherent structure which emerges. This addresses applied issue (A-3) discussed in Section 10.2.

We simplify matters by setting the eastward mean flow and circulation to zero and solve the mean field equations for many values of energy and β. The transition curve in the (E', β)-plane between shear flow and vortex solutions then fits with remarkable accuracy by a parabola of the form

$$E' = (L_R^*)^4 \beta^2 + b\beta + c, \tag{11.19}$$

in which L_R^*, the fourth root of the curvature, is the generalization of the familiar Rhines' scale according to (11.18). In each of the cases we consider, the transition curve is essentially indistinguishable from the best-fit parabola.

It is an interesting question as to which of the two physical length scales in our domain, either the streamwise periodic length, L, or the spanwise channel width, H, determines the generalized Rhines' scale. In fact, we find that the spanwise length is the only relevant length scale here and that L_R^* is in direct proportion to H, for $H \geq L$.

We can best show this by using non-dimensional units where the length scale is fixed by the channel width so that $H = 1$. A change in channel aspect ratio, $AR = H/L = 1/L$, causes no change in the energy density, $E' = E/(HL) = E/L$, of the most probable shear flow solution, since it is translationally invariant in the streamwise direction. This provides a natural basis of comparison for the emergence of vortices in channels with different aspect ratios, and, as we shall see later in this subsection, yields a constant value for the generalize Rhines' length, L_R^*, throughout the range of L that we consider. For a channel of arbitrary width, it is a simple exercise with the above non-dimensionalization to show that the generalized Rhines' scale is directly proportional to the channel width, H, in the problem.

Vortex streets in the dilute PV limit

In the previous subsection we determined the positions in the (E, β)-plane, where vortex streets emerged from the underlying shear flows. We found that, for each value of β, there is a critical energy with coherent vortex streets as the most probable states for larger energies and shear flows for lower energies. In this subsection we consider the precise shape of these transition curves for a wider range of energies and beta-parameters on periodic channels with varying aspect ratio.

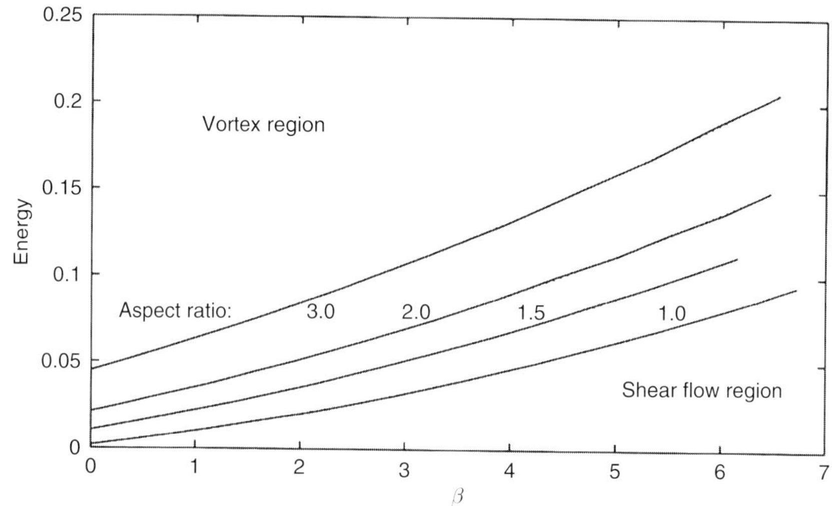

Figure 11.8 Plot of transition curves from shear flow region to vortex region predicted by the dilute PV theory in the (β, E')-plane for the aspect ratios equal 1.0, 1.5, 2.0, and 3.0, with $\alpha = 2$, $V = \Gamma = 0$.

In Figure 11.8 we show four such transition curves in the (β, E')-plane for the aspect ratios $AR = 1.0, 1.5, 2.0$, and 3.0. For simplicity the mean flow, V, and circulation, Γ, vanish. For each of these cases the vortex street solutions exist, and maximize the entropy, in the regions that lie above and to the right of the solid curve. The shear flow solutions exist in all regions of the diagram; however, they are the maximum entropy solutions only in the regions that lie below and to the left of the transition curve. This implies that, for a fixed value of energy, increasing the beta-parameter destroys the vortices, which are re-absorbed into a background shear flow.

The best-fit parabolas, which take the form $E' = (L_R^*)^{\frac{1}{4}}\beta^2 + b\beta + c$, are plotted in dotted lines. The fit is essentially exact and it is difficult to distinguish between the dotted- and solid-line curves. According to our discussion at the beginning of the section, the generalized Rhines' scale is determined from the curvature of these parabolas. Indeed, accounting for the variation in the curvature of the parabolas across the range of aspect ratios considered, we derive a single value for L_R^* to within a 2% error

$$L_R^* = a^{\frac{1}{4}} = 0.180 \pm 0.002, \tag{11.20}$$

which suggests that the empirical length scale is universal with respect to the appearance of vortex streets in the channel. For a channel of arbitrary size with $H \geq L$, we therefore have $L_R^* = (0.180 \pm 0.002)H$. Thus, we have established a quantitative link between the empirical Rhines' scale and the transition between shear flows and coherent vortices predicted by the statistical theory.

Langevin monopole

In previous subsections we showed that, for high-energy and predominantly single-signed vorticity, the monopole vortex is predicted by the Langevin theory to be the most probable structure in a channel with positive beta-parameter. In fact, for any given value of β, there is a critical value of the energy at which these monopole vortices emerge from shear flows. Here in this subsection we determine the precise shape of the transition curve from shear flow solutions to coherent vortices in the (β, E')-plane.

In Figure 11.9 we show the transition curve for a case in which the circulation, Γ, and the vorticity extrema parameters, Q and $\overline{Q}$, are each unity. For any given value of β there is, due to the finite vorticity range $[0, 2]$, a maximum value of energy above which no solutions exist. This region lies above the uppermost curve in the diagram. The lowermost curve of the two curves is the transition boundary. Vortex solutions exist and maximize the entropy in the shaded region above this curve; shear flows, which exist in all parts of the accessible parameter space, maximize the entropy in the region below.

We have plotted the best-fit parabola in dotted lines, which is a nearly indistinguishable solid-line curve representing the experimental data. Here, the generalized Rhines' length is calculated to be $L_R = a^{\frac{1}{4}} = 0.17$, where a is the curvature of the best-fit parabola. This is very close to the value established in the previous

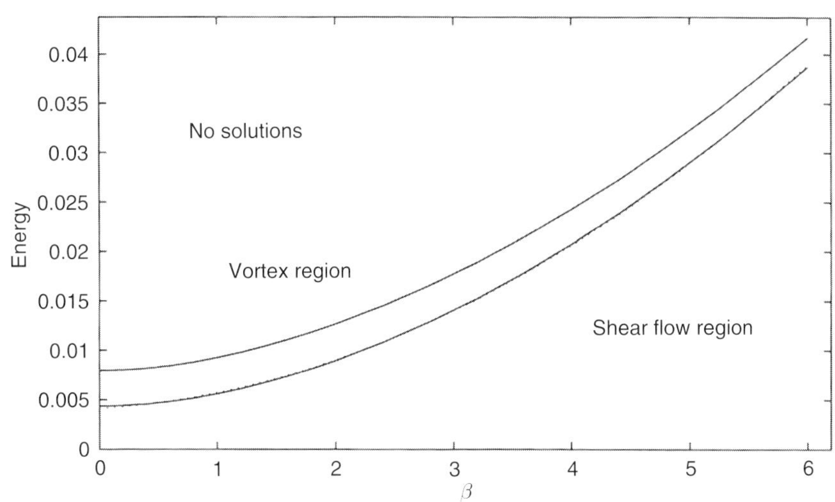

Figure 11.9 Plot of transition curves from shear flow region to vortex region predicted by the Langevin theory in the (β, E')-plane with $\Gamma = \overline{Q} = Q = 1$.

subsection for the emergence of vortex streets in the dilute PV theory. Since we non-dimensionalized with the channel width, H, we have for Langevin monopoles, $L_R^* = 0.17H$. This establishes the quantitative link between a generalized Rhines' scale and the appearance of the monopole vortex as the most probable states in the Langevin statistical theory.

11.3 Statistical sharpness of statistical theories with few constraints

Many different theories based on equilibrium statistical mechanics have been developed to predict large-scale coherent structures in two-dimensional geophysical flows (see Chapters 6, 8, and 9 and the previous section). These statistical theories range from the simple energy–enstrophy theory based on two conserved quantities (EEST) to the maximum entropy principle based on the statistical mechanics of point vortices (PVST, Section 9.3), and most recently to statistical theories that incorporate an infinite number of constraints involving conservation of generalized potential enstrophies in addition to the energy conservation (ESTMC, Section 9.4). Our goal in this section is to clarify the relationship between the predictions of the simpler statistical theories, based on a few constraints, such as energy and circulation, and the more elaborate statistical theories, based on an infinite number of constraints. In other words, we answer the question of how much additional information is actually contained in the predictions of the ESTMC at large scales compared with the predictions of the simpler EEST and PVST, which is one of the theoretical issues from Chapter 10. We will establish the important fact that the most probable mean states predicted by the simple energy–enstrophy theory or point-vortex theory are automatically statistically sharp microstates within the more recent infinite constraint statistical theories. Nevertheless, the family of most probable mean states for the statistical theories with infinitely many constraints (ESTMC) are richer than the family of most probable mean states for the simpler few-constraint statistical theories in general, as we shall demonstrate in the next section via systematic self-consistent asymptotic expansion. This provides answers to the theoretical issues (T-2) and (T-1) from Chapter 10.

For simplicity we consider here the barotropic equations with topography but without beta-effect ($\beta = 0$) and large-scale mean flow ($V \equiv 0$)

$$\frac{\partial q}{\partial t} + \nabla^{\perp}\psi \cdot \nabla q = 0; \quad q = \Delta\psi + h = \omega + h, \tag{11.21}$$

where ψ is the stream function, ω is the relative vorticity, q is the potential vorticity, and $h(x, y)$ represents the normalized topographic height. The incompressible fluid velocity $\vec{v}$ is given by $\vec{v} = \nabla^{\perp}\psi = (-\frac{\partial\psi}{\partial y}, \frac{\partial\psi}{\partial x})$.

Recall from Chapter 1 that (11.21) possesses infinitely many conserved quantities including the energy E and all generalized enstrophies

$$E = -\frac{1}{2} \fint \psi\omega, \quad \mathcal{G}(q) = \fint G(q), \text{ for all functions } G. \tag{11.22}$$

In particular, the extrema of the potential vorticity are conserved. Thus, if the initial data q_0 are bounded, say $q_0^- \le q_0 \le q_0^+$, then future evolutions of the potential vorticity $q(\vec{x}, t)$ must also satisfy $q_0^- \le q(\vec{x}, t) \le q_0^+$. Such an assumption is assumed in this section for simplicity.

11.3.1 Statistical sharpness and generalized selective decay principle

Recall from Section 9.4 that the most probable one-point statistics $\rho(\vec{x}, \lambda)$ associated with the microstate (initial data) q_0 in the ESTMC theory must satisfy

$$\fint \rho(\vec{x}, \lambda) d\vec{x} = \Pi_{q_0}(\lambda) \text{ for any } \lambda \in \text{supp } \Pi_{q_0}, \tag{11.23}$$

where Π_{q_0} is the density function of the distribution associated with q_0 defined in (9.83) or (10.5). Equation (11.23) is equivalent to

$$\int \fint G(\lambda)\rho(\vec{x}, \lambda)\, d\vec{x}\, d\lambda = \int G(\lambda)\Pi_{q_0}(\lambda)\, d\lambda \tag{11.24}$$

for all functions $G(\lambda)$.

Next we introduce the concept of statistically sharp microstates. Roughly speaking, a microstate q is statistically sharp if the the most probable one-point statistics, predicted by ESTMC with q as initial data, contain the same amount of statistical information as q. More precisely, we have the following definition:

Definition 11.1 We say that a microstate, $q(\vec{x})$, is *statistically sharp* for the ESTMC provided that the most probable one-point statistics predicted by ESTMC are Dirac delta functions, i.e. there exists a $\tilde{q}(\vec{x})$ such that

$$\rho(\vec{x}, \lambda) = \delta_{\tilde{q}(\vec{x})}. \tag{11.25}$$

The microstate $q(\vec{x})$ is called *statistically sharp in the strong (pointwise) sense* if the most probable one-point statistics $\rho(\vec{x}, \lambda)$ associated with $q(\vec{x})$ are Dirac delta functions concentrated at $q(\vec{x})$, i.e.

$$\rho(\vec{x}, \lambda) = \delta_{q(\vec{x})}. \tag{11.26}$$

Recall that the coarse-grained mean statistical macrostate (mean field) of $\rho(\vec{x}, \lambda)$ is defined by

$$\bar{q}(\vec{x}) = \int \lambda\rho(\vec{x}, \lambda)\, d\lambda. \tag{11.27}$$

Notice for strongly (pointwise) statistically sharp states, $q(\vec{x})$, the one-point statistics ρ is completely characterized by the pointwise values of $q(\vec{x})$; in particular for statistically sharp states $q(\vec{x})$, it follows from (11.26) and (11.27) that

$$\bar{q}(\vec{x}) = \int \lambda \rho(\vec{x}, \lambda)\, d\lambda = q(\vec{x}), \quad \text{if } q \text{ is statistically sharp in the strong sense.} \tag{11.28}$$

Thus, if the microstate $q(\vec{x})$ is statistically sharp in the strong sense, then all information (statistical and pointwise) of $q(\vec{x})$ is encoded in the mean field of the associated macrostate (one-point statistics).

Now if $q(\vec{x})$ is statistically sharp as defined in (11.25), the most probable one-point statistics ρ may not encode pointwise information of $q(\vec{x})$. However, it encodes all statistical information of $q(\vec{x})$ in the sense that $q(\vec{x})$ and $\bar{q}(\vec{x})$ have the same distribution function. Indeed, combining (11.25) and (11.27) we have

$$\bar{q}(\vec{x}) = \tilde{q}(\vec{x}), \quad \text{if } q \text{ is statistically sharp in the weak sense.} \tag{11.29}$$

Thus, utilizing (11.24), (11.25), and (11.29), we deduce

$$\int G(\lambda)\Pi_q(\lambda) d\lambda = \int ⨍ G(\lambda)\rho(\vec{x}, \lambda) d\vec{x} d\lambda = ⨍ G(\bar{q}(\vec{x})) d\vec{x}. \tag{11.30}$$

Therefore, the distribution functions of $q(\vec{x})$ and $\bar{q}(\vec{x}) = \tilde{q}(\vec{x})$ are the same. Consequently, the coarse-grained macrostate $\bar{q}(\vec{x})$ associated with $q(\vec{x})$ contains all statistical information of $q(\vec{x})$, and $q(\vec{x})$ is statistically sharp in this precise sense.

Next, we consider an arbitrary given microstate, $q_0(\vec{x})$. A natural question is how many of the infinite number of conserved quantities in (11.22) are satisfied by the associated coarse-grained mean statistical macrostate, $\bar{q}(\vec{x})$.

We first observe that since for each $\vec{x}$, $\rho(\vec{x}, \lambda)$ is a probability density function on the line with support contained inside the closed interval of λ with $q_0^- \le \lambda \le q_0^+$ with $q_0^- = \min q_0(\vec{x})$ and $q_0^+ = \max q_0(\vec{x})$, it follows from (11.22) and (11.27) that

$$E(\bar{q}) = E(q_0); \quad q_0^- \le \bar{q}(\vec{x}) \le q_0^+ \tag{11.31}$$

since the energy constraint is imposed on the mean field. Furthermore, if we use the identity in (11.24) with the specific function, $G(\lambda) = \lambda$, it follows by the definition (11.27) that $\bar{q}$ satisfies the same circulation constraint as q_0, i.e.

$$\int \bar{q}(\vec{x}) d\vec{x} = \int q_0(\vec{x}) d\vec{x}. \tag{11.32}$$

We claim that, in general, an infinite number of the constraints in (11.22) are violated and, furthermore, if only one of these particular constraints is satisfied, then, automatically, the mean macrostate $\bar{q}(\vec{x})$ is a rearrangement of q_0, i.e. $\bar{q}(\vec{x})$

and q_0 have the same distribution density function $\Pi_{q_0}(\lambda)$ defined in (9.83) from Chapter 9.

We recall that a smooth function, $G(\lambda)$, is *strictly convex* on the interval $[q_0^-, q_0^+]$ provided that the second derivative, $G''(\lambda)$, satisfies

$$\min_{q_0^- \leq \lambda \leq q_0^+} G''(\lambda) > c_0 > 0. \tag{11.33}$$

We have the following general principle which is the key ingredient in all further developments in this section:

Proposition 11.1 Alternative for relaxed constraints: *Either the coarse-grained mean statistical macrostate, $\bar{q}(\vec{x})$, satisfies the strict inequalities*

$$\int_\Omega G(\bar{q}(\vec{x}))d\vec{x} < \int_\Omega G(q_0(\vec{x}))d\vec{x} \tag{11.34}$$

for all strictly convex functions $G(\lambda)$ on the interval $[q_0^-, q_0^+]$ or $\bar{q}(\vec{x})$ is necessarily a rearrangement of $q_0(\vec{x})$. In other words, if there is equality in (11.34) for only one strictly convex function $G(\lambda)$, then the statistical macrostate $\bar{q}(\vec{x})$ is automatically a rearrangement of the microstate $q_0(\vec{x})$ and the associated most probable one-point statistics satisfy $\rho(\vec{x}, \lambda) = \delta_{\bar{q}(\vec{x})}$, i.e. q_0 is statistically sharp in the sense of (11.25).

In particular, with $G(\lambda) = \frac{1}{2}\lambda^2$, we see that either the enstrophy of the mean macrostate, $\bar{q}(\vec{x})$, is strictly lower than the enstrophy of the microstate $q_0(\vec{x})$, or $\bar{q}(\vec{x})$ is a rearrangement of $q_0(\vec{x})$ and all the moments of the enstrophy coincide for $\bar{q}$ and q_0, and q_0 is statistically sharp.

The mathematical proof of Proposition 11.1 is straightforward given the following proposition which is a strong form of Jensen's inequality (Rudin, 1987):

Proposition 11.2 *Assume that $\nu(\lambda)$ is a probability density function on the line with the support of $\nu(\lambda)$ contained in the interval, $[q_0^-, q_0^+]$, and consider a strictly convex function, $G(\lambda)$, on this interval. Then either*

$$G\left(\int \lambda\nu(\lambda)d\lambda\right) < \int G(\lambda)\nu(\lambda)d\lambda \tag{11.35}$$

or there is some point q with $q_0^- \leq q \leq q_0^+$ so that the probability density function, ν, is the Dirac delta function at q, i.e.

$$\nu = \delta_q \tag{11.36}$$

and equality occurs in (11.35) for any function $G(\lambda)$.

Next, we utilize Proposition 11.2 to verify Proposition 11.1 in a straightforward fashion.

Proof: We consider a strictly convex function, $G(\lambda)$, satisfying (11.33) and apply Proposition 11.2 to the probability density function $\rho(\vec{x}, \lambda)$ defined in (11.23) pointwise for each $\vec{x}$ in the domain Ω. Thus, either $\rho(\vec{x}, \lambda)$ is a Dirac delta function at $\vec{x}$ or we have

$$G(\bar{q}(\vec{x})) = G\left(\int \lambda\rho(\vec{x}, \lambda)d\lambda\right) < \int G(\lambda)\rho(\vec{x}, \lambda)d\lambda. \tag{11.37}$$

Therefore, if $\rho(\vec{x}, \lambda)$ is not a Dirac delta function for a set of points $\vec{x}$ with positive area in Ω, by integrating (11.37) and utilizing (11.24) we obtain

$$⨍ G(\bar{q}(\vec{x}))d\vec{x} < ⨍ \int G(\lambda)\rho(\vec{x}, \lambda)d\lambda\, d\vec{x} = \int G(\lambda)\Pi_{q_0}(\lambda)d\lambda = ⨍ G(q_0(\vec{x}))d\vec{x}. \tag{11.38}$$

This is the first part of Proposition 11.1. On the other hand, assume that there is equality in (11.34) for one strictly convex function, $G(\lambda)$, i.e.

$$\int_\Omega G\left(\int \lambda\rho(\vec{x}, \lambda)d\lambda\right) d\vec{x} = \int_\Omega \int G(\lambda)\rho(\vec{x}, \lambda)d\lambda d\vec{x}.$$

Then, according to Jensen's inequality (Proposition 11.2), necessarily $\rho(\vec{x}, \lambda)$ is a Dirac delta function for all $\vec{x}$ belonging to Ω except for a set with zero area, i.e. (11.25) holds for some $\tilde{q}(\vec{x})$. This further implies that $q(\vec{x})$ is statistically sharp according to Definition 11.1. Combining this with (11.29) we see that the probability measure $\rho(\vec{x}, \lambda)d\lambda$ must be centered at its mean state $\bar{q}$, i.e.

$$\rho(\vec{x}, \lambda) = \delta_{\bar{q}(\vec{x})}. \tag{11.39}$$

This completes our verification of Proposition 11.1.

Next, we utilize Proposition 11.1 to discuss statistically sharp microstates.

Generalized selective decay principles and statistically sharp microstates

Here we assume that a microstate, $q_*(\vec{x})$, is determined by a generalized "selective decay" principle, i.e. q_* is the *unique* solution of the variational principle

$$\int G(q_*)d\vec{x} = \min_{q\in\mathcal{C}} \int G(q)d\vec{x}, \tag{11.40}$$

where $G(\lambda)$ is some strictly convex function and the set $\mathcal{C}$ is given by

$$\mathcal{C} = \left\{ q \;\middle|\; E(q) = E(q_*) \text{ and } \int q\, d\vec{x} = \int q_* d\vec{x} \right\}. \tag{11.41}$$

The reader is referred to Section 4.5 for more on the generalized selective decay principle and the non-linear stability of selective decay states.

Under the above hypothesis we make the following claim which is equivalent to Proposition 11.1.

Proposition 11.3 Statistical sharpness for selective decay states. *q_* is statistically sharp if it is determined by the selective decay principle in (11.40). If q_* is the unique minimizer determined by the selective decay principle in (11.40), the microstate, q_*, is statistically sharp in the strong (pointwise) sense.*

The proof of proposition 11.3 is straightforward. Indeed, let $\bar{q}(\vec{x})$ be the mean field of the macrostate associated with the microstate, $q_*(\vec{x})$, then from (11.31) and (11.32) we have

$$E(\bar{q}) = E(q_*), \quad \int \bar{q} = \int q_*$$

so that $\bar{q}$ belongs to the set $\mathcal{C}$. Thus, from the variational characterization in (11.40) we have, for the strictly convex function G

$$\int G(q_*)d\vec{x} \leq \int G(\bar{q})d\vec{x}. \tag{11.42}$$

The condition in (11.42) and Proposition 11.1 immediately guarantee that

$$\rho(\vec{x}, \lambda) = \delta_{\bar{q}(\vec{x})} \tag{11.43}$$

and hence q_* is statistically sharp. Consequently, we have

$$\int G(\bar{q})d\vec{x} = \int G(q_*)d\vec{x}.$$

Furthermore if the minimizer to the variational principle in (11.40) is unique, we have $\bar{q} = q_*$ and $\rho(\vec{x}, \lambda) = \delta_{q_*(\vec{x})}$. According to (11.26) and (11.28), this guarantees that q_* is a statistically sharp microstate, as claimed in Proposition 11.3.

Below, we use the variational "selective decay" principle elucidated in (11.40) and Proposition 11.3 to verify the statistical sharpness of suitable macroscopic statistical states for both the statistical theories with infinitely many constraints (ESTMC) and various few-constraint statistical theories involving energy/enstrophy principles (EEST) or the statistics of point vortices (PVST).

11.3.2 The statistical sharpness of macrostates from ESTMC

Here we verify the statistical sharpness of suitable macrostates, $\bar{q}(\vec{x})$, of the statistical theory with infinitely many constraints (ESTMC) from Section 9.4. Recall from there that $\bar{q}(\vec{x})$ is determined as the solution of the mean field equation

$$\Delta\bar{\psi} + h = \bar{q} = -\frac{1}{\theta}\frac{\partial}{\partial\psi}\ln(Z(\theta\psi))|_{\bar{\psi}}, \tag{11.44}$$

where the partition function is defined as

$$Z(\psi) = \int e^{-\bar{\gamma}(\lambda)-\lambda\psi}\Pi_{q_0}(\lambda)d\lambda. \tag{11.45}$$

Our strategy in verifying the statistical sharpness of $\bar{q}$ involves establishing under suitable hypothesis, that $\bar{q}$ is automatically the minimum state for a "selective decay" variational procedure in (11.40) so that the general principle in Proposition 11.3 applies to guarantee statistical sharpness. These hypotheses allow for states $\bar{q}$ with any positive temperature θ as well as an interval of negative temperature states θ with $\theta > \theta_{CR} < 0$ (see (11.50) below).

We fix the Lagrange multiplier function $\tilde{\gamma}(\lambda)$ and view the solutions of (11.44), (11.45) as the energy, i.e. θ varies. Recall that the Legendre transform (see for instance Evans, 1998) of a strictly convex function, $\mathcal{W}(\psi)$, is the unique convex function, $g(q)$, given by

$$g(q) = q\psi - W(\psi), \tag{11.46}$$

where ψ is the unique solution to the equation

$$q = W'(\psi).$$

Recall from Section 9.5, the function $\ln(Z(\psi))$ is strictly convex. Let $G(q)$ denote the Legendre transform of $\ln(Z(\psi))$. Then it follows from (11.44) and (11.46) that the mean macrostate $\bar{q}$ satisfies

$$G'(\bar{q}) = \theta\bar{\psi}. \tag{11.47}$$

Now, consider the "selective decay" variational principle defined in (11.40) and (11.41) with this choice of $G(q)$. We claim under suitable hypothesis on the inverse temperature, θ, that $\bar{q}$ satisfies

$$\int G(\bar{q})d\vec{x} = \min_{q\in\mathcal{C}} \int G(q)d\vec{x}. \tag{11.48}$$

Once we know that $\bar{q}$ is the unique minimum satisfying (11.48), it follows automatically that $\bar{q}$ is statistically sharp as a consequence of the general principle in Proposition 11.3.

To check that $\bar{q}$ is actually the strict minimum of (11.48), we set $q = \bar{q} + \delta q$, utilize the energy constraint, and calculate in standard fashion

$$\begin{aligned}
&\int G(\bar{q})d\vec{x} - \int G(\bar{q}+\delta q)d\vec{x} \\
&\quad \theta E(\bar{q}) - \theta E(\bar{q}+\delta q) + \int G(\bar{q})d\vec{x} - \int G(\bar{q}+\delta q)d\vec{x} \\
&\quad \leq -\frac{1}{2}\int (\delta q)^2 d\vec{x}\left[\theta\frac{E(\delta q)}{\int_\Omega(\delta q)^2} + \min_\lambda G''(\lambda)\right].
\end{aligned} \tag{11.49}$$

We observe that the expression in brackets in (11.49) is positive and $\bar{q}$ is guaranteed to be a strict minimum provided that θ satisfies

$$\infty > \theta > \theta_{CR} < 0 \quad \text{with } \theta_{CR} = -(\min_\lambda G''(\lambda))\Lambda_1, \tag{11.50}$$

where Λ_1 is the first (necessarily positive) eigenvalue of $-\Delta$ with either the Dirichlet boundary condition or channel boundary condition or periodic boundary condition with mean zero on the region Ω. This finishes the argument. We also note that this argument yields the non-linear stability of the mean macrostate, $\bar{q}$, within the inviscid non-linear dynamics of (11.21) by the general non-linear stability theory from Section 4.5.

It is an interesting open mathematical problem to decide whether strongly negative temperature states of the statistical theory with infinitely many constraints and $\theta < \theta_{CR} < 0$ are statistically sharp. Such strongly negative temperature states have been utilized by Turkington and Whitaker (1996) recently to develop an interesting new statistical theory of vortex patch merger.

11.3.3 Statistical sharpness of macrostates of ESTP

Energy–enstrophy statistical theories (EEST)

We begin with the well-known *energy–enstrophy statistical theories* (EEST). Recall from Sections 6.6, 6.7 and Chapter 8 that the statistical macrostates (mean fields) coincide with solutions of the "selective decay" principle in (11.40), (11.41) with the strictly convex function $G(q) = \frac{1}{2}q^2$. For simplicity in exposition, we specialize the region Ω to a periodic box T^2, which is 2π-periodic in each direction. In this periodic setting we assume that the mean of the topography, h, vanishes. According to (11.40), the most probable mean field of this two-constraint statistical theory satisfies

$$\Delta\psi_* + h = q_* = \theta\psi_*, \tag{11.51}$$

where θ is the Lagrange multiplier for energy.

First, consider the situation where $h \neq 0$, then the same calculation in (11.49), (11.50) establishes that q_* is a strict minimum for $\theta > \theta_{CR}$ with $\theta_{CR} = -\Lambda_1$. Thus, according to the general principle in Proposition 11.3, we have strong/pointwise statistical sharpness within the infinite constraint statistical theory (ESTMC) of the macrostates q_* from the energy–enstrophy theory provided that θ satisfies

$$-\Lambda_1 < \theta < +\infty. \tag{11.52}$$

We note that, as θ varies over the range in (11.52), the energy, $E(q_*)$ varies over the complete range from zero to infinity if the topography, h, has a non-degenerate spectrum in the sense that the projection of the topography on to the first energy shell is non-trivial (see Sections 6.6 and 6.7). Of course, for the 2π-periodic domain, $\Lambda_1 = 1$. We note that the range of θ in (11.52) with statistical sharpness is exactly the range of admissible macrostates of the energy–enstrophy statistical theory with topography having a non-degenerate spectrum (Chapters 6 and 8).

In the case of topography with a degenerate spectrum, i.e. the projection of the topography on to the ground energy shell vanishes, the energy, $E(q_*)$, varies over a range from zero to E_1 defined in Section 6.6 when θ varies over the range in (11.52).

For the case without topography, from (11.51), we observe that the variational principle yields 2π-periodic eigenfunctions of the Laplacian and recall from Chapter 3 only the ground state eigenfunctions of (11.51) are macrostates of the energy–enstrophy statistical theory, i.e. ψ_* satisfies

$$\Delta\psi_* = -\Lambda_1\psi_* \tag{11.53}$$

and $\Lambda_1 = \theta_{CR}$. See also Sections 6.6 and 6.7. For the 2π-periodic case, these non-unique ground states have the form

$$\psi_* = A\sin x + B\sin y + C\cos x + D\cos y, \tag{11.54}$$

where A, B, C, D are arbitrary parameters and the energy constraint in (11.41) imposes $A^2+B^2+C^2+D^2 = \text{Constant}$. In this case, Proposition 11.3 guarantees that all these states, q_*, are statistically sharp, but not in the strong/pointwise sense.

Point-vortex statistical theories (PVST)

Recall from Section 9.3 that the mean field equation for the statistical theory for point vortices with a single sign in a bounded region, Ω, leads to the well-known equation (Montgomery and Joyce, 1974) for the statistical macrostates

$$\Delta\psi_* + h = q_* = \Gamma\left[\int_\Omega e^{-\beta\psi_*}\right]^{-1} e^{-\beta\psi_*}, \tag{11.55}$$

where Γ is the integral of q_*, i.e. essentially the total circulation.

Recall from Section 9.4, that the appropriate prior distribution for the PVST has a special form given by (9.44). Thus, when combined with (9.47) (with $\alpha = 0$ since no conservation of angular momentum is involved), the minus entropy can be represented in terms of the mean field for the most probable states as

$$\begin{aligned}
-\mathcal{S}(\rho^*, \Pi_0) &= \fint \int \rho^*(\vec{x},\lambda)\ln\frac{\rho^*(\vec{x},\lambda)}{\Pi_0(\lambda)}d\lambda d\vec{x} \\
&= \int \frac{e^{\theta\omega_0\psi_*}}{\int e^{\theta\omega_0\psi_*}d\vec{x}}\left(\theta\omega_0\psi_* - \ln\int e^{\theta\omega_0\psi_*} + 2\ln|\Omega|\right)d\vec{x} \\
&= \int \frac{e^{-\beta\psi_*}}{\int e^{-\beta\psi_*}d\vec{x}}\left(-\beta\psi_* - \ln\int e^{-\beta\psi_*} + 2\ln|\Omega|\right)d\vec{x} \\
&= \frac{1}{\Gamma}\int q_*\ln q_* - \ln\Gamma + 2\ln|\Omega|,
\end{aligned} \tag{11.56}$$

where we used the notation $\beta = -\theta\omega_0$ and utilized (11.55). Thus we have demonstrated that a one-point statistics, $\rho(\vec{x}, \lambda)$, maximizes the Shannon relative entropy if and only if its mean field, q_*, minimizes the following functional

$$G(q) = \begin{cases} q \ln q, & q > 0 \\ 0, & q \leq 0 \end{cases} \tag{11.57}$$

within the constraint set $\mathcal{C}$ defined by

$$\mathcal{C} = \left\{ q \;\middle|\; E(q) = E(q_*), \int q d\vec{x} = \Gamma > 0 \text{ and } q \geq 0 \right\}. \tag{11.58}$$

Since the G given by (11.57) is strictly convex in the region $[0, \infty)$, the entropy-maximizing macrostate must be statistically sharp in the sense of (11.25) according to Proposition 11.3. These arguments can be made rigorous. For instance, for the case without topography, Caglioti *et al.* (1995) have proved that there is a range of β with $0 > \beta_{CR} < \beta < +\infty$ so that the macrostates of the statistical point-vortex theory can be characterized as the unique solution of the variational principle in (11.40) with the constraints in (11.58). Thus, by the general principle developed in Proposition 11.3, we are guaranteed that the macrostates in (11.55) for the statistical point-vortex theory are automatically statistically sharp microstates of the statistical theory with infinitely many constraints. We note that strictly speaking, Proposition 11.3 needs to be modified to allow convex functions with $G''(q) = \frac{1}{|q|}$ near $q = 0$, but this we leave as an exercise for the interested reader.

11.4 The limit of many-constraint theory (ESTMC) with small amplitude potential vorticity

In the previous section we presented results on the statistical sharpness of macrostates for a few constraint equilibrium statistical theories (EEST and PVST). We saw that, under a wide range of circumstances, the most probable macrostates predicted by ESTP are statistically sharp within the more elaborate ESTMC. Thus, we might inquire into the issue of whether the macrostates of the statistical theory with infinitely many constraints (ESTMC) are actually more plentiful than those of the simpler few constraint statistical theories in regimes without extreme negative values of the inverse temperature, β (we use θ in the sequel). Here we consider the macrostates of these statistical theories in a regime with topography and small amplitude potential vorticity and show, through self-consistent asymptotic expansions, the fashion in which these more elaborate statistical theories generate a richer family of statistical macrostates. In fact, we establish below that, in this regime, the *most probable mean fields of the many constraint statistical theories (ESTMC) solve similar equations as* in (11.51) for *the most probable mean fields* *of the simpler energy–enstrophy statistical theories (EEST) with the important difference that the original topography, h, is replaced by a renormalized effective topography, h_{eff}, which depends on the higher moments of the distribution function for potential vorticity* (see (11.75), (11.76) below). This partially addresses the theoretical issues (T-1) and (T-2) from Chapter 10.

To fix ideas, we consider a family of microstates for potential vorticity, $q_0(\vec{x})$, with a common distribution density function, $\Pi_0(\lambda)$, defined in (9.83). We consider the microstates with small amplitude potential vorticity given by

$$aq_0(\vec{x}), \quad a \ll 1. \tag{11.59}$$

We easily calculate that the distribution associated with these microstates is given by

$$\Pi_0^a(\lambda) = a^{-1}\Pi_0\left(\frac{\lambda}{a}\right) \quad a \ll 1. \tag{11.60}$$

According to the statistical theory with many constraints (ESTMC) from Section 9.4, there is a Lagrange multiplier function $\tilde{\gamma}^a(\lambda)$ so that for a fixed inverse temperature, θ, the most probable mean field, $\bar{q}^a$, satisfies

$$\Delta\bar{\psi}^a + h = \bar{q}^a = \frac{1}{\theta}\frac{d}{d\psi}\log\left(Z^a(\theta\psi)\right)\Big|_{\bar{\psi}^a}. \tag{11.61}$$

It is easy to check, after rescaling the variable, that the partition function is given by

$$Z^a(\psi) = \int e^{-\tilde{\gamma}^a(a\lambda)-a\lambda\psi}\Pi_0(\lambda)d\lambda. \tag{11.62}$$

The requirements in (11.23) become, in this case

$$\int_\Omega \frac{e^{-\tilde{\gamma}^a(a\lambda)-\theta a\bar{\psi}^a(\vec{x})}}{Z^a(\theta\bar{\psi}^a(\vec{x}))}d\vec{x} = 1 \tag{11.63}$$

for all λ in the support of $\Pi_0(\lambda)$. We note that the non-linearities in (11.61) and (11.63) are weak for $a \ll 1$, because the partition function, $Z^a(\psi)$, admits a regular asymptotic expansion in this small parameter.

11.4.1 The asymptotic expansion

In the theory, we need to select the Lagrange multiplier functions, $\tilde{\gamma}^a(\lambda)$, so that the constraints in (11.63) are satisfied exactly. Here we make the following ansatz for suitable approximate Lagrange multiplier functions, $\tilde{\gamma}_N^a(\lambda)$, for each positive integer, N

$$\tilde{\gamma}_N^a(\lambda) = \sum_{\substack{i+j\leq N\\ 0\leq i,1\leq j}} \alpha_{ij}a^i\lambda^j. \tag{11.64}$$

We choose the coefficients, α_{ij}, so that the requirements in (11.63) are satisfied within order $o(a^{N+1})$, i.e.

$$\int_\Omega \frac{e^{-\tilde{\gamma}_N^a(a\lambda)-\theta a\bar{\psi}^a(\vec{x})}}{Z^a(\theta\bar{\psi}^a(\vec{x}))}d\vec{x} = 1 + o(a^{N+1}). \tag{11.65}$$

The interested reader can readily verify that with $\tilde{\gamma}_N^a(\lambda)$ chosen to satisfy (11.65), Z_N^a defined as in (11.62) with $\tilde{\gamma}_N^a$, and

$$\rho_N^a(\vec{x},\lambda) = \frac{e^{-\tilde{\gamma}_a^N(\lambda)-\theta\lambda\bar{\psi}^a(\vec{x})}}{Z_N^a(\theta\bar{\psi})}\Pi_0^a(\lambda), \tag{11.66}$$

the constraint requirements from (11.24)

$$\int\int G(\lambda)\rho_N^a(\vec{x},\lambda)d\vec{x}\,d\lambda = \int G(\lambda)\Pi_0^a(\lambda)d\lambda \tag{11.67}$$

are automatically satisfied within the small error, $o(a^{N+1})$, for any function $G(\lambda)$.

It is a tedious but straightforward procedure (Holen, 1995) to expand the left-hand side of (11.65) using (11.62) for $a \ll 1$ and to determine the values of the Lagrange multiplier coefficients, α_{ij}, in (11.64) by the requirement in (11.65). Here we only recall the results of this expansion for $N = 3$. First we recall the spatial average of a function

$$\langle f(\vec{x})\rangle = \fint f(\vec{x})d\vec{x} \tag{11.68}$$

and also recall the notation for the average of a function, $G(q_0)$, with respect to the probability measure $\Pi_0(\lambda)$ by

$$\langle G(q_0)\rangle_0 = \int G(\lambda)\Pi_0(\lambda)d\lambda. \tag{11.69}$$

Now, in order to satisfy (11.65) to order $o(a^4)$, the non-zero coefficients for α_{ij} should satisfy (for the details see Holen, 1995)

$$\begin{aligned}
\alpha_{01} &= \theta\langle\bar{\psi}^a\rangle, \quad \alpha_{02} = \frac{1}{2}\theta^2\left\langle\left(\bar{\psi}^a - \langle\bar{\psi}^a\rangle\right)^2\right\rangle,\\
\alpha_{11} &= -\langle q_0\rangle_0\,\theta^2\left\langle\left(\bar{\psi}^a - \langle\bar{\psi}^a\rangle\right)^2\right\rangle\\
\alpha_{21} &= \left(\langle q_0\rangle_0^2 - \frac{1}{2}\langle q_0^2\rangle_0\right)\theta^3\left\langle\left(\bar{\psi}^a - \langle\bar{\psi}^a\rangle\right)^3\right\rangle\\
\alpha_{12} &= -\frac{1}{2}\langle q_0\rangle_0\,\theta^3\left\langle\left(\bar{\psi}^a - \langle\bar{\psi}^a\rangle\right)^3\right\rangle, \quad \alpha_{03}\frac{1}{6}\theta^3\left\langle\left(\bar{\psi}^a - \langle\bar{\psi}^a\rangle\right)^3\right\rangle
\end{aligned} \tag{11.70}$$

Next, we insert the explicit formulas for the Lagrange multiplier coefficients in (11.70) into $Z^a(\psi)$ in (11.62) and then Taylor expand the right-hand side of (11.61) for $a \ll 1$. We obtain within errors of order $o(a^5)$ the following:

Asymptotic equation for statistical mean macrostates: $\Delta\bar{\psi}^a + h = \bar{q}^a$ with $\bar{q}^a$ given by

$$\begin{aligned}\bar{q}^a(\vec{x}) &= a\langle q\rangle_0 + a^2\theta\left(\bar{\psi}^a(\vec{x}) - \langle\bar{\psi}\rangle\right)\left[\left\langle(q_0 - \langle q_0\rangle_0)^2\right\rangle_0\right] \\ &+ a^3\frac{1}{2}\theta^2\left[\left(\bar{\psi}^a(\vec{x}) - \langle\bar{\psi}^a\rangle\right)^2 - \left\langle\left(\bar{\psi}^a - \langle\bar{\psi}^a\rangle\right)^2\right\rangle\right]\left\langle(q_0 - \langle q_0\rangle_0)^3\right\rangle_0 \\ &+ a^4\theta^3\left\{\left[\left(\bar{\psi}^a(\vec{x}) - \langle\bar{\psi}^a\rangle\right)^3 - \left\langle\left(\bar{\psi}^a - \langle\bar{\psi}^a\rangle\right)^3\right\rangle\right]\right. \\ &\times\left(\frac{1}{6}\left\langle q_0^4\right\rangle_0 - \frac{2}{3}\langle q_0\rangle_0\left\langle q_0^3\right\rangle_0 + 2\langle q_0\rangle_0^2\left\langle q_0^2\right\rangle_0 - \frac{1}{2}\left\langle q_0^2\right\rangle_0^2 - \langle q_0\rangle_0^4\right) \\ &+ \left\langle\left(\bar{\psi}^a(\vec{x}) - \langle\bar{\psi}^a\rangle\right)^2\right\rangle\left(\bar{\psi}^a(\vec{x}) - \langle\bar{\psi}^a\rangle\right) \\ &\times\left.\left(2\langle q_0\rangle_0\left\langle q_0^3\right\rangle_0 - \frac{1}{2}\left\langle q_0^4\right\rangle_0 - \frac{1}{2}\left\langle q_0^2\right\rangle_0^2 - \left\langle q_0^2\right\rangle_0\langle q_0\rangle_0^2\right)\right\}\end{aligned} \tag{11.71}$$

In the periodic setting, we automatically have $\langle q_0\rangle_0 = 0$ and $\langle\bar{\psi}^a\rangle = 0$, so the formula for $\bar{q}^a$ simplifies to

$$\begin{aligned}\bar{q}^a(\vec{x}) &= a^2\theta\bar{\psi}^a(\vec{x})\left\langle q_0^2\right\rangle_0 + a^3\frac{1}{2}\theta^2\left[\bar{\psi}^a(\vec{x})^2 - \left\langle(\bar{\psi}^a)^2\right\rangle\right]\left\langle q_0^3\right\rangle_0 \\ &+ a^4\theta^3\left\{\left[\bar{\psi}^a(\vec{x})^3 - \left\langle(\bar{\psi}^a)^3\right\rangle\right]\left(\frac{1}{6}\left\langle q_0^4\right\rangle_0 - \frac{1}{2}\left\langle q_0^2\right\rangle_0^2\right)\right. \\ &\left. - \left\langle(\bar{\psi}^a)^2\right\rangle\bar{\psi}^a(\vec{x})\left(\frac{1}{2}\left\langle q_0^4\right\rangle_0 + \frac{1}{2}\left\langle q_0^2\right\rangle_0^2\right)\right\}\end{aligned} \tag{11.72}$$

11.4.2 Interpretation of the asymptotic equation through renormalized topography

For simplicity of exposition, we only study the equation for statistical macrostates from (11.72) in the periodic setting, where automatically $\langle q_0\rangle_0 = 0$ and $\langle\bar{\psi}^a\rangle = 0$. First, we remark that within terms of order, $o(a^3)$, the asymptotic equation for statistical macrostates from (11.71) and (11.72) reduces to the equation for $\bar{\psi}_2^a$

$$\Delta\bar{\psi}_2^a + h = \theta a^2\left\langle q_0^2\right\rangle\bar{\psi}_2^a. \tag{11.73}$$

This leading-order statistical equation for $a \ll 1$ is exactly the equation for macrostates (11.51) of the energy–enstrophy statistical theory with topography (EEST, Chapters 6 and 8). Thus, we recover the simplest two constraint theory for statistical macrostates at the first non-trivial order in the asymptotic expansion at small amplitudes of the statistical theory with many constraints. We note that self-consistently the coefficient, $a^2\langle q^2\rangle$ in (11.73), is twice the enstrophy of the microstate defined by the measure Π_0^a, i.e.

$$\frac{1}{2}a^2\langle q_0^2\rangle = \int \frac{1}{2}\lambda^2\Pi_0^a(\lambda)d\lambda. \tag{11.74}$$

The function $\bar{\psi}_2^a$ in (11.73) is determined explicitly through the solution of a linear equation; furthermore, this function $\bar{\psi}_2^a$ agrees with the solution $\bar{\psi}^a$ of (11.71), (11.72) within $o(a^3)$. Thus, in a fashion completely consistent with the asymptotic expansion, we can replace $\bar{\psi}^a$ by $\bar{\psi}_2^a$ in the terms with order $o(a^3)$, $o(a^4)$ on the right-hand side of (11.72). Thus we are led to a new renormalized *effective topography*, h_{eff}, given explicitly by

$$\begin{aligned} h_{\mathrm{eff}} = h &- a^3\frac{1}{2}\theta^2\left[(\bar{\psi}_2^a(\vec{x}))^2 - \langle(\bar{\psi}_2^a(\vec{x}))^2\rangle\right]\langle q_0^3\rangle_0 \\ &- a^4\theta^3\left[(\bar{\psi}_2^a(\vec{x}))^3 - \langle(\bar{\psi}_2^a)^3\rangle\right]\left(\frac{1}{6}\langle q_0^4\rangle_0 - \frac{1}{2}\langle q_0^2\rangle_0^2\right) \\ &+ a^4\theta^3\langle(\bar{\psi}_2^a)^2\rangle\bar{\psi}_2^a(\vec{x})\left(\frac{1}{2}\langle q_0^4\rangle_0 + \frac{1}{2}\langle q_0^2\rangle_0^2\right) \end{aligned} \tag{11.75}$$

so that the approximate mean statistical macrostate $\bar{\psi}^a$ satisfies the equation

$$\Delta\bar{\psi}^a + h_{\mathrm{eff}} = \theta a^2\langle q_0^2\rangle\bar{\psi}^a \tag{11.76}$$

within $o(a^5)$.

The renormalized effective topography is a function of both the third and fourth moments of the microstate distribution function, Π_0^a since

$$a^3\langle q_0^3\rangle_0 = \int \lambda^3\Pi_0^a(\lambda)d\lambda, \quad a^4\langle q_0^4\rangle_0 = \int \lambda^4\Pi_0^a(\lambda)\,d\lambda. \tag{11.77}$$

Thus, even within this asymptotic approximation at small amplitudes, the macrostates in the statistical theory with many constraints encode the richer microstate statistical information from (11.77) involving higher moments of Π_0^a beyond the information in the energy–enstrophy statistical macrostates in (11.51). However, for small amplitudes of the potential vorticity, the energy–enstrophy macrostates emerge at leading order in the asymptotics as developed above in (11.73), (11.74), while the contributions from the third and fourth moments are higher-order effects. In applications where the microstate is not known exactly, such higher moments are more difficult to obtain exactly, and we might hope that crude reduced dynamics utilizing the two-constraint statistical theory is accurate for non-equilibrium flows with topography. We explore these issues in the next chapter.

Finally, we remark that the formal asymptotic expansions presented in (11.64)–(11.76) can be justified with full mathematical rigor in a straightforward fashion (Holen, 1995).

References

DiBattista, M. T., Majda, A. J., and Turkington, B. (1998), Prototype geophysical vortex structures via large-scale statistical theory. *Geophys. Astrophys. Fluid Dynam.* **89** (3–4), 235–283.

Caglioti, E., Lions, P. L., Marchioro, C., and Pulvirenti, M. (1992), A special class of stationary flows for two-dimensional Euler equations: A statistical mechanics description. *Commun. Math. Phys.* **143**, 501–525.

Caglioti, E., Lions, P. L., Marchioro, C., and Pulvirenti, M. (1995), A special class of stationary flows for two-dimensional Euler equations: a statistical mechanics description. *Commun. Math. Phys.* **174**, 229–260.

Evans, L. C. (1998), *Partial Differential Equations*. Providence, Rhode Island: AMS.

Flierl, G. R. (1987), Isolated eddy models in geophysics. *Ann. Rev. Fluid Mech.* **19**, 493–530.

Holen, M. (1995), Comparison of statistical theories for large scale two dimensional flows. Doctoral Dissertation, Princeton University.

Majda, A. and Holen, M. (1997), Dissipation, topography, and statistical theories for large-scale coherent structure. *Comm. Pure Appl. Math.* **50** (12), 1183–1234.

Montgomery, D. and Joyce, G. (1974), Statistical mechanics of negative temperature states. *Phys. Fluids.* **17**, 1139–1145.

Rhines, P. (1975), Waves and turbulence on a beta plane. *J. Fluid Mech.* **69**, 417–443.

Rhines, P. (1979), Geostrophic turbulence. *Ann. Rev. Fluid Mech.* **11**, 404–441.

Rudin, W. K. (1987), *Real and Complex Analysis*, 3rd edn. McGraw-Hill.

Turkington, B. (1999), Statistical equilibrium measures and coherent states in two-dimensional turbulence. *Comm. Pure Appl. Math.* **52** (7), 781–809.

Turkington, B. and Whitaker, N. (1996), Statistical equilibrium computations of coherent structures in turbulent shear layers. *SIAM J. Sci. Comput.* **17**, 1414–1433.

12

Equilibrium statistical theories and dynamical modeling of flows with forcing and dissipation

12.1 Introduction

In the previous chapters we discussed various equilibrium statistical theories. These statistical theories are developed for the idealized inviscid unforced geophysical flows. However, as we have discussed in Section 10.4, virtually all practical geophysical flows are subject to both forcing and dissipation. For instance, the earth's atmosphere is subject to intense random small-scale forcing from convective storms, and the ocean is subject to forcing from unresolved baroclinic instability processes on a small length scale. Thus, a natural question to ask is whether the equilibrium statistical theories can be applied in a forced and damped environment. The purpose of this chapter is to address this question. More precisely, we want to provide answers to the applied issue (A-4) and theoretical issue (T-5) from Chapter 10.

As was discussed in Section 10.4, equilibrium statistical theories will not be able to approximate geophysical flows in a statistical sense for all forcing. This is not a surprise, since intuitively we could only expect equilibrium theories to succeed when the flow is near equilibrium. What we are interested in here are external forcing which is random and small scale in space and kicks in time. This kind of forcing mimics the small-scale random forcing in the atmosphere and ocean as discussed above and in Chapter 10. The "quasi-equilibrium" state of the geophysical flow is achieved if the inverse cascade of energy from the small scales, where the external forcing occurs, to the large scales (the scale characterized by the equilibrium statistical theory) is sufficiently weak.

Another issue is the choice of statistical theories. Based on the discussion in Chapter 10 and the predictions and comparisons in Chapter 11, it seems that empirical statistical theory with a prior distribution (ESTP) is a good choice between the simple (linear) theory, based on energy and enstrophy (EEST) and the complex theory, based on infinite conserved quantities (ESTMC), which we cannot keep track of in practice.

There are two ways to utilize ESTP to approximate in a statistical sense the geophysical flows governed by barotropic quasi-geostrophic equations (10.1). First, we can verify the consistency (meta-stability) of ESTP in a non-equilibrium regime. This is done by matching the instantaneous exact energy, E, and circulation, Γ (and other bulk quantities needed in the statistical theories). Thus given the exact values E and Γ etc., we can compare the most probable mean state predicted by the statistical theory with the exact solution. This can be interpreted as checking the meta-stability of the statistical large structures with dissipation and small-scale random forcing. This is the topic of Section 12.2. This section follows closely the work of DiBattista, Majda, and Grote (2001). Second, we could put the statistical theory to a more stringent test; namely, we will calculate the bulk quantities such as the energy, E, and circulation, Γ, etc. via a crude closure algorithm which depends only on the statistical theory, the form of dissipation and external small-scale random forcing, and does not rely on the exact dynamics of the geophysical flows. Surprising enough, the crude closure algorithm performs extremely well under a wide range of circumstances interesting in geophysical situations. This is the topic of Section 12.3 and the material relies heavily on Majda and Holen (1997) and Grote and Majda (1997). In the last section, Section 12.4, we give partial mathematical justification of the success of the crude closure algorithm in the case of spin-up from rest. Here we demonstrate that the Dirichlet quotient of the flow is bounded above, independent of the small-scale random bombardment. Since the Dirichlet quotient controls the small scales in the flow, this means that the small scales cannot grow without bound, even under persistent small-scale forcing.

12.2 Meta-stability of equilibrium statistical structures with dissipation and small-scale forcing

In Section 11.2 we have shown that the mean field predictions of the ESTP with either the Langevin (10.7) or dilute vortex (10.9) prior distributions include a wide range of interesting geophysical phenomena which are not accessible by the predictions of the simpler EEST or selective decay. On the other hand, in Section 10.4 we have emphasized the importance of understanding the significance of the equilibrium statistical predictions which are based solely on considerations from inviscid dynamics in the content of flows with forcing and dissipation.

Here we apply the equilibrium statistical theory to *evolving* damped and driven geophysical flows. Below, we describe an approximate dynamics that calculates the most probable states of the equilibrium theory, where energy, circulation, and other essential quantities match the instantaneous values of the evolving flow.

Thus, by tracking the values of only a few key quantities, we approximate the state of the evolving flow with the most probable solutions of equilibrium statistical mechanics. There are two goals:

- to demonstrate the meta-stability of the most probable states of equilibrium statistical theory, i.e. vortex monopoles and vortex dipole streets, under a wide range of numerical experiments involving small-scale forcing and dissipation. This study includes both direct and inverse cascade regimes – flow regimes in which energy transfers from large to small scales where dissipation is more efficient, and flow regimes in which strong small-scale forcing near the scales of resolution allows substantial energy transfer to the largest scales. However, our resolution is too crude to compute detailed spectra for these cascades and these are not the quantities of primary interest here. Here we establish the suitability of the solutions of the statistical theory as basic models of atmospheric and oceanographic structures when dominated by suitable random small-scale forcing. This is a key idea motivating the use of ESTP in the application for Jupiter discussed in Chapter 13.
- to evaluate quantitatively the accuracy of the equilibrium statistical theory – how well do the predictions of the Langevin and dilute-vortex theory track the evolution of large-scale coherent structures in a strongly damped and driven flow? In this respect this section serves as a validation study of the approximate dynamics.

These goals are realized through a series of numerical runs initialized by monopole vortices and dipole vortex streets that are the most probable states produced by the equilibrium theory. According to Figures 11.8 and 11.9 from our discussion in Section 11.2, these equilibrium statistical theories predict topological transitions from shear flows to either monopole or vortex dipoles as the energy changes through dissipation and forcing. Such experiments are a stringent test of the viability of the statistical theories. The numerical experiments include cases of pure decay in which only the dissipation is active, and cases with both dissipation and forcing in which small-scale vortices – mimicking the bombardment of the flow by localized convective storms – are randomly placed in the channel domain. Additionally, in a more stringent test of the theory, we compare the end state that arises from an inverse cascade – induced by single-signed, small-scale forcing of an initially quiescent channel – to the most probable states of the equilibrium theory. The numerical results in this section are due to DiBattista, Majda, and Grote (2001).

12.2.1 The numerical model

Following the basic theme of this book, we consider one of the simplest models for geophysical flow at midlatitudes, the barotropic quasi-geostrophic (QG) fluid in a beta-plane channel. Specifically, we consider a QG fluid that is both damped and driven in a beta-plane channel $\Omega = [0, 2\pi] \times [-\pi, \pi]$ periodic in x. We recall that the dynamic equations for this barotropic quasi-geostrophic model in terms of the potential vorticity q are

$$\begin{aligned} \frac{\partial q}{\partial t} + \nabla^{\perp}\Psi \cdot \nabla q &= \mathcal{D}(\Delta)\omega + \mathcal{F}, \\ q = \Delta\Psi + \beta y = \omega + \beta y, \quad \Psi &= -Vy + \psi, \end{aligned} \tag{12.1}$$

together with the boundary conditions that all quantities are periodic in x with period 2π and the small-scale stream function, ψ, vanishes at the walls $(y = \pm\pi)$ of the channel. Here, the corresponding fluid velocity is given by

$$\vec{v} = \nabla^{\perp}\Psi = \begin{pmatrix} V \\ 0 \end{pmatrix} + \begin{pmatrix} -\frac{\partial \psi}{\partial y} \\ \frac{\partial \psi}{\partial x} \end{pmatrix} \tag{12.2}$$

In equations (12.1), $\mathcal{D}$ represents the dissipation

$$\mathcal{D}(\Delta) \equiv \sum_{i=0}^{3} d_i (-1)^i \Delta^i, \; d_i \geq 0, \tag{12.3}$$

and $\mathcal{F}$ represents forcing. For the dissipation operator, $\mathcal{D}$, the coefficient, d_0, represents the Ekman drag, which models the effect of the boundary layer on large-scale geophysical flows, d_1 represents Newtonian viscosity, which models molecular diffusion, and d_3 represents hyper-viscosity, which ensures numerical stability in flows with vanishing Newtonian viscosity. The hyper-viscosity coefficient, d_2, vanishes for all numerical experiments treated here.

We solve the quasi-geostrophic equations in (12.1) numerically, using a pseudo-spectral method, which calculates the gradients $\nabla^{\perp}\psi$ and $\nabla\omega$ in Fourier space, and the non-linear product $\nabla^{\perp}\psi \cdot \nabla\omega$ at the mesh points in physical space. The resolution of the numerical grid varies from 64×64 to 256×256 for the numerical experiments. The interested reader can consult DiBattista *et al.* (2001) for further details.

The forcing in the QG equation in (12.1) is designated by the variable $\mathcal{F}$, which represents a bombardment of the flow by hypothetical small-scale storms in the vorticity field. These small vortices, which have finite support and constant amplitude, are placed at random mesh points in the flow at fixed time intervals (kicks), and take the particular form

$$\mathcal{F}(\vec{x}, t) = \sum_{j=1}^{\infty} \delta(t - j\Delta t)\omega_r(\vec{x} - \vec{x}_j), \tag{12.4}$$

where the jth vortex is centered at $\vec{x}_j$. Each of these localized vortices has support in a small disk of radius, r_j, given by

$$\omega_r(\vec{x}) = \begin{cases} Amp\left(1 - |\vec{x} - \vec{x}_j|^2/r_j^2\right)^2, & |\vec{x} - \vec{x}_j|^2 \leq r_j^2, \\ 0, & |\vec{x} - \vec{x}_j|^2 > r_j^2. \end{cases} \tag{12.5}$$

Here, the parameter Amp represents the strength of the vortex and may be either positive or negative, corresponding to cyclonic or anti-cyclonic forcing.

The forcing term, $\mathcal{F}$, in (12.4) is determined by four parameters: the vortex strength, Amp, the vortex radius, r_j, the vortex position, $\vec{x}_j$, and the time interval, Δt. In each of the experiments described below the random locations, $\vec{x}_j$, are uniformly distributed among the mesh points inside the channel domain, excluding a narrow margin along the channel boundaries. The radius, r_j, of the small-scale vortices is fixed at 3/64. The time interval between forcings is fixed at $\Delta t = 0.1$, which in all cases is at least ten times larger than the time step used for the numerical integration of the (non-linear) advection terms – if the simulation is run up to $T = 1000$, precisely 10 000 small-scale vortices are added to the evolving flow, while at least 100 000 time steps are performed by the numerical method. The main parameter that varies from experiment to experiment is therefore the vortex strength, Amp, which is usually expressed as a proportion of the maximum initial potential vorticity in the flow.

Quantitative comparisons between the evolving flow and the most probable states of the equilibrium theory are evaluated by means of the correlation between two functions, i.e.

$$\mathrm{Corr}(f, g) = \frac{(f, g)}{||f||\,||g||}, \tag{12.6}$$

in which (f, g) denotes the L^2-inner product on the channel domain and $||f||$ its corresponding L^2-norm. For convenience of numerical comparison, we consider the following quantity, Corr-d, that we call correlation deficit

$$\text{Corr-d}(f, g) = 1 - \frac{(f, g)^2}{||f||^2||g||^2} = 1 - \mathrm{Corr}^2(f, g), \tag{12.7}$$

The correlation deficit satisfies $0 \leq \text{Corr-d}(f, g) \leq 1$ and the correlation deficit is zero if and only if the functions f and g are collinear, i.e.

$$f = \mu g,$$

where μ is a constant.

In addition we have two different error measures that quantify the accuracy of the equilibrium theory in reproducing the evolving flow: the relative L^2-error in the velocity

$$\mathrm{Err}(\vec{v}, \vec{v}^*) = \frac{||\vec{v} - \vec{v}^*||_2}{||\vec{v}||_2}, \tag{12.8}$$

and the relative L^2-error in the vorticity

$$\mathrm{Err}(\omega, \omega^*) = \frac{||\omega - \omega^*||_2}{||\omega||_2}, \tag{12.9}$$

in which the unstarred quantities represent the exact solution and starred quantities refer to the maximum entropy state determined by an approximate dynamics to be described below. In general we expect that the relative error in the vorticity is a more severe test than the corresponding error in the velocity, since the velocity field is usually "smoother" than the vorticity field. Mathematically, the L^2 norm of the velocity is controlled by the L^2 norm of the vorticity.

12.2.2 Approximate dynamics for Langevin and dilute PV theory

We check the accuracy and consistency of predictions from the equilibrium statistical theory for evolving damped and driven flow by an algorithm that we call the *approximate dynamics*. We extract key quantities from snapshots of the evolving flow, including the mean field energy and circulation, produce the most probable states in the equilibrium statistical theory based on this limited information, and calculate the accuracy of the fit based on the statistical measures of the velocity and vorticity fields given in (12.8) and (12.9).

Before we provide the algorithm for the approximate dynamics, we must deal with one subtlety of the equilibrium theory. The approximate dynamics must be based entirely on *mean field* quantities, i.e., the mean field potential vorticity, $\overline{q}$, or stream function, $\overline{\psi}$, since these are predicted by the equilibrium theory and observed in the evolving flow. As described in Chapters 6, 8, and 9 above, the mean field equations from (ESTP) are derived from "coarse graining" the information that appears in the smallest scales.

Quantities such as the vorticity extrema, Q_+ and Q_-, in the Langevin theory (see (11.5)) and the absolute vorticity, A, in the dilute-vortex theory (see (11.6)), must be estimated from the mean field data. Indeed, one of the aims of this section is to produce effective mean field surrogates in the approximate dynamics for these small-scale quantities.

For the Langevin theory, this is quite simple. The potential vorticity maximum, Q_+, in the flow is estimated from the maximum potential vorticity in the pre-existing flow and the additional potential vorticity provided by the forcing, which yields

$$Q_{+} = \max(\overline{q}) + Amp + \text{small amount}, \tag{12.10}$$

where Amp is the amplitude of the forcing discussed in Subsection 12.2.1 and the "small amount" is usually a few percent of maximum in the mean field potential vorticity. The addition of a "small amount" tacitly acknowledges the possible insufficiency of estimating small-scale information from the large-scale field; however, as a practical matter we provide ample evidence that the best fit between the evolving flow and the equilibrium statistical theory is given by $\overline{q}$ and Amp alone. We may define a similar relation for the vorticity minima, Q_{-}, but in all the numerical experiments run with predominantly single-signed vorticity in Subsection 12.2.3, we assume the quantity vanishes, i.e. $Q_{-} = 0$, with no practical loss in accuracy of the equilibrium theory.

We cannot measure the mean absolute vorticity, A, in the evolving flow, since this quantity depends on the details of the small-scale potential vorticity field. Therefore, we must find a related mean field quantity that we can measure from mean field data and that is in one-to-one correspondence with A for all equilibrium solutions produced by the dilute-vortex theory. The simplest such quantity is the related "mean field mean absolute vorticity," $\overline{A}$:

$$\overline{A} = \int_{\Omega} |\overline{q}| \, dx \, dy. \tag{12.11}$$

The quantity, $\overline{A}$, in (12.11) is unambiguously determined from the mean field potential vorticity field.

Algorithm for the approximate dynamics

The equilibrium statistical theory for the damped and driven quasi-geostrophic flow in (12.1) is evaluated quantitatively via an approximate dynamics based on the mean field equations for the Langevin theory or the dilute-vortex theory depending on the context. The form of the algorithm for the approximate dynamics is identical for both theories and may be summarized as follows:

(i) For each numerical experiment, we specify the dissipation, $\mathcal{D}$, the forcing, $\mathcal{F}$, and the initial vorticity field and evolve the flow by the quasi-geostrophic dynamics in (12.1).

(ii) At fixed time intervals, we calculate the instantaneous values of energy, E, and circulation, Γ.

For the algorithm based on the Langevin theory we calculate the potential vorticity extrema, Q_{+}, in (12.10).

For the algorithm based on the dilute-vortex theory we calculate the mean field absolute vorticity, $\overline{A}$, in (12.11).

(iii) The values of the Lagrange multipliers θ, γ, and possibly α for these statistical theories (see Section 11.2) are determined to ensure that the energy, circulation, and possibly the mean field absolute vorticity of the most probable states match the values calculated from the evolving snapshots.
(iv) We calculate the correlation, and relative errors in the vorticity and velocity fields between the instantaneous snapshots and the predictions made by the equilibrium theory.

Thus, we have defined an unambiguous algorithm for the approximate dynamics for the Langevin and dilute-vortex equilibrium statistical theories described in Section 11.2. This algorithm is based entirely on mean field quantities that are easily calculated from snapshots of the evolving flow.

12.2.3 Statistical consistency of freely decaying vortex states

We turn now to the numerical experiments that establish the meta-stability of monopole and dipole vortices through both direct and inverse cascades and the quantitative accuracy of the approximate dynamics described in Subsection 12.2.2. We concentrate on two kinds of experiments: (1) the free decay of vortical flows, which is treated in this section, and (2) the damping and driving of large-scale vortices by small-scale forcing, which is discussed below in Subsection 12.2.4.

The numerical experiments are initialized with monopole and dipole vortices produced as the most probable states of the equilibrium statistical theories described earlier in Section 11.2 of Chapter 11. Here, the initial vortices freely decay and we show that a fluid initialized with an equilibrium state will evolve "close" to other such equilibrium solutions, even as the energy substantially decays and the streamlines exhibit topological transitions as vortical flows decay into shear flows with decreasing energy. The results of this section show that the equilibrium statistical theory is quite successful at tracking the topological transition from vortex to shear flow. These simple laminar tests of the consistency of the statistical theory also provide a point of comparison for the numerical experiments run with small-scale forcings that are described afterwards.

Free decay of vortex monopoles

To illustrate the free decay of vortex monopoles, we initialize the flow with one of two different cases, shown in Figures 12.1(a) and 12.1(b). The monopole in Figure 12.1(a) is "peaked," and the potential vorticity is gently distributed throughout the channel. The monopole in 12.1(b) is "flat," and the potential vorticity is sharply divided into two patch-like regions. These vortices, which are both most probable states in the Langevin statistical theory, have identical energy, $E = 0.4$, but different circulations – the circulation in the peaked example

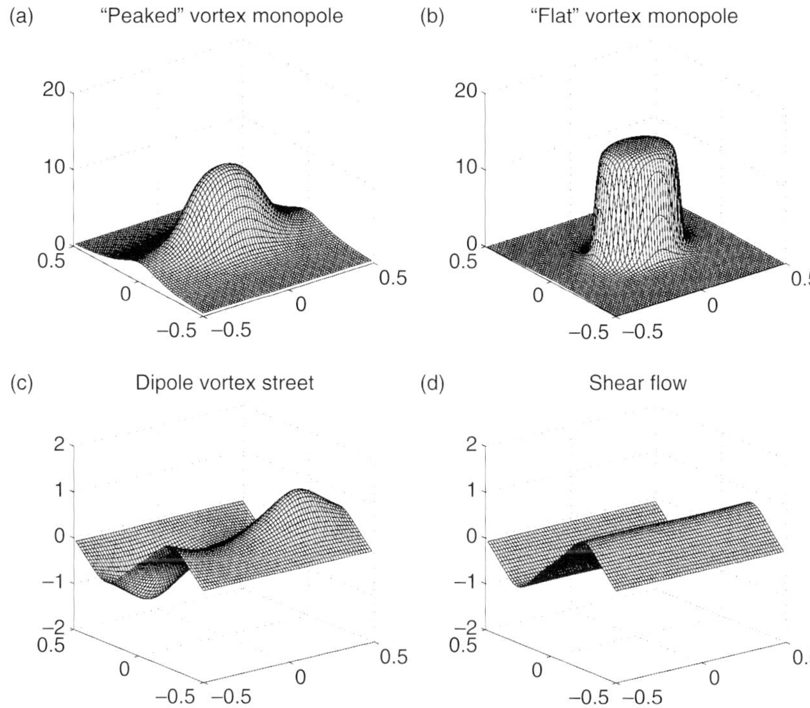

Figure 12.1 Initial potential vorticity surfaces for the numerical experiments run. Each surface is the most probable state produced by the statistical theory. (a) the "peaked" monopole vortex with energy, $E = 0.4$, circulation, $\Gamma = 2.225$, and $\beta = 0.5$. (b) the "flat" monopole vortex with energy, $E = 0.4$, circulation, $\Gamma = 1.85$, and $\beta = 0.5$. (c) a dipole-vortex street with energy, $E = 0.0034$, mean absolute vorticity, $\overline{A} = 0.42$, and $\beta = 0.1$. (d) a shear flow with energy, $E = 0.0034$, mean absolute vorticity, $\overline{A} = 0.42$, and $\beta = 0.1$.

is $\Gamma = 2.225$, and the circulation in the flat example is $\Gamma = 1.85$. The turnaround time for both vortices is near $t = 3$; this is estimated by dividing the mean circumference of the vortex, $d = 2\pi r = \pi/2$, by the "average" velocity in the initial flow, which is estimated at approximately half the peak velocity, $v = 1.2/2$, so that we have

$$\text{vortex turnaround time} = \frac{\pi}{2 \cdot 0.6} \sim 3. \tag{12.12}$$

For the experiments initialized with a peaked vortex and dissipated by Ekman drag and hyper-viscosity, the total run time is usually $T = 1000$, so that the run lasts for more that 300 vortex turnaround times. For the experiments initialized with a flat vortex undergoing Newtonian viscosity, the Reynolds's number is near $Re \sim 2500$, and the total run time is $T = 50$.

In our first numerical experiment the peaked monopole, depicted in Figure 12.1(a), freely decays, dissipated by a combination of Ekman drag, $d_0 = 10^{-3}$, and hyper-viscosity, $d_3 = 10^{-7}$. The geophysical beta-parameter is $\beta = 0.5$. The total run time is $T = 1000$, which is sufficiently long for the initial vortex to decay into a shear flow, which occurs near time $t = 630$. The energy and circulation decrease uniformly throughout the run and the maximum of the potential vorticity increases for a short time after initialization due to the hyper-viscosity, which is shown in Figures 12.2(a) and 12.2(c). Also, the maximum velocity, in the flow, v_{max}

$$v_{max} = max|\vec{v}| \tag{12.13}$$

decreases in the evolution of the flow.

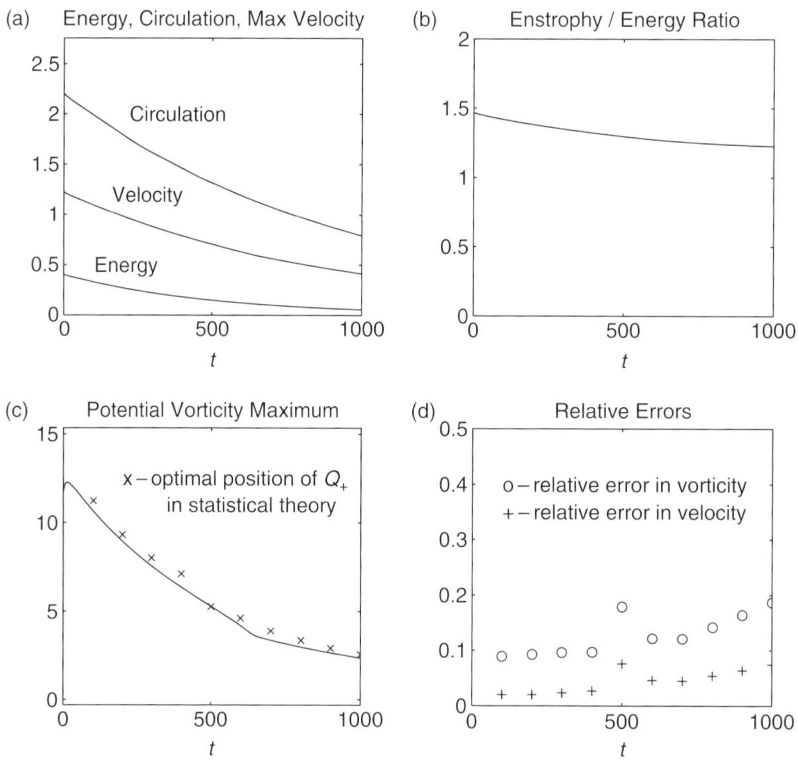

Figure 12.2 Evolving quantities for free decay of peaked monopole shown in Figure 12.1(a), damped by a combination of Ekman drag, $d_0 = 10^{-3}$, and hyper-viscosity, $d_3 = 10^{-7}$. (a) The energy, circulation and maximum velocity in the flow. (b) The enstrophy–energy ratio normalized by π^2. (c) The maximum potential vorticity in the flow and the optimal positions produced by the statistical theory. (d) The relative errors with respect to vorticity field, marked by o's, and with respect to the velocity field, marked by x's.

Three snapshots of the decaying flow are shown in the left-hand column of Figure 12.3 at times $t = 100$, 400, and 700, for the contour lines of the potential vorticity. The equilibrium solutions of the approximate dynamics, which are produced by the algorithm described in Subsection 12.2.2, appear at the corresponding

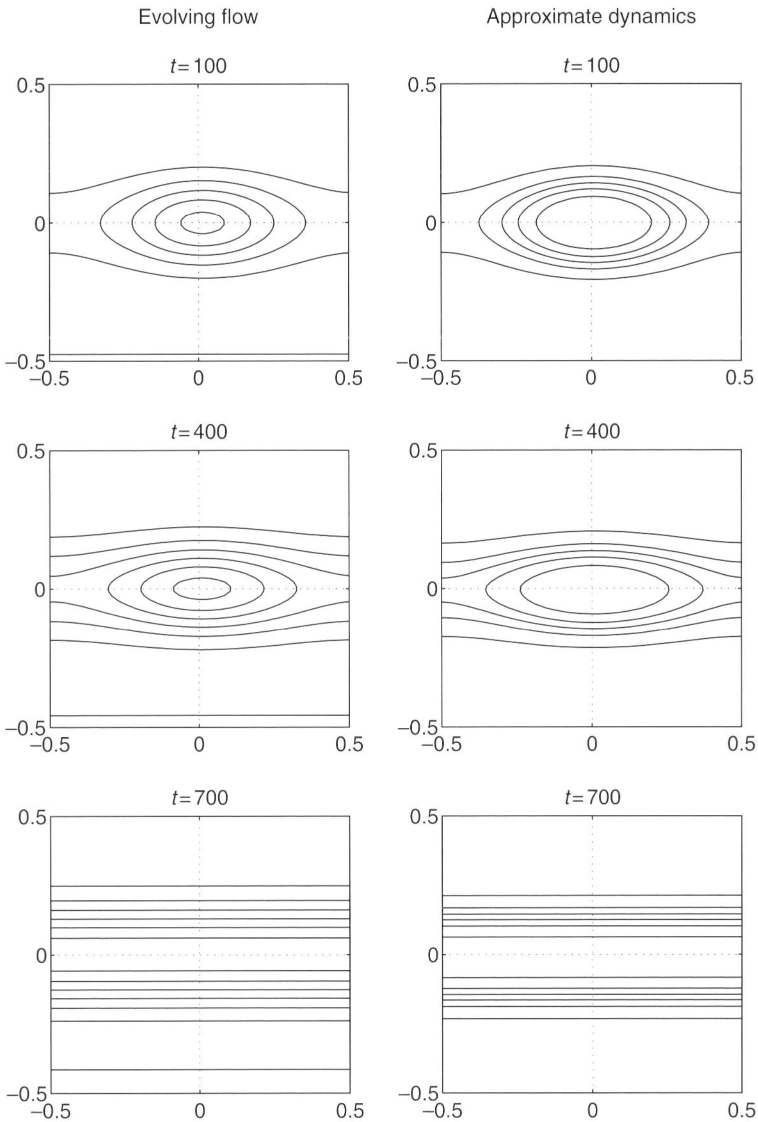

Figure 12.3 Contour lines of the potential vorticity fields at times $t = 100$, 400, and 700 for the numerical experiment described in Figure 12.2. Left-hand column: snapshots of the evolving flow, right-hand column: most probable states of the statistical theory.

positions in the right-hand column of Figure 12.3. The equilibrium solutions are calculated from the instantaneous values of the evolving energy, circulation, and potential vorticity maximum that appear in Figure 12.2. During the numerical experiment the initial vortex slowly weakens and decays to a zonal shear flow at time near $t = 630$ – therefore, two of the snapshots depict vortex flows and one depicts a zonal shear. The qualitative structure of the decaying flow in the left-hand column of Figure 12.3 is clearly captured by the approximate dynamics, which also pinpoints the time at which the evolving flow crosses from vortex to shear flow.

Each of the most probable solutions in Figure 12.3 is very near the *optimal*, in the sense of correlations defined in (12.7), that the approximate dynamics can produce. At each time we extract the instantaneous values of the energy and circulation, and solve the Langevin mean field equations for a range of potential vorticity extrema, Q_+. The values of the potential vorticity maximum that yield the smallest correlations for each snapshot are plotted with x in Figure 12.2(c) – notice that almost all lie just above the maximum PV in the mean field. This supports the use of (12.10) in the approximate dynamics.

The quantitative comparisons, as measured by the relative errors in the velocity and vorticity fields between the evolving and approximate dynamics are shown in Figure 12.2(d). The errors are roughly constant in the early stages of the experiment for times $t < 500$, where the topological structure is a strong monopole. The relative errors in the velocity field are near 2% and relative errors in the vorticity field near 9%. As the monopole vortex weakens and approaches the crossover point to zonal shear flow, there is a bump in the quantitative data, after which the errors slowly increase throughout the run, with relative errors in the velocity field approximately 3–8% and relative errors in the vorticity field approximately 9–20%.

Free decay of vortex streets

To illustrate the free decay of vortex streets, we initialize the flow by the most probable state shown in Figure 12.1(c) and evolve under the quasi-geostrophic dynamics in (12.1) with $\mathcal{F} = 0$. The dissipation, $\mathcal{D}$, is a combination of Ekman drag, $d_0 = 10^{-3}$, and hyper-viscosity, $d_3 = 10^{-7}$. The turnaround time for the vortices is near $t = 12$; this is estimated by dividing the mean circumference of a vortex, $d = 2\pi r = \pi/4$, by the "average" velocity in the flow, which is approximately $v = 0.12/2$, so that

$$\text{vortex turnaround time} = \frac{\pi}{4 \cdot 0.06} \sim 12. \tag{12.14}$$

Since the total run time for the numerical experiments in this section is $T = 750$, each run lasts more than 60 vortex turnaround times.

As the flow evolves, the individual vortices that comprise the initial street weaken and ultimately decay into a zonal shear flow. The energy, E, and absolute vorticity, $\overline{A}$, uniformly decrease, along with the maximum value of the velocity in the flow, which is shown in Figures 12.4(a) and 12.4(c). The circulation, Γ, is identically zero, since the vortex street is spanwise anti-symmetric with respect to the channel, and the quasi-geostrophic dynamics preserves this symmetry (see Subsection 11.2.1 for more details).

A few snapshots of the evolving velocity field are provided in the left-hand column of Figure 12.5, at times $t = 100$, 400, and 700. The weakening vortex

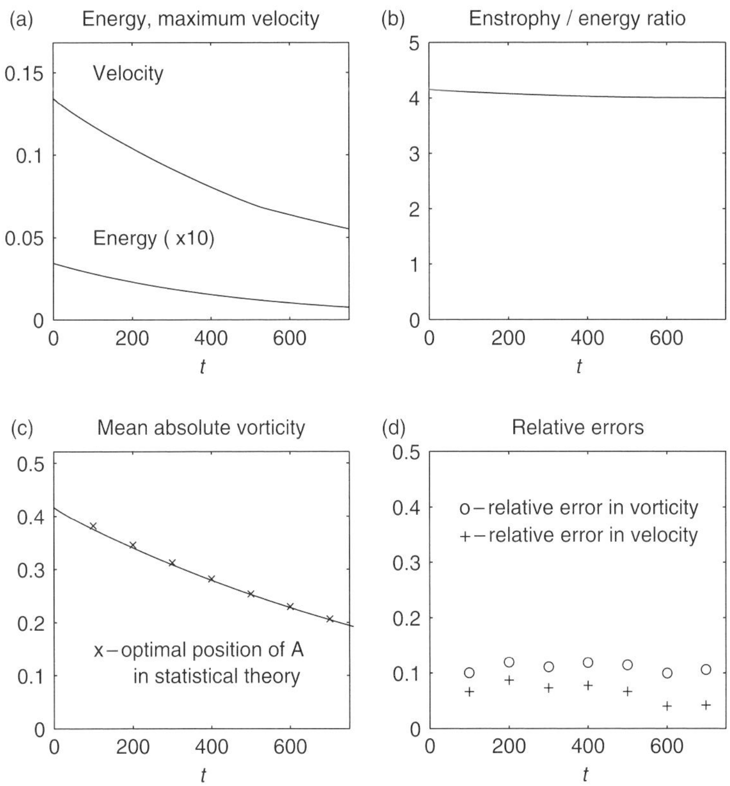

Figure 12.4 Evolving quantities for the free decay of the vortex street shown in Figure 12.1(c), damped by a combination of Ekman drag, $d_0 = 10^{-3}$, and hyperviscosity, $d_3 = 10^{-7}$. (a) The energy and maximum velocity in the flow. (b) The enstrophy–energy ratio, which is normalized by π^2. (c) The mean absolute vorticity in the flow and the optimal positions produced by the statistical theory. (d) The relative errors with respect to vorticity field, marked by o's, and with respect to the velocity field, marked by x's.

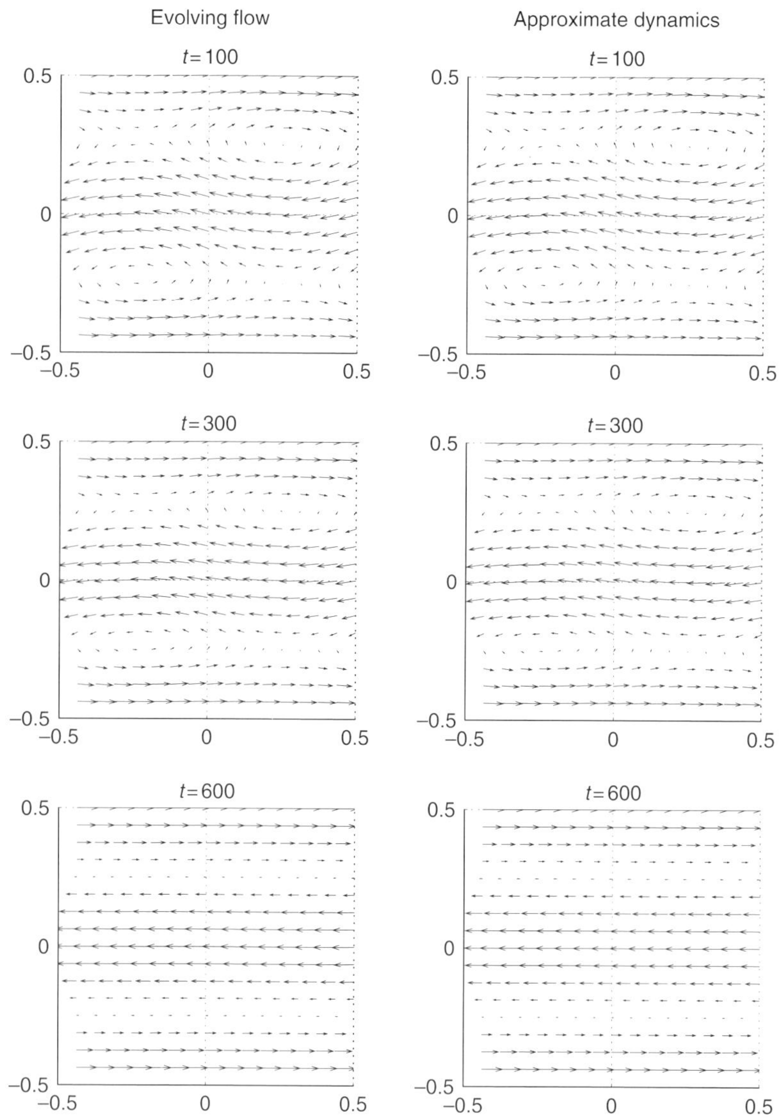

Figure 12.5 Velocity field at times $t = 100$, 300, and 600 for the numerical experiment described in Figure 12.4. Left-hand column: snapshots of the evolving flow, right-hand column: most probable states of the statistical theory.

street is seen in the upper two diagrams, as the arrows indicating the direction and strength of the velocity field straighten and the pattern around the vortices becomes more elliptical. In the diagram shown in the bottom left-hand corner, the flow is a zonal shear and each of the arrows lies parallel to the channel boundaries. The predictions of the equilibrium theory, produced by the algorithm discussed above, are shown in the right-hand column of Figure 12.5. Each of the most probable states has identical energy, and absolute mean field potential vorticity (the circulation very nearly vanishes) as the corresponding snapshot of the evolving flow. The matching velocity fields are indistinguishable by eye. The transition point from vortex to shear flow, which occurs near time $t = 520$, is also pinpointed by the approximate dynamics, although we do not provide a figure to show this.

The relative errors in the velocity and vorticity fields are shown in Figure 12.4(d). The magnitude of the errors between the evolving flow and the equilibrium solutions is very low; approximately 5–8% in the velocity fields and 10–12% in the vorticity fields. We have also calculated the values of the mean field absolute vorticity, $\overline{A}$, that produce most probable states that are optimal, in the sense of correlation deficits defined in (12.7); these values are plotted with x in Figure 12.4(c). For these optimal fits, we match the energy and circulation of the evolving flow, but select the mean field absolute vorticity that produces the lowest correlation deficit with respect to the evolving flow. Notice that the positions of the optimal fit are quite close to the evolving mean field absolute vorticity at all times, which gives strong support for the algorithm for approximate dynamics. There is no practical difference between these "optimal" solutions and those produced by matching the energy, circulation and mean field absolute vorticity in the evolving flow.

12.2.4 Statistical consistency of damped and driven vortex states

In the previous subsection, we demonstrated the meta-stability of freely decaying monopole and dipole vortices initialized by most probable states of an equilibrium statistical theory. Here, we provide more stringent tests on the approximate dynamics – in addition to damping by dissipation that includes both Ekman drag and hyper-viscosity, the evolving flow is driven by small-scale vortices placed randomly in the channel, which mimics a forcing provided by small-scale storms in geophysical flow. We run a number of numerical experiments, including (1) the buildup of a large-scale shear flow via an inverse cascade driven by single-signed vortices, (2) the maintenance of vortex streets with dissipation and small-scale vortex forcing and (3) the emergence of vortex streets from an initial shear flow via an inverse cascade driven by small-scale vortices of both signs.

The key parameter in these experiments is the amplitude of the localized forcing, *Amp*. In the numerical runs initialized by a dipole-vortex street and forced with vortices of both signs, the quantitative measures generally degrade with increased forcing strength. For a strong forcing with $Amp = 3/10$, the peak initial potential vorticity in the flow, relative errors in the velocity measure are usually 20% and relative errors in the vorticity measure are usually 40%.

However, the small-scale forcing helps to maintain the vortex structure of the flow and the qualitative structure is usually captured by the equilibrium statistical theory; larger errors usually occur at higher forcing strengths. The degradation of the relative errors for stronger forcing is undoubtedly related to the dimples in the potential vorticity field due to small-scale vortices not yet strained by the large-scale flow. These corrugations in the potential vorticity surface naturally exert a greater effect on the relative error in the vorticity measure.

Inverse cascade from small-scale forcing of single-signed vortices

A stringent test of the statistical theory is to build a large-scale structure by small-scale forcing of an initially quiescent flow. In order to encourage the formation of coherent structures, we eliminate the Ekman drag, which damps both large and small scales, from the dissipation, i.e., $d_0 = 0$ in the operator, $\mathcal{D}$, defined in (12.3), and damp the evolving flow solely with hyper-viscosity, $d_3 = 10^{-10}$. The flow, computed here at 256×256 resolution, is forced by localized single-signed vortices with radius, $r = 12/256$, $(= 3/64)$, and amplitude, $Amp = 0.1$, placed at random locations in the channel at time intervals, $\Delta t = 0.1$. Thus, by the end of the numerical experiment at time $T = 1000$, we have bombarded the evolving flow with 10 000 localized vortices. Also, for simplicity in discussion we set $\beta = 0$ in these experiments. The emerging structures in these experiments are shear flows – accurately tracked by the approximate dynamics with errors in the velocity field approaching 3% after nearly 10 000 forcings.

The circulation, maximum velocity, and square root of the energy shown in Figure 12.6(a) all grow linearly in this numerical run, which is a consequence of the relatively weak dissipation. The potential vorticity maxima, shown in Figure 12.6(c), also grow linearly and exhibit a transition from a faster to slower rate of growth, which appears near time $t = 200$. The vorticity minima decrease slightly in the run due to the effects of hyper-viscosity. In contrast to the smooth lines in Figure 12.6(a), the potential vorticity extrema depicted in Figure 12.6(c) are jagged, showing the effects of random forcing.

In the left-hand column of Figure 12.7 we show three snapshots of the potential vorticity surface of the evolving flow, at times $t = 100$, 500, and 1000. The jaggedness of the surfaces clearly show the effects of the small-scale forcing. At time $t = 100$ individual vortices have begun to cluster, forming the zonal shear flow; by time $t = 1000$ the shear flow is strong, although it exhibits a curious rippling in the potential vorticity surface in the center of the channel. The most probable states of the equilibrium theory are provided in the matching diagrams

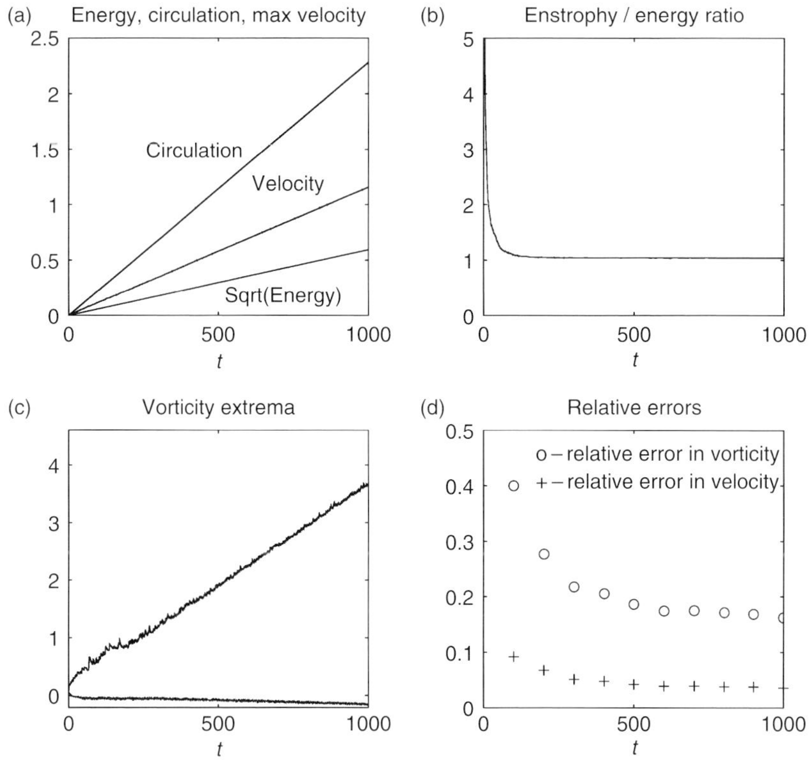

Figure 12.6 As in Figure 12.2, evolving quantities for initially quiescent flow in a channel. Here, the flow is only damped by hyper-viscosity, $d_3 = 10^{-10}$, and forced by localized vortices with amplitude, $Amp = 0.1$.

in the right-hand column of Figure 12.7. Each of these solutions is formed from extracting the energy, E, circulation, Γ, and potential vorticity maximum, Q_+, from the evolving snapshots, and solving the Langevin mean field equation for these values.

Naturally, each of the steady most probable solutions shown in the right-hand column of Figure 12.7 are smooth, even for early times in the numerical run. We might reasonably expect that the quantitative fit between the evolving snapshots and the most probable states of the equilibrium theory improves at longer times. This is shown in Figure 12.6(d), in which the relative errors in the vorticity field uniformly decrease from nearly 40% at time $t = 100$ to 16% at time $t = 1000$ and the relative errors in the velocity field uniformly decrease from 9% to 3% in the same time range. As in other experiments with small-scale forcing, the errors in the velocity field are much lower than the errors in the vorticity field. Indeed, the strongly forced evolving flow quickly approaches the ground state, which is indicated in Figure 12.6(b) by the rapid approach of enstrophy–energy ratio

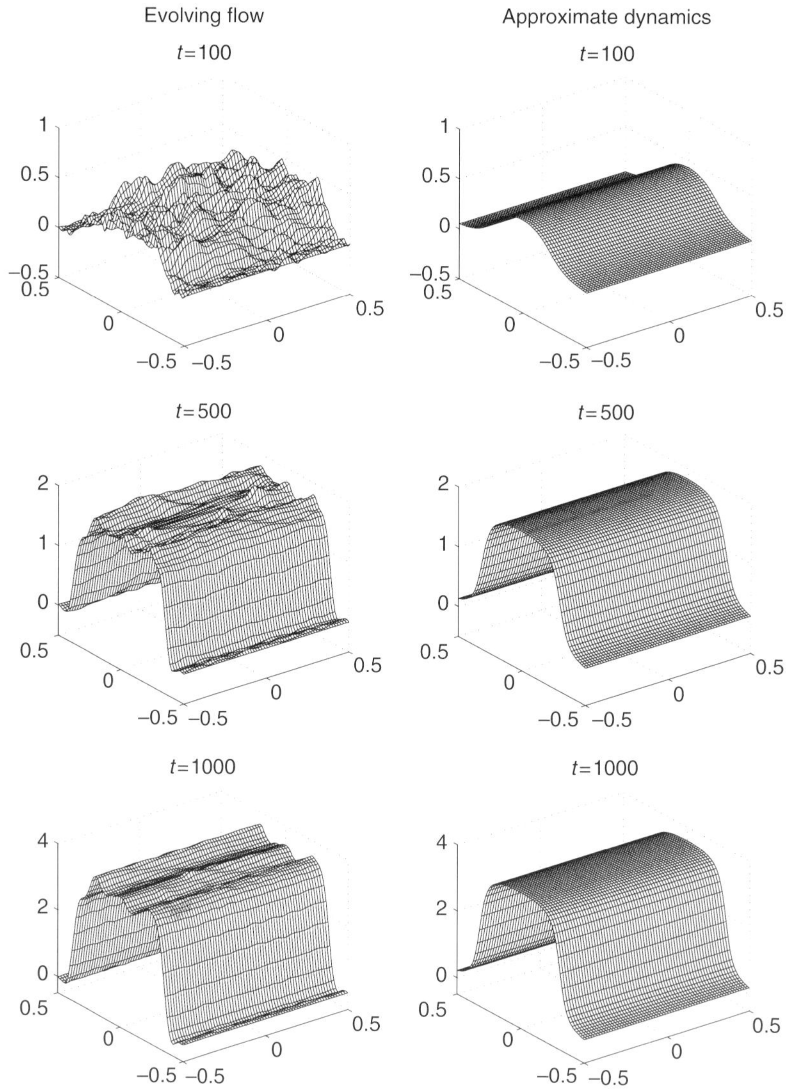

Figure 12.7 Potential vorticity surfaces at times $t = 100$, 500, and 1000 for the numerical experiment described in Figure 12.6. Left-hand column: snapshots of the evolving flow, right-hand column: most probable states of the statistical theory.

to its minimum value, $\Lambda = \pi$. A direct comparison of the most probable states produced by the statistical theory to the linear ground state solutions, which are the eigenfunctions of the Laplacian operator for the channel domain, shows that the two are closely correlated throughout the run, decreasing from a difference of 3% at time $t = 100$ to 2% at time $t = 1000$.

Maintenance of vortex streets by small-scale, double-signed forcing

In Subsection 12.2.3, we showed that freely decaying vortex streets are accurately tracked by the approximate dynamics; here we repeat these numerical experiments with the identical initial conditions and dissipation, adding small-scale forcing by localized vortices of both signs. In general, the forcing delays the transition from vortex to shear-flow vortex streets, even as the vortices that make up the large-scale street are bombarded by localized vortices of the opposite sign.

We repeat the numerical experiment described earlier in Subsection 12.2.3, retaining the level of dissipation and the geophysical beta-parameter but adding small-scale vortices of amplitude, $Amp = 0.1$ and 0.3, in two separate runs. As before the small-scale forcings are placed at random grid points on the channel at regular intervals, $\Delta t = 0.1$, and each such vortex has identical radius, $r = 3$ grid points, where the full channel width is 64 grid points.

For the weaker case with $Amp = 0.1$, the topological transition from vortex street to shear flow is delayed by the small-scale forcing, which helps to maintain the vortical structure in the flow. Indeed, the evolving flow never quite reverts to a zonal shear, but retains a slight signature of the dipolar pattern in the velocity field. The approximate dynamics tends to slightly underpredict the strength of the vortex street, predicting shear flows where the evolving flow contains weak vortex street. Quantitatively, the approximate dynamics performs quite well in this case, with relative errors in the velocity field between 8 and 20% and relative errors in the vorticity field between 15 and 35% throughout the run.

For the stronger case with $Amp = 0.3$, a strong vortex street is maintained throughout the run. The strength of the forcing can be seen in the jaggedness of the key quantities in the flow shown in Figure 12.8, including energy, E, and mean field absolute vorticity, $\overline{A}$. The enstrophy–energy ratio, which is depicted in Figure 12.8(b), holds roughly constant throughout the initial stages of the run, and increases towards the end, which suggests that energy introduced by the localized forcing is entering into the smaller scales. As the largest scales continue to decay – the energy is decreasing throughout the run – the forcing, whose amplitude is held fixed, becomes relatively stronger, and helps to maintain the vortex street.

A few snapshots of the evolving velocity field are provided in the left-hand column of Figure 12.9, at times $t = 100$, 400, and 700. The vortex street remains quite strong throughout the run, with well-formed vortices and a clear meander to the jet that traverses the center of the channel. The predictions of the equilibrium theory are shown in right-hand column of Figure 12.9. Each of these most probable states has identical energy, circulation, and absolute mean field potential vorticity as the corresponding snapshot of the evolving flow.

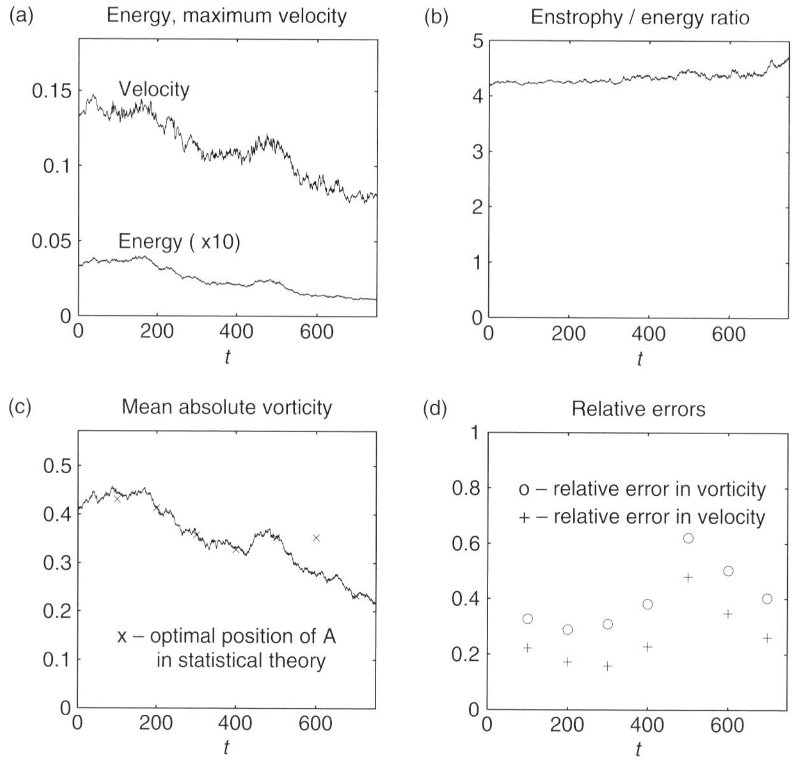

Figure 12.8 As in Figure 12.2, evolving quantities for the vortex street shown in Figure 12.1(c) undergoing damping and driving. The flow is damped by a combination of Ekman drag, $d_0 = 10^{-3}$, and hyper-viscosity, $d_3 = 10^{-7}$, and strongly forced by small-scale vortices with amplitude, $|Amp| = 0.3$.

The quantitative diagnostics for the experiment, the correlation and relative errors in the velocity and vorticity fields, are shown in Figure 12.8(d). The errors in both the velocity and vorticity fields have increased at the higher forcing amplitude: the errors in the velocity field are approximately 20%, excluding the time period between $t = 500$ and $t = 600$, where the approximate dynamics fails, and the errors in the vorticity field are approximately 40%. For the interval of time between $t = 500$ and $t = 600$ the relative velocity errors in the approximate dynamics reach 50%. In this narrow interval of time the equilibrium statistical theory incorrectly produces a shear flow as the most probable state based on the evolving energy, circulation, and absolute vorticity. The rapid increase in the relative errors in the velocity and vorticity fields shows the importance of predicting the correct topological structure in the evolving flow.

Although the relative errors in the velocity field have increased to nearly 20% in this run, the approximate dynamics clearly captures the qualitative structure

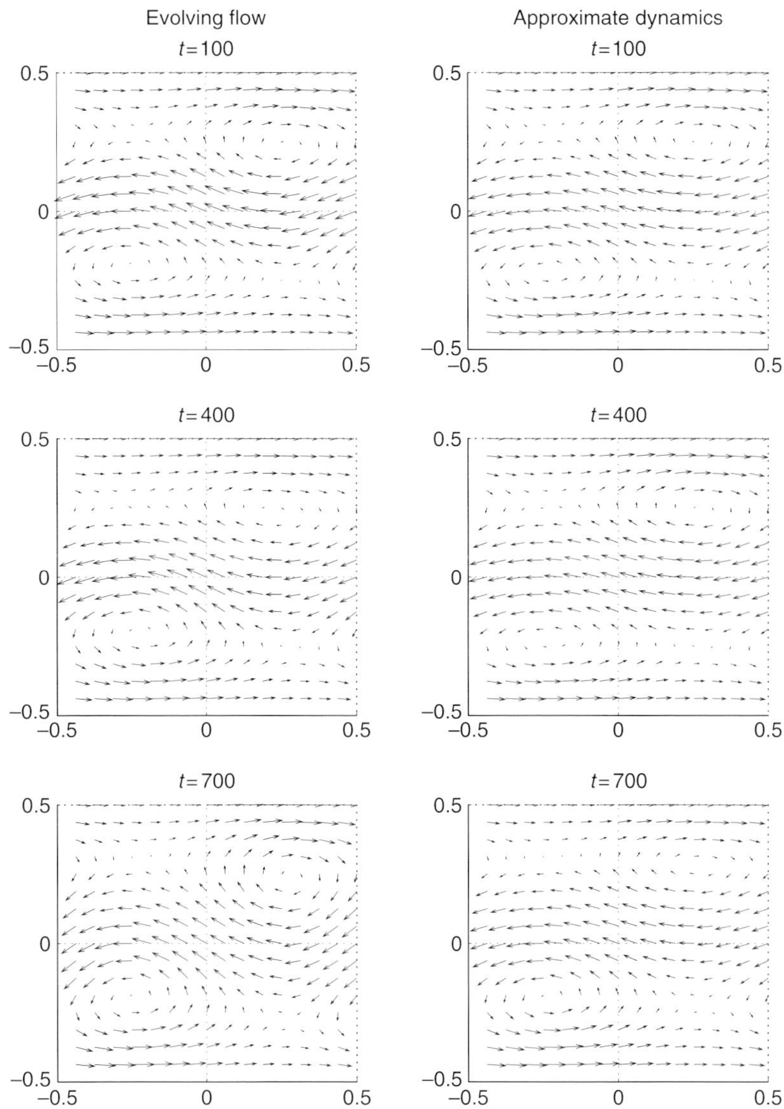

Figure 12.9 Velocity field at times $t = 100$, 400, and 700 for the numerical experiment described in Figure 12.8. Left-hand column: snapshots of the evolving flow, right-hand column: most probable states of the statistical theory.

of the velocity field in the evolving flow, as is demonstrated in Figure 12.9. The relative errors in the potential vorticity field have increased to nearly 40%; however, the basic structure of the vortex street is also seen in the potential vorticity surfaces produced by the approximate dynamics. In the left-hand column of Figure 12.10 we provide the evolving PV surfaces at times $t = 100$, 400, and 700. Here, the strong small-scale forcing, which includes localized vortices of

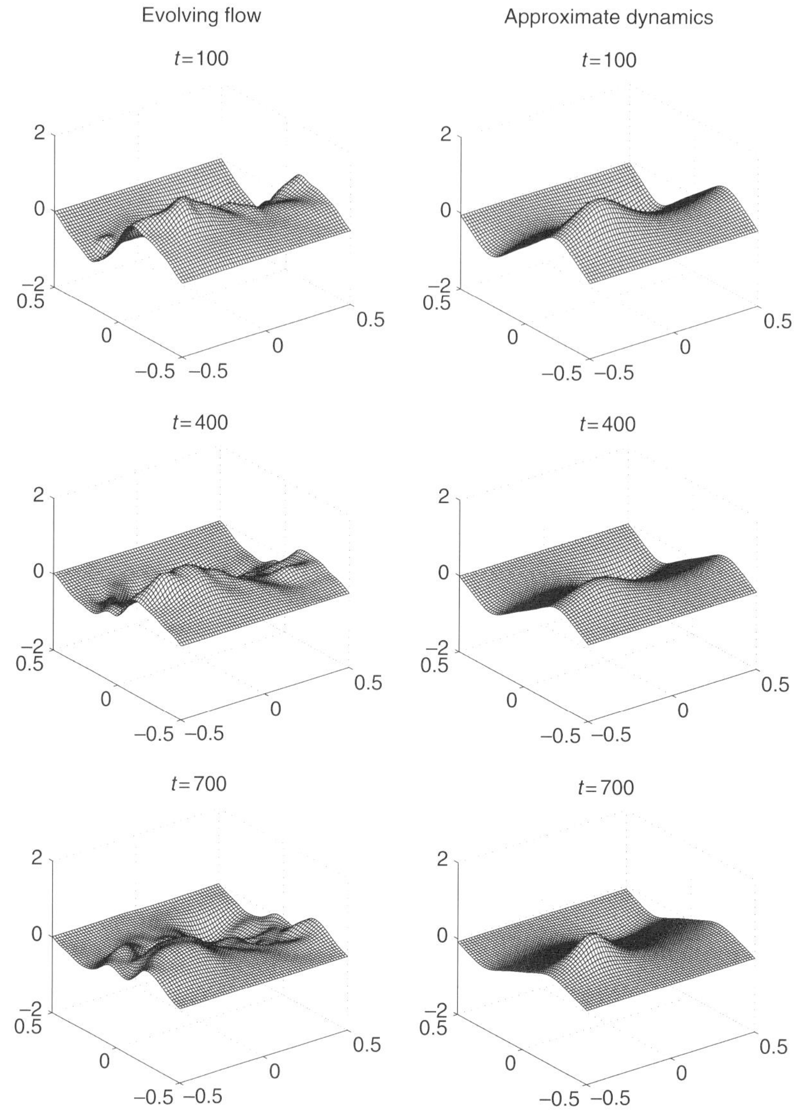

Figure 12.10 Potential vorticity field at times $t = 100$, 400, and 700 for the numerical experiment described in Figure 12.8. Left-hand column: snapshots of the evolving flow, right-hand column: most probable states of the statistical theory.

both signs, introduces large ripples into the potential vorticity surfaces of the evolving flow. The most probable states of the equilibrium theory are shown in the right-hand column of Figure 12.10. The equilibrium most probable states are much smoother than the evolving flow field; nevertheless, the strong cores of the damped and driven vortex streets are clearly reproduced. The degradation of the relative errors in the vorticity field is predominantly caused by the large dimples in the PV field due to localized vortices that have not yet been strained by the large-scale street. A comparison between the velocity fields in Figure 12.9, which are largely smooth even in the evolving snapshots, and the potential vorticity fields in Figure 12.10, demonstrate that the error in the relative vorticity is a much tougher measure for the approximate dynamics.

Vortex coalescence in strongly forced shear flow

In previous subsections we showed that the approximate dynamics performs reasonably well in tracking the evolution of vortex streets bombarded by localized vortices, even at forcing amplitudes at nearly 3/10 the peak vorticity in the initial flow. Here, we test the statistical theory under far more stringent conditions: we repeat the damped and driven numerical experiments run in previous subsections – damped by a combination of Ekman drag and hyper-viscosity and driven by small-scale vortices of both signs with the large amplitude, $|Amp| = 0.3$ – and replace the initial vortex street with the initial shear flow shown in Figure 12.1(d). During the numerical run, the localized forcing eventually creates and maintains a vortex street in the channel.

The evolution of key quantities in the flow, which includes the energy, absolute vorticity, and maximum velocity, are shown in Figures 12.11(a) and 12.11(c). The energy, absolute vorticity, and maximum velocity generally decrease during the length of the numerical experiment. The jaggedness in these curves is caused by the localized vortices added to the potential vorticity field.

The main effect of the small-scale forcing is to induce and maintain a vortex street in the evolving flow. Although the initial flow field is a zonal shear, the small-scale forcings quickly coalesce into the counter-rotating vortices characteristic of the street. A few snapshots of the evolving velocity field are provided in the left-hand column of Figure 12.12, at times $t = 100$, 300, and 640. A weak vortex street is seen in the upper diagram; it gradually strengthens as time evolves, as is evident in the lower diagrams in the column.

The predictions of the equilibrium theory, which are calculated by the approximate dynamics described in Subsection 12.2.2, are shown in the right-hand column of Figure 12.12. Each of these most probable states has identical energy, circulation, and absolute mean field potential vorticity as the corresponding snapshot of the evolving flow. Here, the approximate dynamics tend to predict shear flows – see times $t = 100$ and $t = 300$ – for the weaker vortex streets in the evolving flow. At time $t = 640$, where the strength of the evolving street has strengthened as indicated by the increasing meander in the central jet, the equilibrium theory produces a vortex street, although weaker than the evolving flow. Throughout the

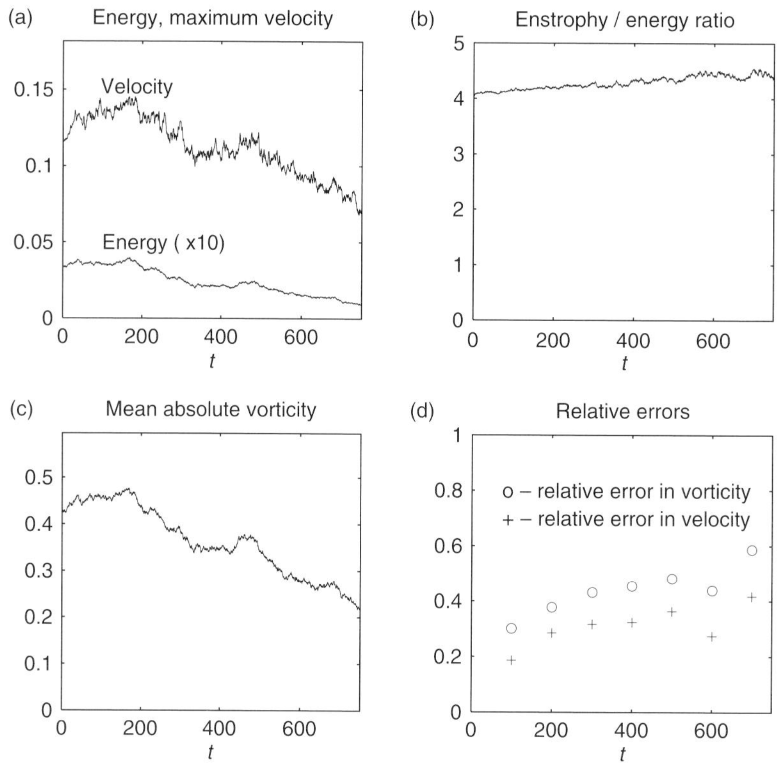

Figure 12.11 As in Figure 12.2, evolving quantities for an initial shear flow shown in Figure 12.1(d) undergoing damping and driving. The flow is damped by a combination of Ekman drag, $d_0 = 10^{-3}$, and hyper-viscosity, $d_3 = 10^{-7}$, and strongly forced by small-scale vortices with amplitude, $|Amp| = 0.3$.

run the approximate dynamics tends to qualitatively underpredict in this fashion, producing zonal shears for weak streets and weak vortex streets for strong ones.

Quantitatively, the relative errors, which are shown in Figure 12.11(d), are fairly high, usually between 20 and 40% in the velocity field and between 30 and 60% in the vorticity field. This is not unexpected given the stringency of the test – tracking a vortex flow induced by small-scale forcing in an initial shear flow. For the times when the most probable state achieves the correct topological structure, the relative errors, especially in the velocity field, are in the lower portions of these ranges. Indeed, for the strong vortex street that appears at time, $t = 640$, which is shown in the bottom row of Figure 12.12, the errors in the velocity field fall to 16%. Thus, the approximate dynamics is capable of tracking the transition from zonal shear to vortex flow induced by small-scale forcing, although it tends to underpredict the strength of the vortex street and frequently predicts zonal shears if the vortex street is sufficiently weak.

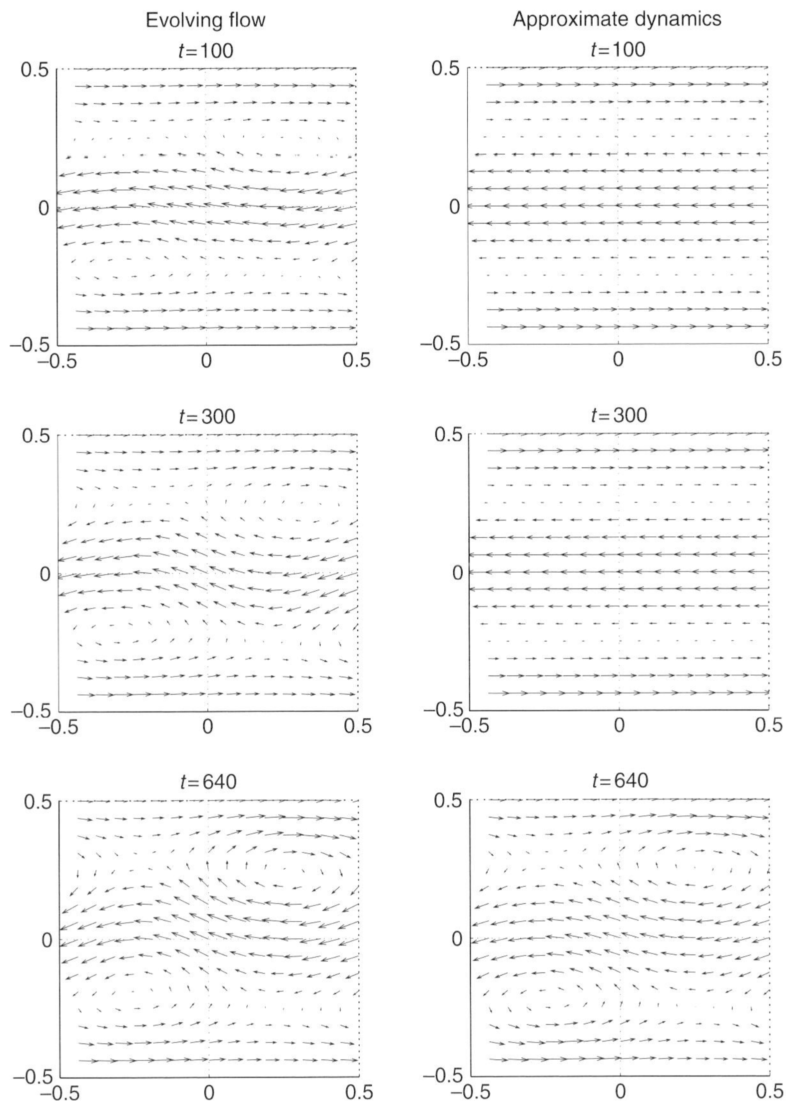

Figure 12.12 Velocity field at times $t = 100$, 300, and 640 for the numerical experiment described in Figure 12.11. Left-hand column: snapshots of the evolving flow, right-hand column: most probable states of the statistical theory.

12.3 Crude closure for two-dimensional flows

We continue our investigation of the applicability of equilibrium statistical theories, especially ESTP, in the environment of forcing and damping. In the previous section, we demonstrated the meta-stability of the equilibrium statistical structures in the environment of damping and small-scale random forcing. There, the quantities such as the energy and circulation needed in the equilibrium statistical theory are computed from the numerical solutions of the exact dynamics of the barotropic quasi-geostrophic equations. Here we perform a much more stringent test. Namely, we will compute the quantities such as the energy and circulation needed in the ESTP theory via a crude closure problem which does not depend on the exact fluid flow. The crude closure problems can be solved very efficiently. Again, we consider environment with dissipation and small-scale forcing. The crude closure procedure performed amazingly well under a wide range of circumstances. This further addresses the basic issues (A-4) and (T-5) discussed in Chapter 10. The numerical results in this section are due mostly to Majda and Holen (1997) and Grote and Majda (1997).

12.3.1 The equation and the problem

For simplicity we consider here a special case of the barotropic quasi-geostrophic equation without geophysical effects ($h \equiv 0, \beta = 0$), but with damping and forcing in a two-dimensional box $Q = [0, \pi] \times [0, \pi]$, equipped with stress-free boundary conditions. The basic dynamic equation is

$$
\begin{aligned}
\frac{\partial \omega}{\partial t} + \vec{v} \cdot \nabla \omega &= \sum_{j=1}^{2} d_j (-\Delta)^j \psi + \mathcal{F}, \\
\omega &= \Delta \psi, \\
\vec{v} = \nabla^{\perp} \psi &= \left(-\frac{\partial \psi}{\partial y}, \frac{\partial \psi}{\partial x} \right),
\end{aligned}
\tag{12.15}
$$

where the dissipation operate, $\mathcal{D}$, and external random small-scale forcing, $\mathcal{F}$, are discussed in Subsection 12.2.1 on meta-stability of equilibrium statistical structures.

The stress-free boundary condition is equivalent to the swimming pool geometry (see Chapter 1), i.e. the stream function takes the form, in terms of Fourier expansion

$$
\psi(x, y, t) = \sum_{k_1, k_2 \geq 1} a_{k_1, k_2}(t) \sin(k_1 x) \sin(k_2 y). \tag{12.16}
$$

At any time, t, the energy, E, and the circulation, Γ, are given by

$$
\begin{aligned}
E(t) &= \frac{1}{2} \int_{\Omega} |\vec{v}|^2 = -\frac{1}{2} \int_{\Omega} \omega \psi, \\
\Gamma(t) &= \int_{\Omega} \omega.
\end{aligned}
\tag{12.17}
$$

Recall from Chapter 9 that the most probable mean field predicted by various equilibrium statistical theories, with few external constraints such as the energy and the circulation, satisfies the following mean field equation

$$\omega^* = f(\psi^*; E, \Gamma), \tag{12.18}$$

where $f = f(\psi^*; E)$ in the case of the energy–enstrophy theory.

In particular, we recall from Chapters 6 and 8 that the energy–enstrophy theory (EEST) produces the ground states as the most probable mean field which takes the form, in our special swimming pool symmetry (12.16)

$$\omega^* = A \sin x \sin y, \tag{12.19}$$

where A is any parameter. Here there are two ground states for a given energy with coefficient $\pm A$ and the uniqueness is enforced by the sign of the circulation. The statistical theory of point vortices (PVST) in the periodic setting leads to the familiar Sinh–Poisson equation

$$\Delta \psi^*_{c,b} = -c \sinh(b\psi^*_{c,b}), \tag{12.20}$$

where the parameters b and c are Lagrange multipliers and are uniquely determined by E and Γ. The analytic expressions for the solution of the Sinh–Poisson equation in this special periodic setting are described in Section 3.7 from Chapter 3.

12.3.2 Description of the crude closure dynamics

In this subsection we investigate whether the empirical statistical theories with a prior distribution (ESTP), especially the point-vortex theory (PVST) and the energy–enstrophy theory (EEST) can be used to define crude closure dynamics for time-dependent flows in a non-equilibrium forced regime. In other words we will study the possibility of constructing an algorithm which predicts the temporal evolution of bulk features of the flow, such as the energy and the circulation, using only the statistical macrostates and without solving the flow equations. For this purpose we postulate two main assumptions on designing these crude closure algorithms

1. All dissipation of energy and circulation occurs in the large-scale statistical flow.
2. The effect of small-scale forcing is to instantaneously change the large-scale statistical structure to a new large-scale statistical structure with an adjusted energy and circulation.

Let us assume that the flow is initially at statistical equilibrium, which is completely defined by its energy and possibly its circulation. As time progresses, the flow evolves under the constant action of small-scale forcing and dissipation. This is where the two assumptions we made become crucial. With the addition of every new localized vortex, the energy and circulation slightly change, thereby forcing the large-scale flow into a new statistical state with adjusted energy and circulation. Next, due to dissipation, the energy and circulation decrease; again we estimate that decline by assuming that all dissipation occurs in the large scales of the flow. This strategy will enable us to devise a crude closure algorithm, which captures the large-scale behavior of the damped and driven flow only by tracking the energy and possibly the circulation, without actually solving the fluid equations.

As the flow evolves under the constant bombardment with localized vortices, it alternates between two extreme flow regimes: a phase of applied forcing and a phase of pure decay. We begin with the addition of a single vortex, ω, at $t=\bar{t}$ to the current approximate state, $\bar{\omega}$, of the flow. Let $\bar{t}_-$ denote the instant just before and $\bar{t}_+$ the instant right after the event. Then

$$\psi_{\bar{t}_+} = \bar{\psi}_{\bar{t}_-} + \psi, \quad \omega_{\bar{t}_+} = \bar{\omega}_{\bar{t}_-} + \omega. \tag{12.21}$$

The change of the energy, E, and circulation, Γ, can be calculated easily using these equations and (12.17). We immediately obtain

$$\begin{aligned} E(\bar{t}_+) &= E(\bar{t}_-) - \int_\Omega \bar{\psi}_{\bar{t}_-}\,\omega - \frac{1}{2}\int_\Omega \psi\omega, \\ \Gamma(\bar{t}_+) &= \Gamma(\bar{t}_-) + \int_\Omega \omega. \end{aligned} \tag{12.22}$$

Since $\bar{\psi}_{\bar{t}_-}$, ω, and ψ are explicitly known, we can evaluate the right-hand sides of (12.22) to update the energy and the circulation. These new values of E and Γ define the new state $\bar{\omega}$ of the approximate solution via (12.18), which depends on the statistical theory considered.

Next, we let Δt denote the time interval between forcing by two successive random vortices. Given the approximate solution $\bar{\omega}$ at $t=\bar{t}_+$, we need to determine the evolution of the approximate dynamics during the decay phase, $\bar{t}_+ \le t < \bar{t}+\Delta t$, up to the instant just before adding the next localized vortex, $t=\bar{t}+\Delta t$. To do so, we must determine the change in energy and circulation in a situation without forcing.

In the case of Ekman drag, i.e. $\nu = 0, d > 0$, it is straightforward to derive exact expressions for the evolution of the energy and the circulation in a situation of pure decay

$$\begin{aligned} E(t) &= e^{-2d\Delta t} E(\bar{t}_+), \\ \Gamma(t) &= e^{-d\Delta t}\Gamma(\bar{t}_+), \quad \bar{t}_+ \le t < \bar{t}+\Delta t. \end{aligned} \tag{12.23}$$

By applying these equations at $t = (\bar{t} + \Delta t)_{-}$, we obtain the new values of E and Γ. This defines the new approximate dynamics solution $\bar{\omega}$ through (12.18) and completes the crude dynamics closure for Ekman drag.

In the case of Newtonian dissipation, i.e. $\nu > 0$, $d = 0$, we do not have simple exact expressions that describe the evolution of E and Γ. Thus we need to estimate the change in E and Γ over the decay phase, $\bar{t}_{+} \le t < \bar{t} + \Delta t$ by using an approximate solution of (12.15) with $\mathcal{F} = 0$. To do so, we approximate the vorticity by the first-order term in the Taylor series expansion

$$\tilde{\omega} = \bar{\omega} + (t - \bar{t}) \left. \frac{\partial \bar{\omega}}{\partial t} \right|_{\bar{t}_{+}} . \tag{12.24}$$

We remark that higher order or Padé approximations also can be used, but our results below in Subsection 12.3.3 indicate that these more sophisticated approximations are not needed. Since $\bar{\psi}$ solves (12.18), the non-linear terms in (12.15) vanish at $t = \bar{t}_{+}$, and we obtain

$$\frac{\partial \bar{\omega}}{\partial t} = \nu \Delta f(\bar{\psi}; \bar{E}, \bar{\Gamma}), \quad t = \bar{t}_{+}. \tag{12.25}$$

Here $\bar{E}$ and $\bar{\Gamma}$ denote the (fixed) energy and circulation associated with $\bar{\omega}$ at $t = \bar{t}_{+}$ via (12.17). We introduce the above equation into (12.24), which yields

$$\tilde{\omega} = \bar{\omega} + \nu(t - \bar{t}) \Delta f(\bar{\psi}; \bar{E}, \bar{\Gamma}), \quad \bar{t}_{+} \le t < \bar{t} + \Delta t. \tag{12.26}$$

The accuracy of this approximation depends mainly on the size of $\nu(t - \bar{t})$. By using $\tilde{\omega}$ from (12.26), we calculate the approximate change in $\bar{E}$ and $\bar{\Gamma}$. The approximate new value of the circulation, $\bar{\Gamma}(t)$, is

$$\bar{\Gamma}(t) = \Gamma(\bar{t}_{+}) + \nu(t - \bar{t}) \int_{\Omega} \Delta f(\bar{\psi}; \bar{E}, \bar{\Gamma}), \quad \bar{t}_{+} \le t < \bar{t} + \Delta t. \tag{12.27}$$

To compute the change in E, we recall that

$$\frac{dE(t)}{dt} = -\nu \int_{\Omega} |\nabla v|^2 = -\nu \int_{\Omega} \omega^2. \tag{12.28}$$

Next we integrate (12.28) in time over $[\bar{t}_{+}, \bar{t} + \Delta t)$, with ω replaced by $\tilde{\omega}$, and use (12.26) to evaluate the integral explicitly. This yields the approximate new value of the energy, $\bar{E}(t)$,

$$\begin{aligned} \bar{E}(t) = \bar{E} - \nu(t - \bar{t}) \int_{\Omega} \bar{\omega}^2 - \nu(t - \bar{t})^2 \int_{\Omega} \bar{\omega} \Delta f(\bar{\psi}; \bar{E}, \bar{\Gamma}) \\ - \frac{\nu^2 (t - \bar{t})^3}{3} \int_{\Omega} [\Delta f(\bar{\psi}; \bar{E}, \bar{\Gamma})]^2. \end{aligned} \tag{12.29}$$

Finally, we set $t = (\bar{t} + \Delta t)_-$ in (12.29) to update $\bar{E}$ and $\bar{\Gamma}$, which completes the crude dynamics closure for the barotropic fluid equations with Newtonian dissipation.

We emphasize that the formulas derived above, which define the crude closure dynamics, track the energy and circulation only approximately. Indeed, whether for Ekman drag or for Newtonian dissipation, the interaction between the current state of the flow $\bar{\omega}$ and the newly added localized vortex ω is computed only approximately; it cannot be calculated exactly without explicit knowledge of the exact solution to the flow equations. The case utilizing explicit knowledge of exact solutions for equilibrium statistical theories was discussed in the Section 12.2. However, in the case of Ekman drag *without external forcing*, and only in this special case, the crude dynamics follow the evolution of E and Γ exactly. For the sake of clarity, we summarize the crude closure algorithm below.

Crude closure algorithm

Step 0. At $t = 0$ match $\bar{E}$ and $\bar{\Gamma}$ with the energy and circulation of the initial conditions. By (12.18) this defines the initial approximate statistical state $\bar{\omega}$.

Step 1. Let $\bar{E}_-$ and $\bar{\Gamma}_-$ denote the current approximate values of the energy and the circulation, which define the statistical state $\bar{\omega} = f(\bar{\psi}; \bar{E}_-, \bar{\Gamma}_-)$ at $t = \bar{t}_-$. Compute the change in $\bar{E}$ and $\bar{\Gamma}$ due to the addition of a localized vortex ω

$$\bar{E}_+ = \bar{E}_- - \int_\Omega \bar{\psi}\omega - \frac{1}{2}\int_\Omega, \quad \bar{\Gamma}_+ = \bar{\Gamma}_- + \int_\Omega \omega. \tag{12.30}$$

This defines the new statistical state $\bar{\omega}$ at $t = \bar{t}_+$ via $\bar{\omega} = f(\bar{\psi}; \bar{E}_+, \bar{\Gamma}_+)$.

Step 2. Compute the change in $\bar{E}$ and $\bar{\Gamma}$ during pure decay, $\bar{t} < t < t + \Delta t$.
For Ekman drag

$$\bar{E} = e^{-2d\Delta t}\bar{E}_+, \quad \bar{\Gamma} = e^{-d\Delta t}\bar{\Gamma}_+. \tag{12.31}$$

For Newtonian dissipation

$$\begin{aligned} \bar{E} &= \bar{E}_+ - \nu\Delta t \int_\Omega \bar{\omega}^2 - \nu(\Delta t)^2 \int_\Omega \bar{\omega}\Delta f(\bar{\psi}) - \frac{\nu^2(\Delta t)^3}{3}\int_\Omega [\Delta f(\bar{\psi})]^2, \\ \bar{\Gamma} &= \bar{\Gamma}_+ + \nu\Delta t \int_\Omega \Delta f(\bar{\psi}). \end{aligned} \tag{12.32}$$

This defines the new statistical state $\bar{\omega}$ at $t = \bar{t} + \Delta t$.

Step 3. Return to 1.

We emphasize that for given initial values of the energy and the circulation, the crude closure algorithm marches in time independently and without any knowledge of the solution of (12.15). It only requires knowledge of the applied external forcing and of the dissipation coefficient.

Now we have proposed the crude closure algorithm, our main task is to investigate how close the statistical states are to the exact solution and how well the closure algorithm tracks the energy and the circulation of the time-dependent evolving flow. For this purpose we shall compare the exact solution with two types of approximate statistical states. On the one hand, we shall verify the *consistency* of the equilibrium statistical theories in a non-equilibrium regime simply by matching the instantaneous exact energy, E, and circulation, Γ, of the flow. Thus, given the exact computed values of E and Γ, we can compare the statistical state $\omega^* = f(\psi^*; E, \Gamma)$ with the exact solution ω. This was done in Section 12.2. On the other hand, we shall measure the *accuracy* of the crude dynamics closure by comparing the approximate statistical state $\bar{\omega}$, defined above by the crude dynamics algorithm, with the exact solution ω. Although we might initially be led to believe that the statistical macrostates obtained by matching E and Γ to those of the exact solution would give much better results, we will provide ample numerical evidence that this is not necessarily so. On the contrary, because the evolution of bulk features in the flow, such as the energy and circulation, will prove to be remarkably well-predicted, even under harshest conditions, both the accuracy of the crude closure and the consistency procedure tend to agree extremely well; in fact, it is usually quite difficult to distinguish between them in any given situation.

12.3.3 Numerical results with Newtonian dissipation

The numerical method used to compute the solution to the dynamic equations (12.15) is the same as the one used in Section 12.2. The interested reader is referred to Majda and Holen (1997), Grote and Majda (1997) for more details.

We shall now submit the crude dynamics closure derived in Subsection 12.3.2 for the Navier–Stokes equations, i.e. (12.15) with $d_2 = \nu > 0$, $d_1 = d = 0$, to a series of successively more stringent tests. The large-scale Reynolds number, Re, is defined as

$$Re = \frac{\pi |\vec{v}|_{\max}}{\nu}. \tag{12.33}$$

Decaying flow

We begin with a purely decaying situation and set $\mathcal{F} = 0$ in (12.15). We set the initial data to be a perturbed state of a solution to the sinh-Poisson equation, which respects the symmetries (12.16). The perturbation in each quadrant consists of six random vortices of amplitude equal to 50% of the maximum vorticity of the underlying sinh–Poisson flow. The initial maximal velocity and vorticity are $|\vec{v}|_{\max} \simeq 3.8$ and $|\omega|_{\max} \simeq 21.1$, respectively. Six random vortex patches, three clockwise and three counter-clockwise, with fixed amplitude $|A| = |\omega|_{\max}/2$, are superimposed over the initial state. With the viscosity set to $\nu = 0.001$, the initial Reynolds number is about $Re = 12\,500$. Figure 12.13 depicts selected snapshots

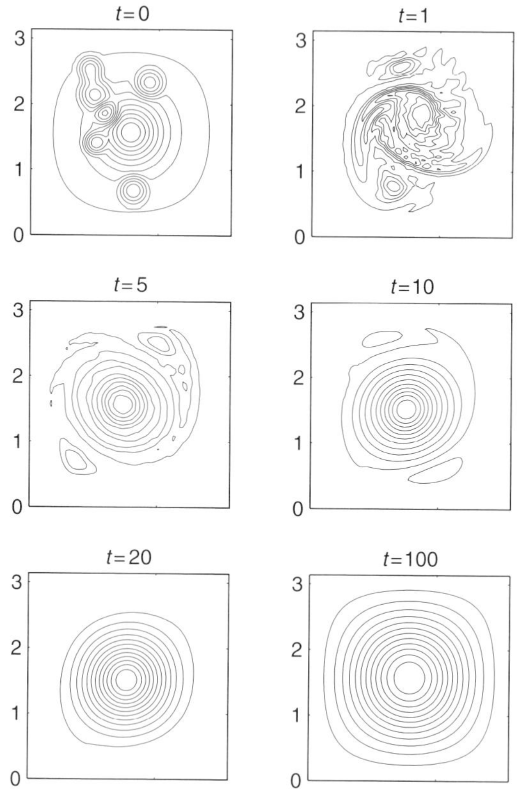

Figure 12.13 Snapshots of the vorticity at $t = 0, 1, 5, 10, 20, 100$ during pure decay, with Newtonian dissipation $\nu = 0.001$.

of the vorticity at times $t = 0, 1, 5, 10, 20, 100$. The small-scale vortices, entrained by the large background vortex, interact and merge rapidly to form a large vortex, which spreads to the boundary of the domain. Hence Figure 12.13 confirms qualitatively the theory of selective decay (see Chapter 3), which predicts the emergence of a Taylor vortex as the long-time solution of the 2D Navier–Stokes equations in a periodic geometry with symmetries.

In Figure 12.14, we compare the numerical solution with the approximate solutions defined by statistical theories. The upper two frames show the relative errors in the macrostates, ω^*, obtained by matching the values of the energy and circulation with those of the exact solution. The sinh–Poisson statistical states greatly outperform the Taylor vortices, with relative errors in the velocity of about 3% for the former and 15% for the latter. As expected, the errors in the vorticity are substantially larger, about twice as large as the relative errors in $\vec{v}$. The graph of the correlation deficit immediately supports this evidence, as the exact solution correlates very rapidly with a sinh–Poisson macrostate; the correlation

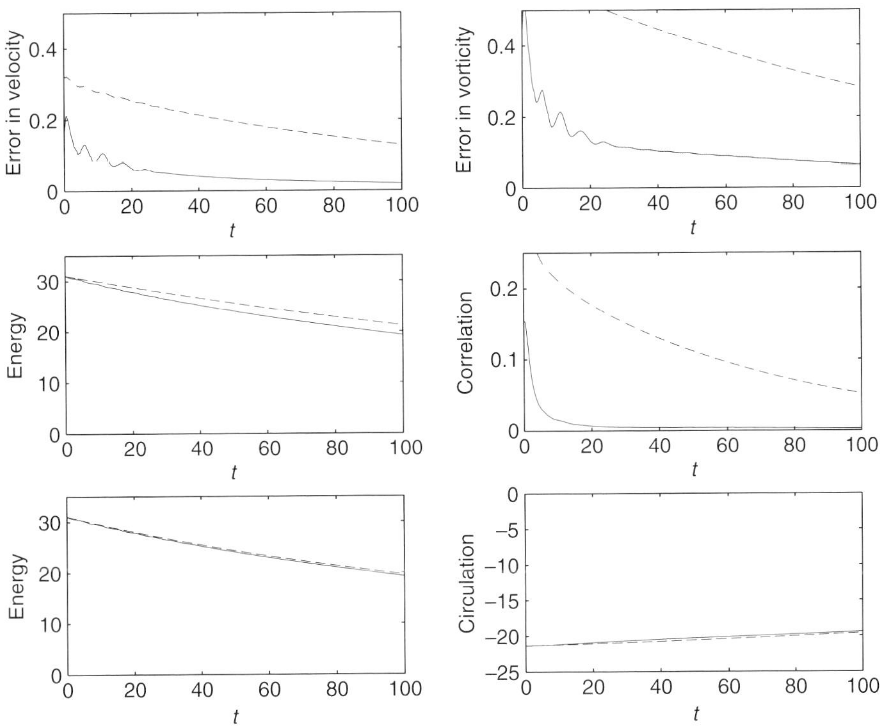

Figure 12.14 Comparison with crude closure dynamics during pure decay, with Newtonian dissipation $\nu = 0.001$. Top left: error in velocity $\mathrm{Err}(\vec{v}, \vec{v}^*)$ with the point-vortex theory (plain) and the energy–enstrophy theory (dashed); top right: error in vorticity $\mathrm{Err}(\omega, \omega^*)$ with the point-vortex theory (plain) and the energy–enstrophy theory (dashed); middle left: energy E (plain) and approximate energy $\bar{E}$ predicted by the energy–enstrophy theory (dashed); middle right: correlation deficit $\text{Corr-d}(\omega, \sinh(b\psi))$ (plain) and $\text{Corr-d}(\omega, \psi)$ (dashed); bottom left: energy E (plain) and approximate energy $\bar{E}$ predicted by the point-vortex theory (dashed); bottom right: circulation Γ (plain) and approximate circulation $\bar{\Gamma}$ predicted by the point-vortex theory (dashed).

deficit with a Taylor vortex, however, is fairly poor and around 0.1, which is at least one order of magnitude larger than the correlation deficit with the sinh–Poisson solution. Next, we explore the accuracy of the approximate solution, $\bar{\omega}$, determined by the crude closure algorithm. In the lower two frames, the exact and approximate values for the energy, E and $\bar{E}$, and for the circulation, Γ and $\bar{\Gamma}$, are shown. The two curves in each frame can barely be distinguished from each other, which demonstrates that the crude closure dynamics follow the evolution of the exact solution remarkably well. We recall that the crude closure dynamics evolve independently and without knowledge of the numerical solution, once the energy and circulation of the initial state have been matched at $t = 0$. Surprisingly the energy of the statistical states, determined by the crude closure algorithm using the energy–enstrophy theory, also follows the energy of the exact solution with an error of at most 5%, despite the rather poor correlation.

Since the statistical states are completely determined by the values of $\bar{E}$ and $\bar{\Gamma}$, which track the exact energy and circulation within just a few percent, the relative L^2-errors of the macrostates obtained via the crude closure dynamics essentially coincide with the curves shown in the top two frames, and hence we omit the figure here. Thus, solving the flow equations numerically provides little extra information for bulk features of the flow.

Spin-up of large vortex

Here we consider a numerical experiment radically different from the previous one. Initially at rest, the flow is bombarded with localized vortices of like negative sign and constant amplitude $|A| = 1$. The interesting question is to determine the statistical quasi-equilibrium state for long times with damping and driving.

In our first calculation we set $\nu = 0.01$, so that equilibrium was reached at $t \simeq 200$, with a final Reynolds number close to 4500. Initially we simply set both $\bar{E}$ and $\bar{\Gamma}$ to zero. Figure 12.15 depicts snapshots of the vorticity at initial and later stages in the computation, $t = 1, 5, 10, 25, 50, 250$. The initial stages of the flow are strongly perturbed and incoherent, yet with time a vortical super-structure emerges from the turbulent flow with the vorticity concentrated in the center of the square. In Figure 12.15 we see the huge increase in energy from 0 up to approximately 130, as the flow accelerates, due to the incessant bombardment with small vortices. The circulation, negative since the vortex rotates clockwise, decreases from 0 down to -50. The two middle frames in Figure 12.16 display the error in velocity and vorticity using the macrostates obtained by matching the energy and enstrophy of the exact numerical solution. The relative errors, quite large during the initial stages, rapidly decrease and reach remarkably small values, 3% for the velocity and 6% for the vorticity. In contrast to the computation performed in the previous calculation, the point-vortex and the energy–enstrophy theories yield comparable results in this situation, with long-time correlation deficits of about 0.001 and 0.002, respectively.

Next we address the remaining crucial question of whether the crude dynamics algorithm is able to bridge over the initial chaotic stages of the flow and predict the bulk features of the emerging super-vortex. Surprisingly, the answer to this question is, yes. Indeed in the lower two frames in Figure 12.16, the errors in velocity and vorticity, obtained with the crude closure algorithm using the point-vortex statistical theory, are barely larger than those obtained from the simple matching procedure. The behavior of the crude closure dynamics using

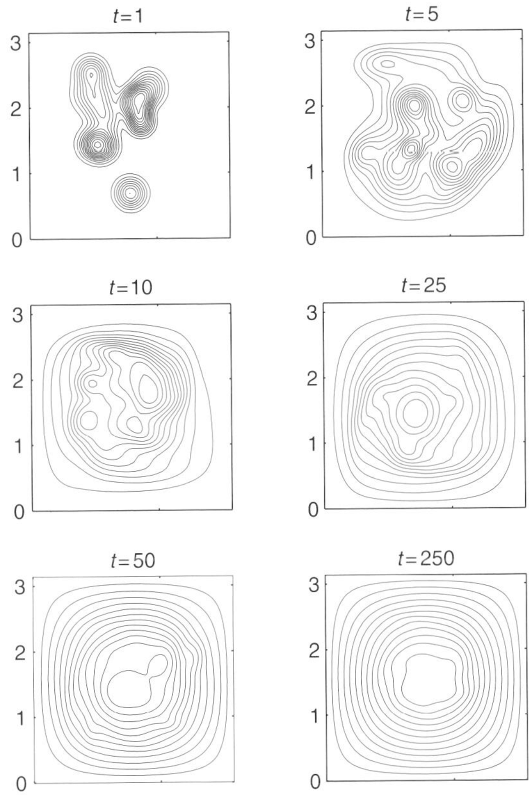

Figure 12.15 Snapshots of the vorticity at $t = 1, 5, 10, 25, 50, 250$, during vortex build-up from zero, with Newtonian dissipation $\nu = 0.01$.

the energy–enstrophy theory is almost identical to that of the point-vortex theory with identical relative L^2-errors.

To test the dependence of the results on the Reynolds number, we set $\nu = 0.001$ and run the simulation up to $t = 1000$ – until the time 10 000 random vortices have been added. Again under the constant bombardment with random-like signed vortices, the flow undergoes a transient incoherent stage until a large vortex emerges and completely fills the square. In Figure 12.17 we see the huge increase in the vorticity up to 125, while the large-scale Reynolds number at $t = 1000$ is close to 365 000. Beyond the initial perturbed flow configurations, the emerging vortex correlates extremely well – the correlation deficit is smaller than 0.003 – with both statistical states. In the lower two frames of Figure 12.17, we see the remarkable performance of the crude closure algorithm for the point-vortex statistical theory. The relative errors are again about 3% for the velocity and 6% for the vorticity. The errors obtained using the sinh–Poisson statistical states with the energy and circulation matched with the exact solution are about half these

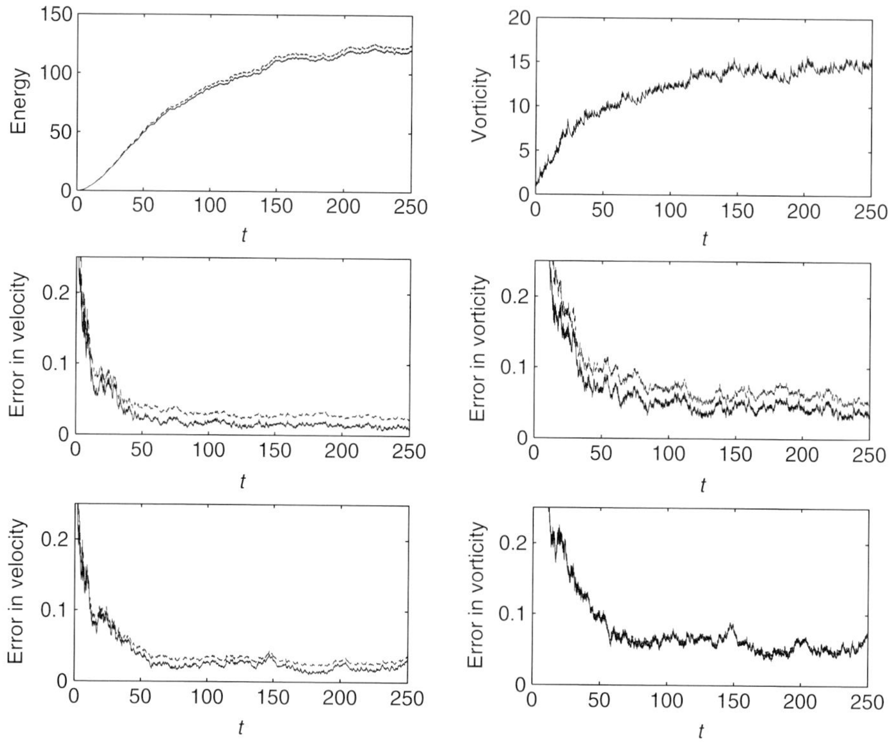

Figure 12.16 Comparison with crude closure dynamics during vortex build-up from zero, with Newtonian dissipation $\nu = 0.01$. Top left: energy E (plain) and approximate energy $\bar{E}$ predicted by the point-vortex theory (dashed); top right: maximal vorticity $|\omega|_{\max}$; middle left: error in velocity using the point-vortex theory, $\mathrm{Err}(\vec{v}, \vec{v}^*)$ (plain) and $\mathrm{Err}(\vec{v}, \bar{\vec{v}})$ (dashed, upper curve); middle right: error in vorticity using the point-vortex theory, $\mathrm{Err}(\omega, \omega^*)$ (plain) and $\mathrm{Err}(\omega, \bar{\omega})$ (dashed, upper curve); bottom left: error in velocity using the energy–enstrophy theory, $\mathrm{Err}(\vec{v}, \vec{v}^*)$ (plain) and $\mathrm{Err}(\vec{v}, \bar{\vec{v}})$ (dashed, upper curve); bottom right: error in vorticity using the energy–enstrophy theory, $\mathrm{Err}(\omega, \omega^*)$ (plain) and $\mathrm{Err}(\omega, \bar{\omega})$ (dashed).

values. The energy–enstrophy theory again performs similarly to the point-vortex theory. Although the flow has not reached statistical equilibrium yet, we observe a constant decrease in the relative errors with increasing time.

In the next section we will provide partial explanation to the numerical results for the spin-up problem.

Forcing by alternating or opposite signed vortices

We attempt to explore the limits of the statistical theory by increasing the intensity of the external forcing. First, we will bombard the large-scale vortex with localized

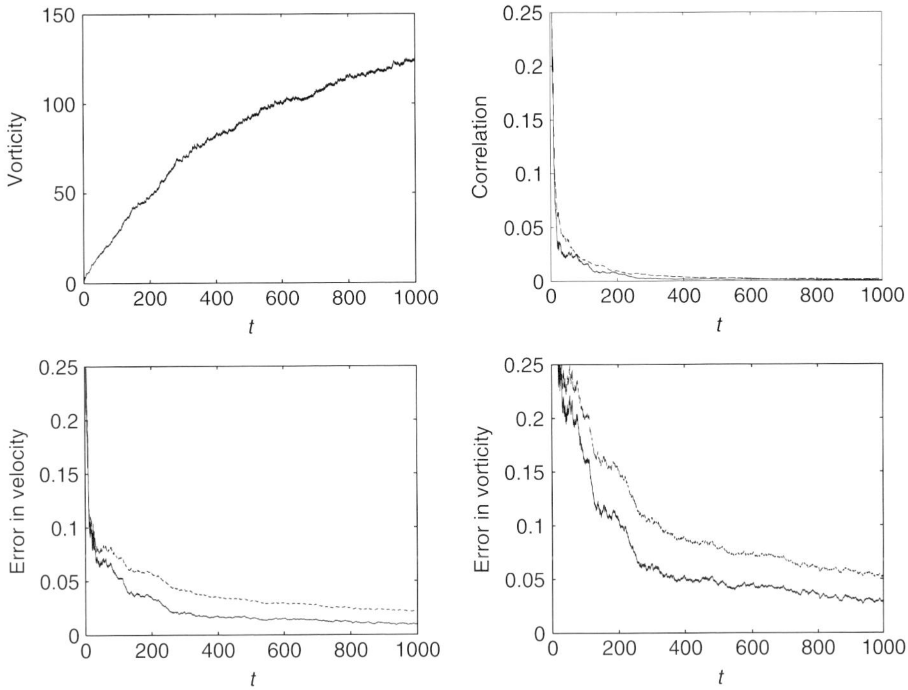

Figure 12.17 Comparison with crude closure dynamics during vortex build-up from zero, with Newtonian dissipation $\nu = 0.001$. Top left: maximal vorticity $|\omega|_{\max}$; top right: correlation deficit Corr-d$(\omega, \sinh(b\psi))$ (plain) and Corr-d(ω, ψ) (dashed, upper curve); bottom left: error in velocity using the point-vortex theory, Err$(\vec{v}, \vec{v}^*)$ (plain) and Err$(\vec{v}, \bar{\vec{v}})$ (dashed, upper curve); bottom right: error in vorticity using the point-vortex theory, Err(ω, ω^*) (plain) and Err$(\omega, \bar{\omega})$ (dashed, upper curve).

vortices of opposite sign to destroy the initially coherent structure and force the flow into a transient incoherent regime. Again we will see that the crude closure algorithm is able to bridge this chaotic state and predict the emergence of the new vortical super-structure with opposite sign. Second, we will force the flow out of its statistical state by strongly perturbing it with random localized vortices of alternating sign, and test the performance of the crude closure algorithm as we increase the intensity in the bombardment.

In Figure 12.18 we display snapshots at selected times $t = 0, 10, 20, 30, 40, 50$, of the vorticity, as we bombard the initial sinh–Poisson state with small vortices of opposite sign and constant amplitude $|A| = 1$. Here the initial maximal velocity was about 3.8 and $\nu = 0.01$, leading to a Reynolds number $Re \simeq 2500$. The large vortex structure, initially rotating clockwise, begins to disintegrate and is completely destroyed by $t = 20$. Under the constant action of the external forcing,

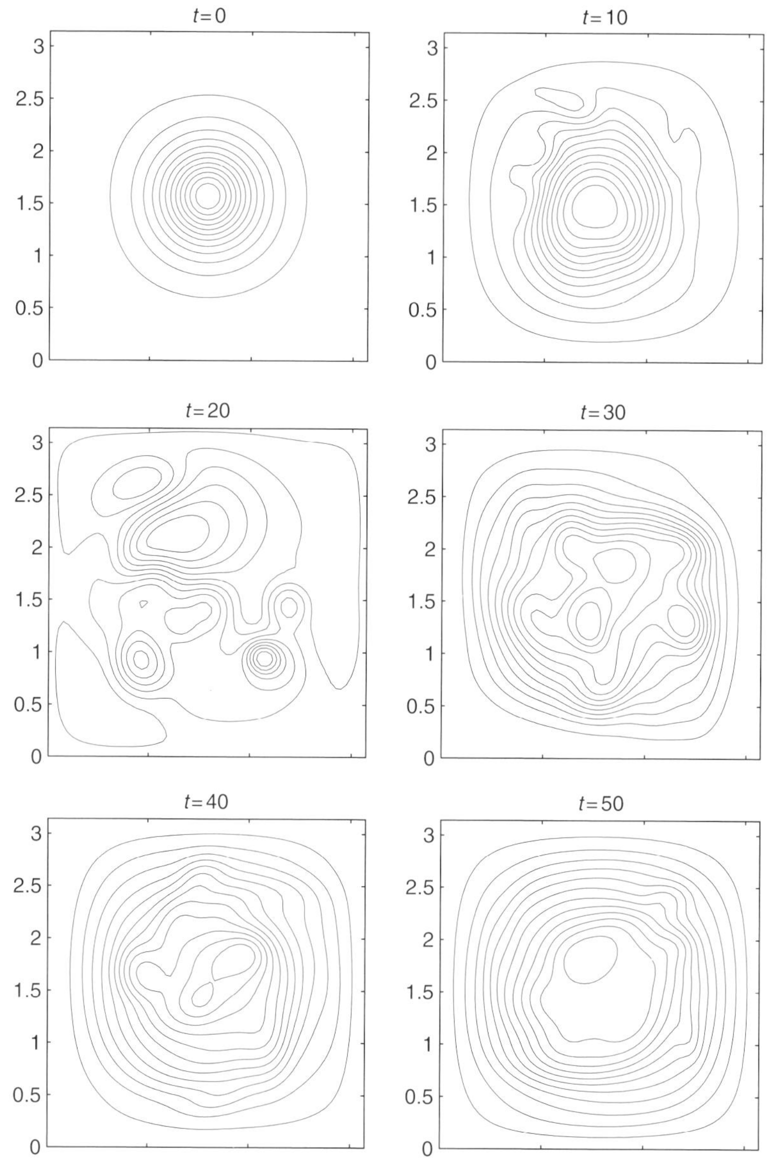

Figure 12.18 Snapshots of the vorticity at $t = 0, 10, 20, 30, 40, 50$, during flow reversal, with Newtonian dissipation $\nu = 0.01$.

the flow lingers in an incoherent intermediate state, reverses its direction, and slowly starts to rebuild a large vortical structure with opposite sign. The graph of the circulation, shown in Figure 12.19, clearly demonstrates the change of direction in rotation as the circulation goes through zero. Since the macrostates of the energy–enstrophy theory are only defined up to a sign, we had to impose the

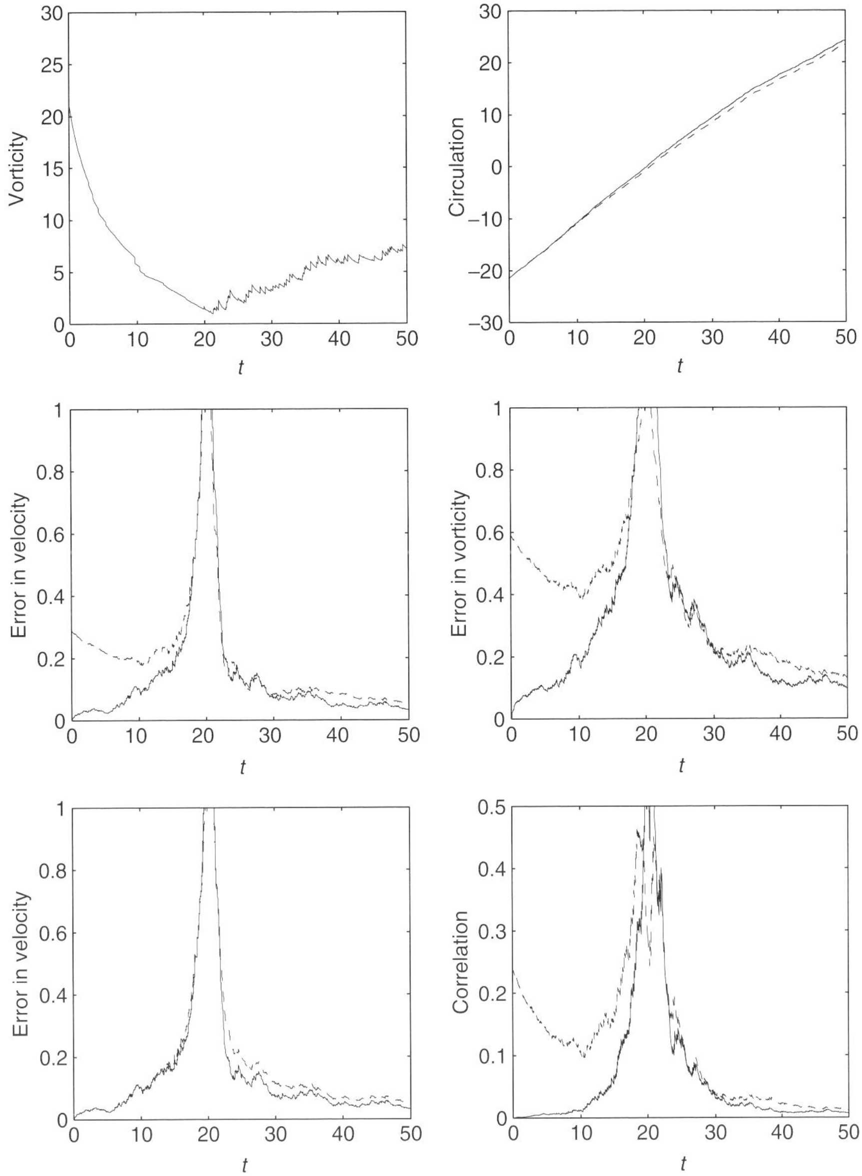

Figure 12.19 Comparison with crude closure dynamics during flow reversal, with Newtonian dissipation $\nu = 0.01$. Top left: maximal vorticity $|\omega|_{\max}$; top right: circulation Γ (plain) and approximate circulation $\bar{\Gamma}$ predicted by the point-vortex theory (dashed); middle left: error in velocity $\mathrm{Err}(\vec{v}, \vec{v^*})$ using the point-vortex (plain) or the energy–enstrophy theory (dashed); middle right: error in vorticity $\mathrm{Err}(\omega, \omega^*)$ using the point–vortex (plain) or the energy–enstrophy theory (dashed); bottom left: error in velocity using the point–vortex theory, $\mathrm{Err}(\vec{v}, \vec{v^*})$ (plain) and $\mathrm{Err}(\vec{v}, \vec{\bar{v}})$ (dashed); bottom right: correlation deficit Corr-d$(\omega, \sinh(b\psi))$ (plain) and Corr-d(ω, ψ) (dashed).

change in sign of the circulation in the exact solution in (12.20). In that same frame we display the evolution of the circulation predicted by the crude closure algorithm using the point-vortex statistical theory: it is difficult to distinguish between the exact and the predicted values, which demonstrates again the remarkable accuracy in the prediction of the crude closure dynamics. In the lower four frames in Figure 12.19, all various error measures of either the velocity or the vorticity display relative errors beyond 100% during the transient stages, where the flow velocity is close to zero and totally incoherent. However, once past the transient stages, the errors in velocity and vorticity are about 5% and 10%, respectively, for either statistical theory, whether we use the matched values of E and Γ from the exact solution or the values $\bar{E}$ and $\bar{\Gamma}$ predicted by the crude closure dynamics. We note the qualitative difference in the initial vortex structure, with a tightly concentrated center, and the later vortex, with tame values in the vorticity spread out up to the boundaries of the square. Thus due to its higher versatility in adapting to various situations, the sinh–Poisson solution outperforms the Taylor vortices in the initial tightly packed vortex configuration. The comparable performance of both statistical theories in the reconstruction phase corroborates the results in the previous section during the build-up from zero.

Next, we consider the same initial sinh–Poisson solution, but strongly perturb the flow by vortices with alternating signs and constant amplitude equal to 5% of the initial maximal vorticity. Here the viscous coefficient was set to $\nu = 0.001$, which led to a Reynolds number close to 25 000. Figure 12.20 depicts the evolution of the flow with snapshots of the vorticity at selected times $t = 0, 20, 40, 60, 80, 100$. In contrast to the behavior during pure decay or build-up from zero, here the vorticity distribution remains tightly packed in the center of the square. In Figure 12.21 the correlation deficit with the Taylor vortex is poor, around 0.1, versus a correlation deficit around 0.02, obtained using the sinh–Poisson solution. As a consequence, the errors in velocity obtained with the energy–enstrophy statistical theory are quite large, slightly above 20%, when compared with the excellent performance of the crude closure based on the point-vortex theory, which yields relative errors of only 6%. We omit showing the comparison between the exact energy and the evolution predicted by the crude dynamics in the context of the point-vortex theory, as the curves are barely distinguishable. Instead, we graph the evolution of the energy predicted by the crude closure based on the energy–enstrophy theory which, despite its poorer performance, tracks the energy within relative errors below 6%.

Finally, we attempt to push the statistical theory to its limits by repeating the above numerical experiment, yet with the magnitude of the localized vortices augmented to 20% of the maximal initial vorticity. In Figure 12.22 we display contours of the vorticity at selected times $t = 0, 20, 40, 60, 80, 100$, which

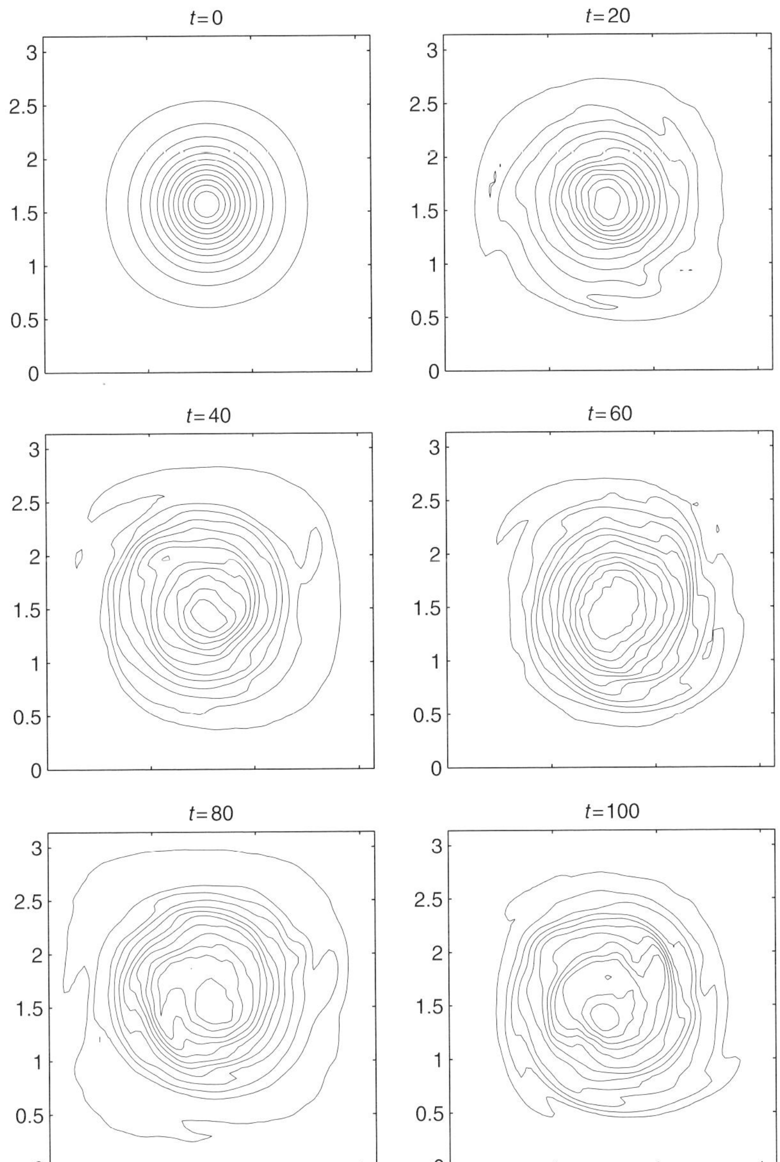

Figure 12.20 Snapshots of the vorticity at $t = 0$, 20, 40, 60, 80, 100, during bombardment with localized vortices of alternating sign and amplitude equal to 5% of initial maximal vorticity $|\omega|_{\max} \simeq 21.1$, with Newtonian dissipation $\nu = 0.001$.

show the perturbed state of the evolving flow as the large-scale vortex leaves the center of the square and navigates about the domain while shedding small vortical structures. As to be expected in this violent regime, both statistical theories correlate rather poorly with the evolving flow, and lead to errors over 50% in the

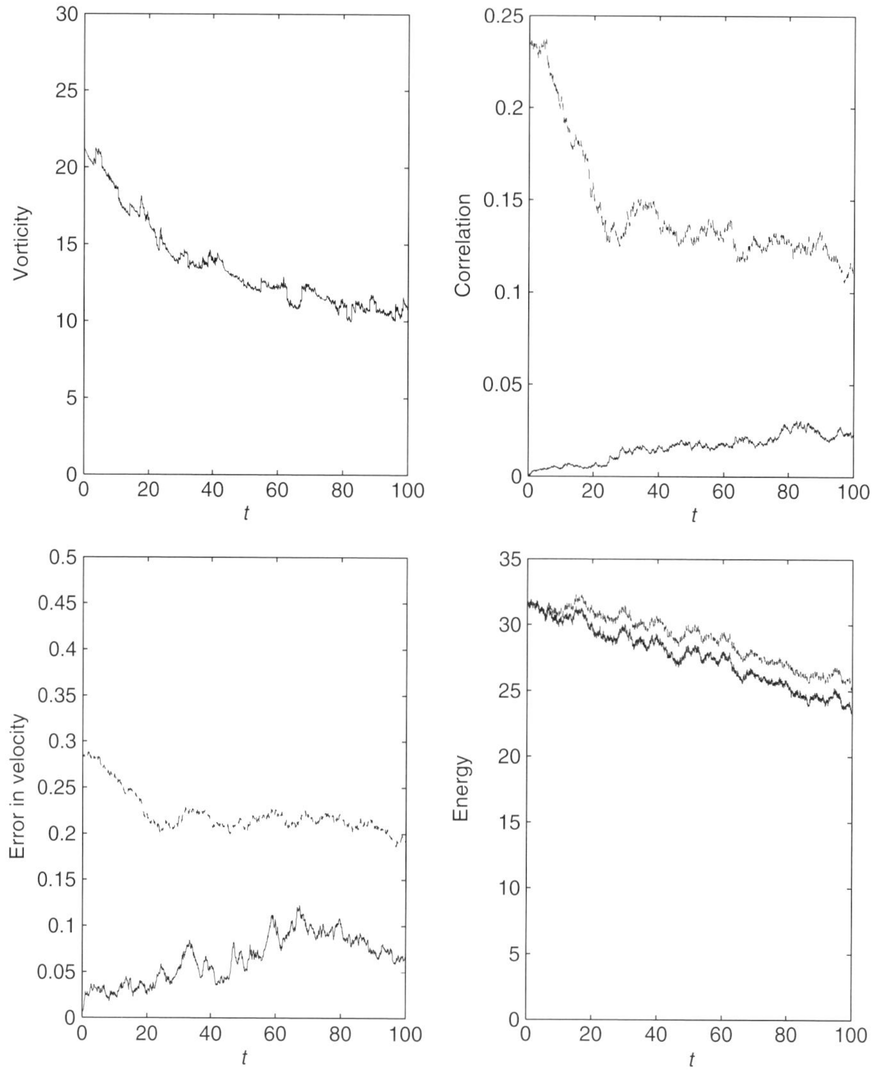

Figure 12.21 Comparison with crude closure dynamics during bombardment with localized vortices of alternating sign and amplitude equal to 5% of initial maximal vorticity $|\omega|_{\max} \simeq 21.1$, with Newtonian dissipation $\nu = 0.001$. Top left: maximal vorticity $|\omega|_{\max}$; top right: correlation deficit Corr-d$(\omega, \sinh(b\psi))$ (plain) and Corr-d(ω, ψ) (dashed, upper curve); bottom left: error in velocity $\mathrm{Err}(\vec{v}, \vec{v^*})$ using the point-vortex (plain) or the energy–enstrophy theory (dashed, upper curve); bottom right: energy E (plain) and approximate energy $\bar{E}$ predicted by the energy–enstrophy theory (dashed, upper curve).

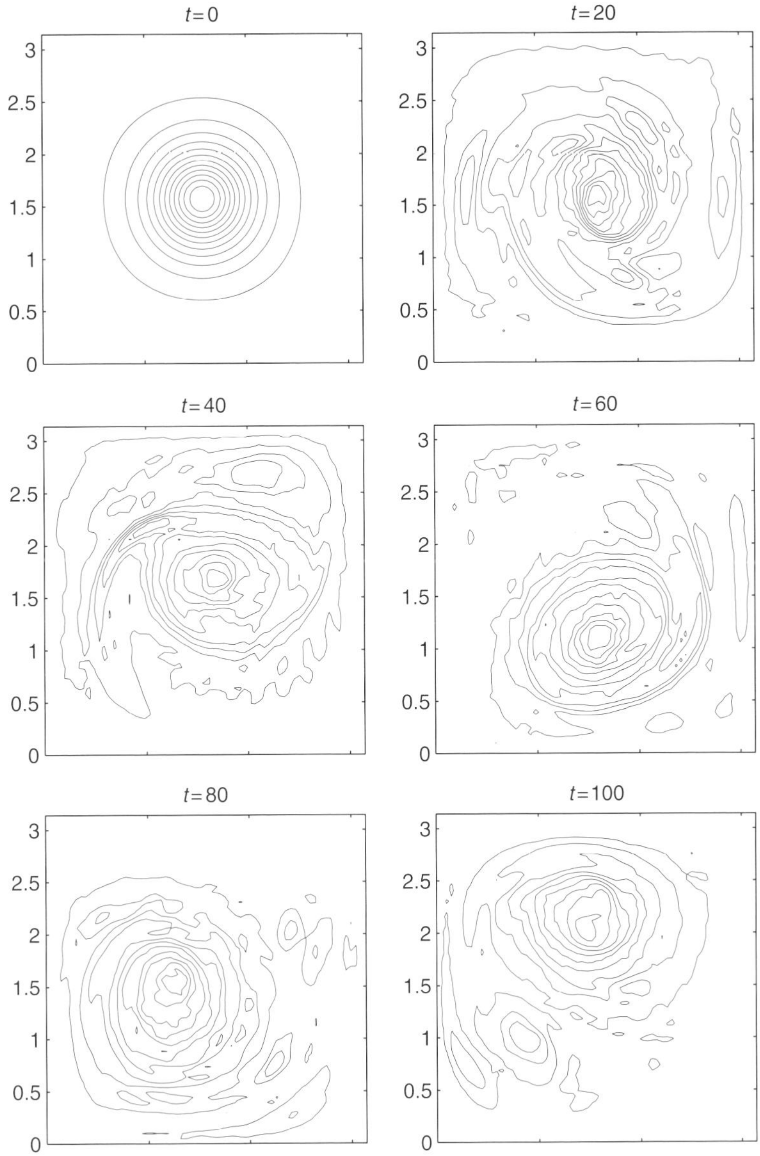

Figure 12.22 Snapshots of the vorticity at $t = 0$, 20, 40, 60, 80, 100, during bombardment with localized vortices of alternating sign and amplitude equal to 20% of initial maximal vorticity $|\omega|_{\max} \simeq 21.1$, with Newtonian dissipation $\nu = 0.001$.

velocity and the vorticity. However, the surprising result in Figure 12.23 is the ability of the crude closure dynamics to track bulk features of the flow, such as the energy and the circulation, quite accurately despite the poor correlation with the statistical equilibrium states.

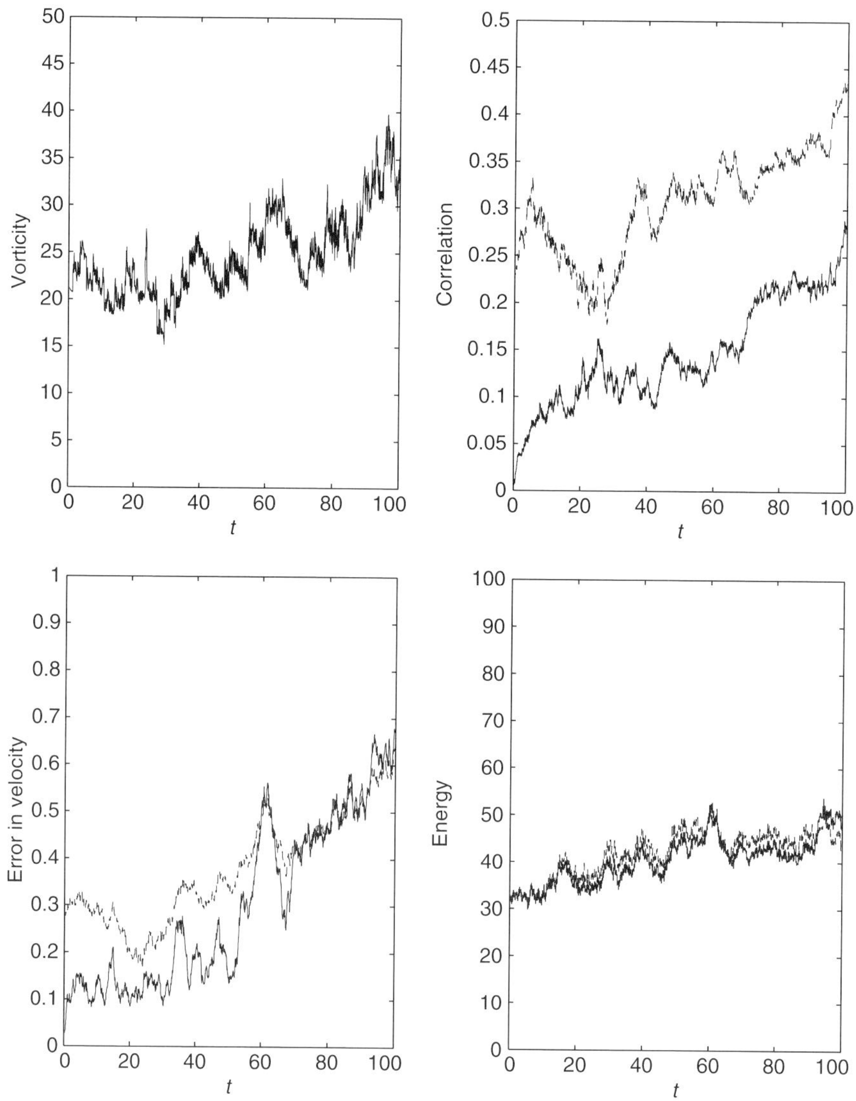

Figure 12.23 Comparison with crude closure dynamics during bombardment with localized vortices of alternating sign and amplitude equal to 20% of initial maximal vorticity $|\omega|_{\max} \simeq 21.1$, with Newtonian dissipation $\nu = 0.001$. Top left: maximal vorticity $|\omega|_{\max}$; top right: correlation deficit Corr-d$(\omega, \sinh(b\psi))$ (plain) and Corr-d(ω, ψ) (dashed, upper curve); bottom left: error in velocity Err$(\vec{v}, \vec{v^*})$ using the point-vortex (plain) or the energy–enstrophy theory (dashed, upper curve); bottom right: energy E (plain) and approximate energy $\bar{E}$ predicted by the point-vortex theory (dashed, upper curve).

Finally we remark on the case with Ekman damping. The crucial difference between Ekman damping and eddy viscosity (or hyper-viscosity) is that Ekman drag dissipates energy at all scales equally, whereas Newtonian viscosity dissipates much more strongly at the smaller scales. Thus we should not expect the crude closure to work since the initial assumptions on designing the crude closure algorithms are violated. This is supported by numerical evidence, and the interested reader is referred to Grote and Majda (2000) for more details.

12.4 Remarks on the mathematical justifications of crude closure

In this section we offer partial justification of the numerical evidence of the emergence of large-scale vortex under random small-scale vortex bombardment with the same sign as presented in Subsection 12.3.3, i.e. the spin-up of large vortex. A completely satisfactory justification would be to establish that the long time behavior is a quasi-equilibrium state in the form of ground state (or a large coherent structure close to the ground state) plus small fluctuations. A somewhat satisfactory explanation would be to establish that the long time average of the flow is a ground state or a suitable large-scale vortex. Such an emergence of large-scale structure fits very well with the law of large numbers, and technically can be derived via proving the existence and uniqueness of certain invariant measures with appropriate supports. This approach has been carried out recently with full mathematical details by the authors Majda and Wang (2005). Indeed, the small-scale random bombardments can be decomposed into a large-scale deterministic part and a random fluctuation with zero mean. Under appropriate scaling of the time step between kicks and the amplitude of the bombardment, the forcing can be approximated by a deterministic large-scale steady forcing plus a small white noise in time. This is basically the infinite-dimensional version of Donsker's invariance principle (see Reed and Simon, 1980). We can then imagine that the main feature of the flow is determined by the large-scale steady state forcing, which produces a large coherent structure. The small white noise forcing produces small (in some appropriate sense) fluctuations from the large-scale coherent structure. Moreover, we can show that the large-scale coherent structure is highly correlated with the ground state, which is consistent with our numerical evidence (Majda and Wang, 2005).

Here, we take an alternative approach. Recall from Chapters 1 and 3 that the Dirichlet quotient controls the small scales in the flow. Indeed, the Dirichlet quotient attains its minimum, the first eigenvalue of the Laplace operator allowed by the geometry, if and only if the flow attains the maximum scale structure allowed by the geometry, i.e. the ground states. Moreover, flows with predominantly small structures are characterized by a large Dirichlet quotient, while flows with predominant large structures are characterized by a small Dirichlet quotient. Therefore, an upper bound on the Dirichlet quotient for the flow is a partial justification for the emergence of the large-scale structure. Such an upper bound on the Dirichlet quotient is the goal of this section.

In general, a flow governed by the quasi-geostrophic dynamics may not be able to maintain a large-scale structure under random small-scale bombardments. Indeed, it is easy to construct a flow consists of minus a given vortex plus very small perturbation of higher Dirichlet quotient (small structure). The bombardment of this flow by the given vortex results in the cancellation of the relatively large structure and leads to the extremely small structure having a much higher Dirichlet quotient. This is supported by the numerical results from Subsection 12.3.3. However, due to the special setting of the spin-up problem, we will see that the vorticity field will be non-negative for all time due to a maximum principle (see for instance Evans, 1998). Such a positive vorticity field cannot be annihilated by the positive small-scale bombardment and this is the main ingredient in the success in deriving an upper bound for the Dirichlet quotient.

There are two stages in the dynamics: a free decay stage and an instantaneous forcing stage. Recall from Chapter 3 that the Dirichlet quotient is a monotonic decreasing function of time in the free decay stage. Thus it is only necessary to establish that the instantaneous forcing stage cannot increase the Dirichlet quotient without bound.

We first recall the equations and consider a slightly more general case with an arbitrary initial vorticity as long as it is positive

$$\begin{gathered}\frac{\partial \omega}{\partial t} + \nabla^{\perp}\psi \cdot \nabla\omega = \nu\Delta\omega + \mathcal{F}, \\ \omega|_{t=0} = \omega_0 \geq 0 \text{ on } Q,\end{gathered} \tag{12.34}$$

where the stream function, ψ, satisfies the free stress boundary condition at the edges of the square $Q = [0, \pi] \times [0, \pi]$, which is equivalent to having a special type of Fourier expansion listed in (12.16). In particular this implies the following boundary condition for the stream function, ψ, and the vorticity, ω

$$\psi|_{\partial Q} = \omega|_{\partial Q} = 0. \tag{12.35}$$

The external forcing takes the form specified in (12.4) with the small-scale vortex given in (12.5). One special feature of the external forcing is the positivity. When this positivity is combined with the positivity of the initial vorticity, we obtain the positivity of future vorticity via a simple maximum principle argument as we may view (12.34) as a convection-diffusion equation for the vorticity, ω. Hence we have

$$\omega(\vec{x}, t) \geq 0, \text{ for all } t > 0. \tag{12.36}$$

(see for instance Evans, 1998). As for the stream function, it can be solved via the Poisson equation together with the zero boundary condition specified in (12.35). The maximum principle for Poisson equation (see for instance Evans (1998)) implies that the stream function is strictly negative inside the box Q unless $\omega \equiv 0$

$$\psi(\vec{x}, t) < 0, \text{ for all } t > 0. \tag{12.37}$$

Next we look into the evolution of the energy, enstrophy, and circulation. It is easy to derive, after multiplying the quasi-geostrophic (Navier–Stokes) equations (12.34) by $-\psi$ or ω or 1 and integrating over the square Q

$$\begin{aligned} \frac{dE}{dt} &= -\nu \int_Q |\omega|^2 - \int_Q \psi \mathcal{F}, \\ \frac{d\mathcal{E}}{dt} &= -\nu \int_Q |\nabla\omega|^2 + \int_Q \omega \mathcal{F}, \\ \frac{d\Gamma}{dt} &= \nu \int_{\partial Q} \frac{\partial \omega}{\partial n}, + \int_Q \mathcal{F}, \end{aligned} \tag{12.38}$$

where $\frac{\partial\omega}{\partial n}$ represents the normal derivative of the vorticity with respect to the unit outer normal at the boundary of the box Q. Since the vorticity is positive inside the box (12.36), the normal derivative of the vorticity is negative, i.e.

$$\frac{\partial \omega}{\partial n} \leq 0, \quad \text{at } \partial Q. \tag{12.39}$$

This implies that the Newtonian dissipation decreases the circulation. This is of course consistent with the basic idea of dissipation. A simple consequence of the equations (12.38) is the fact that the positive external forcing, $\mathcal{F}$, increases energy, enstrophy, and circulation. This partially justifies the notation of spin-up.

The dynamics of the Dirichlet quotient $\Lambda(t) = \frac{\mathcal{E}(t)}{E(t)}$ can be calculated easily using the dynamics of the energy and enstrophy. Indeed we have (see for instance Chapter 3)

$$\begin{aligned} \frac{d\Lambda(t)}{dt} &= \frac{1}{E(t)}(\dot{\mathcal{E}}(t) - \Lambda(t)\dot{E}(t)) \\ &= -\frac{\nu\| -\Delta\vec{v} - \Lambda(t)\vec{v}\|_0^2}{E(t)} + \frac{\int_Q(\omega + \Lambda(t)\psi)\mathcal{F}}{E(t)}. \end{aligned} \tag{12.40}$$

Since the sign of $\omega + \Lambda(t)\psi$ is not definite, we are not sure if the positive external forcing increases or decreases the Dirichlet quotient. In fact numerical evidence from the previous section suggests that it could be both ways.

Our goal here is to derive a time uniform bound on the Dirichlet quotient. The bound would imply that not many small-scale structures are created, even though the system is under constant bombardment by small-scale vortices.

We introduce the notation

$$\bar{\omega}_j = \omega_r(\vec{x} - \vec{x}_j)$$

and

$$\bar{E}_j = -\frac{1}{2}\int_Q \bar{\psi}_j \bar{\omega}_j, \qquad \bar{\mathcal{E}}_j = \frac{1}{2}\int_Q \bar{\omega}_j^2. \tag{12.41}$$

We see immediately, thanks to (12.21), the positivity of ω and $\bar{\omega}_j$ and the negativity of ψ and $\bar{\psi}_j$

$$\begin{aligned} 2E(t_j^-) + 2\bar{E}_j \geq E(t_j^+) \geq E(t_j^-) + \bar{E}_j, \\ 2\mathcal{E}(t_j^-) + 2\bar{\mathcal{E}}_j \geq \mathcal{E}(t_j^+) \geq \mathcal{E}(t_j^-) + \bar{\mathcal{E}}_j, \end{aligned} \tag{12.42}$$

where $t_j = j\ \Delta t$. We also have, by the definition of Dirichlet quotient and the instantaneous forcing effect (12.21)

$$\Lambda(t_j^+) = \frac{\mathcal{E}(t_j^-) + \int_Q \omega(t_j^-)\bar{\omega}_j + \bar{\mathcal{E}}_j}{E(t_j^-) - \int_Q \psi(t_j^-)\bar{\omega}_j + \bar{E}_j}. \tag{12.43}$$

It is clear from our choice of $\bar{\omega}_j$ in (12.5) that there exists a constant λ_1 such that

$$\frac{\bar{\mathcal{E}}_j}{\bar{E}_j} \leq \lambda_1, \forall j. \tag{12.44}$$

In order to derive a uniform in time bound on the Dirichlet quotient, it is enough to prove the following claim:

Claim: There exists a constant λ_2 such that

$$\bar{\omega}_j \leq -\lambda_2 \bar{\psi}_j, \text{ for all } j. \tag{12.45}$$

We observe that with the validity of the claim we have

$$\begin{aligned} \mathcal{E}(t_j^-) + \int_Q \omega(t_j^-)\bar{\omega}_j + \bar{\mathcal{E}}_j &\leq \mathcal{E}(t_j^-) - \lambda_2 \int_Q \omega(t_j^-)\bar{\psi}_j + \lambda_1 \bar{E}_j \\ &\leq \Lambda_{t_j^-} E(t_j^-) - \lambda_2 \int_Q \omega(t_j^-)\bar{\psi}_j + \lambda_1 \bar{E}_j \\ &\leq \bar{\lambda}_j (E(t_j^-) - \int_Q \omega(t_j^-)\bar{\psi}_j + \bar{E}_j), \end{aligned}$$

where

$$\bar{\lambda}_j = \max\{\Lambda_{t_j^-}, \lambda_2, \lambda_1\}, \tag{12.46}$$

and hence

$$\Lambda_{t_j^+} \leq \bar{\lambda}_j. \tag{12.47}$$

Notice that during the pure decay stage $t_{j-1} < t < t_{j-1} + \Delta t = t_j$ the Dirichlet quotient is monotonically decreasing (see Chapter 3 and (12.40)) we have

$$\Lambda_{t_j^-} \leq \Lambda_{t_{j-1}^+}.$$

When this combined with (12.47), (12.46) and a simple iteration we deduce

$$\Lambda_{t_j^+} \leq \bar{\lambda}_1 \leq \bar{\lambda}_0 \overset{def}{=} \max\{\Lambda_0, \lambda_2, \lambda_1\} \tag{12.48}$$

where

$$\Lambda_0 = \frac{\mathcal{E}(\omega_0)}{E(\omega_0)}. \tag{12.49}$$

This proves a uniform in time bound on the Dirichlet quotient.

It remains to prove the claim (12.45). By the special choice of our random forcing (12.4) and the small-scale vortex (12.5), we see that the support of $\bar{\omega}_j$ always overlap with an interior region $Q_{in} = (\Delta x, \pi - \Delta x) \times (\Delta x, \pi - \Delta x)$ of Q since the center of $\bar{\omega}_j$ lies in this subregion Q_{in}. This implies that there exist finitely many boxes $B_i, i = 1, \ldots, N$, such that $B_i \subset Q_{in}$ for all i, there exists a constant A and for each $\bar{\omega}_j$ there exists a B_i satisfying

$$\bar{\omega}_j \geq A\chi_i, \tag{12.50}$$

where

$$\chi_i(\vec{x}) = \begin{cases} 1 & \text{if } \vec{x} \in B_i, \\ 0 & \text{otherwise} \end{cases} \tag{12.51}$$

is the indicator function of the set B_i.

Let ψ_i be the solution of

$$\Delta\psi_i = \chi_i, \qquad \psi_i|_{\partial Q} = 0. \tag{12.52}$$

By a standard comparison principle we have

$$-\bar{\psi}_j \geq -A\psi_i. \tag{12.53}$$

By the Hopf's strong maximum principle (Evans, 1998) we have

$$\frac{\partial\psi_i}{\partial n} > 0, \tag{12.54}$$

at the boundary ∂Q away from the four corners. More precisely, there exists a constant B such that

$$\frac{\partial\psi_i}{\partial n} > B, \tag{12.55}$$

provided

$$
\begin{aligned}
|\vec{x}-(0,0)| &\geq \frac{1}{4}\Delta x, \quad |\vec{x}-(0,\pi)| \geq \frac{1}{4}\Delta x, \\
|\vec{x}-(\pi,0)| &\geq \frac{1}{4}\Delta x, \quad |\vec{x}-(\pi,\pi)| \geq \frac{1}{4}\Delta x.
\end{aligned}
\tag{12.56}
$$

When this combined with the negativity of ψ_i in the interior of the box Q and the choice of $\bar{\omega}_j$ (12.5), we have

$$-\psi_i \geq C\bar{\omega}_j, \quad \text{for all } i, \text{ and } j. \tag{12.57}$$

This combined with (12.53) yields the claim with

$$\lambda_2 = A\ C. \tag{12.58}$$

References

DiBattista, M. T., Majda, A. J. and Grote, M. J. (2001), Meta-stability of equilibrium statistical structures for prototype geophysical flows with damping and driving. *Phys. D* **151** (2–4), 271–304.

Evans, L. C. (1998), Partial Differential Equations. Rhode Island: AMS.

Grote, M. J. and Majda, A. J. (1997), Crude closure dynamics through large scale statistical theories. *Phys. Fluids* **9** (11), 3431–3442.

Grote, M. J. and Majda, A. J. (2000), Crude closure for flow with topography through large-scale statistical theory. *Nonlinearity* **13** (3), 569–600.

Majda, A. and Bertozzi, A. (2001), Vorticity and Incompressible Flow. Cambridge: Cambridge University Press.

Majda, A. J. and Holen, M. (1997), Dissipation, topography, and statistical theories for large-scale coherent structure. *Comm. Pure Appl. Math.* **50** (12), 1183–1234.

Majda, A. J. and Wang, X. (2005), The emergence of large scale coherent structure under small scale random bombardment. *Comm. Pure Appl. Math.*, in press.

Montgomery, D., Matthaeus, W. H., Martinez, D. and Oughton, S. (1992), Relaxation in two dimensions and sinh–Poisson equation. *Phys. Fluids* A **4**, 3–6.

Reed, M. and Simon, B. (1980), Methods of Modern Mathematical Physics. New York: Academic Press.

Rhines, P. (1975), Waves and turbulence on a beta plane. *J. Fluid Mech.* **69**, 417–443.

13

Predicting the jets and spots on Jupiter by equilibrium statistical mechanics

13.1 Introduction

Prominent examples of long-lived large-scale vortices in geophysical flows are those such as the Great Red Spot (GRS) on Jupiter (Dowling, 1995; Marcus, 1993; Rogers, 1995). The emergence and persistence of such coherent structures at specific latitudes such as 22.4° S for the GRS in a background zonal shear flow that seems to violate all of the standard stability criteria (Dowling, 1995) are a genuine puzzle. Here we show how to utilize the equilibrium statistical theory with a suitable prior distribution, ESTP, introduced in Section 9.2 and discussed in Chapters 10, 11, and 12 to predict the actual coherent structures on Jupiter in a fashion which is consistent with the known observational record. As discussed in Section 9.2, the statistical theory is based on a few judiciously chosen conserved quantities for the inviscid dynamics such as energy and circulation and does not involve any detailed resolution of the fluid dynamics. The key ingredient of the ESTP is the prior distribution, $\Pi(\lambda)$, for the one-point statistics of the potential vorticity which parameterizes the unresolved small-scale turbulent eddies that produce the large-scale coherent structures. Below, we show how the observational record of Jupiter from the recent Galileo mission suggests a special structure for such a prior distribution resulting from intense small-scale forcing. Recall from the studies discussed in Chapter 12 (DiBattista, Majda, and Grote, 2001) that ESTP can potentially describe the meta-stable large-scale coherent structures occurring from strong small-scale forcing, provided that the flux of energy to large scales is sufficiently weak. With the confidence gained from this fact, the ESTP with a special prior distribution is utilized to predict the large-scale coherent structures on Jupiter; these predicted structures are in agreement with the observational record from the weather layer of Jupiter as inferred from the Voyager mission.

The approach in making the theoretical prediction in this chapter is inherently multi-disciplinary and has several ingredients:

(A) The observational record for Jupiter.
(B) The quasi-geostrophic model for interpreting observations and predictions.
(C) Fitting the zonal shear profile from observations, the Limaye profile from the Voyager mission (Rogers, 1995), to the quasi-geostrophic model (Dowling, 1995).
(D) The reinterpretation of (C) using the energy–enstrophy statistical theory.
(E) The observational and physical motivation for the prior distribution for ESTP and a convenient mathematical framework for interpreting the ESTP with this family of prior distributions.
(F) Equilibrium statistical predictions from ESTP for the coherent structures on Jupiter and comparison with observations.

(13.1)

Our presentation of this material in this chapter closely follows the recent work of Turkington, Majda, Haven, and DiBattista (Turkington *et al.*, 2001; Haven *et al.*, 2002), and we systematically follow the steps in (13.1) in our presentation. In the remainder of this introduction, we present an overview of the observational record and the model in (B) and (C), and we briefly compare the equilibrium statistical predictions of the ESTP for the large-scale coherent structures on Jupiter with the observational record. The details in (B)–(F) are presented systematically below in the remainder of the chapter.

13.1.1 The observational record for Jupiter and the quasi-geostrophic model

There were two space missions to observe Jupiter, the Voyager mission of the 1970s and the Galileo mission of 1995–1996 (Rogers, 1995). The Voyager mission observed the "weather layer" on Jupiter. This is the upper part of Jupiter's atmosphere, which is like either the stratosphere or the upper layer of the ocean on earth with the important differences that, unlike the stratosphere, radiative forcing from the sun is very weak, and, unlike the upper layer of the ocean, wind forcing is negligible. Thus, the simplest models for interpreting observations on Jupiter are the quasi-geostrophic and one-half layer models derived in (1.159)–(1.163). The main observations of the Voyager mission for the weather layer can be summarized as follows (Rogers, 1995):

(A) *The Limaye profile*: The zonal shear flow that results from averaging of the velocity profile over the east–west at a fixed latitude is a remarkably complex sequence of alternating jets. In Figures 13.1 and 13.2 the centered Limaye profile for the two regions from 36.6° S to 13.7° S and 23.1° N to 42.5° N are displayed. Notice the complex alternating jets in each figure.

(B) *Coherent vortices in the southern hemisphere*: The Great Red Spot centered at 22.4° S and the White Ovals (WO) centered at 32° S are large-scale coherent structures. These structures are all *anti-cyclonic vortex structures*. An anti-cyclonic vortex structure has vorticity which is opposite in sign to the planetary rotation. Thus, an anti-cyclone has counter-clockwise rotation in the southern hemisphere.

(C) *No coherent vortices in the northern hemisphere*: There are very few coherent vortices in the northern hemisphere of Jupiter. There are *no large-scale coherent vortices* in the northern hemisphere between the latitudes 23.1° N and 42.5° N depicted in Figure 13.2.

(13.2)

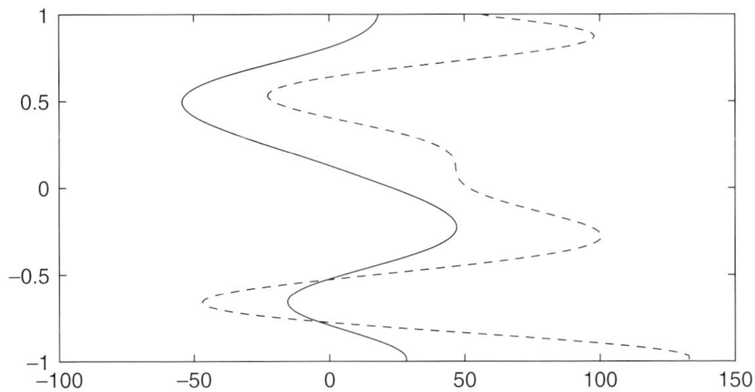

Figure 13.1 The Limaye zonal mean velocity profile for the southern hemisphere band from 36.6° S and 13.7° S (solid), and the zonal velocity profile for the lower layer (dashed) inferred by the Dowling procedure.

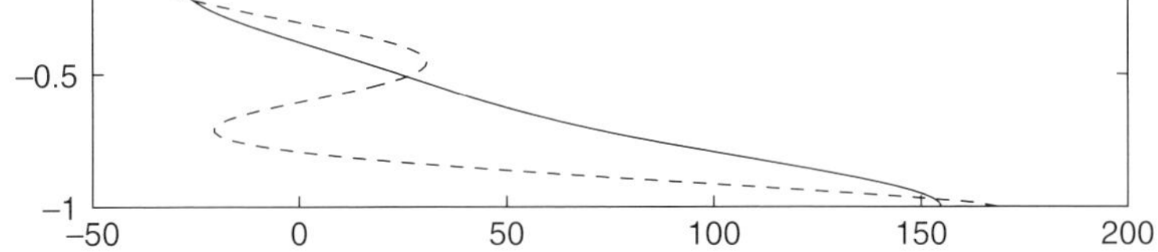

Figure 13.2 The Limaye zonal mean velocity profile for the northern hemisphere band from 23.1° N to 42.5° N (solid), and the zonal velocity profile for the lower layer (dashed) inferred by the Dowling procedure.

The structures in (A) and (B) are observed to be quasi-steady within the human observation time over 200 years.

There is more limited knowledge about the vertical structure of Jupiter's atmosphere (Rogers, 1995). The Galileo probe surveyed the vertical structure of Jupiter's atmosphere to a certain depth. Ingersoll *et al.* (2000) have recently interpreted the data from the Galileo mission as providing evidence that the weather layer is forced by small-scale convective towers. Such convective towers have an analogue on earth associated with the deep penetrative convection that occurs in the tropics (Emanuel, 1994). These clouds from deep moist convection rise to the bottom of the stratosphere where they reach their level of neutral buoyancy, they then spread horizontally, i.e. the horizontal velocity $\vec{v}_H$, has positive divergence, $\operatorname{div}\vec{v}_H > 0$. Recall that the weather layer is the analogue for Jupiter of the stratosphere on earth. On a rotating planet, the curl of the horizontal velocity equations, $\omega = \operatorname{curl}\vec{v}_H$, satisfies (Pedlosky, 1979)

$$\frac{D\omega}{Dt} = -f \operatorname{div}\vec{v}_H, \tag{13.3}$$

where f is the planetary rotation rate. Thus, the intense small-scale forcing by convective towers on Jupiter generates anti-cyclonic vorticity, i.e. ω with the sign opposite to the planetary f. With this interpretation (Ingersoll *et al.*, 2000), the *observational data* from the *Galileo probe* suggest

> The weather layer on Jupiter is subjected to small-scale forcing with the intense small scale forcing having a preponderance of anti-cyclones. (13.4)

There is another significant observational event for the weather layer on Jupiter. This is the collision of comet Shoemaker-Levy with Jupiter. In a fashion much like a rock splashing in a pond yields surface waves, the gravity wave speed for the weather layer on Jupiter has been measured from this collision as $c = 454ms^{-1}$. This value will be utilized in the models below.

13.1.2 Predictions of the ESTP with a suitable prior and the observational record

We show below in steps (C)–(D) from (13.1) that Dowling's fitting procedure for the observed Limaye profile is equivalent to utilizing the energy–enstrophy statistical theory, which has a Gaussian prior distribution for small-scale fluctuations of potential vorticity. Statistically, such a prior distribution has an equal number of cyclones and anti-cyclones in the small-scale probability distribution. To incorporate the observational information from the Galileo probe summarized in (13.4), we use a *prior probability distribution* with the same mean and variance as the Gaussian prior but with an *anti-cyclonic skewness* expressed by the normalized third moment. Below, we use a suitable family of centered Gamma distributions to achieve this (Turkington *et al.*, 2001) so that there is a preponderance of intense anti-cyclones in the prior distribution consistent with (13.4). The numerical valuations of the ESTP with these distributions has the following four important features:

(A) The coherent vortices which emerge from ESTP with the above family of prior distributions are anti-cyclones if and only if the prior distribution for small-scale potential vorticity fluctuations has anti-cyclonic skewness.
(B) For such a prior distribution with anti-cyclonic skewness, coherent large-scale monopolar vortices, such as the GRS, emerge at the appropriate latitudes within the zonal mean velocity profile, such as the Limaye profile derived from Voyager data (see Figure 13.3 below). The emergence of GRS in the ESTP predictions hardly alters the zonal mean Limaye profile (Figure 13.4).
(C) In agreement with the observational record in (13.2), no coherent vortices emerge in the northern hemisphere between 23.1° N and 42.5° N when the predictions of ESTP with an anti-cyclonic skewed prior are calculated.
(D) All steady flows arising as the large-scale predictions of ESTP are non-linearly stable (see Chapter 4), including zonal shear flows that contain prograde and retrograde jets and embedded vortices.

(13.5)

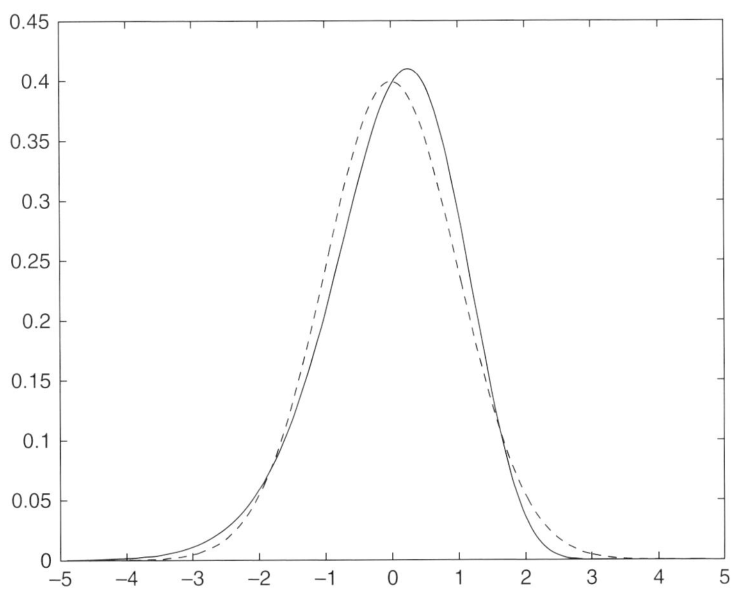

Figure 13.3 Centered Gamma distribution with mean zero, unit variance and anti-cyclonic skewness −0.5 (solid line), and Gaussian distribution with mean zero and unit variance (dashed line)

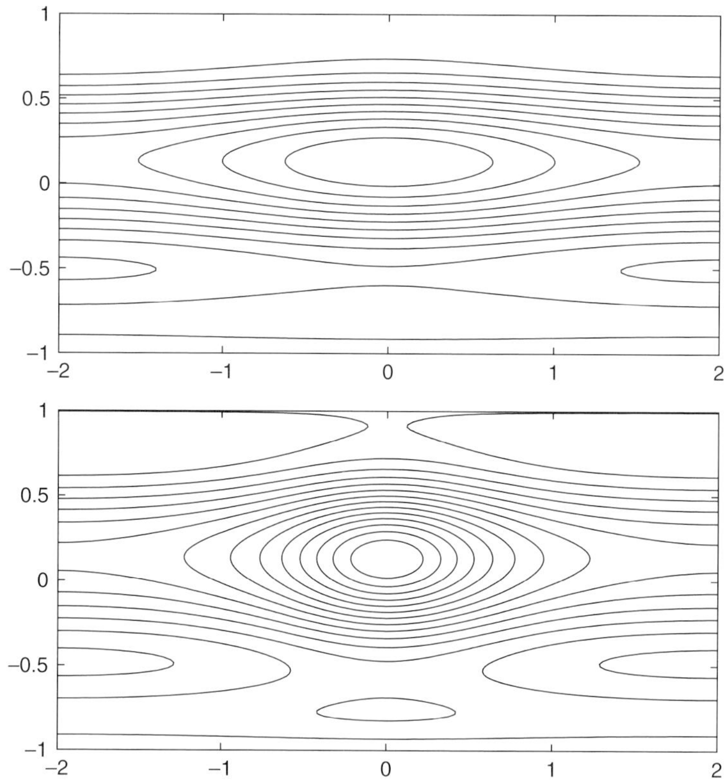

Figure 13.4 Mean streamline plots for the equilibrium states over the effective zonal topography in Figure 13.1, with skewness parameter $\varepsilon = -0.02$ (above) and $\varepsilon = -0.035$ (below), and the same energy and circulation as the Limaye zonal flow. The length scale is $L = 14,000\,\text{km}$.

Looking back at (13.2), we see that the *ESTP with suitable anti-cyclonic prior distribution consistent with the Galileo probe data in (13.4) automatically predicts coherent structures in the weather layer consistent with all the observations in (13.2) from the Voyager mission.* The important fact regarding non-linear stability of the equilibrium steady states is not a simple application of the ideas discussed in Chapter 4 and applied in Section 9.5 to the large-scale predictions of equilibrium statistical theories. Significant additional mathematical ideas beyond Section 9.5 are needed and have been developed recently by Ellis, Haven, and Turkington (2002) in an important paper. The application to Jupiter is discussed elsewhere (Haven *et al.*, 2002). Other applications of ESTP with a judicious physically and observationally motivated choice for a prior distribution have been developed recently for parameterizing the cold water mass in deep ocean convection (DiBattista and Majda, 2000; DiBattista *et al.*, 2002).

13.2 The quasi-geostrophic model for interpreting observations and predictions for the weather layer of Jupiter

As mentioned in the introduction, the weather layer of Jupiter is a comparatively shallow layer of fluid which is visible to observations over a deep internal layer where much less is known about its vertical structure. Under these circumstances, the simplest quasi-geostrophic model to interpret observations on Jupiter in the weather layer is the one-and-one-half layer model derived in (1.159)–(1.163). In this model, the shallow upper layer satisfies

$$\frac{\partial}{\partial t}(\Delta\psi - F_1(\psi-\psi_2)) + J(\psi, \Delta\psi - F_1(\psi-\psi_2)) + \beta\frac{\partial}{\partial x}\psi = 0 \tag{13.6}$$

while the lower layer satisfies the barotropic potential vorticity equation

$$\frac{\partial}{\partial t}\Delta\psi_2 + J(\psi_2, \Delta\psi_2) + \beta\frac{\partial}{\partial x}\psi_2 = 0. \tag{13.7}$$

For simplicity in exposition, we have eliminated the effects of dissipation and forcing in both (13.6) and (13.7). The simplest exact solutions for the lower layer equations in (13.7), which are assumed here, is a zonal shear flow

$$\psi_2 = \psi_2(y) \tag{13.8}$$

so that (13.6) becomes the F-plane potential vorticity equations with a zonal topography. With these simplifications, the quasi-geostrophic model in (13.6) becomes

$$\frac{\partial q}{\partial t} + J(\psi, q) = 0, \quad q = \Delta\psi - \lambda^{-2}(\psi-\psi_2) + \beta y, \tag{13.9}$$

where $q = q(x, y, t)$ is the potential vorticity and $\psi = \psi(x, y, t)$ is the stream function or, equivalently, the normalized height perturbation for the upper layer. The positive parameters $\lambda(= 1/\sqrt{F_1})$ and β are non-dimensionalized Rossby deformation radius and the gradient of the Coriolis parameter, respectively. The choice of λ, β, and $\psi_2(y)$ is discussed below. The velocity fields of the upper and lower layers are non-divergent, and are given by $(u, v) = (-\psi_y, \psi_x)$ and $(u_2, 0) = (-\psi_{2y}, 0)$, respectively.

For all the theory and computations reported below the flow domain is the zonal channel

$$\Omega = \{(x, y) : |x| < 2, |y| < 1\}.$$

The boundary conditions for flow in the channel are achieved by setting $\psi = 0$ on the walls $y = \pm 1$ and imposing periodicity in x.

As in Chapter 1, solutions of (13.9) in a channel domain conserve both the energy and circulation

$$E = \frac{1}{2}\int_{\Omega}(\psi_x^2 + \psi_y^2 + \lambda^{-2}\psi^2)\,dxdy,$$
$$\Gamma = \int_{\Omega}(\psi_{xx} + \psi_{yy} - \lambda^{-2}\psi)\,dxdy. \tag{13.10}$$

We leave this verification as a simple exercise for the reader. The impulse is also conserved by the dynamics in (13.9) but for simplicity in exposition, we remove this constraint here by a choice of reference frame in the subsequent discussion.

13.2.1 Fitting the non-dimensional model with the dimensional parameters of Jupiter

Here we show how to fit the dimensionless parameters and fields in the dynamical equations in (13.9) to their dimensional counterparts for the Jovian atmosphere, which are distinguished by asterisk superscripts. Let Ω^* be the flow domain that lies between the latitudes Θ_- and Θ_+, and let u^* be the zonally averaged velocity profile for this domain of Jupiter determined by the Limaye profile. A centered Limaye profile, $\tilde{u}^*$, obtained by subtracting from u^* its average over the domain, is introduced so that the zonal mean flow satisfies the boundary conditions $\tilde{\psi}^* = 0$ at the channel walls. With this adjustment of frame of reference, no constraint associated with the conservation of linear impulse is required. The characteristic length scale, L, is defined to be half the channel width, and the characteristic velocity scale, U, is defined to be the r.m.s. velocity of the centered Limaye profile; namely

$$L = \frac{(\Theta_+ - \Theta_-)r_0}{2}, \qquad U^2 = \frac{1}{2L}\int (\tilde{u}^*)^2\,dy^*,$$

where $r_0 = 7 \times 10^7\,m$ is used for the radius of Jupiter. The dimensional constants β^* and λ^* are computed from the Coriolis parameter $f(\Theta) = 2\Omega\sin\Theta$, with $\Omega = 1.76 \times 10^{-4}\,s^{-1}$, at the center latitude Θ_0 of the channel; specifically

$$\beta^* = \frac{f'(\Theta_0)}{r_0}, \qquad \lambda^* = \frac{c}{f(\Theta_0)},$$

where $c = 454\,ms^{-1}$ is the estimated gravity wave speed for Jupiter from the collision with comet Shoemaker-Levy. The dimensionless model parameters are then $\beta = L^2U^{-1}\beta^*$ and $\lambda = L^{-1}\lambda^*$. The fields q and ψ are similarly non-dimensionalized by L and U; in particular, the dimensionless centered Limaye stream function is

$$\tilde{\psi} = -\int_{-1}^{y}\tilde{u}(y')\,dy', \quad \text{with} \quad \tilde{u} = U^{-1}\tilde{u}^*.$$

13.2.2 Fitting the lower layer topography

There are no available data on the zonal velocity profile of the lower layer, and consequently there is no direct way to deduce ψ_2. Dowling (1995) has noted that a good fit for the observational data is obtained by imposing the linear relation

$$\tilde{q} = \theta\tilde{\psi} - \gamma \quad \text{with} \quad \theta \approx -\lambda^{-2}. \tag{13.11}$$

This relation combined with the second equation in (13.9) then determines the effective zonal topography

$$\psi_2 = (\lambda^2\theta + 1)\tilde{\psi} + \lambda^2\left(\frac{d\tilde{u}}{dy} - \beta y - \gamma\right). \tag{13.12}$$

In the computations presented here, the parameter θ in (13.12) is selected to be $\theta = -\lambda^{-2} - (\pi/2)^2$, the first eigenvalue of $\partial^2/\partial x^2 + \partial^2/\partial y^2 - \lambda^{-2}$ on Ω. When ψ_2 is fixed with this value, which is close to Dowling's approximate value, $\tilde{\psi}$ is a bifurcation point for the energy–enstrophy theory (see Chapters 6 and 8 above). Other selection criteria for θ are compatible with the analysis of observed data in (Dowling, 1995) and result in small changes in the effective zonal topography; those criteria are discussed elsewhere (Haven *et al.*, 2002). For any given value of θ, the parameter γ in (13.11) is chosen so that $\int_{-1}^{+1} \psi_2\, dy = 0$.

Figures 13.1 and 13.2 show the centered Limaye mean zero profile, the dark curve, and the lower layer zonal topography, the dashed curve, inferred by Dowling's procedure for two different bands of latitudes. Figure 13.1 is the band in the southern hemisphere from 13.7° S to 36.6° S, while Figure 13.2 is the band in the northern hemisphere from 23.1° N to 42.5° N. In these two figures, the velocity in the lower layer is often faster than the velocity on the upper layer. This aspect of the fitting has been controversial in some circles; however, recent observations (Atkinson *et al.*, 1997) of the deep winds of Jupiter through the Galileo probe show that the lower layer has faster velocities. This fact lends observational support to Dowling's fitting procedure. The results of the ESTP theory reported below, which utilize this fitting procedure as a starting point and lead automatically to the large-scale predictions consistent with the observations in (13.5)(A)–(D) for the weather layer, provide further support for Dowling's fitting procedure.

13.3 The ESTP with physically motivated prior distribution

Dowling's fitting procedure is based on the linear relation in (13.11) for the potential vorticity and stream function. Recall from Chapters 6 and 8 and Section 9.2 that the linear equation (13.11) arises as the most probable coarse-grained mean state equilibrium statistical prediction of ESTP with Gaussian prior distribution for small-scale fluctuations of potential vorticity. Through Dowling's fitting procedure, this equilibrium statistical theory provides an excellent fit with the zonal averaged Limaye velocity profile, the jets on Jupiter. On the other hand, energy–enstrophy statistical theory applied here can predict only zonal mean flows on Jupiter in contrast to the observations of coherent vortices such as the Great Red Spot and White Ovals in the weather layer (see (13.2)). Furthermore, the use of a Gaussian prior distribution in deriving (13.11) implies equal statistically symmetric forcing at small-scales by cyclones and anti-cyclones; this implied symmetric small scale forcing from the EEST model contradicts the observational data from the Galileo probe summarized in (13.4), where there is a preponderance of strong anti-cyclones in the small-scale forcing of the weather layer. Our strategy here in addressing these physical discrepancies is to set up the ESTP with a prior distribution $\Pi(\lambda)$ that has the identical first and second moments as the Gaussian distribution, but has a non-vanishing third moment which skews this distribution toward measuring a preponderance of strong anti-cyclones in the small-scale fluctuations of potential vorticity. Recall the intuition from Chapter 12 which demonstrates in several simplified contexts that ESTP can be utilized to represent the large-scale consequences of small-scale forcing, provided that the flux of energy to large scales is sufficiently weak. In this fashion the ESTP theory is set up to predict the large-scale features emerging from this random statistical forcing at small scales, consistent with the observational record from the Galileo probe.

13.3.1 The ESTP with a family of prior distributions with anti-cyclonic skewness

First we briefly review the ESTP described in Section 9.2 with a general prior distribution $\Pi(\lambda)$ adapted to the present setting with constraints on energy and circulation in (13.10). Following the calculations in Section 9.2, the entropy-maximization procedure of ESTP with the constraints in (13.10) leads to the mean field equation

$$\bar{q}(x, y) = \bar{\psi}_{xx} + \bar{\psi}_{yy} - \lambda^{-2}(\bar{\psi} - \psi_2) + \beta y, \tag{13.13}$$

where

$$\bar{q} = \frac{\int \lambda e^{(\theta\bar{\psi}-\gamma)\lambda}\Pi(\lambda)d\lambda}{\int e^{(\theta\bar{\psi}-\gamma)\lambda}\Pi(\lambda)d\lambda}. \tag{13.14}$$

The parameters θ and γ are the Lagrange multipliers for the energy and circulation constraints in (13.10), respectively, and play the role of the "inverse temperature" and "chemical potential." It is instructive to write (13.14) in its variational form

$$\bar{q} = G'(\theta\bar{\psi} - \gamma), \tag{13.15}$$

where

$$G(s) = \ln \int e^{s\lambda}\, \Pi(\lambda)\, d\lambda$$

is the cumulant generating function for the probability distribution Π. Notice that the cumulant generating function is simply the log of the moment generating function and is related to the partition function in statistical mechanics (see Chapters 6 and 9) through $\ln \mathcal{Z}(\psi) = G(\theta\psi - \gamma)$. The properties of the coherent structure determined by the solution to (13.15) depend on the prior distribution Π through the properties of the real function $G' = dG/ds$.

An important property of the cumulant generating function, easily verified by the reader, is that the first three derivatives of $G(s)$ at the origin measure the first three mean-centered moments of $\Pi(\lambda)$, i.e.

$$\left.\frac{d^p G(s)}{ds^p}\right|_{s=0} = G^{(p)}(0) = \int (\lambda - \langle\lambda\rangle)^p \Pi(\lambda) d\lambda,\ p = 2, 3 \tag{13.16}$$

with

$$\frac{dG}{ds}(0) = \langle\lambda\rangle = \int \lambda \Pi(\lambda) d\lambda. \tag{13.17}$$

For a given prior distribution $\Pi(\lambda)$ with zero mean, $G^{(2)}(0) = \mathrm{Var}(\Pi)$, the variance of $\Pi(\lambda)$, while the skewness of Π is given by

$$\mathrm{Skew}\,\Pi = \frac{G^{(3)}(0)}{(\mathrm{Var}\,\Pi)^{3/2}} = \frac{(\lambda - \langle\lambda\rangle)^3 \Pi(\lambda) d\lambda}{(\mathrm{Var}\,\Pi)^{3/2}}. \tag{13.18}$$

Obviously, a Gaussian distribution has zero skewness.

Now consider a family of prior probability distributions $\Pi_\varepsilon(\lambda)$ normalized with zero mean and unit variance so that

$$G'_\varepsilon(0) = 0, \quad G''_\varepsilon(0) = 1, \tag{13.19}$$

while the skewness is given by

$$G^{(3)}_\varepsilon(0) = \mathrm{Skew}\,\Pi_\varepsilon = 2\varepsilon \tag{13.20}$$

with ε sufficiently small. The Taylor approximation

$$G'_\varepsilon(s) \cong G'_\varepsilon(0) + sG''_\varepsilon(0) + \frac{s^2}{2} G^{(3)}_\varepsilon(0) \tag{13.21}$$

substituted into the mean field equation with the normalization in (13.19) and (13.20) yields the approximate mean field equation

$$\bar{q} = \theta\bar{\psi} - \gamma + \varepsilon(\theta\bar{\psi} - \gamma)^2. \tag{13.22}$$

The formula in (13.22) strongly suggests that for the prior distribution $\Pi_\varepsilon(\lambda)$ normalized in (13.19), (13.20), the skewness of the prior distribution is the leading-order contribution to the non-linearity of the original coarse-grained mean field equation. Furthermore, the approximate mean field equation in (13.22) suggests that

> An anti-cyclonically skewed prior distribution, i.e. $\varepsilon < 0$, induces anti-cyclonic, i.e. $\bar{q} \leq 0$, corrections to the predictions of the ESST. (13.23)

This can be understood intuitively by building an approximate solution of (13.22) for $|\varepsilon| \ll 1$ in the form

$$\bar{q} = \tilde{Q} + \Delta\psi_1 - \lambda^{-2}\psi_1,$$

where $\tilde{Q}$ satisfies the mean field equation of the energy–enstrophy theory in (13.11), while the correction satisfies

$$\Delta\psi_1 - \lambda^{-2}\psi_1 = \varepsilon(\theta\tilde{\psi} - \gamma)^2,$$

so that anti-cyclones (resp. cyclones) are generated in the perturbation to the leading order for $\varepsilon < 0$ (resp. $\varepsilon > 0$).

Recall that the observational data summarized in (13.2) at large scales in the weather layer have coherent structures such as the Great Red Spot with anti-cyclonic behavior, while the observations from Galileo probe in (13.4) suggest a prior distribution $\Pi_\varepsilon(\lambda)$ with anti-cyclonic skewness. Thus, the large-scale predictions in (13.22) of the approximate mean field equation in (13.22) indicate that the equilibrium statistical theory predicts a direct link between anti-cyclonic small-scale forcing and anti-cyclonic large-scale coherent structures.

13.3.2 The centered Gamma distribution as a family of skewed prior distributions for ESTP

With all of the motivation just presented, we need to find a family of prior probability distributions measuring skewness and also a higher probability of large amplitude events going in the same direction with this skewness. An attractive choice of the prior distribution fitting these requirements is the centered gamma distribution

$$\Pi_\varepsilon(\lambda) = |\varepsilon|^{-1}R(\varepsilon^{-1}[\lambda + \varepsilon^{-1}];\, \varepsilon^{-2}), \tag{13.24}$$

where $R(z;\, a) = \Gamma(a)^{-1}z^{a-1}e^{-z}(z \geq 0)$, $R(z;\, a) = 0(z < 0)$ denotes the standard gamma density. This family of distributions has a one-sided exponential tail with compact support at the other side. The scaling of Π_ε is chosen so that the normalizations in (13.19) and (13.20) are satisfied. It is easy to calculate that $R(z;\, a)$ has the explicit cumulant generating function $-a\ln(1 - s)$. Thus, the mean field equation (13.15) associated with the prior distribution (13.24) takes the explicit form

$$\bar{q} = G'_{\varepsilon}(\theta\bar{\psi} - \gamma) = \frac{\theta\bar{\psi} - \gamma}{1 - \varepsilon(\theta\bar{\psi} - \gamma)}. \tag{13.25}$$

As expected from the formal arguments presented above, in the limit as the skewness parameter $\varepsilon \to 0$, Π_{ε} converges to the standard Gaussian distribution, and the corresponding statistical equilibrium theory reduces to the standard energy–enstrophy theory.

As an illustration, the skewed prior distribution determined by the centered Gamma distribution in (13.24) is plotted for the anti-cyclonic negative skewness value $\varepsilon = -0.5$ in Figure 13.4. Not only is this distribution skewed to negative values but has an exponential tail for large negative values compared with bounded support for large positive values. Thus, there is a preponderance of large negative value events in this prior probability distribution, even though it has zero mean so that positive and negative events are equally likely on average.

13.4 Equilibrium statistical predictions for the jets and spots on Jupiter

All of the ingredients are now in place for an ESTP with a prior $\Pi_{\varepsilon}(\lambda)$ motivated by the small-scale forcing in (13.4) observed for the weather layer. Namely, we should utilize the ESTP with the prior distribution given by the centered Gamma distribution in (13.24) with a small anti-cyclonic skewness, $\varepsilon < 0$ and the lower layer topography determined by the Dowling procedure described in (13.12). What are our goals? We hope that the large-scale predictions of ESTP with these anti-cyclonic skewed prior distributions automatically produce the large-scale observational features of the weather layer on Jupiter summarized in (13.2). As we shall see below, these expectations are confirmed in a spectacular fashion.

The large-scale mean field predictions for the ESTP presented below involve a similar numerical algorithm as utilized in Section 11.2. The external values for the energy, E, and circulation, Γ, are fixed as given by those from the centered Limaye profile, $\tilde{\psi}$, in all the computations presented below. Consequently, the solution corresponding to $\varepsilon = 0$ is a zonal shear flow that coincides with the centered Limaye profile on the channel domain. Branches of solutions are therefore computed by varying the skewness parameter, ε, in the prior distribution (13.24), holding the bottom topography, ψ_2, energy, E, and circulation, Γ, fixed. Since virtually all of the large coherent vortices on Jupiter have anti-cyclonic relative vorticity, it is natural to expect that they emerge as equilibrium states in a statistical theory with a prior distribution having anti-cyclonic skewness. This conjecture is tested for particular channel domains in the southern and the northern hemispheres. Each domain contains five alternating zonal jets of the Limaye profile.

13.4.1 The southern hemisphere domain

The large-scale coherent structures predicted by the statistical equilibrium theory are computed for a channel domain between 36.6° S and 13.7° S in latitude. (For the sake of clarity, the usual northern hemisphere conventions are used in the discussion of southern hemisphere flows.) Recall that Figure 13.1 displays the centered Limaye profile, $\tilde{u}^*$, for the upper layer together with velocity profile, u_2^*, for the lower layer determined by the Dowling procedure. By construction, the computed solution, $\bar{u}^*$, for $\varepsilon = 0$ coincides with the centered Limaye profile, $\tilde{u}^*$. In Figure 13.3 and the subsequent figures, the zonal profiles of computed equilibrium states are plotted with respect to the reference frame of the uncentered Limaye profile. The streamline plots displayed in Figure 13.4 correspond to solutions with the same values of energy, E, and circulation, Γ, as in Figure 13.1, but with anti-cyclonic skewness $\varepsilon = -0.02, -0.035$. These plots clearly show that a large coherent vortex forms in the zonal shear flow and becomes stronger and more concentrated as the skewness parameter increases. This large vortex is an anti-cyclone centered in latitude at approximately 23° S, the latitude of the Great Red Spot. Furthermore, the vortex for $\varepsilon = -0.035$ has nearly the same size and shape as the GRS. A weaker anti-cyclonic vortex also forms south of the large vortex at 32° S, in the zone that corresponds to the location of the White Ovals. In the belt between those zones and in the belt north of the GRS, there are regions of recirculating cyclonic flow. All these predictions of the statistical equilibrium theory are consistent with the observed large-scale features of the Jovian southern hemisphere summarized in (13.2) (Rogers, 1995). In Figure 13.5, the zonal mean velocity profiles for the upper layer flows corresponding to $\varepsilon = -0.02, -0.035$ are plotted together with the Limaye profile. It is noteworthy that, even though the

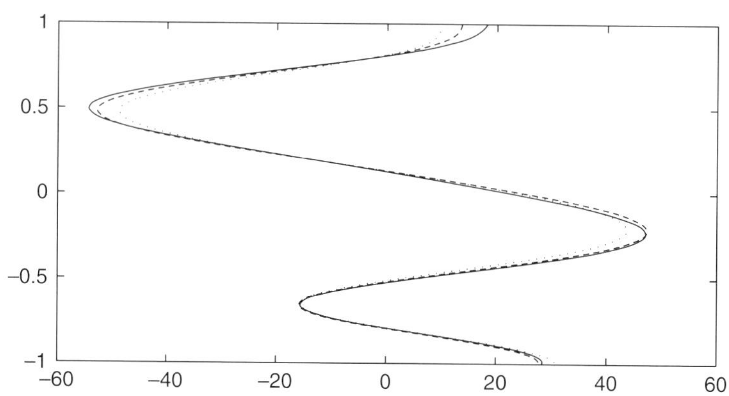

Figure 13.5 The Limaye profile (solid) in the southern hemisphere band, and the zonally averaged velocity profiles for the equilibrium states with $\varepsilon = -0.02$ (dashed) and $\varepsilon = -0.035$ (dotted).

computed solution with $\varepsilon = -0.035$ contains a strong vortex, its zonally averaged profile departs only slightly from the Limaye profile in a self-consistent fashion with the features in (13.2)(A), (B) for the observational record.

13.4.2 The northern hemisphere domain

In analogy to the southern hemisphere computation, the theory is applied to a channel domain from 23.1° N to 42.5° N in latitude. This implementation provides an interesting test of the statistical theory because no large-scale coherent vortices are observed in this domain of Jupiter. Recall that Figure 13.2 displays the centered Limaye profile, $\tilde{u}^*$, and the zonal velocity profile, u_2^*, for the lower layer determined by the Dowling procedure from Subsection 13.2.2. The solutions are then computed corresponding to skewness parameter $\varepsilon = -0.02$, -0.032 and having the same energy and circulation as the Limaye solution with $\varepsilon = 0$. In striking contrast to the southern hemisphere computation, these equilibrium states are purely zonal shear flows in agreement with the observations summarized in (13.2). The zonal velocity profiles for these solutions are plotted together in Figure 13.6. In fact, all the solutions corresponding to $0 > \varepsilon > -0.032$ are zonal, and for $\varepsilon < -0.033$ solutions fail to exist. Even though no coherent vortices emerge, it is evident from Figure 13.6 that the zonal jet structure of these solutions is modified significantly as the anti-cyclonic skewness increases.

As a final note, we remark that these calculations of ESTP were done on a non-dimensional channel domain with length which is twice the height. The horizontal scale of such a channel domain is not long enough to completely circle

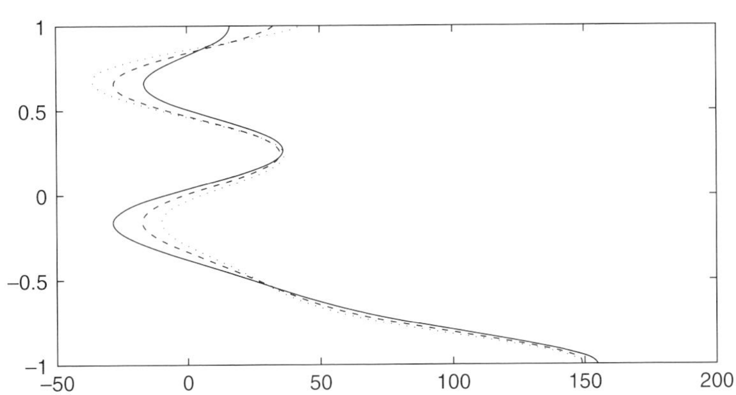

Figure 13.6 The Limaye profile (solid) in the northern hemisphere band, and the zonal velocity profiles for the equilibrium states with $\varepsilon = -0.02$ (dashed) and $\varepsilon = -0.032$ (dotted) and the same energy and circulation as the Limaye zonal flow; each of these computed solutions is purely zonal.

the globe on Jupiter; this was done only for computational convenience. However, if we recall the results from Section 11.2 on the predicted ESTP Rhines' scale, it is clear that the height of the domain controls the vortex structures of the ESTP utilized here for domains with length larger than height. Thus, we expect very little difference in out results, qualitatively, for longer domains. This fact is confirmed in Haven *et al.* (2002).

References

Atkinson, D. H., Ingersoll, A. P., and Seiff, A. (1997), Deep winds on Jupiter as measured by the Galileo probe. *Nature* **388**, 649–650.

DiBattista, M. T. and Majda, A. J. (2000), An equilibrium statistical theory for large-scale features of open-ocean convection. *J. Phys. Oceanogr.* **30** (6), 1325–1353.

DiBattista, M. T., Majda, A. J., and Marshall, J. (2002), A statistical theory for the "patchiness" of open-ocean deep convection: the effect of p-reconditioning. *J. Phys. Oceanogr.* **32**, 599–626.

Dowling, T. E. (1995), Dynamics of Julian atmosphere. *Annual Rev. Fluid Mech.* 27: 293–334.

Ellis, R., Haven, K., and Turkington, B. (2002), Nonequivalent statistical ensembles and refined stability theorem for most probable flows. *Nonlinearity* **15** (2), 239–255.

Emanuel, K. (1994), *Atmospheric Convection*. New York: Oxford University Press.

Haven, K., Majda, A., and Turkington, B. (2003), Predicting the jets and spots on Jupiter by equilibrium statistical mechanics. In preparation for *J. Atoms. Sci.*

Ingersoll, A. P., Geirasch, P. J., Banfield, D., Vasavada, A. R., and the Galileo Imaging Team (2000), Observations of moist convection in Jupiter's atmosphere. *Nature* **403**, 630–632.

Marcus, P. S. (1993), Jupiter's Great Red Spot and other vortices. *Annual Rev. Astrophys.* **31**, 523–573.

Pedlosky, J. (1979) *Geophysical Fluid Dynamics*. New York: Springer.

Rogers, J. H. (1995), *The Giant Planet Jupiter*. Cambridge: Cambridge University Press.

Turkington, B., Majda, A., Haven, K., and DiBbatista, M. (2001), Statistical equilibrium predictions of the jets and spots on Jupiter. *Proc. Nat. Acad. Sci.* **98**, 12346–12350.

14

The statistical relevance of additional conserved quantities for truncated geophysical flows

14.1 Introduction

A major theme of Chapters 6–13 of this book is to illustrate that systematic application of ideas from equilibrium statistical mechanics leads to novel promising strategies for assessing the unresolved scales of motion in geophysical flows. The various theories range from the simplest energy–enstrophy statistical theory (EEST) discussed in Chapters 6 and 8 to the empirical statistical theories discussed in Section 9.4 attempting to encode all the conserved quantities (ESTMC), to point vortex statistical theories discussed in Section 9.3, and, finally, to the empirical statistical theories with a few large-scale constraints and a judicious small-scale prior distribution (ESTP) formulated in Section 9.2. It was established in Chapters 11, 12, and 13 that the ESTP theories have a wide range of applicability in predicting large-scale behavior in damped and driven flows, as well as for observations such as the Great Red Spot of Jupiter. The ESTP formulation also includes the predictions of the energy–enstrophy statistical theory for the mean flow from Chapters 6 and 8 by utilizing a simple Gaussian prior probability distribution for potential vorticity fluctuations.

As discussed earlier, in Chapters 8, 9, and 10, the different equilibrium statistical theories all attempt to predict the coarse-grained behavior at large scales through the use of some of the formally infinite list of conserved quantities for idealized geophysical flows derived in Chapter 1. As discussed earlier in Chapter 10, the natural question arises as to which of the formally infinite list of conserved quantities for geophysical flows is actually statistically relevant for the coarse-grained large-scale statistical features of the flow field? A related question is the suitability of an equilibrium statistical theory such as ESTP to approximate this statistical behavior at large scales. In this chapter these important issues are addressed by utilizing suitable discrete numerical approximations for geophysical flows with many conserved quantities as a numerical laboratory. Related issues are also discussed briefly below in the simpler context of the truncated Burgers–Hopf equations described in Chapter 7. The material in the present chapter is based on Abramov and Majda (2003a, b) and Abramov *et al.* (2003).

In this chapter the results of numerical experiments are presented for truncated geophysical flows with topography in a suitable regime, which establish that the integrated third power of potential vorticity besides the familiar constraints of energy, circulation, and enstrophy is statistically relevant for the coarse-grained equilibrium statistical behavior at large scales. Furthermore, in this regime of fluid motion the integrated higher powers of potential vorticity larger than three are statistically irrelevant for the large-scale equilibrium statistical behavior. Also, numerical evidence is presented here which strongly supports the use of the ESTP theories as arising from the statistical dynamics of the truncated quasi-geostrophic flow with many invariants; these results include the same strategy utilized in Chapter 13 for Jupiter where it is shown below that a skewed prior distribution for small-scale potential vorticity fluctuations predicts the coarse-grained large-scale response. The simplest geophysical model utilized here is barotropic flow with topography in periodic geometry, which is described by the equations

$$q = \Delta\psi + h; \quad \vec{v} = \nabla^{\perp}\psi = \begin{pmatrix} -\psi_y \\ \psi_x \end{pmatrix}; \quad \frac{\partial q}{\partial t} + \vec{v}\cdot\nabla q = 0. \tag{14.1}$$

In (14.1), q is the potential vorticity, $\omega = \Delta\psi$ is the relative vorticity, $\vec{v}$ is the incompressible fluid velocity, ψ is the stream function, and h is the prescribed topography. When h is zero, (14.1) become the equations for 2D incompressible flow. As discussed in Chapter 1, the equations in (14.1) conserve kinetic energy

$$E = \frac{1}{2}\int |\vec{v}|^2 = -\frac{1}{2}\int \psi\omega == -\frac{1}{2}\int \psi(q-h), \tag{14.2}$$

as well as the infinite number of conserved quantities

$$Q_p(q) = \int q^p, \quad p = 1, 2, 3, \ldots . \tag{14.3}$$

In (14.2), (14.3), $\int \cdot$ denotes integration over the period domain in periodic geometry, where the total circulation, Q_1, satisfies $Q_1 \equiv 0$. As discussed in Chapters 6, and 8, the quadratic conserved quantity, the enstrophy

$$\mathcal{E} = \frac{1}{2}Q_2(q), \tag{14.4}$$

is singled out in some statistical theories for large-scale flow as having special significance, while the higher generalized enstrophies, $Q_p(q)$, for $p \geq 3$ are ignored in these theories. Since EEST postulates the least additional statistical information, it is reasonable first to summarize results established in Chapter 8 for this statistical theory as a benchmark null hypothesis for the statistical role of the higher invariants in (14.3).

14.1.1 The traditional spectral truncation and equilibrium statistical theory

The potential vorticity, q, in (14.1) is truncated by projection on to $(2M+1)\times(2M+1)$ Fourier modes by the projection operator

$$P_\varepsilon q = \sum_{|k_1|,|k_2|\le M} \hat{q}_{\mathbf{k}} e^{ix\mathbf{k}}, \quad \varepsilon = \frac{2\pi}{2M+1}. \tag{14.5}$$

As discussed in Chapter 8, the traditional aliased truncation of the quasi-geostrophic equations in (14.1) for these Fourier coefficients $\hat{q}_{\mathbf{k}}$, $|k_1| \le M$, $|k_2| \le M$, is given by

$$\frac{\mathrm{d}\hat{q}_{\mathbf{k}}}{\mathrm{d}t} = \sum_{|k'_1|,|k'_2|\le M} \frac{\mathbf{k}\times\mathbf{k}'}{|\mathbf{k}'|^2} \hat{q}_{\mathbf{k}+\mathbf{k}'}(\hat{q}_{-\mathbf{k}'} - \hat{h}_{-\mathbf{k}'}). \tag{14.6}$$

In (14.6) the finite domain of the Fourier coefficients is extended to $(4M+1)\times(4M+1)$ by using the periodicity rule $\hat{q}_{\mathbf{k}} = \hat{q}_{\mathbf{k}+(2M+1,0)} = \hat{q}_{\mathbf{k}+(0,2M+1)}$. As established in Chapter 8, the traditional spectral truncation in (14.6) conserves energy, circulation, and enstrophy from (14.2)–(14.4) but in general none of the higher-order invariants in (14.3) for $p \ge 3$; this last fact has been demonstrated in recent numerical experiments by Abramov and Majda (2003b), also from Chapter 8. The energy–enstrophy statistical theory which utilizes only these two conserved quantities for the truncation in (14.6) predicts a Gaussian probability measure for the equilibrium statistical behavior of the Fourier modes. Recall that Gaussian measures are uniquely characterized by their mean and variance. Given externally prescribed mean values for the energy and enstrophy, the mean of this probability distribution for $\hat{q}_{\mathbf{k}}$ is determined by the linear equation

$$\bar{q}_\mu = \Delta\bar{\psi}_\mu + h = \mu\bar{\psi}_\mu, \tag{14.7}$$

while as shown in Section 8.5 there is equipartition of pseudo-energy of perturbations around the mean state in (14.7) with a constant variance $\alpha > 0$. Here the pseudo-energy Fourier coefficients, $\hat{p}_{\mathbf{k}}$, are given by

$$\hat{p}_{\mathbf{k}} = \left(1 + \frac{\mu}{|\mathbf{k}|^2}\right)^{1/2} \hat{\omega}_{\mathbf{k}}, \tag{14.8}$$

with $\hat{\omega}_{\mathbf{k}}$ being the Fourier coefficients of relative vorticity. The constants, μ and α, completely characterizing the Gaussian measure, are determined by the values of E and $\mathcal{E}$ as illustrated in Sections 8.4 and 8.5 above. A wide variety of numerical experiments with the truncation in (14.6) confirm the predictions of the equilibrium statistical theory from (14.7) and (14.8). However, these results cannot address the fundamental issue for the statistical relevance of the higher-order invariants in (14.3) for $p \ge 3$, since these fail to be conserved by the approximation in (14.6) over long time intervals of integration. Here, the equilibrium statistical predictions through (14.7) and (14.8) are regarded as a null hypothesis in the tests reported below for the role of the higher-order invariants.

14.2 A numerical laboratory for the role of higher-order invariants

14.2.1 The spectral truncation with many conserved quantities

Here, following the important observation by Zeitlin (1991), we consider the sine-bracket truncation as an approximation to the quasi-geostrophic dynamics in (14.1) through the spectral approximation in (14.6). This finite-dimensional set of equations for the Fourier coefficients is given by

$$\frac{\mathrm{d}\hat{q}_{\mathbf{k}}}{\mathrm{d}t} = \sum_{k'_1,k'_2=-M}^{M} \frac{\sin(\varepsilon \mathbf{k}\times\mathbf{k}')}{\varepsilon|\mathbf{k}'|^2}\hat{q}_{\mathbf{k}+\mathbf{k}'}(\hat{q}_{-\mathbf{k}'}-\hat{h}_{-\mathbf{k}'});\quad \varepsilon=\frac{2\pi}{2M+1}, \tag{14.9}$$

with the same convention of periodicity as in (14.6). The truncation in (14.9) conserves the energy in (14.2) as the Hamiltonian. In addition, the sine-bracket truncation in (14.9) conserves $2M$ invariants of the form

$$C_N = \sum_{Z^N} \hat{q}_{\mathbf{i}_1}\dots\hat{q}_{\mathbf{i}_N}\cos[\varepsilon A(\mathbf{i}_1,\dots,\mathbf{i}_N)],$$

$$Z^N = \left\{(\mathbf{i}_1,\dots,\mathbf{i}_N),\ \sum_{j=1}^{N}\mathbf{i}_j=\mathbf{0}\right\},\quad 1\le N\le 2M,$$

$$A(\mathbf{i}_1,\dots,\mathbf{i}_N) = \mathbf{i}_2\times\mathbf{i}_1+\mathbf{i}_3\times(\mathbf{i}_1+\mathbf{i}_2)+\dots+\mathbf{i}_N\times(\mathbf{i}_1+\dots+\mathbf{i}_{N-1}). \tag{14.10}$$

The Casimir invariant C_2 is a multiple of the enstrophy since $A(\mathbf{i},-\mathbf{i})=0$. However, the higher Casimir invariants C_N for $3\le N\le 2M$ are suitable regularizations of those in (14.3) for $3\le p\le 2M$. A pedagogical detailed discussion of (14.9), (14.10) can be found in the Appendix to the present chapter, while this and related important numerical issues are found in Abramov and Majda (2003b). In particular, to guarantee conservation of the invariants in (14.10) within roundoff error under time discretization, McLachlan's (1993) symplectic integrator is combined with a suitable version of second-order accurate Strang splitting in time (Abramov and Majda, 2003b). This last modification of the basic algorithm in McLachlan (1993) is crucial for accurate energy conservation and results in a 10^{-4} decrease in relative errors for the energy with the time steps used below for only twice the computational expense. We note that as the number of modes, M, increases

$$\frac{\sin(\varepsilon\mathbf{k}\times\mathbf{k}')}{\varepsilon} = \vec{k}^{\perp}\cdot\vec{k}'+\mathcal{O}(\varepsilon^2)$$

so that by comparing and the spectral representation in (8.2), the sine-bracket truncation is formally consistent with the quasi-geostrophic flow in (14.1) as

$M \to \infty$. In fact, a rigorous proof of convergence is presented in Abramov and Majda (2003b).

14.2.2 Numerical experiments demonstrating the statistical relevance of $C_3(q)$ at large scales

In the numerical experiments reported here, the prescribed values of the initial data were chosen so that the energy and enstrophy were fixed at $E = 7$ and $\mathcal{E} = 20$, while the value of the third invariant, $C_3(q)$ from (14.10), normalized through $\hat{C}_3(q) = C_3(q)/\mathcal{E}^{3/2}$, varied through the four values, $\hat{C}_3 = 0, 2, 4, 6$. A standard constrained optimization problem was set up to find initial data with many Fourier modes satisfying these constraints (Abramov, 2002). The higher-order invariants, C_N, for $4 \leq N \leq 2M$ were not specified by pre-determined values for the initial data, although they are automatically conserved within roundoff error for the algorithm in (14.9). The values of energy and enstrophy, $E = 7$, $\mathcal{E} = 20$, guarantee a suitable "negative temperature regime" for the energy–enstrophy equilibrium statistical theory of Chapter 8 summarized in (14.7), (14.8) with $\mu \cong -0.9$, $\alpha \cong 20$. The numerical experiments utilizing (14.9) reported below use $M = 11$, so that there are 528 active Fourier modes and 22 non-trivial conserved quantities in the dynamics. In all numerical experiments reported below, the equations in (14.9) were calculated for times of order 10^4 with a time step, $\Delta t = 0.01$ with a maximum relative error in energy of 1.2×10^{-6}. Statistics were gathered through long-time averaging of individual solutions with

$$\langle f \rangle_\tau = \frac{1}{\tau} \int_{T_0}^{T_0+\tau} f(t)\, \mathrm{d}t \tag{14.11}$$

denoting the averaging. After several initial experiments, it was determined that initial time T_0 with $T_0 \geq 10^3$ were sufficiently long to begin the time-averaging procedure. Three different cases of barotropic geophysical flow from (14.1) with the discretization from (14.9) were studied with the above parameters for the initial data: no topography, large-scale random topography with non-zero Fourier modes at $|\mathbf{k}|^2 = 2, 4$, and the deterministic layered topography, $h = 0.2\cos(x) + 0.4\cos(2x)$. This last case has the advantage in that the energy–enstrophy theory in (14.7), (14.8) predicts a large-scale mean flow which is also layered, so $\bar{\psi}(x)$ is a function of the x-variable alone, for any value of μ; thus, significant departures from such a layered structure in x in stream function plots provide strong visual evidence for departures in the large-scale statistical predictions beyond the energy–enstrophy theory. Most of the results are reported below for this case to take advantage of this fact.

Also, we note that the same layered topography and identical values of α and μ where utilized together with the standard pseudo-spectral code for the numerical results presented earlier in (8.4), (8.5) confirming the predictions of the standard energy–enstrophy theory. Since the focus here is on the statistical relevance of the higher-order invariants in (14.3) or (14.10) for equilibrium statistical theories, a higher-order invariant is called statistically relevant at large scales if it affects single-point spatial statistics such as the large-scale mean, and probability distribution function (PDF) centered about the mean. Below we also study the effect of the higher-order invariants for (14.9) on two-point spatial correlations by computing the energy spectrum and looking for systematic departures from equipartition of pseudo-energy. However, a systematic study of other multi-point, multi-time statistics is not presented here. With all of this background, the next results are reported below for the mixing properties of the system, the energy spectrum, the mean stream function, the mean potential vorticity, and the PDF for potential vorticity.

14.2.3 Mixing and decay of temporal correlations

The normalized temporal correlation functions averaged over the two groups of Fourier modes $|\mathbf{k}|^2 = 1$ and $|\mathbf{k}|^2 = 2$ are depicted in Figure 14.1 for the four cases with varying $\hat{C}_3$ for layered topography. These are the correlation functions with the slowest decay rate among all the groups of 528 Fourier modes with significantly faster decay for time correlations of the higher modes not depicted here. Obviously, the longest tails for the correlations occur for $|\mathbf{k}|^2 = 1$. The two graphs in Figure 14.1 each have four curves with the cases $\hat{C}_3 = 0$ demonstrating the most rapid decay of correlations, $\hat{C}_3 = 2, 4$ exhibiting clear decay of correlations, and the case $\hat{C}_3 = 6$ with $|\mathbf{k}|^2 = 1$ showing marginal decay of correlations to below 15 % of the value with zero lag at time lags of 10^3. The correlation function in this case eventually decays to zero with time lags of order $2 \cdot 10^3$ but the similar plot for $|\mathbf{k}|^2 = 1$ with $\hat{C}_3 = 8$, not depicted here, exhibits no decay of correlations. Thus the value $\hat{C}_3 = 6$ is near the boundary of the parameter regime with decay of temporal correlations. As discussed in Chapter 7, the decay of correlations of Fourier modes is the most primitive test for mixing in phase space of an individual solution and justifies the use of the time averages in (14.11) in calculating statistics of the mean flow and energy spectrum. The other cases with no topography and random topography have similar behavior in the decay of temporal correlations (Abramov, 2002).

14.2.4 The large-scale mean flow

Here the large-scale mean flow is calculated from the numerical solution by processing the time average of the stream function, $\bar{\psi} = \langle \psi \rangle_T$ and $\bar{q} = \langle q \rangle_T$ at

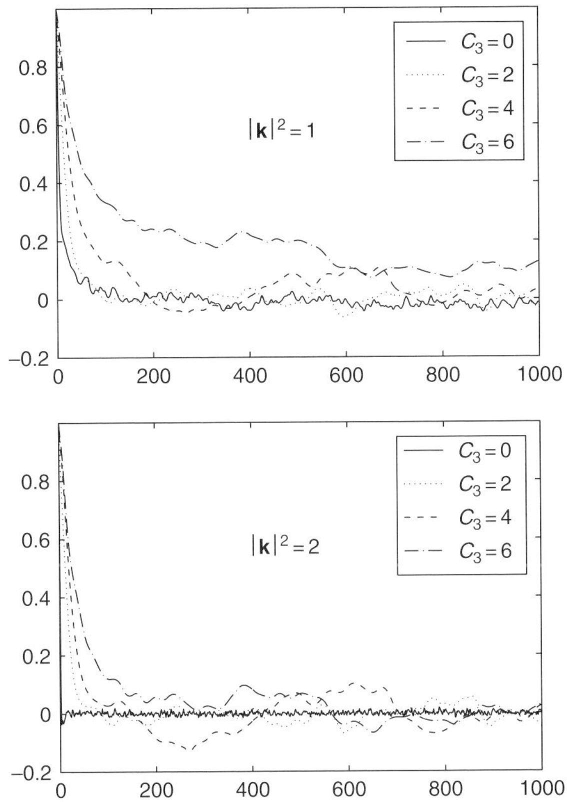

Figure 14.1 Averaged temporal correlations for $|\mathbf{k}|^2 = 1$ and $|\mathbf{k}|^2 = 2$ for varying values of skewness, $\hat{C}_3 = 0, 2, 4, 6$.

individual spatial points and comparing the results. If the energy and enstrophy are the only important conserved quantities, then the energy–enstrophy statistical theory in (14.7) predicts that $\langle\psi\rangle_\tau$ and $\langle q\rangle_\tau$ should become increasingly collinear as τ increases. This is quantified by measuring the graphical correlation between $\langle\psi\rangle_\tau$ and $\langle q\rangle_\tau$, defined by

$$C(\langle\psi\rangle_\tau, \langle q\rangle_\tau) = \frac{(\langle\psi\rangle_\tau, \langle q\rangle_\tau)_0}{\|\langle\psi\rangle_\tau\| \|\langle q\rangle_\tau\|} \tag{14.12}$$

with $(f, g)_0 = \int_{T^2} fg$, $\|f\| = (f, f)_0^{1/2}$. Note that solutions of the energy–enstrophy statistical theory in (14.7) with $\mu < 0$ are collinear and satisfy $C(\bar{\psi}_\mu, \bar{q}_\mu) = -1$. In Figure 14.2, the behavior of the graphical correlation $C(\langle\psi\rangle_\tau, \langle q\rangle_\tau)$ is depicted as a function of the averaging window, τ, in (14.11) for $10^3 \leq \tau \leq 10^4$ for the four cases with $\hat{C}_3 = 0, 2, 4, 6$ for the layered topography. The graphical correlation

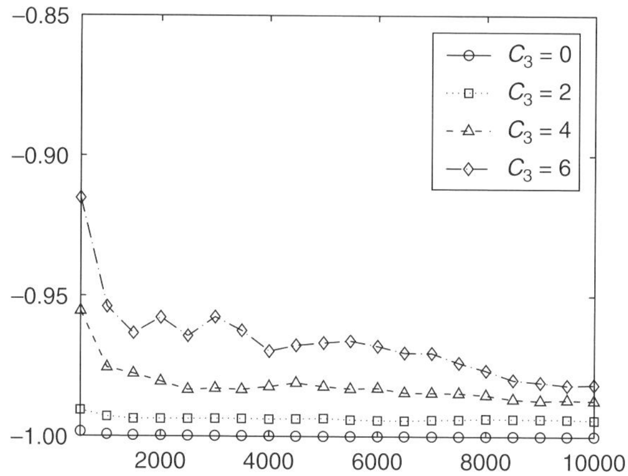

Figure 14.2 The graphical correlation, $C(\langle\psi\rangle_\tau, \langle q\rangle_\tau)$, as a function of the averaging window, τ, $\hat{C}_3 = 0, 2, 4, 6$.

for the case with $\hat{C}_3 = 0$ clearly converges to the value -1, while the graphical correlation for the other cases with non-zero third invariant clearly level off for the largest values of τ and clearly do not asymptote to -1; increasing values of $\hat{C}_3$ lead to increasing values of the graphical correlation. Figure 14.2 provides quantitative evidence that the coarse-grained large-scale mean flows do not satisfy the predictions of the energy–enstrophy statistical theory, unless the invariant $\hat{C}_3$ vanishes identically. To confirm this, in Figure 14.3 we present scatter plots of the large-scale mean stream function $\langle\psi\rangle_\tau$ and mean potential vorticity, $\langle q\rangle_\tau$, for the largest averaging window, $\tau = 10^4$ from Figure 14.2 for the four cases with varying $\hat{C}_3$. For the case with $\hat{C}_3 = 0$ in Figure 14.3, the scatter plot shows collinear behavior, while the scatter plots for the cases with increasing skewness display a coherent non-linear relationship between $\bar{\psi}$ and $\bar{q}$ with increasing curvature as $\hat{C}_3$ increases through the values 2, 4, 6. Figure 14.3 provides powerful evidence for the non-trivial effect of the third invariant, $\hat{C}_3$, on the large-scale coarse-grained statistical mean flow. This non-trivial effect of $\hat{C}_3$ is clearly evident in the contour plots for the coarse-grained stream functions depicted in Figure 14.4, plotted in each case over one period interval in x and y. For $\hat{C}_3 = 0$ the stream function is clearly layered in y as predicted by the energy–enstrophy statistical theory. This information from Figure 14.4 combined with that in Figure 14.2 and Figure 14.3 for the case with $\hat{C}_3 = 0$ indicates that as regards the coarse-grained mean flow, the energy–enstrophy statistical theory predicts the behavior; thus, with $\hat{C}_3 = 0$, the other 19 conserved quantities, $\hat{C}_N$, $4 \le N \le 22$ in (14.10) are statistically irrelevant for predicting the large-scale mean flow. On the other hand, the streamline contours in 14.4(b),(c),(d) indicate stronger more

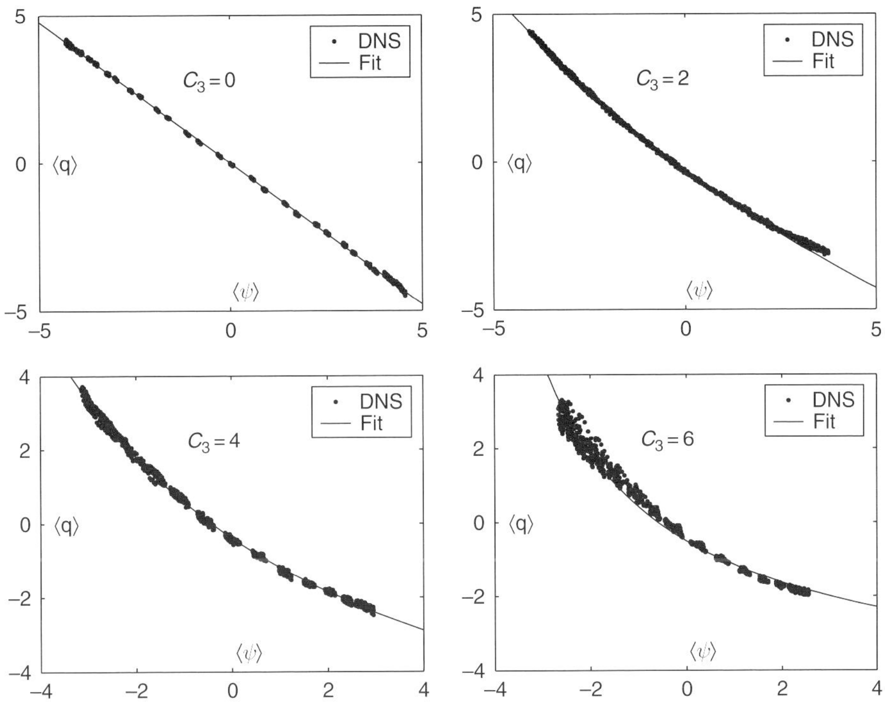

Figure 14.3 The scatterplots $\bar{q}$ vs $\bar{\psi}$ for the 23×23 sine-bracket truncation, layered topography, $\hat{C}_3 = 0, 2, 4, 6$.

localized regions of closed stream lines associated with negative values of the stream function for the cases with $\hat{C}_3 = 2, 4, 6$. Recall that closed stream lines with negative stream function (pressure) correspond to positive (cyclonic) relative vorticity. Thus, the stream line contours in Figure 14.4 indicate the presence of coherent structures through positive cyclonic vortices for positive values of the regularized third moment $\hat{C}_3$; for the case with $\hat{C}_3 = 2$, there are both cyclonic and anti-cyclonic vortices, while for the cases with $\hat{C}_3 = 4, 6$, there are only cyclonic vortices in the mean stream function with increasingly weaker anti-cyclonic flow without coherent vortices. This is further strong evidence demonstrating the statistical relevance of $\hat{C}_3$ in determining the coarse-grained large-scale mean flow.

14.2.5 The energy spectrum

Recall from Section 8.5 that the energy–enstrophy theory predicts the equipartition of the pseudo-energy variables $\hat{p}_{\mathbf{k}}$ defined in (14.8) for perturbations of the mean flow in (14.7). In Figure 14.5 the energy spectrum of the pseudo-energy

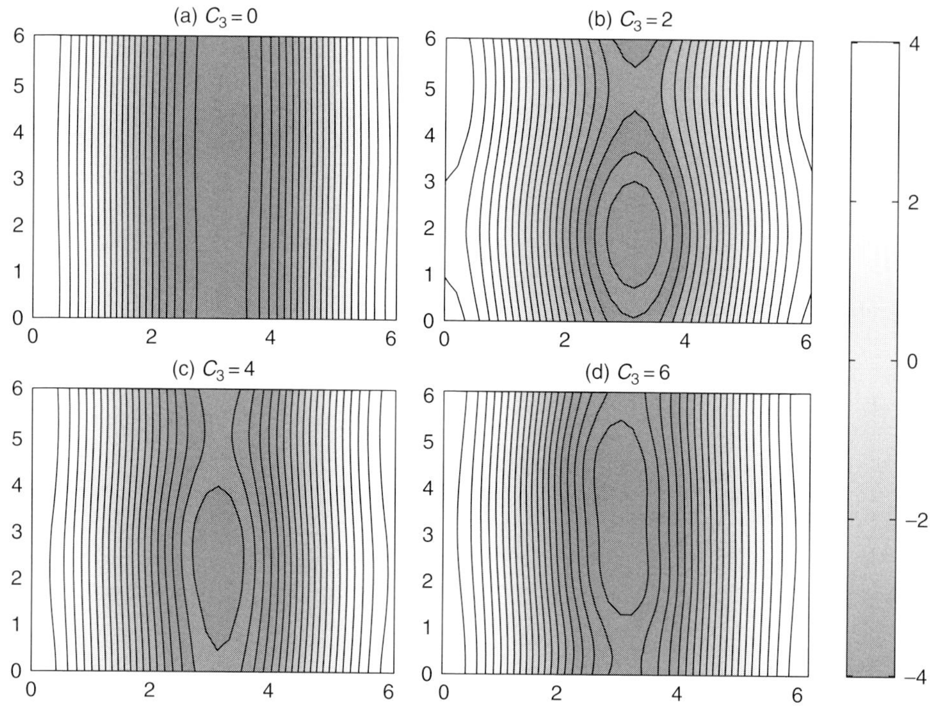

Figure 14.4 The contour plots of the mean stream function, 23×23 sine-bracket truncation, layered topography, $\hat{C}_3 = 0, 2, 4, 6$.

variables $\hat{p}_{\mathbf{k}}$ from (14.8) is plotted for perturbations of the mean flow calculated from the numerical output as in Figure 14.2, Figure 14.3, and Figure 14.4 above for the three cases with no topography, random topography, and layered topography with $\hat{C}_3$ varying for 0, 4, 6. It is remarkable that this energy spectrum is virtually identical for the three different large-scale topographies and fixed values of $\hat{C}_3$ with $E = 7$ and $\mathcal{E} = 20$; this indicates a potentially universal feature of the spectrum for fluctuations, which depends on $\hat{C}_3$ but is independent of the large-scale mean flow. In particular, for the case in Figure 14.5 with $\hat{C}_3 = 0$, equipartition of pseudo-energy is confirmed as predicted by the energy–enstrophy statistical theory; this is a further numerical confirmation of the statistical irrelevance of the 19 higher-order invariants $\hat{C}_N$, $4 \le N \le 22$, for the energy spectrum. The cases with $\hat{C}_3 = 4, 6$ in Figure 14.5 clearly demonstrate the statistical relevance of this invariant in determining the energy spectrum with more pseudo-energy concentrated at small wave numbers and lower pseudo-energy at large wave numbers in a statistically identical fashion for the three cases with varying topography.

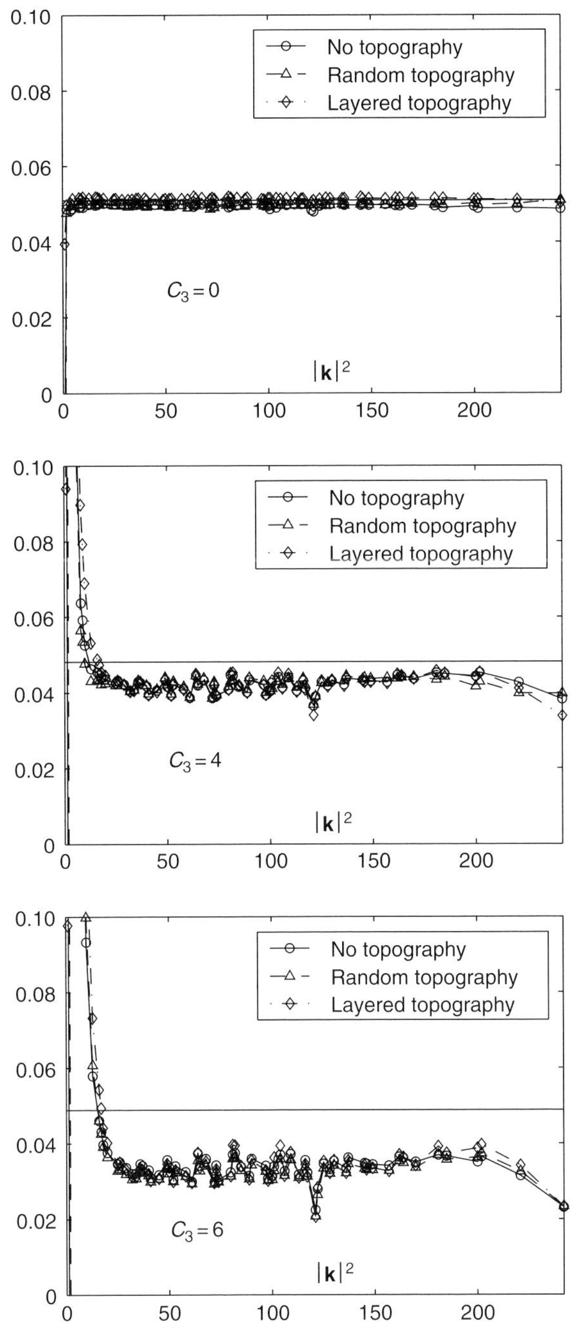

Figure 14.5 The pseudo-energy spectrum, 23×23 sine-bracket truncation, different topographies, $\hat{C}_3 = 0, 4, 6$. The pseudo-energy spectrum does not depend on topography.

14.3 Comparison with equilibrium statistical predictions with a judicious prior

The results in Figures 14.3 and 14.4 show a definite non-linear relationship between coarse-grained stream functions and potential vorticity at large scales for $\hat{C}_3 > 0$, as well as the appearance of intense large-scale cyclones in the stream function. With these results, it is natural to inquire whether the small-scale potential vorticity fluctuations have a cyclonic skewness for $\hat{C}_3 > 0$ and whether the non-linear features in Figures 14.3 are consistent with the ESTP with such a prior. Below, we show that such results are true for the numerical simulations. These results provide powerful evidence for the ESTP theories of Section 9.2 as well as for similar models used in Chapter 13 for studying Jupiter.

14.3.1 The probability distribution function of potential vorticity

While the mean stream function emphasizes the role of $\hat{C}_3$ in determining the coarse-grained large-scale flow, the probability distribution function (PDF) for potential vorticity highlights the effect of $\hat{C}_3$ on the small-scale fluctuations. Here this PDF is determined by a standard bin counting algorithm as time evolves with evaluation at a grid of spatial points. The PDF of potential vorticity is depicted in Figure 14.6 for the case with layered topography and $\hat{C}_3 = 0, 2, 4, 6$. The variance and skewness of this PDF for varying $\hat{C}_3$ are recorded in Table 14.1. The main trends apparent from Figure 14.6 and Table 14.1 is the non-trivial increase in the positive skewness of the PDF for potential vorticity as $\hat{C}_3$ increases, coupled with the nearly Gaussian PDF for potential vorticity for $\hat{C}_3 = 0$, which is predicted

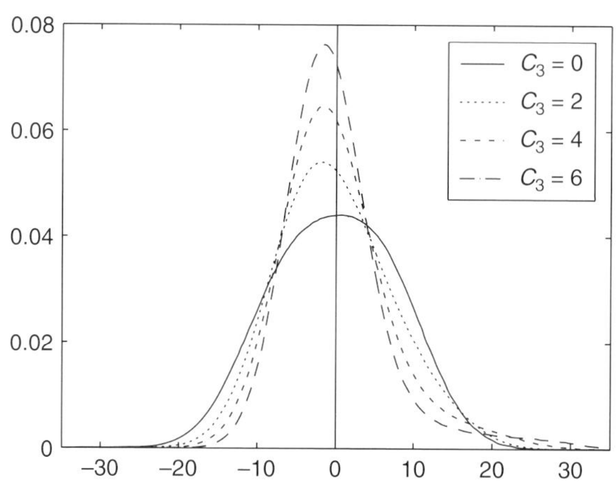

Figure 14.6 The PDFs of the potential vorticity for $\hat{C}_3 = 0, 2, 4, 6$. Note the increasing skewness in PDFs as $\hat{C}_3$ increases.

Table 14.1 *Variance and skewness of PDF of the potential vorticity, $\hat{C}_3 = 0, 2, 4, 6$*

$\hat{C}_3$	Variance	Skewness
0	65.67	$-1.37 \cdot 10^{-2}$
2	56.14	0.3114
4	50.82	0.7752
6	48.08	1.441

from the energy–enstrophy statistical theory. Once again this last result suggests the statistical irrelevance of the invariants $\hat{C}_N, 4 \leq N \leq 22$ in the regime with $\hat{C}_3 = 0$ for $E = 7, \mathcal{E} = 20$.

14.3.2 Equilibrium statistical predictions of the non-linear mean state

The graphs in Figure 13.3 and Figure 13.6 indicate that increasing the third invariant, $\hat{C}_3$, systematically results simultaneously in an increasing skewness in the PDF for potential vorticity, which represents small-scale fluctuations, and in increasing non-linearity in the scatter plot of mean stream function versus mean potential vorticity reflecting large-scale behavior. Recall from Section 9.2 that the ESTP begins with the PDF for the potential vorticity reflecting small-scale fluctuations as a prior distribution and predicts the equilibrium coarse-grained large scale mean flow in the continuum limit through a specific large-scale functional relation, $\bar{q} = G(\bar{\psi})$. Here the non-linear function G depends on the large-scale energy and circulation constraints in a specific fashion. In particular, in Section 13.3 it is shown that a skewed prior distribution for small-scale potential vorticity fluctuations given by the centered Gamma distribution with mean zero, variance, σ, and skewness, ε, yields the equilibrium statistical prediction for the large-scale mean flow,

$$\bar{q} = \frac{\sigma(\theta\bar{\psi} - \gamma)}{4\pi^2 - \varepsilon\sigma^{1/2}(\theta\bar{\psi} - \gamma)}. \tag{14.13}$$

In (14.13), θ and γ are Lagrange multipliers to satisfy the energy and zero circulation constraints. As shown in Chapter 13, this modeling strategy successfully predicts the large-scale coherent spots on Jupiter from the small-scale Gamma prior distribution motivated by recent observations from the Galileo mission. We note the strong visual evidence, comparing positively skewed PDFs for potential vorticity fluctuations in Figure 13.6 with the (negatively) skewed PDFs for the centered Gamma distribution in Figure 13.3. Here we address the much more limited issue: if we approximate the skewed PDFs for potential vorticity in Figure 13.6 by the centered Gamma distributions with the same skewness and flatness in Table 14.1 as $\hat{C}_3$ varies, are there values of the Lagrange multipliers, θ, γ, so that the non-linear large-scale $\bar{q} \leftrightarrow \bar{\psi}$ relation in (14.13) fits the data in Figure 13.3? The answer is yes, and the results are the curved lines overlayed on Figure 13.3. While these results are encouraging, the reader is warned that they are not a confirmation of the equilibrium statistical theory directly: if θ and γ are utilized as Lagrange multipliers to match the energy, $E = 7$, and zero circulation, the vorticity-stream relations show significantly less curvature and the stream functions are significantly non-linear but without closed stream lines as depicted in Figure 13.4. This is not surprising, since simulations with 23×23 Fourier modes are extremely far from the continuum limit, which requires strict spatial scale separation between the large-scale flow and the small-scale potential vorticity fluctuations. Nevertheless, such trends are already strongly evident in Figures 13.3 and 13.6, Table 14.1, and equation (14.13), and, with the above caveats of limited resolution, confirm the approach in the ESTP statistical theories discussed above in Chapters 9, 10, 11, 12, and 13.

14.4 Statistically relevant conserved quantities for the truncated Burgers–Hopf equation

In Section 7.3 we studied equilibrium statistical predictions for the TBH equation

$$(u_\Lambda)_t + \frac{1}{2} P_\Lambda (u_\Lambda^2)_x = 0, \tag{14.14}$$

where P_Λ is the projection on the Fourier modes, $|k| \le \Lambda$, and u_Λ is represented by such a finite Fourier expansion. There we demonstrated that the equations in (14.14) have three conserved quantities, the momentum, M, energy, E, and Hamiltonian, H, given by

$$M = \int_\Omega u_\Lambda, \quad E = \frac{1}{2} \int_\Omega u_\Lambda^2, \quad H = \int_\Omega u_\Lambda^3. \tag{14.15}$$

The verification that H is the Hamiltonian for (14.14) is found in Abramov *et al.* (2003). The momentum, M, is a trivial conserved quantity from the statistical viewpoint and in Section 7.3 we always set $M = 0$. On the other hand, we showed in Section 7.3 that (14.14) has the Liouville property with a non-trivial Gibbs measure associated with the energy, E, alone; in particular, the integral of the third power of u_Λ defining H provided an example of a conserved quantity which is not normalizable in the Gibbs measure. On the other hand, as shown in Section 7.3, numerical simulations of (14.14) with suitable random initial data as defined earlier (7.46) strongly confirm the equilibrium statistical predictions for

(14.14), even though the effect of the Hamiltonian, H, is not taken into account in the Gibbs measure. Thus, the question arises in this setting regarding the statistical significance of the conserved quantity, H, beyond the statistical significance of E. Clearly, this is a simplified context for some issues which we discussed in detail earlier in this chapter for the more complex truncated quasi-geostrophic equations. The advantage of such a simpler model is that detailed equilibrium statistical predictions beyond those utilized in Section 7.3 and involving both E and H can be made and compared directly with numerical simulations. This is done in Abramov *et al.* (2003). Below is a brief summary of the main results and the reader is referred to that paper as well as Section 5 of Abramov and Majda (2003b) for further details.

First, an appropriate equilibrium statistical theory is developed which includes both energy and Hamiltonian by restricting to fixed energy surfaces in a microcanonical ensemble. Then a convergent Monte-Carlo algorithm is developed for computing equilibrium statistical distributions. The probability distribution of the Hamiltonian on a microcanonical energy surface is studied through the Monte-Carlo algorithm and leads to the concept of statistically relevant and irrelevant values for the Hamiltonian. Empirical numerical estimates and simple analysis are combined to demonstrate that the statistically relevant values of the Hamiltonian have vanishingly small measures as the number of degrees of freedom increases with fixed mean energy. The predictions of the theory for relevant and irrelevant values of the Hamiltonian are confirmed through systematic numerical simulations. In particular, for the statistical initial data (7.46), the Hamiltonian, H, is statistically irrelevant, for $\Lambda = 100$, as expected from the results in Section 7.3. For statistically relevant values of the Hamiltonian, these simulations show a surprising spectral tilt rather than equipartition of energy. This spectral title is predicted and confirmed independently by Monte-Carlo simulations based on equilibrium statistical mechanics together with a heuristic formula for the tilt. On the other hand, the theoretically predicted correlation scaling law is satisfied both for statistically relevant and irrelevant values of the Hamiltonian with excellent accuracy. In particular, Monte-Carlo simulations of the equilibrium statistical distribution and simple theory (Abramov *et al.*, 2003) predict a formula for the tilt in the spectrum for the statistical energy distribution for TBH, which reflects the additional conserved quantity, H. This formula is

$$\frac{1}{2}\langle|\hat{u}_k|^2\rangle = \frac{E}{\Lambda}\left(1 + \frac{8}{5}\frac{H^2}{E^3}\frac{(\Lambda+1)/2 - k}{\Lambda}\right). \tag{14.16}$$

Notice that $\frac{H^2}{E^3}$ is a non-dimensional number reflecting the importance of the conserved quantity, H, relative to E and gives the slope of the spectral tilt beyond equipartition of energy; when this number is very small, essentially equipartition of energy is predicted. Many numerical experiments confirming the striking prediction in (14.16) are developed in Abramov *et al.* (2003) and Abramov and Majda (2003b). In particular, Abramov and Majda (2003b) show that the breakdown of (14.16) roughly coincides with the failure of mixing, i.e. non-decay of correlations at large scales, for very large values of H. A list of accessible open mathematical problems regarding statistically relevant conserved quantities for TBH are presented at the end of the paper by Abramov *et al.* (2003).

References

Abramov, R. (2002), Ph.D., thesis, Rensselaer Polytechnic Institute, Supervised by A. Majda and G. Kovacic.

Abramov, R. and Majda, A. (2003a), *Statistically Relevant Conserved Quantities for Truncated Quasi-geostrophic Flow*. Proc. Natl. Acad. Sci., 100(7), 3841–3846, April.

Abramov, R. and Majda, A. (2003b), Discrete approximations with additional conserved quantities: Deterministic and statistical behavior. *Methods and Applications of Analysis* **10**, 151–190.

Abramov, R., Kovacic, G., and Majda, A. (2003), Hamiltonian structure and statistically relevant conserved quantities for the truncated Burgers–Hopf equation. *Comm. Pure Appl. Math.* **56**(1), 1–46.

Ellis, E., Haven, K., and Turkington, B. (2003), Analysis of statistical equilibrium models of geostrophic turbulence. *J. Appl. Math. Stoch. Anal.*

Majda, A. and Bertozzi, A. (2001), Vorticity and Incompressible Flow. Cambridge: Cambridge University Press.

McLachlan, R. (1993), Explicit Lie–Poisson integration and the Euler equations. *Phys. Rev. Lett.* **71**, 3043–3046.

Morrison, P. J. (1998), *Hamiltonian description of the ideal fluid. Reviews of Modern Physics* **70**, 467–521, American Physical Society, College Park.

Salmon, R. (1998), *Lectures on Geophysical Fluid Dynamics*. New York: Oxford Press.

Turkington, B. (1999), Statistical equilibrium measures and coherent states in two-dimensional turbulence. *Comm. Pure Appl. Math.* **52**, 781–809.

Zeitlin, V. (1991), Finite-mode analogues of 2D ideal hydrodynamics: Coadjoint orbits and local canonical structure. *Physica D* **49**, 353–362.

Appendix 1 Spectral truncations of quasi-geostrophic flow with additional conserved quantities

A.1.1 Some basic facts about Hamiltonian systems

We consider the equations for a finite-dimensional Hamiltonian system

$$\frac{\mathrm{d}\mathbf{y}}{\mathrm{d}t} = \mathbf{J}(\mathbf{y})\nabla H(\mathbf{y}), \tag{14.17}$$

where $\mathbf{y}$ is the column vector of dynamical variables, H is a *Hamiltonian* function, and $\mathbf{J}$ is a *symplectic matrix* which is skew-symmetric and satisfies the Jacobi identity

$$\sum_l (J_{pl}\partial_l J_{qr} + J_{rl}\partial_l J_{pq} + J_{ql}\partial_l J_{rp}) = 0 \quad \forall p, q, r \leq N, \tag{14.18}$$

where N is the dimension of the system. The Jacobi identity essentially tells us that the system can be represented in its canonical action-angle variables (Morrison, 1998, and references therein). We will verify the Jacobi identity here for a few cases below; however, as the reader will see, in this work we never really take any advantage of the Jacobi identity in the numerical methods.

We denote the *Poisson bracket* for $F(\mathbf{y})$ and $G(\mathbf{y})$ by

$$\{F, G\} = (\nabla F, \mathbf{J}\nabla G), \tag{14.19}$$

where $(\cdot, \cdot)$ denotes the scalar product of two vectors. Due to the skew-symmetry of $\mathbf{J}$ we have

$$\{F, G\} = -\{G, F\}. \tag{14.20}$$

In this notation the time evolution for any $F(\mathbf{y})$ can be written in the form

$$\frac{\mathrm{d}F}{\mathrm{d}t} = (\nabla F, \dot{\mathbf{y}}) = (\nabla F, \mathbf{J}\nabla H) = \{F, H\}, \tag{14.21}$$

thus from (14.20) and (14.21) we have

$$\frac{\mathrm{d}H}{\mathrm{d}t} = \{H, H\} = -\{H, H\} = 0, \tag{14.22}$$

i.e. H is a *conserved quantity* of the system.

Generally, the conserved quantities of the system (14.17) depend on both $\mathbf{J}$ and H. Let us take the basis of the kernel $\ker(\mathbf{J})$ to be spanned by $\mathbf{K}_j(\mathbf{y}), 0 < j \leq \dim(\ker(\mathbf{J}))$. If for some $\mathbf{K}_j(\mathbf{y})$ (let us denote it as $\mathbf{K}_{j_k}(\mathbf{y})$) we can find $C_k(\mathbf{y})$ such that

$$\mathbf{K}_{jk}(\mathbf{y}) = \nabla C_k(\mathbf{y}), \tag{14.23}$$

then

$$\mathbf{J}\nabla C_k(\mathbf{y}) \equiv 0.$$

This condition automatically guarantees

$$\frac{\mathrm{d}C_k}{\mathrm{d}t} = \{C_k, H\} = -\{H, C_k\} = (\nabla H, \mathbf{J}\nabla C_k) = 0,$$

i.e. C_k is conserved in time regardless of the Hamiltonian and, therefore, is a property of the symplectic structure $\mathbf{J}$ alone. These conserved quantities are called the *Casimir invariants*. For simple Poisson brackets the Casimir invariants are easy to find, and, if we are looking for conserved quantities of a system, then the Casimir invariants are the best candidates to be found quickly.

A.1.2 The equations for Barotropic flow in Fourier space

The two-dimensional vorticity equation in (14.1) on a double-periodic plane transformed through two-dimensional Fourier series takes the form of an infinite system of ODEs

$$\frac{\mathrm{d}\hat{q}_{\mathbf{k}}}{\mathrm{d}t} = \sum_{k'_1,k'_2=-\infty}^{\infty} \frac{\mathbf{k}\times\mathbf{k}'}{|\mathbf{k}'|^2}\hat{q}_{\mathbf{k}+\mathbf{k}'}(\hat{q}_{-\mathbf{k}'}-\hat{h}_{-\mathbf{k}'}). \tag{14.24}$$

Here $\hat{q}_{\mathbf{k}}$ denotes the Fourier coefficient associated with the two-dimensional wave vector $\mathbf{k}$, whose components are integers. The skew-symmetric scalar product $\mathbf{k}\times\mathbf{k}'$ is $k_1k'_2 - k_2k'_1$, and the norm $|\mathbf{k}|$ is $\sqrt{k_1^2+k_2^2}$. Since q is real, then $\hat{q}^*_{\mathbf{k}} = \hat{q}_{-\mathbf{k}}$. In order to derive (14.24), first we look at the equation in (14.1) and conclude that in Fourier space it becomes

$$\hat{q}_{\mathbf{k}} = -|\mathbf{k}|^2\hat{\psi}_{\mathbf{k}} + \hat{h}_{\mathbf{k}}, \tag{14.25}$$

so that

$$\hat{\psi}_{\mathbf{k}} = -\frac{1}{|\mathbf{k}|^2}(\hat{q}_{\mathbf{k}} - \hat{h}_{\mathbf{k}}). \tag{14.26}$$

Now, let us write ψ_x, ψ_y, q_x, q_y in terms of Fourier coefficients:

$$\psi_x = \sum_{\mathbf{k}}(ik_1)\hat{\psi}_{\mathbf{k}}e^{i(k_1x+k_2y)}, \qquad \psi_y = \sum_{\mathbf{k}}(ik_2)\hat{\psi}_{\mathbf{k}}e^{i(k_1x+k_2y)},$$

$$q_x = \sum_{\mathbf{k}}(ik_1)\hat{q}_{\mathbf{k}}e^{i(k_1x+k_2y)}, \qquad q_y = \sum_{\mathbf{k}}(ik_2)\hat{q}_{\mathbf{k}}e^{i(k_1x+k_2y)}.$$

Then we write the dynamic equation (14.1) in terms of Fourier coefficients,

$$\frac{\mathrm{d}}{\mathrm{d}t}\sum_{\mathbf{k}}\hat{q}_{\mathbf{k}}e^{i\mathbf{k}\mathbf{x}} = \sum_{\mathbf{k}'}\sum_{\mathbf{k}''}(k'_1k''_2 - k'_2k''_1)\hat{\psi}_{\mathbf{k}'}\hat{q}_{\mathbf{k}''}e^{i(\mathbf{k}'+\mathbf{k}'')\mathbf{x}}.$$

Now we substitute $\mathbf{k} = \mathbf{k}' + \mathbf{k}''$ and collect terms with equal exponents:

$$\frac{\mathrm{d}}{\mathrm{d}t}\hat{q}_{\mathbf{k}} = -\sum_{\mathbf{k}'}(k'_2k_1 - k'_1k_2)\hat{\psi}_{\mathbf{k}'}\hat{q}_{\mathbf{k}-\mathbf{k}'}.$$

Changing $\mathbf{k}' \to -\mathbf{k}'$ yields

$$\frac{\mathrm{d}}{\mathrm{d}t}\hat{q}_{\mathbf{k}} = -\sum_{\mathbf{k}'}(\mathbf{k}\times\mathbf{k}')\hat{q}_{\mathbf{k}+\mathbf{k}'}\hat{\psi}_{-\mathbf{k}'},$$

and, substituting the expression in (14.26) for $\hat{\psi}_{\mathbf{k}}$, we obtain (14.24).

It is well-known that the system governed by the quasi-geostrophic equations in (14.1) is a Hamiltonian system in physical space with the Hamiltonian given by the energy in (14.2), while the higher-order enstrophies in (14.3) are Casimir invariants (Morrison, 1998; Salmon, 1998). Here we record the corresponding formal Hamiltonian structure in Fourier space in order to motivate the finite-dimensional spectral approximations developed below. The infinite-dimensional equation (14.24) has the Hamiltonian structure

$$\frac{\mathrm{d}}{\mathrm{d}t}\hat{q}_{\mathbf{k}} = \mathbf{J}_{\mathbf{k}\mathbf{k}'}\frac{\partial H}{\partial \hat{q}_{\mathbf{k}}} \tag{14.27}$$

with

$$\mathbf{J}_{\mathbf{k}\mathbf{k}'} = -(\mathbf{k}\times\mathbf{k}')\hat{q}_{\mathbf{k}+\mathbf{k}'}. \tag{14.28}$$

The Hamiltonian H is given by the energy since

$$H = E = \frac{1}{2}\sum_{\mathbf{k}}|\mathbf{k}|^2|\hat{\psi}_{\mathbf{k}}|^2 = \frac{1}{2}\sum_{\mathbf{k}}\frac{|\hat{q}_{\mathbf{k}}-\hat{h}_{\mathbf{k}}|^2}{|\mathbf{k}|^2}, \tag{14.29}$$

$$\frac{\partial H}{\partial \hat{q}_{\mathbf{k}}} = \frac{1}{|\mathbf{k}|^2}(\hat{q}_{-\mathbf{k}}-\hat{h}_{-\mathbf{k}}), \tag{14.30}$$

The fact that $\mathbf{J}_{\mathbf{k}\mathbf{k}'}$ is skew-symmetric follows from $\mathbf{J}_{\mathbf{k}\mathbf{k}'} = -\mathbf{J}_{\mathbf{k}'\mathbf{k}}$. The formal proof of the Jacobi identity follows directly from the definition in (14.18).

$$\begin{aligned}
&\sum_{\mathbf{l}}(\mathbf{J}_{\mathbf{il}}\partial_{\mathbf{l}}\mathbf{J}_{\mathbf{jk}} + \mathbf{J}_{\mathbf{kl}}\partial_{\mathbf{l}}\mathbf{J}_{\mathbf{ij}} + \mathbf{J}_{\mathbf{jl}}\partial_{\mathbf{l}}\mathbf{J}_{\mathbf{ki}})\\
&\quad= \sum_{\mathbf{l}=\mathbf{j}+\mathbf{k}}(\mathbf{i}\times\mathbf{l})(\mathbf{j}\times\mathbf{k})\hat{q}_{\mathbf{i}+\mathbf{j}+\mathbf{k}} + \sum_{\mathbf{l}=\mathbf{i}+\mathbf{j}}(\mathbf{k}\times\mathbf{l})(\mathbf{i}\times\mathbf{j})\hat{q}_{\mathbf{i}+\mathbf{j}+\mathbf{k}}\\
&\qquad+ \sum_{\mathbf{l}=\mathbf{k}+\mathbf{i}}(\mathbf{j}\times\mathbf{l})(\mathbf{k}\times\mathbf{i})\hat{q}_{\mathbf{i}+\mathbf{j}+\mathbf{k}}\\
&\quad= \sum_{\mathbf{l}=\mathbf{j}+\mathbf{k}}(\mathbf{i}\times(\mathbf{j}+\mathbf{k}))(\mathbf{j}\times\mathbf{k})\hat{q}_{\mathbf{i}+\mathbf{j}+\mathbf{k}} + \sum_{\mathbf{l}=\mathbf{i}+\mathbf{j}}(\mathbf{k}\times(\mathbf{i}+\mathbf{j}))(\mathbf{i}\times\mathbf{j})\hat{q}_{\mathbf{i}+\mathbf{j}+\mathbf{k}}\\
&\qquad+ \sum_{\mathbf{l}=\mathbf{k}+\mathbf{i}}(\mathbf{j}\times(\mathbf{k}+\mathbf{i}))(\mathbf{k}\times\mathbf{i})\hat{q}_{\mathbf{i}+\mathbf{j}+\mathbf{k}}\\
&\quad= \sum_{\mathbf{l}=\mathbf{j}+\mathbf{k}}(\mathbf{i}\times\mathbf{j})(\mathbf{j}\times\mathbf{k}) - (\mathbf{k}\times\mathbf{i})(\mathbf{j}\times\mathbf{k})\hat{q}_{\mathbf{i}+\mathbf{j}+\mathbf{k}}\\
&\qquad+ \sum_{\mathbf{l}=\mathbf{i}+\mathbf{j}}(\mathbf{k}\times\mathbf{i})(\mathbf{i}\times\mathbf{j}) - (\mathbf{j}\times\mathbf{k})(\mathbf{i}\times\mathbf{j})\hat{q}_{\mathbf{i}+\mathbf{j}+\mathbf{k}}\\
&\qquad+ \sum_{\mathbf{l}=\mathbf{k}+\mathbf{i}}(\mathbf{j}\times\mathbf{k})(\mathbf{k}\times\mathbf{i}) - (\mathbf{i}\times\mathbf{j})(\mathbf{k}\times\mathbf{i})\hat{q}_{\mathbf{i}+\mathbf{j}+\mathbf{k}}\\
&\quad= 0.
\end{aligned} \tag{14.31}$$

The system (14.24) also has Casimir invariants of the form

$$\hat{C}_N = \sum_{\sum_{j=1}^{N} \mathbf{k}_j = 0} \left(\prod_{j=1}^{N} \hat{q}_{\mathbf{k}_j} \right), \quad \text{for arbitrary } N, \tag{14.32}$$

which are the Fourier space analogues of (14.3). The formal proof of conservation is also a direct calculation

$$\begin{aligned}
\sum_{\mathbf{j}} \mathbf{J}_{\mathbf{jk}} \frac{\partial C_N}{\partial \hat{q}_{\mathbf{j}}} &= \sum_{\mathbf{j}} \mathbf{J}_{\mathbf{jk}} \left(N \sum_{\sum_l^{N-1} \mathbf{m}_l = -\mathbf{j}} \prod_l^{N-1} \hat{q}_{\mathbf{m}_l} \right) \\
&= N \sum_{\mathbf{j}} \sum_{\sum_l^{N-1} \mathbf{m}_l = -\mathbf{j}} (\mathbf{j} \times \mathbf{k}) \hat{q}_{\mathbf{j}+\mathbf{k}} \prod_l^{N-1} \hat{q}_{\mathbf{m}_l} \\
&= N \sum_{\mathbf{j}} \sum_{\sum_l^{N-1} \mathbf{m}_l = -\mathbf{j}} ((\mathbf{j}+\mathbf{k}) \times \mathbf{k}) \hat{q}_{\mathbf{j}+\mathbf{k}} \prod_l^{N-1} \hat{q}_{\mathbf{m}_l} \\
&= N \sum_{\mathbf{i}} \sum_{\sum_l^{N-1} \mathbf{m}_l = \mathbf{k}-\mathbf{i}} (\mathbf{i} \times \mathbf{k}) \hat{q}_{\mathbf{i}} \prod_l^{N-1} \hat{q}_{\mathbf{m}_l} \\
&= N! \sum_{\sum_l^{N} \mathbf{m}_l = \mathbf{0}} (\mathbf{k} \times \mathbf{k}) \prod_l^{N} \hat{q}_{\mathbf{m}_l} = \mathbf{0}
\end{aligned}$$

since $\mathbf{k} \times \mathbf{k} \equiv 0$. The conservation of potential vorticity (which is the first Casimir invariant) now becomes trivial since

$$\frac{\mathrm{d}Q_1(q)}{\mathrm{d}t} = \frac{\mathrm{d}\hat{q}_{\mathbf{0}}}{\mathrm{d}t} = 0.$$

The enstrophy is essentially the second Casimir invariant

$$\mathcal{E} = \frac{1}{2} \sum_{\mathbf{k}} |\hat{q}_{\mathbf{k}}|^2.$$

A.1.3 The sine-bracket truncation with many additional conserved quantities

Here we consider the sine-bracket truncation as an approximation to the quasi-geostrophic dynamics in (14.1) through the spectral representation in Section A.1.2. This finite-dimensional set of equations for the Fourier coefficients is given by

$$\frac{\mathrm{d}}{\mathrm{d}t} \hat{q}_{\mathbf{k}} = \sum_{k_1', k_2' = -M}^{M} \frac{1}{\varepsilon} \frac{\sin(\varepsilon \mathbf{k} \times \mathbf{k}')}{|\mathbf{k}'|^2} \hat{q}_{\mathbf{k}+\mathbf{k}'} (\hat{q}_{-\mathbf{k}'} - \hat{h}_{-\mathbf{k}'}), \quad \varepsilon = \frac{2\pi}{2M+1}, \tag{14.33}$$

so that the summation occurs on the $(2M+1)\times(2M+1)$ domain of the Fourier coefficients, and, as in Section 14.1.1 above, the coefficients, $\hat{q}_{\mathbf{k}}$, have period $2M+1$ in $\mathbf{k}$ in each direction. The name of the truncation follows, first, from the presence of a sine function, and, second, due to the fact that the truncation can be written in its "bracket" form as a commutator defined on the group of unitary $(2M+1)\times(2M+1)$ matrices, which we will show later. Note that for $M\to\infty$ and given $\mathbf{k}$ and $\mathbf{k}'$, $\varepsilon^{-1}\sin(\varepsilon\mathbf{k}\times\mathbf{k}')=\mathbf{k}\times\mathbf{k}'+O(\varepsilon^2)$, so that the equations in (14.33) are formally consistent with the spectral form of the quasi-geostrophic equations discussed in A.1.2. This truncation possesses a Hamiltonian structure with the symplectic operator

$$\mathbf{J}_{\mathbf{k}\mathbf{k}'} = -\frac{1}{\varepsilon}\sin(\varepsilon\mathbf{k}\times\mathbf{k}')\hat{q}_{\mathbf{k}+\mathbf{k}'} \tag{14.34}$$

and Hamiltonian

$$H = E = \frac{1}{2}\sum_{\mathbf{k}}|\mathbf{k}|^2|\hat{\psi}_{\mathbf{k}}|^2 = \frac{1}{2}\sum_{\mathbf{k}}\frac{|\hat{q}_{\mathbf{k}}-\hat{h}_{\mathbf{k}}|^2}{|\mathbf{k}|^2}, \tag{14.35}$$

$$\frac{\partial H}{\partial\hat{q}_{\mathbf{k}}} = \frac{1}{|\mathbf{k}|^2}(\hat{q}_{-\mathbf{k}}-\hat{h}_{-\mathbf{k}}). \tag{14.36}$$

There is no need to demonstrate the direct proof of the energy conservation in the sine-bracket truncation (14.33), since the energy *is* the Hamiltonian for (14.33), and the conservation of the Hamiltonian is the general property of a Hamiltonian system, which is shown in (14.22).

Due to the sine function in (14.34) the symplectic matrix is skew-symmetric and the Jacobi identity also holds: $\forall\mathbf{i},\mathbf{j},\mathbf{k}$ we have

$$\begin{aligned}
&\sum(\mathbf{J}_{\mathbf{il}}\partial_{\mathbf{l}}\mathbf{J}_{\mathbf{jk}}+\mathbf{J}_{\mathbf{kl}}\partial_{\mathbf{l}}\mathbf{J}_{\mathbf{ij}}+\mathbf{J}_{\mathbf{jl}}\partial_{\mathbf{l}}\mathbf{J}_{\mathbf{ki}})\\
&\quad=\sum[\sin(\varepsilon\mathbf{i}\times(\mathbf{j}+\mathbf{k}))\sin(\varepsilon\mathbf{j}\times\mathbf{k})+\sin(\varepsilon\mathbf{k}\times(\mathbf{i}+\mathbf{j}))\sin(\varepsilon\mathbf{i}\times\mathbf{j})\\
&\qquad+\sin(\varepsilon\mathbf{j}\times(\mathbf{k}+\mathbf{i}))\sin(\varepsilon\mathbf{k}\times\mathbf{i})]\hat{q}_{\mathbf{i}+\mathbf{j}+\mathbf{k}}\\
&\quad=[\sin(\varepsilon\mathbf{i}\times\mathbf{j})\cos(\varepsilon\mathbf{i}\times\mathbf{k})\sin(\varepsilon\mathbf{j}\times\mathbf{k})-\cos(\varepsilon\mathbf{i}\times\mathbf{j})\sin(\varepsilon\mathbf{k}\times\mathbf{i})\sin(\varepsilon\mathbf{j}\times\mathbf{k})\\
&\qquad+\sin(\varepsilon\mathbf{k}\times\mathbf{i})\cos(\varepsilon\mathbf{k}\times\mathbf{j})\sin(\varepsilon\mathbf{i}\times\mathbf{j})-\cos(\varepsilon\mathbf{k}\times\mathbf{i})\sin(\varepsilon\mathbf{j}\times\mathbf{k})\sin(\varepsilon\mathbf{i}\times\mathbf{j})\\
&\qquad+\sin(\varepsilon\mathbf{j}\times\mathbf{k})\cos(\varepsilon\mathbf{j}\times\mathbf{i})\sin(\varepsilon\mathbf{k}\times\mathbf{i})\\
&\qquad-\cos(\varepsilon\mathbf{j}\times\mathbf{k})\sin(\varepsilon\mathbf{i}\times\mathbf{j})\sin(\varepsilon\mathbf{k}\times\mathbf{i})]\hat{q}_{\mathbf{i}+\mathbf{j}+\mathbf{k}}\\
&\quad=0.
\end{aligned}$$

As shown below, the sine-bracket truncation in (14.33) conserves $2M$ independent Casimir invariants of the form

$$C_N = \sum_{Z^N} \hat{q}_{\mathbf{i}_1} \ldots \hat{q}_{\mathbf{i}_N} \cos[\varepsilon A(\mathbf{i}_1, \ldots, \mathbf{i}_N)],$$
$$Z^N = \left\{ (\mathbf{i}_1, \ldots, \mathbf{i}_N), \sum_{j=1}^{N} \mathbf{i}_j = \mathbf{0} \right\}, \quad 1 \le N \le 2M, \tag{14.37}$$

where

$$A(\mathbf{i}_1, \ldots, \mathbf{i}_N) = \mathbf{i}_2 \times \mathbf{i}_1 + \mathbf{i}_3 \times (\mathbf{i}_1 + \mathbf{i}_2) + \ldots + \mathbf{i}_N \times (\mathbf{i}_1 + \ldots + \mathbf{i}_{N-1}).$$

The Casimir invariant C_2 is a multiple of the enstrophy since $A(\mathbf{i}, -\mathbf{i}) = 0$. However, the higher Casimir invariants C_N for $3 \le N \le 2M$ are suitable regularizations of those in (14.24), or, equivalently, (14.3).

In order to prove the existence of the Casimir invariants in (14.37), let us consider the family of $(2M+1) \times (2M+1)$ matrices $\mathbf{D_n}$ which have the form

$$\mathbf{D_n} = \sigma^{n_1 n_2/2} g^{n_1} h^{n_2},$$

where the unitary $(2M+1) \times (2M+1)$ matrices h and g satisfy

$$hg = \sigma g h \quad \text{and} \quad g^{2M+1} = h^{2M+1} = I. \tag{14.38}$$

We choose $(2M+1)$ to be prime and

$$\sigma = \exp\left(\frac{4\pi i}{2M+1}\right).$$

For our choice of σ the matrices g and h are

$$g = \begin{pmatrix} 1 & 0 & 0 & \cdots & 0 \\ 0 & \sigma & 0 & \cdots & 0 \\ 0 & 0 & \sigma^2 & \cdots & 0 \\ \vdots & \vdots & \vdots & \ddots & \vdots \\ 0 & 0 & 0 & \cdots & \sigma^{2M} \end{pmatrix}, \quad h = \begin{pmatrix} 0 & 1 & 0 & \cdots & 0 \\ 0 & 0 & 1 & \cdots & 0 \\ \vdots & \vdots & \vdots & \ddots & \vdots \\ 0 & 0 & 0 & \cdots & 1 \\ 1 & 0 & 0 & \cdots & 0 \end{pmatrix}.$$

We can see that, if h is applied to the given matrix from the left, it shifts the given matrix upwards, and if it is applied from the right, it shifts the given matrix to the right. Taking this into account, (14.38) can be easily proven. First, $g^{2M+1} = I$ is obvious for our choice of σ. Second, $h^{2M+1} = I$ because h shifts itself $2M$ times into the identity position. Then

$$hg = \begin{pmatrix} 0 & \sigma & 0 & \cdots & 0 \\ 0 & 0 & \sigma^2 & \cdots & 0 \\ \vdots & \vdots & \vdots & \ddots & \vdots \\ 0 & 0 & 0 & \cdots & \sigma^{2M} \\ 1 & 0 & 0 & \cdots & 0 \end{pmatrix}, \sigma gh = \begin{pmatrix} 0 & \sigma & 0 & \cdots & 0 \\ 0 & 0 & \sigma^2 & \cdots & 0 \\ \vdots & \vdots & \vdots & \ddots & \vdots \\ 0 & 0 & 0 & \cdots & \sigma^{2M} \\ \sigma^{2M+1} & 0 & 0 & \cdots & 0 \end{pmatrix}$$

and since $\sigma^{2M+1} = 1$ we obtain $hg = \sigma gh$.

The set of $\mathbf{D_n}$s is the group of unitary matrices which satisfy the following relations

$$\mathbf{D}_{\mathbf{n}}^{\dagger} = \mathbf{D}_{-\mathbf{n}} = (\mathbf{D}_{\mathbf{n}})^{-1}, \tag{14.39}$$

$$\mathbf{D}_{\mathbf{n}}\mathbf{D}_{\mathbf{m}} = \exp(i\varepsilon(\mathbf{m}\times\mathbf{n}))\mathbf{D}_{\mathbf{n}+\mathbf{m}}, \tag{14.40}$$

$$\frac{1}{2M+1}\mathrm{Tr}(\mathbf{D}_{\mathbf{m}}) = \delta_{\mathbf{m}}^{\mathbf{0}}, \tag{14.41}$$

where $\varepsilon = 2\pi/(2M+1)$. We present the proof for the three statements in (14.39) below.

- Proof of (14.39):

$$\begin{aligned}
\mathbf{D}_{\mathbf{n}}^{-1} &= \exp\left(-\frac{2\pi i n_1 n_2}{2M+1}\right)(g^{n_1}h^{n_2})^{-1} = \exp\left(-\frac{2\pi i n_1 n_2}{2M+1}\right)(h^{n_2})^{-1}(g^{n_1})^{-1} \\
&= \exp\left(-\frac{2\pi i n_1 n_2}{2M+1}\right)(h^{n_2})^{\dagger}(g^{n_1})^{\dagger} = \exp\left(-\frac{2\pi i n_1 n_2}{2M+1}\right)(g^{n_1}h^{n_2})^{\dagger} = \mathbf{D}_{\mathbf{n}}^{\dagger}. \\
\mathbf{D}_{\mathbf{n}}^{-1} &= \exp\left(-\frac{2\pi i n_1 n_2}{2M+1}\right)(g^{n_1}h^{n_2})^{-1} = \exp\left(\frac{2\pi i n_1 n_2}{2M+1}\right)\sigma^{-n_1 n_2}h^{-n_2}g^{-n_1} \\
&= \exp\left(\frac{2\pi i n_1 n_2}{2M+1}\right)g^{-n_1}h^{-n_2} = \mathbf{D}_{-\mathbf{n}}.
\end{aligned}$$

- Proof of (14.40):

$$\begin{aligned}
&\mathbf{D}_{\mathbf{n}}\mathbf{D}_{\mathbf{m}} \\
&= \exp\left(\frac{2\pi i(n_1 n_2 + m_1 m_2)}{2M+1}\right) g^{n_1}h^{n_2}g^{m_1}h^{m_2} \\
&= \exp\left(\frac{2\pi i(n_1 n_2 + m_1 m_2)}{2M+1}\right) \sigma^{m_1 k}g^{n_1}h^{n_2-k}g^{m_1}h^{m_2+k} \\
&= \exp\left(\frac{2\pi i(n_1 n_2 + m_1 m_2 + 2m_1 n_2)}{2M+1}\right) g^{m_1+n_1}h^{m_2+n_2}
\end{aligned}$$

$$= \exp\left(\frac{2\pi i[(m_1+m_2)(n_1+n_2)+(m_1n_2-m_2n_1)]}{2M+1}\right) g^{m_1+n_1}h^{m_2+n_2}$$

$$= \exp(i\varepsilon(\mathbf{m}\times\mathbf{n}))\mathbf{D}_{\mathbf{n}+\mathbf{m}}.$$

- Proof of (14.41): first note that

$$\mathrm{Tr}(\mathbf{D}_{\mathbf{k}}) = \sigma^{k_1k_2/2}\mathrm{Tr}(g^{k_1}h^{k_2}).$$

Then, $\mathrm{Tr}(g^j) = 0$ for $j \neq (2M+1)s_1$, $s_1 \in \mathcal{Z}$, because

$$\sum_{p=0}^{2M} \exp\left(\frac{4\pi ijp}{2M+1}\right) = 0, \quad j \neq (2M+1)s_1.$$

Now we can see that $\mathrm{Tr}(g^jh^k) = 0$ for $k \neq (2M+1)s_2$, $s_2 \in \mathcal{Z}$, because h^k shifts non-zero entries of g^j off the diagonal position. Otherwise, we obtain $\mathrm{Tr}(g^jh^k) = \mathrm{Tr}(I) = 2M+1$ for $j = (2M+1)s_1$ and $k = (2M+1)s_2$. Taking into account that our indices are between $-M$ and M we have

$$\mathrm{Tr}(\mathbf{D}_{\mathbf{k}}) = \begin{cases} 2M+1, & \mathbf{k}=\mathbf{0}, \\ 0, & \mathbf{k}\neq\mathbf{0}, \end{cases}$$

which gives the third statement of (14.39).

From (14.39) immediately follows

$$\begin{aligned} [\mathbf{D}_{\mathbf{n}}, \mathbf{D}_{\mathbf{m}}] &= 2i\sin(\varepsilon(\mathbf{m}\times\mathbf{n}))\mathbf{D}_{\mathbf{n}+\mathbf{m}}, \\ [\mathbf{D}_{\mathbf{n}}, \mathbf{D}_{\mathbf{m}}]_+ &= 2\cos(\varepsilon(\mathbf{m}\times\mathbf{n}))\mathbf{D}_{\mathbf{n}+\mathbf{m}}, \end{aligned} \tag{14.42}$$

where $[\cdot,\cdot]$ and $[\cdot,\cdot]_+$ are the commutation and anti-commutation operators, respectively, defined as follows

$$[x, y] = xy - yx, \quad [x, y]_+ = xy + yx.$$

Now we will show how to write the sine-bracket equation in its "bracket" form via the commutation rule (14.42). We denote

$$Q = \sum_{\mathbf{n}} \hat{q}_{\mathbf{n}}\mathbf{D}_{\mathbf{n}} \quad \text{and} \quad F = \sum_{\mathbf{n}} f_{\mathbf{n}}\mathbf{D}_{\mathbf{n}},$$

where

$$f_{\mathbf{k}} = \frac{\partial H}{\partial \hat{q}_{\mathbf{k}}},$$

and H is an arbitrary Hamiltonian. We write the analogue of the sine-bracket truncation (14.33) with the arbitrary Hamiltonian in the form

$$\dot{Q} = \frac{i}{2\varepsilon}[Q, F]. \tag{14.43}$$

In order to show that (14.43) is true, we first write down the commutator $[Q, F]$

$$\begin{aligned}\frac{i}{2\varepsilon}[Q,F] &= \frac{i}{2\varepsilon}\left(\sum_{\mathbf{n}}\hat{q}_{\mathbf{n}}\mathbf{D}_{\mathbf{n}}\right)\left(\sum_{\mathbf{m}}f_{\mathbf{m}}\mathbf{D}_{\mathbf{m}}\right)-\left(\sum_{\mathbf{m}}f_{\mathbf{m}}\mathbf{D}_{\mathbf{m}}\right)\left(\sum_{\mathbf{n}}\hat{q}_{\mathbf{n}}\mathbf{D}_{\mathbf{n}}\right)\\
&= \frac{i}{2\varepsilon}\sum_{\mathbf{n},\mathbf{m}}\hat{q}_{\mathbf{n}}f_{\mathbf{m}}(\mathbf{D}_{\mathbf{n}}\mathbf{D}_{\mathbf{m}}-\mathbf{D}_{\mathbf{m}}\mathbf{D}_{\mathbf{n}})\\
&= \frac{1}{\varepsilon}\sum_{\mathbf{n},\mathbf{m}}\sin(\varepsilon(\mathbf{m}\times\mathbf{n}))\hat{q}_{\mathbf{n}}f_{\mathbf{m}}\mathbf{D}_{\mathbf{m}+\mathbf{n}}\\
&= -\frac{1}{\varepsilon}\sum_{\mathbf{k},\mathbf{m}}\sin(\varepsilon(\mathbf{k}\times\mathbf{m}))\hat{q}_{\mathbf{k}+\mathbf{m}}f_{-\mathbf{m}}\mathbf{D}_{\mathbf{k}}.\end{aligned}$$

Here we denote $\mathbf{k}=\mathbf{m}+\mathbf{n}$ and then change $\mathbf{m}$ to $-\mathbf{m}$. Now we write $\dot{Q}$ as

$$\dot{Q}=\sum_{\mathbf{k}}\dot{\hat{q}}_{\mathbf{k}}\mathbf{D}_{\mathbf{k}},$$

and (14.43) becomes

$$\sum_{\mathbf{k}}\dot{\hat{q}}_{\mathbf{k}}\mathbf{D}_{\mathbf{k}} = -\frac{1}{\varepsilon}\sum_{\mathbf{k},\mathbf{m}}\sin(\varepsilon(\mathbf{k}\times\mathbf{m}))\hat{q}_{\mathbf{k}+\mathbf{m}}f_{-\mathbf{m}}\mathbf{D}_{\mathbf{k}}. \tag{14.44}$$

The matrices $\mathbf{D}_{\mathbf{k}}$ are linearly independent; therefore the equality (14.44) must hold separately for each $\mathbf{D}_{\mathbf{k}}$, hence (14.43) is equivalent to (14.33).

We expect Casimirs to be traces of Q^k. To show this, first note that, from the commutation rule in (14.42), we can see that $\mathrm{Tr}[\mathbf{D}_{\mathbf{n}},\mathbf{D}_{\mathbf{m}}]=0$. From (14.39), Q^k is a linear combination of $\mathbf{D}$s so that the time derivative of Q^k is a set of $\mathbf{D}$-commutators, therefore the traces, $\mathrm{Tr}(Q^k)$, are conserved in time with an arbitrary Hamiltonian. Explicitly

$$\begin{aligned}\frac{1}{2M+1}\mathrm{Tr}(Q^k) &= \frac{1}{2M+1}\mathrm{Tr}\left(\sum_{\mathbf{i}}\hat{q}_{\mathbf{i}}\mathbf{D}_{\mathbf{i}}\right)^k\\
&= \frac{1}{2M+1}\sum_{\mathbf{i}_1,\ldots,\mathbf{i}_k}\hat{q}_{\mathbf{i}_1},\ldots,\hat{q}_{\mathbf{i}_k}\mathrm{Tr}(\mathbf{D}_{\mathbf{i}_1},\ldots,\mathbf{D}_{\mathbf{i}_k})\\
&= \sum_{\sum_j^k \mathbf{i}_j=0}\hat{q}_{\mathbf{i}_1},\ldots,\hat{q}_{\mathbf{i}_k}\exp\{i\varepsilon[\mathbf{i}_2\times\mathbf{i}_1+\mathbf{i}_3\times(\mathbf{i}_1+\mathbf{i}_2)\\
&\qquad +\cdots+\mathbf{i}_{k-1}\times(\mathbf{i}_1+\cdots+\mathbf{i}_{k-2})]\}\\
&= \sum_{\sum_j^k \mathbf{i}_j=0}\hat{q}_{\mathbf{i}_1},\ldots,\hat{q}_{\mathbf{i}_k}\cos\{\varepsilon[\mathbf{i}_2\times\mathbf{i}_1+\mathbf{i}_3\times(\mathbf{i}_1+\mathbf{i}_2)\\
&\qquad +\cdots+\mathbf{i}_{k-1}\times(\mathbf{i}_1+\cdots+\mathbf{i}_{k-2})]\},\end{aligned} \tag{14.45}$$

i.e. (14.45) coincides with (14.37).

15

A mathematical framework for quantifying predictability utilizing relative entropy

15.1 Ensemble prediction and relative entropy as a measure of predictability

Complex systems with many spatial degrees of freedom arise in diverse context, such as atmosphere/ocean general circulation models (GCMs) for climate or weather prediction, pollution models, and the models for the spread of hazardous biological, chemical, or nuclear plumes. These non-linear models are intrinsically chaotic over many time scales with sensitive dependence on initial conditions. In this chapter, as in Chapters 7 and 8 earlier, such models are represented discretely as a large system of ODEs for a vector $\vec{X} \in \mathcal{R}^N$ given by

$$\frac{d\vec{X}}{dt} = \vec{F}(\vec{X}, t), \vec{X} \in \mathcal{R}^N, N \gg 1. \tag{15.1}$$

Given both the uncertainty in a deterministic initial condition, $\vec{X}_0$, as well as the intrinsic chaos in solutions of (15.1), it is natural instead to consider an ensemble of initial data characterized by a probability density, $p_0(\vec{X})$, satisfying $p_0 \geq 0$, $\int p_0 = 1$ and with mean given by $\vec{X}_0$, i.e.

$$\int \vec{X} p_0(\vec{X}) d\vec{X} = \vec{X}_0. \tag{15.2}$$

The idea is to utilize the ensemble of solutions of (15.1) drawn from the initial data $p_0(\vec{X})$ to quantify the uncertainty and measure the confidence interval and predictive power of the deterministic solution beginning at $\vec{X}_0$. For example, consider $\vec{X}_0$, and an associated probability density consisting of small random perturbations of these initial data. This probability density measures the uncertainty in the measurement of the initial data. For instance, the random initial data might be sampled from a Gaussian probability distribution centered about $\vec{X}_0$ (see Section 6.4)

$$p_0(\vec{X}) = (2\pi)^{-N/2} \epsilon^{-N/2} e^{-\frac{|\vec{X}-\vec{X}_0|^2}{2\epsilon}}, \epsilon \ll 1, \tag{15.3}$$

where $\epsilon \ll 1$ measures the variance in each component. In other words, the ensemble of initial data satisfies

$$\int \vec{X} p_0(\vec{X}) d\vec{X} = \vec{X}_0,$$
$$\int (X_i - X_{0,i})(X_j - X_{0,j}) p_0(\vec{X}) d\vec{X} = \epsilon \delta_{ij}, \tag{15.4}$$

where δ_{ij} are the Kronecker symbols and take the value 1 if $i = j$ and 0 otherwise. In general, here we assume an initial probability measure $p_0(\vec{X})$, reflecting an ensemble of initial conditions. Theoretically, this is straight forward because it is easy to establish, following the discussion in Chapter 7, that the initial probability density, $p_0(\vec{X})$, evolves to a new probability density, $p_t(\vec{X})$, at later times, which satisfies the Liouville equation

$$\frac{\partial}{\partial t} p_t(\vec{X}) + \mathrm{div}_{\vec{X}}(\vec{F}(\vec{X}, t) p_t(\vec{X})) = 0,$$
$$p_t(\vec{X})\Big|_{t=0} = p_0(\vec{X}). \tag{15.5}$$

The Liouville equation is a linear PDE in a very large spatial dimension, $\mathbb{R}^N$, with $N \gg 1$, and is impractical to solve directly. In practice, instead a finite ensemble of individual solutions of (15.1), $\{\vec{X}_t^r, 1 \leq r \leq R\}$, is constructed by sampling the initial distribution, $p_0(\vec{X})$. Then $p_t(\vec{X})$ is approximated by the empirical probability distribution for ensemble prediction

$$p_t^E = \frac{1}{R} \sum_{r=1}^{R} \delta_{\vec{X}_t^r}(\vec{X}), \tag{15.6}$$

where $\delta_{\vec{X}_0}(\vec{X})$ is the Dirac delta measure at $\vec{X}_0$. The probability density in (15.6) is an explicit solution of the Liouville equation in (15.5).

The empirical ensemble prediction in (15.6) is the central probability measure of interest for practical ensemble prediction. How can we quantify and estimate the uncertainty in such a prediction? In the remainder of this introduction, some of the important scientific issues associated with empirical ensemble prediction are discussed as well as the potential use of information theory as developed in Chapter 6 in quantifying many aspects of uncertainty.

The first practical issue to confront is that the empirical ensemble size, R, in (15.6) is often very small in environmental science, on the order of $R \leq 50$ for weather prediction and short-term climate prediction in GCMs (Toth and Kalnay, 1993; Palmer, 2000), while, for example, $R = 500$ for intermediate models for predicting El Niño (Kleeman and Moore, 1997, 1999). This is due to the computational cost in these extremely complex systems in an operational setting requiring real time prediction. Furthermore, the idealized ensemble distribution, $p_0(\vec{X})$, is also unknown in practice. For theoretical studies, in the perfect predictability scenario, the initial distribution, $p_0(\vec{X})$ from (15.5), is known exactly and the approximation by an ensemble $R \gg 1$ is the main emphasis. These studies in idealized systems are very useful in determining the features governing predictability in a given system (Kleeman, 2002; Kleeman and Majda, 2004). Both viewpoints are emphasized in Sections 15.3 and 15.4 below.

What is available to quantify the uncertainty from the empirical ensemble prediction p_t^E and readily evaluated are some of the moments

$$\int_{\mathbb{R}^N} (\vec{X} - \bar{\vec{X}})^\alpha p_t^E(\vec{X})\,\mathrm{d}\vec{X} = \frac{1}{R}\sum_{r=1}^{R} (\vec{X}_t^r - \bar{\vec{X}}^r)^\alpha, \quad 0 \le |\alpha| \le 2L, \tag{15.7}$$

with $\vec{X}^\alpha = X_1^{\alpha_1} \ldots X_N^{\alpha_N}$ (Kalnay, 2003), $|\alpha| = \sum \alpha_i$. Of course, in practice only the low-order moments, where $L = 2$ or 4, are utilized for a judicious restricted set of variables

$$\vec{X}_i,\ 1 \le i \le M. \tag{15.8}$$

15.1.1 Practical and mathematical issues for predictability

Some of the main important issues in quantifying the uncertainty of an ensemble prediction are listed in (15.9)

(A) How much lack of information is contained in a given prediction strategy, for example, measuring only the second moments in (15.7), compared with p_t^E?
(B) When does the ensemble prediction distribution exhibit bimodality in some variable so that there are at least two different scenarios of significant change predicted in a given variable for times of interest? How can this be quantified in a computationally tractable fashion?
(C) In a given dynamical system in (15.1), which subsets of variables, $X_1 \ldots X_M$, are more predictable than the others? How can this be quantified?
(D) For long-term climate prediction, when is an ensemble prediction useful at all beyond the historical climate record? How can we estimate the lack of information in the climate record beyond the ensemble prediction? What features control the variation of this predictive utility in a given dynamical system?
(E) What features characterize the rare ensemble predictions with more than typical information beyond the climate record?

(15.9)

All of the above issues involve the lack of information in one probability distribution, $\Pi(\vec{X})$, compared with another probability distribution, $p(\vec{X})$, over some subset of variables, $\vec{X}_1 \ldots X_M$. As discussed extensively in Chapter 6, on information theory, this lack of information content is quantified by the relative entropy

$$P(p, \Pi) = \int_{\mathbb{R}^M} p \ln\left(\frac{p}{\Pi}\right). \tag{15.10}$$

As used earlier in the book, we will call Π the prior distribution. Recently, in an important paper, Kleeman (2002) has suggested the use of relative entropy to address the issue in (15.9(D)) for long-term climate prediction. In this case, Π is the historical record climate distribution on a given set of variables, while $p = p_t^E$ is the empirical ensemble prediction, so $P(p_t^E, \Pi)$ should be estimated. The main issues in (15.9(D)) were raised earlier by Anderson and Stern (1996). In the situation in (15.9(A)), monitoring only the first and second moments of the prediction as is usually done in practice (Toth and Kalnay, 1993; Kalnay, 2003) leads to the Gaussian prediction, p_G; so in this case $P(p_t^E, p_G)$ should be estimated. Recently, Roulston and Smith (2002) have suggested a similar use of information theory as a "score" or predictability measure for ensemble forecasts. Below we show how to develop rigorous computationally feasible tests for bimodality to address (15.9(B)) through non-Gaussian estimators for the empirical prediction ensemble and apply them in Section 15.3 to the Lorenz 96 model. As regards (15.9(C)), for the special case when both p and Π are Gaussian distributions, Schneider and Griffies (1999) have developed the idea of principal predictability components utilizing the entropy difference; this interesting concept has been generalized by Majda, Kleeman, and Cai (2004) to one typical practical situation, where Π is Gaussian, while p is non-Gaussian, with (15.10) as the more precise measure of lack of information in Π (see Section 15.2 below). Finally, to address the important issue in (15.9(E)) for long-term prediction, as well as the key factors that determine variability in predictive utility, Kleeman (2002) has introduced and applied the concept of the signal/dispersion decomposition for the special case when both Π and p are Gaussian distributions. Recently, Kleeman, Majda, and Timofeyev (2002) have determined the controlling factors of this facet of predictability in a simple model with statistical features of the atmosphere in a Gaussian setting, the truncated Burgers–Hopf model discussed in Chapter 7. This work is discussed below in Section 15.4. This decomposition for suitable non-Gaussian measures, p and Π, has been developed by Majda, Kleeman, and Cai (2004), and Abramov and Majda (2004) developed a new, more precise, decomposition into a signal, dispersion, and cross term. Both the signal and cross term are directly and cheaply evaluated from the moments of p. Thus, they are readily available in determining variations in utility in a given dynamical system.

With the background presented above, once we adopt the relative entropy in (15.10) as the theoretical tool for addressing the important issues in (15.9), the following practical issues arise:

(A) How can we numerically compute or approximate the relative entropy, $P(p, \Pi)$, for a complex system with many degrees of freedom and a subset of variables with $M \gg 1$?
(B) Are there computationally feasible and mathematically rigorous strategies to estimate $P(p, \Pi)$ from below, i.e. $P(p, \Pi) \geq P(p^*, \Pi)$, where $P(p^*, \Pi)$ can be evaluated either analytically or by a rapid numerical procedure?

(15.11)

15.1.2 Gaussian prior distribution for predictability

Often a Gaussian measure is utilized as an estimate of an ensemble prediction as in the applications in 15.9(A). We show below, how a rigorous lower bound estimator $P(p_4^*, p_G^*)$ as discussed in Section 6.4.5 can be utilized to estimate non-Gaussianity as in (15.9(B)). Furthermore, in many applications, the climate distribution is Gaussian on a subset of variables and thus completely determined by the mean and covariance. For such a Gaussian distribution

$$\Pi_0(X_1, \ldots, X_M) = (2\pi)^{-M/2}(\det \mathcal{C})^{-1/2} \exp(-\frac{1}{2}((\vec{X} - \vec{X}_0), \mathcal{C}^{-1}(\vec{X} - \vec{X}_0))), \tag{15.12}$$

where $\vec{X}_0 \in \mathcal{R}^M$ is the mean

$$\int X_j \Pi_0 = X_{0,j}, \quad 1 \leq j \leq M \tag{15.13}$$

and $\mathcal{C} = (C_{ij})$ is the positive definite, $\mathcal{C} > 0$, symmetric, $M \times M$ covariance matrix

$$\int (X_i - X_{0,i})(X_j - X_{0,j})\Pi_0 = C_{ij}. \tag{15.14}$$

For example, the truncated Burgers–Hopf (TBH) model and the inviscid unforced Lorenz 96 (IUL96) model from Chapter 7 both have Gaussian climates which are realized through direct numerical simulation. The spectrally truncated quasi-geostrophic (TQG) equations discussed extensively in Chapter 8 provide another very rich family of examples with a predicted Gaussian climate, which can be realized in various regimes through direct numerical simulation (see Chapter 7 and Majda *et al* 2001). The practical and mathematical issues for predictability listed in (15.1.1) are most readily developed for the special situation where the prior distribution $\Pi(X_1, \ldots, X_M)$ is Gaussian. The theory for this important special case is developed in Section 15.2 below and is applied to the examples from Chapter 7 in Sections 15.3 and 15.4. The final section of this chapter, Section 15.5, discusses other recent developments in ensemble prediction and information theory.

15.1.3 Invariance of the predictability measure under a general change of coordinates

Clearly, a good predictability measure for features of a given dynamical system should not depend on the coordinate system used to describe the underlying dynamics. While this seems obvious intuitively, many *ad-hoc* predictability measures utilized in practice fail to have this property and often depend on the choice of metric (energy, enstrophy, geopotential height etc.); Schneider and Griffies (1999) have emphasized this point in their work. Here we show that the predictability measure, $P(p, \Pi)$, defined in (15.10) by the relative entropy is invariant under a general non-linear change of coordinates.

Consider a smooth invertible transformation from $\mathcal{R}^M$ to $\mathcal{R}^M$ defined by

$$\vec{Y} \to \Phi(\vec{Y}) = \vec{X}.$$

Any smooth probability density, $p(\vec{X})$, determines a smooth probability density $p_\Phi(\vec{Y})$ via the formula

$$p_\Phi(\vec{Y}) = p(\Phi(\vec{Y})) \left| \det\left(\frac{d\Phi}{d\vec{Y}}\right) \right|, \tag{15.15}$$

where $\frac{d\Phi}{d\vec{Y}}$ denotes shorthand for the $M \times M$ Jacobian matrix of the mapping, $\frac{d\Phi}{d\vec{Y}} = \left(\frac{\partial \Phi_i}{\partial Y_j}\right)$ and det is the determinant.

The standard change of variable formula shows that, if $p(\vec{X})$ is a probability density, then $p_\Phi(\vec{Y})$ is also a probability density; the factor $\det\left(\frac{d\Phi}{d\vec{Y}}\right)$ is needed to guarantee this. The invariance of the relative entropy predictability measure under change of coordinates is the statement that

$$\int p \ln\left(\frac{p}{\Pi}\right) = P(p, \Pi) = P(p_\Phi, \Pi_\Phi) = \int p_\Phi \ln\left(\frac{p_\Phi}{\Pi_\Phi}\right) \tag{15.16}$$

for any smooth invertible transformation $\Phi(\vec{Y})$.

However, it follows immediately from (15.15) and the change of variables formula that (15.16) is true.

15.1.4 Canonical form for a Gaussian climate: EOF basis

As a simple application of the principle in (15.16), we show how measuring predictability for a Gaussian climate can be simplified through the special use of a linear change of coordinates. For the Gaussian climate defined in (15.12), consider the change of variables

$$\vec{X} = \vec{X}_0 + \mathcal{C}^{1/2}\vec{Y}. \tag{15.17}$$

In (15.17), $\mathcal{C}^{1/2}$, is the square root of the positive definite matrix $\mathcal{C}$ (Lax, 1996); the matrix $\mathcal{C}^{1/2}$ is also symmetric and positive definite, commutes with $\mathcal{C}$, and satisfies $\mathcal{C}^{1/2}\ \mathcal{C}^{-1}\ \mathcal{C}^{1/2} = I$. Since $\mathcal{C}$ is positive definite symmetric, there is an orthonormal basis, $\{\vec{e}_{i=1}^{M}\}$, called empirical orthogonal functions (EOFs) in the atmosphere/ocean community and corresponding positive eigenvalues, $\lambda_1 \geq \lambda_2 \geq \cdots \geq \lambda_M > 0$ so that the covariance matrix $\mathcal{C}$ is diagonalized, i.e.

$$\mathcal{C}\vec{f} = \sum_i \lambda_i \vec{e}_i(\vec{f}, \vec{e}_i). \tag{15.18}$$

The eigenvectors, $\vec{e}_1$, is called the first EOF, etc. With (15.18), the matrix $\mathcal{C}^{1/2}$ is defined by

$$\mathcal{C}^{1/2}\vec{f} = \sum_i \lambda_i^{1/2} \vec{e}_i(\vec{f}, \vec{e}_i). \tag{15.19}$$

From (15.15) and (15.12), it follows that in the coordinate system defined by (15.17), the Gaussian climate distribution has the simplified form

$$\Pi_0(\vec{Y}) = (2\pi)^{-M/2} e^{-\frac{1}{2}|\vec{Y}|^2}. \tag{15.20}$$

The TBH and IUL96 models from Chapter 7 already have a Gaussian climate with the diagonal form in (15.20), while the pseudo-energy metric discussed in Section 8.4 achieves the same result for the TQG models discussed in Chapter 8. The only minor difference is that in (15.20) we have rescaled the modes, $\vec{Y}$, to be uncorrelated with unit variance, while TBH, IUL96, and TQG models have a natural Gaussian climate represented by uncorrelated variables with a constant covariance not set to unity. With (15.15) and (15.17), given $p(\vec{X})$, we get

$$p_\Phi(\vec{Y}) = p(\vec{X}_0 + \mathcal{C}^{1/2}\vec{Y}) \det(\mathcal{C}^{1/2}). \tag{15.21}$$

For a general probability density, p, the covariance matrix of p, cov$p$, is defined by the centered second moments in (15.14) where Π_0 is replaced by p. Given the probability density $p(\vec{X})$ with mean $\bar{\vec{X}}_0$ and covariance matrix, Cov$p$, it follows easily from (15.17), (15.19) that the mean and covariance of the transformed probability density, p_Φ, are given by

$$\begin{aligned} \text{Mean } p_\Phi &= \bar{\vec{Y}} = \mathcal{C}^{-1/2}(\bar{\vec{X}} - \bar{\vec{X}}_0), \\ \text{Cov}\, p_\Phi &= \mathcal{C}^{-1/2} \text{Cov}\, p\, \mathcal{C}^{-1/2}. \end{aligned} \tag{15.22}$$

Of course, the same considerations apply to a general Gaussian prior distribution in general applications. Here this change of variables is called the "whitening" transformation.

15.2 Quantifying predictability for a Gaussian prior distribution

Here we address the practical and mathematical issues elucidated in Subsection 15.1.1 regarding predictive information content for the special case of a Gaussian prior distribution. Here as shown above in (15.17)–(15.21) without loss of generality we assume $\vec{X} = (X_1, \ldots, X_M)$ and

$$\Pi_0(\vec{X}) = (2\pi)^{-M/2} e^{-\frac{1}{2}|\vec{X}|^2}. \tag{15.23}$$

First we calculate the relative entropy of $p(\vec{X})$, which measures the additional information in p beyond Π_0. For simplicity in notation, here we suppress the preliminary transformation Φ that leads to the canonical form in (15.23). Recall from Sections 6.3 and 6.4, the entropy of a probability density, p, is given by

$$\mathcal{S}(p) = -\int_{\mathcal{R}^M} p \ln p, \tag{15.24}$$

which measures the average lack of information in p.

With (15.23), we calculate that

$$\begin{aligned} P(p, \Pi_0) &= \int p \ln p - \int p \ln \Pi_0 \\ &= -\mathcal{S}(p) + \frac{M}{2}\ln(2\pi) + \int_{\mathcal{R}^M} \frac{|\vec{X}|^2}{2} p, \end{aligned} \tag{15.25}$$

while for the Gaussian distribution in (15.23) we have the explicit formula

$$\mathcal{S}(\Pi_0) = \frac{M}{2}\ln(2\pi) + \frac{M}{2}. \tag{15.26}$$

Thus, we obtain the *relative entropy identity for a normalized Gaussian prior in* $\mathcal{R}^M$

$$P(p, \Pi_0) = \mathcal{S}(\Pi_0) - \mathcal{S}(p) + \int_{\mathcal{R}^M} \frac{|\vec{X}|^2}{2} p - \frac{M}{2}. \tag{15.27}$$

The elementary identity in (15.27) states that except for an interesting correction involving second moments, the measure of predictability through relative entropy for a Gaussian prior is directly related to the entropy difference. In fact, an immediate corollary of (15.27) is the following

Assume the ensemble prediction probability density p satisfies $\int |\vec{X}|^2 p = M = \int |\vec{X}|^2 \Pi_0$, then $P(p, \Pi_0) = \mathcal{S}(\Pi_0) - \mathcal{S}(p)$. (15.28)

Thus, if the trace of the second moments of the probability density p coincides with that of the climate, the relative entropy is exactly the entropy difference. This generalizes the discussion in Subsections 6.4.2 and 6.4.4 to the multi-dimensional case.

15.2.1 The signal and dispersion decomposition for a Gaussian prior

For the Gaussian prior in (15.23), here we develop a simple decomposition of the predictability measure $P(p, \Pi_0)$ into signal and dispersion components. This provides a theoretical framework for addressing the issues in (15.9A). Roughly speaking, the signal measures the contribution of the mean to the information content of the prediction beyond the climate while the dispersion measures the information content of the variance and other high moments for the prediction. Given the fact that the climate has zero mean, the knowledge that the prediction ensemble has a non-zero mean has potentially significant information content beyond the normalized variance of a prediction. Measures of predictability involving only the variance alone were the usual concept in atmosphere/ocean science (Schneider and Griffies, 1999) until the important paper of Kleeman (2002) who introduced the signal–dispersion decomposition for the special case when both $p(\vec{X})$ and $\Pi_0(\vec{X})$ are Gaussian distributions. In that paper, Kleeman (2002) also demonstrated the practical utility of the signal component in predictions for a series of stochastic and deterministic models for El Nino, while Kleeman, Majda, and Timofeyev (2002) demonstrated the surprising importance of the signal component in determining predictive utility for the TBH models from Chapter 7. These last results will be discussed briefly in Section 15.3.

For a Gaussian climate, the signal–dispersion decomposition is an immediate consequence of the formula in (15.27) and the elementary identity

$$\int |\vec{X}|^2 p = \sum_{i=1}^{M} \left(\int |\bar{X}_i|^2 p + \int (X_i - \bar{X}_i)^2 p \right) = |\text{Mean } p|^2 + \text{tr}(\text{Cov } p), \quad (15.29)$$

where given an $M \times M$ matrix $A = (A_{ij})$, the trace of A, $\text{Tr}(A) = \sum_{1 \le i \le M} A_{ii}$. With the definition of the mean of covariance, the proof of (15.29) is left as a simple exercise for the reader. With (15.27) and (15.29), we have the following:

Proposition 15.1 The signal dispersion decomposition for a Gaussian climate *Under these circumstances $P(p, \Pi_0)$ admits the signal–dispersion decomposition*

$$P(p, \Pi_0) = S + D, \quad (15.30)$$

where the signal S is given by

$$S = \frac{1}{2} |\text{Mean } p|^2 \quad (15.31)$$

and the dispersion, D, is given by

$$D(p, \Pi_0) = \mathcal{S}(\Pi_0) - \mathcal{S}(p) + \frac{1}{2}(\mathrm{tr}(\mathrm{Cov}\, p) - \mathrm{tr}(\mathrm{Cov}\, \Pi_0)). \tag{15.32}$$

The formulas in (15.30), (15.31), and (15.32) are an immediate consequence of (15.27), (15.29) and the identity, $\mathrm{tr}(\mathrm{Cov}\, \Pi_0) = M$. Clearly, only a non-zero mean value of p contributes to the information content of the prediction as measured by the signal. On the other hand, only higher moments of p contribute to the information content in the dispersion, D, as defined in (15.32). Note that both $\mathcal{S}(p)$ and $\mathrm{tr}(\mathrm{Cov}\, p)$ have identical values for p and any translated distribution with an arbitrary mean, $p_{\vec{\tau}} = p(\vec{X} - \vec{\tau})$ for a constant vector $\vec{\tau}$; thus, for the dispersion, D

$$D(p_{\vec{\tau}}, \Pi_0) = D(p, \Pi_0) \quad \text{for any } \vec{\tau} \tag{15.33}$$

and the dispersion in (15.32) measures the information content in p beyond Π_0 which is independent of the value of the mean of p. These facts provide the intuition behind the signal–dispersion decomposition in Proposition 15.1. This decomposition is generalized to non-Gaussian prior distribution in Majda *et al.* (2004) and Abramov and Majda (2004).

15.2.2 Rigorous lower bounds on predictive information content with a Gaussian prior

As noted in (15.7), in practical ensemble prediction, the mean and a few moments up to order typically at most four of the predictive probability density, $p(\vec{X})$, are known with significant accuracy. Here we provide a rigorous mathematical framework for estimating the predictive information content by quantities which can be readily computed. Thus, we address the issues (15.9(A),(B)) for the special case of a Gaussian prior distribution as in (15.23). Significant generalizations for a non-Gaussian prior are presented in Majda *et al.* (2004), while Subsection 6.4.5 already contains more explicit details of the procedure sketched below for the special case of univariate probability densities.

Motivated by (15.7), we assume that we know the mean and a finite number of the higher moments for $p(\vec{X})$, i.e.

$$\begin{aligned} \bar{X}_i &= \int X_i\, p, \quad 1 \le i \le M, \\ M_{i,j}^{\alpha,\beta} &= \int (X_i - \bar{X}_i)^{\alpha} (X_j - \bar{X}_j)^{\beta}\, p, \\ &\qquad 0 \le \alpha + \beta \le 2L, \quad 1 \le i, j \le M, \text{ with } L > 1 \text{ fixed.} \end{aligned} \tag{15.34}$$

The issues we address here are how can we estimate and rigorously bound $P(p, \Pi_0)$ in a systematic and practically significant fashion?

As background information, recall that we established in Subsection 6.4.2 that the Gaussian distribution maximizes the entropy among all probability measures with specified first and second moments. Recall from (15.30)–(15.32) that

$$P(p, \Pi_0) = -\mathcal{S}(p) + R(p, \Pi_0), \tag{15.35}$$

where R depends solely on the first and second moments of p, i.e.

$$R(p, \Pi_0) = \mathcal{S}(\Pi_0) + \frac{1}{2}(\mathrm{tr}(\mathrm{Cov}\, p) - \mathrm{tr}(\mathrm{Cov}\, \Pi_0)) + \frac{1}{2}(\mathrm{Mean}\, p)^2. \tag{15.36}$$

From (15.5) it follows that if we pick a probability density, p^*, which maximizes the entropy, $\mathcal{S}(p)$, subject to some of the constraints with $L \geq 1$ so that $R(p, \Pi_0)$ remains constant, then automatically $P(p^*, \Pi_0)$ provides a rigorous and potentially practical lower bound on the information content in the ensemble prediction beyond the prior distribution, Π_0. Next, we formalize the statements above in a precise fashion.

For any fixed $\tilde{L}$ with $1 \leq \tilde{L} \leq L$, define a set , $\mathcal{C}_{\tilde{L}}$, of linear constraints on probability measures through all the moment constraints in (15.34) for $\alpha + \beta \leq 2\tilde{L}$. By definition, the prediction ensemble, p, belongs to $\mathcal{C}_{\tilde{L}}$ so this set is non-empty and convex, while the entropy is a concave function on this set. Thus, we can always find $p^*_{2\tilde{L}}$ which satisfies the *maximum entropy principle*

$$p^*_{2\tilde{L}} \in \mathcal{C}_{\tilde{L}}, \quad \mathcal{S}(p^*_{2\tilde{L}}) = \max_{p \in \mathcal{C}_{\tilde{L}}} \mathcal{S}(p). \tag{15.37}$$

Clearly, we have

$$\mathcal{S}(p^*_{2L_1}) \geq \mathcal{S}(p^*_{2L_2}) \geq \mathcal{S}(p) \quad \text{for any } 1 \leq L_1 \leq L_2 \leq L. \tag{15.38}$$

Furthermore, because $\tilde{L} \geq 1$, and $R(p, \Pi_0)$ only involves moments up to order two

$$R(p^*_{2\tilde{L}}, \Pi_0) = R(p, \Pi_0) \quad \text{for any } \tilde{L}, \text{with } 1 \leq \tilde{L} \leq L. \tag{15.39}$$

Also, as developed in Subsection 6.4.2

$$\begin{array}{c} p^*_2 = p^*_G \text{ is a Gaussian distribution with the same mean and} \\ \text{variance matrix as } p;\ \text{thus } p^*_G \text{ is given by explicit formulas} \\ \text{as in (15.12)–(15.14).} \end{array} \tag{15.40}$$

With (15.35)–(15.40), we have the following:

Proposition 15.2 (Estimating the information content in an ensemble prediction for a Gaussian prior) *For any fixed number of prediction moments, $2L$, with $L \geq 1$ with a Gaussian prior, we have the rigorous lower bounds*

$$P(p, \Pi_0) \geq P(p^*_{2L}, \Pi_0) \geq P(p^*_{2\tilde{L}}, \Pi_0) \geq P(p^*_G, \Pi_0), \quad \textit{for } L \geq \tilde{L} \geq 1. \tag{15.41}$$

In fact, the same chain of estimates in (15.41) also applies to the dispersion $D(p, \Pi_0)$, *alone, i.e.*

$$D(p, \Pi_0) \geq D(p^*_{2L}, \Pi_0) \geq D(p^*_{2\tilde{L}}, \Pi_0) \geq D(p^*_G, \Pi_0), \quad \textit{for } L \geq \tilde{L} \geq 1. \quad (15.42)$$

The estimates in (15.41) are generalized to a non-Gaussian prior in Majda *et al.* (2004) and specific estimates for the information loss are contained in our earlier discussion in Subsection 6.4.5 for univariate distributions. Kleeman (2002) introduced and applied the relative entropy, $P(p, \Pi_0)$, for measuring the information content where he assumes that both p and Π_0 were Gaussian distributions. Here in Proposition 15.2, we have shown that the relative entropy of the Gaussian distribution, $P(p^*_G, \Pi_0)$ is automatically a rigorous lower bound on the information content of any prediction density, p, with the same first and second moments. With the formula in (15.32) for the dispersion and (15.12), for the Gaussian estimator $P(p^*_G, \Pi_0)$, we have

$$\mathcal{S}(\Pi_0) - \mathcal{S}(p^*_G) = \ln(\det(\operatorname{Cov} p)^{-1/2}) \quad (15.43)$$

so that

$$\begin{aligned} D(p^*_G, \Pi_0) &= \ln(\det(\operatorname{Cov} p)^{-1/2}) + \frac{1}{2}(\operatorname{tr}(\operatorname{Cov} p) - M), \\ P(p^*_G, \Pi_0) &= D(p^*_G, \Pi_0) + \frac{1}{2}|\text{Mean } p|^2. \end{aligned} \quad (15.44)$$

Of course $D(p^*_G, \Pi_0)$ is just the relative entropy of the Gaussian measure with the same variance as P^*_G but with zero mean.

Finally we mention that the practical advantage for utilizing the maximum entropy principle in (15.37) with the explicit moment information in (15.34) is that standard constrained optimization numerical methods, as mentioned in Chapter 11, can be utilized to find $p^*_{2\tilde{L}}$ accurately. Here an even number of moments are utilized to guarantee that the optimization problem has a solution.

15.2.3 Choosing reduced variables to order the predictive information content

Here we describe a nice construction for Gaussian prediction distribution, p^*_G, essentially due to Schneider and Griffies (1999) which organizes the variables $\vec{X}$ into subspaces with a hierarchy of predictive information content. Recall from Section 15.1.3 that through the change of variables in (15.17), a general Gaussian climate assumes the canonical form in (15.20), while the relative entropy of a general prediction remains invariant; the covariant matrix for a general prediction distribution in the new coordinates is given by the formula in (15.22). For estimating predictive information content, it is very natural to introduce a second change of coordinates, which diagonalizes the covariant matrix of p. Since $\mathrm{Cov}\, p$ is a symmetric positive definite $M \times M$ matrix, there exists a rotation matrix, $\mathcal{O}$, with $\mathcal{O}^T = \mathcal{O}^{-1}$, so that

$$\mathcal{O}^{-1}\mathrm{Cov}\, p\, \mathcal{O} = D = \begin{pmatrix} \gamma_1 & 0 & \cdots & 0 \\ 0 & \gamma_2 & \cdots & 0 \\ \vdots & \vdots & \vdots & \vdots \\ 0 & 0 & 0 & \gamma_M \end{pmatrix} > 0, \tag{15.45}$$

where D is the positive diagonal matrix with non-zero diagonal entries, $\gamma_i > 0$. Consider the new variable

$$\vec{X} = \mathcal{O}\,\vec{Z}. \tag{15.46}$$

Since $\mathcal{O}$ is a rotation matrix, it follows that the Gaussian prior measure $\Pi_0(\vec{Z})$ retains the same normalized form in (15.20).

In this coordinate system, as a consequence of (15.45), the Gaussian distribution $p^*_G(\vec{Z})$ with the same variance and mean as p, assumes the factored form

$$p^*_G(\vec{Z}) = \prod_{i=1}^{M}(2\pi\gamma_i)^{-1/2}e^{-\frac{(Z_i-\bar{Z}_i)^2}{2\gamma_i}} \stackrel{\mathrm{def}}{\equiv} \prod_{i=1}^{M} p^*_i(Z_i), \tag{15.47}$$

while the Gaussian prior distribution $\Pi_0(\vec{Z})$ retains the factored form

$$\Pi_0(\vec{Z}) = \prod_{i=1}^{M}(2\pi)^{-1/2}e^{-\frac{Z_i^2}{2}} = \prod_{i=1}^{M}\Pi_{0,i}(Z_i). \tag{15.48}$$

In (15.47), $\bar{Z}_i$ is the mean of $p(\vec{Z})$ in the ith coordinate. In the Z variables, with (15.47) and (15.48), the Gaussian estimator for the predictability factors into a sum of one-dimensional principal predictability factors

$$P(p, \Pi_0) \geq P(p^*_G, \Pi_0) = \sum_{i=1}^{M} P(p^*_i, \Pi_{0,i}) = \sum_{i=1}^{M}\frac{1}{2}[\ln\gamma_i^{-1} + \gamma_i - 1] + \sum_{i=1}^{M}\frac{1}{2}[\bar{Z}_i^2]. \tag{15.49}$$

Clearly, the first term in the summation in (15.49) is the relative entropy contribution to the dispersion from the one-dimensional Gaussian distributions p^*_i in (15.47), while the second term arises from the contribution of p^*_i to the signal. Thus, these variables can obviously be organized into groups with higher predictive information content in the dispersion, the signal, or the total combination, depending on the nature of the application. These variables are called the *principal predictability components*. The change of coordinates in (15.46) has been utilized in Majda *et al.* (2004) to get improved estimators beyond Proposition 15.2 for multi-variable distributions in terms of mutual information of joint distributions.

15.2.4 The relative entropy and the entropy difference for quantifying predictive information content

For a normalized Gaussian climate, Π_0, and a Gaussian prediction, p_G, Schneider and Griffies (1999) advocate the entropy difference in (15.43)

$$\Delta\mathcal{S}(p_G, \Pi_0) = \mathcal{S}(\Pi_0) - \mathcal{S}(p_G) = \ln(\det(\text{Cov}p_G)^{-1/2}), \tag{15.50}$$

as a measure of predictive information under the tacit assumption that distribution, p_G, of interest for prediction satisfies $\Delta\mathcal{S}(p_G, \Pi_0) \geq 0$. First, consider the special case of (15.50) for one-dimensional probability distributions for quantities with variance for p_G given by σ^2. Clearly, from (15.50) we have

$$\Delta\mathcal{S}(p_G, \Pi_0) < 0 \quad \text{if and only if } \sigma^2 > 1. \tag{15.51}$$

Thus, the entropy difference does not measure information content for distributions with variance $\sigma^2 > 1$. On the other hand, the fact that the variance of a predicted quantity at a short fixed time, such as the temperature at a fixed location, is predicted to be higher than that of the historical climate record has obvious information content ignored in (15.50). Since $P(p, \Pi_0)$ is non-negative, the relative entropy predictive measure always assigns non-trivial information content to these events. Furthermore, as mentioned earlier in Subsection 15.2.1, the entropy difference in (15.50) cannot measure the information content expressed in the fact that the mean of the prediction density, p_G, can be non-zero and different from that of Π_0, while the signal contribution to the relative entropy measure of predictive information content developed in Subsection 6.4.1 does this precisely. Finally, as developed in Subsection 15.2.1, the relative entropy is invariant under a general change of coordinates, while the entropy difference in (15.50) is only unchanged by linear transformation. Quite often quantities such as humidity can be complex non-linear functions of other variables in the atmosphere/ocean system, so a measure of information content which is unchanged by non-linear transformations is clearly superior to one which is not. For all of these three reasons, we claim that, even for Gaussian predictions and climate, the relative entropy measure is superior to the entropy difference in quantifying predictive information content. For general Markov processes, such as a finite Markov chain with a non-uniform invariant measure, there are additional compelling dynamic reasons to favor the relative entropy over entropy difference related to the decrease of information as equilibrium approaches (Cover and Thomas, 1991); however, an additional discussion here is beyond the scope of the material as presented here.

15.3 Non-Gaussian ensemble predictions in the Lorenz 96 model

Here the development of non-Gaussian bimodal and skewed distributions in ensemble predictions is studied within the context of the damped forced L96 model discussed in detail in Section 7.4. Recall that this elementary model with 40 degrees of freedom has features mimicking the baroclinic wave packets in the real atmosphere associated with actual weather systems. The robust dynamical regime is used with constant forcing $F = 8$ and damping coefficient $d = 1$. Below we examine the non-Gaussian information content in ensemble predictions with fairly small sample sizes for the three Fourier modes, $k = 0, k = 3$, and $k = 8$. Recall from the discussion in Section 7.4 that the mode $k = 0$ defines the climatology, while $k = 8$ is the most unstable mode in the model and $k = 3$ is a linearly stable but chaotic large-scale mode.

The numerical set-up for the experiments with the Lorenz '96 model is the following: first, a long (10 000 time units) "climatological" time series of a single solution is generated. Then an instantaneous snapshot of this solution is recorded at the end of the series. The statistical ensembles of various sizes are then generated around this single recorded snapshot, by perturbing each gridpoint via a narrow Gaussian probability with small variance (10^{-5} fraction of the climatological variance). The ensembles then propagate further, as their time series are being recorded.

Four different ensemble sizes with 25, 50, 100, and 200 members are employed in the current work to evaluate and compare the efficiency of the methodologies for measuring information content for different sample sizes. Since an ensemble provides a single value of the relative entropy at any given time, a super-ensemble (ensemble of ensembles) has to be generated for each of the four sample sizes in order to study meaningful statistical properties of non-Gaussianity in different regimes. For each sample size, a super-ensemble consisting of 20 ensembles is employed.

For the real part of each of the three Fourier modes, $k = 0, 3, 8$, the non-Gaussian information content is estimated from below for each ensemble prediction as time evolves by $P(p_4^*, p_G^*)$, where the probability density, p, is the empirical ensemble measure given in (15.6). Thus, we apply the theory developed in Proposition 15.2 in its simplest univariate form as developed in more detail in Subsection 6.4.5 to the empirical ensemble prediction distribution, p. The reader might want to re-examine Figure 6.2 and Table 6.1 to check the wide variety of behavior allowed for $P(p_4^*, p_G^*)$ for varying families of known probability distributions as well as the actual magnitude of $P(p_4^*, p_G^*)$ associated with distributions with strong bimodal or skewed behavior. Of course, this theory as used here assumes that the empirical distribution is the actual prediction distribution; this assumption is an excellent one for large ensemble sizes like $R = 200$, but is unrealistic for small

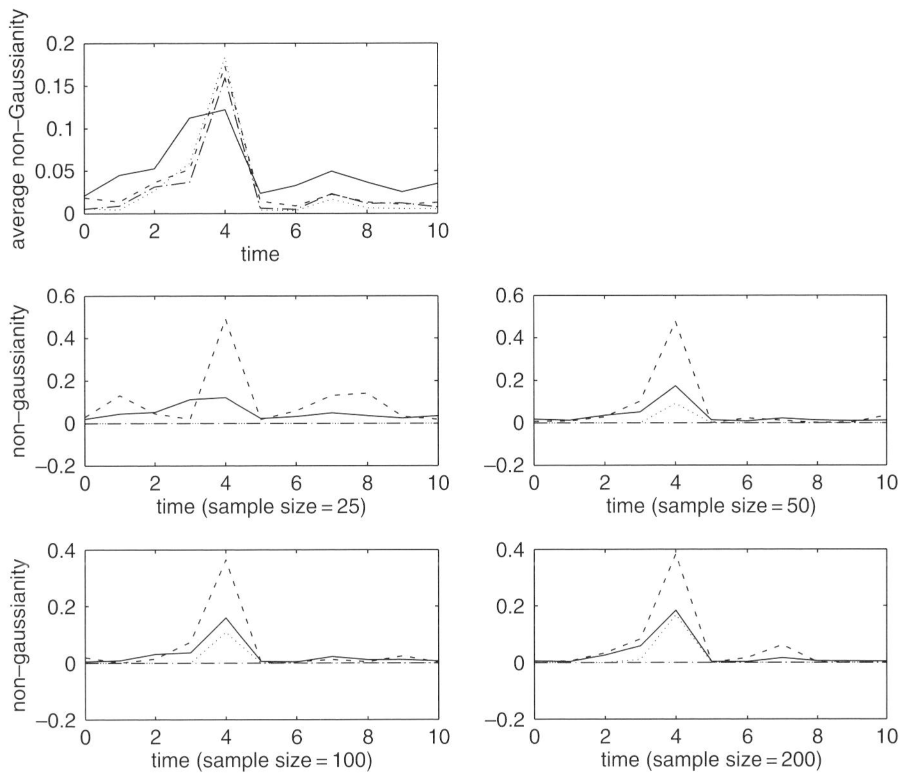

Figure 15.1 Fourier mode: $\mathbf{k} = 0$. Top panel: shows the average non-Gaussianity, over 20 ensembles, with sample sizes 25 (solid), 50 (dashed), 100 (dot–dashed), and 200 (dotted). Panels 2–4: The solid curve shows the average non-Gaussianity for the indicated sample size. The dashed curve shows the ensemble with the largest non-Gaussianity at time $t = 4$. The dot–dashed curve shows the adjusted moment technique applied to this ensemble. The dotted curve shows the central limit technique applied to this ensemble.

ensemble sizes like $R = 25$. Of course, the detection of non-Gaussianity with high confidence in an ensemble prediction with small ensemble size is an important practical topic. Thus, in the plots below we also include two statistical estimating procedures for finite sample size, called the central limit and adjusted moments methodologies, developed by Haven, Majda, and Abramov (2004), which provide 95% confidence intervals on lower bounds on non-Gaussian information content. The reader is referred to that paper for details of those algorithms.

The average over the super-ensemble of the sample non-Gaussianities of the Fourier mode $k = 0$ as measured by $P(p_4^*, p_G^*)$ are plotted in the first panel of Figure 15.1, with ensemble sizes 25 (solid), 50 (dashed), 100 (dot-dashed), and 200 (dotted). In general, the average of the sample non-Gaussianities is not equal to the true non-Gaussianity, but here it is assumed. The graphs show the averages

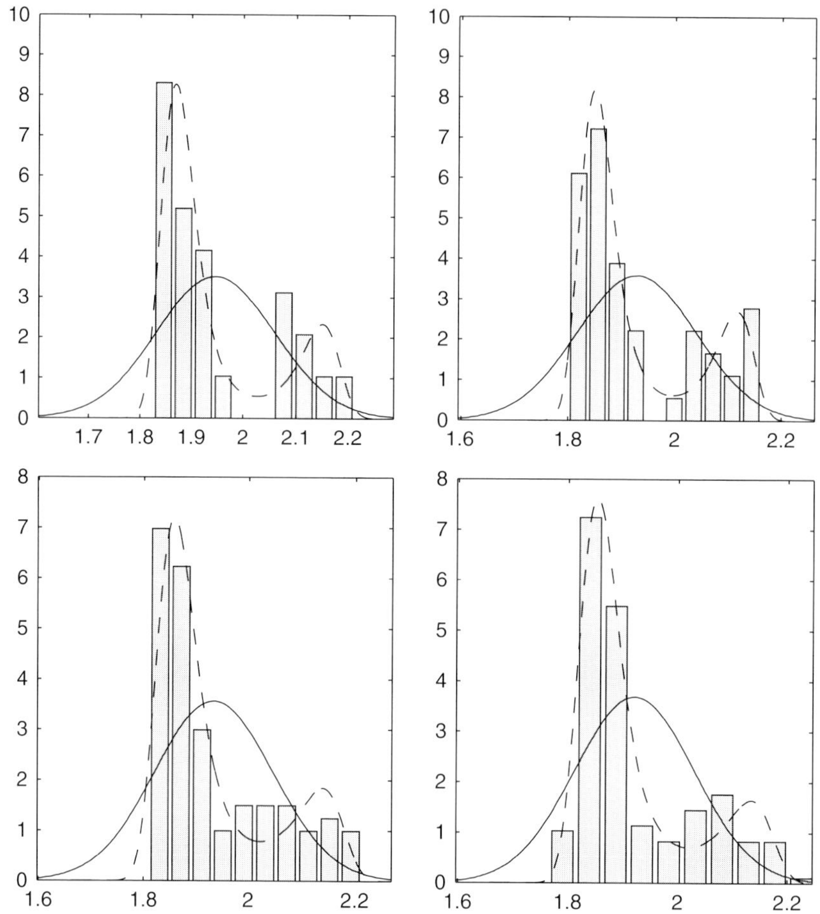

Figure 15.2 Fourier mode: $\mathbf{k} = \mathbf{0}$. Each graph shows a histogram of the data from panels 2–4 in Figure 15.1 at time $t = 4$. The Gaussian and four-moment fits are overlayed.

decreasing with increasing ensemble size, except for a spike in the non-Gaussianity at time $t = 4$, and unless the super-ensemble size is very large.

The second panel of Figure 15.1 shows the average non-Gaussianity for ensemble size 25 as a solid curve. The dashed curve represents the ensemble with the largest non-Gaussianity at time $t = 4$. The dot–dashed and dotted curves, which overlap on this graph, show the adjusted moment and central limit methodologies, respectively, applied to this ensemble. Similar graphs are shown for each ensemble size in panels three, four, and five. In each panel, the solid curve represents the average sample non-Gaussianities, while the dashed, dot–dashed, and dotted curves represent the non-Gaussianity and, adjusted moment, and central limit statistical methodologies applied to the ensemble with the largest non-Gaussianity

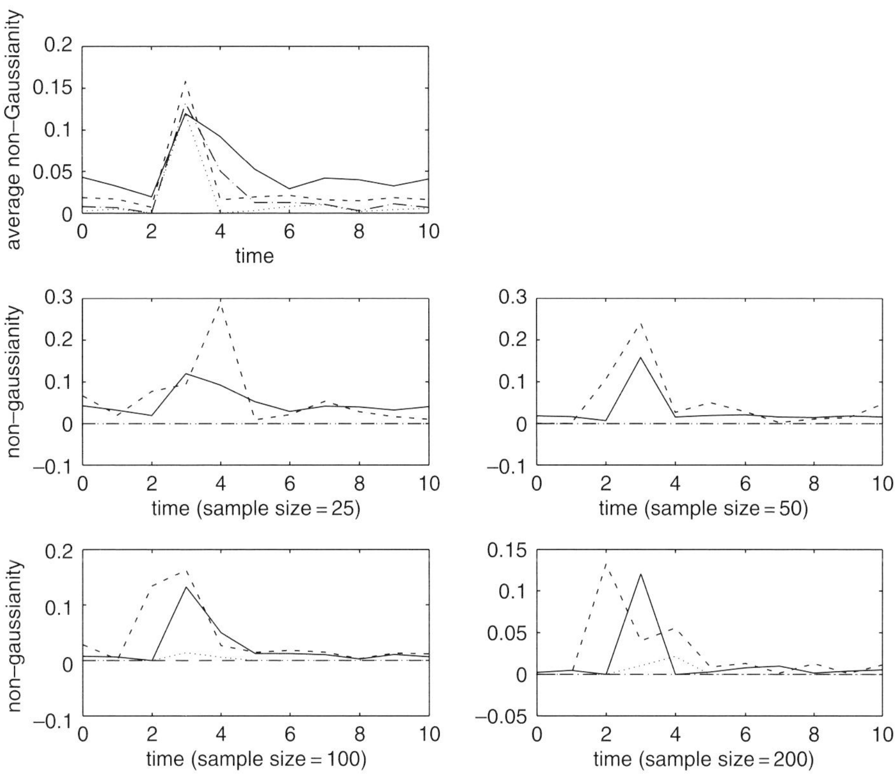

Figure 15.3 Fourier mode: $\mathbf{k} = 3$. Top panel: shows the average non-Gaussianity, over 20 ensembles, with sample sizes 25 (solid), 50 (dashed), 100 (dot–dashed), and 200 (dotted). Panels 2–4: The solid curve shows the average non-Gaussianity for the indicated sample size. The dashed curve shows the ensemble with the largest non-Gaussianity in the $k = 0$ mode at time $t = 3$. The dot–dashed curve shows the adjusted moment technique applied to this ensemble. The dotted curve shows the central limit technique applied to this ensemble.

at time $t = 4$. In addition, the plots in Figure 15.1 using the adjusted moment methodology demonstrate that evidence of non-Gaussian behavior of the prediction density can be detected, with 95% confidence, at ensemble sizes as small as 50. Figure 15.2 shows the histograms of the data, at each ensemble size and time $t = 4$, for the ensemble with the maximum non-Gaussianity at time $t = 4$. The solid curve shows the Gaussian density, p_G^*, while the dashed curve shows the four-moment density p_4^*. Clearly, the four-moment density fits the data better than the Gaussian density, and tracks bimodality in the ensemble prediction extremely well.

Figures 15.3–15.6 show the same plots as Figures 15.1 and 15.2 for Fourier modes $k = 3$ and $k = 8$. In each case, the observed ensemble is the one with

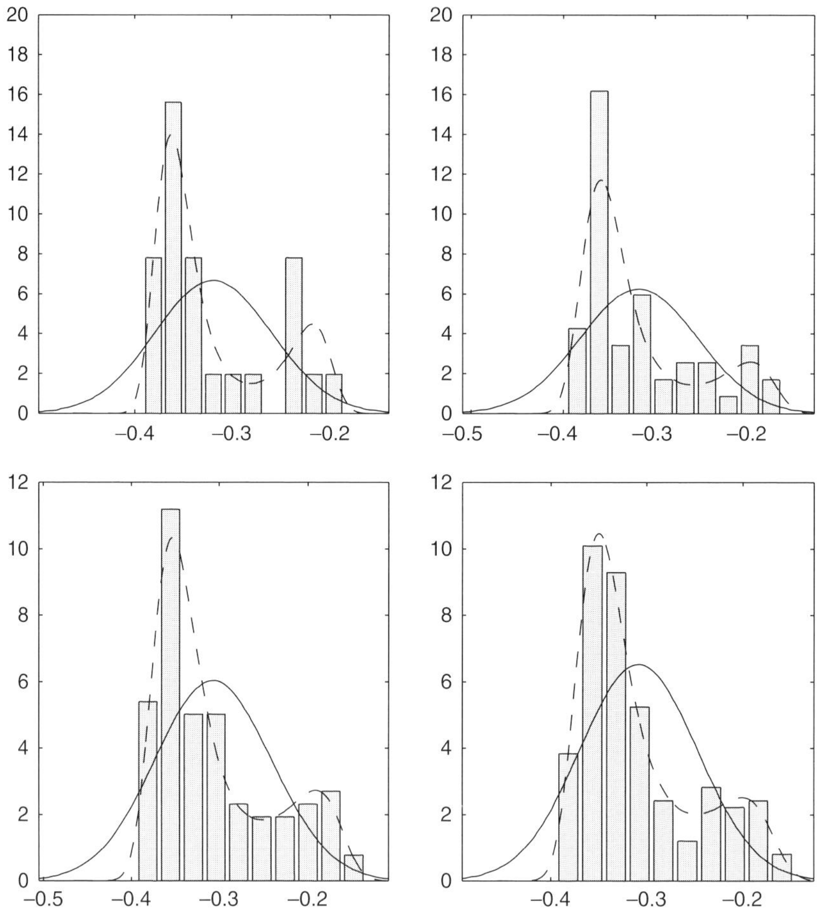

Figure 15.4 Fourier mode: $\mathbf{k} = 3$. Each graph shows a histogram of the data from panels 2–4 in Fig. 15.3 at time $t = 3$. The Gaussian and four-moment fits are overlayed.

maximum non-Gaussianity in Fourier mode $k = 0$. For these figures, the histograms are plotted for the time $t = 3$, where the largest peak of non-Gaussianity seems to appear. Once again bimodality is detected and tracked very well by p_4^* for the mode $k = 3$ and, the central limit statistical methodology detects non-Gaussian behavior of the prediction density, with 95% confidence.

However, an interesting example of a false indication of bimodality for the most unstable mode, $k = 8$ can be seen in Figure 15.6 for small ensemble sizes. In the first panel, at sample size 25, there seems to be a strong indication of non-Gaussianity. Visually, the data seem bimodal, and the corresponding estimate for $P(p_4^*, p_G^*)$, the dashed line in panel two of Figure 15.6 at $t = 3$, is

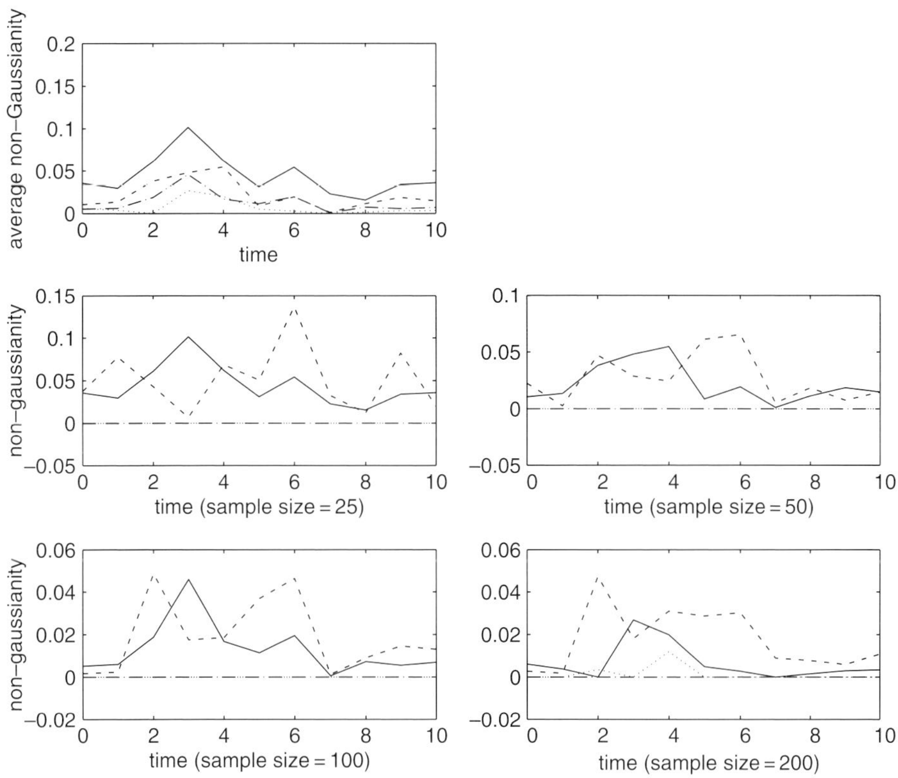

Figure 15.5 Fourier mode: $\mathbf{k} = 8$. Top panel: shows the average non-Gaussianity, over 20 ensembles, with sample sizes 25 (solid), 50 (dashed), 100 (dot–dashed), and 200 (dotted). Panels 2–4: The solid curve shows the average non-Gaussianity for the indicated sample size. The dashed curve shows the ensemble with the largest non-Gaussianity in the $k = 0$ mode at time $t = 3$. The dot–dashed curve shows the adjusted moment technique applied to this ensemble. The dotted curve shows the central limit technique applied to this ensemble.

approximately 0.5. However, at a sample size of 200, in the fourth panel, the bimodality is almost completely gone, both visually and in the estimate from $P(p_4^*, p_G^*)$. Interestingly, the central limit statistical methodology detects little or no non-Gaussian behavior for all sample sizes, and this is a highly desirable feature of such a sample estimator.

The above results suggest, with 95% confidence, that non-Gaussian tendencies exist for some modes of the prediction densities at various times. That the higher moment information is detected, with confidence and for small sample sizes, is evidence that a strictly Gaussian prediction strategy will sometimes miss substantial levels of predictability.

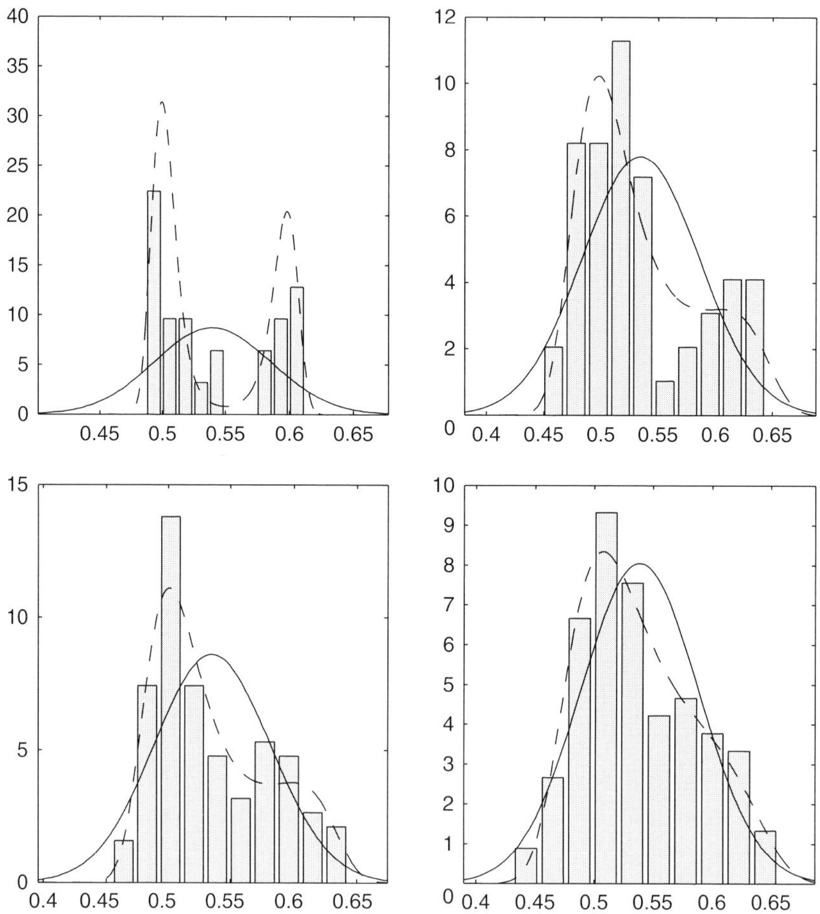

Figure 15.6 Fourier mode: $\mathbf{k} = 8$. Each graph shows a histogram of the data from panels 2–4 in Figure 15.5 at time $t = 3$. The Gaussian and four-moment fits are overlayed.

15.4 Information content beyond the climatology in ensemble predictions for the truncated Burgers–Hopf model

Here we study ensemble predictions in the truncated Burgers–Hopf (TBH) model discussed extensively earlier in Section 7.3. As established there, this model exhibits equipartition of energy for statistically irrelevant values of the Hamiltonian with a Gaussian climatological distribution; the model also has the interesting property discussed in Subsection 7.3.5 that the de-correlation time scale of individual Fourier components is inversely proportional to their wave number. In other words, large-scale structures have low frequency variability and are much more persistent than the "weather modes" at smaller scales. Such a statistical property is a well-known feature of the atmosphere and many other large-dimensional dynamical systems of physical interest. Furthermore, as shown in Subsection 7.3.5, this scaling behavior in TBH is predicted by elementary theory and confirmed by numerical simulations. Without further comment, we use the same numerical methods for TBH as described in Section 7.3 with the same values as used there with $\Lambda = 50$ and a variance of 0.1 for each individual Fourier mode in the Gaussian climatological distribution from (7.44), (7.45). Thus, there are 100 total modes in the results presented below.

15.4.1 Relaxation of ensemble predictions to the climate distribution

Statistical prediction at longer times may be viewed as the relaxation of a relatively tight probability distribution at the initial time towards an equilibrium distribution which can be considered the climatological distribution. The initial time p.d.f. can be considered as the uncertainty in the initial specification of the systems state vector. In general we would expect the mean of the initial condition distribution to be drawn according to a p.d.f. identical to that of climatology, since this is the historical distribution under the assumption of ergodicity. We adopt such an approach here in generating a super-ensemble.

Additionally, we assume that the initial condition p.d.f. of each ensemble prediction is Gaussian with a mean distributed as just discussed and a variance four orders of magnitude smaller than that of the equilibrium p.d.f., which as noted above is also Gaussian for the TBH model. This choice for the initial condition variance is somewhat arbitrary and is intended to represent the realistic scenario where uncertainty in the initial conditions state vector is much less than the historical spread in the same vector. Prediction ensembles of 500 members are used here with initial conditions drawn according to the initial condition p.d.f. discussed above. With these large ensemble sizes, we are in the perfect predictability regime in the present study in contrast to that in Section 15.3. The relaxation behavior for a typical ensemble of initial conditions is displayed in Figure 15.7, which shows the evolution of the mean and standard deviation of the Fourier components for a particular set of initial conditions. As can be seen from Figure 15.7 the smaller-scale modes converge much more rapidly towards equilibrium for both first and second moments of the p.d.fs. Notice that the first moment can sometimes exhibit some oscillatory-like behavior as it converges to zero.

In general the distributions governing the prediction p.d.fs appear to be approximately Gaussian at all lags. Figure 15.8 shows the distributions for five of the modes at various prediction lags. The modes are chosen to be representative of the various spatial scales of the model. This degree of Gaussian behavior

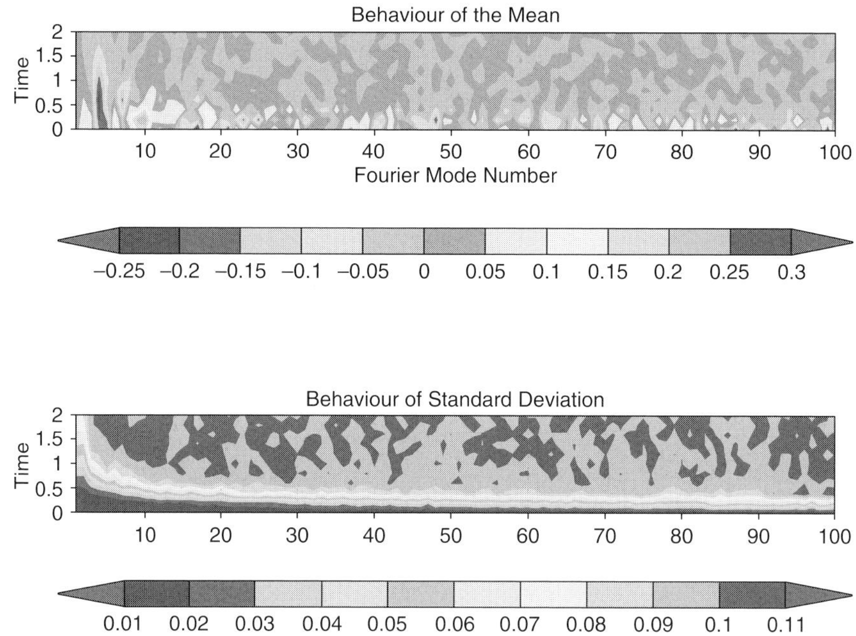

Figure 15.7 Convergence of the first and second moments for a particular ensemble of predictions. The initial conditions are drawn from a Gaussian distribution with means drawn from the equilibrium distribution which has a variance of 0.1 for each mode (see text).

is rather surprising given the significant non-linearity operating in the TBH model. Statistical significance tests reported in Kleeman (2002) confirm this high degree of Gaussianity in ensemble predictions for the TBH model. This contrasts very strongly with the ensemble prediction results in Section 15.3 for the Lorenz 96 model, which displays a large degree on non-Gaussianity including bimodality.

15.4.2 The signal, dispersion, and variations in predictive utility

Given the fact that the climatological distribution is essentially Gaussian and, as just established in Subsection 15.4.1, the ensemble prediction is also Gaussian with large statistical significance; below we utilize the Gaussian formulas for the signal and dispersion decomposition described in Proposition 15.1 in Section 15.2.1. Here we are interested in deciding quantitatively what controls the variations in predictive utility as measured by the relative entropy, $P(p, \Pi)$, where Π is the Gaussian climatological distribution and p is the Gaussian ensemble prediction. Recall that $P(p, \Pi)$ in the present context measures the

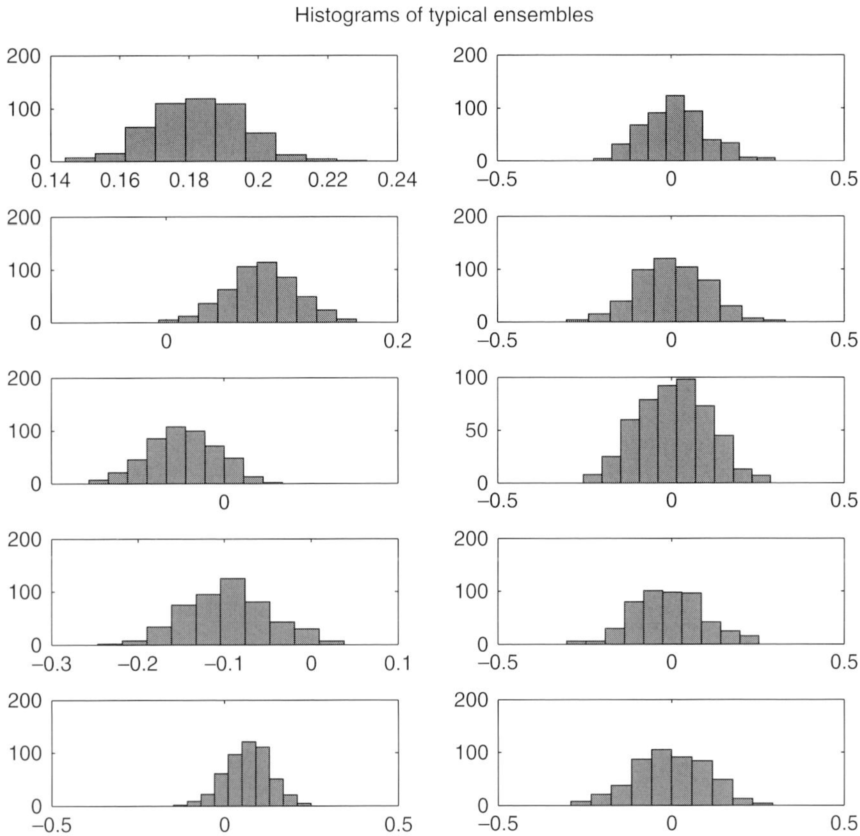

Figure 15.8 Histograms of mode distributions of particular ensembles at particular times. The left-hand panels are for $t = 0.2$ while the right-hand panels are for $t = 0.8$. Each row refers to a different Fourier mode from the model. From the top these are modes 5, 15, 25, 45, and 65.

lack of information in the climatological distribution relative to the ensemble prediction and thus measures quantitatively if the prediction distribution has information content beyond the climatological distribution and is useful at all. According to Proposition 15.1, $P(p, \Pi)$, the relative information content splits into the sum of the signal, an ensemble mean contribution to the information content, and the dispersion, a non-linear function of the ensemble variance relative to the climatology. Following the important paper of Kleeman (2002), we discuss here which of these quantities controls the variations in utility at various prediction times for various sets of modes. To gather meaningful results, we consider a 100 member super-ensemble of 500 member ensemble predictions as discussed at the beginning of Subsection 15.4.1. Majda and

Timofeyev (2000) showed that for $\Lambda = 50$ with variance 0.1 for individual Fourier modes, large-scale initial data form a shock and transfer energy to small scales at roughly the time 0.5. This is consistent with the statistical relaxation process described earlier in Subsection 15.4.1. Below, we call ensemble prediction times, shorter than this characteristic time scale, short-range predictions and, conversely, longer time ensemble predictions, long-range predictions.

In Figure 15.9 we see that for short-range predictions, the signal, and dispersion are of roughly equal importance in determining utility variation with initial condition but for longer ranges the signal is somewhat more important in deter-

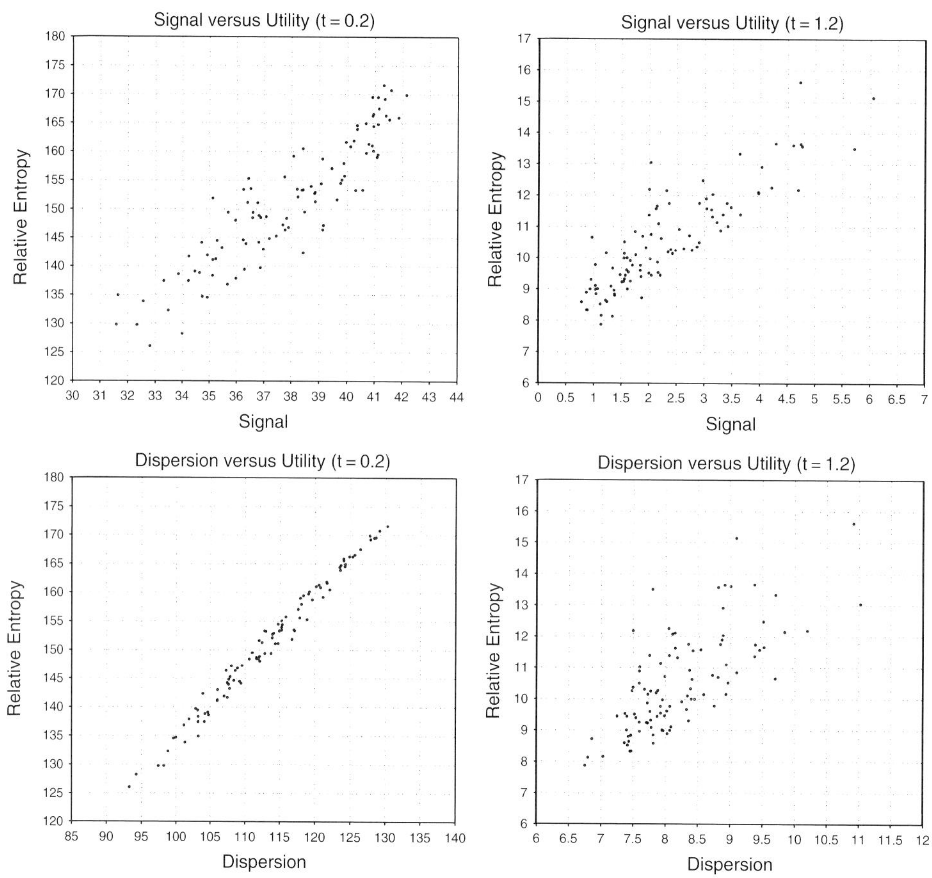

Figure 15.9 A scatter plot for the 100 ensembles considered, of signal and dispersion versus relative entropy. The latter was calculated under the assumption that distributions are Gaussian and so is the sum of signal and dispersion (see the text).

mining utility variability. It is clear that the signal is an important determinant of prediction utility variability for the TBH model. It is worth emphasizing that we are interested here in which parameter determines *variation* of utility with initial condition, not the absolute value of the particular parameter. This latter quantity is somewhat arbitrary, since it depends on the assumption we make about the tightness of the initial condition p.d.f. (a tighter value implies a higher absolute value of the dispersion and conversely).

In general, in prediction scenarios for climate, we are interested in determining the large-scale component of the flow with low frequency variability. This separation of scales is the motivation for the climate-oriented stochastic modeling often utilized in studies of atmospheric dynamics (Majda, Timofeyev, and Vanden Eijnden, 2001). It is clear in the TBH model that the large-scale modes are much more predictable than the small-scale ones. This may be seen in Figure 15.10 for one particular ensemble initial condition set. Plotted is the evolution of utility $P(p, \Pi)$ as a one-dimensional function of each Fourier mode and it is evident that the utility of the large-scale modes remains for a considerably longer period than the same quantity for the small-scale modes.

In order to examine the stochastic "climate" scenario we calculated the utility of the first ten (large-scale) spectral modes. Figure 15.11 shows the role of signal and dispersion in determining total large-scale utility. Rather strikingly, the signal component completely dominates utility variation at all prediction times. Similar results (not shown) were also found when even 20 and 40 modes were retained. This result indicates that in general the signal is the main determinant of prediction utility in the TBH model. Abramov and Majda (2004) have established a similar behavior for predictive utility for statistically relevant values of the Hamiltonian (see Chapter 14) for the TBH model.

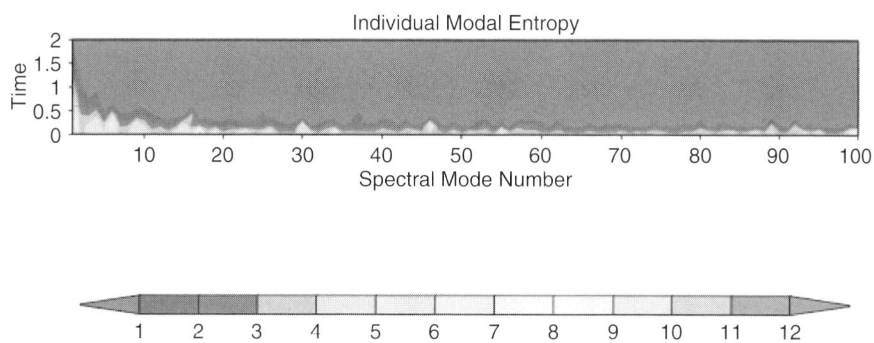

Figure 15.10 The evolution of (univariate Gaussian) relative entropy with time for various Fourier modes and for a particular initial condition ensemble.

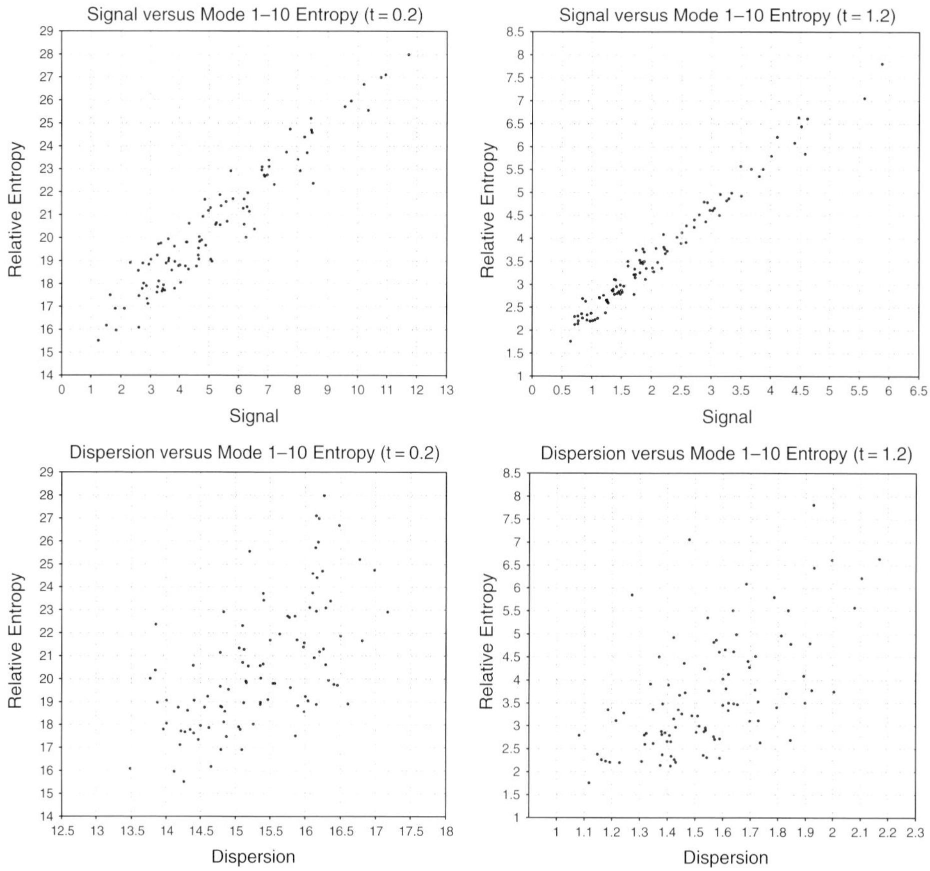

Figure 15.11 The same as Figure 15.10 but for multivariate Gaussian relative entropy of the first ten Fourier modes (see text). These modes represent the large-scale slowly evolving part of the flow.

15.5 Further developments in ensemble predictions and information theory

The applicability of information theory for weather or climate prediction has been studied previously by Carnevale and Holloway (1982), Schneider and Griffies (1999), Roulston and Smith (2002), and Leung and North (1990). Recently, as discussed earlier in this chapter, Kleeman (2002) has suggested the relative entropy as an estimate of predictive utility in an ensemble forecast relative to the climatological record, as well as the Gaussian signal-dispersion decomposition. Majda, Kleeman, and Cai (2004) developed a more sophisticated framework of predictability through relative entropy for non-Gaussian probability density functions, which includes a hierarchy of rigorous lower bounds on relative entropy through the statistical moments beyond the mean and covariance through maximum entropy optimization as illustrated in Subsection 6.4.5 and Section 15.2 in simple cases. Abramov and Majda (2004) converted the non-Gaussian predictability framework into a practical tool through the hierarchy of lower bounds and a rapid numerical optimization algorithm. Recently, Cai, Haven, and Majda (2004) exhaustively tested several facets of the non-Gaussian information theoretic predictability framework in a simple chaotic mapping model with an explicit attractor ranging from Gaussian to fractal as parameters are varied. Kleeman and Majda (2004) have quantified the loss of information in coarse-grained ensemble estimators and applied these ideas to geophysical turbulence. Different applications of relative entropy as a predictability tool were developed in Abramov and Majda (2004); besides a straightforward measure of lack of information in the climate relative to the prediction ensemble, the relative entropy can be used in estimating lack of information in a forecast ensemble relative to the actual events (Roulston and Smith, 2002), in evaluating additional information content in the skewness and higher-order forecast moments (non-Gaussianity) as illustrated in Section 15.3, and the information flow between different subsets of phase space in an ensemble forecast (statistical correlation between different large-scale phenomena). In Abramov and Majda (2004) all of these facets were demonstrated for the Lorenz '96 model (Lorenz and Emanuel, 1998), including highly non-Gaussian behavior and also the use of principal predictability decomposition in Subsection 15.2.3. Finally, Abramov, Majda, and Kleeman (2005) successfully applied the relative entropy framework to the simplest midlatitude atmospheric climate model, barotropic T21 spherical truncation with realistic orography in two different dynamical regimes, with each regime mimicking the behavior of the atmosphere at a certain height. In particular, information flow was found to be responsible for correlated switches in large-scale structures like the Arctic Oscillation, North Atlantic Oscillation, and Pacific/North American pattern. All of the above work demonstrates many practical facets of quantifying uncertainty in ensemble forecasts through the relative entropy; however, all of the work in idealized settings described above utilized large ensemble sizes. The work of Haven *et al.* (2004) addresses the central issue of quantifying the uncertainty of an ensemble forecast with limited ensemble size in the non-Gaussian framework of information theory and relative entropy.

Perhaps, the most sophisticated contemporary uses of ensemble predictions in complex systems with small ensemble size occur in weather and climate predictions. Studies of predictions with forecast ensembles were performed, among others, by Anderson and Stern (1996), Toth and Kalnay (1993), and Palmer (2000). Practice shows that with current computational facilities, in order to perform real-time forecasts with large atmospheric GCMs, the size of the prediction ensemble has to be small, about ten to 50 members, depending on spatial resolution, with the possibility of 50 to 100 member ensembles in the near future. Due to limited forecast ensemble size, certain complications arise concerning the credibility of information provided by such ensemble. The common strategy of dealing with small forecast ensembles is to maximize the information provided by the limited sample size, via generating a forecast ensemble in a very specific way. In particular, Ehrendorfer and Tribbia (1997) show that, for correct error growth reconstruction, the fastest-growing directions of the phase space have to be sampled. Two efficient methods of ensemble generation are usually used in practical ensemble forecasts: local Lyapunov vectors (Toth and Kalnay, 1993; Kalnay, 2003) and singular vectors (Palmer *et al.*, 1993; Reynolds and Palmer, 1998). The efficiency of prediction depending on ensemble size has been studied previously by Buizza and Palmer (1998). The book of Kalnay is strongly recommended for more reading on these topics.

References

Abramov, R. and Majda, A. (2004), Quantifying uncertainty for non-Gaussian ensembles in complex systems. *SIAM J. Sci. Stat.* Comp. **26**(2), 411–447.

Abramov, R., Kovacic, G., and Majda, A. (2003), Hamiltonian structure and statistically relevant conserved quantities for the truncated Burgers-Hopf equation. *Comm. Pure Appl. Math.* **56**(1), 1–46.

Abramov, R., Majda, A., and Kleeman, R. (2005), Information theory and predictability for low frequency varaibility. *J. Atmos. Sci.* **62**(1), 65–87.

Anderson, J. and Stern, W. (1996), Evaluating the potential predictive utility of ensemble forecasts. *J. Clim.* **9**, 260–269.

Buizza, R. and Palmer, T. (1998), Impact of ensemble size on ensemble prediction. *Mon. Wea. Rev.* **126**, 2503–2518.

Cai, D., Haven, K., and Majda, A. (2004), Quantifying predictability in a simple model with complex features. *Stoch. Dynam.* **4**(4), 547–569.

Carnevale, G. and Holloway, G. (1982), Information decay and the predictability of turbulent flows. *J. Fluid Mech.* **116**, 115–121.

Cover, T.M. and Thomas, J.A. (1991), *Elements of Information Theory*. New York: Wiley.

Ehrendorfer, M. and Tribbia, J. (1997), Optimal prediction of forecast error covariances through singular vectors. *J. Atmos. Sci.* **54**, 286–313.

Haven, K., Majda, A., and Abramov, R. (2004), Predictability through information theory: small sample estimation in a non-Gaussian framework. Submitted to *J. Comp. Phys.* **206**, 334–362.

Kalnay, E. (2003), *Atmospheric Modeling, Data Assimilation and Predictability*. New York: Cambridge University Press.

Kleeman, R. (2002), Measuring dynamical prediction utility using relative entropy. *J. Atmos. Sci.* **59**, 2057–2072.

Kleeman, R. and Majda, A. J. (2004), Predictability in a model of geophysical turbulence. *J. Atmos. Sci.*

Kleeman, R., Majda, A. J. and Timofeyev, I. (2002), Quantifying predictability in a model with statistical features of the atmosphere. *Proc. Natl. Acad. Sci. USA* **99**(24), 15291–15296.

Kleeman, R. and Moore, A. (1997), A theory for the limitation of ENSO predictability due to stochastic atmospheric transients. *J. Atmos. Sci.* **54**, 753–767.

Kleeman, R. and Moore, A. (1999), A new method for determining the reliability of dynamical ENSO predictions. *Mon. Wea. Rev.* **127**, 694–705.

Lasota, A. and Mackey, M. (1994), Chaos, fractals, and noise. *Applied Math. Sciences* **97**, New York: Springer-Verlag.

Lax, P. D. (1996), *Linear Algebra*. New York: Wiley.

Leung, L.-Y. and North, G. (1990), Information theory and climate prediction. *J. Clim.* **3**, 5–14.

Lorenz, E. and Emanuel, K. (1998), Optimal sites for supplementary weather observations. *J. Atmos. Sci.* **55**, 399–414.

Majda, A. J., Kleeman, R. and Cai, D. (2002), A mathematical framework for quantifying predictability through relative entropy. *Methods and Applications of Analysis* **9**(3), 425–444.

Majda, A. J. and Timofeyev, I. (2000), Remarkable statistical behavior for truncated Burgers–Hopf dynamics. *Proc. Nat. Acad. Sci.* **97**(23), 12413–12417.

Majda, A. J., Timofeyev, I. and Vanden Eijinden, E. (2001), A mathematical framework for stochastic climate model. *Comm. Pure Appl. Math.* **54**(8), 891–974.

Palmer, T., 2000, Predicting Uncertainty in Forecasts of Weather and Climate, Rep. Prog. Phys, 63, 71–116, 2000.

Palmer, T., Molteni, F., Mureau, R., Buizza, R., Chapelet, P. and Tribbia, J. (1993), Ensemble prediction, Proc. validation of models over Europe, vol. 1, 21–66.

Reynolds, C. and Palmer, T. (1998), Decaying singular vectors and their impact on analysis and forecast correction. *J. Atmos. Sci.* **55**, 2576–2596.

Roulston, M. and Smith, L. (2002), Evaluating probabilistic forecasts using information theory. *Mon. Wea. Rev.* **130**, 1653–1660.

Schneider, T. and Griffies (1999), A conceptual framework for predictability studies. *J. Clim.* **12**, 3133–3155.

Toth, Z. and Kalnay, E. (1993), Operational ensemble prediction at the national meteorological center: practical aspects, *Bull. Amer. Meteor. Soc.* **74**, 2317–2330.

16

Barotropic quasi-geostrophic equations on the sphere

16.1 Introduction

It is apparent that the earth is not flat, at least not on large scale. However, we have completely neglected the spherical geometry of the earth except the beta-plane approximation in the study of geophysical flows in the last 15 chapters. In this chapter, we will focus on geophysical flows on the sphere. In particular, we are interested in clarifying those properties that are parallel to the flat geometry case and those properties that are unique to the spherical geometry. The study of geophysical flows on the sphere is of great importance due to the geometry of the earth and other planets. We will see that most of the phenomena on the sphere have their counterparts for the flat geometry. However, there are certain features unique to the sphere due to the special geometry.

Unlike the periodic or channel (flat) geometry case, there are several useful and natural coordinate systems on the unit sphere S^2. We will utilize two coordinate systems in this chapter. The first coordinate system is an intrinsic coordinate system, the surface spherical coordinates

$$(\phi, \theta), \ 0 \leq \phi < 2\pi, \ -\frac{\pi}{2} < \theta < \frac{\pi}{2}, \tag{16.1}$$

with ϕ being the longitude and θ being the latitude (see Figure 16.1). The second coordinate system is an extrinsic one with the latitudal variable θ replaced by $z = \sin\theta$

$$(\phi, z), 0 \leq \phi < 2\pi, \ -1 < z < 1. \tag{16.2}$$

The two coordinate systems are related through the following relation

$$z = \sin\theta, \quad -1 < z < 1, \quad \frac{\partial}{\partial z} = \frac{1}{\cos\theta}\frac{\partial}{\partial\theta}, \tag{16.3}$$

and we will work back and forth between the two coordinate systems depending on the convenience.

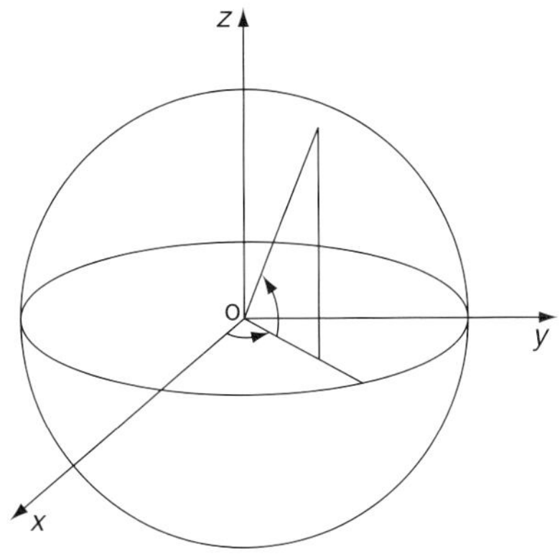

Figure 16.1 Spherical coordinates.

We first recall the basic dynamics equations, the **barotropic quasi-geostrophic equations on the unit sphere** S^2

$$\begin{aligned} &\frac{\partial q}{\partial t} + J(\psi, q) = \mathcal{D}(\Delta)\psi + \mathcal{F}, \\ &q = \Delta\psi + 2\Omega \sin\theta + h = \Delta\psi + 2\Omega z + h, \end{aligned} \tag{16.4}$$

where q is the potential vorticity with $2\Omega \sin\theta = 2\Omega z$ corresponding to the Coriolis force and Ω being the angular velocity of the sphere, h is the bottom topography, ψ is the stream function, Δ is the Laplace operator on scalars on the unit sphere defined as

$$\begin{aligned} \Delta\psi &= \frac{1}{\cos\theta}\left\{\frac{1}{\cos\theta}\frac{\partial^2\psi}{\partial\phi^2} + \frac{\partial}{\partial\theta}\left(\cos\theta\frac{\partial\psi}{\partial\theta}\right)\right\} \\ &= \frac{1}{1-z^2}\frac{\partial^2\psi}{\partial\phi^2} + \frac{\partial}{\partial z}\left((1-z^2)\frac{\partial\psi}{\partial z}\right), \end{aligned} \tag{16.5}$$

the Jacobian J is defined as

$$J(f, g) = \frac{1}{\cos\theta}\left(\frac{\partial f}{\partial\phi}\frac{\partial g}{\partial\theta} - \frac{\partial f}{\partial\theta}\frac{\partial g}{\partial\phi}\right) = \frac{\partial f}{\partial\phi}\frac{\partial g}{\partial z} - \frac{\partial f}{\partial z}\frac{\partial g}{\partial\phi} = \nabla^{\perp} f \cdot \nabla g, \tag{16.6}$$

the dissipation operator $\mathcal{D}$ takes the form

$$\mathcal{D}(\Delta) = \sum_{j=1}^{k} d_j(-\Delta)^j, \quad d_j \geq 0, \tag{16.7}$$

and $\mathcal{F}$ represents external forcing. The interested reader may find a derivation of this system from a primitive system in many classical treatises on geophysical fluid dynamics, such as Pedlosky (1987).

Notice that there is no boundary condition needed for the Laplacian, since the sphere has no boundary. The reader is referred to the next subsection for a quick summary of various differential operators and integration by parts formulas on the sphere.

It is easy to see that the mean of the stream function $\int_{S^2}\psi$ does not play any role in the dynamics just as in the flat geometry with periodic boundary condition (see Section 1.1). Hence without loss of generality we may assume that the stream function has zero mean, i.e.

$$\int_{S^2}\psi = 0. \tag{16.8}$$

Thus we may recover the stream function from the relative vorticity ω or the potential vorticity q by solving the Laplace equation

$$\Delta\psi = \omega = q - 2\Omega\sin\theta - h,$$

together with the zero mean condition (16.8) on ψ.

We observe that the barotropic quasi-geostrophic equation on the sphere (16.4) in the (ϕ, z) coordinates has the same structure as that in the flat geometry case with ϕ playing the role of x and z playing the role of y (see equation (1.1)); the Fourier series which corresponds to eigenfunction expansion for the Laplacian with flat geometry and periodic boundary condition should be replaced by the eigenfunction expansion for the operator Δ on the sphere, which are associated with the Legendre functions as we shall introduce below. Thus we expect resemblance of solutions for the two different cases. However due to the difference in symmetry groups, we will also witness differences in solution behaviors.

The chapter is organized as follows.

In the remaining part of the first section we recall several useful differential operators and integration by parts formula in this spherical geometry. We also recall the Legendre functions and their relation to the surface spherical harmonics as well as their relation to the eigenfunctions of the Laplace operator on the sphere.

In Section 2 we study conserved quantities under various conditions, introduce some special exact solutions, and study their stabilities. This is parallel to Chapters 1 and 4. We discover that both energy and all generalized potential enstrophies are conserved quantities and the subtle issue related to the beta-plane disappear here, since the term corresponding to the Coriolis forcing is a smooth function in this spherical geometry case, in comparison to the flat geometry case, where the beta plane effect is not a smooth function (not periodic in the longitude direction, see Subsection 1.3.4). In the special case of topography living on the ground energy shell, we discover that the energy and enstrophy of the ground state modes are conserved. This further implies that the energy and enstrophy of the high modes (modes on the second energy shell or higher) are conserved. However high moments of the high modes are not conserved in general. The special exact solutions that we introduced in Section 2 include those simple steady state solutions with a linear potential vorticity–stream function relation, or exact solution with generalized Kolmogorov forcing (see Section 1.2 for the counter part in flat geometry). We also discover an interesting fact unique to spherical geometry that the ground state modes have their independent (of the high modes) linear dynamics, assuming the topography lives on the ground energy shell. This independent dynamics of the ground modes has implications on the non-linear stability, response to large-scale forcing, selective decay and statistical theories. The ground state modes and any other energy shell form an invariant subspace of the non-linear dynamics provided the external forcing lives on the ground energy shell and that chosen higher energy shell as well, and the topography lives on the ground energy shell. The motion on the ground energy shell is non-linearly stable if we have ground state topography. This is in comparison to the marginal stability of ground shell motion in the flat geometry case (see Section 4.2). The motion on the first two shells are marginally stable and unstable in general. Stability of steady states ranging from simple ones induced by the ansatz of the linear relation between the potential vorticity and the streamfunction to more general steady states induced by more general functional relationship between the potential vorticity and the streamfunction are discussed following the same methodology as Chapter 4.

Then in Section 3 we study response to large-scale forcing. Instead of allowing the largest scale (ground state modes) external forcing only as in the flat geometry case (see Chapter 2), we can allow the external forcing acting on both the ground energy shell and the second energy shell. This is due to the independent dynamics of the ground state modes. This provides an interesting case of having arbitrary large amplitude external forcing with some variation in structure (two energy shells), but having very regular motion in the long run.

In Section 4, we study selective decay on the sphere. The result and technique here are parallel to those for the flat geometry case (see Chapter 3). However, we are able to explicitly characterize those initial data whose Dirichlet quotient tends to the first eigenvalue of the Laplacian in this spherical case, due to the independent dynamics of the ground states. Selective decay of the high modes and selective decay with symmetry are also discussed.

In Section 5 we apply the empirical energy–enstrophy statistical theory and the complete statistical mechanics to the barotropic quasi-geostrophic equations on the sphere (16.1). This is parallel to Chapters 6, 7, and 8 for the flat geometry case. The predicted mean fields of these statistical theories are steady states with linear potential vorticity-stream function relation whose stability are studied in Subsection 16.2.3. In the case with ground state topography, we have to consider the statistics of the high modes (away from the ground state modes), since the ground state modes do not possess enough mixing. The statistical theory is applicable since the energy and enstrophy for the high modes are conserved as we have established in Subsection 16.2.2. The predictions of the statistical theories here resemble the predictions of the statistical theories in the flat geometry case without geophysical effects.

In the last section, i.e. Section 6, we apply other statistical theories to the barotropic quasi-geostrophic flows on the sphere. The statistical theories include the energy–circulation theory, the point-vortex theory, and the empirical statistical theories with infinitely many constraints. This is parallel to Chapter 9. In the energy–circulation theory and point-vortex theory we need to impose odd symmetry, since we need to utilize the circulation constraint which is trivial in the whole sphere case. For the energy–circulation theory we also need to consider the statistics of the high modes in the case of ground state topography, due to the trivial dynamics of the ground state modes. The predictions are similar to the flat geometry case without topography. For the point-vortex theory, we can only consider the statistics of the high modes, since the dynamics of the ground states are trivial under the assumption of no geophysical effects. The empirical statistical theory with many constraints (ESTMC) can only be applied in the general case, not to the high modes, since the high moments of the high modes are not necessarily conserved even in the case with ground state topography.

In Appendix 1, we elaborate on the invariant dynamics of the first two energy shells. Numerical evidence suggests that the motion on the first two energy shells could be complicated and might include quasi-periodic motion. Appendix 2 deals with a technical detail related to the proof of the response to large-scale forcing in Section 16.3.

Most of the results here are parallel to the flat geometry case, and the techniques are similar too. We often refer the reader to the corresponding material in the flat geometry case for more details. In a sense, this chapter can be viewed as a review of the basic theories, both dynamics and statistics, for barotropic quasi-geostrophic flows, with an added twist of the spherical geometry.

16.1.1 Common differential operators and integration by parts formulas on the sphere

For the sake of convenience, we list here some of the commonly used differential operators and integration by parts formulas on the sphere. These are standard materials and can be found in any standard textbook on differential geometry.

The surface element on a sphere with radius a is

$$ds = a^2 \cos\theta \, d\phi d\theta = a \, d\phi dz, \quad (z = a \sin\theta). \tag{16.9}$$

The L^2 inner product of two functions f, g on the unit sphere (radius equal to 1) is defined as

$$\begin{aligned}(f, g)_{L^2(S^2)} &= \fint_{S^2} f(\mathbf{s})\, g(\mathbf{s})\, d\mathbf{s} = \frac{1}{4\pi} \int_{-\pi/2}^{\pi/2} \int_0^{2\pi} f\, g \cos\theta d\phi d\theta \\ &= \frac{1}{4\pi} \int_{-1}^{1} \int_0^{2\pi} f\, g\, d\phi dz.\end{aligned} \tag{16.10}$$

At each point of the space (outside $\theta = \pm\pi/2$ or $z = \pm 1$), we define the usual local orthonormal system, $\{\vec{e}_\phi, \vec{e}_\theta\}$, corresponding to increasing values of ϕ and θ. We define $\vec{n} = \vec{e}_r$ as the unit vector corresponding to the increasing values of the radius.

Let ψ be a scalar field on S^2 and $\vec{v} = v_\phi \vec{e}_\phi + v_\theta \vec{e}_\theta$ be a tangent vector field on S^2. We recall several common differential operators and integration by parts formulas on the sphere.

The **gradient operator** on a **scalar field** is defined as

$$\begin{aligned}\nabla\psi = \operatorname{grad}\psi &= \frac{1}{a\cos\theta} \frac{\partial\psi}{\partial\phi} \vec{e}_\phi + \frac{1}{a} \frac{\partial\psi}{\partial\theta} \vec{e}_\theta \\ &= \frac{1}{\sqrt{a^2 - z^2}} \frac{\partial\psi}{\partial\phi} \vec{e}_\phi + \frac{\sqrt{a^2 - z^2}}{a^2} \frac{\partial\psi}{\partial z} \vec{e}_\theta.\end{aligned} \tag{16.11}$$

The **perpendicular gradient operator**, or the **curl operator** on a **scalar field** is defined as

$$\begin{aligned}\nabla^\perp\psi = \operatorname{curl}\psi = \nabla\psi \times \vec{n} &= \frac{1}{a} \frac{\partial\psi}{\partial\theta} \vec{e}_\phi - \frac{1}{a\cos\theta} \frac{\partial\psi}{\partial\phi} \vec{e}_\theta \\ &= \frac{\sqrt{a^2 - z^2}}{a^2} \frac{\partial\psi}{\partial z} \vec{e}_\phi - \frac{1}{\sqrt{a^2 - z^2}} \frac{\partial\psi}{\partial\phi} \vec{e}_\theta.\end{aligned} \tag{16.12}$$

The **divergence operator** on a **vector field** is defined as

$$\begin{aligned}\operatorname{div}\vec{v} = \nabla\cdot\vec{v} &= \frac{1}{a\cos\theta} \frac{\partial v_\phi}{\partial\phi} + \frac{1}{a\cos\theta} \frac{\partial(v_\theta \cos\theta)}{\partial\theta} \\ &= \frac{1}{\sqrt{a^2 - z^2}} \frac{\partial v_\phi}{\partial\phi} + \frac{1}{a} \frac{\partial(v_\theta \sqrt{a^2 - z^2})}{\partial z}.\end{aligned} \tag{16.13}$$

The **curl** of a **vector field** is defined as

$$\begin{aligned}\text{curl } \vec{v} = \text{div}\,(\vec{v} \times \vec{n}) &= \frac{1}{a\cos\theta}\frac{\partial v_\theta}{\partial\phi} - \frac{1}{a\cos\theta}\frac{\partial(v_\phi \cos\theta)}{\partial\theta} \\ &= \frac{1}{\sqrt{a^2-z^2}}\frac{\partial v_\theta}{\partial\phi} - \frac{1}{a}\frac{\partial(v_\phi\sqrt{a^2-z^2})}{\partial z}.\end{aligned} \tag{16.14}$$

The **Laplacian operator** on a **scalar field** is defined as

$$\begin{aligned}\Delta\psi &= \text{div grad } \psi = -\text{curl curl } \psi \\ &= \frac{1}{a^2\cos\theta}\left\{\frac{1}{\cos\theta}\frac{\partial^2\psi}{\partial\phi^2} + \frac{\partial}{\partial\theta}\left(\cos\theta\frac{\partial\psi}{\partial\theta}\right)\right\} \\ &= \frac{1}{a^2-z^2}\frac{\partial^2\psi}{\partial\phi^2} + \frac{\partial}{\partial z}\left((a^2-z^2)\frac{\partial\psi}{\partial z}\right).\end{aligned} \tag{16.15}$$

We recall here one of the possible definitions of $\Delta\vec{v}$, i.e. the **Laplace–Beltrami operator** for **tangential vector fields** on a sphere with radius a

$$\begin{aligned}\Delta\vec{v} &= \text{grad div } \vec{v} - \text{curl curl}\vec{v} \\ &= \left\{\Delta v_\phi - \frac{2\sin\theta}{a^2\cos^2\theta}\frac{\partial v_\theta}{\partial\phi} - \frac{v_\phi}{a^2\cos^2\theta}\right\}\vec{e}_\phi \\ &\quad + \left\{\Delta v_\theta + \frac{2\sin\theta}{a^2\cos^2\theta}\frac{\partial v_\phi}{\partial\phi} - \frac{v_\theta}{a^2\cos^2\theta}\right\}\vec{e}_\theta \\ &= \left\{\Delta v_\phi - \frac{2z}{a(a^2-z^2)}\frac{\partial v_\theta}{\partial\phi} - \frac{v_\phi}{a^2-z^2}\right\}\vec{e}_\phi \\ &\quad + \left\{\Delta v_\theta + \frac{2z}{a(a^2-z^2)}\frac{\partial v_\phi}{\partial\phi} - \frac{v_\theta}{a^2-z^2}\right\}\vec{e}_\theta.\end{aligned} \tag{16.16}$$

We also recall the **Stokes formulas**

$$\begin{aligned}&\int_{S^2} \text{div } \vec{v}\ \psi d\mathbf{s} = -\int_{S^2} \vec{v}\cdot\nabla\psi d\mathbf{s}, \\ &\int_{S^2} \text{curl } \vec{v}\ \psi d\mathbf{s} = \int_{S^2} \vec{v}\cdot\text{curl } \psi d\mathbf{s}, \\ &-\int_{S^2} \Delta\psi\tilde{\psi} d\mathbf{s} = -\int_{S^2} \psi\Delta\tilde{\psi} d\mathbf{s} = \int_{S^2} \nabla\psi\cdot\nabla\tilde{\psi} d\mathbf{s}, \\ &-\int_{S^2} \Delta\vec{v}\cdot\tilde{\vec{v}} d\mathbf{s} = \int_{S^2} (\nabla\vec{v}\cdot\nabla\tilde{\vec{v}} + \frac{1}{a^2}\vec{v}\cdot\tilde{\vec{v}}) d\mathbf{s}.\end{aligned} \tag{16.17}$$

Lastly, we would like to emphasize that both the intrinsic coordinate system (ϕ, θ) and the extrinsic coordinate system (ϕ, z) have singularity at the poles, $\theta = \pm\pi/2$ or $z = \pm 1$. Thus, differentiable functions on the sphere are not mere differentiable functions in terms of coordinates. For instance the function $\sin\phi$ is a smooth function of the variables in either of the coordinate systems. However, this is not a smooth function on the sphere, since it has irremovable singularities at the poles.

16.1.2 Eigenfunctions of the Laplace operator and the Legendre functions

As in the flat geometry case, the Laplace operator will play an important role in studying the behavior of solutions to the barotropic quasi-geostrophic equations on the unit sphere (16.4). For the doubly periodic domain we have witnessed that the eigenfunctions of the Laplacian operator, the trigonometric functions, played an essential role in our analysis of the behavior of solutions. Applying the technique of separation of variables to the Laplacian operator in the (ϕ, z) coordinates we may derive that the **eigenvalues of the Laplacian** are the set

$$\Omega = \left\{-n(n+1) = -\Lambda_n \quad | \quad n \in \mathcal{Z}^+\right\} \tag{16.18}$$

and for each eigenvalue $-n(n+1) = -\Lambda_n$ the associated **eigenfunctions** are **surface spherical harmonics** given by

$$w_{nm} = P_{nm}(\sin\theta)\exp(im\phi) = P_{nm}(z)\exp(im\phi), \quad \text{for } 0 \le |m| \le n, \tag{16.19}$$

with the convention that

$$P_{n,-m} = P_{n,m}. \tag{16.20}$$

The P_{nm} are the so-called **Legendre functions** and are given by

$$P_{nm}(z) = \frac{(1-z^2)^{\frac{m}{2}}}{2^n n!}\frac{d^{m+n}(z^2-1)^n}{dz^{n+m}}, \quad \text{for } 0 \le m \le n. \tag{16.21}$$

These functions enjoy the following *orthogonal property*:

$$\int_{-1}^{1} P_{nm}(z)P_{n'm}(z)dz = \frac{2}{2n+1}\frac{(n+m)!}{(n-m)!}\delta_{nn'}. \tag{16.22}$$

The interested reader is referred to Courant and Hilbert (1962) for a detailed discussion.

These eigenfunctions are *mutually orthogonal* with respect to the inner product defined in (16.10), and play the role of trigonometric functions in the flat geometry case. The normalized form of these eigenfunctions are given by

$$Y_{nm} = N_{nm}P_{nm}(\sin\theta)\exp(im\phi) = N_{nm}P_{nm}(z)\exp(im\phi), \tag{16.23}$$

where

$$N_{nm} = \left(\frac{(2n+1)(n-|m|)!}{(n+|m|)!}\right)^{\frac{1}{2}}, \tag{16.24}$$

or their real-valued form (not normalized)

$$\begin{aligned} w_n &= P_n(\sin\theta) = P_n(z), \\ w_{anm} &= P_{nm}(\sin\theta)\cos(m\phi) = P_{nm}(z)\cos(m\phi), \\ w_{bnm} &= P_{nm}(\sin\theta)\sin(m\phi) = P_{nm}(z)\sin(m\phi) \end{aligned} \tag{16.25}$$

for $1 \leq m \leq n$.

These *surface spherical harmonics* form a complete orthogonal basis for the function space of square integrable functions on the sphere ($L^2(S^2)$). Their gradients are mutually orthogonal as well just, as in the flat geometry case with the Fourier series. Roughly speaking, these functions play the same role in the spherical geometry as the trigonometric functions in the flat geometry.

16.2 Exact solutions, conserved quantities, and non-linear stability

Parallel to the flat geometry case, conserved quantities and special exact solutions as well as their non-linear stabilities are central to our understanding of the basic dynamics and statistics of the barotropic quasi-geostrophic flows. The purpose of this section is to study conserved quantities, special exact solutions and their non-linear stabilities to the barotropic quasi-geostrophic equations on the sphere (16.4). The techniques and most of the results are similar to the flat geometry case. However, we also have distinctive features associated with the spherical geometry as we shall demonstrate below.

16.2.1 Some special exact solutions

We first derive several exact solutions for the barotropic quasi-geostrophic equations (16.4) on the sphere. Parallel to the flat geometry case (see Section 1.2 from Chapter 1), we have simple steady state solutions, where the potential vorticity is a function of the stream function, and exact solutions with generalized Kolmogorov forcing. Due to the special spherical geometry, we also have special exact solutions that have no obvious analogy in the flat geometry case. For instance, the dynamics of the ground state modes will be independent of the dynamics of the higher modes, if the topography lives on the ground energy shell. This also has implications for the conserved quantities and non-linear stability to be discussed later in the section and statistical theories to be discussed later in the chapter. We will also discover that the space spanned by the ground energy shell and another arbitrary energy shell will be invariant under the quasi-geostrophic dynamics provided that the topography lives on the ground energy shell. This is unique to the spherical geometry.

Independent linear dynamics on the ground energy shell

An interesting phenomenon in the spherical geometry case is that the dynamics of the ground state modes are independent of motions of the higher modes at least in the case when there is no topography (or more generally when the topography lives on the ground energy shell). This means that the higher modes have no influence on the motion of the ground modes even though the ground state modes do have impact on the dynamics of the high modes. This fact has no analogy in the flat geometry case and it has consequences for the conserved quantities, on the non-linear stability problem in the marginal case (see Subsection 16.2.3 below) and the statistical theories.

To derive this property we first notice that there is a simple correspondence between the Cartesian coordinates and the canonical base of ground energy shell

$$z = \sin\theta,\ \ x = \cos\theta\cos\phi,\ \ y = \cos\theta\sin\phi, \tag{16.26}$$

and hence

$$\int_{S^2} J(\psi, \Delta\psi)z = -\int_{S^2} J(\psi, z)\Delta\psi = -\int_{S^2} \frac{\partial\psi}{\partial\phi}\Delta\psi = 0. \tag{16.27}$$

This implies, by rotation invariance of the Jacobian and the Laplacian

$$\begin{aligned} \int_{S^2} J(\psi, \Delta\psi)x &= \int_{S^2} J(\psi, \Delta\psi)\cos\theta\cos\phi = 0, \\ \int_{S^2} J(\psi, \Delta\psi)y &= \int_{S^2} J(\psi, \Delta\psi)\cos\theta\sin\phi = 0. \end{aligned} \tag{16.28}$$

This further implies

$$P_1 J(\psi, \Delta\psi) = 0, \tag{16.29}$$

where P_1 is the orthogonal projection from $L^2(S^2)$ on to the ground energy shell

$$W_1 = \operatorname{span}\{x, y, z\} = \operatorname{span}\{\sin\theta, \cos\theta\cos\phi, \cos\theta\sin\phi\}, \tag{16.30}$$

and P_1 is given explicitly as

$$P_1\psi = 3\left(⨍_{S^2} \psi x\right)x + 3\left(⨍_{S^2} \psi y\right)y + 3\left(⨍_{S^2} \psi z\right)z. \tag{16.31}$$

We also notice that

$$P_1 J(\psi, z) = P_1\frac{\partial\psi}{\partial\phi} = \frac{\partial P_1\psi}{\partial\phi} = J(P_1\psi, z), \tag{16.32}$$

and by symmetry

$$P_1 J(\psi, x) = J(P_1\psi, x), \quad P_1 J(\psi, y) = J(P_1\psi, y). \tag{16.33}$$

Thus, after applying P_1 to the barotropic quasi-geostrophic equations (16.4) on the sphere we deduce

$$-2\frac{\partial P_1\psi}{\partial t}+J(P_1\psi, 2\Omega z)+P_1J(\psi, h)=\mathcal{D}(-2)P_1\psi+P_1\mathcal{F}. \qquad (16.34)$$

In the special case of the topography, h, living on the ground energy shell $W_1 = \text{span}\{x, y, z\}$ we have, utilizing (16.32)–(16.33)

$$P_1J(\psi, h)=J(P_1\psi, h), \quad \text{if } P_1h=h. \qquad (16.35)$$

Thus we further deduce, after combining with (16.34), the following
Dynamics of the ground state modes when $h\in W_1$

$$-2\frac{\partial P_1\psi}{\partial t}+J(P_1\psi, 2\Omega z+h)=\mathcal{D}(-2)P_1\psi+P_1\mathcal{F}, \text{ if } h\in W_1. \qquad (16.36)$$

We then observe that the *the dynamics on the ground energy shell is linear and independent of motions of other modes* provided that topography lives on the ground energy shell. This is in contrast to the flat geometry case where we have either the case that the large-scale mean flow is driven by topographic stress and hence influenced by motion of high modes, or the case that the large-scale mean flow is a constant in the case of a zonal topography (see Subsection 1.3.2).

The dynamics of the ground state modes can be represented in terms of the coefficients of the surface spherical harmonics. Indeed, if we denote a_1, a_{11}, b_{11} the coefficients of ψ for $w_1=\sin\theta=z$, $w_{a11}=\cos\theta\cos\phi=x$ and $w_{b11}= \cos\theta\sin\phi=y$ respectively, i.e.

$$\begin{aligned}\psi(\phi, \theta, t)&=a_1(t)\sin\theta+a_{11}(t)\cos\theta\cos\phi+b_{11}(t)\cos\theta\sin\phi+\psi'(\phi, \theta, t)\\&=a_1(t)z+a_{11}(t)x+b_{11}(t)y+\psi',\end{aligned} \qquad (16.37)$$

and the topography h is represented as

$$\begin{aligned}h&=h_1z+h_{a11}x+h_{b11}y \qquad (16.38)\\&=h_1\sin\theta+h_{a11}\cos\theta\cos\phi+h_{b11}\cos\theta\sin\phi,\end{aligned}$$

we have

$$\frac{d\vec{a}}{dt}=A\vec{a}-\frac{1}{2}(F_1, F_{a11}, F_{b11})^{tr}, \quad \vec{a}=(a_1, a_{11}, b_{11})^{tr}, \qquad (16.39)$$

where the matrix A is given by

$$A=\begin{pmatrix}-\mathcal{D}(-2)/2 & h_{b11}/2 & -h_{a11}/2\\ -h_{b11}/2 & -\mathcal{D}(-2)/2 & \Omega+h_1/2\\ h_{a11}/2 & -\Omega-h_1/2 & -\mathcal{D}(-2)/2\end{pmatrix}, \qquad (16.40)$$

where we have utilized the dynamics of the ground state modes given in (16.36) and the fact

$$
\begin{aligned}
P_1 J(\psi, z) &= P_1 \frac{\partial \psi}{\partial \phi} = -a_{11} y + b_{11} x, \\
P_1 J(\psi, x) &= J(P_1 \psi, x) = a_1 y - b_{11} z, \\
P_1 J(\psi, y) &= J(P_1 \psi, y) = -a_1 x + a_{11} z.
\end{aligned}
\tag{16.41}
$$

In the very special case of no topography ($h = 0$), no damping ($\mathcal{D} = 0$) and no forcing ($\mathcal{F} = 0$), we have

$$
A = \begin{pmatrix} 0 & 0 & 0 \\ 0 & 0 & \Omega \\ 0 & -\Omega & 0 \end{pmatrix}
\tag{16.42}
$$

and the solution takes the form

$$
\begin{aligned}
a_1(t) &= a_1(0), \\
a_{11}(t) &= \cos(\Omega t) a_{11}(0) + \sin(\Omega t) b_{11}(0), \\
b_{11}(t) &= -\sin(\Omega t) a_{11}(0) + \cos(\Omega t) b_{11}(0),
\end{aligned}
\tag{16.43}
$$

i.e. a solid body rotation plus two spherical Rossby waves with phase velocity $-\Omega$. (The two spherical Rossby waves can be combined into one.) Exact solutions on the nth energy shell will be a combination of a zonal flow plus $2n$ Rossby waves, as we shall see later in this section for exact solutions with Kolmogorov forcing. These exact solutions resemble those Rossby waves of the flat geometry case (see Section 1.2).

The general case (the case of $h \in W_1$) looks exactly like the case without topography ($h = 0$). In fact we notice that A is anti-symmetric and hence a normal matrix. This implies that A is diagonalizable with a (complex) unitary matrix. Notice the three eigenvalues of A are 0 and αi, and $-\alpha i$, where α is a real number. Thus we may reorganize the complex unitary matrix into a real unitary matrix U such that (see for instance Lax, 1997)

$$
U^T A U = \begin{pmatrix} 0 & 0 & 0 \\ 0 & 0 & \alpha \\ 0 & -\alpha & 0 \end{pmatrix}.
$$

We further notice this unitary matrix is a change of coordinates of R^3 (and hence on S^2 when restricted) via rotation and reflection. Thus, after we perform some rotation and reflections, the problem reduces to that with a new Coriolis parameter (α) and no topography.

If we need to figure out the exact influence of the topography (separate from the Coriolis force), it may still be worthwhile for us to work out the details for general A and carry out the dynamics on the first eight eigenfunctions (see subsections below).

Exact interesting dynamics of the ground state modes and another energy shell

In this subsection we assume that the topography lives on the ground energy shell, i.e. $h \in W_1$. It is known (see for instance Section 1.3) that each energy shell, i.e. all eigenfunctions associated with one fixed eigenvalue of the Laplace operator, is invariant under the barotropic quasi-geostrophic equation without forcing in the flat geometry. This is also true for the spherical geometry (see the next subsection). Moreover we are able to deduce more invariance due to the special sphere geometry of the sphere.

We observe from the computation above (see also below) that the interaction (through the non-linear advection term) of $z = \sin\theta$ with other eigenfunctions is a rotation of 90° about the z axis combined with a scalar multiplication. Similar results hold for $x = \cos\theta\cos\phi$ and $y = \cos\theta\sin\phi$. This is easily generalized to eigenfunctions corresponding to higher eigenvalues ($-\Lambda_n, n \geq 2$). And, if we consider the dynamics on the space spanned by all eigenfunctions corresponding to the eigenvalues $-\Lambda_1$ and $-\Lambda_n$ (for fixed n), we will see that the dynamics is an *exact nonlinear dynamics* of the quasi-geostrophic equations, if we have ground state topography. These exact dynamics exhibit *periodic* and possibly *quasi-periodic* motions. This phenomena is not present for the flat geometry, even if we make the analogy between the ground state modes in the spherical geometry case to the large mean motion in the flat geometry case. We also observe that we always have non-linear dynamics of the ground state modes plus another energy shell for the spherical geometry case, while we have non-linear dynamics for the flat geometry case only if there is non-trivial topography.

To derive the exact dynamics of the ground state modes and another energy shell, say the nth, we start with the following simple observation based on the explicit form of the Jacobian given in (16.6) and the explicit form of the surface spherical harmonics given by (16.19)

$$J(w_{nm}, z) = im\ w_{nm}, \tag{16.44}$$

Denoting W_n the nth energy shell, i.e.

$$W_n = \text{span}\{w_n, w_{anm}, w_{bnm},\ 1 \leq m \leq n\} = \text{span}\{w_{nm},\ 1 \leq |m| \leq n\}, \tag{16.45}$$

we have

$$J(\psi_n, z) \in W_n, \text{ if } \psi_n \in W_n. \tag{16.46}$$

Thanks to the rotation symmetry of the Jacobian and the nth energy shell W_n, the same result holds with z replaced by x or y, i.e.

$$J(\psi_n, x) \in W_n, \ \ J(\psi_n, y) \in W_n, \ \text{if} \ \psi_n \in W_n. \tag{16.47}$$

This implies

$$J(\psi_n, \psi_1) \in W_n, \ \text{if} \ \psi_n \in W_n, \ \ \psi_1 \in W_1. \tag{16.48}$$

Now let

$$\psi = \psi_1 + \psi_n \in W_1 + W_n, \ \text{with} \ \psi_1 \in W_1, \psi_n \in W_n, \tag{16.49}$$

we have

$$\begin{aligned} J(\psi, \Delta\psi) &= J(\psi_1 + \psi_n, -2\psi_1 - n(n+1)\psi_n) \\ &= -n(n+1)J(\psi_1, \psi_n) - 2J(\psi_n, \psi_1) \\ &= (n(n+1) - 2)J(\psi_n, \psi_1) \\ &\in W_n. \end{aligned} \tag{16.50}$$

This proves that the dynamics on $W_1 + W_n$ is an exact dynamics for the original equation (in the absence of topography). In particular, the dynamics of the two energy shells are given by:

Exact dynamics of the ground state modes and the nth energy shell with $h \in W_1$

$$\begin{aligned} &\frac{\partial \Delta\psi}{\partial t} + (n(n+1) - 2)J(P_n\psi, P_1\psi) + J(\psi, 2\Omega z + h) \\ &= \mathcal{D}(-2)P_1\psi + \mathcal{D}(-n(n+1))P_n\psi + \mathcal{F}, \\ &\text{if } \psi(0) = P_1\psi(0) + P_n\psi(0), \quad \mathcal{F} = P_1\mathcal{F} + P_n\mathcal{F}, \ \text{and } h \in W_1. \end{aligned} \tag{16.51}$$

In Appendix 1 we will elaborate on the special case of $n = 2$, i.e. the motion on the first eight eigenfunctions of the Laplace operator. We will see that the motion on the first two energy shells can be periodic and possibly quasi-periodic.

Exact solutions with generalized Kolmogorov forcing

Here we consider a special case of the previous subsection, namely we consider solutions of the barotropic quasi-geostrophic equations (16.4) living on one single energy shell. This is possible, i.e. one energy shell is invariant, provided we have zero topography ($h = 0$)(the case with ground state topography can be treated similarly with a rotation and reflection) and the external forcing term, $\mathcal{F}$, takes the form of a generalized Kolmogorov forcing, i.e.

$$\mathcal{F} = \sum_{|m| \le n} \hat{F}_{nm}(t) P_{nm}(\sin\theta) \exp(im\phi) = \sum_{|m| \le n} \hat{F}_{nm}(t) P_{nm}(z) \exp(im\phi). \tag{16.52}$$

The material here is analogous to Subsections 1.2.2 and 1.2.3 from Chapter 1.

Since the external forcing lives on the nth energy shell W_n, it makes sense to look for solution that lives on W_n as well. Now suppose that ψ has the following expansion in terms of the surface spherical harmonics (see Figure 16.2)

$$\psi = \sum_{|m|\le n} \hat{\psi}_{nm}(t) P_{nm}(\sin\theta)\exp(im\phi) = \sum_{|m|\le n} \hat{\psi}_{nm}(t)P_{nm}(z)\exp(im\phi). \quad (16.53)$$

We then have

$$\Delta\psi = -n(n+1)\psi$$

and hence

$$J(\psi, \Delta\psi) = 0.$$

It follows then that the coefficients satisfy the following equations

$$-n(n+1)\frac{\partial\hat{\psi}_{nm}}{\partial t} + 2\Omega mi\hat{\psi}_{nm} = \mathcal{D}(-n(n+1))\hat{\psi}_{nm} + \hat{F}_{nm}. \quad (16.54)$$

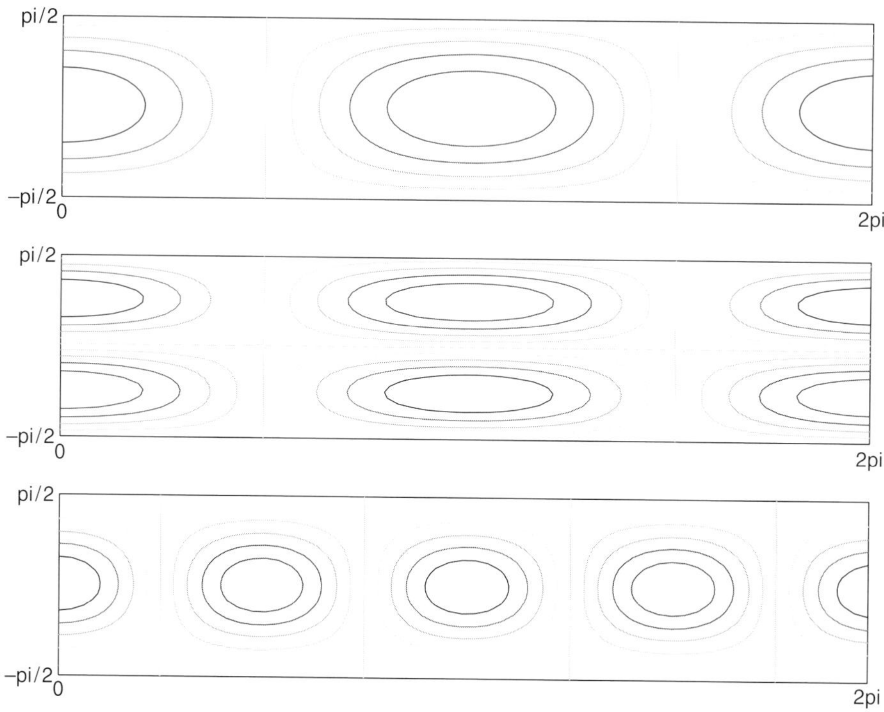

Figure 16.2 Surface spherical harmonics corresponding to spherical Rossby waves in moving frames. Top: $\cos\theta\cos\phi, n = m = 1$; middle: $\cos\theta\sin\theta\cos\phi, n = 2, m = 1$; bottom:$\cos^2\theta\cos(2\phi), n = m = 2$.

The solutions take the form

$$\hat{\psi}_{nm}(t) = \hat{\psi}_{nm}(0)e^{-\frac{\mathcal{D}(-n(n+1))-2m\Omega i}{n(n+1)}t} - \int_0^t e^{-\frac{\mathcal{D}(-n(n+1))-2m\Omega i}{n(n+1)}(t-s)} \frac{\hat{F}_{nm}(s)}{n(n+1)}\, ds. \quad (16.55)$$

In the special case of no external forcing ($\mathcal{F} = 0$) and no damping ($\mathcal{D} = 0$), the stream function takes the form

$$\psi = \hat{\psi}_{n0}(0)P_n(\sin\theta) + \sum_{1\le|m|\le n} \hat{\psi}_{nm}(0)P_{nm}(\sin\theta)\exp\left(im\left(\phi + \frac{2\Omega}{n(n+1)}t\right)\right). \quad (16.56)$$

It is then easy to recognize that the flow is the superposition of a zonal shear flow ($P_n(\sin\theta)$) and $2n$ spherical Rossby waves $\left(P_{nm}(\sin\theta)\exp\left(im\left(\phi+\frac{2\Omega}{n(n+1)}t\right)\right)\right)$ with phase velocity $\omega = -\frac{2\Omega}{n(n+1)}$. These spherical Rossby waves are dispersive, since waves with different wave numbers travel at different speeds. See Subsection 1.2.3 for discussion on dispersive waves.

Steady state solutions

Lastly, we consider steady state solutions of the barotropic quasi-geostrophic equations on the sphere (16.4) in the case of no damping ($\mathcal{D} = 0$), and no external forcing ($\mathcal{F} = 0$). Similar to the flat geometry case (Section 1.2 from Chapter 1), there are exact solutions to (16.62), where the potential vorticity q is a function of the stream function ψ, i.e.

$$q = g(\psi) \quad (16.57)$$

since the non-linear term (the Jacobian) vanishes when such a relation holds.

If the function g is a linear function, i.e.

$$q(= \Delta\psi + 2\Omega\sin\theta + h) = \mu\psi, \quad (16.58)$$

then the steady state solutions can be explicitly written as

$$\begin{aligned}
\psi =& \frac{2\Omega + \hat{h}_{10}}{2+\mu}\sin\theta + \sum_{m=1,-1} \frac{\hat{h}_{1m}}{2+\mu} P_{1m}(\sin\theta)\exp(im\phi) \\
&+ \sum_{n\ge 2}\sum_{|m|\le n} \frac{\hat{h}_{nm}}{n(n+1)+\mu} P_{nm}(\sin\theta)\exp(im\phi) \\
=& \frac{2\Omega + \hat{h}_{10}}{2+\mu} z + \sum_{m=1,-1} \frac{\hat{h}_{1m}}{2+\mu} P_{1m}(z)\exp(im\phi) \\
&+ \sum_{n\ge 2}\sum_{|m|\le n} \frac{\hat{h}_{nm}}{n(n+1)+\mu} P_{nm}(z)\exp(im\phi),
\end{aligned} \quad (16.59)$$

provided that $\mu \ne -n(n+1)$.

In the case of

$$\mu = -n(n+1), \tag{16.60}$$

such steady state exists only if

$$P_n(2\Omega \sin\theta + h) = P_n(2\Omega z + h) = 0, \tag{16.61}$$

where P_n is the projection on to

$$W_n = \text{span}\{P_{nm}(\sin\theta)\exp(im\phi)\} = \text{span}\{w_{nm}\}$$

by Fredholm's alternative.

16.2.2 Conserved quantities

Just as for the flat geometry case, conserved quantities are essential to the understanding and analysis of the barotropic quasi-geostrophic dynamics on the sphere. In general, the conserved quantities here are similar to the flat geometry case as those discussed in Section 1.3 with the difference that the subtlety related to the beta-plane effect disappears, since the Coriolis force here is a smooth function on the sphere. This, of course, implies that we always have infinitely many conserved quantities. However, for the case of zero topography ($h = 0$), or more generally topography living on the ground energy shell ($P_1 h = h$), we see from Subsection 16.2.1 that the dynamics of the ground state modes are independent of the higher modes. This implies that we have extra conserved quantities that are special to the spherical geometry.

The basic dynamics equation is the inviscid unforced barotropic quasi-geostrophic equations on the unit sphere

$$\frac{\partial q}{\partial t} + J(\psi, q) = 0, \quad q = \Delta\psi + 2\Omega z + h. \tag{16.62}$$

In the next two subsubsections we will discuss the conserved quantities of (16.62) in the general case and the extra conserved quantities in the special case of ground state topography as well as extra symmetry.

Conserved quantities with general topography

As we mentioned earlier, the situation here is similar to the flat geometry case. However the subtle issue related to the beta-plane is no longer present since the Coriolis force term $\Omega \sin\theta = \Omega z$ is a smooth function and the sphere has no boundary. It is easy to check that we have both the energy, E, and all the generalized enstrophy, $\mathcal{G}$ (see below), conserved in time.

Indeed let

$$E = \frac{1}{2} \fint_{S^2} |\nabla^\perp \psi|^2 = -\frac{1}{2} \fint_{S^2} \psi \Delta \psi \tag{16.63}$$

be the *kinetic energy*. It is easy to see that

$$\begin{aligned}\frac{dE}{dt} &= -\fint_{S^2} \psi \frac{d\Delta\psi}{dt} = \fint_{S^2} J(\psi, q)\psi = \fint_{S^2} \nabla^\perp \psi \cdot \nabla q \; \psi \\ &= \frac{1}{2} \fint_{S^2} \nabla^\perp (\psi^2) \cdot \nabla q = 0.\end{aligned} \tag{16.64}$$

Thus the energy is conserved.

Observe that the barotropic quasi-geostrophic equation can be written in the form

$$\frac{\partial q}{\partial t} + \nabla^\perp \psi \cdot \nabla q = 0.$$

Thus all *generalized potential enstrophy* $\mathcal{G}(q) = \fint_{S^2} G(q)$ are conserved for all sufficiently smooth G. When $G = \frac{1}{2} q^2$ we have that the *potential enstrophy* $\mathcal{E}_p = \frac{1}{2} \fint_{S^2} q^2$ is conserved. Just like the flat geometry case, we may check that the enstrophy $\mathcal{E} = \frac{1}{2} \fint_{S^2} (\Delta\psi)^2$ is not necessarily conserved in general.

For the spherical geometry, a quantity of apparent importance is the relative *angular momentum* parallel to the earth's axis of rotation. Since the radius is set to one, and $v_\phi = \frac{\partial\psi}{\partial\theta}$, the mean angular momentum is given by

$$M = \fint_{S^2} \cos\theta \frac{\partial \psi}{\partial \theta} \tag{16.65}$$

Simple integration by parts yields

$$M = 2 \fint_{S^2} \psi \sin\theta = 2 \fint_{S^2} \psi z = \frac{2}{3} a_1, \tag{16.66}$$

where a_1 is the coefficient of $z = \sin\theta$ in the surface spherical harmonics expansion of ψ, i.e.

$$\psi(\phi, \theta, t) = a_1(t) \sin\theta + \psi'(\phi, \theta, t), \quad \text{such that } \psi' \perp \sin\theta.$$

Observe that $z = \sin\theta$ corresponds to *solid body rotation* of the flow. Thus (16.66) implies that the relative angular momentum is linearly proportional to the rate of solid body rotation of the flow.

The natural question now is how the angular momentum or the solid body rotation evolves in time, i.e. the dynamics of a_1 or M. For this purpose we take the inner product of the inviscid unforced equation (16.62) with $-z = -\sin\theta$ and obtain, thanks to the orthogonality property of various surface spherical harmonics

$$\begin{aligned}\frac{2}{3}\frac{da_1}{dt} &= \int_{S^2} J(\psi, q)z = -\int_{S^2} J(\psi, z)\,q = -\int_{S^2}\frac{\partial\psi}{\partial\phi}(\Delta\psi + 2\Omega z + h)\\ &= \int_{S^2}\nabla\frac{\partial\psi}{\partial\phi}\cdot\nabla\psi - \int_{S^2}\frac{\partial\psi}{\partial\phi}h = -\int_{S^2}\frac{\partial\psi}{\partial\phi}h\\ &= \int_{S^2}\psi\frac{\partial h}{\partial\phi},\end{aligned}$$

and hence

$$\frac{da_1(t)}{dt} = \frac{3}{2}\int_{S^2}\psi\frac{\partial h}{\partial\phi}. \tag{16.67}$$

In terms of the angular momentum parallel to the earth's rotation axis, we have

$$\frac{dM}{dt} = \int_{S^2}\psi\frac{\partial h}{\partial\phi}. \tag{16.68}$$

This may be interpreted as the topographic stress driving the solid body rotation parallel to the earth's rotation axis, or the angular momentum. This resembles the case in the flat geometry, where the large-scale mean flow is driven by topographic stress. The difference here is that such a relationship comes naturally in the spherical geometry while it is a consequence of the imposed conservation of energy in the flat geometry (see Subsection 1.3.2).

When there is no topography, i.e. $h = 0$ (or more generally, topography independent of the longitude variable ϕ), we see that the *solid body rotation*, or the *angular momentum* parallel to the axis of rotation of the earth is conserved. The existence of extra conserved quantities in the spherical case is no surprise, since we have more symmetry here than the flat geometry case.

Conserved quantities with topography living on the ground energy shell

In this case of topography living on the ground energy shell ($h \in W_1$), the ground state modes have their independent dynamics given by (16.36). We may then easily check, after multiplying (16.36) by $P_1\psi$ and integrating over S^2, utilizing the fact that there is no damping ($\mathcal{D} = 0$) and no external forcing ($\mathcal{F} = 0$), that the energy of the ground state modes

$$E_1 = -\frac{1}{2}\int_{S^2} P_1\psi\,\Delta P_1\psi = \int_{S^2}(P_1\psi)^2 \tag{16.69}$$

is conserved. This further implies that the enstrophy of the ground state modes

$$\mathcal{E}_1 = \frac{1}{2}\fint_{S^2} (\Delta P_1\psi)^2 = 2\fint_{S^2} (P_1\psi)^2 \tag{16.70}$$

is also conserved.

As we mentioned earlier, the enstrophy $\mathcal{E} = \frac{1}{2}\fint_{S^2} (\Delta\psi)^2$ is not necessarily conserved for general topography. However, as we shall demonstrate below, the enstrophy is conserved provided that the topography lives on the ground energy shell ($h \in W_1$). Indeed, thanks to (16.27) and (16.28), we have

$$\fint_{S^2} J(\psi, 2\Omega z + h)\Delta\psi = -\fint_{S^2} J(\psi, \Delta\psi)(2\Omega z + h) = 0 \quad \text{if } h \in W_1. \tag{16.71}$$

Thus multiplying the inviscid unforced barotropic quasi-geostrophic equations (16.62) by $\Delta\psi$ and integrating over the unit sphere S^2 we have that the enstrophy

$$\mathcal{E} = \frac{1}{2}\fint_{S^2} (\Delta\psi)^2 \tag{16.72}$$

is conserved in time if $h \in W_1$.

A consequence of the conservation of the total energy, E, and the energy of the ground state modes E_1 and the Pythagorean identity is that the energy of the higher modes

$$E' = -\frac{1}{2}\fint_{S^2} (Id - P_1)\psi\, \Delta(Id - P_1)\psi \tag{16.73}$$

is conserved in time for the inviscid unforced barotropic quasi-geostrophic equations (16.62) provided that the topography lives on the ground energy shell, i.e. $h \in W_1$.

Likewise, the enstrophy of the higher modes

$$\mathcal{E}' = \frac{1}{2}\fint_{S^2} (\Delta(Id - P_1)\psi)^2 \tag{16.74}$$

is also conserved in time if $h \in W_1$ since the total enstrophy $\mathcal{E}$ and the enstrophy of the ground state modes $\mathcal{E}_1$ are conserved in this case.

In general, we cannot expect high moments of the higher modes to be conserved. Indeed, it is a straightforward exercise to verify, utilizing relationships developed in Subsection 16.2.1, that the high modes, $\psi' = (Id - P_1)\psi = \psi - P_1\psi = \psi - \psi_1$ satisfy the following equations (with general dissipation and forcing)

$$\frac{\partial \Delta\psi'}{\partial t} + J(\psi_1, \Delta\psi' + 2\psi') + J(\psi', \Delta\psi' + 2\Omega z + h) = \mathcal{D}(\Delta)\psi' + (Id - P_1)\mathcal{F}. \tag{16.75}$$

It is then easy to see that high moments of $\Delta\psi'$ are not conserved in general in the inviscid unforced environment. This is similar to the case when the beta-effect is the only geophysical effect in the periodic geometry we discussed in Subsection 1.3.5.

Conserved quantities in the presence of odd symmetry in z

Occasionally it is interesting to consider flows with symmetry on the sphere just as in the flat geometry case. A symmetry of obvious interest is the odd symmetry in z (or equivalently θ). Utilizing the explicit form of the Jacobian given in (16.6), it is easy to see that the odd symmetry in z is preserved under the barotropic quasi-geostrophic dynamics on the sphere (16.4) provided that the topography, h, and the external forcing, $\mathcal{F}$, also respect this odd symmetry in z.

For systems with such symmetry, it is natural to consider quantities on each hemisphere, instead of the whole sphere. We consider the northern hemisphere

$$\begin{aligned} S_N^2 &= \{(\phi, \theta) \,|\, 0 \le \phi < 2\pi,\ 0 < \theta < \pi/2\} \\ &= \{(\phi, z) \,|\, 0 \le \phi < 2\pi,\ 0 < z < 1\}. \end{aligned} \tag{16.76}$$

Since both the stream function ψ, the relative vorticity $\omega = \Delta\psi$ and the potential vorticity q are odd functions in z, we see that the energy in the northern hemisphere

$$E_N = -\frac{1}{2} \fint_{S_N^2} \psi\Delta\psi \tag{16.77}$$

and all the generalized potential enstrophy in the northern hemisphere

$$\mathcal{G}_N = \fint_{S_N^2} G(q) \tag{16.78}$$

are conserved, which can be verified directly utilizing equation (16.62) and the fact that $h = \psi = \omega = q = 0$ at $\theta = 0 (z = 0)$.

The special choice of $G(q) = q$ in (16.78) implies the conservation of circulation in the northern hemisphere

$$\Gamma_N = \fint_{S_N^2} q. \tag{16.79}$$

This circulation is not necessarily zero, which is in contrast to the circulation on the whole sphere, which is identically zero, since we imposed the zero mean condition.

Conserved quantities in the presence of odd symmetry in z and topography living on the ground energy shell

Here we consider conserved quantities with the assumptions on odd symmetry in z and topography living on the ground energy shell ($h \in W_1$). The odd symmetry in z and $h \in W_1$ implies that the topography is a constant multiply of z, i.e. there exists a constant c_h such that

$$h(\phi, z) = c_h z, \tag{16.80}$$

since $x = \cos\theta\cos\phi$ and $y = \cos\theta\sin\phi$ are not odd in z (θ). In particular, this means that the topography is a zonal one and hence the angular momentum (rate of solid body rotation) is conserved.

Another immediate consequence of the assumptions on the odd symmetry in z and the topography is that the enstrophy in the northern hemisphere

$$\mathcal{E}_N = \frac{1}{2}\fint_{S_N^2} (\Delta\psi)^2 \tag{16.81}$$

is conserved, since the total enstrophy is conserved by the assumption on topography (see (16.72)), and the relative vorticity $\omega = \Delta\psi$ is odd in z by the odd symmetry assumption on ψ.

It is a straightforward exercise to verify that the energy of the high modes in the northern hemisphere

$$E'_N = -\frac{1}{2}\fint_{S_N^2} \psi'\,\Delta\psi',\ \psi' = (Id - P_1)\psi \tag{16.82}$$

and the enstrophy in the high modes in the northern hemisphere

$$\mathcal{E}'_N = \frac{1}{2}\fint_{S_N^2} (\Delta\psi')^2,\ \psi' = (Id - P_1)\psi \tag{16.83}$$

are conserved in time by utilizing the conservation of the total energy in the northern hemisphere, the conservation of the energy of the ground state modes in the northern hemisphere, the conservation of the total enstrophy in the northern hemisphere, and the conservation of the enstrophy of the ground state modes in the northern hemisphere.

Also, for the circulation of the high modes on the northern hemisphere, we notice that

$$\fint_{S_N^2} x = \fint_{S_N^2} y = 0,$$

and hence

$$\fint_{S_N^2} P_1\psi = \fint_{S_N^2} 3z \cdot \fint_{S^2} P_1\psi\, z = 3M\fint_{S_N^2} z \tag{16.84}$$

is conserved in time since the angular moment M is conserved in time. Thus, the circulation of the high modes in the northern hemisphere

$$\Gamma'_N = \fint_{S_N^2} (Id - P_1)\Delta\psi = \fint_{S_N^2} \Delta\psi - \fint_{S_N^2} P_1\Delta\psi = \fint_{S_N^2} \Delta\psi + 2\fint_{S_N^2} P_1\psi \tag{16.85}$$

is also conserved in time since both $\fint_{S^2_N} \Delta\psi$ and $\fint_{S^2_N} P_1\psi$ are conserved according to (16.79) and (16.84).

Summary of conserved quantities

Here we summarise the conserved quantities under various assumptions.

General conserved quantities

(i) Total energy $E = -\frac{1}{2}\fint_{S^2} \psi\,\Delta\psi$
(ii) All generalized potential enstrophy $\mathcal{G} = \fint_{S^2} G(q)$, for all G
(iii) Extrema of the potential vorticity max q and min q (a consequence of (ii))
(iv) Angular momentum $M = \fint_{S^2} \psi z$ if the topography is zonal, i.e. $h = h(z)$.

(16.86)

Conserved quantities with odd symmetry in z

(i) All conserved quantities from (16.86)
(ii) Energy in the northern hemisphere $E_N = -\frac{1}{2}\fint_{S^2_N} \psi\,\Delta\psi$
(iii) All generalized potential enstrophy in the northern hemisphere $\mathcal{G}_N = \fint_{S^2_N} G(q)$, for all G
(iv) Circulation of the potential vorticity in the northern hemisphere $\Gamma_N = \fint_{S^2_N} q$ (a consequence of (iii))

(16.87)

Conserved quantities with $h \in W_1$

(i) All conserved quantities from (16.86)
(ii) Total enstrophy $\mathcal{E} = \frac{1}{2}\fint_{S^2} (\Delta\psi)^2$
(iii) Energy in ground state modes $E_1 = -\frac{1}{2}\fint_{S^2} P_1\psi\,\Delta P_1\psi$
(iv) Energy in high modes $E' = -\frac{1}{2}\fint_{S^2} \psi'\,\Delta\psi'$, $\psi' = \psi - P_1\psi$ (a consequence of the conservation of energy and (iii))
(v) Enstrophy in ground state modes $\mathcal{E}_1 = \frac{1}{2}\fint_{S^2} (\Delta P_1\psi)^2$ (same as (iii))
(vi) Enstrophy in high modes $\mathcal{E}' = \frac{1}{2}\fint_{S^2} (\Delta\psi')^2$, $\psi' = \psi - P_1\psi$ (a consequence of (ii) and (v))

(16.88)

Conserved quantities with $h \in W_1$ and odd symmetry in z

(i) All conserved quantities from (16.86), (16.87) and (16.88)

(ii) Energy in ground state modes in the northern hemisphere $E_{1N} = -\frac{1}{2}\int_{S_N^2} P_1\psi\, \Delta P_1\psi$

(iii) Energy in high modes in the northern hemisphere $E'_N = -\frac{1}{2}\int_{S_N^2} \psi'\,\Delta\psi',\ \psi' = \psi - P_1\psi$ (a consequence of the conservation of energy and (ii))

(iv) Enstrophy in ground state modes in the northern hemisphere $\mathcal{E}_{1N} = \frac{1}{2}\int_{S_N^2} (\Delta P_1\psi)^2$ (same as (ii))

(v) Enstrophy in high modes in the northern hemisphere $\mathcal{E}'_N = \frac{1}{2}\int_{S_N^2} (\Delta\psi')^2,\ \psi' = \psi - P_1\psi$ (a consequence of the conservation of enstrophy and (iv))

(vi) Circulation of the high modes in the northern hemisphere $\Gamma'_N = \int_{S_N^2} \Delta\psi',\ \psi' = \psi - P_1\psi$

(16.89)

16.2.3 Non-linear stability of exact solutions

The purpose of this subsection is to study the non-linear stability of some of the special exact solutions that we introduced in Subsection 16.2.1. The ideas are similar to the flat geometry case from Chapter 4, namely, utilize conserved quantities to study non-linear stability.

Non-linear stability of steady states

Here we discuss the stability of the exact steady state solutions introduced in Subsection 16.2.1. The non-linear stability of the steady state solutions are studied in exactly the same way as in Sections 4.2 and 4.3. Since almost everything is parallel to the case of flat geometry we only single out the interesting case of ground state topography. In this case, the ground state modes have their independent linear dynamics and thus we can allow the parameter μ to extend to the second eigenvalue of the Laplacian. More precisely we have:

Proposition 16.1 (Stability of simple steady states). *Let $\overline{q} = \mu\overline{\psi}$ be the steady state solutions of the inviscid unforced barotropic quasi-geostrophic equations (16.62) on the unit sphere:*

1. *If $(Id - P_1)h \neq 0$, then the steady state $\overline{q} = \mu\overline{\psi}$ is non-linearly stable if $\mu > -2$. It is marginally stable if $\mu = -2$.*
2. *If $(Id - P_1)h \equiv 0$, then the steady state $\overline{q} = \mu\overline{\psi}$ is non-linearly stable if $\mu > -6$, and marginally stable if $\mu = -6$.*

Proof: We only prove the second part.

Since we have ground state topography, the ground state modes have their independent linear dynamics given by (16.36) with $\mathcal{D}=0$ and $\mathcal{F}=0$. Thus let $\delta\psi$ be a perturbation of $\overline{\psi}$, we have, by the linearity of (16.36), $\delta\psi$ satisfies (16.36) with $\mathcal{D}=0$ and $\mathcal{F}=0$. In particular, we have

$$\|P_1\delta\psi\|_0^2 = \|P_1\delta\psi_0\|_0^2, \tag{16.90}$$

which implies the nonlinear stability of the ground state modes.

We need to prove the stability of the high modes (motion off the ground state modes) for the given range of μ. For this purpose, we follow the strategy of Section 4.2, utilize the conserved quantities for the high modes from (16.88) to form a positive definite quadratic form for the high modes of the perturbation

$$\begin{aligned}
\mathcal{W}'_\mu(\delta\psi) &= \{\mathcal{E}'(\overline{\psi}+\delta\psi)-\mathcal{E}'(\overline{\psi})\}+\mu\{E'(\overline{\psi}+\delta\psi)-E'(\overline{\psi})\}\\
&= \frac{1}{2}⨍_{S^2}|\delta q'|^2 - \frac{\mu}{2}⨍_{S^2}\delta q'\Delta^{-1}\delta q'\\
&= \frac{1}{2}\sum_{n\geq 2}\left(1+\frac{\mu}{\Lambda_n}\right)|\delta\hat{q}_{nm}|^2.
\end{aligned}$$

For the case of $\mu > -6$ we notice that

$$\mathcal{W}'_\mu(\delta\psi) \geq \frac{1}{2}\left(1-\frac{|\mu|}{6}\right)\sum_{n\geq 2}|\delta\hat{q}_{nm}|^2 = \left(1-\frac{|\mu|}{6}\right)\|\delta q'\|_0^2,$$

$$\mathcal{W}'_\mu(\delta\psi) \leq \frac{1}{2}\left(1+\frac{|\mu|}{6}\right)\sum_{n\geq 2}|\delta\hat{q}_{nm}|^2 = \left(1+\frac{|\mu|}{6}\right)\|\delta q'\|_0^2,$$

and hence the high modes ($\delta q' = \Delta\delta\psi' = \Delta\psi - P_1\Delta\psi$) are non-linearly stable. This implies the non-linear stability when combined with the non-linear stability of the ground state modes ensured by (16.90).

For the case of $\mu = -6$ we have

$$\mathcal{W}'_\mu(\delta\psi) \geq \frac{1}{2}\left(1+\frac{\mu}{12}\right)\sum_{n\geq 3}|\delta\hat{q}_{nm}|^2 = \left(1+\frac{\mu}{12}\right)\|(Id-P_1-P_2)\delta q\|_0^2,$$

$$\mathcal{W}'_\mu(\delta\psi) \leq \frac{1}{2}\left(1+\frac{|\mu|}{12}\right)\sum_{n\geq 3}|\delta\hat{q}_{nm}|^2 = \left(1+\frac{|\mu|}{12}\right)\|(Id-P_1-P_2)\delta q\|_0^2,$$

which implies the desired marginal stability.

This completes the proof.

Likewise we have a result on the non-linear stability of more general steady states:

Proposition 16.2 (Sufficient condition for nonlinear stability) *Let $\overline{q}$ be a given bounded steady state*

$$\overline{\psi} = G(\overline{q}), \quad \overline{q} = \Delta\overline{\psi} + h + 2\Omega\sin\theta, \quad \overline{a} \leq \overline{q} \leq \overline{b}.$$

Assume there is a bounded initial perturbation $\delta\omega_0$ added so that

$$q|_{t=0} = \overline{q} + \delta\omega_0, \qquad \alpha \leq \delta\omega_0 \leq \beta.$$

We further assume there are constants c_1, c_2, c_3, c_4 such that either
Case (1)

$$0 < c_1 \leq G'(r) \leq c_2, \qquad a \leq r \leq b$$

or Case (2)

$$-c_3 \leq G'(r) \leq -c_4 \leq -\Lambda_1^{-1}, \qquad a \leq r \leq b.$$

Then either Case (1) or (2) implies the non-linear stability of the steady state, i.e.

$$\int |\delta\omega(t)|^2 \leq C \int |\delta\omega_0|^2,$$

for some $C = C(c_1, c_2, c_3, c_4)$, where

$$a = \overline{a} + \alpha, b = \overline{b} + \beta.$$

The proofs are the same as those in Sections 4.2 and 4.3. We omit the details here.

Restricted stability of motion on the first two energy shells

In this subsection we will study the stability of those exact solutions on the first eight eigenfunctions, i.e. the first two energy shells. We will see that these solutions are stable in the restricted sense. But most of them are unstable due to the interaction between the first three eigenfunctions (ground energy shell) and the next five eigenfunctions (second energy shell). More precisely, we will show that the solutions are unstable even for perturbations within the first three eigenfunctions, unless the motion occurs on the ground energy shell only, a situation discussed in the previous subsection. This is similar to the case of flat geometry where the dynamics on the ground energy shell are stable in the restricted sense, but unstable due to resonance. The difference is in the spherical geometry case we have a rigorous proof of the non-linear instability while in the flat geometry case we need to resort to truncated systems.

The restricted stability analysis is a slight variation of the discussion in the previous subsection, which goes back to Sections 4.2 and 4.3. We recall the main steps here for the sake of completeness.

Proposition 16.3 (Restricted stability of motion on the first two energy shells) *Let $\bar{q}(t)$ be an exact solution of the inviscid unforced quasi-geostrophic equation on the sphere (16.62). Moreover, we assume that $\bar{q}(t)$ lives on the first two energy shells in the sense that*

$$(P_1 + P_2)\bar{q} = \bar{q},$$

and that the topography lives on the ground energy shell ($h \in W_1$). Then $\bar{q}(t)$ is stable in the restricted sense. More precisely, there exists a constant C such that for any initial perturbation δq_0, we have

$$\|(Id - P_1 - P_2)\delta q\|_0^2(t) \le C\|(Id - P_1 - P_2)\delta q_0\|_0^2, \tag{16.91}$$

and $\bar{q}(t)$ is unstable in general.

Proof: If $\bar{q}$ is a steady state, then the marginal stability of it is discussed in Proposition 16.1. Here we focus on the general case that $\bar{q}$ is time dependent.

We consider a small perturbation of $\bar{q}$ of the form

$$\begin{aligned} q|_{t=0} &= \bar{q}_0 + \delta q_0, \\ \psi(t) &= \bar{\psi}(t) + \delta\psi(t), \\ q(\phi, \theta, t) &= \bar{q}(\phi, \theta, t) + \delta q(\phi, \theta, t). \end{aligned}$$

Since $\bar{q}$ lives on the first two energy shells, we have

$$\bar{\omega} = \Delta\bar{\psi} = \Delta P_1\bar{\psi} + \Delta P_2\bar{\psi} = -6P_2\bar{\psi} - 2P_1\bar{\psi}. \tag{16.92}$$

Following the ideas from Chapter 4, we consider a linear combination of the conserved quantities, the energy, E, and the enstrophy, $\mathcal{E}$, in order to produce a positive definite quadratic form of δq

$$\begin{aligned} \mathcal{W}(\delta q) &\overset{def}{=} \mathcal{E}(\bar{q} + \delta q) - 6E(\bar{q} + \delta q) - (\mathcal{E}(\bar{q}) - 6E(\bar{q})) \qquad (16.93) \\ &= (\bar{\omega}, \delta q) + \mathcal{E}(\delta q) + 6(\bar{\psi}, \delta q) - 6E(\delta q) \\ &= 4(P_1\bar{\psi}, \delta q) + \mathcal{E}(\delta q) - 6E(\delta q) \\ &= 4(P_1\bar{\psi}, P_1\delta q) + \frac{1}{2}\sum_{n,m}\left(1 - \frac{6}{n(n+1)}\right)|\delta\hat{q}_{nm}|^2 \\ &= 4(P_1\bar{\psi}, P_1\delta q) - \|P_1\delta q\|_0^2 + \frac{1}{2}\sum_{n\ge 3}\left(1 - \frac{6}{n(n+1)}\right)|\delta\hat{q}_{nm}|^2 \\ &= -2(P_1\bar{\omega}, P_1\delta q) - \|P_1\delta q\|_0^2 + \frac{1}{2}\sum_{n\ge 3}\left(1 - \frac{6}{n(n+1)}\right)|\delta\hat{q}_{nm}|^2 \\ &= -\|P_1(\bar{\omega} + \delta q)\|_0^2 + \|P_1\bar{\omega}\|_0^2 + \frac{1}{2}\sum_{n\ge 3}\left(1 - \frac{6}{n(n+1)}\right)|\delta\hat{q}_{nm}|^2. \end{aligned}$$

We also recall from (16.2.2) that the energy in the ground state modes is a conserved quantities, and hence

$$\|P_1(\bar{\omega}+\delta q)\|_0^2, \quad \|P_1\bar{\omega}\|_0^2$$

are conserved quantities. Therefore we deduce that

$$\frac{1}{2}\sum_{n\geq 3}\left(1-\frac{6}{n(n+1)}\right)|\delta\hat{q}_{nm}|^2 = \mathcal{W}(\delta q) - \|P_1\bar{\omega}\|_0^2 + \|P_1(\bar{\omega}+\delta q)\|_0^2 \tag{16.94}$$

is a conserved quantity. This implies restricted stability since

$$\begin{aligned}\sum_{n\geq 3}|\delta\hat{q}_{n,m}|^2(t) &\leq 2\sum_{n\geq 3}\left(1-\frac{6}{n(n+1)}\right)|\delta\hat{q}_{nm}|^2(t) \\ &= 2\sum_{n\geq 3}\left(1-\frac{6}{n(n+1)}\right)|\delta\hat{q}_{nm}|^2(0) \\ &\leq 2\sum_{n\geq 3}|\delta\hat{q}_{nm}|^2(0) \\ &= 2\|(Id-P_1-P_2)\delta q_0\|_0^2.\end{aligned} \tag{16.95}$$

In general however, these motions are unstable, even to perturbations within the first two energy shells W_1+W_2. Here are two examples:

Example 1: (A periodic solution with $h=0$) Let

$$\bar{\psi}(t) \equiv a\cos(\Omega t/3)w_{a21} - a\sin(\Omega t/3)w_{b21} = \frac{1}{2}a\sin(2\theta)\cos(\Omega t/3+\phi).$$

Consider a small perturbation at time zero of the form

$$\delta\psi_0 = \epsilon w_1 = \epsilon\sin\theta,$$

then we have, thanks to the exact solution formula from Appendix 1

$$\begin{aligned}&a_1(t)\equiv\epsilon,\ a_{11}(t)\equiv 0,\ b_{11}(t)\equiv 0,\ a_{22}(t)\equiv 0,\ b_{22}(t)\equiv 0,\\ &a_{21}(t) = a\cos(\Omega t/3+2\epsilon t/3),\ b_{21}(t) = -a\sin(\Omega t/3+2\epsilon t/3),\end{aligned}$$

and

$$\begin{aligned}\psi(t) &= \epsilon w_1 + a\cos(\Omega t/3+2\epsilon t/3)w_{a21} - a\sin(\Omega t/3+2\epsilon t/3)w_{b21} \\ &= \epsilon\sin\theta + \frac{1}{2}a\sin(2\theta)\cos((\Omega+2\epsilon)t/3+\phi).\end{aligned}$$

Compare this to

$$\bar{\psi}(t) = \frac{1}{2}a\sin(2\theta)\cos(\Omega t/3+\phi)$$

we immediately see that $\bar{\psi}$ is unstable since small perturbation changes its frequency while keeping the amplitude.

Example 2: (A steady state solution with $h=0,\ \Omega=0$) For simplicity we let $\Omega=0$. The calculation above is still valid except that $\bar{\psi}$ is a steady state now. A small perturbation changes it to a periodic orbit with amplitude a. Thus $\bar{\psi}$ is unstable.

This completes the proof of Proposition 16.3.

16.3 The response to large-scale forcing

Recall that in the flat geometry case without topographic effects ($h\equiv 0$), the long time behavior of the solutions of the barotropic quasi-geostrophic equations with damping and large-scale (ground state modes) forcing is the unique exact solution on the ground state modes determined by the external large-scale forcing (see Chapter 2). This remains true in the spherical geometry case. Once again due to the special geometry we are able to relax the assumption on the external forcing a little bit: instead of requiring the external forcing acting on the largest scale (ground state modes) only, we can allow external forcing acting on both the ground energy shell and the second energy shell. A key step in our proof is to utilize the fact that the dynamics on the ground energy shell is independent of motions on other modes (see Subsection 16.2.1).

We consider a force on the first two energy shells, i.e.

$$\mathcal{F}=P_1\mathcal{F}+P_2\mathcal{F}\in W_1+W_2, \tag{16.96}$$

where

$$P_n\mathcal{F}\in W_n,\ W_n=\operatorname{span}\{Y_{nm},\ |m|\le n\}$$

(See Subsections 16.1.2 and 16.2.1 for notation).

It follows from our discussion on exact dynamics on the first two energy shells for the quasi-geostrophic equation in the absence of topography that the space W_1+W_2 is an invariant subspace of the dynamical system generated by this damped driven equation. See equation (16.51) for more details.

Our goal here is to prove that the long time dynamics is that of the one on the first two energy shells (first eight eigenfunctions) with the initial value the projection of the original initial value onto this subspace. In the case of a steady state forcing (or quasi-periodic), then the exact dynamics on the subspace is also steady state (or quasi-periodic), which further implies that there is *no turbulence* regardless of the size of the generalized Grashoff number (and hence the Reynolds number). This generalizes earlier results on 2D Navier–Stokes equations on a torus where Yudovitch, Marchioro, Constantin-Foias-Temam proved the absence of turbulence under the assumption that the forcing is on the ground energy shell, i.e. $\mathcal{F} \in W_1$ (see Chapter 2 for more details).

The main result of this section is:

Theorem 16.1 (Response to large-scale forcing) *Consider the barotropic quasi-geostrophic equations on the unit sphere (16.4) and consider the special case of having genuine dissipation, i.e.* $d \overset{def}{=} \sum_{j=1}^{k} 12^{j-1} d^j > 0$, *external forcing acting on the lowest two energy shells, i.e.* $\mathcal{F} = P_1\mathcal{F} + P_2\mathcal{F} \in W_1 + W_2$ *or (16.96) holds, and topography living on the ground energy shell, i.e.* $h \in W_1$. *Then any solution* ψ *of the equation converges to the exact solution on the first two energy shells with the projected initial data, i.e.*

$$\psi - \psi_{\text{exact}} \to 0, \quad \textit{as } t \to \infty, \tag{16.97}$$

where ψ_{exact} *is the solution to*

$$\begin{aligned} \frac{\partial}{\partial t}\Delta\psi_{\text{exact}} + J(\psi_{\text{exact}}, \Delta\psi_{\text{exact}} + 2\Omega z + h) = \mathcal{D}(\Delta)\psi_{\text{exact}} + \mathcal{F} \\ \psi_{\text{exact}}|_{t=0} = (P_1 + P_2)\psi(0). \end{aligned} \tag{16.98}$$

Proof: We follow the same strategy as in Chapter 2: we first prove that the dynamics outside the subspace is trivial, in fact decay exponentially fast in time. This is done via an appropriate combination of the energy and potential enstrophy and the fact that the ground energy shell has its independent dynamics. In the second step we argue that the dynamics on the first two energy shells converge to the solution of the exact dynamics with the projected initial value as prescribed by (16.98). This is done by proving that the dynamics on the ground energy shell are the same and the difference on the second energy shell satisfies a linear equation (with the motion on the ground energy shell as given) with a forcing term which converges to zero exponentially fast in time.

Since the proof is similar to those in Section 2.2 where we explained everything in detail, we will only give a sketch here.

Now we move on to the first step, namely to prove that the projection outside the first two shells of the solution decay exponentially fast in time. Recall that P_n denotes the projection on to W_n and Id denotes the identity operator. By an analogy of conservation of energy, i.e. multiplying (16.62) by $-\psi$ and integrating over S^2, we have

$$\frac{1}{2}\frac{d}{dt}\|\nabla\psi\|_0^2+\sum_{j=1}^{k} d_j\|(-\Delta)^{\frac{j-1}{2}}\nabla\psi\|_0^2=-(\mathcal{F},\psi)$$

$$=-(P_1\mathcal{F},\psi)-(P_2\mathcal{F},\psi). \quad (16.99)$$

Similarly by an analogy of the conservation of enstrophy, i.e. multiplying (16.62) by $\Delta\psi$ and integrating over S^2, we have

$$\frac{1}{2}\frac{d}{dt}\|\Delta\psi\|_0^2+\sum_{j=1}^{k} d_j\|(-\Delta)^{\frac{j-1}{2}}\Delta\psi\|_0^2=(\mathcal{F},\Delta\psi)=(\Delta\mathcal{F},\psi)$$

$$=-2(P_1\mathcal{F},\psi)-6(P_2\mathcal{F},\psi). \quad (16.100)$$

The idea is again to form appropriate difference of these two equations so that the terms involving the external forcing drop out. However, since the external forcing is involved in two terms of the equations we are only able to eliminate one term by forming the difference of the second equation (16.100) and six times the first equation (16.99)

$$\frac{1}{2}\frac{d}{dt}(\|\Delta\psi\|_0^2-6\|\nabla\psi\|_0^2)+\sum_{j=1}^{k} d_j(\|(-\Delta)^{\frac{j-1}{2}}\Delta\psi\|_0^2-6\|(-\Delta)^{\frac{j-1}{2}}\nabla\psi\|_0^2)$$

$$=4(P_1\mathcal{F},\psi). \quad (16.101)$$

In this way we have eliminated the term involving the projection of the external forcing onto the second energy shell.

As for the remaining term involving the projection of the external forcing onto the ground energy shell, we turn to the exact dynamics on the ground energy shell (16.36) for help. For this purpose we take the inner product of (16.36) with $P_1\psi$ and deduce

$$\frac{1}{2}\frac{d}{dt}\|\nabla P_1\psi\|_0^2+\sum_{j=1}^{k} d_j\|(-\Delta)^{\frac{j-1}{2}}\nabla P_1\psi\|_0^2=-(\mathcal{F},P_1\psi)=-(P_1\mathcal{F},\psi). \quad (16.102)$$

To completely eliminate terms involving the external forcing we take the sum of (16.101) and four times the equation (16.102) and we obtain

$$\frac{1}{2}\frac{d}{dt}(\|(Id-P_1-P_2)\Delta\psi\|_0^2-6\|(Id-P_1-P_2)\nabla\psi\|_0^2)=$$

$$-\sum_{j=1}^{k} d_j(\|(Id-P_1-P_2)(-\Delta)^{\frac{j-1}{2}}\Delta\psi\|_0^2$$

$$-6\|(Id-P_1-P_2)(-\Delta)^{\frac{j-1}{2}}\nabla\psi\|_0^2), \quad (16.103)$$

where we used the identity

$$\|(P_1+P_2)(-\Delta)^{\frac{j}{2}}\Delta\psi\|_0^2-6\|(P_1+P_2)(-\Delta)^{\frac{j}{2}}\nabla\psi\|_0^2 \\ +4\|P_1(-\Delta)^{\frac{j}{2}}\nabla\psi\|_0^2=0, \tag{16.104}$$

for each $j\geq 0$ and the Pythagorean theorem. This identity can be verified easily using the surface spherical harmonics representation

$$\psi(t)=\sum_{n,m}\hat{\psi}_{nm}(t)Y_{nm}=\sum_{n,m}\frac{-1}{n(n+1)}\hat{\omega}_{nm}(t)Y_{nm}, \tag{16.105}$$

and the fact that the Y_{nm} are normalized eigenfunctions of the Laplace operator associated with the eigenvalue $-n(n+1)$. Moreover, we have

$$\begin{aligned}
&\sum_j d_j(\|(Id-P_1-P_2)(-\Delta)^{\frac{j-1}{2}}\Delta\psi\|_0^2-6\|(Id-P_1-P_2)(-\Delta)^{\frac{j-1}{2}}\nabla\psi\|_0^2)\\
&=\sum_j d_j\sum_{n\geq 3}\sum_{|m|\leq n}\left(\Lambda_n^{j+1}|\hat{\psi}_{nm}|^2-6\Lambda_n^j|\hat{\psi}_{nm}|^2\right)\\
&=\sum_j d_j\sum_{n\geq 3}\sum_{|m|\leq n}\Lambda_n^j(\Lambda_n-6)|\hat{\psi}_{nm}|^2\\
&\geq\sum_j d_j\Lambda_3^{j-1}\sum_{n\geq 3}\sum_{|m|\leq n}(\Lambda_n^2-6\Lambda_n)|\hat{\psi}_{nm}|^2\\
&=d(\|(Id-P_1-P_2)\Delta\psi\|_0^2-\|\nabla(Id-P_1-P_2)\psi\|_0^2) \qquad (16.106)\\
&\geq\frac{d}{2}\sum_{n\geq 3}\sum_{|m|\leq n}\Lambda_n^2|\hat{\psi}_{nm}|^2\\
&=\frac{d}{2}\|(Id-P_1-P_2)\Delta\psi\|_0^2, \qquad (16.107)
\end{aligned}$$

where

$$d=\sum_{j=1}^{k}12^{j-1}d_j=\sum_{j=1}^{k}\Lambda_3^{j-1}d_j. \tag{16.108}$$

Substituting this one into the previous equation (16.103) and integrating in time we have

$$\|(Id-P_1-P_2)\Delta\psi\|_0^2\leq 2\left(\|(Id-P_1-P_2)\Delta\psi\|_0^2-\|\nabla(Id-P_1-P_2)\psi\|_0^2\right) \\ \leq\kappa\exp(-dt), \tag{16.109}$$

where κ is a generic constant independent of time (in this case it is $2(\|(Id-P_1-P_2)\Delta\psi(0)\|_0^2-6\|(Id-P_1-P_2)\nabla\psi(0)\|_0^2))$. This implies the restricted stability and the claim that the long time dynamics is that on the first eight eigenfunctions (or the first two energy shells).

We now turn to the second task, namely to prove that the dynamics on the first two energy shells converges to the exact solution given by (16.98).

We first recall that the ground state modes have their independent dynamics, depending only on the projection of the initial data $P_1\psi(0)$ and the projection of the forcing $P_1\mathcal{F}$ (see equation (16.36)). Since the ψ and ψ_{exact} have the same initial data and are subject to the same external forcing after projection onto the ground energy shell, we conclude that

$$P_1\psi = P_1\psi_{\text{exact}}. \tag{16.110}$$

Thus it only remains to prove

$$P_2(\psi(t) - \psi_{\text{exact}}(t)) \to 0, \text{ as } t \to \infty. \tag{16.111}$$

The information on the dynamics of the ground energy shell and the decay of high modes can be summarized as

$$P_1\psi = P_1\psi_{\text{exact}}, \qquad \sum_{n\geq 3,m} n^2(n+1)^2|\hat{\psi}_{nm}|^2(t) \leq \kappa \exp(-dt). \tag{16.112}$$

The key to derive dynamics on the second energy shell is to write the non-linear term in an appropriate form which is amenable to analysis. For this purpose we notice

$$\begin{aligned}
J(\psi, \Delta\psi) &= J(P_1\psi + P_2\psi + (Id - P_1 - P_2)\psi, \\
&\qquad -2P_1\psi - 6P_2\psi + \Delta(Id - P_1 - P_2)\psi) \\
&= -6J(P_1\psi, P_2\psi) - 2J(P_2\psi, P_1\psi) \\
&\quad + J(P_1\psi, \Delta(Id - P_1 - P_2)\psi) - 2J((Id - P_1 - P_2)\psi, P_1\psi) \\
&\quad + J(P_2\psi, \Delta(Id - P_1 - P_2)\psi) - 6J((Id - P_1 - P_2)\psi, P_2\psi) \\
&\quad + J((I - P_1 - P_2)\psi, \Delta(I - P_1 - P_2)\psi) \\
&= -4J(P_1\psi, P_2\psi) \\
&\quad + J(P_1\psi, \Delta(Id - P_1 - P_2)\psi) + 2J(P_1\psi, (Id - P_1 - P_2)\psi) \\
&\quad + J(P_2\psi, \Delta(I - P_1 - P_2)\psi) + 6J(P_2\psi, (Id - P_1 - P_2)\psi) \\
&\quad + J((Id - P_1 - P_2)\psi, \Delta(Id - P_1 - P_2)\psi).
\end{aligned}$$

Recall that

$$J(P_1\psi, P_n\tilde{\psi}) \in W_n.$$

Thus we have

$$\begin{aligned} P_2J(\psi, \Delta\psi) &= -4J(P_1\psi, P_2\psi) + P_2J(P_2\psi, \Delta(Id - P_1 - P_2)\psi) \\ &\quad +6P_2J(P_2\psi, (Id - P_1 - P_2)\psi) \\ &\quad +P_2J((Id - P_1 - P_2)\psi, \Delta(Id - P_1 - P_2)\psi) \\ &= -4J(P_1\psi, P_2\psi) \\ &\quad + \sum_{n\geq 3}(6 - n(n+1))\hat{\psi}_{nm}P_2J(P_2\psi, Y_{nm}) \\ &\quad - \sum_{n\geq 3, n'\geq 3} n(n+1)\hat{\psi}_{nm}\hat{\psi}_{n'm'}P_2J(Y_{n'm'}, Y_{nm}). \end{aligned} \tag{16.113}$$

Denoting

$$\begin{aligned} \tilde{F} &= \sum_{n\geq 3}(n(n+1) - 6)\hat{\psi}_{nm}P_2J(P_2\psi, Y_{nm}) \\ &\quad + \sum_{n\geq 3, n'\geq 3} n(n+1)\hat{\psi}_{nm}\hat{\psi}_{n'm'}P_2J(Y_{n'm'}, Y_{nm}), \end{aligned} \tag{16.114}$$

we have

$$-6\frac{d}{dt}P_2\psi(t) - J(P_2\psi, 4P_1\psi + 2\Omega z + h) - \mathcal{D}(-6)P_2\psi = P_2\mathcal{F} + \tilde{F}, \tag{16.115}$$

since $P_2J(\psi, \sin\theta) = P_2J(\psi, z) = J(P_2\psi, z)$.

On the other hand, after projecting the (16.98) on to the second energy shell, we have

$$-6\frac{d}{dt}P_2\psi_{\text{exact}}(t) - J(P_2\psi_{\text{exact}}, 4P_1\psi + 2\Omega z + h) - \mathcal{D}(-6)P_2\psi_{\text{exact}} = P_2\mathcal{F}. \tag{16.116}$$

Thus denoting

$$\eta = P_2\psi - P_2\psi_{\text{exact}}, \tag{16.117}$$

we have, since $P_1\psi = P_1\psi_{\text{exact}}$ (see (16.110)),

$$-6\frac{d}{dt}\eta(t) - J(\eta, 4P_1\psi + 2\Omega z + h) - \mathcal{D}(-6)\eta = \tilde{F}. \tag{16.118}$$

and hence, by a usual energy method,

$$3\frac{d}{dt}\|\eta(t)\|_0^2 + \mathcal{D}(-6)\|\eta(t)\|_0^2 \leq \kappa\|\eta(t)\|_0\|\tilde{F}\|_0. \tag{16.119}$$

Now it is clear that our result relies on proving the smallness of $\tilde{F}$. We first make the claim:

Claim: $\tilde{F}$ decay exponentially with rate d, i.e. there exists positive constants κ, such that

$$\|\tilde{F}\|_0^2(t) \le \kappa \exp(-d\,t). \tag{16.120}$$

We leave the proof of the claim to the second Appendix.

Now combining (16.119) and (16.120) and applying Gronwall's inequality, we deduce

$$\|\eta(t)\|_0 \to 0, \text{ as } t \to \infty. \tag{16.121}$$

This completes the proof of (16.111).

16.4 Selective decay on the sphere

In this section we study the selective decay phenomena associated with the decaying barotropic quasi-geostrophic equations on the unit sphere (16.4) in the presence of Newtonian viscosity or hyper-viscosity or their combination, i.e. $\sum_{j=2}^{k} d_j > 0$, the absence of external forcing ($\mathcal{F} \equiv 0$) and ground state topography ($h \in W_1$). The reason that we can allow ground state topography is that a ground state topography merely changes the Coriolis force, so that it aligns with possibly another axis and with another magnitude (see Subsection 16.2.1). Thus all the identities that we need for the analysis and proof of selective decay remain unchanged, since the Coriolis forcing term is skew symmetric (see Chapter 3 and the argument below). We refrain from studying the case with general topography even though the general situation can be analyzed as well.

The result and approach here is very much similar to those in Chapter 3 for the flat geometry case. Thus we will omit most of the details and just state the result and sketch the main steps in the proof. A noticeable difference in the spherical geometry when compared with the flat geometry is that we are able to identify those initial data whose Dirichlet quotient approaches the first eigenvalue of the Laplace operator. This is again due to the independent dynamics of the ground state modes.

We follow the structure of Chapter 3 in this section. We first recall the basic dynamics equation. After that we calculate the physicist's selective decay states. We then formulate the main results regarding selective decay on the sphere. Next we sketch the proof of the weak form of selective decay. We demonstrate that the Dirichlet quotient approaches 2, the first eigenvalue of the Laplace operator, if and only if the projection of the initial data to the ground energy shell is non-trivial. This immediately implies the instability of selective decay states corresponding to higher eigenvalues of the Laplace operator. Selective decay with odd symmetry in z as well as selective decay for high modes (modes away from the ground energy shell) will also be discussed.

16.4.1 The equation

We recall first the governing equation which can be written in the form

$$\begin{aligned}&\frac{\partial \Delta\psi}{\partial t}+J(\psi,\Delta\psi+2\Omega z+h)=\mathcal{D}(\Delta)\psi,\\ &q=\Delta\psi+2\Omega+h,\qquad h\in W_1,\end{aligned} \tag{16.122}$$

where we assume that the total effect of Newtonian viscosity and hyper-viscosity is non-trivial, i.e.

$$\sum_{j=2}^{k} d_j > 0. \tag{16.123}$$

16.4.2 Physicist's selective decay states

As for the flat geometry case (Section 3.2), the associated mean energy and mean enstrophy are given by

$$E(t)=\frac{1}{2}\int_{S^2}|\text{curl }\psi|^2 d\mathbf{s},$$

and

$$\mathcal{E}(t)=\frac{1}{2}\int_{S^2}|\Delta\psi|^2 d\mathbf{s}$$

respectively.

Here we use enstrophy instead of potential enstrophy, since we assumed ground state topography which is equivalent to no topography but with a new Coriolis forcing aligned with a possibly different axis and with different magnitude (see Subsection 16.2.1).

The physicist's selective decay states are those which minimize the enstrophy for fixed energy. A simple calculation as in Section 3.2 yields that any selective decay state, $\psi_\star$, must be an eigenfunction of the Laplacian on the sphere, or surface spherical harmonics, i.e.

$$\psi_\star(\phi,\theta)=\sum_{|m|\le n}\hat{\psi}_{nm}P_{nm}(\sin\theta)\exp(im\phi)=\sum_{|m|\le n}\hat{\psi}_{nm}P_{nm}(z)\exp(im\phi),$$

where the $P_{nm}(z)$s are the associated Legendre functions of order n (see Subsection 16.1.2).

Notice that these surface spherical harmonics (after normalization) form an orthonormal basis in $L^2(S^2)$ and their gradients are orthogonal to each other.

It is also easy to check that they are invariant under the dynamics (16.122) in the sense that starting with a selective decay state, the future states must be selective states as well.

16.4.3 Formulation of the selective decay principle

Now we formulate the main result for the selective decay principle on the sphere. We first formulate the weak version.

Theorem 16.2 (Weak form of selective decay principle) *Consider the decaying barotropic quasi-geostrophic equations on the sphere (16.122) with ground state topography ($h \in W_1$). Assume the existence of Newtonian viscosity or hyper-viscosity or both and no external forcing.*

(i) *(The general case) For any initial data $\psi(0) \in H^1(S^2)$, there exist an eigenvalue Λ_n of the Laplacian on the sphere so that the Dirichlet quotient of the solution converges to Λ_n, i.e.*

$$\lim_{t\to\infty} \frac{\mathcal{E}(\psi(t))}{E(\psi(t))} = \Lambda_n, \tag{16.124}$$

and $n = 1$ if and only if $P_1\psi(0) \neq 0$. Moreover, for any increasing sequence of times $\{t_j\}_{j=1}^{\infty}$ such that $t_j \to \infty$, there exists a subsequence $\{t_{j_l}\}_{l=1}^{\infty}$ and a selective decay state $\psi^* \in H^1(S^2)$ so that

$$\left\| \nabla \frac{\psi(t_{j_l})}{\|\nabla\psi(t_{j_l})\|_0} - \nabla\psi^* \right\|_0 \to 0 \text{ as } l \to \infty \tag{16.125}$$

and ψ^* satisfies

$$\Delta\psi^* = -\Lambda_n\psi^*.$$

(ii) *(Selective decay with odd symmetry) Assume that the topography h is odd in z. Then for any initial data $\psi(0) \in H^1(S^2)$ which is odd in z, there exists an eigenvalue Λ_n of the Laplacian on the sphere so that (16.124) and (6.125) holds. Moreover, $n = 1$ if and only if $P_1\psi(0) \neq 0$, and, in this case, there exists a constant c_∞ such that $\psi^* = c_\infty z$.*

(iii) *(Selective decay of high modes) For any initial data $\psi(0) \in H^1(S^2)$ satisfying*

$$(Id - P_1)\psi(0) = \psi(0) - P_1\psi(0) \neq 0$$

there exists an eigenvalue Λ_n, $n \geq 2$ of the Laplacian on the sphere so that the Dirichlet quotient of the high modes of the solution converges to Λ_n, i.e.

$$\lim_{t\to\infty} \frac{\mathcal{E}((\psi - P_1\psi)(t))}{E((\psi - P_1\psi)(t))} = \Lambda_n. \tag{16.126}$$

More over (16.125) holds with ψ replaced by the high modes $\psi' = \psi - P_1\psi$, and the energy shell with $n \geq 3$ are unstable under perturbation from lower energy shells.

Parallel to the flat geometry case, we also have a strong version of the selective decay principle.

Theorem 16.3 (Mathematical form of selective decay) *Consider the decaying barotropic quasi-geostrophic equations on the sphere (16.122) with ground state topography ($h \in W_1$). Assume the existence of Newtonian viscosity or hyper-viscosity or both and no external forcing.*

(i) (The general case) For any initial data $\psi(0) \in H^1(S^2)$, the results for the general case (part 1) of the weak form (Theorem 16.2) hold. Moreover, we have a strong form of (16.125). Namely, there exists a solution $\psi_n \in W_n$ (where the n corresponds to the n in (16.124)) to the linear inviscid unforced barotropic quasi-geostrophic equation

$$\frac{\partial \Delta\psi_n}{\partial t} + J(\psi_n, 2\Omega z + h) = 0 \tag{16.127}$$

such that the solution $\psi(t)$ converges time asymptotically in $H^1(S^2)$ to ψ_n after renormalization. More precisely, we have

$$\left\| \frac{\nabla\psi(t)}{\|\nabla\psi(t)\|_0} - \nabla\psi_n \right\|_0 \to 0 \text{ as } t \to \infty. \tag{16.128}$$

In particular, $P_1\psi(0) \neq 0$ if and only if and $n = 1$.

(ii) (Selective decay with odd symmetry) Assume that the topography h is odd in z. Then for any initial data $\psi(0) \in H^1(S^2)$ which is odd in z, the results from the general case hold. Moreover, $n = 1$ if and only if $P_1\psi(0) \neq 0$, and, in this case, there exists a constant c_∞ such that $\psi_1 = c_\infty z$, i.e. there exists a unique stable selective decay state: the solid body rotation.

(iii) (Selective decay of high modes) For any initial data $\psi(0) \in H^1(S^2)$ satisfying

$$(Id - P_1)\psi(0) = \psi(0) - P_1\psi(0) \neq 0$$

there exists a solution ψ_n to the linear inviscid unforced barotropic quasi-geostrophic equation (16.127) such that the normalized high modes converges to ψ_n, i.e.

$$\left\| \frac{\nabla(\psi - P_1\psi)(t)}{\|\nabla(\psi - P_1\psi)(t)\|_0} - \nabla\psi_n \right\|_0 \to 0 \text{ as } t \to \infty. \tag{16.129}$$

Remark: It is easy to see that every energy shell W_n is invariant under the linear inviscid unforced barotropic quasi-geostrophic dynamics (16.127), since we assumed ground state topography (see Subsection 16.2.1 for the non-linear case). Thus (16.127) possesses solutions living on a single energy shell W_n, and the solutions must be linear combination of selective decay states on this energy shell at any fixed time. Thus this theorem is consistent with the physicist's selective decay principle. However, the long time asymptotics, i.e. ψ_n, is not necessarily the combination of a zonal flow and n spherical Rossby waves since the ground state topography effectively changes the Coriolis force and hence the rotation axis.

Remark: The condition $\sum_{j=1}^{k} d_j > 0$ is necessary for selective decay to be true. The Dirichlet quotient would be a constant of motion if only Ekman drag is present.

16.4.4 Sketch of the proof

Recall from Chapter 3 that the proof of the weak form of selective decay relies on energy type estimates mostly. Hence the proof for the flat geometry case readily translates to this spherical geometry case except the new phenomena, where we can characterize those initial data whose solution has a Dirichlet quotient converging to the first eigenvalue $\Lambda_1 = 2$.

We skip the part of the proof which is the same as the flat geometry case and focus on the part that is unique to the spherical geometry, i.e. focus on the characterization of the set of initial data whose Dirichlet quotient converges to 2. We will also briefly discuss why selective decay with odd symmetry and selective decay for high modes (part 2 and 3 of the theorem) are true.

The heuristic idea of the characterization of those initial data whose Dirichlet quotient converges to 2 is the following. Since the ground state modes have their independent dynamics, their decay rate is determined by the first eigenvalue, i.e. $\mathcal{D}(-2)$. On the other hand, the high modes should have a decay rate of at least $\mathcal{D}(-\Lambda_2) = \mathcal{D}(-6)$. This separation of decay rate yields that the Dirichlet quotient is the first eigenvalue, $\Lambda_1 = 2$, if and only if the projection of the initial data on to the ground energy shell is non-trivial.

To verify our heuristic idea, we apply a simple energy estimate to the exact dynamics of the ground state modes (16.36) and we deduce

$$\frac{d}{dt}\|P_1\psi\|_0^2 = -\mathcal{D}(-2)\|P_1\psi\|_0^2. \tag{16.130}$$

Thus

$$\|P_1\psi(t)\|_0^2 = \|P_1\psi(0)\|_0^2 e^{-\mathcal{D}(-2)t}, \tag{16.131}$$

where

$$\mathcal{D}(-2) = \sum_{j=1}^{k} 2^j d_j.$$

On the other hand, multiplying the freely decaying barotropic quasi-geostrophic equations (16.122) by ψ and integrating over S^2 we have

$$\begin{aligned}
\frac{d}{dt}\|\nabla\psi\|_0^2 - 2 &= \sum_{j=1}^{k} d_j \|(-\Delta)^{j/2}\psi\|_0^2 \\
&= -2\sum_{j=1}^{k} d_j \left(\sum_{n=1} \Lambda_1^j |\hat{\psi}_{nm}|^2 + \sum_{n\geq 2} \Lambda_n^j |\hat{\psi}_{nm}|^2 \right) \\
&\leq -2\mathcal{D}(-2)\|P_1\psi(t)\|_0^2 - 2\left(\sum_{j=1}^{k} d_j \Lambda_2^{j-1} \sum_{n\geq 2} \Lambda_n |\hat{\psi}_{nm}|^2 \right) \\
-2 &= \mathcal{D}(-2)\|P_1\psi(t)\|_0^2 - \frac{2}{\Lambda_2}\mathcal{D}(-\Lambda_2)\|\nabla(Id-P_1)\psi\|_0^2. \qquad (16.132)
\end{aligned}$$

The left-hand side of (16.132) can be written as, thanks to Pythagorean identity and (11.131)

$$\begin{aligned}
\frac{d}{dt}\|\nabla\psi\|_0^2 &= \frac{d}{dt}\|\nabla P_1\psi\|_0^2 + \frac{d}{dt}\|\nabla(Id-P_1)\psi\|_0^2 \\
&= 2\frac{d}{dt}\|P_1\psi\|_0^2 + \frac{d}{dt}\|\nabla(Id-P_1)\psi\|_0^2 \\
&= -2\mathcal{D}(-2)\|P_1\psi\|_0^2 + \frac{d}{dt}\|\nabla(Id-P_1)\psi\|_0^2. \qquad (16.133)
\end{aligned}$$

Combining (16.132) and (16.133) we deduce

$$\begin{aligned}
\frac{d}{dt}\|\nabla(Id-P_1)\psi\|_0^2 &\leq -\frac{2}{\Lambda_2}\mathcal{D}(-\Lambda_2)\|\nabla(Id-P_1)\psi\|_0^2 \\
&= -\frac{\mathcal{D}(-6)}{3}\|\nabla(Id-P_1)\psi\|_0^2, \qquad (16.134)
\end{aligned}$$

and hence

$$\|\nabla(Id-P_1)\psi(t)\|_0^2 \leq \|\nabla(Id-P_1)\psi(0)\|_0^2 \exp\left(-\frac{\mathcal{D}(-6)}{3}t\right). \qquad (16.135)$$

Similarly, utilizing the enstrophy estimates for the exact dynamics of the ground state modes (16.36) and the freely decaying barotropic quasi-geostrophic equations (16.122) we deduce

$$\|\Delta(Id-P_1)\psi(t)\|_0^2 \leq \|\Delta(Id-P_1)\psi(0)\|_0^2 \exp\left(-\frac{\mathcal{D}(-6)}{3}t\right). \qquad (16.136)$$

Notice that

$$\frac{\mathcal{D}(-6)}{3} = \frac{1}{3}\sum_j 6^j d_j > \sum_j 2^j d_j = \mathcal{D}(-2) \qquad (16.137)$$

provided that there are genuine dissipation, i.e. $\sum_{j\geq 2} d_j > 0$. Thus if

$$\|P_1\psi(0)\|_0 \neq 0, \tag{16.138}$$

we have, thanks to (16.131)

$$\|P_1\psi(t)\|_0 \neq 0, \quad \text{for all } t. \tag{16.139}$$

Therefore

$$\begin{aligned} E(t) &= \frac{1}{2}\|\nabla P_1\psi(t)\|_0^2 + \frac{1}{2}\|\nabla(Id - P_1)\psi(t)\|_0^2 \\ &= \|P_1\psi(t)\|_0^2 + \mathbf{o}(e^{-\mathcal{D}(-2)t}) \\ &= \|P_1\psi(0)\|_0^2 e^{-\mathcal{D}(-2)t} + \mathbf{o}(e^{-\mathcal{D}(-2)t}). \end{aligned} \tag{16.140}$$

Likewise

$$\begin{aligned} \mathcal{E}(t) &= \frac{1}{2}\|\Delta P_1\psi(t)\|_0^2 + \frac{1}{2}\|\Delta(Id - P_1)\psi(t)\|_0^2 \\ &= 2\|P_1\psi(t)\|_0^2 + \mathbf{o}(e^{-\mathcal{D}(-2)t}) \\ &= 2\|P_1\psi(0)\|_0^2 e^{-\mathcal{D}(-2)t} + \mathbf{o}(e^{-\mathcal{D}(-2)t}). \end{aligned} \tag{16.141}$$

Consequently

$$\Lambda(t) = \frac{\mathcal{E}(t)}{E(t)} \sim \frac{2\|P_1\psi(0)\|_0^2 e^{-\mathcal{D}(-2)t}}{\|P_1\psi(0)\|_0^2 e^{-\mathcal{D}(-2)t}} = 2, \quad \text{for large } t. \tag{16.142}$$

On the other hand if

$$\|P_1\psi(0)\|_0 = 0, \tag{16.143}$$

we then have, thanks to (16.131)

$$\|P_1\psi(t)\|_0 = 0, \quad \text{for all } t. \tag{16.144}$$

Thus

$$\begin{aligned} \Lambda(t) &= \frac{\mathcal{E}(t)}{E(t)} = \frac{\|\Delta\psi(t)\|_0^2}{\|\nabla\psi(t)\|_0^2} = \frac{\|\Delta(Id - P_1)\psi(t)\|_0^2}{\|\nabla(Id - P_1)\psi(t)\|_0^2} \\ &= \frac{\sum_{n\geq 2}\Lambda_n^2\|P_n\psi(t)\|_0^2}{\sum_{n\geq 2}\Lambda_n\|P_n\psi(t)\|_0^2} \geq \Lambda_2 = 6. \end{aligned} \tag{16.145}$$

This completes the characterization of those initial data whose Dirichlet quotient converges to the first eigenvalue (2).

It is easy to see, utilizing the explicit form of the Jacobian given in (16.6), that odd symmetry in z is preserved under the decaying barotropic quasi-geostrophic dynamics (16.122), provided that the topography also respects this symmetry.

Thus it makes sense to discuss selective decay with odd symmetry in z. Since the only ground states that are odd in z are multiples of z, we see that the normalized flow must converge to a constant multiple of the solid body rotation if $n = 1$ as claimed in part (ii) of the theorem.

Finally we consider selective decay of the high modes ψ' defined as

$$\psi' = \psi - P_1\psi.$$

It is easy to check, after applying $I - P_1$ to the decaying barotropic quasi-geostrophic equations (16.122) and utilizing (16.29), (16.32), and (16.33), that the high modes satisfy the following equation

$$\frac{\partial \Delta\psi'}{\partial t} + J(\psi', \Delta\psi' + 2\Omega z + h) + J(P_1\psi, \Delta\psi' + 2\psi') = \mathcal{D}(\Delta)\psi'.$$

This is the same as the original decaying barotropic quasi-geostrophic equations (16.122) except the extra term $J(P_1\psi, \Delta\psi' + 2\psi')$. This term is skew symmetric in the sense that

$$\fint_{S^2} J(P_1\psi, \Delta\psi' + 2\psi')\psi' = 0, \qquad \fint_{S^2} J(P_1\psi, \Delta\psi' + 2\psi')\Delta\psi' = 0$$

after a straightforward calculation similar to (16.27) together with the rotation invariance. This implies all the machinery from the flat geometry case is readily translated to this case, since the arguments rely on energy estimates only. The condition $\psi'(0) \neq 0$ is necessary to ensure we do not start with the zero solution.

We may also speculate that the long time dynamics is independent of this extra term, since it decays at a rate faster than the lowest eigenvalue of the high modes which is $\Lambda_2 = 6$. This is why the long time asymptotics of the high modes is independent of the dynamics on the ground energy shell as stated in the strong version.

This ends the proof of the theorem.

Overview of equilibrium statistical theories for the sphere

In the next two sections we apply equilibrium statistical theories that we developed in Chapters 6, 7, 8, 9 to inviscid unforced barotropic quasi-geostrophic flows on the unit sphere. The dynamics of such flows are governed by (16.62).

Recall that the equilibrium statistical theories rely mostly on the conserved quantities utilized in the theory. Thus, we expect that the predictions of equilibrium statistical theories are very much the same for the spherical and for the flat geometry, with the trigonometric functions replaced by the surface spherical harmonics. However, there is one important distinction for the spherical geometry. When the topography lives on the ground energy shell, i.e. $h \in W_1$, the ground state modes have their independent linear dynamics according to Subsection 16.2.1. This implies that there will not be enough mixing for the ground state modes and hence the predictions of the equilibrium statistical theory is unlikely to be realizable. Thus it makes sense to consider the statistics of the high modes (motion away from the ground energy shell) in the case of ground state topography. Recall from Subsection 16.2.2 that we only have the conservation of energy and enstrophy (plus circulation in the case with odd symmetry) for the high modes. Therefore we should only consider energy–enstrophy theory (EEST), energy–circulation theory with prior distribution (a special case of ESTP) and point vortex theory (PVST), but not the statistical theory with infinitely many constraints (ESTMC). We will observe that the predictions of the statistics of the high modes in the case of ground state topography resembles the prediction of the corresponding statistics theory in the flat geometry without geophysical effects.

16.5 Energy enstrophy statistical theory on the unit sphere

In this section, we apply statistical theories with two conserved quantities, energy and potential enstrophy, both the empirical theory introduced in Chapter 6 and the complete statistical mechanics theory in Chapter 8, to the inviscid unforced barotropic quasi-geostrophic equation on the sphere (16.62). The mean field of the most probable state turns out to be those exact steady state solutions of the equation that we introduced in Subsection 16.2.1 and hence their stability is discussed in Subsection 16.2.3. The results here are parallel to the flat geometry case and the argument here is very much the same as in Chapters 6 and 8, since they rely mostly on conserved quantities. Nevertheless, there is an important special case of ground state topography. In this special case, the ground state modes have their independent linear dynamics and thus no sufficient mixing on the ground energy shell. Thus, we need to consider the statistics of the high modes (motion off the ground energy shell) in the case of $h \in W_1$.

For empirical statistics theory we will postulate the one-point statistics, define the Shannon entropy and then apply the maximum entropy principle to derive the most probable state. For complete statistical mechanics theory, we introduce various truncations, verify the Liouville property as well as the conservation of the truncated energy and potential enstrophy, apply the equilibrium statistical mechanics theory for ODEs from Chapter 7 to the truncated system, compute the mean state, and then take the continuum limit. We will only give a sketch and refer to the counterpart for the flat geometry case when appropriate.

16.5.1 Energy–enstrophy theory with topography – empirical statistical theory

In this subsection we derive the most probable state under the classical empirical statistical theory with energy and enstrophy as the two conserved quantities. This is parallel to Section 6.7 and the reader is referred to that section for more details.

As mentioned earlier, we need to consider two cases: the case with general topography and the case with ground state topography.

Empirical energy–enstrophy theory with general topography

Here we consider the case with general topography.

We first introduce the one-point statistics $\rho(\mathbf{s}, \lambda)$ for $\mathbf{s} \in S^2$ and $\lambda \in R^1$. We assume the following property on ρ

$$
\begin{aligned}
&\rho(\mathbf{s}, \lambda) \in \mathcal{PM}(S^2 \times R^1), \rho \geq 0,\\
&⨍_{S^2} \int \rho(\mathbf{s}, \lambda) d\lambda d\mathbf{s} = 1,\\
&\mathbf{s} \to \rho(\mathbf{s}, \lambda) \in \mathcal{PM}(R^1), \text{ for all } \mathbf{s} \in S^2,\\
&\int \rho(\mathbf{s}, \lambda) d\lambda = 1, \ a.e. \ \mathbf{s} \in S^2,\\
&\int_{q-}^{q+} \rho(\mathbf{s}, \lambda) = \text{Prob}\{q- \leq q(\mathbf{s}) < q+\}.
\end{aligned}
\tag{16.146}
$$

As before, the pointwise and overall average with respect to the one-point statistics are given by

$$\bar{F}_\rho(\mathbf{s}) = \int F(\mathbf{s}, \lambda)\rho(\mathbf{s}, \lambda) d\lambda = \langle F(\mathbf{s}, \cdot)\rangle_\rho,$$

$$\bar{F} = \langle F\rangle_\rho = ⨍_{S^2} \bar{F}_\rho(\mathbf{s}) d\mathbf{s} = ⨍_{S^2} \int F(\mathbf{s}, \lambda)\rho(\mathbf{s}, \lambda) d\lambda d\mathbf{s}.$$

In particular the mean potential vorticity is given by

$$\bar{q}(\mathbf{s}) = \int \lambda \rho(\mathbf{s}, \lambda) d\lambda,$$

and the pointwise variance is given by

$$\sigma_\rho^2(\mathbf{s}) = \langle (q - \bar{q})^2(\mathbf{s})\rangle_\rho = \int (\lambda - \bar{q})^2 \rho(\mathbf{s}, \lambda) d\lambda = \langle q^2\rangle_\rho(\mathbf{s}) - \bar{q}^2(\mathbf{s}).$$

The corresponding Shannon entropy is defined as

$$\mathcal{S}(\rho) = -⨍_{S^2} \int \rho(\mathbf{s}, \lambda) \ln \rho(\mathbf{s}, \lambda) d\lambda d\mathbf{s}. \tag{16.147}$$

The constraints on the one point statistics are

$$\mathcal{C} = \mathcal{C}^{(0)} \cap \mathcal{C}^{(\mathcal{E})} \cap \mathcal{C}^{(E)}, \tag{16.148}$$

where

$$\begin{aligned}
\mathcal{C}^{(0)} &= \left\{\rho \middle| \int \rho(\mathbf{s}, \lambda) d\lambda = 1, \text{ for all } \mathbf{s} \in S^2 \right\}, \\
\mathcal{C}^{(\mathcal{E})} &= \left\{\rho \middle| \mathcal{E}_0 = \mathcal{E}(\rho) = \frac{1}{2} \fint_{S^2} \int \lambda^2 \rho(\mathbf{s}, \lambda) d\lambda d\mathbf{s} \right\}, \\
\mathcal{C}^{(E)} &= \left\{\rho \middle| E_0 = E(\rho) = -\frac{1}{2} \fint_{S^2} \bar{\psi} \bar{\omega} d\mathbf{s} \right\},
\end{aligned} \tag{16.149}$$

where the mean stream function is solved via

$$\bar{q}(\mathbf{s}) = \int \lambda \rho d\lambda = \Delta \bar{\psi} + 2\Omega \sin\theta + h = \bar{\omega} + 2\Omega z + h. \tag{16.150}$$

Here $\mathcal{C}^{(\mathcal{E})}$ is the potential enstrophy constraint, $\mathcal{C}^{(E)}$ is the energy constraint and $\mathcal{C}^{(0)}$ is the constraint which ensures that ρ is a probability density function for each $\mathbf{s}$.

The maximum entropy principle states that the least biased (most probable) probability measure

$$\rho^\star \in \mathcal{PM}(S^2 \times R^1)$$

is the one that maximizes the Shannon entropy (16.147).

A straightforward Lagrangian multiplier calculation as in Subsection 6.7.3 leads to

$$\rho^\star(\mathbf{s}, \lambda) = \frac{\sqrt{\alpha}}{\sqrt{2\pi}} \exp\left(-\frac{\alpha}{2}\left(\lambda - \frac{\beta}{\alpha}\bar{\psi}^\star\right)^2\right) \tag{16.151}$$

with α and β being the Lagrange multipliers for the enstrophy and energy respectively, and $\bar{\psi}^*$ is the stream function of the most probable mean field.

Note that $\rho^\star$ is a Gaussian for each $\mathbf{s}$ with the property

$$\int \rho^\star d\lambda = 1, \quad \int \lambda \rho^\star d\lambda = \frac{\beta}{\alpha}\bar{\psi}^\star, \quad \int \lambda^2 \rho^\star d\lambda = \left(\frac{\beta}{\alpha}\right)^2 (\bar{\psi}^\star)^2 + \alpha^{-1}.$$

Therefore the most probable mean field $\bar{\psi}^\star = \psi_\mu$ must satisfy the following **mean field equation**

$$\Delta \psi_\mu + 2\Omega \sin\theta + h = \mu \psi_\mu, \tag{16.152}$$

where $\mu = \frac{\beta}{\alpha}$. Here we used ψ_μ to denote the most probable mean field, since the mean field equation depends only on the parameter μ according to (16.152).

This mean field equation takes the form of potential vorticity linearly proportional to the stream function and hence it can be solved explicitly as (see Subsection 16.2.1)

$$\psi_\mu = \begin{cases} \dfrac{2\Omega/\sqrt{3}+\hat{h}_{10}}{\mu+2}Y_{10} + \sum\limits_{(n,m)\neq(1,0)} \dfrac{\hat{h}_{nm}}{n(n+1)+\mu}Y_{nm}, \\ \qquad \text{if } \mu \neq -n(n+1), \\ \dfrac{2\Omega/\sqrt{3}+\hat{h}_{10}}{\mu+2}Y_{10} + \sum\limits_{(n,m)\neq(1,0),n\neq n_0} \dfrac{\hat{h}_{nm}}{n(n+1)+\mu}Y_{nm} + \sum\limits_{|m|\leq n_0} c_{n_0m}Y_{n_0m}, \\ \qquad \text{if } \mu = -(n_0+1)n_0 \text{ and } P_{n_0}(2\Omega z+h)=0, \end{cases} \tag{16.153}$$

where c_{nm} are arbitrary constants satisfying the reality condition and the topography has the surface spherical harmonics expansion

$$h = \sum_{n,m} \hat{h}_{nm}Y_{nm}.$$

The energy of the most probable mean field takes the form, in the case $\mu \neq -\Lambda_n$, for all n

$$E(q_\mu) = \left(\frac{2\Omega/\sqrt{3}+\hat{h}_{10}}{\mu+2}\right)^2 + \sum_{(n,m)\neq(1,0)} \frac{\Lambda_n}{2}\left(\frac{\hat{h}_{nm}}{\mu+\Lambda_n}\right)^2. \tag{16.154}$$

The Lagrange multiplier α, β can be determined via enforcing the energy and enstrophy constraints similar to the flat geometry case treated in Section 6.7. In particular, we have:

(i) *Generic case*:

$$\hat{h}_{10} \neq -2\Omega/\sqrt{3}, \text{ or } \hat{h}_{11} \neq 0. \tag{16.155}$$

Then there exists a unique $\mu \in (-2,\infty)$ such that the energy constraint is satisfied, i.e. $E(q_\mu) = E_0$. This completely determines the unique most probable mean field q_μ, which is non-linearly stable according to Proposition 16.1. The Lagrange multiplier α can then be determined as

$$\alpha = (2(\mathcal{E}_0 - \mathcal{E}(q_\mu)))^{-1}, \quad \beta = \mu\alpha. \tag{16.156}$$

(ii) *Degenerate case*:

$$\hat{h}_{10} = -2\Omega/\sqrt{3}, \text{ and } \hat{h}_{11} = 0. \tag{16.157}$$

There are two subcases depending on the energy level:

- Subcase of low energy: $E_0 < E_2 = \frac{1}{2}\sum_{n\geq 2} \frac{n(n+1)|\hat{h}_{nm}|^2}{(n(n+1)-2)^2}$.
In this case, for each energy level E_0, the mean field equation (16.152) has only one solution ψ_μ given by the first case of (16.153) with $\mu = \mu(E_0) \in (-2,\infty)$. This

is the unique mean field corresponding to the most probable state and is non-linear stable according to Proposition 16.1.

- Subcase of high energy: $E_0 > E_2$.
  In this case, for each energy level E_0, there are infinitely many solutions to the mean field equation (16.152) given by the second case of (16.153) with $\mu = -2(n_0 = 1)$ and the c_{1m}s satisfying the energy constraint. These are the (non-unique) most probable mean fields. They are marginally stable according to Proposition 16.1.

The interested reader may recover the details via following the argument for the flat geometry case in Subsections 6.6.4 and 6.7.4.

Empirical energy–enstrophy theory with ground state topography

Here we consider the case when the topography lives on the ground energy shell, i.e. $P_1 h = h$. In this case, the ground state modes have their independent linear dynamics. Thus there is no sufficient mixing on the ground energy shell which makes the prediction from the previous subsection unrealizable. Hence we consider the statistics of the high modes, $\psi' = \psi - P_1\psi$ instead. Recall that the energy and enstrophy of the high modes are conserved for this case of ground state topography (see (16.88). Thus the empirical statistical theory with energy and enstrophy as constraints can be applied. The procedure is exactly the same as for the general topography case but with ρ' representing the one-point statistics of the vorticity of the high modes ($\Delta\psi' = q'$), i.e.

$$\int_a^b \rho'(\mathbf{s}, \lambda) = \text{Prob}\{a \le q'(\mathbf{s}) < b\}. \tag{16.158}$$

Thus the mean field is given by

$$\bar{q}'(\mathbf{s}) = \int \lambda\rho'(\mathbf{s}, \lambda)\, d\lambda = \Delta\bar{\psi}'(\mathbf{s}).$$

Following the same argument as for the general topography case (or the flat geometry case from Chapter 6), we deduce that the most probable mean field, $\bar{\psi}'_\mu$, must satisfy the following mean field equation

$$\Delta\bar{\psi}'_\mu = \mu\bar{\psi}'_\mu, \tag{16.159}$$

where $\mu = \frac{\beta}{\alpha}$ with α and β being the Lagrange multipliers for the enstrophy and energy respectively.

This is similar to the degenerate subcase in the case with general topography. In particular, there is no unique most probable mean field. Indeed, for a given energy level E'_0, the set of most probable mean fields consist of all states in the second energy shell with this given energy level. These most probable mean fields are marginally stable according to Proposition 16.1 and unstable in general according to Proposition 16.3. The Lagrange multipliers are determined as

$$\mu = -6, \quad \alpha = \left(2\left(\mathcal{E}_0' - \mathcal{E}'\left(q_\mu'\right)\right)\right)^{-1}, \quad \beta = \mu\alpha.$$

16.5.2 Complete statistical mechanics.

We now consider the complete statistical mechanics for the inviscid unforced barotropic quasi-geostrophic equation on the unit sphere (16.62). The approach here is parallel to the flat geometry case and the reader is referred to Chapter 8 for more details. As usual we first consider a finite-dimensional truncated system of the quasi-geostrophic equations. Derive the most probable state for the truncated system and then take the continuum limit.

Again we have to separate the case with general topography and the case with ground state topography, where the statistics of the high modes must be considered. We will treat here the case with general topography and comment on the case of ground state topography.

The situation on the sphere with general topography is roughly treated in the work of Frederiksen and Sawford (1980). But we will supply more details including the Liouville property and the non-extensive thermodynamic limit (a topic which was neglected by Frederiksen and Sawford).

Unlike the flat geometry case, there are several common truncations to the whole system in the spherical case. We will recall three here. It seems convenient to use the complex (in ϕ) notation in this section (see Section 16.1).

We recall that $z = \sin\theta$, and introduce the expansion in terms of the complex surface spherical harmonics

$$\psi = \sum_{n,m} \hat{\psi}_{nm}(t) Y_{nm}, \quad h = \sum_{n,m} \hat{h}_{nm}(t) Y_{nm}, \tag{16.160}$$

where $\hat{\psi}_{nm}$ and $\hat{h}_{nm}$ satisfies the following realty property

$$\hat{\psi}_{n,-m} = \hat{\psi}^{\star}_{nm}, \; \hat{h}_{n,-m} = \hat{h}^{\star}_{nm},$$

where $\star$ denote the complex conjugation operation.

Rhomboidal truncation: For fixed Λ

$$\psi_\Lambda(\phi, z, t) = \sum_{m=-\Lambda}^{\Lambda} \sum_{n=|m|}^{|m|+\Lambda} \hat{\psi}_{nm}(t) Y_{nm}. \tag{16.161}$$

Triangular truncation:

$$\psi_\Lambda(\phi, z, t) = \sum_{m=-\Lambda}^{\Lambda} \sum_{n=|m|}^{\Lambda} \hat{\psi}_{nm}(t) Y_{nm}. \tag{16.162}$$

Parallelogrammic truncation: For fixed Λ, Λ'

$$\psi_{\Lambda,\Lambda'}(\phi, z, t) = \sum_{m=-\Lambda}^{\Lambda} \sum_{n=|m|}^{|m|+\Lambda'} \hat{\psi}_{nm}(t) Y_{nm}. \tag{16.163}$$

In any case the *truncated dynamics* takes the following form of:

Truncated Dynamics of the Barotropic Quasi-Geostrophic Equations on the unit sphere

$$\frac{\partial \Delta\psi_\Lambda}{\partial t} + P_\Lambda \; J(\psi_\Lambda, \Delta\psi_\Lambda + 2\Omega z + h_\Lambda) = 0, \tag{16.164}$$

where ψ_Λ and h_Λ are the truncated stream function and topography, and P_Λ is the projection onto the finite-dimensional space, specified by the truncation procedure. The truncated dynamics can be also formulated in terms of the coefficients of the surface spherical harmonics

$$\begin{aligned} -n(n+1)\frac{d\hat{\psi}_{nm}(t)}{dt} &= \frac{i}{2}\sum_{k,l,r,s} K_{nls}^{mkr}(l(l+1) - s(s+1))\hat{\psi}_{lk}(t)\hat{\psi}_{sr}(t) \\ &\quad + i\sum_{k,l,r,s} K_{nls}^{mkr}\hat{\psi}_{sr}(t)\hat{h}_{lk}(t) - 2\Omega m i\hat{\psi}_{nm}(t), \end{aligned} \tag{16.165}$$

where

$$K_{nls}^{mkr} = N_{nm}N_{lk}N_{sr}\int_{-1}^{1} P_{nm}(z)(kP_{lk}(z)\frac{dP_{sr}(z)}{dz} - rP_{sr}(z)\frac{dP_{lk}(z)}{dz})dz, \tag{16.166}$$

with the following restriction on the summation indices

$$m = k + r, \quad |l - s| < n < l + s, \quad n + l + s = \text{odd integer}, \tag{16.167}$$

see for instance Jones (1985, pages 1–19), plus additional restriction on the indices m, n, l, k, s, r dictated by the chosen truncation.

We need to check the conservation of energy and potential enstrophy and the Liouville property of the truncated system before we can proceed to derive the most probable states for the truncated system and take the continuum limit.

Conservation of energy and potential enstrophy

We first check that the truncated system conserves truncated energy and potential enstrophy. This follows easily from the following identities

$$\begin{aligned} \int_{S^2} P_\Lambda(J(\psi_\Lambda, q_\Lambda))\psi_\Lambda d\mathbf{s} &= \int_{S^2} J(\psi_\Lambda, q_\Lambda)\psi_\Lambda d\mathbf{s} = 0, \\ \int_{S^2} P_\Lambda(J(\psi_\Lambda, q_\Lambda))q_\Lambda d\mathbf{s} &= \int_{S^2} J(\psi_\Lambda, q_\Lambda)q_\Lambda d\mathbf{s} = 0. \end{aligned} \tag{16.168}$$

Therefore

$$E_\Lambda = \frac{1}{2}\sum_{n,m}\Lambda_n|\hat{\psi}_{nm}|^2 = \frac{1}{2}\fint_{S^2}|\nabla\psi_\Lambda|^2,$$

$$\mathcal{E}_\Lambda = \frac{1}{2}\sum_{(n,m)\neq(1,0)}|-\Lambda_n\hat{\psi}_{nm}+\hat{h}_{nm}|^2+|-2\hat{\psi}_{10}+2\Omega+\hat{h}_{10}|^2 = \frac{1}{2}\fint_{S^2}q_\Lambda^2, \tag{16.169}$$

are conserved in time. The proof is exactly the same as in the flat geometry case (see Section 8.3).

The Liouville property

Next we check the Liouville property. Notice that the finite-dimensional truncated system may be viewed as a system of ODEs with the variables

$$\frac{d\vec{X}}{dt} = \vec{F}(\vec{X}),\ \ \vec{X}|_{t=0} = \vec{X}_0, \tag{16.170}$$

where

$$\vec{X} = (a_{n0}, a_{nm}, b_{nm}),\, m > 0$$

is the real determining variables and

$$\hat{\psi}_{nm} = a_{nm} + ib_{nm}.$$

The vector $\vec{F}$ has the following properties: F_n^0 is independent of a_{n0}; F_{nm}^a is independent of a_{nm}; F_{nm}^b is independent of b_{nm}.

To see why this is true, we consider the equation for $\hat{\psi}_{nm}$ and we just list the terms on the right-hand side that might involve either $\hat{\psi}_{nm}$ or $\hat{\psi}_{n,-m}$. We have

$$\begin{aligned}
&\frac{-i}{n(n+1)}\sum_s(n(n+1)-s(s+1))K_{nns}^{mm0}\hat{\psi}_{nm}\hat{\psi}_{s0}\\
&+\frac{-i}{n(n+1)}\sum_s(n(n+1)-s(s+1))K_{nns}^{m,-m,2m}\hat{\psi}_{n,-m}\hat{\psi}_{s,2m}\\
&+\frac{-i}{n(n+1)}\sum_s K_{nns}^{mm0}\psi_{nm}(t)h_{s0}\\
&+\frac{-i}{n(n+1)}\sum_s K_{nns}^{m,-m,2m}\hat{\psi}_{n,-m}h_{s,2m}\frac{2\Omega mi}{n(n+1)}\hat{\psi}_{nm},
\end{aligned} \tag{16.171}$$

according to the restrictions on the summation indices.

Notice that $\hat{\psi}_{s0}$ and $\hat{h}_{s0}$ are real and $K_{nns}^{m,-m,2m} = 0$ by a simple integration by parts. This implies that the dynamic equation of a_{nm} is independent of a_{nm}, the dynamic equation of b_{nm} is independent of b_{nm}. Thus we have the detailed Liouville property

$$\frac{\partial F^a_{nm}}{\partial a_{nm}} = \frac{\partial F^b_{nm}}{\partial b_{nm}} = 0, \tag{16.172}$$

which further implies the Liouville property

$$\operatorname{div} \vec{F} = \sum \left(\frac{\partial F^a_{nm}}{\partial a_{nm}} + \frac{\partial F^b_{nm}}{\partial b_{nm}} \right) + \sum \frac{\partial F^0_n}{\partial a_{n0}} = 0. \tag{16.173}$$

Most probable state for the truncated system

We can now apply the general theory of equilibrium statistical mechanics for ODEs from Chapter 7 to the truncated barotropic quasi-geostrophic equations on the sphere. This is parallel to the flat geometry case treated in Section 8.3 from Chapter 8.

According to Section 7.2, the most probable measure, i.e. the Gibbs measure, for the truncated system is given by

$$\mathcal{G} = C^{-1} \exp(-\beta E_\Lambda - \alpha \mathcal{E}_\Lambda), \tag{16.174}$$

which can be rewritten as

$$\mathcal{G} = C^{-1} \exp\left(-\alpha \sum_{n,m} \Lambda_n(\mu + \Lambda_n) |\hat{\psi}_{nm} - \bar{\hat{\psi}}_{nm}|^2 \right), \tag{16.175}$$

where α is the Lagrange multiplier for the potential enstrophy and β is the Lagrange multiplier for the energy, $\mu = \frac{\beta}{\alpha}$, and the mean states are defined as

$$\begin{aligned} \bar{\hat{\psi}}_{nm} &= \frac{\hat{h}_{nm}}{\mu + \Lambda_n}, \quad \text{for } (n, m) \neq (1, 0), \\ \bar{\hat{\psi}}_{10} &= \frac{2\Omega/\sqrt{3} + \hat{h}_{10}}{\mu + 2}, \end{aligned} \tag{16.176}$$

or, in terms of the real valued surface spherical harmonics

$$\begin{aligned} \bar{a}_{nm} &= \frac{ha_{nm}}{\mu + \Lambda_n}, \quad \bar{b}_{nm} = \frac{hb_{nm}}{\mu + \Lambda_n}, \quad \text{for } (n, m) \neq (1, 0), \\ \bar{a}_{10} &= \frac{2\Omega/\sqrt{3} + ha_{10}}{\mu + 2}, \; \bar{b}_{10} = 0. \end{aligned} \tag{16.177}$$

It is clear that $\mathcal{G}$ is a product of Gaussian functions of one variable. In order to ensure that it is a probability density function, i.e. to ensure it is *realizable*, we must require that the Gaussian functions decay at infinity, i.e. $\beta + \alpha \Lambda_n > 0$ for all n such that $\Lambda_n \leq \Lambda$.

Case 1: $\alpha > 0$. The ratio $\mu = \beta/\alpha$ must be larger than the maximum of $-\Lambda_n$, i.e.

$$\mu = \frac{\beta}{\alpha} > -\Lambda_1 = -2, \tag{16.178}$$

Case 2: $\alpha < 0$.

$$\mu < -\Lambda. \tag{16.179}$$

The second case is physically uninteresting, since we are not able to take the limit of $\Lambda \to \infty$ in this scenario as we shall see below.

We may determine the Lagrange multipliers α, β by using the probability density constraint, and the energy, potential enstrophy constraints.

It is also possible to verify that $\mathcal{G}$ corresponds to a maximum of the Shannon entropy $\mathcal{S}(P)$ simply by the fact that the entropy is concave and the first-order variation vanishes at the critical point $\mathcal{G}$.

Now we compute the mean of the most probable mean state of the truncated dynamic system. It is easy to see, since the Gibbs measure is a product of Gaussians, that

$$\begin{aligned}
\langle \hat{\psi}_{nm} \rangle &= \int_{R^N} \psi_{nm} \mathcal{G}(\vec{X}) d\vec{X} = \bar{\hat{\psi}}_{nm} = \frac{\hat{h}_{nm}}{\mu + \Lambda_n}, \quad \text{for } (n, m) \neq (1, 0), \\
\langle \hat{\psi}_{10} \rangle &= \bar{\hat{\psi}}_{10} = \frac{2\Omega/\sqrt{3} + \hat{h}_{10}}{\mu + 2}.
\end{aligned} \tag{16.180}$$

Therefore the mean of the most probable stream function $\bar{\psi}_\Lambda$ is given by

$$\bar{\psi}_\Lambda = \sum_{n,m} \bar{\hat{\psi}}_{nm} Y_{nm} \tag{16.181}$$

with the restriction on the summation index governed by the truncation. We observe that the most probable mean state solves the equation

$$\Delta \bar{\psi}_\Lambda + h_\Lambda + 2\Omega \sin\theta = \mu \bar{\psi}_\Lambda. \tag{16.182}$$

This is the truncated version of the exact solution studied in Subsection 16.2.1. These solutions are non-linearly stable in the case $\mu > -2$ via an argument similar to the untruncated case (see Proposition 16.2.3.).

The continuum limit

The problem of taking continuum limit here is very much the same as in the flat geometry case in Section 8.6 from Chapter 8. The reader is referred to that chapter for more details.

We first observe that there is no limit if $\mu (= \mu_\Lambda) < -\Lambda$ (or the limit is zero).

As in the flat geometry case, the finite-dimensional energy and potential enstrophy have two parts: a mean part and a fluctuation part. More precisely

$$E_\Lambda = \langle E \rangle_{\mathcal{G}} = \bar{E}_\Lambda + E'_\Lambda, \tag{16.183}$$

where

$$\begin{aligned} \bar{E}_\Lambda &= \frac{1}{2}\left(\sum_{(n,m)\neq(1,0)} \frac{\Lambda_n |\hat{h}_{nm}|^2}{(\mu+\Lambda_n)^2} + \frac{2|2\Omega/\sqrt{3}+\hat{h}_{10}|^2}{(\mu+2)^2} \right), \\ E'_\Lambda &= \frac{1}{4\alpha} \sum_{n,m} \frac{1}{\mu+\Lambda_n}. \end{aligned} \tag{16.184}$$

The potential enstrophy has an analogous decomposition

$$\mathcal{E}_\Lambda = \langle \mathcal{E} \rangle_{\mathcal{G}} = \bar{\mathcal{E}}_\Lambda + \mathcal{E}'_\Lambda, \tag{16.185}$$

where

$$\begin{aligned} \bar{\mathcal{E}}_\Lambda &= \frac{1}{2}\left(\sum_{(n,m)\neq(1,0)} \frac{\mu^2 |\hat{h}_{nm}|^2}{(\mu+\Lambda_n)^2} + \frac{\mu^2|2\Omega/\sqrt{3}+\hat{h}_{10}|^2}{(\mu+2)^2} \right), \\ \mathcal{E}'_\Lambda &= \frac{1}{4\alpha} \sum_{n,m} \frac{\Lambda_n}{\mu+\Lambda_n}. \end{aligned} \tag{16.186}$$

Similar to the flat geometry case which is treated in Chapter 8, we may take the so-called *non-extensive thermodynamic limit* as follows.
Case with $2\Omega/\sqrt{3}+\hat{h}_{10} \neq 0$ or $\hat{h}_{11} \neq 0$, the procedure is the following:

1. (Choice of $\mu = \mu_\Lambda$) We choose μ_Λ to be the largest μ such that the energy constraint is met by the mean state only, i.e. for given E^0, pick a unique $\mu_\Lambda \in (-2, \infty)$ such that $E(\bar{q}_{\mu_\Lambda}) = E^0$.
2. (Choice of $\alpha = \alpha_\Lambda$) We choose α_Λ so that the enstrophy constraint is met exactly, i.e. for given $\mathcal{E}^0$, pick α_Λ so that $\mathcal{E}^0 = \bar{\mathcal{E}}_\Lambda + \mathcal{E}'_\Lambda$.
3. (The limit of the mean states) Notice that the μ_Λs picked in 1 are monotonically decreasing as Λ increases

   $$\lim_{\Lambda\to\infty} \mu_\Lambda = \mu_\infty,$$

   where μ_∞ is the unique number in $(-2, \infty)$ such that $E(\psi_{\mu_\infty}) = E_0$. It is then straightforward to verify that

   $$\lim_{\Lambda\to\infty} \psi_{\mu_\Lambda} = \psi_{\mu_\infty},$$

   and

   $$\Delta\psi_{\mu_\infty} + h + 2\Omega\sin\theta = \mu\psi_{\mu_\infty}.$$

   ψ_μ is non-linearly stable according to Proposition 16.1.

For the degenerate case ($P_1(2\Omega\sin\theta + h) = 0$), taking the continuum limit is more involved and the fluctuation part could persist if the energy is high enough. The reader is referred to Chapter 8 for more details for the flat geometry case. The spherical geometry case can be treated similarly.

Comment on the case with ground state topography

In the case with ground state topography, we need to consider the statistics of the high modes, $\psi' = \psi - P_1\psi$, since the statistics of the ground state modes are trivial and the EEST theory is applicable to the high modes due to the conservation of energy and enstrophy for the high modes (see Subsection 16.2.1 and (16.2.2)).

The truncated high modes, ψ'_Λ, satisfies the following equations, thanks to (16.75) which governs the dynamics of the high modes

$$\frac{\partial \Delta\psi'_\Lambda}{\partial t} + J(P_1\psi, \Delta\psi'_\Lambda + 2\psi'_\Lambda) + P'_\Lambda J(\psi'_\Lambda, \Delta\psi'_\Lambda + 2\Omega z + h) = 0, \qquad (16.187)$$

where P'_Λ is the projection from the high modes to the appropriate finite-dimensional subspace dictated by the specific truncation.

It is easy to verify that the statistical mechanics theory from Chapter 7 applies to this system of ODEs with the Gibbs measure taking the form

$$\mathcal{G} = C^{-1}\exp\left(-\alpha \sum_{n\geq 2, |m|\leq n} \Lambda(\mu+\Lambda)|\hat{\psi}_{nm}|^2\right). \qquad (16.188)$$

This is the same situation as the flat geometry case without geophysical effect that we treated in Subsection 8.6.3. In particular, we realize that the mean state is identically zero and hence all energy and enstrophy must reside in the fluctuation part. The continuum limit can be carried out as follows:

(i) (Choice of μ_Λ and α_Λ) We choose μ_Λ and α_Λ so that all energy resides in the lowest energy shell (in this case it is the second energy shell) and the enstrophy constraint is satisfied as well, i.e.

$$E'_0 = \frac{1}{4\alpha_\Lambda}\sum_{|m|\leq 2}\frac{1}{\mu_\Lambda + \Lambda_2} = \frac{5}{4\alpha_\Lambda(\mu_\Lambda + 6)},$$

$$\mathcal{E}'_0 = \mathcal{E}_*(E'_0) + \mathcal{E}'_{\Lambda,2} = 6E'_0 + \frac{1}{4\alpha_\Lambda}\sum_{|m|\leq n, n\geq 3}\frac{\Lambda_n}{\mu_\Lambda + \Lambda_n}.$$

It is easy to see that α_Λ approaches infinity as Λ approaches infinity and $\mu_\Lambda \to -6$ as Λ approaches infinity.

(ii) (Limit of the Gibbs measure) It is easy to see that the Gibbs measure converges in this case to the product of Gaussian measures on the lowest modes (second energy shell) and Dirac measures centered at zero for higher modes (third energy shell and higher). Indeed, we have

$$\alpha_\Lambda \Lambda_n(\mu_\Lambda + \Lambda_n) \to \infty, \text{ as } \Lambda \to \infty, \text{if } n \geq 3,$$

which implies the convergence of to the Dirac measures for modes on the 3rd energy shell or higher; and

$$\alpha_\Lambda \Lambda_2(\mu_\Lambda + \Lambda_2) \to \frac{5}{2E_0'}, \text{ as } \Lambda \to \infty,$$

which implies the convergence to Gaussians for the modes on the 2nd energy shell.

The interested reader may fill in the details via consulting Subsection 8.6.3.

16.6 Statistical theories with a few constraints and statistical theories with many constraints on the unit sphere

The energy–enstrophy theory is basically a linear theory which is evident from the mean field equations. Parallel to the flat geometry case, we may consider equilibrium statistical theories for barotropic quasi-geostrophic flows on the sphere, which produce non-linear mean field equations. These include the empirical statistical theories with prior distribution (ESTP), point vortex statistical theory (PVST), and the empirical statistical theories with many constraints (ESTMC) (see Chapter 9). Once again we need to distinguish the case with general topography and the case with ground state topography. However the ESTMC is only applicable in the general topography case since the high moments for the high modes in the ground state topography case are not necessarily conserved as we discussed in Subsection 16.2.2.

A few constraint theory with prior distribution

Here we consider equilibrium statistical theories, which takes into consideration a few judiciously chosen conserved quantities and appropriate prior distribution depending on the applications. The material here is similar to Sections 9.2 and 9.3 from Chapter 9. The reader is referred to that chapter for more details while we only provide a sketch here.

Since we need to take into consideration the robust conserved quantity circulation into our statistical theory, and, since the circulation is identically zero on the whole sphere, we will impose odd symmetry in z for the stream function and the topography. As we have checked in Subsection 16.2.2, the odd symmetry is preserved under the barotropic quasi-geostrophic dynamics and the energy and circulation in the northern hemisphere are conserved in time. Hence our statistical theory will apply to flows on the northern hemisphere. This is similar to the flat geometry case where we considered the channel geometry or swimming pool geometry (see Chapters 1 and 9 for details).

Energy–circulation theory

This is the parallel of Subsection 9.2.1 for the spherical geometry.

Here we consider two conserved quantities, the energy E_N and the circulation Γ_N in the northern hemisphere (see (16.2.2))

$$E_N = -\frac{1}{2}\fint_{S_N^2} \psi\omega, \quad \Gamma_N = \fint_{S_N^2} q, \tag{16.189}$$

with a prior distribution $\Pi_0(\mathbf{s}, \lambda)$

$$\fint_{S_N^2} \int \Pi_0(\mathbf{s}, \lambda) d\lambda d\mathbf{s} = 1. \tag{16.190}$$

As usual we postulate the one-point statistics $\rho(\mathbf{s}, \lambda)$

$$\begin{aligned} &\rho(\mathbf{s}, \lambda) \in \mathcal{PM}(S_N^2 \times R^1), \rho \geq 0, \fint_{S_N^2} \int \rho(\mathbf{s}, \lambda) d\lambda d\mathbf{s} = 1, \\ &\mathbf{s} \to \rho(\mathbf{s}, \lambda) \in \mathcal{PM}(R^1), \text{ for all } \mathbf{s} \in S_N^2, \\ &\int \rho(\mathbf{s}, \lambda) d\lambda = 1, a.e.\ \mathbf{s} \in S_N^2, \\ &\int_{q-}^{q+} \rho(\mathbf{s}, \lambda) = \text{Prob}\{q- \leq q(\mathbf{s}) \leq q+\}. \end{aligned} \tag{16.191}$$

We define the relative Shannon entropy $\mathcal{S}$ as

$$\mathcal{S}(\rho, \Pi_0) = -\fint_{S_N^2} \int \rho \ln\left(\frac{\rho}{\Pi_0}\right) d\lambda d\mathbf{s}. \tag{16.192}$$

We may represent the energy and circulation as well as the mean field in terms of the one-point statistics as

$$\begin{aligned} &\bar{q}(\mathbf{s}) = \int \lambda\rho(\mathbf{s}, \lambda) d\lambda, \quad \Gamma_N = \fint_{S_N^2} \int \lambda\rho(\mathbf{s}, \lambda) d\lambda d\mathbf{s}, \\ &E(\rho) = E(\bar{q}) = -\frac{1}{2}\fint_{S_N^2} \bar{\psi}\bar{\omega}. \end{aligned} \tag{16.193}$$

We then proceed to search for the most probable states with given constraints on the energy and circulation.

Going through a standard Lagrangian multiplier argument procedure (see Section 9.2.1 for more details), we deduce that the most probable one-point statistics ρ^* must satisfy

$$\rho^\star(\mathbf{s},\lambda) = \frac{\exp(\beta\psi^\star(\mathbf{s})\lambda - \gamma\lambda)\Pi_0(\mathbf{s},\lambda)}{\int \exp(\beta\psi^\star(\mathbf{s})\lambda - \gamma\lambda)\Pi_0(\mathbf{s},\lambda)d\lambda}, \tag{16.194}$$

where β and γ are the Lagrange multipliers for the energy and circulation, respectively, and ψ^* is the most probable mean stream function.

As usual we introduce the *partition function*

$$\mathcal{Z}(\beta,\psi,\mathbf{s}) = \log\left(\int \exp((\beta\psi(\mathbf{s}) - \gamma)\lambda)\Pi_0(\mathbf{s},\lambda)d\lambda\right). \tag{16.195}$$

We then have

$$q^\star = \int \lambda\rho^\star(\mathbf{s},\lambda)d\lambda = \frac{1}{\beta}\left(\frac{\partial\mathcal{Z}}{\partial\psi}\right)\bigg|_{\psi^\star}.$$

Hence the most probable mean state satisfies the following equation

$$\Delta\psi^\star + 2\Omega\sin\theta + h = \frac{1}{\beta}\left(\frac{\partial\mathcal{Z}}{\partial\psi}\right)\Big|_{\psi^\star}. \tag{16.196}$$

For the special choice of prior distribution as a uniform distribution on an given interval, i.e.

$$\Pi_0(\mathbf{s},\lambda) = \frac{1}{q_0^+ - q_0^-} I_{q_0^-}^{q_0^+},$$

we recover the Langevin theory. For another special choice of prior distribution as a Gaussian

$$\Pi_0(\lambda) = c\exp\left(-\frac{\beta\lambda^2}{2}\right),$$

we then recover the previous theory with energy and enstrophy conservation. See Subsection 9.2.1 for more details.

In the case with ground state topography, we need to consider the statistics of the high modes with odd symmetry. The energy–circulation theory still applies, since they are conserved for the high modes in the northern hemisphere (see (16.89)). The predictions are very much the same except there is no geophysical effects for the predictions of the high modes. We leave the detail to the interested reader.

The point vortex theory

Here we consider equilibrium statistical theories for point vortices. This is parallel to Section 9.3.1 for the flat geometry case.

We exclude geophysical effects here and hence

$$h \equiv 0, \quad \Omega \equiv 0. \tag{16.197}$$

Since this implies that we have ground state topography and odd symmetry may be imposed, we will consider statistics for the high modes (modes in the second energy shell or higher) in the northern hemisphere with odd symmetry.

Recall that the energy and circulation for the high modes in the northern hemisphere are conserved in this case (see Section 16.2.2) and they can be represented in terms of the one-point statistics ρ' for the high modes ψ' as

$$\begin{aligned} E'_N &= -\frac{1}{2} \fint_{S_N^2} \bar{\psi}' \bar{\omega}' = E_0, \\ \Gamma'_N &= \fint_{S_N^2} \bar{\omega}' = \fint_{S_N^2} \int \lambda \rho'(\mathbf{s}, \lambda) d\lambda d\mathbf{s} = \Gamma_0, \\ \bar{\omega}' &= \int \lambda \rho'(\mathbf{s}, \lambda) d\lambda. \end{aligned} \tag{16.198}$$

We also consider a prior distribution Π_0 on the northern hemisphere

$$\fint_{S_N^2} \int \Pi_0(\mathbf{s}, \lambda) d\lambda d\mathbf{s} = 1. \tag{16.199}$$

In order to accommodate point vortices, we do not enforce the one-point statistics to be probability density at each point on the northern hemisphere. We instead postulate

$$\fint_{S_N^2} \int \rho'(\mathbf{s}, \lambda) d\lambda d\mathbf{s} = 1. \tag{16.200}$$

As usual we define the relative Shannon entropy $\mathcal{S}$ as

$$\mathcal{S}(\rho', \Pi_0) = -\fint_{S_N^2} \int \rho' \ln\left(\frac{\rho'}{\Pi_0}\right) d\lambda d\mathbf{s}. \tag{16.201}$$

Going through a standard Lagrangian multiplier argument procedure we have that the most probable one-point statistics $\rho*$ must satisfy

$$\rho'^{\star}(\mathbf{s}, \lambda) = \frac{\exp((\beta\psi'^{\star}(\mathbf{s}) - \gamma)\lambda)\Pi_0(\mathbf{s}, \lambda)}{\fint_{S_N^2} \int \exp((\beta\psi'^{\star}(\mathbf{s}) - \gamma)\lambda)\Pi_0(\mathbf{s}, \lambda) d\lambda d\mathbf{s}}, \tag{16.202}$$

where β and γ are Lagrange multipliers for the energy and circulation respectively, and $\psi'*$ is the most probable mean stream function.

This implies, since

$$q'^{\star} = \omega'^{\star} = \int \lambda \rho'^{\star}(\mathbf{s}, \lambda) d\lambda,$$

that we have the mean field equation

$$\Delta\psi'^{\star} = \omega'^{\star} = \frac{\int \lambda \exp((\beta\psi'^{\star}(\mathbf{s}) - \gamma)\lambda)\Pi_0(\mathbf{s}, \lambda)ds}{\fint_{S_N^2} \int \exp((\beta\psi'^{\star}(\mathbf{s}) - \gamma)\lambda)\Pi_0(\mathbf{s}, \lambda)d\lambda d\mathbf{s}}. \tag{16.203}$$

For the special choice of prior distribution

$$\Pi_0(\mathbf{s}, \lambda) = \frac{1}{2\pi}\delta_{\omega_0}, \tag{16.204}$$

we have

$$\Delta\psi'^{\star} = \frac{\omega_0 \exp(\beta\omega_0\psi'^{\star})}{\fint_{S_N^2} \exp(\beta\omega_0\psi'^{\star})d\mathbf{s}}. \tag{16.205}$$

Note that β can be determined by the energy constraint and ω_0 can be determined by the circulation constraint

$$\omega_0 = \frac{\Gamma_0}{2\pi}.$$

The reader is referred to Chapter 9 for more details.

16.6.1 Infinitely many constraint statistical theory

In this last subsection of this chapter we discuss the empirical statistical theory with infinitely many constraints consists of all generalize potential enstrophy for the inviscid unforced barotropic quasi-geostrophic equations on the unit sphere (16.62). The result and argument here are very similar to those for the flat geometry, which we treated in detail in Section 9.4. The reader is referred to that section for more ingredients and motivations. One catch here is that we cannot apply the empirical statistics with many conserved quantities (ESTMC) to the high modes even if the topography lives on the ground energy shell, since high moments of the high modes are not necessarily conserved (see Subsection 16.2.2 for more details). This raises the question of realizability of the predictions of the ESTMC in the case of ground state topography, since the ground state modes do not possess enough mixing, a required ingredient of the physical realizability of the most probable states.

As in the flat geometry case (see Section 9.4), we avoid postulating infinitely many constraints in the Lagrange multiplier method by using the distribution function for the initial condition.

Let P_q denote the distribution function of q. Thus if q_0 is the initial potential enstrophy and q its evolution, the conservation of generalized enstrophy is equivalent to, utilizing the distribution functions

$$dP_q(\lambda) = dP_{q_0}(\lambda), \quad \text{for all } \lambda \in supp(dP_{q_0}). \tag{16.206}$$

This can be reformulated, in terms of the one-point statistics $\rho(\mathbf{s}, \lambda)$ of the potential vorticity q as

$$\fint_{S^2} \rho(\mathbf{s}, \lambda) d\mathbf{s}\, d\lambda = dP_q(\lambda) = dP_{q_0}(\lambda), \quad \text{for all } \lambda \in supp(dP_{q_0}). \tag{16.207}$$

Assumption: For convenience in exposition, we consider the special case that $dP_{q_0}(\lambda)$ is absolutely continuous with respect to the Lebesgue measure, i.e.

$$dP_{q_0}(\lambda) = \Pi_{q_0}(\lambda) d\lambda, \tag{16.208}$$

for some integrable function $\Pi_{q_0}(\lambda)$.

As usual we define the relative entropy with prior distribution Π_{q_0} as

$$\mathcal{S}(\rho, P_{q_0}) = -\fint_{S^2} \int \ln\left(\frac{\rho}{\Pi_{q_0}}\right) \rho d\lambda d\mathbf{s}. \tag{16.209}$$

We recall the constraints on the one-point statistics ρ are

$$\begin{gathered} E(\rho) = -\frac{1}{2} \fint_{S^2} \bar{\psi}\bar{\omega} = E_0, \int \rho(\mathbf{s}, \lambda) d\lambda = 1, \text{ for all } \mathbf{s} \in S^2, \\ \fint_{S^2} \rho(\mathbf{s}, \lambda) d\mathbf{s} = \Pi_{q_0}(\lambda), \text{ for all } \lambda \in supp(\Pi_{q_0}). \end{gathered} \tag{16.210}$$

The usual Lagrange multiplier method implies that the most probable state $\rho^\star$ must satisfy

$$\rho^\star(\mathbf{s}, \lambda) = \frac{\exp(\beta\psi^\star(\mathbf{s})\lambda - \gamma(\lambda))\Pi_{q_0}(\lambda)}{\int \exp(\beta\psi^\star(\mathbf{s})\lambda - \gamma(\lambda))\Pi_{q_0}(\lambda) d\lambda}, \tag{16.211}$$

where β and γ are Lagrange multipliers for the energy and generalized enstrophy constraint respectively, and $\psi*$ is the stream function for the most probable mean field $q* = \int \lambda\rho^* \, d\lambda$.

Again we introduce the partition function

$$\mathcal{Z}(\psi, \beta, \gamma(\lambda)) = \log\left(\int \exp(\beta\psi\lambda - \gamma(\lambda)\right) \Pi_{q_0}(\lambda) d\lambda). \tag{16.212}$$

We have

$$q^\star = \int \lambda\rho^\star(\mathbf{s}, \lambda) d\lambda = \frac{1}{\beta} \frac{\partial \mathcal{Z}}{\partial \psi}\bigg|_{\psi = \psi^\star}.$$

and hence the mean field equation takes the form

$$\Delta\psi^\star + 2\Omega \sin\theta + h = \frac{1}{\beta}\frac{\partial \mathcal{Z}}{\partial \psi}\bigg|_{\psi=\psi^\star}. \tag{16.213}$$

The Lagrange multipliers β, γ are determined by the constraints

$$E(q^\star(\beta,\gamma(\lambda))) = E_0, \quad \fint_{S^2} \rho^\star d\mathbf{s} = \Pi_{q_0}, \quad \text{for all } \lambda \in \text{supp}(\Pi_{q_0}). \tag{16.214}$$

The stability of the most probable state we just derived can be discussed using the stability results we derived in Subsection 16.2.3. This is very similar to the flat geometry case (see for instance Section 9.5). We formulate the result here for the sake of completeness.

Theorem 16.4 *The most probable mean field given by (16.213) are exact steady state solutions to the inviscid unforced barotropic quasi-geostrophic equation on the sphere (16.62). These steady state solutions are non-linearly stable provided one of the following conditions holds:*

(A) $\beta > 0$ *and* $supp(\Pi_{q_0})$ *is bounded;*
(B) $\beta < 0$, $supp(\Pi_{q_0}) \subset [-a, a]$, $-\beta a^2 < 2$.

This is an application of Proposition 16.2 and the interested reader is referred to Section 9.5 for details.

References

Courant, R. and Hilbert D. (1962), *Methods of mathematical physics.* Vols. I and II New York-London: Interscience Publishers (a division of John Wiley & Sons).

Frederiksen, J. S. and Sawford, B. L. (1980), Statistical dynamics of two-dimensional inviscid flow on a sphere. *J. Atmospheric Sci.* **37**(4), 717–732.

Frederiksen, J. S. and Sawford, B. L. (1981), Topographic waves in nonlinear and linear spherical barotropic models. *J. Atmospheric Sci.* **38**, 69–86.

Jones, M. N. (1985), *Spherical Harmonics and Tensor for Classical Field Theory.* John Wiley & Sons Inc.

Lax, P. D. (1997), *Linear Algebra.* New York: John Wiley and Sons.

Pedlosky, J. (1987), *Geophysical Fluid Dynamics*, 2nd edn., New York: Spring-Verlag.

Sawford, B. L. and Frederiksen, J. S. (1983), Mountain torque and angular momentum in barotropic planetary flows: equilibrium solutions. *Quart. J. R. Met. Soc.* **109**, 309–324.

Appendix 1

The purpose of this Appendix is to elaborate on the invariant dynamics on the first two energy shells. This is a special case of the exact dynamics of the ground state modes and the nth energy shell discussed in Subsection 16.2.1 with $n = 2$.

For this purpose let us recall the following tables:

Eigenfunctions corresponding to $-\Lambda_1 = -2 (n = 1)$

$$\begin{array}{ccc} \sin\theta & \cos\theta\cos\phi & \cos\theta\sin\phi \\ z & x & y \\ w_1 & w_{a11} & w_{b11} \end{array} \tag{16.215}$$

Eigenfunctions corresponding to $-\Lambda_2 = -6 (n = 2)$

$$\begin{array}{ccccc} 1 - 3\sin^2\theta & \cos\theta\sin\theta & \cos\theta\sin\theta & \cos^2\theta & \cos^2\theta\times \\ & \times\cos\phi & \times\sin\phi & \times\cos(2\phi) & \sin(2\phi) \\ 1 - 3z^2 & zx & yz & x^2 - y^2 & 2xy \\ w_2 & w_{a21} & w_{b21} & w_{a22} & w_{b22} \end{array} \tag{16.216}$$

It is easy to see, utilizing (16.6)

$$\begin{aligned} J(w_2, z) &= 0, & J(1 - 3z^2, z) &= 0, \\ J(w_{a21}, z) &= -w_{b21}, & J(zx, z) &= -yz, \\ J(w_{b21}, z) &= w_{a21}, & J(yz, z) &= zx, \\ J(w_{a22}, z) &= -2w_{b22}, & J(x^2 - y^2, z) &= -2(2xy), \\ J(w_{b22}, z) &= 2w_{a22}, & J(2xy, z) &= 2(x^2 - y^2). \end{aligned}$$

This implies, since $x^2 = 0.5(x^2 - y^2) + 0.5(1 - z^2)$, that

$$J(x^2, z) = -2xy; \quad J(y^2, z) = 2xy.$$

This further implies, thanks to the rotation symmetry of the sphere and the eigenfunctions (which are homogeneous polynomials of degree 1 and 2),

$$\begin{gathered} J(w_2, x) = J(-3z^2, x) = -3(2yz) = -6w_{b21}, \\ J(w_{a21}, x) = J(zx, x) = xy = 0.5w_{b22}, \\ J(w_{b21}, x) = J(yz, x) = y^2 - z^2 = 0.5w_2 - 0.5w_{a22}, \\ J(w_{a22}, x) = J(-y^2, x) = 2yz = 2w_{b21}, \\ J(w_{b22}, x) = J(2xy, x) = -2zx = -2w_{a21}, \\ J(w_2, y) = J(-3z^2, y) = 6xz = 6w_{a21}, \\ J(w_{a21}, y) = J(zx, y) = z^2 - x^2 = -0.5w_{a22} - 0.5w_2, \\ J(w_{b21}, y) = J(yz, y) = -xy = -0.5w_{b22}, \\ J(w_{a22}, y) = J(x^2, y) = 2xz = 2w_{a21}, \\ J(w_{b22}, y) = J(2xy, y) = 2yz = 2w_{b21}. \end{gathered}$$

Thus for

$$\psi(t) = \sum_{n=1}^{2} a_n(t) w_n + \sum_{n=1}^{2} \sum_{m=1}^{n} (a_{nm}(t) w_{anm} + b_{nm}(t) w_{bnm}),$$

we have

$$\begin{aligned}
J(\psi, \Delta\psi + 2\Omega z) &= 2\Omega J(\psi, z) + J(\psi, \Delta\psi) \\
&= 2\Omega J(\psi, z) + 4J(\psi_2, \psi_1) \\
&= 2\Omega \textstyle\sum_{n=1}^{2} \sum_{m=1}^{n} (-m a_{nm}(t) w_{bnm} + m b_{nm}(t) w_{anm}) \\
&\quad + 4a_1 J(\psi_2, z) + 4a_{11} J(\psi_2, x) + 4b_{11} J(\psi_2, y).
\end{aligned}$$

We observe

$$\begin{aligned}
J(\psi_2, z) &= -a_{21} w_{b21} + b_{21} w_{a21} - 2a_{22} w_{b22} + 2b_{22} w_{a22} \\
&= b_{21} w_{a21} - a_{21} w_{b21} + 2b_{22} w_{a22} - 2a_{22} w_{b22}, \\
J(\psi_2, x) &= -6a_2 w_{b21} + 0.5a_{21} w_{b22} + 0.5b_{21} w_2 - 0.5b_{21} w_{a22} + 2a_{22} w_{b21} \\
&\quad - 2b_{22} w_{a21} \\
&= 0.5b_{21} w_2 - 2b_{22} w_{a21} + (2a_{22} - 6a_2) w_{b21} - 0.5b_{21} w_{a22} \\
&\quad + 0.5a_{21} w_{b22}, \\
J(\psi_2, y) &= 6a_2 w_{a21} - 0.5a_{21} w_{a22} - 0.5a_{21} w_2 - 0.5b_{21} w_{b22} \\
&\quad + 2a_{22} w_{a21} + 2b_{22} w_{b21} \\
&= -0.5a_{21} w_2 + (6a_2 + 2a_{22}) w_{a21} + 2b_{22} w_{b21} \\
&\quad - 0.5a_{21} w_{a22} - 0.5b_{21} w_{b22}.
\end{aligned}$$

Hence the dynamics of the coefficients are given by, for the simple case of no topography ($h = 0$), no damping ($\mathcal{D} = 0$) and no external forcing ($\mathcal{F} = 0$)

$$\begin{aligned}
\frac{da_2}{dt} &= -\frac{1}{3} b_{11} a_{21} + \frac{1}{3} a_{11} b_{21}, \\
\frac{da_{21}}{dt} &= \frac{1}{3} \Omega b_{21} + \frac{2}{3} a_1 b_{21} - \frac{4}{3} a_{11} b_{22} + 4b_{11} a_2 + \frac{4}{3} b_{11} a_{22}, \\
\frac{db_{21}}{dt} &= -\frac{1}{3} \Omega a_{21} - \frac{2}{3} a_1 a_{21} + \frac{4}{3} a_{11} a_{22} - 4a_{11} a_2 + \frac{4}{3} b_{11} b_{22}, \\
\frac{da_{22}}{dt} &= \frac{2}{3} \Omega b_{22} + \frac{4}{3} a_1 b_{22} - \frac{1}{3} a_{11} b_{21} - \frac{1}{3} b_{11} a_{21}, \\
\frac{db_{22}}{dt} &= -\frac{2}{3} \Omega a_{22} - \frac{4}{3} a_1 a_{22} + \frac{1}{3} a_{11} a_{21} - \frac{1}{3} b_{11} b_{21}.
\end{aligned}$$

This system of linear equations can be written in a compact form

$$\frac{d\vec{a}}{dt} = A\vec{a}, \quad \vec{a} = (a_2, a_{21}, b_{21}, a_{22}, b_{22})^{tr}, \tag{16.217}$$

where the matrix A is given by

$$A = \begin{pmatrix} 0 & -\frac{1}{3}b_{11} & \frac{1}{3}a_{11} & 0 & 0 \\ 4b_{11} & 0 & \frac{1}{3}\Omega + \frac{2}{3}a_1 & \frac{4}{3}b_{11} & -\frac{4}{3}a_{11} \\ -4a_{11} & -\frac{1}{3}\Omega - \frac{2}{3}a_1 & 0 & \frac{4}{3}a_{11} & \frac{4}{3}b_{11} \\ 0 & -\frac{1}{3}b_{11} & -\frac{1}{3}a_{11} & 0 & \frac{2}{3}\Omega + \frac{4}{3}a_1 \\ 0 & \frac{1}{3}a_{11} & -\frac{1}{3}b_{11} & -\frac{2}{3}\Omega - \frac{4}{3}a_1 & 0 \end{pmatrix} \tag{16.218}$$

In general the above matrix A is periodic in t with period π/Ω (since a_{11}, b_{11} are periodic in t with period π/Ω) and Floquet theory is needed to solve such system.

Special Case 1: $a_{11} = a$; $b_{11}(0) = 0$, $\Omega + 2a_1 = ac$.
This implies

$$a_{11} = a\cos(\Omega t), b_{11} = a\sin(\Omega t),$$

and the system reduce to

$$A = \frac{a}{3}\begin{pmatrix} 0 & -\sin(\Omega t) & \cos(\Omega t) & 0 & 0 \\ 12\sin(\Omega t) & 0 & c & 4\sin(\Omega t) & -4\cos(\Omega t) \\ -12\cos(\Omega t) & -c & 0 & 4\cos(\Omega t) & 4\sin(\Omega t) \\ 0 & -\sin(\Omega t) & -\cos(\Omega t) & 0 & 2c \\ 0 & \cos(\Omega t) & -\sin(\Omega t) & -2c & 0 \end{pmatrix}$$

The eigenvalues of this matrix are independent of time and are given by

$$0, \frac{a}{3}\sqrt{4+c^2}i, \ -\frac{a}{3}\sqrt{4+c^2}i,$$
$$\frac{2a}{3}\sqrt{4+c^2}i, \ -\frac{2a}{3}\sqrt{4+c^2}i.$$

Numerical evidence indicates aperiodic motion for certain choices of parameters, for instance $a = 1$, $c = \sqrt{2}$.

Special Case 2: $\Omega = 0$ (no Coriolis force), the matrix is a constant matrix.

$$A = \begin{pmatrix} 0 & -\frac{1}{3}b_{11} & \frac{1}{3}a_{11} & 0 & 0 \\ 4b_{11} & 0 & \frac{2}{3}a_1 & \frac{4}{3}b_{11} & -\frac{4}{3}a_{11} \\ -4a_{11} & -\frac{2}{3}a_1 & 0 & \frac{4}{3}a_{11} & \frac{4}{3}b_{11} \\ 0 & -\frac{1}{3}b_{11} & -\frac{1}{3}a_{11} & 0 & \frac{4}{3}a_1 \\ 0 & \frac{1}{3}a_{11} & -\frac{1}{3}b_{11} & -\frac{4}{3}a_1 & 0 \end{pmatrix}$$

The eigenvalues of this matrix can be calculated explicitly as

$$0, \quad \frac{2}{3}\sqrt{a_1^2+a_{11}^2+b_{11}^2}\,i, \quad -\frac{2}{3}\sqrt{a_1^2+a_{11}^2+b_{11}^2}\,i,$$
$$\frac{4}{3}\sqrt{a_1^2+a_{11}^2+b_{11}^2}\,i, \quad -\frac{4}{3}\sqrt{a_1^2+a_{11}^2+b_{11}^2}\,i.$$

The eigenvalues must be zero or purely imaginary which is consistent with the fact that the system conserves energy. Thus the motion is periodic in time. However the period can be any real number. This gives us an indication of the complexity of the dynamics.

Appendix 2

The purpose of this Appendix is to prove claim (16.120) from section 16.3.

We need to postulate the following assumption:

Assumptions: $d>0$ and $\|P_2\psi\|_0$ is uniformly bounded in time.

Remark: The first condition is equivalent to saying that there is real dissipation in the system. The second condition is satisfied provided that $\|\mathcal{F}\|_0$ is uniformly bounded in time.

In order to prove the smallness of $\tilde{F}$ we have to deal with the interaction term between different surface spherical harmonics. This is not a trivial task. We recall from Jones (1985, page 180), formula (46), that the Euler's equation on the unit sphere can be written in the form in terms of the coefficients of the surface spherical harmonics

$$\frac{d\hat{\omega}_{n''m''}}{dt}+\sum_{n,m,n',m'}\left[\frac{n(n+1)}{n''(n''+1)n'(n'+1)}\right]^{\frac{1}{2}}K_{n''n'n}^{m''m'm}\hat{\omega}_{n'm'}\hat{\omega}_{nm}=0, \qquad (16.219)$$

where (see Jones, 1985, page 175, equation (30))

$$K_{n''n'n}^{m''m'm}=\frac{i}{2}(-1)^{m''}\left(\frac{(2n''+1)(2n'+1)(2n+1)}{4\pi n''(n''+1)n'(n'+1)n(n+1)}\right)^{\frac{1}{2}}$$
$$\times((n'+n-n'')(n'+n''-n)(-n'+n''+n+1)(n+n'+n''+1))^{\frac{1}{2}}$$
$$\times\begin{pmatrix} n'' & n & n'-1\\ 0 & 0 & 0\end{pmatrix}\begin{pmatrix} n'' & n' & n\\ -m'' & m' & m\end{pmatrix} \qquad (16.220)$$

$\omega = \Delta\psi$ is the vorticity[†], and there are restrictions on the summation indices which must satisfy (see Jones, 1985 page 193)

$$\begin{aligned} m'' &= m' + m, \ |m''| \leq n'', \ |m'| \leq n', \ |m| \leq n, \\ |n' - n| &< n'' < n' + n. \end{aligned} \tag{16.221}$$

Since the non-linear term in Euler's equation is $J(\psi, \Delta\psi)$, we must have

$$\begin{aligned} P_2 J(\psi, \Delta\psi) &= \sum_{m'',n',m',n,m} \left[\frac{n(n+1)}{6n'(n'+1)}\right]^{\frac{1}{2}} K_{2n'n}^{m''m'm} \hat{\omega}_{n'm'} \hat{\omega}_{nm} Y_{2m''} \\ &= \sum_{m'',n',m',n,m} n(n+1) \left[\frac{n(n+1)n'(n'+1)}{6}\right]^{\frac{1}{2}} K_{2n'n}^{m''m'm} \hat{\psi}_{n'm'} \hat{\psi}_{nm} Y_{2m''}. \end{aligned} \tag{16.222}$$

We may then identify the contribution from different sources

(I) Contribution from the interaction between the first shell and the second shell. This must correspond to

$$-4J(P_1\psi, P_2\psi).$$

(II) Contribution from the interaction between the second shell and higher modes

$$\sum_{m'',m',m} \sum_{n \geq 3} (n(n+1))^{\frac{2}{3}} K_{22n}^{m''m'm} \hat{\psi}_{2m'} \hat{\psi}_{nm} Y_{2m''}.$$

(III) Contribution from the interaction between higher modes (≥ 3)

$$\sum_{m'',m',m,n\geq 3,n'\geq 3} n(n+1) \left[\frac{n(n+1)n'(n'+1)}{6}\right]^{\frac{1}{2}} K_{2n'n}^{m''m'm} \hat{\psi}_{n'm'} \hat{\psi}_{nm} Y_{2m''}.$$

(IV) Contribution from the interaction of the ground energy shell and higher modes ($n \geq 3$) is zero since

$$J(P_1\psi, P_n\psi) \in W_n.$$

We observe that the combination of contributions of type II and III is the total effect of $\tilde{F}$.

To estimate the contribution from type II terms, we notice the summation restriction (16.221) in this special case is

$$|2 - n| < 2, \ \ n \geq 3$$

which leads to

$$n = 3.$$

[†] Notice Jones used ζ instead of ω to denote the vorticity and he didn't write down all the indices.

Therefore

$$
\begin{aligned}
&| \sum_{m'',m',m} \sum_{n\geq 3} (n(n+1))^{\frac{3}{2}} K_{22n}^{m''m'm} \hat{\psi}_{2m'} \hat{\psi}_{nm} Y_{2m''} |_0^2 \\
&\quad = \sum_{m''} \sum_{m',m} 12^3 (K_{223}^{m''m'm})^2 |\hat{\psi}_{2m'} \hat{\psi}_{3m}|^2 \\
&\quad \leq \kappa \|P_2\psi\|_0^2 \|P_3\psi\|_0^2 \\
&\qquad \text{(Thanks to the restricted stability (16.109))} \\
&\quad \leq \kappa \exp(-dt) \|P_2\psi\|_0^2 \\
&\qquad \text{(By our assumption)} \\
&\quad \leq \kappa \exp(-dt). \qquad (16.223)
\end{aligned}
$$

To estimate the contribution from type III terms we have, thanks to the summation restriction on the indices in this case

$$
|n-n'| < 2, \quad n \geq 3, \quad n' \geq 3,
$$

and hence n and n' are equivalent and we may exchange them freely at the expenses of introducing a uniformly bounded coefficient. We also notice, thanks to the explicit three-symbol formula (see Jones, 1985 pages 193–195)

$$
\begin{pmatrix} 2 & n & n'-1 \\ 0 & 0 & 0 \end{pmatrix} \sim n^{-\frac{1}{2}}, \quad \begin{pmatrix} 2 & n' & n \\ -m'' & m' & m \end{pmatrix} \sim n^{-\frac{1}{2}}.
$$

We also notice that

$$
\left(\frac{5(2n'+1)(2n+1)}{24\pi n'(n'+1)n(n+1)} \right)^{\frac{1}{2}} \quad ((n'+n-2)(n'+2-n)
$$
$$
(-n'+n+3)(n+n'+3))^{\frac{1}{2}} \sim 1.
$$

Combining the above three relations together with the formula for K (16.220) we deduce

$$
|K_{2n'n}^{m''m'm}| \leq \kappa \frac{1}{n}, \qquad (16.224)
$$

where κ is a generic constant independent of the indices. This further implies, for the contribution from type III

$$
\left| \sum_{m,m',n\geq 3,n'\geq 3} n(n+1) \left[\frac{n(n+1)n'(n'+1)}{6} \right]^{\frac{1}{2}} K_{2n'n}^{m''m'm} \hat{\psi}_{n'm'} \hat{\psi}_{nm} \right|^2
$$

(Since n and n' are equivalent and the estimate on K.)

$$\leq \kappa \left| \sum_{m,m',n\geq 3,n'\geq 3} n^2(n+1)^2 \hat{\psi}_{n'm'} \hat{\psi}_{nm} \right|^2$$

$$\leq \kappa \left(\sum_{m,m',n\geq 3,n'\geq 3} n^2(n+1)^2 |\hat{\psi}_{nm}|^2 \right) \left(\sum_{m,m',n\geq 3,n'\geq 3} n'^2(n'+1)^2 |\hat{\psi}_{n'm'}|^2 \right)$$

(By the summation restriction)

$$\leq \kappa \left(\sum_{m,n\geq 3} n^2(n+1)^2 |\hat{\psi}_{nm}|^2 \right) \left(\sum_{m',n'\geq 3} n'^2(n'+1)^2 |\hat{\psi}_{n'm'}|^2 \right)$$

(Thanks to the restricted stability (16.109))

$$\leq \kappa \exp(-2dt). \tag{16.225}$$

Combining the estimates on contributions from type II and type III terms and their relationship with $\tilde{F}$ we deduce the claim (16.120).

Index

advective derivative, 2
auto-correlation function, 252

barotropic beta-plane equations, 19
barotropic one-layer model, 54
barotropic quasi-geostrophic equations, 2, 59
barotropic quasi-geostrophic equations in a channel with topographic stress, 46
barotropic quasi-geostrophic equations on the unit sphere, 483
beta plane effect, 2, 17
beta plane equations with mean shear flow, 28

Casimir invariants, 443
Cauchy–Schwarz inequality, 216, 315
channel geometry, 45, 49
circulation, 48, 50
complete statistical mechanics, 256, 303, 529
conserved quantities, 33, 504
continuously stratified quasi-geostrophic equations, 53
continuum limit, 270
Coriolis force, 2
correlation deficit, 365
correlation function, 25, 238
correlation time, 238
crude closure algorithm, 390

dilute PV theory, 332
Dirichlet quotient, 84
dispersion relation, 20
dispersive wave, 20
dissipation mechanism, 2, 18, 30

EEST, 318
Ekman drag, 5
empirical statistical theory with infinitely many constraints, 309
energy, 22, 33, 42, 61
energy–circulation theory, 292, 537
energy–circulation impulse theory, 297
energy–enstrophy theory, 211, 525
energy spectrum, 435
ensemble prediction, 452, 453
enstrophy, 30, 40, 42, 61
enstrophy identity, 61
EOF basis, 457
equations for large-scale and small-scale flow interaction via topographic stress, 38, 43, 116
ergodicity, 228
ESTMC, 318, 540
ESTP, 318, 536
Euler equations, 6, 12
exact dynamics of the ground state modes and the nth energy shell, 495

flatness, 201
Fourier series representation of barotropic quasi-geostrophic equations, 68
Fourier series tool kit, 10
F-plane equation, 53
free-slip boundary condition, 16

Gamma distribution, 422
Gaussian distribution, 195, 452
generalized enstrophy, 40, 46, 49
geophysical fluid dynamics, 1
geostrophic balance, 4
Gibbs measure, 226
Great Red Spot (GRS), 411
Gronwall inequality, 65
ground state eigenmode, 12
ground state modes dynamics, 492

Hamiltonian system, 442

impulse, 47, 49
information-theoretic entropy, see Shannon entropy
invariant measure, 227
isobars, 4

Jacobi identity, 443
Jacobian determinant, 9

kinetic energy identity, 36
Kolmogorov forcing, 17, 60, 76

Langevin theory, 332
Laplace–Beltrami operator, 488
large-scale enstrophy, 39, 43, 46, 49
latitude, 4
layered topographic equations, 143
Legendre functions, 489
Limaye profile, 412
Liouville equation, 222, 453
Liouville property, 221
longitude, 4
Lorenz 96 model (L96), 239

maximum entropy principle, 188, 196
maximum relative entropy principle, 192, 196, 204
mean field equation, 207, 214
mean flow, 17
mean velocity–impulse difference, 49
mesoscale, 2, 4
meta-stability, 362

Navier–Stokes equations, 6
no-penetration boundary condition, 16
nonlinear stability, 62, 116, 117

one and one-half layer model, 56
one layer model, 53
one-point statistics, 203, 211

Parseval identity, 11
partition function, 189, 293
periodic boundary condition, 6
Poincaré inequality, 11
point-vortex equations, 304
point-vortex theory, 300, 538
Poisson bracket, 443
potential enstrophy, 40
potential vorticity, 2
potential vorticity extrema, 47, 50
predictability, 452, 454, 456
predictive information content, 461, 463, 465
prior distribution, 203
pseudo-energy, 269
PVST, 318

radiative damping, 5
relative entropy, 192, 195, 204, 452, 455
relative entropy identity, 459
relative vorticity, 2
relaxation, 473
response to large-scale forcing, 67, 511
restricted stability of motion on the first two energy shells, 508
Reynolds number, 97
Rhines measure of anisotropy, 96
Rossby waves, 19, 496

selective decay principle, 81, 85, 518, 519
selective decay state, 81
Shannon entropy, 185, 195, 212, 225
shear flow, 13
signal and dispersion decomposition, 460
sine-bracket truncation, 430
skewness, 201, 421
slaving principle, 78
spectrally truncated quasi-geostrophic equations, 259
spin-up, 394
stability of simple steady states, 116, 505
statistical relevance, 427
statistical solution, 219
statistically sharp, 347
Stokes formula, 488
stratification, 53
stream function, 2
surface spherical coordinates, 482
surface spherical harmonics, 489
swimming pool geometry, 41
swirling eddies, 13
symplectic matrix, 443

tangent plane approximation, 4
Taylor vortices, 15
topographic stress, 37
topography, 2, 25
total energy, 43, 46, 49
truncated Burgers–Hopf equation (TBH), 230
two layer model, 54

undamped unforced L-96 model (IL96), 244
utility, 474

variational characterization, 133
variational derivative, 50
velocity, 2
viscosity (Newtonian, eddy, hyper), 5

whitening, 459
White Ovals, 413

zonal flow, 129